Communications in Computer and Information Science

2829

Series Editors

Gang Li ⓘ, *School of Information Technology, Deakin University, Burwood, VIC, Australia*

Joaquim Filipe ⓘ, *Polytechnic Institute of Setúbal, Setúbal, Portugal*

Zhiwei Xu, *Chinese Academy of Sciences, Beijing, China*

Rationale

The CCIS series is devoted to the publication of proceedings of computer science conferences. Its aim is to efficiently disseminate original research results in informatics in printed and electronic form. While the focus is on publication of peer-reviewed full papers presenting mature work, inclusion of reviewed short papers reporting on work in progress is welcome, too. Besides globally relevant meetings with internationally representative program committees guaranteeing a strict peer-reviewing and paper selection process, conferences run by societies or of high regional or national relevance are also considered for publication.

Topics

The topical scope of CCIS spans the entire spectrum of informatics ranging from foundational topics in the theory of computing to information and communications science and technology and a broad variety of interdisciplinary application fields.

Information for Volume Editors and Authors

Publication in CCIS is free of charge. No royalties are paid, however, we offer registered conference participants temporary free access to the online version of the conference proceedings on SpringerLink (http://link.springer.com) by means of an http referrer from the conference website and/or a number of complimentary printed copies, as specified in the official acceptance email of the event.

CCIS proceedings can be published in time for distribution at conferences or as post-proceedings, and delivered in the form of printed books and/or electronically as USBs and/or e-content licenses for accessing proceedings at SpringerLink. Furthermore, CCIS proceedings are included in the CCIS electronic book series hosted in the SpringerLink digital library at http://link.springer.com/bookseries/7899. Conferences publishing in CCIS are allowed to use Online Conference Service (OCS) for managing the whole proceedings lifecycle (from submission and reviewing to preparing for publication) free of charge.

Publication process

The language of publication is exclusively English. Authors publishing in CCIS have to sign the Springer CCIS copyright transfer form, however, they are free to use their material published in CCIS for substantially changed, more elaborate subsequent publications elsewhere. For the preparation of the camera-ready papers/files, authors have to strictly adhere to the Springer CCIS Authors' Instructions and are strongly encouraged to use the CCIS LaTeX style files or templates.

Abstracting/Indexing

CCIS is abstracted/indexed in DBLP, Google Scholar, EI-Compendex, Mathematical Reviews, SCImago, Scopus. CCIS volumes are also submitted for the inclusion in ISI Proceedings.

How to start

To start the evaluation of your proposal for inclusion in the CCIS series, please send an e-mail to ccis@springer.com.

Francesco Marcelloni · Kurosh Madani ·
Niki van Stein · Joaquim Filipe
Editors

Computational Intelligence

17th International Joint Conference, IJCCI 2025
Marbella, Spain, October 22–24, 2025
Proceedings, Part III

 Springer

level is a task that can only be achieved by the collaborative effort of a dedicated and highly competent team.

We hope you all had an exciting and inspiring conference. We hope to have contributed to the development of our research community, and we look forward to having additional research results presented at the next edition of IJCCI, details of which are available at https://ijcci.scitevents.org.

October 2025

Francesco Marcelloni
Kurosh Madani
Niki van Stein
Joaquim Filipe

Organization

Conference Chair

AGENTICS, ECTA, EXPLAINS, FCTA, NCTA

Joaquim Filipe — Polytechnic Institute of Setubal/INSTICC, Portugal

Program Co-chairs

AGENTICS

Francesco Marcelloni — University of Pisa, Italy
Niki van Stein — Leiden University, Netherlands
Kurosh Madani — University of Paris-Est Créteil, France

ECTA

Niki van Stein — Leiden University, Netherlands

EXPLAINS

Francesco Marcelloni — University of Pisa, Italy
Niki van Stcin — Leiden Unıversity, Netherlands
Kurosh Madani — University of Paris-Est Créteil, France

FCTA

Francesco Marcelloni — University of Pisa, Italy

NCTA

Kurosh Madani — University of Paris-Est Créteil, France

AGENTICS Program Committee

Emel Aktas	Cranfield University, UK
Frederic Alexandre	Inria, France
Isabel Alexandre	Instituto Universitário de Lisboa (ISCTE-IUL) and Instituto de Telecomunicações, Portugal
Patrícia Alves	Instituto Superior de Engenharia do Porto - Instituto Politécnico do Porto, Portugal
Luis Anido-Rifon	University of Vigo, Spain
Patricia Anthony	Lincoln University, New Zealand
Nahla Ben Amor	Institut Supérieur de Gestion, Tunisia
Lars Braubach	City University of Hamburg, Germany
Filipe Calegario	Federal University of Pernambuco, Brazil
Luis Camarinha-Matos	New University of Lisbon, Portugal
Olivier Camp	Graduate School of Electronics of the West, France
Dumitru-Clementin Cercel	National University of Science and Technology Politehnica Bucharest, Romania
Flavio Correa da Silva	University of São Paulo, Brazil
Partha Das	Independent Researcher, India
Ajay Dholakia	Lenovo, USA
Khaldoon Dhou	Texas A&M University Central Texas, USA
Hércules do Prado	Catholic University of Brasília, Brazil
Shlomo Dubnov	University of California, San Diego, USA
Sabeur Elkosantini	University of Carthage, Tunisia
Gal Engelberg	University of Haifa, Israel
Edilson Ferneda	Catholic University of Brasilia, Brazil
Raffaella Folgieri	Universita degli Studi di Milano, Italy
Irene Fondón	Universidad de Sevilla, Spain
Stéphane Galland	Université de Technologie de Belfort-Montbéliard, France
Leonardo Garrido	Tecnológico de Monterrey, Campus Monterrey, Mexico
Pablo Gervás	Universidad Complutense de Madrid, Spain
Cinzia Giannetti	Swansea University, UK
Paolo Giudici	University of Pavia, Italy
Giorgio Gosti	Consiglio Nazionale delle Ricerche, Italy
Frederic Guerrero-Solé	Pompeu Fabra University, Spain
Maki Habib	American University in Cairo, Egypt
Allel Hadjali	ISAE-ENSMA, France
Mirsad Hadzikadic	University of North Carolina Charlotte, USA

Hisashi Hayashi	Advanced Institute of Industrial Technology, Japan
Nadia Hocine	University of Mostaganem, Algeria
Pengyu Hong	Brandeis University, USA
Mohd Nor Akmal Khalid	Universiti Kebangsaan Malaysia, Malaysia
Constantine Kotropoulos	Aristotle University of Thessaloniki, Greece
Boris Kovalerchuk	Central Washington University, USA
Carsten Lanquillon	Heilbronn University of Applied Sciences, Germany
Wookey Lee	Inha University, South Korea
Florin Leon	Technical University "Gheorghe Asachi" of Iași, Romania
Kecheng Liu	University of Reading, UK
Sebastian Lobentanzer	Helmholtz Centre Munich, Germany
Martin Ma	New Mexico Tech, USA
Juan Martinez-Miranda	Independent Researcher, Mexico
Xabier Martínez-Rolán	University of Vigo, Spain
Bruno Mermet	Independent Researcher, France
Silvan Mertes	University of Augsburg, Germany
Michele Missikoff	ISTC-CNR, Italy
Jelena Mitrovic	University of Passau, Germany
Farid Mokhati	University of Oum El Bouaghi, Algeria
Mario Molinara	Universita di Cassino e del Lazio Meridionale, Italy
Maxime Morge	Université Lyon 1, France
Gildas Morvan	Université d'Artois, France
Kshirasagar Naik	University of Waterloo, Canada
Sanaz Nikghadam-Hojjati	Nova University Lisbon, Portugal
Shimon Nof	Purdue University, USA
Flavio Oquendo	UMR IRISA (CNRS) - Université Bretagne Sud, France
Ermelinda Oro	National Research Council (CNR), Italy
Yash Patel	Capella University, USA
Milica Petrovic	University of Belgrade, Serbia
Aske Plaat	Leiden University, Netherlands
Xinchi Qiu	University of Cambridge, UK
Sivaramakrishnan Rajaraman	National Library of Medicine, USA
Ana Paula Rocha	LIACC/FEUP, University of Porto, Portugal
Juha Röning	University of Oulu, Finland
Mika Saari	Tampere University, Finland
Luca Sabatucci	ICAR-CNR, Italy
Christophe Sabourin	Université Paris-Est Créteil, LISSI, France

Markus Schatten	University of Zagreb, Croatia
Michael Schumacher	University of Applied Sciences and Arts Western Switzerland (HES-SO), Switzerland
Yuichi Sei	University of Electro-Communications, Japan
Emilio Serrano	Universidad Politécnica de Madrid, Spain
Soharab Shaikh	BML Munjal University, India
Vishakha Sharma	Roche, USA
Marius Silaghi	Florida Institute of Technology, USA
Adrian-Mihail Stoica	Polytechnic University Bucharest, Romania
Junichi Suzuki	University of Massachusetts, Boston, USA
Ridwan Taiwo	Hong Kong Polytechnic University, China
Hristo Tanev	European Commission, Italy
Takao Terano	Chiba University of Commerce, Japan
Michele Tomaiuolo	University of Parma, Italy
Francisco Túnez	Saint Anthony Catholic University of Murcia, Spain
Gregg Vesonder	Stevens Institute of Technology, USA
Marco Volpe	University of Leicester, UK
Hai Wang	Saint Mary's University, Canada
Hayden Wimmer	Georgia Southern University, USA
Levent Yilmaz	Auburn University, USA

AGENTICS Additional Reviewer

Luisa Mich	University of Trento, Italy

ECTA Program Committee

Anca Andreica	Babeș-Bolyai University, Romania
Mohd Ashraf Ahmad	University Malaysia Pahang, Malaysia
Adrian Bekasiewicz	Gdańsk University of Technology, Poland
Fatima Benbouzid-Si Tayeb	Ecole nationale Supérieure d'Informatique, Algeria
Peter Bentley	University College London, UK
Ying Bi	Victoria University of Wellington, New Zealand
Zafer Bingul	Kocaeli University, Turkey
Janos Botzheim	Eötvös Loránd University, Hungary
Alexander Brownlee	University of Stirling, UK
Stefano Cagnoni	University of Parma, Italy
Pedro Angel Castillo Valdivieso	University of Granada, Spain

Paolo Cazzaniga	University of Bergamo, Italy
Francisco Chicano	University of Málaga, Spain
Vincent Cicirello	Stockton University, USA
Vincenzo Conti	Kore University of Enna, Italy
Marisol B. Correia	Universidade do Algarve & CiTUR, Portugal
Erik Cuevas	Universidad de Guadalajara, Mexico
Wellington dos Santos	Federal University of Pernambuco, Brazil
Marc Ebner	Ernst-Moritz-Arndt-Universität Greifswald, Germany
Gusz Eiben	Vrije Universiteit Amsterdam, Netherlands
Anton Eremeev	Sobolev Institute of Mathematics SB RAS, Russian Federation
Márcia Fernandes	Federal University of Uberlândia, Brazil
Paola Festa	University of Naples, Italy
Gianluigi Folino	ICAR-CNR, Italy
Carmen Galé	Universidad de Zaragoza, Spain
Shangce Gao	University of Toyama, Japan
Zong Woo Geem	Gachon University, South Korea
Luis Gomes	UNL/UNINOVA, Portugal
David Greiner	Universidad de Las Palmas de Gran Canaria, Spain
Francesco Grimaccia	Politecnico di Milano, Italy
Aldy Gunawan	Singapore Management University, Singapore
Gareth Howells	University of Kent, UK
Giovanni Iacca	University of Trento, Italy
Nicolas Jozefowiez	University of Lorraine, France
Chih-Chin Lai	National University of Kaohsiung, Taiwan
Nuno Leite	Instituto Superior de Engenharia de Lisboa, Portugal
Wen-Yang Lin	National University of Kaohsiung, Taiwan
Michalis Mavrovouniotis	Cyprus University of Technology, Cyprus
Mohamed Arezki Mellal	M'Hamed Bougara University, Algeria
Carmelo Militello	National Research Council (CNR), Italy
Soumya D. Mohanty	University of Texas Rio Grande Valley, USA
Nasimul Noman	University of Newcastle, Australia
Colm O'Riordan	University of Galway, Ireland
Andrzej Obuchowicz	University of Zielona Góra, Poland
Schütze Oliver	CINVESTAV-IPN, Mexico
Ender Özcan	University of Nottingham, UK
David A. Pelta	University of Granada, Spain
Pupong Pongcharoen	Naresuan University, Thailand

Abdellatif Rahmoun	École supérieure en informatique de Sidi Bel Abbès, and Graduate School of Computer Science, Algeria
Günther Raidl	TU Wien, Austria
Bendib Riad	20 August 1955 University of Skikda, Algeria
Hendrik Richter	Leipzig University of Applied Sciences, Germany
Olympia Roeva	Institute of Biophysics and Biomedical Engineering, Bulgarian Academy of Sciences, Bulgaria
Conor Ryan	University of Limerick, Ireland
Guillaume Sandou	CentraleSupélec, France
Sevil Sen	Hacettepe University, Turkey
Andrea Serani	National Research Council-Institute of Marine Engineering, Italy
Ankit Sharma	Nirma University, India
Arlindo Silva	Polytechnic University of Castelo Branco, Portugal
Andrzej Skowron	University of Warsaw, Poland
Terence Soule	Independent Researcher, USA
Pedro Sousa	University of Minho, Portugal
Stephen Swift	Brunel University London, UK
Gianluca Tempesti	University of York, UK
Takao Terano	Chiba University of Commerce, Japan
Sunita Tiwari	Indian Institute of Technology, Delhi, India
Stephan Winkler	Independent Researcher, Austria
Man Leung Wong	Lingnan University, China
Takeshi Yamada	Kindai University, Japan

ECTA Additional Reviewer

Igor Kulachenko	Sobolev Institute of Mathematics SB RAS, Russian Federation

EXPLAINS Program Committee

Michael Affenzeller	Independent Researcher, Austria
Mojtaba Ahmadieh Khanesar	University of Nottingham, UK
Emel Aktas	Cranfield University, UK
Wudhichai Assawinchaichote	King Mongkut's University of Technology Thonburi, Thailand

Rosangela Ballini	Universidade Estadual de Campinas, Brazil
Fevzi Belli	Izmir Institute of Technology, Turkey
Stefano Bistarelli	University of Perugia, Italy
Ulrich Bodenhofer	University of Applied Sciences Upper Austria, Hagenberg, Austria
Francesco Calimeri	University of Calabria, Italy
Giovanna Castellano	University of Bari "Aldo Moro", Italy
Juan Luis Castro	Universidad de Granada, Spain
Amitava Chatterjee	Jadavpur University, India
Mario G. C. A. Cimino	University of Pisa, Italy
Flavio Correa da Silva	University of São Paulo, Brazil
António Cunha	Universidade de Trás-os-Montes e Alto Douro, Portugal
Pietro Ducange	University of Pisa, Italy
Márcia Fernandes	Federal University of Uberlândia, Brazil
Gianna Figà-Talamanca	University of Perugia, Italy
Romain Giot	Université de Bordeaux, France
Maki Habib	American University in Cairo, Egypt
Gareth Howells	University of Kent, UK
Giovanni Iacca	University of Trento, Italy
Constantine Kotropoulos	Aristotle University of Thessaloniki, Greece
Christophe Labreuche	Thales Research and Technology, France
Jean-Charles Lamirel	LORIA, University of Strasbourg, France
Wookey Lee	Inha University, South Korea
Ahmad Lotfi	Nottingham Trent University, UK
Hichem Maaref	Universite d'Évry, France
Christos Makris	University of Patras, Greece
Stefano Mariani	University of Modena and Reggio Emilia, Italy
Manuel Mazzara	Innopolis University, Russian Federation
Enza Messina	University of Milano-Bicocca, Italy
José Molina	Universidad Carlos III de Madrid, Spain
Colm O'Riordan	University of Galway, Ireland
Lale Özbakir	Erciyes University, Turkey
Dawid Polap	Silesian University of Technology, Poland
Olympia Roeva	Institute of Biophysics and Biomedical Engineering, Bulgarian Academy of Sciences, Bulgaria
Roseli Romero	University of São Paulo, Brazil
Daniel Sánchez	University of Granada, Spain
Catarina Silva	University of Coimbra, CISUC, Portugal
José Silvestre Silva	Military Academy, Portugal
Philippe Thomas	Université de Lorraine, France

Gregg Vesonder — Stevens Institute of Technology, USA

EXPLAINS Additional Reviewers

Marco Cuccarini — University of Perugia, Italy
Corrado Mencar — University of Bari Aldo Moro, Italy

FCTA Program Committee

Mojtaba Ahmadieh Khanesar — University of Nottingham, UK
Mohd Ashraf Ahmad — University Malaysia Pahang, Malaysia
Fernando Bobillo — University of Zaragoza, Spain
Said Broumi — University of Hassan II -Casablanca, Morocco
Zhen-Song Chen — Wuhan University, China
Scott Dick — University of Alberta, Canada
Antonin Dvorak — University of Ostrava, Czech Republic
Stefka Fidanova — Bulgarian Academy of Sciences, Bulgaria
Pier Luigi Gentili — Università degli Studi di Perugia, Italy
Irina Georgescu — Bucharest University of Economics, Romania
Etienne Kerre — Ghent University, Belgium
Szilveszter Kovacs — University of Miskolc, Hungary
Francisco Lupianez — Universidad Complutense de Madrid, Spain
Luis Martinez Lopez — University of Jaén, Spain
Corrado Mencar — University of Bari, Italy
José Molina — Universidad Carlos III de Madrid, Spain
Javier Montero — Complutense University of Madrid, Spain
Claudio Moraga — Technical University of Dortmund, Germany
Vesa Niskanen — University of Helsinki/VM University, Finland
Alessandro Renda — University of Trieste, Italy
Olympia Roeva — Institute of Biophysics and Biomedical Engineering, Bulgarian Academy of Sciences, Bulgaria
Raul-Cristian Roman — Politehnica University of Timisoara, Romania
Jose de Jesus Rubio — Instituto Politécnico Nacional, Mexico
Luciano Stefanini — University of Urbino Carlo Bo, Italy
Andreja Tepavcevic — Mathematical Institute of Serbian Academy of Sciences and Arts, Serbia
Carlos M. Travieso-González — Universidad de Las Palmas de Gran Canaria, Spain

Alexandre Voisin	Université de Lorraine, CNRS, CRAN UMR 7039, France
Slawomir Zadrozny	Polish Academy of Sciences, Poland

FCTA Additional Reviewers

Francesco Marcelloni	University of Pisa, Italy
Fabrizio Ruffini	University of Pisa, Italy

NCTA Program Committee

Salwa Abd-El-Hafiz	Cairo University, Egypt
Mojtaba Ahmadieh Khanesar	University of Nottingham, UK
Rui Araujo	University of Coimbra, Portugal
Monica Bianchini	University of Siena, Italy
Alessandro Biondi	Scuola Superiore Sant'Anna, Italy
Mounir Bouhedda	University of Medea, Algeria
Jose de Jesus Rubio	Instituto Politécnico Nacional, Mexico
Artur Ferreira	Instituto Superior de Engenharia de Lisboa and Instituto de Telecomunicações, Portugal
Abbas Fotouhi	Cranfield University, UK
Petr Hajek	University of Pardubice, Czech Republic
Sarangapani Jagannathan	Missouri University of Science and Technology, USA
Khairul Kasmiran	Universiti Putra Malaysia, Malaysia
Tarek Khadir	University of Badji Mokhtar Annaba, Algeria
Ramanathan Lakshmanan	Independent Researcher, India
Wei-Min Liu	National Chung Cheng University, Taiwan
Brian Nils Lundstrom	Mayo Clinic, USA
Vincenzo Piuri	Università degli Studi di Milano, Italy
Vijayakumar Ponnusamy	SRM IST, Kattankulathur Campus, India
Antonello Rizzi	Università di Roma "La Sapienza", Italy
Gerald Schaefer	Loughborough University, UK
Pankaj Singh	Central Institute of Technology Kokrajhar, India
Andrzej Skowron	University of Warsaw, Poland
Ryszard Tadeusiewicz	AGH University of Science and Technology, Poland
Philippe Thomas	Université de Lorraine, France
Constantin Volosencu	Politehnica University Timisoara, Romania
Hai Wang	Saint Mary's University, Canada

| Salman Yussof | Universiti Tenaga Nasional, Malaysia |
| Cleber Zanchettin | Federal University of Pernambuco, Brazil |

NCTA Additional Reviewer

| Reza Askari-Moghadam | Sorbonne University, France |

Invited Speakers

IJCCI

Luís Paulo Reis	University of Porto, Portugal
Pietro Ducange	University of Pisa, Italy
Fei Liu	City University of Hong Kong, China
Henry Prakken	Utrecht University, Netherlands

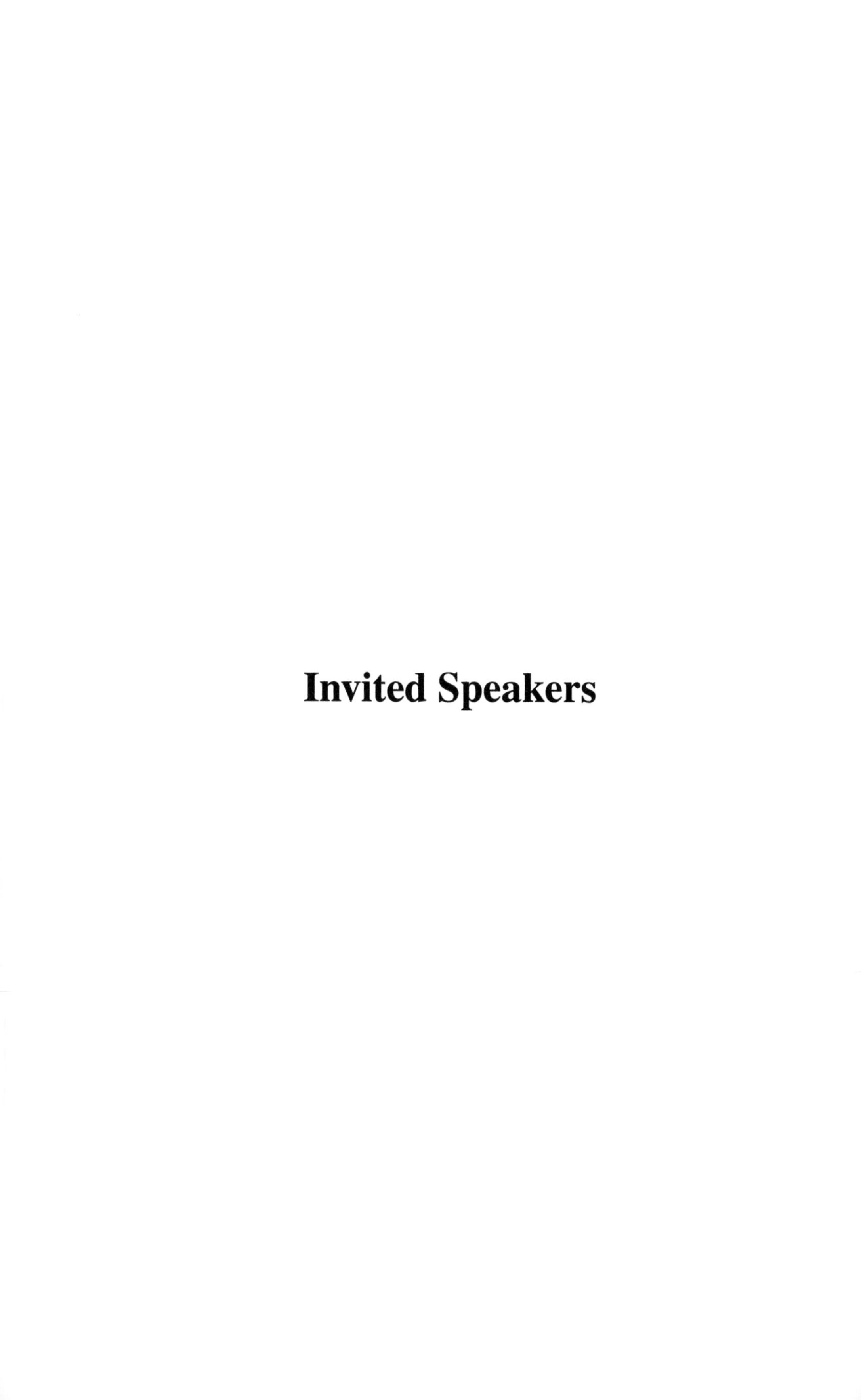

Invited Speakers

Federated Learning (FED) of eXplainable Artificial Intelligence (XAI) Models

Pietro Ducange

University of Pisa, Italy

Abstract. The current era is characterized by an increasing pervasiveness of applications and services based on data processing and often built on Artificial Intelligence (AI) and, in particular, Machine Learning (ML) algorithms. In fact, extracting insights from data is so common in the daily life of individuals, companies, and public entities and so relevant for market players, to become an important matter of interest for institutional organizations. The topic is so important and hot that ad hoc guidelines and regulations for designing trustworthy AI-based applications have also been proposed by the European Union and other national and supra-national bodies. One important aspect is given by the capability of the applications to tackle the data privacy issue. Additionally, depending on the specific application field, paramount importance is given to the possibility for humans to understand why a certain AI/ML-based application is providing that specific output. Trustworthy AI models should be trained with the simultaneous goals of preserving data privacy and ensuring a certain level of explainability of the system. In this talk, we discuss the concept of Federated Learning (FL) of eXplainable AI (XAI) models, in short FED-XAI, purposely designed to address the two requirements simultaneously. We first introduce the motivations at the foundation of FL and XAI, along with their basic concepts. Then, we provide a brief survey regarding approaches, models, results, issues and applications of FED-XAI. Finally, we also show our recently released framework for providing user-friendly support to Federated Learning (FL) of Fuzzy Rule-Based Systems (FRBS) as explainable-by-design models.

From Traditional AI to the Future of Agentic AI and Robotics

Luís Paulo Reis

University of Porto, Portugal

Abstract. This talk analyzes the evolution of AI from symbolic AI to machine learning and the shift from rule-based expert systems to data-driven approaches, the limitations of traditional AI and why machine learning emerged, the role of data in AI, and the rise of deep learning, deep reinforcement learning, large language models (LLMs), and generative AI. It will then analyze agentic AI as the next frontier, moving from predictive models to autonomous, goal-driven AI agents and from the early examples of agents and agentic AI in automation, research, and self-improving systems to the new generation of agents powered by strong LLMs. It will then analyze Deep Reinforcement Learning (DRL) and how DRL enables AI to learn from trial and error without being data-driven with applications in robotics, autonomous systems, and gaming. Then it will focus on the emergence of Large Behavior Models (LBMs), AI models trained on large-scale behavior data, moving beyond text to multimodal inputs (vision, actions, speech) with applications in robotics, industrial automation, and human-AI collaboration. It will conclude with a deep analysis of the future of robotics and AI integration with the shift from reactive robots to proactive and adaptive intelligent robots, combining LLMs, DRL, and LBMs to create more autonomous robotic systems. Will businesses and society be prepared for this AI next generation? What should we do to prepare ourselves?

Automated Algorithm Design with Large Language Model

Fei Liu

City University of Hong Kong, China

Abstract. Algorithm Design (AD) plays a pivotal role in many fields. The emergence of Large Language Models (LLMs) has significantly advanced automation and innovation in this field. We first provide a systematic overview and taxonomy of the literature on algorithm design utilizing large language models. Next, we introduce Evolution of Heuristic (EoH), an evolutionary framework that integrates LLMs with Evolutionary Search to automate algorithm design. Furthermore, we present LLM4AD, an open-source, user-friendly platform to support algorithm development with large language models. This platform aims to provide modularized tools, methodologies, and tasks, fostering research and practical applications in this emerging field.

An Overview of AI and Law Models of Case-Based Reasoning, With Applications in Decision Analysis and Explainable AI

Henry Prakken

Utrecht University, Netherlands

Abstract. In this talk I will give an overview of recent research of myself and my students on formal and computational models of legal case-based reasoning. Legal case-based reasoning is the process of arguing for or against decisions in new cases by drawing analogies to or stressing differences with precedent cases. In the field of AI & law, seminal work on case-based reasoning was by Ashley & Rissland on the HYPO system, followed by Aleven's work on the CATO system. This work primarily focussed on generating case-based debates. More recently, John Horty initiated the formal study of so-called precedential constraint, which addresses a question of a more logical nature, namely, to what extent a decision in a new case is constrained by a body of precedents. In our recent work we have further developed this work and, among other things, developed gradual consistency measures for collections of case-based decisions. We have also studied the application of models of legal case-based reasoning in explainable AI, by exploiting an analogy between the legal notion of precedent and the machine-learning concept of training data.

Contents

**International Conference on Neural Computation Theory and
Applications**

International Conference on Explainable AI for Neural and Symbolic Methods

AutoCausalAIME: A CMA-ES-Driven Framework for Parametric Penalty Tuning in Causal Inverse Explanations

Takafumi Nakanishi$^{(\boxtimes)}$ [ID]

Tokyo University of Technology, 1404-1 Katakura, Hachioji, Tokyo 135-8181, Japan
nakanishitf@stf.teu.ac.jp

Abstract. In recent years, the importance of artificial intelligence (AI) and machine learning model explainability has led to growing interest in Explainable AI (XAI). Specifically, global feature importance (GFI), which identifies the key explanatory variables (features) and their contributions to the target variable, plays a central role from two perspectives: (1) understanding the overall model behavior and (2) discovering true associations. Previous methods, such as LIME and SHapley Additive exPlanations, have primarily addressed the first perspective. To handle both, we introduced approximate inverse model explanations (AIME), which derive GFI in a data-driven manner using algebraic operations centered on the target variable. However, AIME was not fully robust to feature interdependencies, prompting us to develop CausalAIME, which integrates causal structure estimation (via the Peter–Clark algorithm) and the penalty-based suppression of multicollinearity. In this paper, we propose "AutoCausalAIME," a method that eliminates the need for manual adjustment of CausalAIME's hyper-parameters—namely the penalty strength (λ) and causal ratio (α)—by leveraging covariance matrix adaptation evolution strategy (CMA-ES). We compare AIME, AutoCausalAIME, and "worst case" CausalAIME. While the Wilcoxon signed-rank test does not reveal statistically significant differences among them, comparisons across six metrics (descriptive accuracy, sparsity, stability, efficiency, robustness, and completeness) show that AutoCausalAIME achieves high explanatory accuracy, suppressed multicollinearity, and sufficient robustness. Thus, we demonstrate that AutoCausalAIME (1) eliminates the need for manual trial and error, (2) is applicable to large-scale and diverse tasks, and (3) derives interpretable GFI through a causal-structure-based difference penalty.

Keywords: Explainable AI (XAI) · Approximate Inverse Model Explanations (AIME) · Causal Structure Estimation · Covariance Matrix Adaptation Evolution Strategy (CMA-ES) · Multicollinearity Suppression

1 Introduction

In recent years, the use of artificial intelligence (AI) and machine-learning models has become increasingly prevalent, with societal implementation advancing steadily. Consequently, there is a growing need to interpret the predictions made by these models and

F. Marcelloni et al. (Eds.): IJCCI 2025, CCIS 2829, pp. 3–16, 2026.
https://doi.org/10.1007/978-3-032-15638-9_1

understand why these predictions arise. Explainable AI (XAI) [1, 2] has attracted considerable attention as a technical approach to address this demand. XAI techniques are now widely employed across domains such as decision support systems, social implementation, and accountability determination in ethical and regulatory contexts. They have also been used in tasks such as detecting hateful or fake news. Specifically, in fields such as medicine, finance, and public administration, explaining how a model makes predictions is essential.

The global feature importance (GFI) is used as a core XAI methodology to quantify how each feature contributes to the model's outputs. GFI serves two key purposes: (1) to help understand the overall behavior of the model (i.e., model-dependent importance) and (2) to identify the true relationships (true association) between the target variable and its explanatory variables. However, most existing approaches (e.g., SHapley Additive exPlanations (SHAP) [3]) predominantly target the former. To address both perspectives, we previously proposed approximate inverse model explanations (AIME) [4], a data-driven framework that clarifies model behavior and analyzes the true relationships within the data.

Local interpretable model-agnostic explanations (LIME) [5], a widely used XAI technique, focuses on the importance of local features. Given that AI systems often conceal various algorithms from end users, we emphasize model-agnostic methods—such as LIME, SHAP, and AIME—that are not tied to a specific algorithm. Nonetheless, these conventional methods do not consistently account for multicollinearity and feature interdependence. While AIME's formulation offers partial robustness to multicollinearity, our empirical investigations suggest the need for more proactive strategies for addressing feature interdependencies.

To this end, we developed CausalAIME [6], which combines causal structure estimation (using the Peter–Clark (PC) algorithm [7]) with penalty-based suppression of multicollinearity, allowing the derivation of GFI that considers feature interdependencies. However, CausalAIME still requires a manual configuration of hyperparameters, such as penalty strength and causal ratio, leading to trial-and-error tuning for each dataset or task.

Therefore, we propose AutoCausalAIME, which automates the hyperparameter selection process in CausalAIME, eliminating the need for manual adjustment. Specifically, we employ the covariance matrix adaptation evolution strategy (CMA-ES) [8, 9] to solve an optimization problem over penalty strength and causal ratio via cross-validation, aiming to minimize reconstruction error and improve explanatory accuracy. This approach enables the derivation of the GFI while suppressing multicollinearity and feature interdependence, without labor-intensive manual hyperparameter selection.

We compare AutoCausalAIME with AIME [4] and CausalAIME [6] (worst case), where CausalAIME's hyperparameters are deliberately set to unfavorable values. We evaluate these methods using the descriptive accuracy, sparsity, stability, efficiency, robustness, and completeness metrics proposed by Arreche et al. [10] along with the Wilcoxon signed-rank test. Although the Wilcoxon test does not reveal statistically significant differences, the metric comparisons indicate that AutoCausalAIME achieves a good balance of high explanatory accuracy, robustness, and multicollinearity suppression.

The contributions of this study are summarized as follows.

- We propose a new framework, AutoCausalAIME, that integrates CMA-ES into CausalAIME to automatically adjust the hyperparameters.
- Through real-data experiments and statistical verification (via the Wilcoxon test), we demonstrate that this method effectively handles multicollinearity, maintains explanatory accuracy, and provides robustness. While no statistically significant difference is observed, the metric-based qualitative comparison confirms its superiority in practice.

The remainder of this paper is organized as follows: Section 2 reviews related work. Section 3 describes the proposed method, AutoCausalAIME. Section 4 presents the experimental setup and results. Section 5 discusses our findings, and Sect. 5 concludes with final remarks and future prospects.

2 Related Works

As a counterfactual explanation method using causal discovery, Takahashi et al. [11] proposed a framework that estimates an unknown causal graph and uses counterfactual probabilities to explain the behavior of a black-box model. In applications such as credit ratings, the primary focus is on constructing causal graphs and leveraging counterfactual reasoning rather than deriving GFI.

TCEC-FCM [12] targets large-scale fuzzy cognitive maps and employs binary search along with graph traversal to efficiently calculate causal effects, thereby enhancing the interpretability of AI models. However, this work is specialized for fuzzy cognitive maps and does not address applying the method to general black-box models or visualizing global-level feature importance using positive and negative sign information.

Cinquini and Guidotti [13] proposed a local model-dependent explanation method that incorporates causal knowledge into the data to generate stable local explanations. Specifically, they extended LIME by building a local approximation model that accounts for causal structures, aiming to improve fidelity at the local level. However, their approach does not emphasize the importance of global features, which accounts for causality and multicollinearity.

Similarly, Termine et al. [14] learned surrogate causal models in local neighborhoods and generated causal hypothesis-based explanations (MaLESCaMo). Although this method infers feature interactions based on an actual causal structure, it places greater emphasis on local neighborhood analysis rather than on a comprehensive global analysis that addresses the entire model output.

Furthermore, Zecevic et al. [15] introduced a structural causal interpretation (SCI) approach employing Pearl's structural causal model to align a trained model's output with human causal thinking, thus emphasizing causal consistency between machine learning models and their tasks. However, the focus is largely on the consistency between human mental models and causal frameworks, and there is minimal discussion on GFI (including positive/negative contributions) or class-specific details.

The Shapley flow [16] attempts to visualize the interdependencies among features that biased explanation (Shapley) methods often overlook by assigning weights to edges rather than to nodes within a causal graph. Specifically, it calculates how an input's influence propagates to the output at the edge level; however, it continues to emphasize model-internal influence visualization along a causal graph rather than providing overall data-driven importance using true labels.

Finally, Breuer et al. [17] introduced Causality-Aware Shapley Value for Global Explanations (CAGE), which incorporates causal models into Shapley values to yield more causally valid global explanations by applying causal interventions (do operators) and reflecting causal structures during sampling. Nonetheless, it assumes a known causal graph, does not address automatic causal structure learning or data-driven modes using true labels, and does not comprehensively address multicollinearity suppression.

Overall, various forms of integration between causality and XAI have been proposed; however, many studies still face the following issues:

1. a predominant focus on hypothesis-based analysis and local-level (e.g., neighborhood) explanations,
2. an absence of sufficiently detailed GFI that explicitly captures positive/negative contributions and class-specific information,
3. reliance on existing causal structures, without a clear framework for switching between model-dependent and data-driven modes.

In contrast, CausalAIME integrates the PC algorithm into AIME's approximate inverse operator framework, enabling it to estimate causal structures among features while deriving coefficients that suppress multicollinearity and further provide GFI with positive and negative signs as well as class-specific interpretations. Moreover, the same framework allows switching between a "data-driven mode" that uses the true labels Y and a "model-dependent mode" that uses model outputs $\hat{Y}$. The novelty of this work lies in "AutoCausalAIME," which extends CausalAIME by incorporating CMA-ES for automated hyperparameter tuning, thereby eliminating the need for manual trial and error.

Finally, CausalAIME (and consequently AutoCausalAIME) does not rely exclusively on the PC algorithm for causal discovery. In principle, any causal inference method, such as greedy equivalence search (GES) [18], fast causal inference (FCI) [19], linear non-gaussian acyclic model (LiNGAM) [20], or Bayesian structure learning [21], can be integrated, provided it yields a directed or partially directed graph among the features. By replacing the PC algorithm with an alternative approach that aligns with the assumptions and data characteristics at hand (e.g., linearity, non-Gaussianity, or the presence/absence of latent variables), the same penalty-based framework for suppressing multicollinearity can be applied. This flexibility ensures that the overall method can adapt to diverse causal discovery paradigms while retaining its ability to derive signed GFI and class-specific interpretations.

In the global-vs-local perspective, conventional explainers such as SHAP, KernelSHAP, and LIME first compute local feature importance for individual samples and then aggregate them (e.g., via averaging) to approximate global relevance. In contrast, AIME and its extension AutoCausalAIME directly solve for a global inverse operator $A^{\dagger}$ such that $X \approx A^{\dagger}Y$, , thereby providing feature importance without relying on local-to-global averaging. This fundamental difference renders a direct benchmark with SHAP/LIME theoretically unfair; therefore, we restrict our quantitative comparison to methods that also derive global importance in a single step, namely permutation importance (PI) [22] and CAGE [23]. AutoCausalAIME differs fundamentally from PI and CAGE by calculating comprehensive GFI for each class through a single approximate inverse matrix mapping. PI is a sensitivity analysis that sequentially perturbs individual features and measures prediction degradation, incurring inference costs that increase linearly with the number of features d and repetitions R. Additionally, it does not model causal dependencies beyond the base model. CAGE respects causal graphs while mitigating the combination explosion of Shapley values through Monte Carlo approximation; however, its computational complexity still scales with the number of samples S. Additionally, it is theoretically defined as the average of local contributions across all samples; therefore, it does not directly provide class-conditional interpretations. In contrast, the computational complexity of AutoCausalAIME is dominated by regularized matrix operations, ensuring practical scalability even in high-dimensional and multiclass situations without the need to repeat inference iterations. Furthermore, it adopts a modular design that supports easy extension by simply adding regularization terms such as causal penalties or fairness constraints to the objective function, enabling the incorporation of new domain knowledge or constraints with minimal implementation effort. Therefore, AutoCausalAIME demonstrates superiority in three aspects: computational efficiency, granularity of explanation, and extensibility. From the perspective of accurately and efficiently visualizing class-dependent causal feature contributions, it offers a more practical method than existing GFI estimation techniques.

3 Method

3.1 Overview of PC Algorithm

We employed the PC algorithm [7] to estimate the causal structure among the features. Specifically, the following steps are used to first build an undirected graph called the "skeleton" and then determine the directions of the edges to derive a causal graph (a directed acyclic graph, DAG):

Skeleton Initialization. Suppose there are d features. Initially, we consider a complete graph among these d nodes (each feature is a node, and every pair of nodes is connected by an undirected edge).

Conditional Independence Testing Based on Partial Correlations. To construct the skeleton, we check whether node i and node j are conditionally independent given some set of conditioning variables S. If deemed independent, the edges between them are removed. Specifically, the conditional independence test uses partial correlation computed by the function

$$partial\_correlation(x, y, Z).$$

This function regresses x and y onto Z, calculates the correlation between the resulting residuals, and returns the correlation coefficient along with its p-value. If the obtained p-value exceeds the significance level α (e.g., 0.05), i.e., p-value $> \alpha$, x and y are considered independent, and the edge is deleted. This test is repeated up to the maximum order, max_k. . Thus, by incrementally increasing the size of the conditioning set (from zero variables up to max_k), edges that can be definitively removed are identified.

V-Structure Detection and Edge Orientation. After the skeleton is constructed, consider three nodes (i, j, k) such that edges $i - j$ and $j - k$ exist, but $i - k$ does not. If j is not part of the separating set for (i, k), then the V-structure $i \rightarrow j \leftarrow k$ is established (i.e., both arrows converge on j). Subsequently, by sequentially applying additional orientation rules starting from this V-structure, directions are assigned to all edges, resulting in a DAG.

Thus, the PC algorithm iteratively checks the conditional independence based on partial correlation tests to produce a DAG. In our proposed method, the estimated causal structure (including directed and undirected edges) is used within CausalAIME's penalty term to suppress multicollinearity and adjust the feature-importance coefficients. While the present code implements the PC algorithm illustratively, other causal inference methods (e.g., GES [18], FCI [19], LiNGAM [20], or Bayesian structure learning [21]) could be substituted if desired.

3.2 Overview of AIME

AIME [4] provides an approximate inverse operator $A^\dagger$ that maps from the output/target space to the feature space, thereby yielding GFI. Let $X \in \mathbb{R}^{m \times N}$ be the feature matrix (where m is the number of features and N the number of samples) and $Y \in \mathbb{R}^{n \times N}$ be the corresponding output or target matrix (with n output dimensions and the same N samples). We solve the following least-squares problem:

$$\min_{A^\dagger \in \mathbb{R}^{m \times n}} \left\| X - A^\dagger Y_F^2 \right\|, \tag{1}$$

where $\|\cdot\|_F^2$ denotes the Frobenius norm. Intuitively, this objective attempt to reconstruct X as a linear transformation of Y. The matrix $A^\dagger$, which we call the approximate inverse operator, is then interpreted as mapping from the n -dimensional output space back to the m -dimensional feature space.

If YY^T is invertible (Y^T represents the transformation matrix of Y), the closed-form solution is

$$A^\dagger = XY^T \left(YY^T \right)^{-1}. \tag{2}$$

In practice, pseudo-inverse or numerical optimization can be used.

As $A^\dagger$ has dimensions $m \times n$, its j-th row corresponds to feature i, and its j-th column corresponds to output dimension j. The entry $(A^\dagger)_{i,j}$ thus indicates how a change in the j-th output (or class) dimension influences the i-th feature. A large positive value suggests that increasing output dimension j is associated with an increase in feature i, whereas a large negative value indicates a suppressive relationship. Interpreted in a global sense, each column of $A^\dagger$ reveals how all m features relate—positively or negatively—to that particular output dimension.

Furthermore, the AIME can operate in two modes:

- **Model-Dependent Mode.** Uses $Y = \widehat{Y} = f(X)$, the predictions from a black-box model f. This highlights how the model itself transforms outputs into features, revealing its internal "importance" structure.
- **Data-Driven Mode.** Uses Y as the ground-truth labels or target values, Y_{true}. This yields feature importance, reflecting the genuine associations inherent in the dataset rather than the learned behavior of the model.

Although AIME elucidates these global patterns, it does not inherently account for multicollinearity or feature interdependencies. In the next section, we introduce CausalAIME, which incorporates penalties derived from causal structure to stabilize $A^\dagger$ and mitigate multicollinearity.

3.3 Overview of CausalAIME

CausalAIME [6] extends AIME [4] by introducing a causal inference approach, such as the PC algorithm [7], and incorporating the resulting penalty terms into the objective function. This approach enables the suppression of multicollinearity and derivation of feature importance, which accounts for inter-feature dependencies. Specifically, the straightforward least-squares formulation in AIME [4] may produce an unstable solution when strong correlations (multicollinearity) or latent causal relationships exist among the features; for instance, by causing excessively large coefficients for a particular feature. To address this issue, CausalAIME leverages a causal graph (comprising directed edges E_{dir} and undirected edges E_{undir}) estimated from a causal discovery method such as the PC algorithm [7]. Then, it solves an augmented least-squares problem with additional penalty terms.

When the PC algorithm yields a directed edge $i \rightarrow j$, a "difference penalty" is imposed to ensure that the child node j's contribution coefficient does not diverge too far from that of the parent node i. Concretely, for each output dimension k,

$$\left(A^\dagger_{j,k} - \alpha A^\dagger_{i,k} \right) \tag{3}$$

is added to the penalty, where α is a ratio indicating how strongly the child's coefficient should refer to the parents. If the PC algorithm retains an undirected edge $u - v$, it imposes a bidirectional difference penalty to discourage extreme divergences in the contribution coefficients of features u and v. Thus, the penalty controls the degree to which correlated or causally linked features can "spread out" or produce large discrepancies in their coefficients.

Considering these, CausalAIME solves the following extended optimization problem:

$$\min_{A^\dagger \in \mathbb{R}^{m \times n}} \left\| X - A^\dagger Y_F^2 \right\| + \lambda \times P(A^\dagger, E_{dir}, E_{undir}) \tag{4}$$

where λ is a hyperparameter controlling the strength of the penalty, and $P(A^\dagger, E_{dir}, E_{undir})$ is the sum of difference penalties based on directed and undirected edges. Specifically:

- For a directed edge $(i \rightarrow j) \in E_{dir}$, add

$$\sum_{k=1}^{n} \left(A_{j,k}^{\dagger} - \alpha A_{i,k}^{\dagger}\right)^{2}. \tag{5}$$

- For an undirected edge $(u - v) \in E_{undir}$, add

$$\sum_{k=1}^{n} \left[\left(A_{v,k}^{\dagger} - \alpha A_{u,k}^{\dagger}\right)^{2} + \left(A_{u,k}^{\dagger} - \alpha A_{v,k}^{\dagger}\right)^{2}\right]. \tag{6}$$

These difference penalties incorporate the structure of the causal graph into the estimation of $A^{\dagger}$, effectively mitigating extreme coefficient biases among features exhibiting strong correlations or causal relationships.

CausalAIME has two primary hyperparameters: the penalty strength (λ) and the causal ratio (α). A larger λ enforces stronger suppression of multicollinearity but may increase the reconstruction error, whereas α controls "how much the child node's coefficient should refer to its parent node's coefficients." Rather than adjusting these hyperparameters manually, this study proposes to search for them automatically using cross-validation and CMA-ES, yielding the AutoCausalAIME framework.

3.4 AutoCausalAIME

We treat λ and α in CausalAIME as hyperparameters to be optimized via cross-validation, and employ the CMA-ES [8, 9] to perform this search automatically. Concretely, let our data be divided into K folds; in fold k, we designate a training set *traink* and a held-out test set *testk*. For a given pair (λ, α), we train (i.e., solve the penalized least-squares problem in CausalAIME) on each *traink* to obtain a corresponding approximate inverse operator.

$$A^{\dagger}(\lambda, \alpha)^{k} = \arg\min_{A^{\dagger}}\left[X^{(j)}train - A^{\dagger}Y^{(j)}train_{F}^{2} + \lambda \times P(A^{\dagger}, E_{dir}, E_{undir}, \alpha)\right],$$

where $X^{(j)}train$ and $Y^{(j)}train$ collect the training samples (features and labels) in fold k, and $P(A^{\dagger}, E_{dir}, E_{undir}, \alpha)$ is the penalty term defined by the directed and undirected edges from the PC algorithm's causal structure (including α for parent–child referencing). Then, we measure the mean-squared reconstruction error on the test set *testk*, denoted by $MSE^{(k)}(\lambda, \alpha)$. More explicitly, if the test set has $|testk|$ samples indexed by $i \in testk$, each sample having a feature vector x_i and label vector y_i, we define

$$MSE^{(k)}(\lambda, \alpha) = \frac{1}{|testk|} \sum_{i \in testk} \left\|x_i - A^{\dagger}(\lambda, \alpha)^{k}y_i\right\|_{2}^{2}.$$

We then average these mean squared errors (MSEs) over the K folds to obtain our evaluation function

$$CV - MSE(\lambda, \alpha) = \frac{1}{K} \sum_{k=1}^{K} MSE^{(k)}(\lambda, \alpha).$$

To identify (λ, α) that minimize $CV - MSE$, we apply CMA-ES in the two-dimensional space of (λ, α). CMA-ES initializes a mean vector, typically chosen as a guess around $\lambda \approx 0.5$ and $\alpha \approx 0.9$ (or another prior), along with an initial covariance matrix or step size. In each generation, it samples a population of candidate pairs $\{(\lambda, \alpha)\}$ from a multivariate Gaussian, evaluates $CV - MSE(\lambda, \alpha)$ using the above procedure, ranks the candidates, and then updates its mean vector and covariance by emphasizing the top-performing portion of the population. Iterating these steps converges toward a (λ, α) that yield low cross-validated reconstruction error and thereby provide stable GFI for CausalAIME. Once CMA-ES terminates, we fix $\lambda = \lambda$ and $\alpha = \alpha$, retrain on the entire dataset if desired, and thereby obtain an approximate inverse operator $A^{\dagger}$ that (1) suppresses multicollinearity effectively via the causal-structure-based penalty and (2) requires no manual tuning of hyperparameters. Although CMA-ES is used here owing to its well-documented robustness and efficiency in multimodal search, other methods—such as Bayesian optimization, gradient-free evolutionary algorithms like Differential Evolution, or grid/random search—can likewise be employed to minimize $CV - MSE(\lambda, \alpha)$ under computational constraints.

4 Experiment

In this section, we present the experimental setup and the results. Specifically, we compare the following three approaches: AIME and two versions of AutoCausalAIME obtained via CMA-ES parameter exploration—one with the highest (worst) *MSE* and another with the lowest (best) *MSE*. For each approach, we evaluated six metrics proposed by Arreche et al. [10] (descriptive accuracy, sparsity, stability, efficiency, robustness, and completeness). We then conducted a Wilcoxon signed-rank test [24] to determine whether there were any statistically significant differences among the three methods.

4.1 Experimental Environment

In our experiments, we used the Wisconsin Breast Cancer (diagnostic) dataset [25], which is a medium-sized tabular public dataset that quantifies nuclear morphology features from fine-needle aspiration images of breast tumors as 30-dimensional continuous values, including 357 benign cases and 212 malignant cases (total of 569 cases). The sample size and number of features are neither excessively large nor small, reproducing the "thousands of cases and dozens of features" conditions commonly encountered in medical diagnostic tasks. This makes the dataset an effective benchmark where model complexity and computational requirements are less influenced by data scale. Furthermore, since its release in 1993, this dataset has been used as a standard evaluation framework in machine learning and XAI and has been frequently cited in recent research on explainable deep learning and SHAP applications, establishing international consensus for reproducibility and comparability. Furthermore, it has a benign/malignant ratio of approximately 2:1, which is moderately imbalanced, and contains numerous highly correlated morphological features; hence, its statistical diversity is suitable for verifying feature importance estimation methods that consider multicollinearity and class dependency (e.g., AutoCausalAIME). In summary, this dataset is considered a "typical" diagnostic tabular benchmark in terms of public availability, scale, structure, and literature accumulation and is therefore an appropriate evaluation target. All implementations were conducted using Python, version 3.11.11. For AIME, we employed the publicly available library aime-xai 0.1 [26, 27]. CausalAIME was developed using NumPy 1.26.4, pandas 2.2.2, networkx 3.4.2, matplotlib 3.10.0, and seaborn 0.13.2. Additionally, AutoCausalAIME, which requires CMA-ES, was implemented using cma 4.0.0. All experiments were performed using Google Colaboratory Pro +, and the results are presented below.

4.2 Experimental Result

Figure 1 shows the exploration results for λ and α obtained using CMA-ES. $\lambda = 0.1040714257155052$ and $\alpha = 0.08123420618462901$ yield the smallest MSE ($MSE = 0.7430990191409051$), representing the best condition. In the following experiments, we refer to this parameter setting as "AutoCausalAIME (best)." Fig. 1 also indicates that around $\lambda \approx 1.3$ and $\alpha \approx 1.3$, the MSE becomes largest (worst condition). Therefore, we designate this hyperparameter setting as "AutoCausalAIME (worst)."

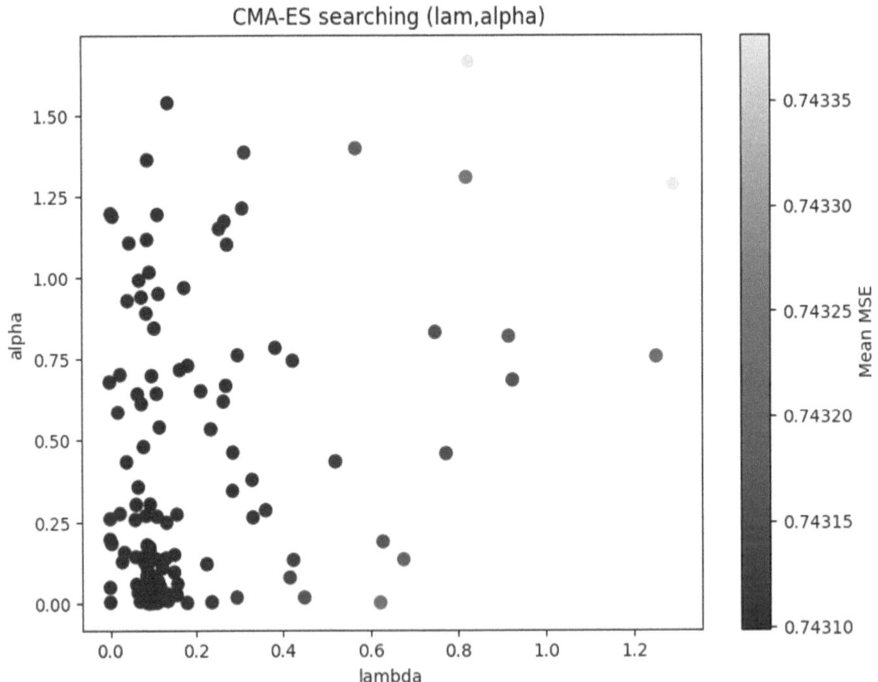

Fig. 1. Results of CMA-ES search for λ and α.

Table 1 compares the three explanation methods—AIME, AutoCausalAIME (best), and AutoCausalAIME (worst)—using the six metrics proposed by Arreche et al. (descriptive accuracy, sparsity, stability, efficiency, robustness, and completeness), while Table 2 summarizes the results of the Wilcoxon signed-rank test at a 5% significance level. Specifically, descriptive accuracy indicates the appropriateness of a feature's importance based on the extent of decline of the model performance when that feature is removed; the higher this value, the more essential the selected top features are. Sparsity measures the degree to which a small number of top features account for the total importance; a higher value suggests that importance is more heavily concentrated in only a few features. Stability refers to the extent to which the top features remain consistent when the process is repeated under the same conditions, with a value of 1.0000 meaning perfect consistency. Efficiency (in seconds) is the computation time required to produce the explanation; a smaller value corresponds to a higher speed. Robustness (Bias/Adv) indicates the susceptibility of the method to biased models or adversarial noise; a value of 0.00 indicates that it is not fooled at all, whereas values closer to 1.00 indicate that it is easily misled. Completeness is the proportion of cases in which manipulating the top features does not change the final model prediction. A value of 0.0000 suggests that altering the top features immediately changes the prediction, signifying that they are essential.

In actual measurements, AutoCausalAIME (best) maintained a descriptive accuracy of approximately 0.91, which is comparable to AIME's high explanatory power. Although its efficiency is slower at 6.97 s compared with AIME's 0.01 s, it is still slightly faster than AutoCausalAIME (worst) at 7.80 s. Robustness shows a moderately balanced resilience of 0.50/0.50 against biased models and adversarial attacks. AIME, which does not explicitly address multicollinearity or causal structures, has a considerably high resistance (0.00) to biased models and extremely short computation time (0.01 s) but cannot visualize causal structures or mitigate multicollinearity; it might be more easily tricked (0.50) by adversarial attacks. According to the Wilcoxon signed-rank test results, no statistically significant differences were found at the 5% level for any of the comparisons: AutoCausalAIME (best) versus AIME, AutoCausalAIME (best) versus AutoCausalAIME (worst), and AIME versus AutoCausalAIME (worst), each with a $p > 0.05$. However, "no significant difference" in this test does not necessarily imply equivalent performance; the sample size, variance in metrics, and other factors can reduce the statistical power to detect differences. AutoCausalAIME (best) preserves high descriptive accuracy and suppresses multicollinearity by incorporating causal-structure-based penalties, delivering slightly better speed and accuracy than AutoCausalAIME (worst). Thus, CMA-ES optimization is effective to some extent. Consequently, although no statistically significant differences were found, larger datasets or different application environments in the future may further underscore the effectiveness of AutoCausalAIME (best) for high-accuracy causal visualization. In medicine and finance, where multicollinearity and causal reasoning are crucial, this method merits consideration.

Table 1. Comparison of AIME, AutoCausalAIME (worst), and AutoCausalAIME(best) on the Breast Cancer Wisconsin Dataset. Min–max ranges are shown with the average value in parentheses. Robustness was evaluated under Bias/Adv conditions. Completeness is the proportion of instances in which altering the top features failed to change the prediction.

	Desc. Acc	Sparsity	Stability	Efficiency (s)	Rob. (B/A)	Compl
AIME	0.55–1.00, 0.91	0.41–0.41, 0.41	1.0000	0.01	0.00 / 0.50	0.0000
AutoCausalAIME(worst)	0.55–1.00, 0.88	0.40–0.40, 0.40	1.0000	7.80	0.50 / 0.50	0.0000
AutoCausalAIME(best)	0.60–1.00, 0.91	0.40–0.41, 0.40	1.0000	6.97	0.50 / 0.50	0.0000

Table 2. Wilcoxon signed-rank test results for pairwise comparisons of GFI.

	W-stat	p-value	Significance (5%)
AIME vs. AutoCausalAIME(best)	741.0	0.277077	not significant
AutoCausalAIME(best) vs. AutoCausalAIME(worst)	801.0	0.316211	not significant
AIME vs. AutoCausalAIME(worst)	723.0	0.372144	not significant

5   Conclusion

We proposed AutoCausalAIME, an extension of CausalAIME that integrates a penalty term and causal structure estimation (via the PC algorithm) into the approximate inverse operator framework for black-box models and further incorporates CMA-ES for automatic hyperparameter optimization. The goal was to derive the GFI while achieving both the mitigation of multicollinearity and the visualization of causal relationships. Specifically, we constructed a framework wherein a difference penalty based on the directed/undirected edges estimated by the PC algorithm is imposed on AIME's approximate inverse operator. This ensures that features with strong correlations or causal dependencies can be handled without causing an excessive bias in their coefficients. Moreover, by automatically searching for hyperparameters, such as λ and α, using CMA-ES, AutoCausalAIME eliminates the need for manual trial and error and can potentially provide more stable feature importance.

In experiments using the Wisconsin Breast Cancer (diagnostic) dataset, we compared AIME with CausalAIME (worst case/best case) and evaluated them using the six indices proposed by Arreche et al., along with the Wilcoxon signed-rank test. Although the results showed no statistically significant difference, AutoCausalAIME (best) achieved a balance of robust multicollinearity suppression, reflecting causal structures, while maintaining a descriptive accuracy comparable to that of AIME. Thus, hyperparameter tuning via CMA-ES functions effectively.

Future work will clarify the effectiveness of AutoCausalAIME by conducting experiments on larger datasets and across broader application domains as well as re-examining statistical power and evaluation metrics. Additionally, it may be possible to further enhance the generality and explanatory power of this method by incorporating other causal inference algorithms, such as LiNGAM or FCI (in place of the PC algorithm), and by improving the flexibility to switch between model-dependent and data-driven modes.

Acknowledgments. In the preparation of this manuscript, we used language proofing and translation tools, including ChatGPT, Open PaperPal, and DeepL, to improve the clarity and readability of the English text. These tools have assisted in language refinement. The full responsibility for the accuracy and integrity of the final content remains with the authors. Any errors, oversights, or misinterpretations are the responsibility of the authors.

Disclosure of Interests. The authors have no competing interests to declare relevant to the content of this article. All aspects of the research, including data collection, analysis, and presentation of the results, were conducted independently and impartially, without external influence or financial support that could affect the integrity of the findings.

References

1. Mahto, M.K.: Explainable artificial intelligence: fundamentals, approaches, challenges, XAI evaluation, and validation. In: Malik, K., Sharma, M., Deswal, S., Gupta, U., Agarwal, D., Al-Shamsi, Y.O.B. (eds.) Explainable Artificial Intelligence for Autonomous Vehicles, pp. 25–49. CRC Press (2025)
2. Ngueajio, M., Aryal, S., Atemkeng, M., Washington, G., Rawat, D.: Decoding fake news and hate speech: a survey of explainable AI techniques. ACM Comput. Surv. **57**(1), 1 (2025). https://doi.org/10.1145/3711123
3. Lundberg, S.M., Lee, S.I.: A unified approach to interpreting model predictions. In: Proceedings of the 31st International Conference on Neural Information Processing Systems, pp. 4768–4777 (2017). https://doi.org/10.5555/3295222.3295230
4. Nakanishi, T.: Approximate inverse model explanations (AIME): unveiling local and global insights in machine learning models. IEEE Access **11**, 101020–101044 (2023). https://doi.org/10.1109/ACCESS.2023.3314336
5. Ribeiro, M.T., Singh, S., Guestrin, C.: "Why should I trust you?": Explaining the predictions of any classifier. In: Proceedings of the 22nd ACM SIGKDD International Conference on Knowledge Discovery and Data Mining, pp. 1135–1144 (2016). https://doi.org/10.1145/2939672.2939778
6. Nakanishi, T.: CausalAIME: leveraging peter-clark algorithms and inverse modeling for unified global feature explanation in healthcare. In: Proceedings of the 3rd World Conference on eXplainable Artificial Intelligence (2025, under review)
7. Spirtes, P., Glymour, C.: An algorithm for fast recovery of sparse causal graphs. Soc. Sci. Comput. Rev. **9**(1), 62–72 (1991). https://doi.org/10.1177/089443939100900106
8. Hansen, N., Ostermeier, A.: Adapting arbitrary normal mutation distributions in evolution strategies: the covariance matrix adaptation. In: Proceedings of the 1996 IEEE International Conference on Evolutionary Computation, pp. 312–317. IEEE Press, Nagoya, Japan (1996). https://doi.org/10.1109/ICEC.1996.542381

9. Hansen, N., Ostermeier, A.: Completely derandomized self-adaptation in evolution strategies. Evolut. Comput. **9**(2), 159–195 (2001). https://doi.org/10.1162/106365601750190398

10. Arreche, O., Guntur, T.R., Roberts, J.W., Abdallah, M.: E-XAI: Evaluating black-box explainable AI frameworks for network intrusion detection. IEEE Access **12**, 23954–23988 (2024). https://doi.org/10.1109/ACCESS.2024.3365140

11. Takahashi, D., Shimizu, S., Tanaka, T.: Counterfactual explanations of black-box machine learning models using causal discovery with applications to credit rating. In: 2024 International Joint Conference on Neural Networks (IJCNN), pp. 1–8 (2024). https://doi.org/10.1109/IJCNN60899.2024.10650130

12. Tyrovolas, M., Kallimanis, N., Stylios, C.: Advancing explainable AI with causal analysis in large-scale fuzzy cognitive maps. ArXiv, abs/2405.09190 (2024). https://doi.org/10.48550/arXiv.2405.09190

13. Cinquini, M., Guidotti, R.: CALIME: causality-aware local interpretable model-agnostic explanations. In: Proceedings of Explainable Artificial Intelligence. xAI 2024. Communications in Computer and Information Science, vol. 2155. Springer, Cham (2024). https://doi.org/10.1007/978-3-031-63800-8_6

14. Termine, A., Antonucci, A., Facchini, A.: Machine learning explanations by surrogate causal models (MaLESCaMo). In: Proceedings of xAI (Late-breaking Work, Demos, Doctoral Consortium), pp. 59–64 (2023)

15. Zecevic, M., Dhami, D., Rothkopf, C., Kersting, K.: XAI establishes a common ground between machine learning and causality. In: Proceedings of xAI (Late-breaking Work, Demos, Doctoral Consortium) (2022)

16. Wang, J., Wiens, J., Lundberg, S.: Shapley flow: a graph-based approach to interpreting model predictions. In: Proceedings of the 24th International Conference on Artificial Intelligence and Statistics, PMLR, vol. 130, pp. 721–729 (2021)

17. Breuer, N.O., Sauter, A., Mohammadi, M., Acar, E.: CAGE: causality-aware shapley value for global explanations. In: Proceedings of Explainable Artificial Intelligence, xAI 2024, Communications in Computer and Information Science, vol. 2155. Springer, Cham (2024). https://doi.org/10.1007/978-3-031-63800-8_8

18. Chickering, D.M.: Optimal structure identification with greedy search. J. Mach. Learn. Res. **3**, 507–554 (2002)

19. Spirtes, P., Meek, C., Richardson, T.: Causal inference in the presence of latent variables and selection bias. In: Besnard, P., Hanks, S. (eds.) Proceedings of the Eleventh Conference on Uncertainty in Artificial Intelligence, pp. 499–506. Morgan Kaufmann, San Francisco (1995)

20. Shimizu, S., Hoyer, P.O., Hyvärinen, A., Kerminen, A.: A linear non-Gaussian acyclic model for causal discovery. J. Mach. Learn. Res. **7**, 2003–2030 (2006)

21. Koller, D., Friedman, N.: Probabilistic Graphical Models: Principles and Techniques. MIT Press, Cambridge (2009)

22. Breiman, L.: Random forests. Mach. Learn. **45**(1), 5–32 (2001)

23. Breuer, N.O., Sauter, A., Mohammadi, M., Acar, E.: CAGE: causality-aware Shapley value for global explanations. arXiv preprint arXiv:2404.11208 (2024)

24. Wilcoxon, F.: Individual comparisons by ranking methods. Biometrics Bull. **1**(6), 80–83 (1945)

25. Wolberg, W., Mangasarian, O., Street, N., Street, W.: Breast Cancer Wisconsin (Diagnostic). UCI Mach. Learn. Repository (1993). https://doi.org/10.24432/C5DW2B

26. AIME: Approximate Inverse Model Explanations,

27. https://github.com/ntakafumi/aime/. Accessed 3 Feb 2024

28. https://pypi.org/project/aime-xai/. Accessed 3 Feb 2024

Leveraging Large Language Models for Generating and Evaluating Natural Language Explanations in XAI: A Comparative Study

Renato Okabayashi Miyaji[(⊠)] and Pedro Luiz Pizzigatti Corrêa

Escola Politécnica, Universidade de São Paulo, Av. Prof. Luciano Gualberto, 380, Butantã, São Paulo - SP, 05508-010, Brazil
{re.miyaji,pedro.correa}@usp.br

Abstract. Making machine learning (ML) model predictions understandable to diverse users is a critical challenge in Explainable Artificial Intelligence (XAI). While XAI techniques such as SHAP and LIME provide feature attributions, their outputs often require technical expertise to interpret. This study investigates the use of Large Language Models (LLMs) to bridge this gap by generating accessible natural language explanations from technical XAI outputs and to evaluate the quality of these explanations. We conduct a comparative analysis employing two distinct LLMs (OpenAI's GPT-4o and o4-mini) and two prompting strategies (Zero-Shot and Few-Shot learning) for explanation generation using the TEXEN dataset. The generated explanations are evaluated using traditional NLP metrics (BLEU, ROUGE, METEOR, BERTScore) against human-authored references, and through an *LLM-as-a-Judge* approach (Prometheus 2) assessing Soundness, Completeness, Fluency, and Context-Awareness. Results indicate that while GPT-4o with Few-Shot prompting achieved higher scores on traditional NLP metrics (e.g., BERTScore F1 of 0.858), the *LLM-as-a-Judge* evaluation revealed that the smaller o4-mini model, particularly with Few-Shot prompting, often matched or surpassed GPT-4o in human-aligned qualitative criteria, achieving, for instance, a fluency score of 4.30 and completeness of 4.10. These findings highlight the nuanced capabilities of different LLMs and the importance of multifaceted evaluation frameworks for developing truly interpretable XAI systems.

Keywords: Explainable Artificial Intelligence · Evaluation · LLM-as-a-Judge

1 Introduction

Making machine learning (ML) model predictions understandable and interpretable to humans is a longstanding and complex challenge in the field [1]. As high-performing models have become increasingly complex—often referred to as *black-box* models, such as Deep Neural Networks (DNN)—the need to explain their decisions has gained significant attention in the literature [2]. Understanding why a model makes a certain prediction is essential not only for user trust but also for debugging, fairness, and ethical deployment in real-world applications [3].

© The Author(s), under exclusive license to Springer Nature Switzerland AG 2026
F. Marcelloni et al. (Eds.): IJCCI 2025, CCIS 2829, pp. 17–29, 2026.
https://doi.org/10.1007/978-3-032-15638-9_2

To address this problem, the field of Explainable Artificial Intelligence (XAI) has emerged with the goal of making model behavior more transparent to users [2]. Among the most widely adopted techniques are SHAP (SHapley Additive exPlanations) [4] and LIME (Local Interpretable Model-agnostic Explanations) [5]. SHAP provides feature attributions by estimating the contribution of each input feature to the model's prediction based on cooperative game theory [4]. LIME, on the other hand, builds a local surrogate model around a specific prediction to approximate the behavior of the original model and explain individual decisions [5].

Despite their popularity, these techniques still suffer from a major limitation: interpreting their outputs requires a certain level of technical knowledge. Users must understand how the underlying XAI method works and be aware of its limitations in order to correctly interpret the explanations [2]. This creates a barrier for domain experts and stakeholders who may not have expertise in machine learning but still need to understand and trust model outputs [3].

To overcome this limitation, recent research has explored the use of Language Models (LMs) to generate natural language explanations based on XAI outputs [6]. These efforts range from developing rule-based explanations [7,8] and fine-tuning models, such as T5 and BART [9], to employing prompt engineering strategies with Large Language Models (LLMs) to translate technical explanations into more accessible narratives [10]. While these methods have shown promise, many open questions remain, particularly around their reliability and effectiveness [6].

One of the most pressing challenges is how to properly evaluate the natural language explanations generated by LMs. This includes both defining appropriate evaluation metrics and determining suitable methods to assess explanation quality [11]. Without reliable evaluation frameworks, it is difficult to measure progress or compare different approaches objectively.

In this work, we investigate the viability of using LLMs to generate and evaluate natural language explanations of ML predictions by adopting the *LLM as a Judge* approach [12]. We compare this with traditional NLP metrics to assess whether automatic evaluation can adequately capture the quality of explanations when compared to human ground truths.

The main contributions of this study are threefold: (1) we evaluate different LLMs (OpenAI's GPT-4o [13] and o4-mini [14]) to generate natural language explanations from the outputs of various XAI methods, comparing Zero-Shot and In-Context Learning with Few-Shot approaches [15]; (2) we evaluate the effectiveness of traditional NLP metrics in assessing explanation quality against human-authored references; and (3) we explore the feasibility of using *LLM-as-a-Judge* approach with state-of-the-art evaluation methods, aligned with criteria established in the literature.

2 Related Works

2.1 Explainable Artificial Intelligence (XAI)

The growing adoption of complex *black-box* models—such as Deep Neural Networks, and large-scale transformers—has raised significant concerns about their interpretability and transparency [1]. These models often achieve state-of-the-art performance but at the cost of being difficult for humans to understand. In response, the research community has proposed several taxonomies and frameworks to better categorize and assess the explainability of machine learning models [2]. These include distinctions between model-specific and model-agnostic approaches, local and global explanations, and post-hoc versus intrinsic interpretability. Such classifications help guide the selection of appropriate techniques depending on the model type, domain constraints, and user needs [2].

Explainable Artificial Intelligence (XAI) emerged as a dedicated area of study aiming to bridge the gap between model performance and interpretability [16]. XAI techniques are classified according to several key dimensions: the explanation scope, the type of explanation, the level of user expertise, the explanation target, and the stage of application [17]. The explanation scope distinguishes between global explanations, which aim to provide an overview of the entire model behavior, and local explanations, which clarify individual predictions [17]. The type of explanation refers to whether the method is intrinsic—where interpretability is built into the model—or post hoc, where explanations are generated after the model is trained [17]. The taxonomy also considers the intended user expertise, differentiating between explanations suited for experts, data scientists, or lay users. Another axis is the explanation target, which can focus on the model, the data, or the individual predictions. Finally, the taxonomy incorporates the application stage, indicating whether the technique is used during model training, validation, or deployment. This multifaceted taxonomy enables a structured understanding of the diverse landscape of XAI methods and facilitates selecting appropriate techniques based on the context and user needs [17].

Among the most prominent XAI techniques is SHapley Additive exPlanations (SHAP) [4], which provides a unified framework for feature attribution based on Shapley values from cooperative game theory. SHAP assigns an importance value to each feature by considering all possible combinations of features and their marginal contributions to the prediction. Its key strengths are consistency and local accuracy, making it suitable for a wide range of models and applications [4].

Local Interpretable Model-agnostic Explanations (LIME) [5] is another widely adopted XAI technique that focuses on explaining individual predictions by learning a local surrogate model around the input instance. This surrogate is usually a simple linear model or decision tree trained on perturbed samples of the input, weighted by their similarity to the original instance. LIME is valued for its flexibility and ease of interpretation, although its reliance on local approximations can sometimes lead to unstable or misleading explanations [5].

2.2 Explainable Artificial Intelligence (XAI) and Large Language Models (LLMs)

To make machine learning (ML) models truly understandable to end-users, recent literature has proposed leveraging language models (LMs) to generate natural language explanations for model predictions [9]. This approach aims to bridge the gap between complex model outputs and human interpretability, facilitating better trust and usability in AI systems.

Early efforts in this domain involved fine-tuning models, such as T5 and BART, using expert-generated explanations [9]. While these methods demonstrated the potential of LMs in producing coherent explanations, they often lacked flexibility, as they were tailored to specific datasets or tasks. Consequently, their applicability across diverse scenarios was limited [6].

More recent studies have shifted towards utilizing Large Language Models (LLMs) for this task [18], capitalizing on their advanced reasoning capabilities and adaptability. However, several open challenges persist that influence the performance of LMs in generating effective explanations. Building on this trend, recent research has outlined four key directions to address the open challenges associated with using LLMs for explainability [6].

The first direction emphasizes the development of rigorous evaluation metrics tailored to natural language explanations. Traditional metrics such as completeness and soundness remain relevant, but new dimensions—such as fluency, and the ability to incorporate contextual knowledge—are crucial to assess the quality and usefulness of narrative explanations produced by LLMs [6, 11].

A second direction focuses on prompt engineering as a critical factor in explanation quality. By systematically varying prompt templates the literature aims to understand how LLMs respond to different instructions and how robustly they can generate coherent and accurate explanations in Zero-Shot or Few-Shot settings [6, 18].

The third direction involves comparative analysis of LLMs with different architectures and scales. Preliminary evidence suggests that larger models, such as GPT-4, generally outperform smaller ones in generating sound and context-aware explanations [6]. A systematic evaluation encompassing variations in model's size, type, and design, is necessary to reveal the trade-offs in explanation fidelity, quality, and computational efficiency.

Lastly, enhancing narrative explanations through targeted training and external knowledge integration has shown promise [6]. Fine-tuning LLMs for a specific task can improve their ability to generate high-quality explanations [9]. Additionally, techniques such as Retrieval-Augmented Generation (RAG) can supplement LLM outputs with relevant background information drawn from training data or user-provided knowledge bases [18]. These strategies aim to produce explanations that are not only accurate and fluent, but also rich in insight and aligned with stakeholders' informational needs.

In summary, we observe that there remains substantial room in the literature to improve the performance of LLMs in generating explanations from Machine Learning model predictions. In particular, it is essential to investigate the impact of using different model architectures and sizes, as well as to advance prompt design strategies and the application of In-Context Learning techniques. Most critically, there is a pressing need to enhance the evaluation process for generated explanations—both in terms of establishing more comprehensive and meaningful evaluation criteria, and in refining assessment methodologies. This includes integrating human evaluation protocols and leveraging emerging automatic methods, such as *LLM-as-a-Judge*. Together, these research efforts can lead to more reliable, informative, and user-aligned explanations in practical XAI applications.

3   Methodology

3.1   Proposed Contributions

The methodological procedure adopted in this study involves three distinct variations within the standard pipeline commonly employed for generating natural language explanations using Language Models (LMs). These variations target different stages of the pipeline to assess the effectiveness and quality of explanations produced for Machine Learning model predictions. The main goal is to investigate how different Large Language Models (LLMs), prompt engineering, and In-Context Learning strategies influence the quality of generated explanations. Moreover, our goal is to define a more robust and scalable evaluation method for the generated explanations.

The standard pipeline for generating natural language explanations typically consists of the following sequential steps: (1) a dataset is used to train or apply a Machine Learning (ML) model; (2) the model generates predictions; (3) an Explainable AI (XAI) method is applied to interpret the predictions; (4) the output of the XAI method is passed as input to a language model; (5) the Language Model (LM) generates a natural language explanation; and finally, (6) the explanation is evaluated either manually or using automatic methods. In this work, we focus on steps 4 and 5, introducing controlled variations to evaluate their impact on explanation quality. Figure 1 presents the pipeline with the proposed contributions.

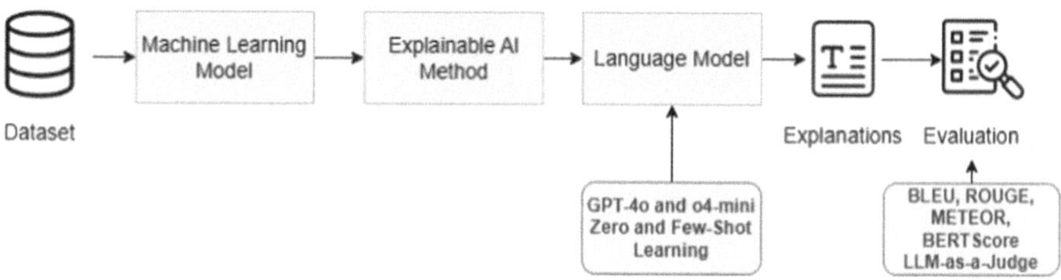

Fig. 1. Pipeline for generating natural language explanations with the proposed contributions.

To investigate the role of the Language Model in the pipeline, we employ two distinct LLMs: GPT-4o [13] and o4-mini [14]. GPT-4o is OpenAI's latest multimodal model designed for high efficiency and faster inference, while maintaining strong performance across reasoning, summarization, and instruction-following tasks. It integrates vision, audio, and text processing, though in this study we focus exclusively on its text capabilities [13]. In contrast, o4-mini is a lightweight model optimized for efficient reasoning in resource-constrained settings. Despite its smaller size, o4-mini is designed to match or exceed reasoning capabilities of larger models in Few-Shot and instruction-based tasks [14].

In addition to evaluating different models, we experiment with two prompt design strategies for generating explanations. The first is a Zero-Shot approach based on the template proposed by Zytek et al. (2024), where the model is provided only with a task description and the XAI output. The second approach uses Few-Shot Learning [15] with five carefully curated examples included in the prompt to guide the model in generating more structured and consistent explanations. These examples were selected to cover diverse reasoning patterns and output structures, aiming to improve the completeness and coherence of generated explanations. Figure 2 presents the prompts used in Zero-shot and Few-shot learning approaches.

Zero-Shot Prompt

Few-Shot Prompt

Fig. 2. Zero-shot and Few-shot learning prompts for generating explanations.

For the evaluation phase, we adopt two complementary strategies. First, we apply traditional Natural Language Processing (NLP) metrics, including BLEU [19], ROUGE [20], METEOR [21], and BERTScore [22]. These metrics are selected to capture a range of textual similarity aspects: BLEU focuses on exact n-gram precision and is traditionally used in machine translation, despite its limitations in handling paraphrastic or semantically equivalent explanations due to its reliance on surface-level overlap. ROUGE complements this by emphasizing recall of n-grams and subsequences, which is useful for identifying missing but expected content. METEOR incorporates stemming and synonym matching, making it more tolerant of lexical variation, which is common in natural explanations. Finally, BERTScore leverages contextual embeddings from pre-trained language models to estimate semantic similarity at a deeper level, addressing some of the shortcomings of purely lexical metrics. Despite their widespread use, these automatic metrics may fall short in capturing higher-order qualities such as factual correctness, explanation coherence, or domain relevance—particularly important in our setting. Therefore, we complement them with human evaluations to provide a more holistic assessment of explanation quality.

To address these limitations, we employ *LLM-as-a-Judge* evaluation, using Prometheus 2 [23], a fine-tuned LLM trained specifically to evaluate natural language outputs. Prometheus 2 has demonstrated state-of-the-art alignment with human judgments in various tasks, showing high correlation with human evaluators across several benchmarks [23]. In our evaluation, we adopt the Direct Assessment method based on the criteria proposed by Zytek et al. (2024), which scores each explanation along four dimensions: Soundness, Completeness, Fluency, and Context-Awareness. Table 1 presents the criteria. Prometheus 2.0 outputs a score for each criterion on a fixed scale, enabling detailed quantitative comparison across models and prompting strategies [23].

Table 1. Evaluation Metrics (Scale: 1–5). Higher scores indicate better performance. Adapted from Zytek et al. (2024).

Metric	1 (Poor)	3 (Fair)	5 (Excellent)
Soundness: Accuracy of Information	Contains significant factual errors or inaccuracies	Includes some misleading or questionable information	Information is completely accurate and free of errors
Fluency: Naturalness of Language	Reads as artificial, confusing, or repetitive; lacks natural flow	The language is understandable but not particularly smooth or engaging	The language is natural, clear, and reads as if written by a fluent human
Completeness: Information Provided	Essential features or contextual details are missing or omitted	All features are mentioned, but specific values or relationships may not be detailed	Provides a comprehensive overview, including all relevant values, relationships, and context
Context-awareness: Depth of Explanation	Provides no relevant external context or explanation	Provides some comparison to typical values or trends	Offers in-depth explanations with appropriate context, clarifying underlying causes

By analyzing performance across these dimensions, our methodology enables a comprehensive understanding of how different LLMs and In-Context Learning affect the generation of explanations in the context of XAI. It also allows us to assess the extent to which traditional NLP metrics correlate with human-aligned evaluation through LLMs. This multifaceted evaluation framework is essential for advancing the development of truly interpretable machine learning systems capable of communicating insights to non-expert users in a trustworthy and accessible manner.

3.2 Dataset

We used the TEXtual Explanation Narratives (TEXEN) dataset [9] to evaluate the proposed variations. More specifically, we employed it to obtain the outputs of the XAI methods, to select examples for Few-Shot Learning, and to retrieve the human-annotated ground truth for comparison with the explanations generated by the LLM. The TEXEN dataset represents a valuable contribution to the field of XAI, addressing the critical need for translating numerical explanations generated by XAI techniques into human-understandable narratives [9]. It pairs outputs from local-level XAI methods, such as LIME [5], SHAP [4], Integrated Gradients (IG) [24], and Layer-Wise Relevance Propagation (LRP) [25], with corresponding textual descriptions crafted by computer science experts. These narratives articulate the key insights and factors influencing a machine learning model's classification decision, making complex feature importance scores accessible to a wider audience [9].

The dataset creation involved a meticulous two-stage process. First, a range of supervised machine learning models—including Decision Trees, Random Forests, Gradient-boosted Machines, and Neural Networks—were trained on diverse tabular datasets drawn from multiple domains such as healthcare, finance, and public policy. Then, XAI techniques were applied to generate feature importance scores for randomly selected test cases, which were visualized as graphs to form the numerical explanations. Subsequently, computer science experts crafted concise and accurate textual narratives based on these graphs. They were tasked with summarizing the model's prediction, highlighting influential features, and explaining the impact of positive and negative contributions [9]. TEXEN facilitates several important tasks in XAI. The primary task is data-to-text generation, where the goal is to automatically produce fluent and accurate textual explanations from the numerical XAI outputs.

To address potential limitations of the dataset size, an augmented version of TEXEN was also created. This augmentation involved re-randomizing the feature and class name placeholders within the original narratives [9]. The augmented version of the dataset is composed of 3,421 records, distributed as follows: train, validation, and test splits contain 3,280, 47, and 94 records, respectively. The TEXEN dataset offers a solid foundation for advancing research in XAI, providing a resource for developing models and algorithms capable of generating high-quality textual explanations. An example of the TEXEN dataset is presented in Fig. 3.

4 Results

The evaluation using traditional Natural Language Processing (NLP) metrics, as detailed in Table 2, reveals distinct performance patterns between the o4-mini and GPT-

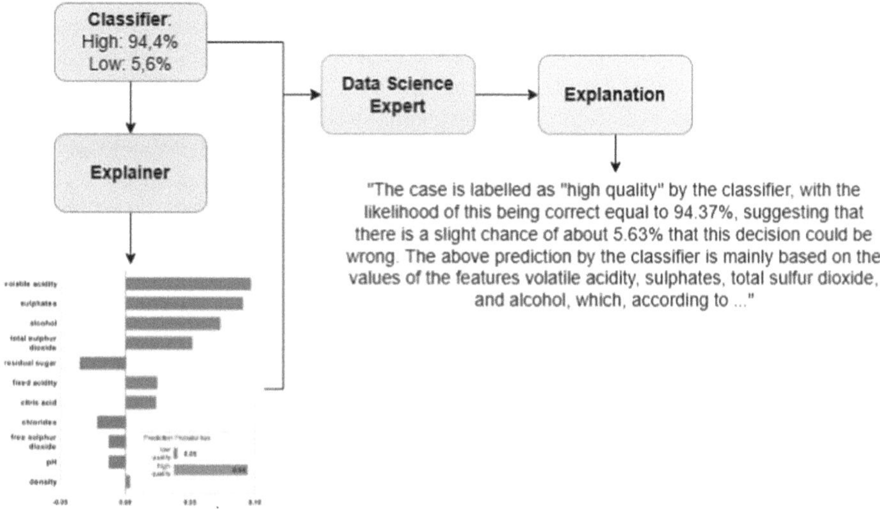

Fig. 3. Example of an explanation provided in TEXEN dataset.

Table 2. Average metrics and standard deviations for each evaluation criterion across models and prompting strategies.

Criterion	o4-mini		GPT-4o	
	Zero-Shot	Few-Shot	Zero-Shot	Few-Shot
BLEU	0.006 ± 0.003	0.007 ± 0.003	0.010 ± 0.009	0.032 ± 0.023
ROUGE-1 (F1)	0.217 ± 0.039	0.253 ± 0.051	0.292 ± 0.042	0.391 ± 0.050
ROUGE-L (F1)	0.126 ± 0.019	0.149 ± 0.024	0.165 ± 0.022	0.211 ± 0.028
METEOR	0.131 ± 0.031	0.148 ± 0.031	0.200 ± 0.035	0.222 ± 0.039
BERTScore (F1)	0.814 ± 0.015	0.827 ± 0.016	0.828 ± 0.015	0.858 ± 0.016

4o models, as well as the impact of Zero-Shot versus Few-Shot prompting strategies. Across all metrics—BLEU, ROUGE-1 (F1), ROUGE-L (F1), METEOR, and BERTScore (F1)—GPT-4o consistently outperformed o4-mini. For instance, in the Few-Shot setting, GPT-4o achieved a ROUGE-1 (F1) score of 0.391, compared to o4-mini's 0.253. Similarly, GPT-4o's BERTScore (F1) reached 0.858 with Few-Shot prompting, surpassing o4-mini's 0.827 under the same condition. This general trend suggests that the larger, more advanced architecture of GPT-4o is more adept at generating explanations that align lexically and semantically with human-authored references.

The introduction of Few-Shot prompting generally led to improved scores for both models compared to their Zero-Shot counterparts. This effect was particularly noticeable for GPT-4o. For example, GPT-4o's BLEU score increased from 0.010 (Zero-Shot) to 0.032 (Few-Shot), and its ROUGE-L (F1) score improved from 0.165 to 0.211. While o4-mini also benefited from Few-Shot examples (e.g., ROUGE-L (F1) increasing from 0.126 to 0.149), the magnitude of improvement was often less pronounced than that observed with GPT-4o. This indicates that providing in-context examples effectively guides the LLMs, especially GPT-4o, towards generating outputs that better match the target style and content of the reference explanations.

BERTScore (F1), which measures semantic similarity, yielded the highest absolute scores among all metrics, underscoring a reasonable degree of semantic alignment between the generated explanations and the ground truth for both models. GPT-4o with Few-Shot prompting achieved the highest BERTScore (0.858), followed by its Zero-Shot performance (0.828). Even o4-mini demonstrated strong semantic similarity, with scores of 0.827 (Few-Shot) and 0.814 (Zero-Shot). The relatively smaller gap in BERTScore improvements from Zero-Shot to Few-Shot, compared to n-gram based metrics, suggests that while Few-Shot prompting refines the output, both models inherently capture significant semantic content even with minimal instruction.

Conversely, BLEU scores remained notably low across all configurations, with the highest score being 0.032 for GPT-4o (Few-Shot). These low values indicate limited exact n-gram overlap between machine-generated explanations and human references. This finding is not entirely unexpected, as natural language explanations often prioritize fluency, rephrasing, and conveying meaning over verbatim replication of specific phrases from a reference set. Therefore, while BLEU is a standard metric, its utility in assessing the nuanced quality of generated explanations for XAI might be limited, as it may penalize valid and fluent rephrasing.

In summary, the traditional NLP metrics suggest that GPT-4o, particularly when employing Few-Shot prompting, produces explanations with greater lexical and semantic similarity to human references compared to o4-mini. The consistent, albeit sometimes modest, improvements with Few-Shot learning highlight its value in guiding LLM outputs. However, the low BLEU scores and the inherent limitations of n-gram based metrics in capturing aspects like factual accuracy or depth of reasoning underscore the importance of complementing these evaluations with qualitative assessments or more semantically-aware metrics, such as the ones measured by BERTScore and the *LLM-as-a-Judge* approach.

Table 3. Average scores and standard deviations for each evaluation criterion across models and prompting strategies.

Criterion	o4-mini		GPT-4o	
	Zero-Shot	Few-Shot	Zero-Shot	Few-Shot
Soundness	4.00 ± 1.08	4.05 ± 0.94	3.85 ± 0.75	3.95 ± 0.89
Completeness	3.90 ± 0.97	4.10 ± 0.55	3.80 ± 0.69	3.75 ± 0.72
Fluency	4.25 ± 0.44	4.30 ± 0.57	3.65 ± 0.49	3.75 ± 0.64
Context-Awareness	3.50 ± 1.10	4.00 ± 0.86	3.60 ± 0.88	3.50 ± 0.69

The evaluation using the *LLM-as-a-Judge* approach (Prometheus 2 [23]), as presented in Table 3, offers insights into the qualitative aspects of the generated explanations across four key criteria: Soundness, Completeness, Fluency, and Context-Awareness [6]. Notably, the o4-mini model demonstrated strong performance, often comparable or even superior to GPT-4o on these human-aligned metrics. For instance, o4-mini achieved the highest average Fluency scores (4.25 for Zero-Shot and 4.30 for Few-Shot), indicating that its generated explanations were perceived as highly natural and clear. In terms of Soundness, o4-mini also scored well, with 4.00 (Zero-Shot) and 4.05 (Few-Shot), suggesting a good level of factual accuracy in its outputs.

When comparing prompting strategies, Few-Shot learning generally enhanced the performance of o4-mini, particularly in Completeness (increasing from 3.90 to 4.10) and Context-Awareness (improving from 3.50 to 4.00). This suggests that providing a few examples helped o4-mini generate more comprehensive and contextually rich explanations. Conversely, the impact of Few-Shot prompting on GPT-4o was more nuanced and, in some cases, slightly detrimental according to the LLM-as-a-Judge. While Soundness (3.85 to 3.95) and Fluency (3.65 to 3.75) saw marginal improvements with Few-Shot for GPT-4o, Completeness slightly decreased (3.80 to 3.75) as did Context-Awareness (3.60 to 3.50).

Across both models, Fluency scores were generally high, indicating that both o4-mini and GPT-4o are capable of producing linguistically coherent text. However, o4-mini consistently outperformed GPT-4o in this dimension. For Soundness, both models performed commendably, with average scores hovering around 3.85 to 4.05, suggesting that the generated explanations were largely accurate. The scores for Completeness and Context-Awareness showed more variability, with o4-mini (Few-Shot) achieving the highest Completeness (4.10) and Context-Awareness (4.00), while GPT-4o (Few-Shot) registered the lowest scores for these criteria among the Few-Shot configurations.

The standard deviations observed in Table 3, while generally moderate, were sometimes notable. This indicates a degree of variability in the *LLM-as-a-Judge's* ratings for certain configurations, which could reflect either inconsistencies in the generated explanations or inherent subjectivity in evaluating these qualitative aspects. Nevertheless, the overall trend suggests that o4-mini, particularly with Few-Shot prompting, can produce explanations that are highly rated on qualitative dimensions crucial for user understanding and trust.

In summary, the *LLM-as-a-Judge* evaluation reveals that the smaller o4-mini model can be highly competitive, and sometimes preferable to the larger GPT-4o, in generating explanations that are sound, complete, fluent, and context-aware. The impact of Few-Shot learning was consistently positive for o4-mini but yielded mixed results for GPT-4o on these qualitative metrics. These findings highlight the importance of evaluating LLM-generated explanations not just with traditional NLP metrics but also with methods that capture human-perceived quality, as these dimensions are critical for the practical utility of XAI systems.

5   Conclusions

This study successfully demonstrated the dual role of LLMs in enhancing XAI, both for generating accessible natural language explanations from technical XAI outputs and for evaluating their quality. Our findings indicate that while larger models such as GPT-4o, particularly with Few-Shot prompting, achieve superior scores on traditional NLP metrics reflecting lexical and semantic similarity to human references, the *LLM-as-a-Judge* approach highlighted the strong performance of the smaller o4-mini model. Specifically, o4-mini, when guided by in-context examples, often matched or surpassed GPT-4o in human-aligned qualitative criteria such as Soundness, Fluency, Completeness, and Context-Awareness, suggesting a potential divergence between traditional metrics and perceived explanation quality. This underscores the value of employing multifaceted evaluation strategies, including *LLM-as-a-Judge*, to gain a more holistic understanding of explanation effectiveness.

Future work should prioritize comprehensive human evaluation studies to directly assess the perceived quality and utility of LLM-generated explanations, and to further validate the *LLM-as-a-Judge* approach. Expanding the investigation to include a broader array of LLM architectures, more sophisticated prompt engineering strategies (e.g., Retrieval-Augmented Generation), and diverse XAI methods and datasets will be crucial. Furthermore, a deeper analysis of the faithfulness, reliability, and errors of these narrative explanations, ensuring they accurately reflect the underlying model's behavior without introducing biases or misinterpretations, remains a critical avenue for research to foster trustworthy and interpretable AI systems.

Acknowledgments and Disclosure of Funding. This study was supported by the Amigos da Poli – project 2025_021. It was also financed in part by the Coordenação de Aperfeiçoamento de Pessoal de Nível Superior – Brasil (CAPES) – Finance Code 001. It was made possible by the Thematic Projects of FAPESP 2017/17047-0 and 2020/15230-5.

References

1. Nguyen, V.B., Schlötterer, J., Seifert, C.: From black boxes to conversations: incorporating XAI in a conversational agent. In: Longo, L. (ed.) xAI 2023. CCIS, vol. 1903, pp. 71–96. Springer, Cham (2023). https://doi.org/10.1007/978-3-031-44070-0_4
2. Arrieta, A.B., et al.: Explainable artificial intelligence (XAI): concepts, taxonomies, opportunities and challenges toward responsible AI. Inf. Fusion **58**(C) (2020)
3. Bhatt, U., et al.: Explainable machine learning in deployment. In: Proceedings of the 2020 Conference on Fairness, Accountability, and Transparency (2020)
4. Lundberg, S.M., Lee, S.: A unified approach to interpreting model predictions. In: Proceedings of the 31st International Conference on Neural Information Processing Systems (2017)
5. Ribeiro, M.T., Singh, S., Guestrin, C.: "Why should I trust you?": Explaining the predictions of any classifier. In: Proceedings of the 22nd ACM SIGKDD International Conference on Knowledge Discovery and Data Mining (2016)
6. Zytek, A., Pidò, S., Veeramachaneni, K.: LLMs for XAI: future directions for explaining explanations. In: Proceedings of the ACM CHI Workshop on Human-Centered Explainable AI (2024)
7. Alonso, J.M., Bugarín, A.: ExpliClas: automatic generation of explanations in natural language for weka classifiers. In: Proceedings of the 2019 IEEE International Conference on Fuzzy Systems (FUZZ-IEEE) (2019)
8. Alonso, J.M., Ducange, P., Pecori, R., Vilas, R.: Building explanations for fuzzy decision trees with the ExpliClas software. In: Proceedings of the 2020 IEEE International Conference on Fuzzy Systems (FUZZ-IEEE) (2020)
9. Burton, J., Al Moubayed, N., Enshaei, A.: Natural language explanations for machine learning classification decisions. In: Proceedings of the 2023 International Joint Conference on Neural Networks (IJCNN) (2023)

10. Wu, X., Zhao, H., Zhu, Y., Shi, Y., Yang, F., et al.: Usable XAI: 10 strategies towards exploiting explainability in the LLM era. arXiv (2024)
11. Zhou, J., Gandomi, A.H., Chen, F., Holzinger, A.: Evaluating the quality of machine learning explanations: a survey on methods and metrics. Electronics **10**(5), 593 (2021)
12. Gu, J., et al.: A survey on LLM-as-a-judge. arXiv (2024)
13. Hello GPT-4o. https://openai.com/index/hello-gpt-4o/. Accessed 26 May 2023
14. Introducing OpenAI o3 and o4-mini. https://openai.com/index/introducing-o3-and-o4-mini/. Accessed 06 May 2023
15. Brown, T.B., Mann, B., Ryder, N., Subbiah, M.: Language models are few-shot learners. In: Proceedings of the 34th International Conference on Neural Information Processing Systems (2020)
16. Mersha, M., Lam, K., Wood, J., AlShami, A., Kalita, J.: Explainable artificial intelligence: a survey of needs, techniques, applications, and future direction. Neurocomputing **599** (2024)
17. Schwalbe, G., Finzel, B.: A comprehensive taxonomy for explainable artificial intelligence: a systematic survey of surveys on methods and concepts. Data Min. Knowl. Disc. **38**, 3043–3101 (2024)
18. Spitzer, P., Celis, S., Martin, D., Kühl, N., Satzger, G.: Looking through the deep glasses: how large language models enhance explainability of deep learning models. IN: Proceedings of Mensch und Computer 2024 (2024)
19. Papineni, K., Roukos, S., Ward, T., Zhu, W.: BLEU: a method for automatic evaluation of machine translation. In: Proceedings of the 40th Annual Meeting on Association for Computational Linguistics (2002)
20. Lin, C.: ROUGE: A Package for Automatic Evaluation of Summaries. Text Summarization Branches Out (2004)
21. Lavie, A., Agarwal, A.: Meteor: an automatic metric for MT evaluation with high levels of correlation with human judgments. In: Proceedings of the Second Workshop on Statistical Machine Translation (2007)
22. Zhang, T., Kishore, V., Wu, F., Weinberger, K.Q., Artzi, Y.: BERTScore: evaluating text generation with BERT. In: Proceedings of the International Conference on Learning Representations (2020)
23. Kim, S., et al.: Prometheus 2: an open source language model specialized in evaluating other language models. In: Proceedings of the 2024 Conference on Empirical Methods in Natural Language Processing (2024)
24. Sundararajan, M., Taly, A., Yan, Q.: Axiomatic attribution for deep networks. In: Proceedings of the 34th International Conference on Machine Learning (2017)
25. Binder, A., Montavon, G., Lapuschkin, S., Müller, K.R., Samek, W.: Layer-wise relevance propagation for neural networks with local renormalization layers. Artif. Neural Netw. Mach. Learn. (2016)

Uncertainty in Deep Model Performance for Radiology: A Case Study of Classifying Maxillary Sinus Appearance

Fara Aninha Fernandes[1,2]([✉]) , Martin Gerdes[1] , Georgi Chaltikyan[2] ,
and Christian W. Omlin[1]

[1] Department of Information and Communication Technology, University of Agder
(UiA), 4879 Grimstad, Norway
[2] Faculty European Campus Rottal-Inn, Deggendorf Institute of Technology (DIT),
84347 Pfarrkirchen, Germany
`fara.fernandes@th-deg.de`

Abstract. Predictions from deep learning models need to be perfect in radiology, where incorrect predictions challenge their trustworthiness. As models of different architectures enter the field of image analysis, it is important to uncover the source and quantify the extent of the uncertainty in the models' predictions. In this study, uncertainty methods are applied to a downstream task in classifying maxillary sinus images using three trained models: convolutional neural network (CNN), vision transformer (ViT), and gated multilayer perceptron (gMLP). The uncertainty is explored through probability distribution, expected calibration error, calibration curves, predictive entropy, and second-order representation. Of the three models, the ViT was the most uncertain in its predictions. Images that affected model confidence were identified. Regardless of the model architecture, most 'clear' sinus images were identified with certainty, while considerable uncertainty in predicting 'opaque' and 'thick' radiographic appearances was noted. These findings represent a novel direction towards the explainability of deep learning models as uncertainty estimation is a viable method for scrutinizing their inner workings.

Keywords: Uncertainty Estimation · Deep Learning Models · Radiology · Maxillary Sinus

1 Introduction

Deep learning is poised to revolutionize the field of radiology. In radiographic image analysis, experimental models offer high levels of performance in classification tasks [6]. Although high performance metrics are often correlated as evidence of a model's superior performance, they can be misleading; such metrics may not fully reflect the nuances of clinical performance and can obscure significant limitations. To fulfill the indispensable need for trustworthiness, the

F. Marcelloni et al. (Eds.): IJCCI 2025, CCIS 2829, pp. 30–49, 2026.
https://doi.org/10.1007/978-3-032-15638-9_3

confidence of the predictions of these systems must be determined. One way to lend credence to predictions is through uncertainty estimation or quantification.

Most deep learning systems use complex models based on the convolutional neural network (CNN), while models based on architectures like the vision transformer (ViT) and gated multilayer perceptron (gMLP) have been experimented with. These models have demonstrated comparable performance in tasks such as classifying radiographic images of the maxillary sinus [6]. Conventionally, these models are evaluated using performance metrics that provide limited insight into the inner workings of the model and do not explore what it does not know, how unsure it is or why it doubts [1]. By measuring uncertainty, it is possible to find limitations in the dataset and the model, thus improving decision-making [18,20] and enhancing model reliability.

The uniqueness and complexity of radiographic images differ significantly from other images due to anatomical variations, pathologies, and imaging conditions. Additionally, the number of channels, dimensions and resolution [4] add further complexity. If uncertainty can detect subtle anomalies in an image [19] and handle variability, it can address some of the stringent requirements of these sensitive tasks. The concept of uncertainty is related to the 'state of not knowing'. In deep learning, it refers to the confidence of the predictions of a neural network [21]. Uncertainty quantification distinguishes predictions that are likely to be correct from those that are uncertain or incorrect. Lambert B. et al. determined that uncertainty quantification can identify uncertain samples for human review, find pitfalls in the model, reveal anomalies in a dataset, and facilitate interaction between the algorithm and the user [10].

The association between uncertainty quantification and explainability is interconnected and complementary, as the estimation of uncertainty provides a more comprehensive understanding of a model's behavior and the confidence of its predictions. When uncertainty techniques are applied to an explanation, they quantify the reliability and stability of a model's decision-making process. If an explanation is highly uncertain, it means that the understanding of 'why' the model made a decision is uncertain. Such explanations are critical since they reveal that the model is focusing on ambiguous or out-of-distribution samples.

Despite the importance of uncertainty on its use, there is limited research in radiology and dental radiographic applications. The goal of this study is to assess the uncertainty of three deep learning models in classifying radiographs of the maxillary sinus. The objectives are as follows:

1. Quantify the uncertainty in both the dataset and model predictions through various measures.
2. Identify data points with the highest levels of uncertainty.
3. Compare the three models through aleatoric and epistemic uncertainty.

2   Background

2.1   Types of Uncertainty Quantification

Uncertainty quantification is categorized into aleatoric and epistemic uncertainty based on the source [9]. Aleatoric uncertainty (irreducible) arises from noise within a dataset and due to errors in the data generation process [19]. In datasets for radiology, aleatoric uncertainty arises from image noise, patient movement, equipment settings, mislabeled data, ambiguous characteristics of cases, differences in brightness and contrast, and the presence of artifacts [4,11]. On the other hand, epistemic uncertainty is theoretically reducible and arises from a lack of knowledge that limits a model's understanding of data distribution [9,19]. In radiology, it may be attributed to limited and poor representation of training data, previously unseen radiographic structures or pathologies, incomplete or missing contexts, and flaws in the model [4,11,15].

2.2 Approaches to Uncertainty Quantification

Conventionally, the uncertainty of a deep learning model is obtained by a probabilistic distribution that estimates the mean and variance [4,9]. Gal Y. et al. suggested using the Monte Carlo dropout technique, where dropout is used during inference time (prediction phase) to enable a stochastic process, introducing randomness into the computation and generating multiple predictions [2,7]. This technique provides a Bayesian approximation to simulate sampling from a distribution over the model's weights [7]. The Monte Carlo dropout technique turns a deterministic neural network into a probabilistic model, thus estimating posterior distribution [8]. The derived metric is the mean of the predictions across all iterations or the average probability of a class in classification tasks. The prediction variance quantifies the uncertainty around the prediction mean; it measures the spread of the distribution of possible outcomes, thus identifying regions of the input space where the model is uncertain.

The calibration error of a probabilistic model is calculated using expected calibration error (ECE) that quantifies the difference between predicted probabilities and observed outcomes across confidence bins [19]. Calibration curves - also called reliability diagrams, serve to assess the calibration of probabilistic models and evaluate how well the predicted probabilities of a model align with the actual outcomes [18]. A perfectly calibrated model is one in which the predicted probabilities represent the true likelihood of each outcome.

Predictive entropy quantifies the amount of uncertainty in the probability distribution over possible outcomes. In classification tasks, predictive entropy is used to assess the uncertainty in predicted class probabilities and detect misclassification due to out-of-distribution images [1,15]. For a second-order representation of uncertainty, aleatoric and epistemic uncertainties are quantified [18]. One approach is based on information theory where Shannon entropy is the sum of the conditional entropy and mutual information [11].

2.3 Applications of Uncertainty Quantification in Medical Imaging

The primary use of uncertainty methods is to ascertain the predictions of models as a means to trustworthiness, improving performance through calibration and uncovering bias [4]. The targeted use of uncertainty quantification includes the improvement of performance metrics, modification of diagnostic protocol and refinement of image quality as evidenced by the following research. Yang S. et al. investigated the factors unknown to and unlearnt by a deep learning model when they explored the relation between model accuracy for radiographic image classification and model optimization with uncertainty quantification methods [20]. Using entropy, the researchers removed the ten most uncertain samples in the dataset, effectively increasing the accuracy, sensitivity and specificity [20]. In an attempt to improve classification performance, Feng M. et al. estimated the uncertainty of predictions at each magnification of whole slide images [5]. By recognizing the uncertainty at different magnifications and consequently using optimal magnifications, the researchers were able to integrate more trusted image features so as to improve the classification task [5].

3 Methods

This study was carried out with permission and in full compliance with the ethical standards of using anonymized data.

3.1 Data Collection and Image Processing

A carefully curated dataset comprising of 1,098 two-dimensional images of the lower third of the maxillary sinuses was used [6]. Minimal data preprocessing was performed with minor adjustments to brightness and contrast to enhance image quality. The final image was devoid of any biased indicators and cropped to a square-shaped region of interest. The exported images in joint photographic experts group format were divided into training, validation and test datasets (825:165:108), measuring 955×955 pixels and with a resolution of 819 dpi. The dataset is categorized into three distinct classes (Fig. 1):

1. Clear: Images showing the coronal view of a clear maxillary sinus, devoid of any sinus lining thickening or presence of radiodense material.
2. Opaque: Images illustrating the coronal view of an opacified maxillary sinus, which may contain discrete pockets of air spaces within the opacification.

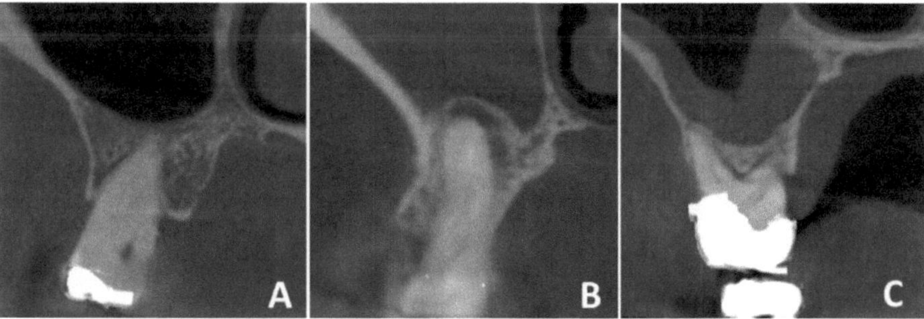

Fig. 1. Classification of radiographic appearance of the maxillary sinus. A. Clear sinus, B. Opaque sinus, C. Thickened sinus lining.

3. Thick: Images depicting the coronal view of a maxillary sinus with thickened lining, appearing as irregular flat, polypoidal, or semispherical thickening ranging from 3 mm to 8 mm.

3.2   Model Training and Evaluation

We trained three deep learning models - convolutional neural network (CNN), vision transformer (ViT) and gated multilayer perceptron (gMLP) and compared their performances. The models were developed and trained using TensorFlow version 2.15.0, Python version 3.10.12 and executed in Google Colaboratory. The input images were resized to 200×200 pixels and followed a sequential data augmentation process (normalization, rotation $= 0.02$ radians, zoom $= 0.2$). Each model was trained independently with optimal parameters to achieve the best performance. The models were compiled and trained for 150 epochs, a learning rate of 0.001, batch size of 32, a weight decay of 0.0001 and with an Adam optimizer. The weights of the most optimal performance of each model were saved and utilized to estimate the uncertainty.

On training the model, performance metrics - sensitivity, specificity, precision, accuracy and f1-score of each of the three models were derived. The CNN had performance metrics ranging from 0.96 to 0.98, the ViT showed metrics of 0.94 to 0.97 and the gMLP displayed metrics that ranged from 0.91 to 0.97. This was done in order to use high-performance models for estimating the uncertainty.

3.3   Estimation of Uncertainty Mean and Variance of Predictions

The Monte Carlo dropout technique was used to estimate the uncertainty of predictions from the three trained deep learning models. It was used on the test dataset consisting of 108 images divided into three classes - 'clear', 'opaque', and 'thick'. For each of the models, the average of the predicted probabilities after 100 stochastic forward passes (mean) and the variance of the predicted probabilities (variance) was calculated. A distribution of predictions for each input image was thus obtained. As output softmax probabilites can be used as a marker of uncertainty [10] or confidence, the logits obtained from the mean predictions were converted to probabilities using a softmax function. An analysis of the difference in variance between two classes was done to investigate the confidence of the model's predictions. The workflow was as follows:

1. Calculation of mean and variance: Implementation of the Monte Carlo dropout technique to obtain the mean and variance after 100 iterations, generating predictions for each image in the test dataset [2].
2. Conversion to probabilities: Conversion of the logits into probabilities to get a set of probabilities between 0 and 1 [10].
3. Plotting of mean prediction representing uncertainty for each image [12].
4. Visualization of class probabilities: Plotting of a heatmap of the probabilities for each class to visualize the distribution of predicted probabilities across the test dataset images.

5. Variance analysis: Calculation of the differences in prediction variances between the three classes (between 'clear' and 'opaque', 'opaque' and 'thick' and 'clear' and 'thick' classes).

3.4   Evaluation of Calibration of Trained Models

The Expected Calibration Error (ECE) was used to evaluate the model's calibration, that is the alignment between the model's predicted probabilities and the actual fraction of true positives. Using binning and iterating over the bins, the fraction of true positives (accuracy of predictions) and average predicted probability (confidence) were calculated for each bin [13]. Subsequently, calibration curves were plotted to evaluate how well the predicted probabilities align with actual outcomes. The curves were plotted by taking the true binary labels and predicted probabilities for a specific class as inputs. By setting the number of bins to 10 and iterating over each class, the calibration curves provided a visual assessment of the model's calibration [12]. The following was the workflow:

1. Calculation of ECE [13]:

$$ECE = \sum_{m=1}^{M} \frac{|B_m|}{n} |\mathrm{acc}(B_m) - \mathrm{conf}(B_m)|$$

  - M is the number of equally spaced bins
  - B represents "bins"
  - m is the bin number
  - n is the number of samples
  - $\mathrm{conf}(B_m)$ represents the average estimated probabilities in bin m
  - $\mathrm{acc}(B_m)$ represents the accuracy per bin m

2. Plotting of calibration curves for each class and each model to visualize the alignment of the predicted probabilities with actual outcomes [12].

3.5   Predictive Entropy

The numerical uncertainty value of each image in the test dataset for all three models was computed as the predictive entropy by summing the negative products of probabilities and their logarithms [12].

In order to investigate whether correct predictions exhibit different levels of uncertainty, the predictive entropy values of the correctly classified images of each model were analyzed. These correct predictions were categorized into two groups based on their entropy values: those with zero entropy (indicating full model confidence) and those with entropy greater than zero (indicating a degree of uncertainty). The non-parametric Mann–Whitney U test was used to assess if the group with non-zero entropy values had a greater and statistically significant uncertainty than the zero values entropy group.

3.6   Calibration of Models

In order to increase the agreement between the estimated probability and the true frequency of a specific class [4], models that appear to require calibration were calibrated. Calibration was done using Platt scaling - transforming scores into probability estimates with logistic regression and isotonic regression (learns bin boundaries from data, thus having variable number of bins depending on the training set) using sigmoid and isotonic methods respectively [14,16]. This was followed by repeat plotting of the calibration curves.

3.7   Second-Order Representation

The total, aleatoric and epistemic uncertainty was calculated as follows [17, 18]: Total Uncertainty (TU): Overall uncertainty in prediction (Y), represented as Shannon entropy of the probabilistic prediction, measuring uncertainty of a random variable (Y).

$$TU = ENT(Y) = ENT(\bar{p}) = ENT\left(\int p_\theta \, dQ(\theta)\right) \tag{1}$$

- ENT(Y) represents Shannon entropy
- $\bar{p}$ represents the average or expected probability distribution over possible outcomes
- Integral $\int ENT(p_\theta), dQ(\theta)$ represents the expected probability distribution over model parameters θ, weighted by distribution $Q(\theta)$

Aleatoric Uncertainty (AU): Conditional entropy of a model's prediction and uncertainty inherent in data due to inherent randomness.

$$AU = ENT(Y \mid \theta) = \int ENT(p_\theta) \, dQ(\theta) \tag{2}$$

- Conditional Entropy $ENT(Y \mid \theta)$ measures the uncertainty in (Y) given a specific model θ
- Integral $\int ENT(p_\theta), dQ(\theta)$ represents the expected entropy over the distribution of model parameters

Epistemic uncertainty (EU): Uncertainty in a model.

$$EU = I(Y,\theta) = ENT(Y) - ENT(Y \mid \theta) \tag{3}$$

- $I(Y,\theta)$ is the mutual information quantifying the amount of information shared between the prediction (Y) and the model parameters θ
- $ENT(Y) - ENT(Y \mid \theta)$ is the difference between the total and the aleatoric uncertainty.

3.8   Analysis of Class Uncertainty

The inherent uncertainty biases of the three models was investigated by comparing the prediction variances between the three classes across all 108 images of the test dataset. This was done to determine if a model is more uncertain about one class than others, regardless of the final prediction. The raw variance outputs for each of the three classes for every single image were used to calculate the measures of central tendency. Levene's test was done to determine if there was a significant difference in the variances among the three classes. Additionally, the Friedman's test was done to identify a difference in the distributions of variance scores across the three classes.

3.9   Analysis of Uncertainty Scores for Correct and Incorrect Predictions

For each type of uncertainty - total, aleatoric, epistemic - a quantitative assessment was done to check if the uncertainty estimates differ between correct and incorrect predictions. The Mann-Whitney U test was applied (independently) to the total uncertainty (TU), aleatoric uncertainty (AU), and epistemic uncertainty (EU) scores. This was done to compare the values of the 'correct' group against the 'incorrect' group for each of the three models. For each model, predictions were labeled as correct or incorrect based on their agreement with the ground truth. Separate Mann–Whitney U tests were conducted for each uncertainty type to assess whether the uncertainties were significantly lower in correct predictions as compared to the incorrect predictions.

4   Results

The following are the results of the uncertainty quantification in CNN, ViT and gMLP models when classifying radiographic images of the maxillary sinus. The calculations were consistent across ten runs.

4.1   Estimation of Uncertainty Mean and Variance Across Predictions

The mean of the predictions and variance across all three models was noted. The logits from the mean predictions were converted to probabilities and plotted. For each model, 108 plots were used to find the images where the models were uncertain about their decisions. The CNN, ViT and gMLP models were uncertain about five, ten and eight images, respectively. It was also noted that the uncertainty for some images overlapped with two or all three models (Table 1). A heatmap was generated where each column represented one class. In each column, each row represented the uncertainty of predicting the class of an image (Fig. 2). These images were then checked by a dental radiologist to investigate the cause of the uncertainty.

Table 1. Images that showed uncertainty and misclassification across the three deep learning models.

Class Label	Clear	Opaque	Thick
	Image number	Image number	Image number
CNN	24, 26	56	77, 90
ViT	10, 13, 14, 24, 25, 26	56, 63	74, 92
gMLP	10	56, 63	74, 78, 90, 92, 93

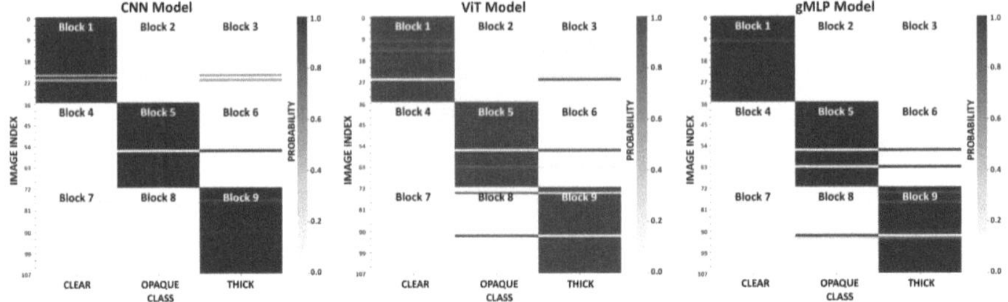

Fig. 2. Heatmaps of class probabilities for all three deep learning models showing correctly classified and misclassified images with their uncertainties. Note: Darker rows represent higher predicted probabilities, indicating greater model confidence. Correct classifications appear in blocks 1, 5, and 9, while darker shaded rows outside these blocks indicate confidently misclassified images.

Among the images, there was one image that was misclassified by all three models with high confidence (Fig. 3). This image (from the 'opaque' class and classified as 'thick') was noted to be an image showing an opaque sinus with evidence of graft material. Two other images with graft material were also misclassified (image 63 was wrongly classified wrongly by the ViT and gMLP and image 78 was wrongly classified by the gMLP). The other images that were classified with uncertainty, particularly the 'clear' images (images 10, 13, 14, 24, 25 and 26) had no plausible reason nor any characteristic finding. Images of the 'thick' category were also characterized with uncertainty (images 74, 77, 90, 92, and 93), revealing an inherent difficulty and uncertainty in classifying images of this class. It must be noted that images of the 'thick' class display varying patterns. The variance analysis reiterated these results by showing that the ViT model was uncertain when distinguishing between the classes as compared to the other models, with the gMLP model showing the least variance in distinguishing between the classes (Table 2).

4.2   Evaluation of Calibration of Trained Models

The ECE quantifying the deviation between the predicted confidence and actual accuracy was computed (Table 3). The highest calibration error was caused by

Table 2. Inter-class variance in model prediction for the CNN, ViT, and gMLP models showing the range of prediction variance when distinguishing between 'clear', 'opaque', and 'thick' image classes. Higher variance indicates greater uncertainty.

Variance between image classes	CNN	ViT	gMLP
'clear' and 'opaque'	0.29–26.96	0.90–124.98	1.05–10.02
'opaque' and 'thick'	0.13–65.21	2.34–268.62	0.22–11.64
'clear' and 'thick'	0.13–89.01	3.57–98.54	0.00 to 9.36

Table 3. Expected Calibration Error (ECE) for each class label across the CNN, ViT, and gMLP models.

Class Label	CNN model	ViT model	gMLP model
Clear	0.0144	0.0132	0.0046
Opaque	0.0092	0.0251	0.0315
Thick	0.0223	0.0359	0.0340

the ViT and gMLP in predicting 'thick' images. The lowest ECE was in predicting 'clear' images by the gMLP. Calibration curves for each class showed that the CNN and gMLP had almost perfect calibration curves (Fig. 4). The calibration curves of the ViT showed that it was not calibrated.

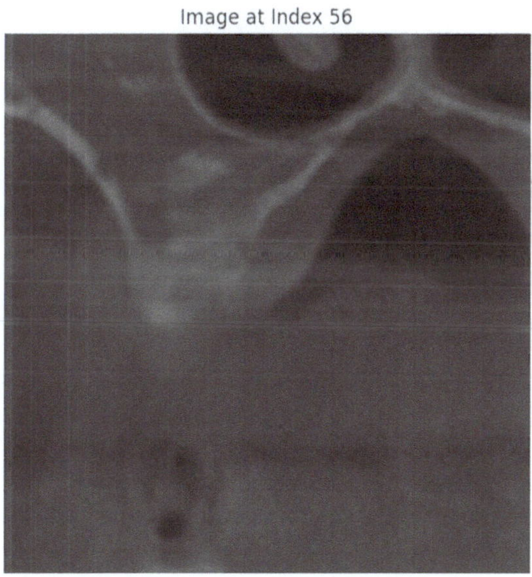

Fig. 3. A misclassified image (by all models) identified through uncertainty measures showing a probable intra-class variation suggestive of an out-of-distribution datapoint.

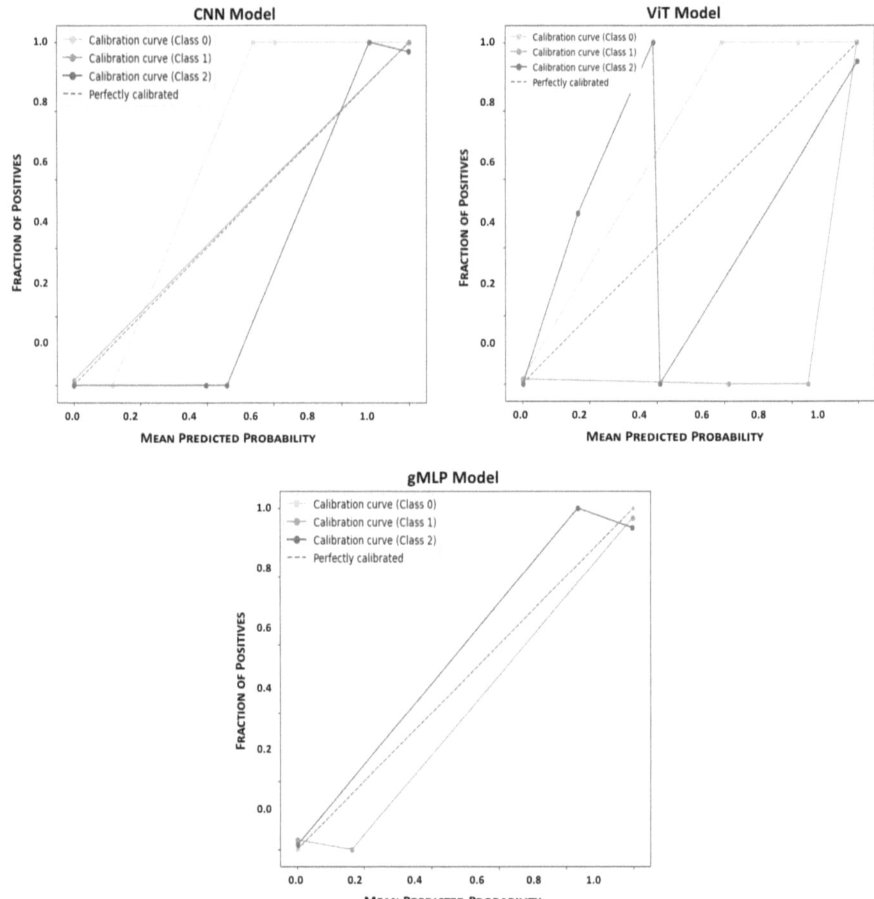

Fig. 4. Uncertainty calibration curves showing the relationship between mean predicted probability and fraction of positives. Note: Curves close to the diagonal line indicate good model calibration.

4.3 Predictive Entropy

Measurement of predictive entropy revealed 295 certain decisions for all 108 images predicted by all three models. Of the 29 uncertain predictions, most were by the ViT (11.11%), followed by the gMLP (8.33%) and CNN (7.40%) (Tables 4, 5, 6). All three models were most uncertain when predicting images of the 'thick' class. The number of correct predictions with zero entropy was compared with the number of non-zero entropy values (CNN model: 100 zero entropy versus 8 non-zero entropy values, ViT model: 96 zero entropy versus 11 non-zero entropy values, gMLP model: 99 zero entropy versus 8 non-zero entropy values). The MannâĂŞWhitney U test revealed differences in entropy distributions between these groups and for every model with p-values effectively zero and U statistics of 800.0, 1056.0, and 792.0. These results confirm that within a model's correct predictions, there is a statistically significant subset of outputs where the model expresses notable uncertainty.

4.4   Calibration of Models

Since the calibration curves revealed that the ViT model was uncalibrated, post-hoc calibration was carried out using Platt scaling and isotonic regression methods. Repeated calibration curves of the ViT model showed almost perfect calibration (Fig. 5). The CNN and the gMLP models were not calibrated, as their calibration curves were almost perfect.

4.5   Second-Order Representation for Further Uncertainty Analysis

The total, aleatoric, and epistemic uncertainty of all 108 images in the test dataset was calculated and compared. It was seen that the gMLP model was the most certain about its predictions, with aleatoric uncertainty ranging from 0.003 to 0.4605 and epistemic uncertainty ranging from 0.000 to 0.529. The CNN model showed low epistemic uncertainty (99% of the values between 0.006 to 0.401) but high aleatoric uncertainty (68% of the values between 0.604 to 0.848). A moderate amount of uncertainty was noted by the ViT model which showed equal amounts of aleatoric and epistemic uncertainty values ranging from 0.061 to 0.604 and 0.095 to 0.525, respectively (Fig. 6).

Table 4. Correctly classified images by the CNN model with non-zero predictive entropy values, indicating uncertainty despite correct predictions.

Image Number	Predicted	Actual	Entropy value
Image 40	Opaque	Opaque	0.2299
Image 45	Opaque	Opaque	0.1340
Image 50	Opaque	Opaque	0.2724
Image 72	Thick	Thick	0.3349
Image 83	Thick	Thick	0.1613
Image 85	Thick	Thick	0.3423
Image 97	Thick	Thick	0.3665
Image 104	Thick	Thick	0.3222

4.6   Analysis of Class Uncertainty

The analysis of central tendency scores of the CNN model showed similar mean and median uncertainty values. It was observed that Levene's test was significant

Table 5. Correctly classified images by the ViT model with non-zero predictive entropy values, indicating uncertainty despite correct predictions.

Image Number	Predicted	Actual	Entropy value
Image 7	Clear	Clear	0.1637
Image 13	Clear	Clear	0.3670
Image 36	Opaque	Opaque	0.1170
Image 41	Opaque	Opaque	0.1903
Image 45	Opaque	Opaque	0.3042
Image 49	Opaque	Opaque	0.3005
Image 50	Opaque	Opaque	0.1914
Image 52	Opaque	Opaque	0.2974
Image 80	Thick	Thick	0.3652
Image 84	Thick	Thick	0.2120
Image 104	Thick	Thick	0.2163

Table 6. Correctly classified images by the gMLP model with non-zero predictive entropy values, indicating uncertainty despite correct predictions.

Image Number	Predicted	Actual	Entropy value
Image 10	Clear	Clear	0.3048
Image 37	Opaque	Opaque	0.3315
Image 62	Opaque	Opaque	0.0141
Image 74	Thick	Thick	0.0846
Image 78	Thick	Thick	0.0983
Image 90	Thick	Thick	0.3545
Image 93	Thick	Thick	0.2611
Image 96	Thick	Thick	0.2976

(p=0.0003). This indicates a variability in the model's uncertainty for Class 2 ('thick') (standard deviation = 22.69), as compared to Class 0 ('clear') (standard deviation = 15.48). The Friedman test for the CNN model showed that it is more confident about Class 0 ('clear') than it is about Class 1 ('opaque') and Class 2 ('thick') (p=0.0001). The higher standard deviation of the CNN model's uncertainty for Class 2 ('thick') suggests low confidence levels for this set of images (Table 7).

In contrast, the ViT model exhibited a notable mean and median uncertainty for Class 1 ('opaque') (100.47, 106.91) as compared to that of Class 0 ('clear') (63.35, 39.70) and Class 2 (34.06, 23.88). This finding was supported by Levene's test (p < 0.0001) and Friedman's test ($p \approx 0$) that confirmed that the observed difference in uncertainty for Class 1 ('opaque') is statistically significant. This

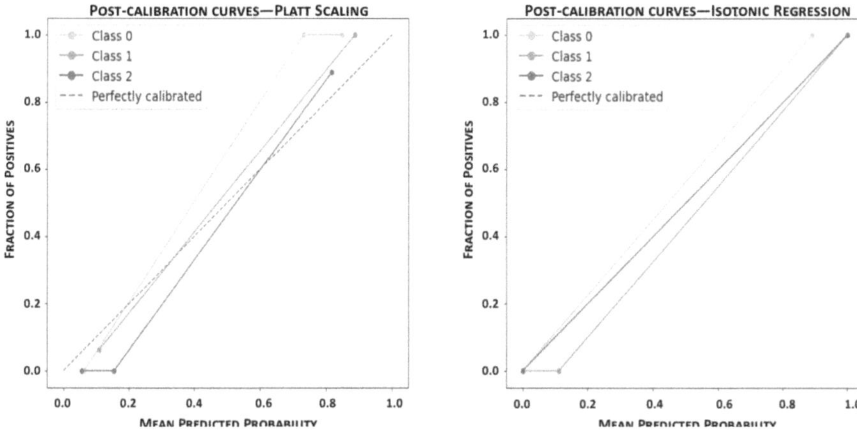

Fig. 5. Post-hoc calibration plots for the initially uncalibrated ViT model. The plots show the relationship between the mean predicted probability and the fraction of positives. The plot on top shows the calibration curve after applying Platt scaling, which provides moderate improvement. The bottom plot shows the result of applying isotonic regression, which yields curves closer to the diagonal indicating a better calibrated model. Note: Calibration curves close to the diagonal line indicate good model calibration.

high uncertainty demonstrates an inherent difficulty and lack of confidence when classifying images of Class 1 ('opaque') (Table 7).

The gMLP model appeared to be the most confident of the three models, with the lowest uncertainty scores. However, the gMLP model also displays an inherent bias in being consistently more uncertain about Class 2 ('thick'). Levene's test (p = 0.0121) corroborates that the uncertainty estimations differ significantly between the classes. Friedman's test indicates that the gMLP model is significantly more uncertain about Class 2 ('thick') where the median = 3.58, than it is about Class 0 ('clear') where the median = 1.61 and Class 1 ('opaque') where the median = 0.91 (Table 7).

4.7 Analysis of Uncertainty Scores for Correct and Incorrect Predictions

The CNN model showed no significant differences in epistemic or total uncertainty between correct and incorrect predictions (p > 0.2), but aleatoric uncertainty was marginally higher in incorrect predictions (p = 0.048). This suggests a possible link to input-level (image) noise and ambiguous data. The analysis of the results revealed markedly different uncertainty characteristics across the three architectures. This indicates the model is just as confident when it is wrong. In contrast, the ViT model revealed significant differences for both total (p = 0.0011) and aleatoric uncertainty (p = 0.0067), indicating that incorrect predictions are characterized by greater model uncertainty. However, its epis-

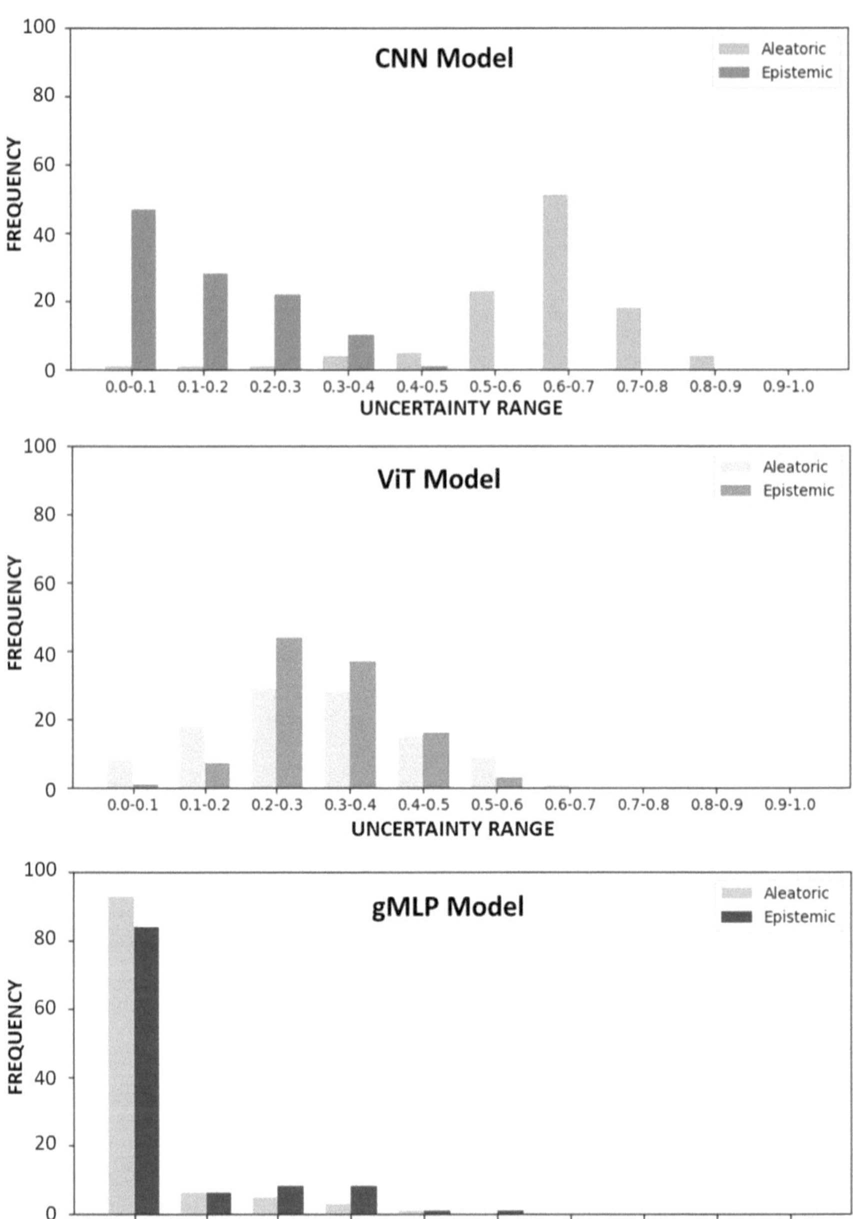

Fig. 6. Frequency distribution of aleatoric and epistemic uncertainty across the test set for the CNN, ViT, and gMLP models. The gMLP model demonstrates the highest overall confidence with predominantly low uncertainty values. In contrast, the CNN model shows high aleatoric but low epistemic uncertainty, while the ViT model exhibits moderate levels of both.

temic uncertainty was not reliable (p = 0.136). The gMLP model showed the most consistent results, with all three uncertainty types (epistemic, total, and aleatoric) significantly higher in incorrect predictions (p < 0.005), implying that the gMLP model's uncertainty estimates effectively differentiate between confident and uncertain decisions. The gMLP was the only model whose epistemic uncertainty (p = 0.003) was noted as a reliable indicator of error.

Table 7. Central tendency values of class-wise prediction uncertainty for the CNN, ViT, and gMLP models presenting the mean, median, and standard deviation of the uncertainty scores of all images in the test dataset.

Model and Class	Mean	Median	Standard Deviation
CNN Model			
Class 'clear'	32.46	29.60	15.48
Class 'opaque'	38.92	34.96	19.16
Class 'thick'	35.44	33.66	22.69
ViT Model			
Class 'clear'	63.35	39.70	53.80
Class 'opaque'	100.47	106.91	72.45
Class 'thick'	34.06	23.88	34.26
gMLP Model			
Class 'clear'	2.19	1.61	1.67
Class 'opaque'	2.37	0.91	2.66
Class 'thick'	4.28	3.58	2.90

5   Discussion

This study was carried out to assess the uncertainty of three deep learning models for a downstream task in dental radiology, using a pragmatic approach to investigate its relevance and utility [10].

The first objective was to quantify the uncertainty in the dataset as well as the predictions of the models. By using the Monte Carlo dropout technique, it is possible to obtain the mean and variance of the model's predictions. The mean of the predictions across multiple stochastic forward passes gives an estimate of the expected output, that is similar to sampling different configurations of the weights of a trained model. This is an effective way to estimate uncertainty. Additionally, the variance provides a distribution of predictions for each sample. This technique encourages a review of each datapoint that is crucial to the performance of the model and its overall trustworthiness [10].

In this study, a possible out-of-distribution image that was noted as a variation of the 'opaque' class was identified. Few images of graft tissue similar to this image were included in the training dataset. When all three models were compared, it was noted that the transformer architecture on which the ViT is based, demonstrated a high level of uncertainty. This may be due to the tendency of ViTs to overfit to data, resulting in unreliable and overconfident predictions [3].

Furthermore, uncertainty methods unveil the images where models are uncertain and their confidence in the resulting prediction by using entropy. A high entropy value signals the uncertain samples. In the present study, the most uncertain category was that of the thickened lining, which could be due to its varied appearances. Hence, it can be inferred that current deep learning models have a limited ability to understand the subtle variability of this anatomical feature, which otherwise is easily understood by trained radiologists. Besides the 'thick' images, the CNN model was also unsure about some 'clear' images, which needs further investigation.

An overall view of the uncertainty of the models was noted through the estimation of aleatoric and epistemic measures. This second-order representation of uncertainty strengthens the explainable aspect of the methods. Images that are out-of-distribution are linked to epistemic uncertainty, while images akin to noisy datapoints are linked to aleatoric uncertainty [10]. This clear depiction of overconfidence in the predictions of the model addresses the missing component of current explainable artificial intelligence methods [15].

The findings also suggest that uncertainty estimates, particularly total and aleatoric uncertainty serve as useful indicators of prediction reliability, with model-specific variations in the manifestation of uncertainty.

As demonstrated in the downstream task used here, uncertainty may be appreciated as an important component of model outcome. The methods are able to detect errors in data and the model, as well as fulfill the assessment of calibration of models. Faghani et al. [4] explain that uncertainty shown by deep learning models is analogous to the confidence of a radiologist and is inversely related to the trustworthiness of model outputs. Just as a radiologist would be less confident in making a decision when there is insufficient information, so also can uncertainty methods identify insufficient information in making reliable decisions. In this regard, Bley F. et al. find that uncertainty prediction can be turned into methods for explainable artificial intelligence (XAI) [1].

This underscores the importance of considering uncertainty metrics alongside accuracy and other explainable methods, as models may deliver correct classifications with varying degrees of confidence, which has implications for reliability and downstream decision-making. Model uncertainty quantification assists in model explainability by providing crucial context about the reliability of a model's predictions. It essentially reveals when and how much to trust the model's output, which is a fundamental aspect of understanding its behavior.

Explainability aims to understand not just what a model predicts, but why. A model that produces a prediction with high uncertainty effectively communicates that there is a deficiency in the system and that it is operating in ambiguous conditions. In high-stakes domains like radiology, a model's prediction is insufficient on its own. For a system to be truly trustworthy, it must communicate its own limitations. This is a core tenet of human-centered artificial intelligence (AI). Uncertainty quantification thus provides a direct mechanism for this communication. By signaling when it is "unsure", a model provides a critical checkpoint, inviting human domain expert oversight. This moves the system from a black-box to a human-AI collaboration, enhancing clarity and building trust, and ensuring safer clinical workflows. This work, therefore, investigates the uncertainty characteristics of different model architectures as a fundamental step toward building explainable systems that can be responsibly integrated into clinical practice. By quantifying epistemic and aleatoric uncertainty, we move beyond a simple prediction to a more insightful output that tells us how to gauge the model's prediction and its deficiencies.

Since the ultimate goal of explainable AI (XAI) in a clinical setting is to foster effective human-AI collaboration built on a foundation of trust, our comparative analysis of uncertainty provides several direct insights into how this can be achieved. As demonstrated through this case study, uncertainty acts as a critical check and a trust-building mechanism that flags confident but incorrect classifications and ambiguous cases with high uncertainty. Using these signals, clinicians are encouraged to apply their own expertise, effectively triaging the decision of the AI system. Such a mechanism allows them to trust the model's confident predictions more, knowing that it will alert them to cases that fall outside its scope. When a model is uncertain, it can serve as a form of first-level explanation, explaining that the input is challenging, novel, or ambiguous (e.g., due to poor image quality, rare pathology, or out-of-distribution images). This is a crucial step before a clinician might use other post-hoc XAI tools (like saliency maps) to ask why the case was difficult.

While uncertainty estimates contribute meaningfully to explainability, their limitations include being challenging to non-expert users. Relating model uncertainty estimates to diagnostic criteria might be difficult for radiologists [19]. The lack of reference standards, ground truths (for uncertainty), mathematical complexity, and limited empirical evaluations further strain its use [4,21]. Though it was possible to identify some images that were misclassified and some that were classified with uncertainty, it is still unknown why the model was uncertain about these images. Another limitation is related to the size of the dataset. While the dataset was carefully curated, its limited size may not fully capture the wide spectrum of anatomical and pathological variations present in the general population as noted in the isolation of out-of-distribution cases. This potentially constrains the generalizability of the deep learning models used in this study.

As the integration of uncertainty methods into explainability frameworks becomes increasingly important, there is a need to prioritize practical and collaborative downstream tasks in medical imaging. This includes the use of second-order representation, the visualization of uncertainty through heatmaps, and rectification of overconfidently misclassified images through accuracy-rejection curves.

6   Conclusion

This study explored the use of uncertainty methods in a downstream task in dental radiology. Uncertainty estimation can enhance the trustworthiness of deep learning models by quantifying both aleatoric and epistemic uncertainty to reveal model confidence. This study highlights how uncertainty contributes to the field of explainable AI by comparing the uncertainty of the predictions by three deep learning models. The integration of uncertainty measures into model evaluation makes it possible to obtain model insights beyond accuracy. This lays the groundwork for future research into the use of ensemble methods and other calibration techniques to further enhance model reliability in medical imaging.

Disclosure of Interests. The authors have no competing interests to declare that are relevant to the content of this article.

References

1. Bley, F., Lapuschkin, S., Samek, W., Montavon, G.: Explaining predictive uncertainty by exposing second-order effects. Pattern Recogn. **160**, 111171 (2024). https://doi.org/10.1016/j.patcog.2024.111171
2. Leibig, C., Allken, V., Ayhan, M.S., Berens P., Wahl, S.: Leveraging uncertainty information from deep neural networks for disease detection. Sci. Rep. **7**, 17816 (2017). https://doi.org/10.1038/s41598-017-17876-z
3. Erick, F.X., Rezaei, M., Müller, J.P., Kainz, B.: Uncertainty-aware vision transformers for medical image analysis. In: Sudre, C.H., Mehta, R., Ouyang, C., Qin, C., Rakic, M., Wells, W.M. (eds.) UNSURE 2024. LNCS, vol. 15167, pp. 171–180. Springer, Cham (2024). https://doi.org/10.1007/978-3-031-73158-7_16
4. Faghani, S., et al.: Quantifying uncertainty in deep learning of radiologic images. Radiology **308**, e222217 (2023). https://doi.org/10.1148/radiol.222217
5. Feng, M., et al.: Trusted multi-scale classification framework for whole slide image. Biomed. Sig. Process. Control **89**, 105790 (2024). https://doi.org/10.1016/j.bspc.2023.105790
6. Fernandes, F., Ge, M., Chaltikyan, G., Gerdes, M., Omlin, C.: Preparing for downstream tasks in artificial intelligence for dental radiology: a baseline performance comparison of deep learning models. Dentomaxillofacial Radiol. **54**, 149–162 (2025). https://doi.org/10.1093/dmfr/twae056
7. Gal, Y., Ghahramani, Z.: Dropout as a Bayesian approximation: representing model uncertainty in deep learning. In: Proceedings of the 33rd International Conference on Machine Learning JMLR W&CP, vol. 48, pp. 1050–1059. PMLR (2016)
8. Goel, P., Chen, L.: On the robustness of Monte Carlo dropout trained with noisy labels. In: Conference on Computer Vision and Pattern Recognition Workshops, pp. 2219–2228 (2021). https://doi.org/10.1109/CVPRW53098.2021.00251
9. Hüllermeier, E., Waegeman, W.: Aleatoric and epistemic uncertainty in machine learning: an introduction to concepts and methods. Mach. Learn. **110**(3), 457–506 (2021). https://doi.org/10.1007/s10994-021-05946-3
10. Lambert, B., Forbes, F., Doyle, S., Dehaene, H., Dojat, M.: Trustworthy clinical AI solutions: a unified review of uncertainty quantification in deep learning models for medical image analysis. Artif. Intell. Med. **150**, 102830 (2024). https://doi.org/10.1016/j.artmed.2024.102830

11. Löhr, T., Ingrisch, M., Hüllermeier, E.: Towards aleatoric and epistemic uncertainty in medical image classification. In: Finkelstein, J., Moskovitch, R., Parimbelli, E. (eds.) AIME 2024, Part II. LNCS, vol. 14845, pp. 145–155. Springer, Cham (2024). https://doi.org/10.1007/978-3-031-66535-6_17
12. Nguyen, H.: mcdrop_classification (2020). https://github.com/huyng/incertae/blob/master/mcdropclassification.ipynb. Accessed 23 Jan 2025
13. Pavlovic, M.: Expected calibration error (ECE): a step-by-step visual explanation (2023). https://towardsdatascience.com/expected-calibration-error-ece-a-step-by-step-visual-explanation-with-python-code-c3e9aa12937d/. Accessed 23 Jan 2025
14. Scikit-learn: Probability calibration of classifiers in scikit learn (2023). https://www.geeksforgeeks.org/probability-calibration-of-classifiers-in-scikit-learn/. Accessed 23 Jan 2025
15. Seoni, S., Jahmunah, V., Salvi, M., Datta Barua, P., Molinari, F., Acharya, U.R.: Application of uncertainty quantification to artificial intelligence in healthcare: a review of last decade (2013–2023). Comput. Biol. Med. **165**, 107441 (2023). https://doi.org/10.1016/j.compbiomed.2023.107441
16. Silva Filho, T., Song, H., Perello-Nieto, M., Santos-Rodriguez, R., Kull, M., Flach, P.: Classifier calibration: a survey on how to assess and improve predicted class probabilities. Mach. Learn. (3), 1–50 (2023). https://doi.org/10.1007/s10994-023-06336-7
17. Smith, L., Gal, Y.: Understanding measures of uncertainty for adversarial AI. In: Proceedings of the Conference on Uncertainty in Artificial Intelligence, Monterey, California, USA (2018)
18. Waegeman, W.: Aleatoric and epistemic uncertainty in statistics and machine learning (2025). [Workshop at NLDL Winter School]
19. Wang, T., et al.: From aleatoric to epistemic: exploring uncertainty quantification techniques in artificial intelligence. Comput. Res Repository abs/2501.03282 (2025). https://doi.org/10.48550/arXiv.2501.03282
20. Yang, S., Fevens, T.: Uncertainty quantification and estimation in medical image classification. In: Farkaš, I., Masulli, P., Otte, S., Wermter, S. (eds.) ICANN 2021. LNCS, vol. 12893, pp. 671–683. Springer, Cham (2021). https://doi.org/10.1007/978-3-030-86365-4_54
21. Zou, K., Chen, Z., Yuan, X., Shen, X., Wang, M., Fu, H.: A review of uncertainty estimation and its application in medical imaging. Meta-Radiol. **1**(1), 100003 (2023). https://doi.org/10.1016/j.metrad.2023.100003

Explain to Gain: Optimising Performance Through Explainable Reinforcement Learning Parameter Investigation

Patrick Capaldo[1], Varniethan Ketheeswaran[1,2], Santiago Quintana-Amate[1(✉)], Delaney Stevens[1], and Mark Hall[1]

[1] Airbus AI Research, Filton, UK
santiago.quintana-amate@airbus.com
[2] University of Bath, Bath, UK

Abstract. Building upon the foundational work of "Explain to Gain: Introspective Reinforcement Learning for Enhanced Performance," this article further investigates methods for leveraging explainable reinforcement learning (XRL) knowledge to enhance the performance of reinforcement learning (RL) agents. While our initial work demonstrated the potential of XRL approaches to guide and optimise RL agent training beyond merely improving interpretability and user trust, this paper extends that exploration. We expand upon the previously introduced introspective analysis framework by incorporating an additional XRL metric parameter into the search space of XRL parameter configurations within the training pipelines of model-free RL algorithms. This refined integration allows for even more nuanced dynamic adjustments of algorithm-specific parameters based on real-time feedback from a broader set of XRL metrics. The proposed methodology is validated across diverse OpenAI Gym environments (CartPole and Taxi). By evaluating both on-policy and off-policy approaches with this expanded parameter space, we demonstrate that incorporating these additional XRL insights leads to further significant improvements in agent performance. The analysis of the results highlights the enhanced benefits of deepened explainability and more finely tuned decision-making. This work contributes to the XRL research area by continuing to align interpretability with actionable performance gains, advancing the development of more reliable, transparent, and effective RL systems for complex, real-world applications.

Keywords: Explainable Reinforcement Learning (XRL) ·
Introspective Reinforcement Learning (IxDRL) · Explainability
Metrics · XRL Parameter Search Space · Dynamic Algorithm
Adjustment · Exploration-Exploitation Trade-off · Explainable
Autonomous Agents

1 Introduction

Reinforcement Learning (RL) algorithms enable AI agents to improve their sequential decision-making by interacting with an environment to optimise for long-term cumulative rewards [30]. This learning framework has achieved promising results in a wide variety of applications, achieving high performance in areas ranging from complex strategic games to robotic control [29,33]. Despite these advances, the implementation of RL solutions is still limited by the requirements of real-world problems. These include excessive data requirements, computationally expensive assets, and the generally "black-box" nature of the decision-making policies produced, leading to serious difficulties in ensuring reliability and interpretability where these are essential [23].

The research community has identified a lack of trust in RL systems by end users [11,23], highlighting the need for explainability techniques that aim to overcome the "black-box" nature of deep reinforcement learning (DRL) models used to solve decision-making problems. The objective of Explainable Reinforcement Learning (XRL) is to address the gap between human comprehension of the rationale underlying an agent's decision-making processes, actions, and policies. In certain domains, such as medical diagnostics or self-driving vehicle technology, the ability to explain AI decisions is crucial, as it can reduce costly errors [14]. The insights provided by the XRL methods offer an opportunity for actionability and can improve the performance and robustness of the agent's decision-making process [23,27].

However, XRL research has focused mainly on explaining an agent's behaviour—with the goal of making its action choices more comprehensible—without paying much attention to the actionability of the insights provided by these XRL methods [6,17].

With the aim of providing means for the actionability of XRL, this work builds on and extends the research presented in [27]. The original paper established a methodology for leveraging metrics derived from agent data, such as agent confidence and riskiness, towards model improvement and demonstrated its application to the Proximal Policy Optimisation (PPO) and Deep Q-Networks (DQN) algorithms. This paper takes the next step by conducting a detailed investigation into the parametric sensitivities of this XRL-enhanced training process. To achieve this, we integrate the XRL provided insights into the training process of model-free RL algorithms like Proximal Policy Optimisation (PPO) and Deep Q-Networks (DQN). This approach bridges the gap in the original research where explainability was largely used for post-hoc analysis and not actively guiding training processes. By incorporating these insights into training loops, we generate agents that not only have high performance but are also interpretable and reliable. This work makes the following contributions:

1. Parametric Investigation of XRL-Integrated Training: We present a detailed experimental analysis of critical parameters within our XRL-integrated training framework [27] for PPO and DQN. This extends previous work by evaluating the effects of varying the start iteration for applying XRL metrics, the number of previous state values used to gauge the agent's progress, and different threshold values used by the equations employed by the presented approach to adjust the agent's exploration coefficient.

2. Empirical Guidelines for Optimising XRL Integration: Through experiments in CartPole and Taxi environments, we provide empirical evidence on how specific choices for these parameters influence agent performance and training dynamics, offering actionable insights for optimising the XRL-guided improvement process.

By aligning explainability with performance enhancement through detailed parametric analysis, this paper contributes to the growing body of research advocating for interpretable and high-performing RL systems.

2   Background

A challenge in the field of reinforcement learning (RL) is determining how to act upon the insights from explainability methods to improve agent performance. Several strategies aiming to bridge this gap by making RL systems more effective and transparent while increasing model performance are described below.

Reward Shaping Methods. Modifying the reward function, known as reward shaping, has been widely adopted to help RL agents learn more efficiently. The work by [20] introduced a potential-based approach that maintains the optimality of learned policies while speeding up the learning process. In this line of work, [5] presented an approach where the shaping reward adapts as the agent interacts with its environment. Further developments by [12] framed the problem as a bi-level optimisation, allowing the agent to discover the most effective way to use these additional rewards. In addition, [9] found that such methods can reduce the amount of experience required for learning, without compromising performance.

Human-in-the-Loop Correction. Explanations generated from reinforcement learning systems can assist individuals in recognising when an agent's choice is suboptimal or when it encounters a particularly significant state or set of states. This understanding enables people to intervene by offering corrective actions or modifying the reward function to guide the agent toward better decisions. For example, [36] investigated methods that incorporate human feedback to identify and repair failures in agent behaviour. However, as noted in recent surveys [3], this human-in-the-loop process can become impractical in environments with very large action spaces, highlighting the need for more scalable, automated approaches.

Policy Refinement. Current research focuses on reducing the need for direct human intervention by using explainability techniques to automatically identify vulnerabilities in the agent's policy. For example, [8] developed a method that identifies relevant states in an agent's trajectory and uses this information to guide targeted updates to the policy. In this context, [2,38] proposed additional approaches where the environment is reset to important states highlighted by explainability methods, intensifying training from these points. As pointed out in a recent survey by [3], these strategies have the potential to accelerate policy refinement but they must be appropriately applied to avoid overfitting to specific states or episodes.

Integration with Advanced Neural Architectures. Recent advances have shown a synergy between explainability and state-of-the-art neural network architectures, leading to more interpretable and effective reinforcement learning systems. For instance, [18] developed a hybrid transformer model that integrates interpretability techniques to identify and better handle risky situations, demonstrating the value of explanation-driven adaptation in complex domains. In the same line of work, [1] developed the Explain to Improve Streaming Learning (ESL) framework, which provides explanations during transformer training to improve learning efficiency and model transparency.

Beyond transformers, recent work has begun to explore the intersection of explainability and selective state space models. In this context, the Mamba architecture [7] has been shown to possess implicit attention mechanisms that can be analysed using existing interpretability tools, offering new insights into the internal dynamics of these models and enabling direct comparisons with transformer-based approaches.

Evaluation and Performance Assessment. A key question in the development of explainable reinforcement learning (XRL) is whether the use of explanations actually translates into improved agent performance. [4] addressed this by evaluating how policy refinement methods, which leverage critical steps identified through explanation techniques, impact agent outcomes. By comparing agent performance before and after applying these explanation-driven modifications, they found that high-quality explanations can significantly improve learning, helping agents overcome local optima and achieve higher rewards. Supporting these findings, other studies [3,28] also demonstrated that explanations can both reveal weaknesses in agent policies and provide actionable guidance for improvement.

Advanced Frameworks and Hybrid Approaches. Recently, several studies have presented different strategies focused on improving the robustness of RL policies. For instance, [15] alternates between human corrections and proxy rewards, using pseudo-labels to reduce the need for human intervention and to deliver a more stable learning process. Of high interest is the study presented in [16], where a framework integrating explainability to improve RL is described. This framework operates at two levels, providing a high-level explanation used to learn a reward function and low-granularity explanations to identify specific state-action pairs that are responsible for the agent's vulnerability, highlighting the practical value of using explainability not just for explaining the agent's behaviour but also as a tool for improving its performance.

Other Notable Strategies. Other relevant approaches include:

- Lazy-MDP [13]: This method delays decisions, allowing the agent to focus on the most relevant moments, showing an improvement in learning and the agent's interpretability.
- Self-imitation learning [22]: In this study, agents are encouraged to repeat successful actions more often, which leads to more effective policies.
- External guidance methods [31, 32, 35]: These approaches use knowledge from pre-trained agents or human demonstrations to help RL agents solve complex problems.

Gaps in Current Research. While progress has been made in this area, much of the current research still focuses on previously gathered knowledge or external input, rather than focusing on the reasons why an agent's policy might fail or on ways to improve how decisions are made [26]. Understanding these underlying factors is important, not only for interpreting an agent's actions, but also for identifying concrete steps to improve its policy. In this work, we aim to address this challenge by presenting a new approach that extends the work in [24] and builds upon the insights of [27], enabling agents to recognise and adapt to their own shortcomings and uncertainties. Our methodology not only delivers a boost in performance; it is also designed to make reinforcement learning systems more transparent, more resilient, and better suited to handle the complexities of real-world problems.

3 Methods

This section details the methodological framework employed to leverage explainable reinforcement learning (XRL) metrics for enhancing the performance of reinforcement learning (RL) agents. Building upon the foundational approach introduced in our precursor study [24], which itself utilised the IxDRL framework [27], the present work conducts a more granular investigation into the parametric sensitivities inherent in this XRL-augmented training process. We begin by recapitulating core RL concepts, including the relevant mathematical formalisms and primary algorithmic classes (Sect. 3.1). Following this, we describe the specific XRL metrics adopted, detailing their operational significance, computational characteristics, and potential contributions to agent performance (Sect. 3.2).

The central aim of this line of research is to harness XRL metrics for addressing the critical exploration-exploitation dilemma in RL. This objective is pursued through dynamic adjustment mechanisms that integrate real-time explainability feedback to modulate exploration strategies, as further elaborated in Sect. 3.3. Subsequently, we present the mathematical formulations that enable these dynamic adjustments, outlining two distinct methodologies for modifying exploration coefficients in response to XRL metric values (Sect. 3.4). These components collectively establish a robust framework for the synergistic integration of explainability into RL training paradigms, targeting the development of more reliable and performant learning systems.

3.1   Reinforcement Learning Fundamentals

Reinforcement learning (RL) is a sub-discipline of machine learning where an agent learns to make optimal sequences of decisions through iterative, trial-and-error interactions within an environment. RL problems are commonly formalised as Markov Decision Processes (MDPs), defined by the tuple $M = (S, A, P, R, \gamma, \rho_0)$, where:

- S: The set of all possible states in the environment.
- A: The set of all actions available to the agent.
- $P(s'|s, a)$: The state transition probability function, denoting the probability of transitioning to state s' after executing action a in state s.
- $R(s, a) \in \mathbb{R}$: The reward function, which provides a scalar feedback signal for taking action a in state s.
- $\gamma \in [0, 1)$: The discount factor, determining the present value of future rewards.
- ρ_0: The initial state distribution.

The primary goal in RL is to learn a policy, denoted $\pi(a|s)$, that maps states to actions in such a way as to maximise the cumulative expected reward. This objective is captured by:

$$\pi^* = \arg\max_{\pi} \mathbb{E} \left[\sum_{t=0}^{\infty} \gamma^t R_t \right] \tag{1}$$

where π^* represents the optimal policy and R_t is the reward at timestep t.

RL algorithms are generally bifurcated into two principal categories: policy gradient methods and value-based methods. Policy gradient approaches focus on directly optimising the policy's parameters to maximise the expected return, using techniques like REINFORCE or actor-critic frameworks. Algorithms such as Proximal Policy Optimisation (PPO) and Soft Actor-Critic (SAC) are prominent examples, demonstrating efficacy in both discrete and continuous action spaces.

Value-based methods, in contrast, concentrate on learning value functions—such as the state-value function $V(s)$ or the action-value function $Q(s, a)$—which estimate the expected return from states or state-action pairs. Policies are then derived by selecting actions that maximise these learned values. Deep Q-Networks (DQN) epitomise this approach, often augmented by techniques like prioritised experience replay to enhance learning efficiency. These methods are particularly effective for problems with bounded action spaces or clearly defined value structures, as detailed further in [24].

3.2   Explainable RL Metrics for Performance Enhancement

In this investigation, aiming to improve agent performance, we utilise selected explainability metrics initially proposed within the IxDRL framework [27]. These metrics—namely *confidence* (execution certainty), *incongruity, goal conduciveness*, and *riskiness*—were originally considered for their proven actionability (i.e., their capacity to meaningfully inform interventions during training [10]) and their computational tractability, which ensures a low overhead suitable for real-time analysis. Each metric, along with its relevance to performance optimisation, is briefly described below:

Execution Certainty or Confidence: This metric quantifies an agent's assurance in its action selections by measuring the entropy of its policy distribution. For discrete actions, as employed in this study, confidence is computed using an evenness index applied to the action probabilities [19]. Low execution certainty often signals hesitant decision-making, highlighting opportunities for targeted retraining or refined guidance. Its computational simplicity allows for continuous real-time monitoring of policy stability. Note, however, that based on findings from [24] where combinations of only the subsequent three metrics yielded the most promising results, *execution certainty is not actively used to modulate agent behaviour in the XRL experiments presented in this paper.*

Incongruity: This metric captures significant deviations between expected and observed outcomes, typically by leveraging the temporal difference (TD) error. High incongruity values can indicate unexpected environmental transitions or reward anomalies, thereby facilitating the rapid detection of environmental stochasticity or inaccuracies within the agent's learned model. This metric is crucial for identifying scenarios that may require policy adaptation or environmental model recalibration. For value-based algorithms like DQN, incongruity was calculated using softmax Q-values, following the methodology described in [37].

Goal Conduciveness: This metric assesses the agent's progress towards its designated task objectives by analysing the first derivative of its value function over time. Positive derivative values suggest that the agent is advancing towards its goals, while negative values may indicate movement along suboptimal paths. This metric serves to guide exploration strategies, encouraging the prioritisation of states that are likely to lead to higher long-term rewards. The specific window of historical state values used for this derivative approximation (i.e., history length) is a key parameter explored in the current study.

Riskiness: This metric evaluates the potential severity of suboptimal actions by comparing the difference between the highest and lowest valued actions (in value-based methods) or the gap between primary and secondary action probabilities (in policy gradient methods). Elevated riskiness scores flag critical decision points where minor errors could precipitate significant performance declines, thus enabling proactive implementation of safeguarding measures.

These XRL metrics are derived directly from the agent's learned models and interaction history, ensuring both computational efficiency and the provision of actionable insights. Their computation relies on capturing various data streams, including rewards, policy distributions (action probabilities), value function estimates (e.g., state and action values), and trajectories of recent interactions. Some metrics, such as incongruity, additionally use TD errors or other measures of prediction deviation.

3.3   Leveraging XRL Metrics for Adaptive Agent Performance

While the literature presents various strategies for enhancing RL agent performance, such as reward function shaping [21] or self-imitation learning techniques [34], the research presented herein specifically continues our focus on tackling the critical exploration-exploitation dilemma [25]. This is achieved by dynamically adjusting the agent's exploration behaviour based on real-time feedback from the aforementioned XRL metrics (excluding Execution Certainty for active modulation in this work).

To implement this, we utilise two distinct mathematical functions (elaborated in Sect. 3.4). These functions use the primary hyperparameter governing the exploration-exploitation trade-off within the specific RL algorithm (e.g., the entropy coefficient for PPO, or ε for DQN), alongside the current values of the selected XRL metrics, to compute an adjusted value for this exploration hyperparameter. This allows for a more adaptive and potentially more efficient exploration strategy. The efficacy and parametric sensitivities of this approach are rigorously tested using Proximal Policy Optimisation (PPO) and Deep Q-Networks (DQN) across two distinct OpenAI Gym environments. This systematic investigation facilitates a detailed examination of the benefits derived from integrating XRL feedback into the training pipeline. The general procedure for adaptively tuning the exploration-exploitation balance, as depicted in Fig. 1, involves the following steps:

1. **Data Capture:** At each training iteration, essential data such as action probabilities (policy), value estimates, rewards, and Temporal Difference (TD) errors are recorded.
2. **Metric Computation:** The selected XRL metrics for active modulation—*goal conduciveness*, *incongruity*, and *riskiness*—are calculated using the captured agent-environment interaction data.
3. **Parameter Update:** One of the two specified equations (see Sect. 3.4) is applied to adjust the exploration coefficient (entropy for PPO or ε for DQN), based on the current metric values and the coefficient's value from the previous iteration.
4. **Clipping:** The newly calculated exploration coefficient is constrained within predefined boundaries (e.g., $\varepsilon \in [0.01, 0.5]$) to maintain training stability and prevent erratic behaviour.

This introspective strategy aims to convert XRL-derived knowledge into direct, implementable policy modifications, thereby seeking a synergistic balance between performance optimisation and enhanced system explainability. By aligning exploration strategies with the agent's dynamically assessed capabilities, we anticipate observing quantifiable improvements in cumulative rewards, as will be further detailed in Sect. 4.

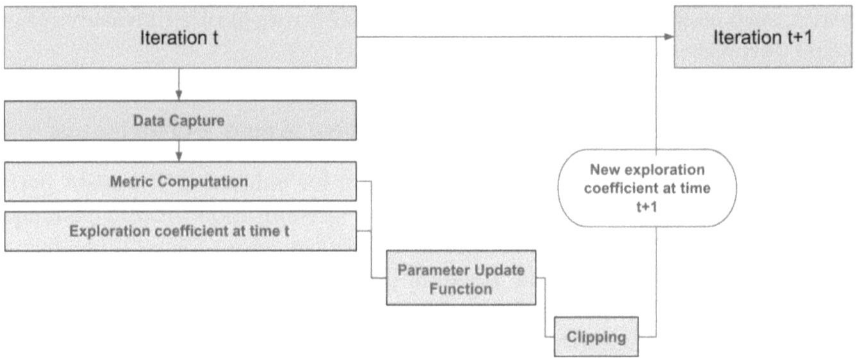

Fig. 1. Method to dynamically adjust agent exploration.

3.4 Functional Mechanisms for Dynamic Exploration Adjustment

In this study, building upon [24], we employ two primary functional forms for the dynamic adjustment of the exploration-exploitation trade-off during agent training. For simplicity, the key variable controlling this balance is referred to as the *exploration coefficient*. The precise nature of this coefficient varies with the RL algorithm. For Proximal Policy Optimisation (PPO), it is the entropy coefficient, governing the stochasticity of action selection. For Deep Q-Networks (DQN), it is the epsilon (ε) value, which dictates the probability of selecting a random action versus one from the learned policy. A core aspect of the current investigation is to analyse the sensitivity of agent performance to variations in how these functions utilise XRL metrics, including the configuration of parameters such as metric-specific thresholds within Function 1.

Function 1: Compounded Threshold-Based Adjustment. The first approach calculates the updated exploration coefficient, $\mathrm{EC}_{\mathrm{new}}$, using Eq. 2:

$$\mathrm{EC}_{\mathrm{new}} = \mathrm{softclip}\left(\mathrm{EC}_{\mathrm{cur}} \prod_m \begin{cases} \alpha_m & \text{if } v_m > \tau_m \\ \beta_m & \text{otherwise} \end{cases}, [\min, \max], \delta\right) \qquad (2)$$

where:

- $\mathrm{EC}_{\mathrm{new}}$: The newly computed exploration coefficient.
- softclip(): A function that constrains its input value to the range $[\min, \max]$ using soft boundaries.
- δ: The parameter defining the width of the soft transition zone for the clipping function.
- $\mathrm{EC}_{\mathrm{cur}}$: The current exploration coefficient before this adjustment.
- m: An individual active XRL metric (e.g., goal conduciveness, incongruity, riskiness).
- α_m: The multiplicative adjustment factor applied if the value of metric m is above its threshold τ_m.

- v_m: The current calculated value of metric m.
- τ_m: The predefined or varied threshold specific to metric m.
- β_m: The multiplicative adjustment factor applied if the value of metric m is at or below its threshold τ_m.

This method involves a dynamic update to the exploration coefficient by comparing the current value of each active XRL metric (v_m) against its specific threshold (τ_m). If v_m exceeds τ_m, the exploration coefficient is scaled by α_m; otherwise, it is scaled by β_m. As detailed in Sect. 3.5, the threshold τ_m for Goal Conduciveness is varied as part of our parametric study, while thresholds for other active metrics (Incongruity, Riskiness) are set to predefined values based on algorithm-environment pairs. The adjustment factors α_m and β_m are also pre-specified. These predefined values for Incongruity and Riskiness thresholds and the adjustment factors were established based on preliminary experiments and analysis of typical metric value ranges to ensure meaningful and stable adjustments. A full tabulation is omitted for brevity, but this approach aims to anchor the adjustment logic to empirically grounded heuristics for each specific learning context.

Following these compounded adjustments, the resulting EC_{new} is passed through a clipping function (e.g., *gradual_clip_positive*) to ensure it remains within a specified stable range, defined by [min, max] and the soft zone δ. This approach offers direct control over how individual metrics influence exploration, allowing, for instance, high "riskiness" to decrease exploration or low "goal conduciveness" to increase it. The ability to define distinct thresholds (fixed or varied) and adjustment factors per metric makes it adaptable to varied scenarios where specific metric states signal opportune moments to shift the exploration-exploitation balance.

Function 2: Compounded Normalisation-Based Adjustment. The second functional approach computes the exploration coefficient as per Eq. 3:

$$\text{EC}_{\text{new}} = \text{softclip}\left(\text{EC}_{\text{cur}} \prod_{m} \frac{\max_{\text{norm\_range}} - \min_{\text{norm\_range}}}{\max_m - \min_m}(v_m - \min_m), [\min, \max], \delta\right) \tag{3}$$

where:

- EC_{new}, softclip(), δ, m, v_m: These terms are defined similarly to their counterparts in Function 1.
- EC_{cur}: The current exploration coefficient before this adjustment.
- norm_range: Specifies the target range (e.g., $[0, 1]$ or other empirically determined bounds) to which each metric value v_m will be normalised.
- $\max_m, \min_m$: The maximum and minimum values observed for metric m during training, used for normalisation.

This method first normalises the value of each active XRL metric (v_m) to a common target range (norm_range), based on the observed extrema ($\max_m, \min_m$) for that metric during training. These normalised metric values are then typically used to proportionally scale the current exploration coefficient EC_{cur}. The direction of scaling depends on the intended relationship between the metric and exploration; for instance, high normalised goal conduciveness might decrease EC_{cur}, while high normalised incongruity might increase it.

The product of these scaled factors across all active metrics is then applied to EC_{cur}, and the result is clipped to maintain stability, similar to Function 1. This approach offers less granular control over individual metric influences compared to Function 1 but excels when a dynamic combination of multiple metrics is desired without reliance on hard-coded thresholds. It is particularly useful in environments where metrics exhibit diverse scales or varying importance, as normalisation ensures a more balanced contribution from each. The inherent flexibility and scalability of Function 2 make it a compelling alternative for managing the exploration-exploitation trade-off in complex RL tasks. Further details on the specific implementation of this function were presented in [24].

3.5   Parametric Search Space

A central contribution of this work is an extensive grid search over a multi-dimensional parameter space to identify optimal XRL configurations. This search space encompasses seven key parameters:

- **RL Algorithm:** PPO or DQN.
- **Environment:** CartPole-v1 or Taxi-v3.
- **Random Seed:** Five distinct seeds (5, 17, 42, 123, 8) are used for each configuration to ensure robust evaluation.
- **Goal Conduciveness (GC) History Length:** A novel parameter for this study, varying the number of historical state values for GC derivative approximation. Values: 3, 5, or 7.
- **XRL Adjustment Start Iteration (AftrItr):** The training iteration at which XRL-based dynamic adjustments commence. Values: 2, 30, or 60. However, and as explained in [24], this logic is directly applicable to PPO but not to DQN due to the nature of the algorithms. Therefore, for DQN, this value is essentially zero at all times.
- **Active XRL Metrics:** Two combinations are explored: (1) Goal Conduciveness, Incongruity (IC), and Riskiness active; (2) Goal Conduciveness and Riskiness active (Execution Certainty is inactive in all XRL experiments).
- **Adjustment Function & Configuration:** Either Function 1 (Compounded Threshold Approach) or Function 2 (Compounded Combination Approach) from Sect. 3.4. For Function 1, the Goal Conduciveness threshold is varied (0.0, 0.2, or 0.4), while thresholds and factors for other active metrics (IC, Riskiness) are set to predefined values specific to each algorithm-environment pair (as detailed by our configuration generation process). Function 2 operates via normalisation without these specific thresholds.

The systematic combination of these parameters results in 1,440 unique XRL-enhanced experimental configurations. In addition, 20 baseline configurations (standard PPO/DQN per environment and seed, without XRL adjustments) are executed for comparative analysis, leading to a total of 1,460 experiments.

4   Results and Discussion

Of the 1,440 unique XRL parameter combinations explored via the extensive grid search, approximately 40 exhibited performance profiles superior to their corresponding baseline configurations. The criteria for "superior" performance encompassed the overall mean reward achieved, the stability of this mean reward, and the average standard deviation of returns. Among these promising configurations, a subset of around 20 demonstrated performance that was significantly better than their baselines, thereby warranting a more detailed analytical focus. The primary implication of these top-performing configurations is that while XRL integration is not a panacea, a well-parameterised setup can yield substantial gains. These setups consistently pointed towards the benefits of using smoother guidance signals (from longer Goal Conduciveness history lengths) and applying this guidance early in the training process to effectively shape the agent's policy. The subsequent sections delve into these findings. To provide a comprehensive picture, some figures present performance averaged across broad categories of configurations, while others highlight specific, top-performing examples from the most successful subset to illustrate the method's full potential.

4.1   Environment-Specific Performance Gains

A primary observation from our extensive experimentation is that significant performance enhancements attributable to the XRL-driven dynamic parameter adjustments were predominantly concentrated in the Taxi environment for both the PPO and DQN algorithms. While experiments were also conducted in the CartPole environment, the most compelling improvements were observed in the more complex Taxi navigation task. This pattern suggests that the utility of the introspective XRL mechanisms, as implemented in this study, is more pronounced in environments characterised by richer state-action spaces and sparser reward signals, where guidance afforded by XRL offers a more substantial advantage.

The CartPole environment, by contrast, is comparatively simple. It features a continuous but low-dimensional state space and provides a dense reward signal at every timestep. An optimal policy can be learned relatively quickly, and the exploration-exploitation trade-off is less challenging. Consequently, the nuanced adjustments provided by the XRL metrics offer marginal benefits over a standard, well-tuned RL algorithm. In the Taxi environment, however, with its larger discrete state space and sparse terminal rewards, the agent faces a much harder exploration problem. Here, the XRL metrics provide crucial intermediate guidance—signaling goal progress, highlighting surprising outcomes, and reinforcing confident decisions—that helps the agent navigate the vast search space more effectively, leading to the observed significant performance gains.

4.2 Comparative Algorithm Performance in the Taxi Environment Based on Rewards

Focusing on the results within the Taxi environment, distinct performance characteristics based on reward trajectories were evident when comparing the PPO and DQN algorithms, particularly when these were augmented with effective XRL configurations. The following subsections explore these differences in sample efficiency and training stability.

Figure 2 provides an overview of the average performance trends for XRL-enhanced PPO and DQN configurations against their respective baselines in the Taxi environment. On average, XRL-DQN configurations (solid blue line with circle markers) outperform the DQN baseline (dashed blue line). Interestingly, in this averaged view across all tested setups, the XRL-PPO configurations (solid orange line) do not show a clear average improvement over the PPO baseline (dashed orange line), suggesting that while some specific XRL-PPO configurations perform exceptionally well (as discussed below), the average impact across all tested XRL-PPO setups was more varied, with some configurations being detrimental.

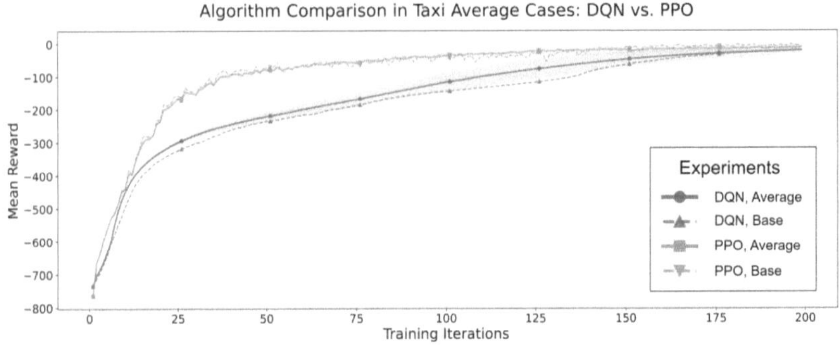

Fig. 2. Average performance comparison of XRL-enhanced DQN vs. PPO configurations, and their respective baselines, in the Taxi environment. Lines represent averages over all tested configurations within each category. (Color figure online)

PPO's Superior Sample Efficiency in Reward Attainment. The PPO algorithm, when paired with optimal XRL parameter settings, consistently demonstrated superior sample efficiency in the Taxi environment in terms of reward attainment. This is particularly evident when examining specific high-performing configurations, as illustrated in Fig. 3. The selected XRL-PPO configuration (solid orange line) achieves near-optimal mean rewards significantly faster than both its baseline (PPO Baseline, dashed orange line) and the XRL-DQN counterpart shown (solid blue line). The learning curve for this XRL-PPO agent shows a steep initial climb. The average case shown in Fig. 2 also supports this trend, with the PPO overall average (solid orange line with square markers) rising more quickly than the DQN overall average (solid blue line with circle markers).

DQN's Enhanced Training Stability in Rewards. While PPO often learned faster, XRL-enhanced DQN configurations tended to exhibit greater training stability in terms of lower variance in mean rewards across training iterations. This characteristic is clearly visible in Fig. 3. The shaded area representing the standard deviation around the mean reward for the XRL-DQN configuration (solid blue line) is markedly narrower than that of the XRL-PPO configuration (solid orange line). Interestingly, while the specific XRL-PPO case in Fig. 3 shows high variance, the averaged XRL-PPO curve in Fig. 2 (solid orange line with square markers) displays a relatively small shaded area. This suggests that while individual PPO runs can be highly variable, these variations might average out across different configurations or seeds.

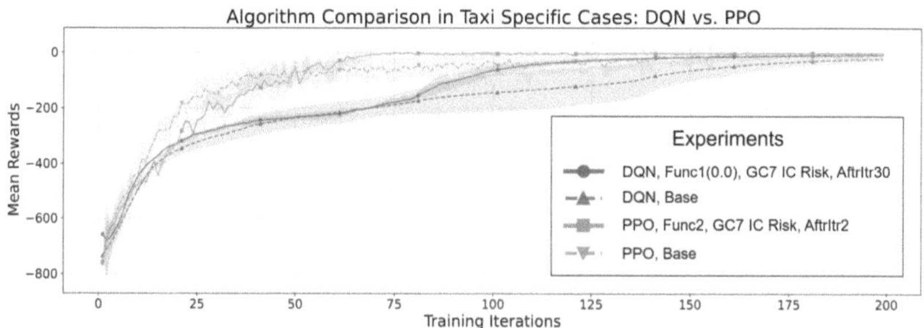

Fig. 3. Performance comparison (mean rewards) of specific, representative top-performing XRL-enhanced DQN vs. PPO configurations, alongside their respective baselines, in the Taxi environment. The legend provides details for each configuration: 'GC' indicates the Goal Conduciveness history length, and 'AftrItr' specifies the start iteration for XRL adjustments. (Color figure online)

4.3 Comparative Efficacy of Adjustment Functions for DQN in Taxi Based on Rewards

A particularly insightful pattern emerged from the analysis of DQN performance in the Taxi environment based on reward curves, specifically concerning the relative effectiveness of the two exploration adjustment mechanisms: Function 1 (Compounded Threshold Approach) and Function 2 (Compounded Combination Approach), as detailed in Sect. 3.4. Our findings indicate a general superiority of Function 1 over Function 2 for DQN within this environmental context.

This trend is first illustrated by an aggregated comparison of average performance (Fig. 4). On average, configurations utilising Function 1 (blue line with circle markers) achieve a slightly better learning trajectory and final mean reward compared to those using Function 2 (orange line with square markers). Both XRL-augmented function types, on average, demonstrate an improvement over the baseline DQN (green dashed line).

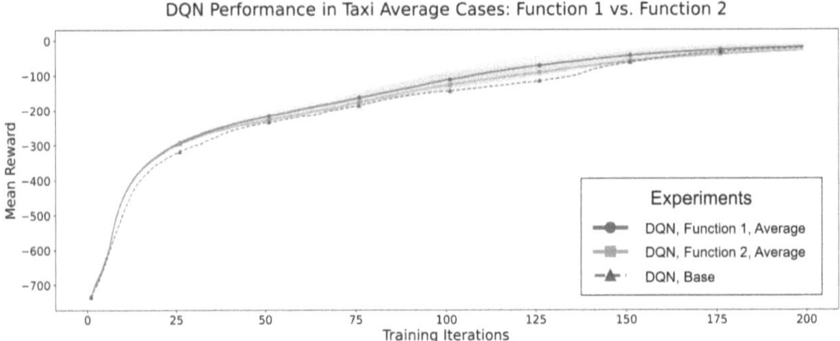

Fig. 4. Average performance comparison (mean rewards) of DQN configurations in the Taxi environment utilising Adjustment Function 1 versus Adjustment Function 2, relative to the DQN baseline. Lines represent the average performance across all DQN configurations using the respective adjustment function. (Color figure online)

The performance distinction becomes even more pronounced when examining specific, individual experimental configurations (Fig. 5). This figure highlights a representative high-performing XRL-DQN configuration utilising Function 1 (blue line) against a representative XRL-DQN configuration employing Function 2 (orange line) and the DQN baseline (green dashed line). The selected Function 1 configuration not only reaches a higher final mean reward more rapidly but also significantly outperforms the illustrated Function 2 case and exhibits substantially greater stability (narrower shaded band). Notably, some specific configurations using Function 2 performed on par with, or marginally worse than, the baseline.

4.4   Linking XRL Metric Dynamics to Performance Outcomes in Taxi

To explain the mechanisms behind the observed performance variations, we analysed the trajectories of the core XRL metrics: Goal Conduciveness (GC), Incongruity (IC), and Riskiness. We compared the metric profiles of the top three and bottom three performing XRL configurations against the baseline for both DQN and PPO in the Taxi environment.

DQN: Metric Signatures of Success and Failure. The reward trajectories for the selected best and worst XRL-DQN configurations, alongside the baseline,

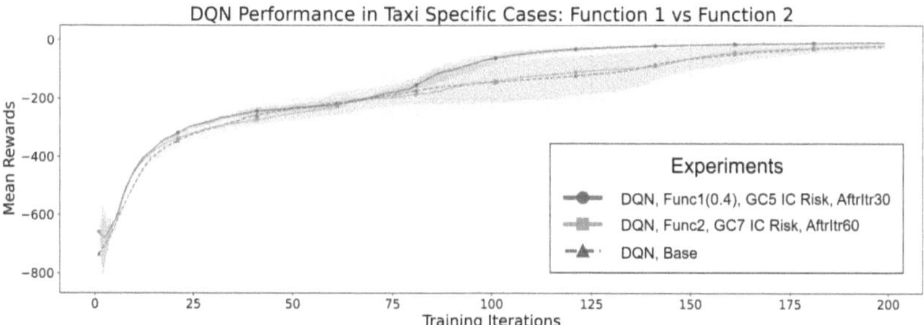

Fig. 5. Performance comparison (mean rewards) of specific, representative top-performing DQN configurations in the Taxi environment: Function 1 vs. Function 2, and baseline. The legend provides configuration details for the XRL-DQN Function 1 (blue line) and Function 2 (orange line) examples shown, where ' (Color figure online)GC' indicates the Goal Conduciveness history length. (Color figure online)

are depicted in Fig. 6. This plot clearly illustrates the substantial performance gap: the top configurations (lines in shades of green) rapidly converge to near-optimal rewards, significantly outperforming the baseline (black line), while the bottom configurations (lines in shades of red and orange) lag considerably or even perform worse.

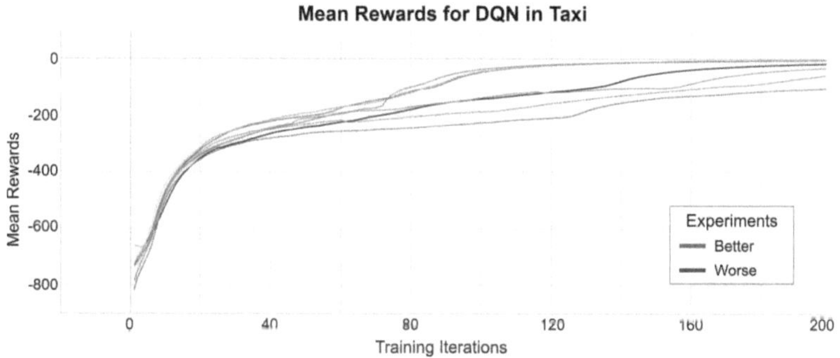

Fig. 6. Mean Rewards for best 3 (lines in shades of green), worst 3 (lines in shades of red and orange) XRL-DQN configurations, and baseline (black) in the Taxi environment, illustrating performance disparities. (Color figure online)

Examining the underlying XRL metrics provides insights into these disparities. As shown in Fig. 7, the best-performing XRL-DQN configurations rapidly achieve and sustain high levels of Mean Goal Conduciveness. Regarding Mean Incongruity (Fig. 8), after an initial learning spike, these top configurations demonstrate a quicker and more pronounced reduction in IC. Finally, Mean Riskiness profiles (Fig. 9) for the best XRL-DQN agents tend to rise to and maintain higher levels. In stark contrast, the worst-performing configurations typically exhibit low and stagnant GC, persistently high or slowly decreasing IC, and low and erratic Riskiness.

Collectively, for DQN, a 'healthy' XRL-enhanced learning trajectory that correlates with high rewards appears characterised by quickly establishing a strong sense of goal progress (high GC), rapidly improving environmental prediction (low IC), and developing decisiveness in action selection (high Riskiness).

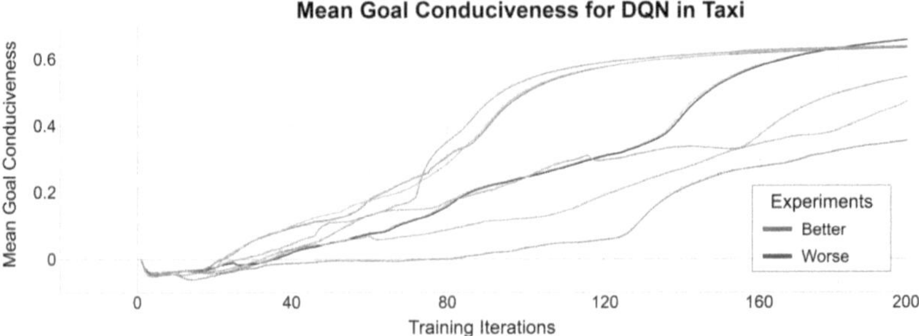

Fig. 7. Mean Goal Conduciveness for best 3 (lines in shades of green), worst 3 (lines in shades of red and orange) XRL-DQN configurations, and baseline (black) in Taxi, corresponding to rewards in Fig. 6 (Color figure online).

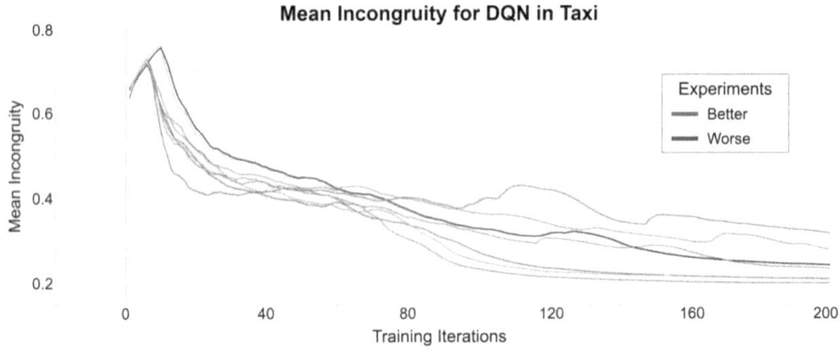

Fig. 8. Mean Incongruity for best 3 (lines in shades of green), worst 3 (lines in shades of red and orange) XRL-DQN configurations, and baseline (black) in Taxi, corresponding to rewards in Fig. 6 (Color figure online).

PPO: Metric Signatures of Success and Failure. A similar analysis was conducted for PPO agents. Figure 10 displays the reward trajectories for the best three and worst three XRL-PPO configurations compared to the PPO baseline. Again, the top configurations (lines in shades of green) demonstrate clear performance advantages in terms of faster learning and higher final rewards over

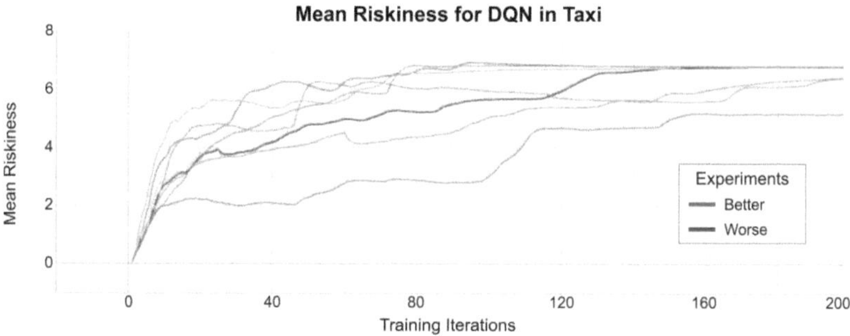

Fig. 9. Mean Riskiness for best 3 (lines in shades of green), worst 3 (lines in shades of red and orange) XRL-DQN configurations, and baseline (black) in Taxi, corresponding to rewards in Fig. 6 (Color figure online).

the baseline (black line) and especially over the poorly performing XRL configurations (lines in shades of red and orange).

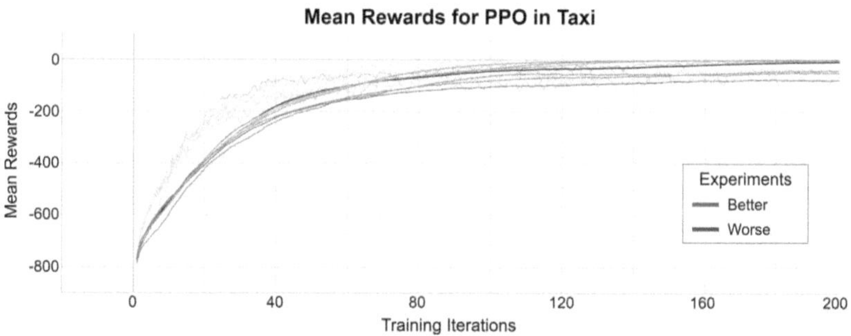

Fig. 10. Mean Rewards for best 3 (lines in shades of green), worst 3 (lines in shades of red and orange) XRL-PPO configurations, and baseline (black) in the Taxi environment, illustrating performance disparities. (Color figure online)

The corresponding metric dynamics for these PPO agents reveal related patterns. The best-performing XRL-PPO configurations (Fig. 11) tend to achieve positive Mean Goal Conduciveness values earlier and more consistently. Mean Incongruity (Fig. 12), after the initial spike, tends to decrease more rapidly and reach more stable negative values for these top configurations. Regarding Mean Riskiness (Fig. 13), the best XRL-PPO configurations generally achieve higher riskiness values sooner and sustain them more effectively. Conversely, the worst XRL-PPO configurations often show negative and erratic GC, slower decrease or higher plateaus in IC, and lower and slower increase in Risk.

Thus, for PPO, successful XRL integration leading to superior rewards appears to foster an early establishment of positive goal-directedness (positive GC), efficient learning of environmental predictability (rapidly decreasing IC to more negative values), and quicker development of policy decisiveness (higher Riskiness).

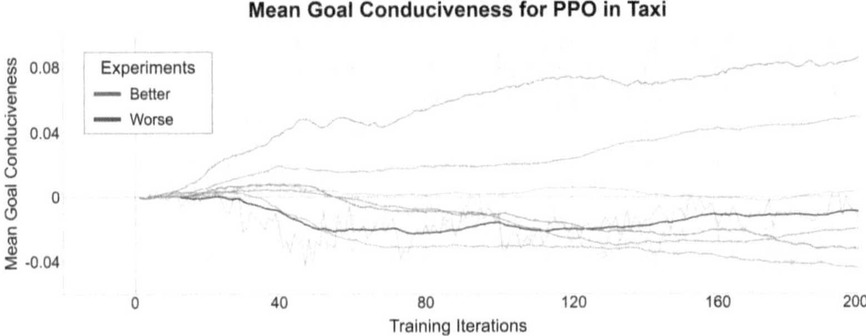

Fig. 11. Mean Goal Conduciveness for best 3 (lines in shades of green), worst 3 (lines in shades of red and orange) XRL-PPO configurations, and baseline (black) in Taxi, corresponding to rewards in Fig. 10 (Color figure online).

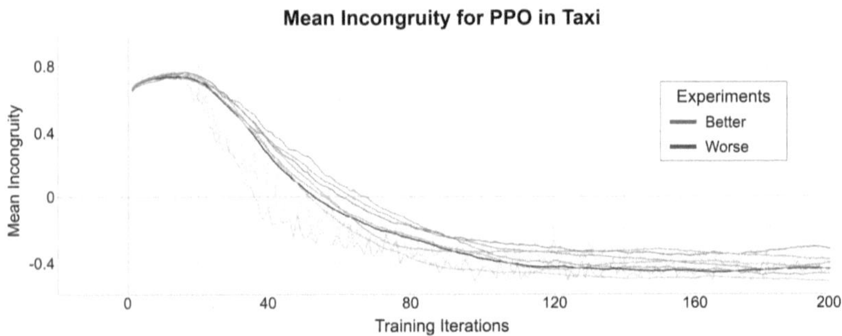

Fig. 12. Mean Incongruity for best 3 (lines in shades of green), worst 3 (lines in shades of red and orange) XRL-PPO configurations, and baseline (black) in Taxi, corresponding to rewards in Fig. 10 (Color figure online).

Translating Metric Dynamics to Agent Behaviour. While the correlation between metric profiles and rewards is clear, it is crucial to understand how these numerical trends translate into observable agent behaviour. In the Taxi environment, for a successful agent:

– **High Goal Conduciveness (GC)** means the agent is consistently choosing actions that reduce the distance to the passenger or the destination. For example, instead of wandering randomly, it learns to prioritise 'move north' or 'move east' actions that directly advance its sub-goal.

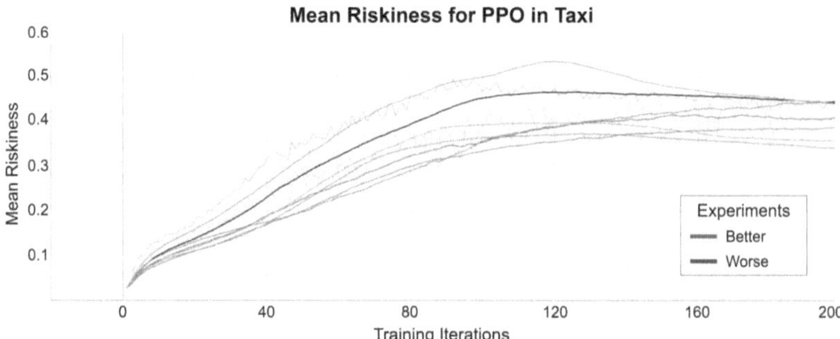

Fig. 13. Mean Riskiness for best 3 (lines in shades of green), worst 3 (lines in shades of red and orange) XRL-PPO configurations, and baseline (black) in Taxi, corresponding to rewards in Fig. 10 (Color figure online).

– **Low Incongruity (IC)** reflects a well-calibrated internal model. The agent is no longer "surprised" by the environment's response. This manifests as efficient, deliberate action sequences. For instance, it successfully navigates a series of turns to reach a passenger without making unexpected moves into walls, which would have generated high TD errors (high IC) earlier in training.
– **High Riskiness** indicates the agent is making confident, high-stakes decisions. When faced with picking up a passenger versus moving to an empty adjacent square, a high-riskiness agent decisively chooses the 'pickup' action because its Q-value for that action is significantly higher than any alternative. This avoids the dithering behaviour seen in early training, where action probabilities or Q-values might be very close.

This qualitative link between XRL metrics and agent behaviour bridges the gap between numerical improvements and an intuitive understanding of why the XRL-guided agent becomes more effective. The pronounced difference in metric profiles between high and low-performing XRL configurations for both algorithms strongly underscores the critical importance of precise XRL parameter tuning.

5   Conclusion

This research has significantly extended our prior investigations [24] by undertaking a comprehensive parametric study focused on optimising reinforcement learning (RL) agent performance through the strategic integration of explainable reinforcement learning (XRL) metrics. Our core objective was to meticulously map the sensitivities of agent learning to a diverse range of XRL configurations. By evaluating Proximal Policy Optimisation (PPO) and Deep Q-Networks (DQN) across 1,440 unique XRL-enhanced setups in the CartPole and Taxi environments, we aimed to generate robust empirical guidelines for more potent XRL integration.

Our extensive experimentation confirmed that while XRL offers substantial potential for performance enhancement, its efficacy is profoundly dependent on precise parameterisation. A relatively small fraction of the explored configurations (20) yielded statistically significant improvements, with these gains being most evident in the more complex Taxi environment. We provided a detailed rationale for this disparity, attributing the lack of improvement in CartPole to its relative simplicity and dense reward structure, which makes the nuanced guidance of XRL less critical. In the Taxi environment, XRL-augmented PPO agents frequently demonstrated superior sample efficiency, while their XRL-DQN counterparts often exhibited enhanced training stability.

Crucially, our analysis of XRL metric trajectories for top and bottom-performing configurations provided deeper insights into these performance differentials. We identified distinct "healthy" metric profiles associated with successful learning—typically involving high Goal Conduciveness, low Incongruity, and high Riskiness—and provided a qualitative explanation of how these metric values translate into more efficient and deliberate agent behaviours. This metric-level analysis not only reinforces the critical need for careful XRL parameter tuning but also illuminates the internal learning dynamics that underpin XRL-driven performance gains. The implications of this work are twofold: it provides empirical evidence that well-configured introspective XRL can lead to demonstrably better-performing RL agents, and it offers a more nuanced understanding of how these improvements manifest at the level of interpretable learning metrics. This contributes to the broader goal of making XRL a truly actionable tool, capable of guiding the development of more effective, reliable, and transparent autonomous systems.

6   Future Work

The findings and limitations of this study pave the way for several exciting avenues of future research, aimed at further unlocking the potential of XRL for optimising RL agent performance.

Firstly, the scope of empirical validation should be expanded. Investigating the generalisability of the identified "healthy" XRL metric profiles and optimal parameter sensitivities across a wider array of environments—including those with continuous action spaces, partial observability, and different reward structures—is paramount. Similarly, applying these introspective XRL techniques to a broader suite of RL algorithms will be crucial.

Secondly, the XRL mechanisms themselves can be evolved. Future work could explore additional XRL metrics or design more sophisticated adaptive adjustment functions. This could involve mechanisms that dynamically select or weight XRL metrics during training based on their reliability, or even adapt XRL parameters (e.g., thresholds) online by monitoring their correlation with performance.

Thirdly, the challenge of efficiently navigating the vast XRL parameter space warrants attention. While our grid search was systematic, more advanced optimisation techniques, such as Bayesian optimisation or evolutionary algorithms, could be employed to more efficiently discover optimal XRL configurations. This would reduce the reliance on extensive manual tuning and represents a key step towards more autonomous, self-optimising RL systems.

Furthermore, a deeper causal understanding of the link between XRL parameters, metric dynamics, and agent behaviour is needed. This involves moving beyond correlational studies to experiments designed to dissect how specific XRL interventions directly influence an agent's internal decision-making processes. Perhaps the observed metric profiles could be detected early to salvage unpromising training runs, thereby saving computational resources.

Finally, the application of these refined XRL-enhanced agents to challenging real-world problems remains a key objective. Demonstrating the practical benefits of these introspective optimisation techniques in domains such as robotics, autonomous navigation, or complex system control would provide compelling validation and undoubtedly uncover new research frontiers at the intersection of performance, explainability, and reliability.

By pursuing these directions, we can continue to refine XRL not just as a tool for interpretation, but as an integral component for building more intelligent, adaptive, and effective autonomous agents.

References

1. Ayyar, M., Benois-Pineau, J., Zemmari, A.: Esl: explain to improve streaming learning for transformers. In: International Conference on Pattern Recognition, pp. 160–175. Springer (2025). https://doi.org/10.1007/978-3-031-78189-6_11
2. Cheng, Z., Wu, X., Yu, J., Sun, W., Guo, W., Xing, X.: Statemask: explaining deep reinforcement learning through state mask. In: Advances in Neural Information Processing Systems (2023)
3. Cheng, Z., Wu, X., Yu, J., Yang, S., Wang, G., Xing, X.: Rice: breaking through the training bottlenecks of reinforcement learning with explanation. arXiv preprint arXiv:2405.03064 (2024)
4. Cheng, Z., Wu, X., Yu, J., Yang, S., Wang, G., Xing, X.: Rice: breaking through the training bottlenecks of reinforcement learning with explanation. In: International Conference on Machine Learning (2024)
5. Devlin, S.M., Kudenko, D.: Dynamic potential-based reward shaping. In: International Conference on Autonomous Agents and Multiagent Systems, pp. 433–440 (2012)
6. Finkelstein, M., Liu, L., Schlot, N., Kolumbus, Y., Parkes, D., Rosenschein, J., Keren, S.: Explainable reinforcement learning via model transforms. In: Proceedings of the AAAI Conference on Artificial Intelligence, vol. 36, pp. 12943–12944 (2022)
7. Gu, A., Dao, T.: Mamba: Linear-time sequence modeling with selective state spaces. arXiv preprint arXiv:2312.00752 (2023)
8. Guo, W., Wu, X., Khan, U., Xing, X.: Edge: explaining deep reinforcement learning policies. In: Advances in Neural Information Processing Systems, vol. 34, pp. 12222–12236 (2021)
9. Gupta, A., Pacchiano, A., Zhai, Y., Kakade, S., Levine, S.: Unpacking reward shaping: Understanding the benefits of reward engineering on sample complexity.

In: Advances in Neural Information Processing Systems, vol. 35, pp. 15281–15295 (2022)

10. He, C., Nachum, O., Theocharous, G., Campbell, M.: Explainable reinforcement learning via model transforms. In: Proceedings of Advances in Neural Information Processing Systems (NeurIPS), vol. 35, pp. 12345–12358 (2022)

11. Heuillet, A., Couthouis, F., Daz-Rodriguez, N.: Explainability in deep reinforcement learning. Knowledge-Based Syst. (2021)

12. Hu, Y., et al.: Learning to utilize shaping rewards: a new approach of reward shaping. In: Advances in Neural Information Processing Systems, vol. 33, pp. 15931–15941 (2020)

13. Jacq, A., Ferret, J., Pietquin, O., Geist, M.: Lazy-mdps: towards interpretable rl by learning when to act. In: International Conference on Autonomous Agents and Multiagent Systems, pp. 669–677 (2022)

14. Jia, Y., Mcdermid, J., Lawton, T., Habli, I.: The role of explainability in assuring safety of machine learning in healthcare. IEEE Trans. Emerging Topics Comput. **10**, 1746–1760 (2021). https://api.semanticscholar.org/CorpusID:237385754

15. Jiang, Z., et al.: Reinforcement learning from imperfect corrective actions and proxy rewards. arXiv preprint arXiv:2410.05782 (2024)

16. Liu, S., Zhu, M.: Utility: utilizing explainable reinforcement learning to improve reinforcement learning. In: The Thirteenth International Conference on Learning Representations (2025)

17. Lu, W., Zhao, X., Fryen, T., Lee, J., Li, M., Magg, S., Wermter, S.: Causal state distillation for explainable reinforcement learning. In: 3rd Conference on Causal Learning and Reasoning (CLeaR) (2024)

18. Mallic, R., et al.: A hybrid transformer with domain adaptation using interpretability techniques for the application to the detection of risk situations. Multimedia Tools Appli. **83**(35), 83339–83356 (2024)

19. Mulder, C., Bazeley-White, E., Dimitrakopoulos, P., Hector, A., Scherer-Lorenzen, M., Schmid, B.: Species evenness and productivity in experimental plant communities. Oikos **107**(1), 50–63 (2004)

20. Ng, A.Y., Harada, D., Russell, S.J.: Policy invariance under reward transformations: theory and application to reward shaping. In: International Conference on Machine Learning, pp. 278–287 (1999)

21. Ng, A., Harada, D., Russell, S.: Policy invariance under reward transformations: theory and application to reward shaping. In: International Conference on Machine Learning, pp. 278–287 (1999)

22. Oh, J., Guo, Y., Singh, S., Lee, H.: Self-imitation learning. In: International Conference on Machine Learning, pp. 3878–3887 (2018)

23. Puiutta, E., Veith, E.: Explainable reinforcement learning: a survey, pp. 77–95 (2020)

24. Quintana-Amate, S., Stevens, D., ketheeswaran, V., Capaldo, P., Sheldon, D., Hall, M.: Explain to gain: Introspective reinforcement learning for enhanced performance. In: Communications in Computer and Information Science, vol. 30. Springer (2025). https://doi.org/10.1007/978-3-032-08324-1_11

25. Retzlaff, C., et al.: Human-in-the-loop reinforcement learning: a survey and position on requirements, challenges, and opportunities. J. Artifi. Intel. Res. **79**, 359–415 (2024). https://api.semanticscholar.org/CorpusID:267448673

26. Sequeira, P., Gervasio, M.: Interestingness elements for explainable reinforcement learning: understanding agents' capabilities and limitations. Artifi. Intell. (2020)

27. Sequeira, P., Gervasio, M.: Ixdrl: a novel explainable deep reinforcement learning toolkit based on analyses of interestingness. In: World Conference on Explainable Artificial Intelligence, pp. 373–396. Springer (2023). https://doi.org/10.1007/978-3-031-44064-9_20

28. Shicheng Liu, M.Z.: Utility: utilizing explainable reinforcement learning to improve reinforcement learning. In: The Thirteenth International Conference on Learning Representations (2025)

29. Silver, D., et al.: A general reinforcement learning algorithm that masters chess, shogi, and go through self-play. Science **362**(6419), 1140–1144 (2018)

30. Sutton, R., Barto, A.: Reinforcement Learning: An Introduction, second edition edn.. MIT Press (2018)

31. Taylor, M.E.: Improving reinforcement learning with human input. In: International Joint Conference on Artificial Intelligence, pp. 5724–5728 (2018)

32. Taylor, M.E., Nissen, N., Wang, Y., Navidi, N.: Improving reinforcement learning with human assistance: an argument for human subject studies with hippo gym. Neural Comput. Appl. **35**(32), 23429–23439 (2023)

33. Vinyals, O., et al.: Grandmaster level in starcraft ii using multi-agent reinforcement learning. Nature **575**(7782), 350–354 (2019)

34. Wang, Z., Taylor, M.: Improving reinforcement learning with confidence-based demonstrations. In: International Joint Conference on Artificial Intelligence, pp. 3027–3033 (2017)

35. Wang, Z., Taylor, M.E.: Improving reinforcement learning with confidence-based demonstrations. In: International Joint Conference on Artificial Intelligence, pp. 3027–3033 (2017)

36. van Waveren, S., Pek, C., Tumova, J., Leite, I.: Correct me if i'm wrong: using non-experts to repair reinforcement learning policies. In: Proceedings of the 2022 ACM/IEEE International Conference on Human-Robot Interaction (HRI), pp. 493–501. IEEE (2022)

37. Xu, M., Chen, X., Wang, J.: Policy correction and state-conditioned action evaluation for few-shot lifelong deep reinforcement learning. IEEE Trans. Neural Netw. Learn. Syst. (2024)

38. Yu, J., Guo, W., Qin, Q., Wang, G., Wang, T., Xing, X.: Airs: Explanation for deep reinforcement learning based security applications. In: Proc. of USENIX Security (2023)

Quantifying Prototype Stability in ProtoPNet Without Manual Part Annotations

Iker Sancho[1,2]([✉]) [ID], Ruben Naranjo[1,2] [ID], Nerea Aranjuelo[1] [ID], and Itsaso Rodríguez-Moreno[2] [ID]

[1] Vicomtech Foundation, Basque Research and Technology Alliance (BRTA), Mikeletegi 57, 20009 Donostia-San Sebastián, Spain
[2] University of the Basque Country (UPV/EHU), Donostia-San Sebastián, Spain
isancho@vicomtech.org

Abstract. Prototype-based networks such as ProtoPNet offer intrinsically interpretable predictions by matching regions of the inferenced image to learned prototypical patches. However, existing stability metrics rely on expensive manual part annotations and are limited to narrow perturbation types. In this work, we introduce the *Structural Stability Score* (S_{ss}), a scalable, annotation-free metric that quantifies prototype stability under a diverse set of visual transformations by comparing prototype activation maps. We evaluate S_{ss} on two ProtoPNet variants (using VGG19 and Resnet34 backbones) trained on the CUB-200-2011 dataset, and assess stability across six distinct perturbations. Our results reveal clear differences in robustness both between models and among different transformations. These findings demonstrate that S_{ss} is a practical tool for highlighting stability variations within and across prototype-based networks, guiding model selection and interpretability analysis.

Keywords: Prototype-based Networks · Interpretability · Stability

1 Introduction

In recent years, the adoption of complex machine learning models, such as deep neural networks, has yielded unprecedented performance across a wide range of tasks, from medicine [12] to autonomous driving [10]. However, the opaqueness of these "black-box" models impedes trust, accountability, and regulatory compliance, especially in high-stakes domains where understanding the rationale behind a decision is critical [4,9,18]. This growing demand for transparency [8] has motivated the development of interpretable machine learning approaches that either explain predictions post-hoc [22] or embed interpretability directly into the model architecture [1].

Prototype-based networks [6,11,24] represent a promising paradigm for intrinsic interpretability. Inspired by human reasoning, which often relies on

© The Author(s), under exclusive license to Springer Nature Switzerland AG 2026
F. Marcelloni et al. (Eds.): IJCCI 2025, CCIS 2829, pp. 74–85, 2026.
https://doi.org/10.1007/978-3-032-15638-9_5

recalling representative examples, ProtoPNet learns a dictionary of prototypical image patches and classifies new inputs by measuring the similarity between image regions and these learned prototypes [6,21]. By displaying the most relevant prototypes next to each prediction, ProtoPNet provides a clear, example-based justification: the model indicates "this part of the image looks like that prototype", facilitating intuitive human inspection without sacrificing performance.

However, the mere visualization of matching prototypes can hide semantic inconsistencies and fragility to slight perturbations in the input [17]. Existing measures of prototype quality, such as consistency and stability scores, are based on manual part annotations, which limits their scalability and objectivity [14,15]. To address these problems, we introduce the Structural Stability Index (S_{ss}), which leverages the Jaccard index and the distance between maximum heatpoints to compare prototype activation maps between original and perturbed images, quantifying robustness across multiple perturbation types without human labeling.

2   Related Work

Interpretable machine learning has become a critical requirement in many application domains [18], as black-box models, especially deep neural networks, can be difficult to trust or audit when making decisions [4,13]. Prototype-based networks address this need by embedding interpretability directly in the model [21]. In particular, ProtoPNet [6] learns a set of prototypical image patches and classifies new images by comparing local regions to these prototypes. This design yields a transparent reasoning process: each prediction is justified by showing the image parts that match learned prototypes, akin to how a human might recognise an object by recalling similar examples. Importantly, ProtoPNet and its variants have been shown to attain accuracy on par with conventional Convolutional Neural Networks (CNNs) while providing such built-in explanations [5,7,11,16,19].

Despite their promise, prototype-based explanations may require extra processes to improve their quality and robustness [2] as well as rigorous evaluations to ensure their meaningfulness and robustness. Visualising a few "matching prototypes" is not enough to guarantee interpretability in general [17].

One approach to enhance the quality of part-prototype explanation is presented in the paper "This Looks Like That, Because... Explaining Prototypes for Interpretable Image Recognition" [19]. During the inference process, rather than simply showing which patches match, this methodology provides quantitative information about the contribution that each visual characteristic (colour, shape, texture, contrast, saturation) makes to each prototype in this process. This allows for a more detailed breakdown of how each prototype encodes knowledge and how much each feature influences the model's decision.

Other recent studies have found that learned prototypes can be fragile or inconsistent. For example, even small perturbations or distortions can cause a prototype to activate on a semantically different part of the image [15,20]. To address this issue, the paper "Evaluation and Improvement of Interpretability for Self-Explainable Part-Prototype Networks" [15] introduces two significant quantitative metrics for assessing prototype quality: the *consistency score* and the *stability score*. These metrics, respectively, measure whether each prototype corresponds to the same object part across different images, and whether that correspondence holds under slight input perturbations. However, computing these metrics requires detailed human-provided part annotations: each prototype's location is mapped to ground-truth object parts in every image, so the stability score (sometimes called S_{sta}) can only be evaluated with manual labelling of parts. Such manual intervention limits scalability and objectivity [3,14].

In view of these limitations, an objective methodology that ensures the correct measurement of the stability of prototypes in Part-Prototype Networks is of utmost importance, specially given the relevance of interpretable machine learning in high-risk applications.

3 Method

In this section, we first review existing metrics and enhanced explainability techniques for prototype-based models, and then, introduce our proposed metric.

3.1 Importance Scores for Image Characteristics

As explained in Sect. 2, in the search for improved explanations, a new approach to measure the importance of different features in the inference of an image is proposed in the paper "This Looks Like That, Because... Explaining Prototypes for Interpretable Image Recognition" [19].

Following the indications mentioned in the cited research, a closed set of transformations is defined, Φ, trying to address as many important features in an image as possible (see Eq. 1). These transformations modify contrast, saturation, hue, shapes (using a wavelet distortion effect), and texture (applying denoising techniques) as seen in Fig. 1.

$$\Phi = \{f_{\text{Contrast}}(x), f_{\text{Saturation}}(x), f_{\text{Hue}}(x), f_{\text{Shape}}(x), f_{\text{Texture}}(x)\}. \tag{1}$$

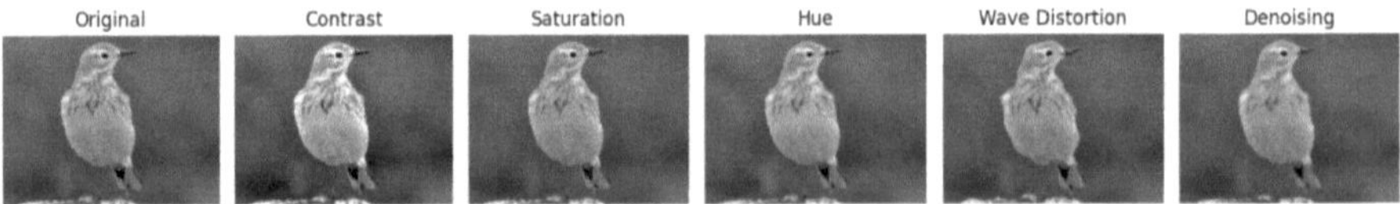

Fig. 1. Transformations from the set Φ applied to a test image.

With the transformations already defined, two distinct metrics are introduced to precisely quantify the effect of specific visual features on the model's decision process: the *local importance score*, which measures the difference in similarity score between a prototype and an original image versus the same prototype and the image modified by a specific visual transformation, and the *global importance score*, which computes a weighted arithmetic mean of these local importances across all training images, weighting each term by its prototype similarity to ensure that higher-similarity images have greater influence.

In this way, it is possible to decide which is the most influential characteristic in the classification, achieving a quantitative and precise measurement.

3.2   Stability Score

Prototype evaluation via qualitative human studies (where multiple annotators score visual characteristics under strict guidelines) is rare and resource-intensive. To address this, Huang et al. (2023) [15] propose two differente scores to evaluate the **interpretability** of Part-Prototype Networks: **(1) Consistency**, which tests whether a prototype activates on the same object part across different images. **(2) Stability**, which tests whether a prototype activates on the same part of an object before and after perturbing an image. In this work, we focus exclusively on the Stability metric (S_{sta}) defined in Eq. 2.

$$S_{\mathrm{sta}} = \frac{1}{M} \sum_{j=1}^{M} \frac{\sum_{x \in \mathcal{I}_{c(j)}} \mathbb{1}\{\mathbf{o}^{p_j}(x) = \mathbf{o}^{p_j}(x + \xi)\}}{|\mathcal{I}_{c(j)}|}. \tag{2}$$

For each prototype p_j, it is observed how many parts of the object remain within the activation performed on the original image, $\mathbf{o}^{p_j}(x)$. Then, the process is repeated on the disturbed image, $\mathbf{o}^{p_j}(x + \xi)$ (where ξ is Gaussian noise) and each part of the object is compared (Fig. 2 is a visual representation of this process). The stability of prototype p_j is then averaged over all test images in its category $c(j)$, and the overall score is the mean over all M prototypes.

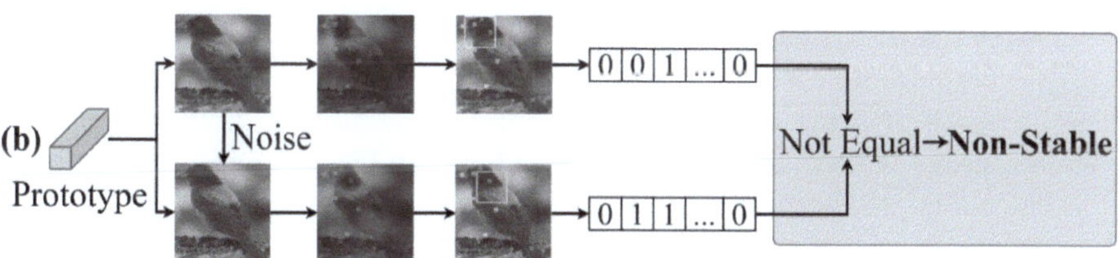

Fig. 2. Visual example of the process of calculating stability (S_{sta}) for a single prototype on one image. Image extracted from Huang et al. (2023) [15].

This example highlights how minor perturbations can affect prototype activation maps and underscores the importance of robustness in interpretability evaluations.

3.3   Towards Enhanced Prototype Interpretability

Since the stability score introduced by Huang et al. (2023) [15] relies on manually identified object parts to assess whether a prototype activation remains anchored to the same region, this process still demands costly annotations and may lead to problems in the scalability and applicability of the metric. To overcome these limitations, we propose the **Structural Stability Score** (S_{ss}) which eliminates manual part labeling, and calculates the Stability with a prototype activation area comparison process. In addition, we propose the use of a closed set of transformations (defined in Sect. 3.1), in order to calculate the stability against different transformations and to determine which is the visual characteristic against which each prototype is less stable.

To better understand our proposal, it is necessary to define the way in which two activations maps are compared and scored. The process is as follows:

1. **Activation Area Comparison:** The activation area comparison process consists of different phases: **(1)** First, let $V_1, \in [0,1]^{H \times W}$ be the activation map of prototype p_j over the image x. **(2)** The activation map is then binarized by assigning a value of 1 to activations above γ, and 0 to activations below γ; $B(\bar{V}_1) = \{\mathbf{1}(\bar{V}_1 \geq \gamma)\ \ \&\ \ \mathbf{0}(\bar{V}_1 < \gamma)\}$. **(3)** Now, let $V_2, \in [0,1]^{H \times W}$ be the activation map of the prototype p_j over the perturbed image x_d. Repeat the previous process to calculate $B(\bar{V}_2)$. **(4)** Finally, we calculate the *Jaccard Index*, also known as Intersection over Union (IoU), on both maps (Eq. 3).

$$\mathrm{J}(B(\bar{V}_1),\, B(\bar{V}_2)) \;=\; \frac{\left| B(\bar{V}_1) \cap B(\bar{V}_2)\right|}{\left| B(\bar{V}_1) \cup B(\bar{V}_2)\right|}. \tag{3}$$

2. **Maximum Activation Distance:** To avoid falling into the analysis of the activation area only, this metric also evaluates the displacement of the maximum peak of the activation after the transformation of the image: **(1)** First we calculate the Euclidean distance between each maximum activation peak, $d(\bar{V}_1, \bar{V}_2) \;=\; \left\| \arg\max \bar{V}_1 - \arg\max \bar{V}_2 \right\|_2$. **(2)** Then, knowing that both activation maps have the same dimensions $(H \times W)$, the maximum possible distance is calculated, $d_{\max} = \sqrt{(H-1)^2 + (W-1)^2}$. **(3)** Finally, the distance between the maximum peaks is normalized (scaled to lie between 0 and 1) and the similarity is calculated as shown in Eq. 4.

$$S_d(\bar{V}_1, \bar{V}_2) = 1 - \frac{d(\bar{V}_1, \bar{V}_2)}{d_{\max}}. \tag{4}$$

3. **Weighted Sum:** The numerical value of the comparison between two activation maps (such as $\bar{V}_1$ and $\bar{V}_2$) will be composed of the two parts previously calculated: the Jaccard's coefficient between the binarized maps, and the similarity of the Euclidean distance between the maximum activation peaks. Finally, the weighted sum of each coefficient is calculated to obtain a single value, as seen in Eq. 5.

$$f_s(\bar{V}_1, \bar{V}_2) = w \cdot \mathrm{J}(B(\bar{V}_1), B(\bar{V}_2)) + (1 - w) \cdot S_d(\bar{V}_1, \bar{V}_2), \qquad w \in [0, 1]. \quad (5)$$

Figure 3 visually represents this stability calculation process. Knowing how two activation maps are compared, the next step is to integrate this process into the equation of the stability calculation. First, as one of the fundamental pillars of this metric is to be able to calculate the stability against different perturbations (and not only against Gaussian noise), we must define a set of transformations with which the images will be perturbed. In this case, we use the same set as the one proposed in the paper "This Looks Like That, Because... Explaining Prototypes for Interpretable Image Recognition" [19] and add the Gaussian noise transformation proposed in the original stability formula (See Eq. 2). Figure 4 and Fig. 5 visually show how different transformations can affect the stability calculation. In addition, we do a comparison between a stable prototype and an unstable prototype.

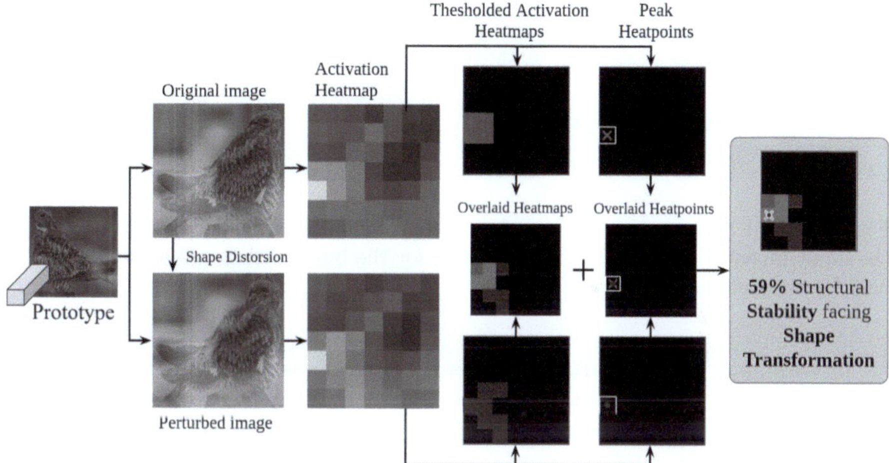

Fig. 3. Visual representation of the process of calculating the stability of a prototype. The process compares the activation map of the prototype on the original image and on the transformed image.

From now on, the expression $\bar{V}^{p_j}(x)$ will refer to the activation map of prototype p_j on image x, and $\phi_i(x)$, will refer to an image x perturbed with the transformation $\phi_i \in \Phi$. So the activation map of a prototype over a perturbed image will be expressed as $\bar{V}^{p_j}(\phi_i(x))$. Equation 7 shows how the Structural Stability Score of a Part-Prototype Network against a ϕ_i transformation is calculated.

$$\Phi = \{f_{\text{Contrast}}(x), f_{\text{Saturation}}(x), f_{\text{Hue}}(x), f_{\text{Shape}}(x), f_{\text{Texture}}(x), f_{\text{Noise}}(x)\}. \quad (6)$$

$$S_{ss}^{\phi_i} = \frac{1}{M} \sum_{j=1}^{M} \frac{\sum_{x \in \mathcal{I}_{c(j)}} f_s\{\bar{V}^{p_j}(x), \bar{V}^{p_j}(\phi_i(x))\}}{\|\mathcal{I}_{c(j)}\|}, \quad \phi_i \in \Phi. \quad (7)$$

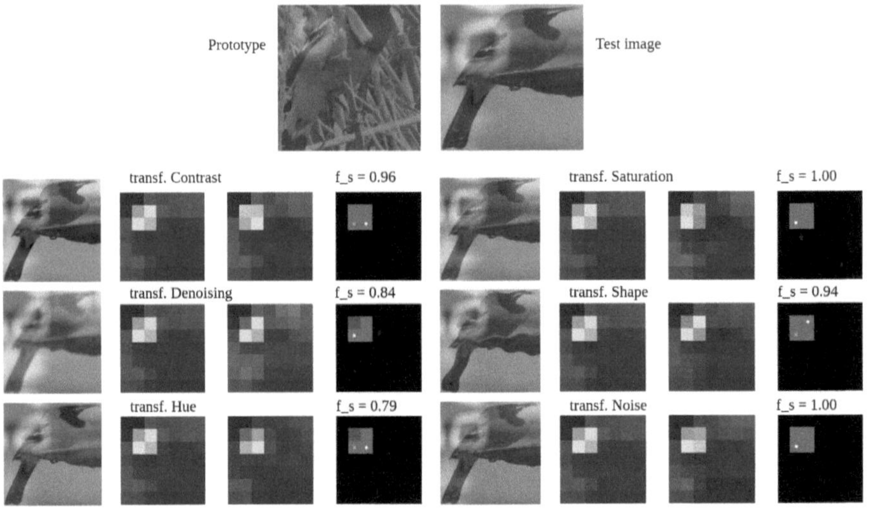

Fig. 4. Calculation of the stability of a prototype against different transformations in the input image. A threshold, γ, of 0.6 is set for the binarization and a coefficient w of 0.65 for the weighted sum.

In summary, this research provides a comprehensive metric for prototype-based models by seamlessly integrating two complementary criteria: the overlap of activation regions (quantified via the Jaccard index) and the normalized Euclidean proximity of activation peaks. This dual assessment not only offers a more faithful spatial consistency measure under a diverse set of perturbations, but also indicates the specific transformation to which each prototype is most sensitive. Furthermore, by relying on a closed set of predefined transformations, our metric achieves both extensibility and scalability, enabling effortless adaptation to emerging perturbation types and large-scale vision tasks.

4   Experiments

In order to validate the efficacy and robustness of our proposed metric, we design a set of experiments that evaluate how prototype-based models behave under various visual perturbations.

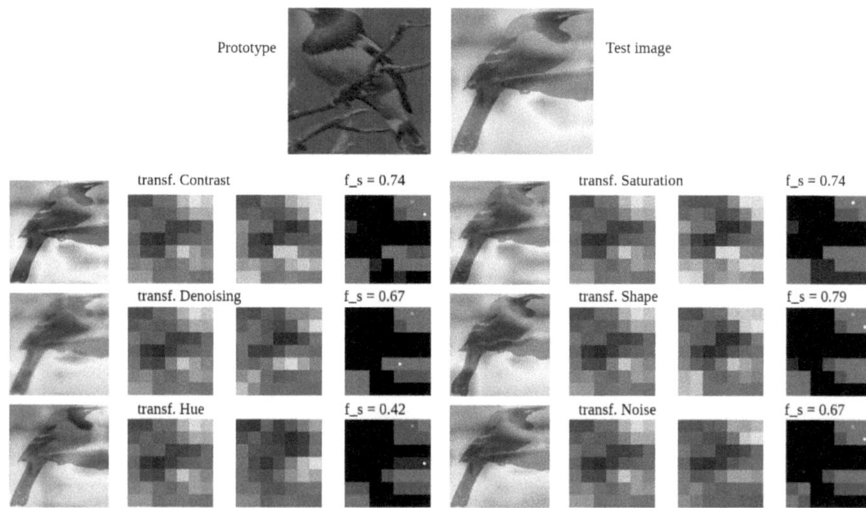

Fig. 5. Calculation of the stability of a unstable prototype against different transformations in the input image. A threshold, γ, of 0.6 is set for the binarization and a coefficient w of 0.65 for the weighted sum.

4.1 Experimental Setting

Datasets. All experiments are conducted on the CUB-200-2011 dataset [23], a fine-grained bird classification benchmark consisting of 11,788 images across 200 species, each annotated with bounding boxes, segmentations, and part-level attributes. Following prior work on prototype-based interpretability [6,15,19], we train and evaluate our models under the same train/test split and preprocessing protocols.

Model Configurations. We instantiate two variants of our prototype-based model differing only in backbone architecture using both *Resnet34* backbone and *VGG19* backbone. Both variants are trained end-to-end, optimizing the classification loss alongside prototype clustering and separation terms, in accordance with the implementation details of [6,19].

Parameters. Regarding the dimensions of the prototype activation maps, we set H, W to be 7 and 7. For the first stability calculation, we set γ and w to be 0,5 both (in Tables 2, 3 and 4, we test with different configurations of these parameters). Our models are trained for 15 epochs with the Adam optimizer (including 5 epochs for warm-up). We set the learning rates of the backbone, the add-on module, and the prototypes to be $1e^{-4}$, $3e^{-3}$, and $3e^{-3}$, respectively. The number of prototypes per category and the dimension of prototypes is set to 10 and 128 for our method.

4.2   Stability Evaluation

We assess model robustness by measuring Structural Stability Score under the six different transformations defined in the Eq. 6: noise, contrast, saturation, hue, shape and texture. For the Gaussian Noise transformation we set σ to 0.2, the rest of the transformations maintain the parameters of the paper "This Looks Like That, Because... Explaining Prototypes for Interpretable Image Recognition" [19]. Table 1 reports these scores for VGG19 and Resnet34 with $\gamma = 0.5$, $w = 0.5$. The results in the table demonstrate how the stability of a model varies based on the transformation applied, allowing a better understanding of how prototypes respond to visual perturbations at the time of inference.

Table 1. Average Stability results for Resnet34 and VGG19, for each different transformation, using $\gamma = 0.5$ and $w = 0.5$

Stability	Noise	Contrast	Saturation	Hue	Shape	Texture	Avg Stability
VGG19	0,7529	0,8157	0,8065	0,7269	0,8307	0,6924	0,7709
Resnet34	0,8087	0,8391	0,8459	0,7406	0,8386	0,7537	0,8044

4.3   Hyperparameter Sensitivity for Stability

We next examine how the weighting parameter w and the binarization threshold γ influence prototype stability. Due to the hard discrimination made by the first metric (labeling as unstable any prototype that fails to activate just on one part of the inferred object), it is necessary to study our sensitivity to slight changes in the hyperparameters. Tables 2, 3 and 4, report Avg Stability under $(w, \gamma) \in \{0.5, 0.65, 0.8\} \times \{0.5, 0.7\}$.

Table 2. Stability results for Resnet34 and VGG19 using $w = 0.5$ and $\gamma = 0.5, 0.7$, for each different transformation.

Stability	$w = 0{,}50$			
	$\gamma = 0{,}5$		$\gamma = 0{,}7$	
Transformation	VGG19	Resnet34	VGG19	Resnet34
Noise	0,7529	0,8087	0,7081	0,7595
Contrast	0,8157	0,8391	0,7776	0,7946
Saturation	0,8065	0,8459	0,7682	0,8022
Hue	0,7269	0,7406	0,6831	0,6833
Shape	0,8307	0,8386	0,7946	0,7938
Texture	0,6924	0,7537	0,6410	0,6956
Avg Stability	**0,7709**	**0,8044**	**0,7288**	**0,7548**

Table 3. Stability results for Resnet34 and VGG19 using $w = 0.65$ and $\gamma = 0.5, 0.7$, for each different transformation.

Stability	$w = 0{,}65$			
	$\gamma = 0{,}5$		$\gamma = 0{,}7$	
Transformation	VGG19	Resnet34	VGG19	Resnet34
Noise	0,7183	0,7802	0,6601	0,7162
Contrast	0,7886	0,8147	0,7391	0,7570
Saturation	0,7781	0,8224	0,7284	0,7656
Hue	0,6869	0,7002	0,6299	0,6257
Shape	0,8058	0,8145	0,7587	0,7563
Texture	0,6473	0,7166	0,5804	0,6412
Avg Stability	**0,7375**	**0,7748**	**0,6828**	**0,7103**

Table 4. Stability results for Resnet34 and VGG19 using $w = 0.5$ and $\gamma = 0.5, 0.7$, for each different transformation.

Stability	$w = 0{,}80$			
	$\gamma = 0{,}5$		$\gamma = 0{,}7$	
Transformation	VGG19	Resnet34	VGG19	Resnet34
Noise	0,6837	0,7904	0,6122	0,6729
Contrast	0,7617	0,7990	0,7006	0,7193
Saturation	0,7497	0,6598	0,6885	0,7291
Hue	0,6469	0,7905	0,5767	0,5681
Shape	0,7809	0,6795	0,7230	0,7187
Texture	0,6022	0,7517	0,5199	0,5867
Avg Stability	**0,7042**	**0,7452**	**0,6368**	**0,6658**

In both backbones, higher w (higher importance to Jaccard's coefficient in the weighted sum) tends to score stability lower, particularly for texture. On the other hand, a larger w (more restrictive binarization) tends to score the stability of the models higher. For our metric, Resnet34 consistently outperforms VGG19 in terms of stability, although this does not always mean better model performance.

5   Conclusions

This work introduces the **Structural Stability Score** (S_{ss}), an automated metric designed to assess the stability of prototype activations in ProtoPNet without requiring manual part annotations. Rather than comparing how a prototype activates on the same part of an object before and after perturbing an image, S_{ss} quantifies the stability by measuring the similarity of two activation maps under a diverse set of visual transformations, including changes in noise, contrast, saturation, hue, shape, and texture.

This metric helps us better understand how interpretable a prototype truly is: stable prototypes suggest that the model has learned reliable and robust associations, while unstable ones reveal fragility in the explanatory process. The S_{ss} metric allows us to probe the explanatory behavior of the model in a systematic and scalable way, making it possible to examine interpretability across large datasets without subjective judgments or annotation costs.

In future work, we aim to expand this methodology in three directions: **(1)** Develop optimization strategies to accelerate the computation of S_{ss}, enabling large-scale analysis; **(2)** Extend this approach to the concept of *consistency* across prototypes, with the goal of achieving a fully automated, quantitative, and objective framework for interpretability assessment; and **(3)** Integrate stability and consistency metrics directly into the training loss, encouraging models to learn not only accurate but also interpretable and robust representations. Through these efforts, we hope to bridge the gap between transparency and performance in prototype-based models.

Acknowledgments. This work has received funding from the European Union's Horizon Programme under Grant 101076754—AIthena Project.

Disclosure of Interests. The authors have no competing interests to declare that are relevant to the content of this article.

References

1. Melis, A.D., Jaakkola, T.: Towards robust interpretability with self-explaining neural networks. In: Bengio, S., et al. (eds.) Advances in Neural Information Processing Systems, vol. 31. Curran Associates, Inc

2. Melis, A.D., Jaakkola, T.: Towards robust interpretability with self-explaining neural networks. In: Bengio, S., et al. (eds.) Advances in Neural Information Processing Systems, vol. 31. Curran Associates, Inc. (2018)

3. Bhagat, P., Choudhary, P.: Image annotation: then and now. Image Vis. Comput. **80**, 1–23 (2018)

4. Brożek, B., Furman, M., Jakubiec, M., Kucharzyk, B.: The black box problem revisited. Real Imaginary Challenges Autom. Legal Decis. Making **32**(2), 427–440

5. Carmichael, Z., Redgrave, T., Cedre, D.G., Scheirer, W.J.: This probably looks exactly like that: an invertible prototypical network

6. Chen, C., et al.: This looks like that: deep learning for interpretable image recognition. In: Wallach, H., Larochelle, H., Beygelzimer, A., Alché-Buc, F.d., Fox, E., Garnett, R. (eds.) Advances in Neural Information Processing Systems, vol. 32. Curran Associates, Inc

7. Donnelly, J., Barnett, A.J., Chen, C.: Deformable ProtoPNet: an interpretable image classifier using deformable prototypes. In: 2022 IEEE/CVF Conference on Computer Vision and Pattern Recognition (CVPR), pp. 10255–10265. ISSN: 2575-7075

8. European Commission. Directorate General for Communications Networks, Content and Technology., High Level Expert Group on Artificial Intelligence.: Ethics guidelines for trustworthy AI. Publications Office
9. Ferreira, J.J., Monteiro, M.: The human-AI relationship in decision-making: AI explanation to support people on justifying their decisions, version Number: 2
10. Fujiyoshi, H., Hirakawa, T., Yamashita, T.: Deep learning-based image recognition for autonomous driving. IATSS Res. **43**(4), 244–252 (2019)
11. Gautam, S., et al.: Protovae: a trustworthy self-explainable prototypical variational model (2022)
12. Goyal, M., Knackstedt, T., Yan, S., Hassanpour, S.: Artificial intelligence-based image classification methods for diagnosis of skin cancer: challenges and opportunities **127**, 104065
13. Guidotti, R., et al.: A survey of methods for explaining black box models. ACM Comput. Surv. **51**(5) (2018)
14. Hanbury, A.: A survey of methods for image annotation. J. Vis. Lang. Comput. **19**(5), 617–627 (2008)
15. Huang, Q., et al.: Evaluation and improvement of interpretability for self-explainable part-prototype networks, pp. 2011–2020
16. Keswani, M., Ramakrishnan, S., Reddy, N., Balasubramanian, V.N.: Proto2proto: Can you recognize the car, the way i do? pp. 10223–10233. IEEE Computer Society
17. Kim, S.S.Y., Meister, N., Ramaswamy, V.V., Fong, R., Russakovsky, O.: Hive: evaluating the human interpretability of visual explanations. In: Avidan, S., Brostow, G., Cissé, M., Farinella, G.M., Hassner, T. (eds.) Computer Vision - ECCV 2022, pp. 280–298. Springer Nature Switzerland, Cham (2022)
18. Molnar, C.: Interpretable Machine Learning. A Guide for Making Black Box Models Explainable. 2nd edn
19. Nauta, M., Jutte, A., Provoost, J., Seifert, C.: This looks like that, because ... explaining prototypes for interpretable image recognition. In: Machine Learning and Principles and Practice of Knowledge Discovery in Databases, pp. 441–456. Springer International Publishing
20. Nauta, M., Seifert, C.: The co-12 recipe for evaluating interpretable part-prototype image classifiers. In: Longo, L. (ed.) Explainable Artificial Intelligence. pp, 397–420. Springer Nature Switzerland
21. Rudin, C.: Stop explaining black box machine learning models for high stakes decisions and use interpretable models instead **1**(5), 206–215
22. Slack, D., Hilgard, A., Singh, S., Lakkaraju, H.: Reliable post hoc explanations: Modeling uncertainty in explainability. In: Ranzato, M., Beygelzimer, A., Dauphin, Y., Liang, P.S., Vaughan, J.W. (eds.) Advances in Neural Information Processing Systems, vol. 34, pp. 9391–9404. Curran Associates, Inc
23. Wah, C., Branson, S., Welinder, P., Perona, P., Belongie, S.: Technical Report CNS-TR-2011-001, California Institute of Technology (2011)
24. Yang, H.M., Zhang, X.Y., Yin, F., Yang, Q., Liu, C.L.: Convolutional prototype network for open set recognition. IEEE Trans. Pattern Anal. Mach. Intell. **44**(5), 2358–2370 (2022)

Interpretable Railway Object Classification Using Part-Prototype Networks

Ruben Naranjo[1,2(✉)] [iD], Iker Sancho[1,2] [iD], Nerea Aranjuelo[1] [iD],
Itsaso Rodríguez-Moreno[2] [iD], and Marcos Nieto[1] [iD]

[1] Vicomtech Foundation, Basque Research and Technology Alliance (BRTA), Mikeletegi 57,
20009 Donostia-San Sebastián, Spain
[2] University of the Basque Country (UPV/EHU), Donostia-San Sebastián, Spain
rnaranjo@vicomtech.org

Abstract. Enhancing safety and operational efficiency in railway systems benefits from robust AI-powered perception, particularly for reliable obstacle and pedestrian detection. However, the prevalent black-box nature of contemporary deep learning models presents significant challenges for verification and trust, especially within safety-critical railway environments characterised by dynamic weather, illumination changes, and high speed, which create undesirable effects such as cluttered backgrounds, and motion blur, which may hinder the performance of computer vision approaches. This paper addresses the need for more transparent models by proposing the application of Prototypical Part Networks (ProtoPNet) for interpretable obstacle and pedestrian classification within the railway domain. Experiments with the OSDaR23 dataset demonstrate that training with a careful selection of data augmentation processes enhances key metrics such as precision, recall and F1-score while yielding transparent results with visually robust prototypes.

Keywords: Object Classification · Interpretability · Railway

1   Introduction

The continuous drive towards enhancing safety and operational efficiency within railway systems has spurred significant interest in AI-powered perception technologies, particularly for the realisation of autonomous trains [29]. Among the myriad of perception tasks, the reliable detection of obstacles and pedestrians on or near the railway track stands as a paramount safety function [19, 25]. While contemporary deep learning models have demonstrated substantial capabilities in object recognition, their prevalent *black-box* nature poses considerable challenges [24].

Current methodologies for detecting obstacles and pedestrians in railway environments frequently lack transparency, making verification of the decision-making processes a complex endeavour [31]. The railway context itself introduces a unique confluence of challenges for computer vision systems. These include, but are not limited to, highly dynamic weather patterns, significant and sudden variations in illumination (e.g. transitions between open tracks and tunnels, day and night operations), visually

© The Author(s), under exclusive license to Springer Nature Switzerland AG 2026
F. Marcelloni et al. (Eds.): IJCCI 2025, CCIS 2829, pp. 86–97, 2026.
https://doi.org/10.1007/978-3-032-15638-9_6

cluttered and complex backgrounds (comprising elements such as catenary lines, signaling equipment and dense vegetation), and the potential for motion blur due to train speed. These factors can adversely affect not only the raw detection performance but also the consistency and reliability of any perception system. Consequently, there is a pressing need to rigorously evaluate the performance and robustness of models within this demanding operational domain.

Explainable AI (XAI), and more specifically, self-explainable or interpretable models such as Prototypical Part Networks (ProtoPNet) [2], present a promising venue towards achieving transparent and trustworthy systems which can elucidate the internal reasoning of the AI models by providing insights into the *how* and *why* a specific prediction is formulated. Thus, enabling an easier analysis of the behaviour of models under different circumstances, which enables better debugging and reduces complexity in certification processes.

In this paper, we propose the adaptation and application of ProtoPNet for the task of interpretable obstacle and pedestrian detection within a railway environment. The principal contributions of this research are articulated as follows:

- The novel application and systematic adaptation of ProtoPNet for the task of obstacle and pedestrian classification within complex railway environments.
- Comparative analysis of the classification performance of several ProtoPNet configurations.
- Qualitative analysis of the prototypes learned by the model, which leads to discovery of issues that would not be noticed in black-box models.

2   Related Work

The application of advanced image classification techniques, particularly deep learning, holds significant promise for enhancing railway safety and operational efficiency [12]. However, the deployment of such complex models in this safety-critical environment necessitates a thorough understanding of their decision-making processes.

2.1   Explainability and Interpretability Methods

The increasing complexity of AI models, particularly deep neural networks used in image classification tasks, often results in *black-box* systems where the internal decision-making processes are completely opaque. This lack of transparency presents significant hurdles, especially in safety-critical domains where understanding the *why* behind a prediction is crucial for trust, accountability, debugging, and safe deployment [11].

To remedy this opaqueness, current research has seen an increasing interest in XAI. Explainability often denotes *post-hoc* techniques applied to opaque *black-box* models to produce understandable explanations without altering model architecture. The key challenge here is *fidelity*, ensuring the explanations accurately reflect the model's internal reasoning [10]. Prominent post-hoc techniques include saliency or gradient-based methods like Grad-CAM [26] and its variants (e.g. [1,18]), which generate heat-maps highlighting important image regions for a given prediction. Model-agnostic perturbation-based methods, such as LIME [22] and SHAP [17], explain predictions by observing output changes in response to input perturbations, offering local (LIME) or theoretically grounded global/local (SHAP) feature importance values. However, these can be computationally expensive and rely on potentially problematic perturbation strategies [11]. Concept-based explanations, like TCAV [13], aim for a higher-level understanding by quantifying model sensitivity to human-defined concepts, but require concept examples and face challenges like concept entanglement [11]. Counterfactual explanations identify minimal input changes needed to alter a prediction, offering insights into decision boundaries but facing difficulties in generating realistic changes [6].

In contrast, self-explainable models or interpretable models typically refer to models that are inherently transparent: they integrate explainability directly into the model's design and training. These methods aim for inherent faithfulness by tying explanations to the model's mechanisms, addressing the fidelity concerns of post-hoc approaches [10]. Attention mechanisms, adapted from Natural Language Processing architectures, allow models to weight input regions, providing focus maps as inherent explanations, although whether attention truly equates to explanation is debated [11]. Prototype-based learning, exemplified by ProtoPNet [2], uses case-based reasoning, comparing input parts to learned prototypes, which are often patches from images of training data. This offers intuitive *"this looks like that"* explanations, but interpretability depends heavily on the quality and clarity of the learned prototypes. Inherently interpretable concept-based architectures, like Concept Bottleneck Models (CBM) [14], force information through an intermediate layer representing explicit concepts, aiming for transparency in the final concept-to-prediction mapping. However, defining adequate concepts, preventing information leakage, and obtaining concept supervision remain significant issues [5].

2.2 Railway Object Classification

Computer vision has become increasingly vital for enhancing railway safety, maintenance and efficiency [12]. Automated vision systems are replacing traditional manual inspections for tasks like track inspection, obstacle detection, and component monitoring. The systems must be robust to challenging railway conditions and often require real-time processing capabilities [23].

Methodologically, deep learning has become the standard approach, outperforming traditional computer vision techniques. Convolutional Neural Networks (CNN) of various architectures (e.g. ResNet [9], VGG [27]) are widely used for feature extraction and classification [23].

For tasks requiring localisation, object detection frameworks like Faster R-CNN [21], SSD [15], and YOLO variants [20] are common, with YOLO often favoured for real-time applications.

More recently, Transformer architectures such as Vision Transformer [7] and Swin Transformer [16] are being applied, sometimes in hybrid models with CNNs, leveraging their ability to capture global image dependencies [23].

Traditional computer vision methods involving hand-crafted features such as Histogram of Oriented Gradients [4], and Haar-like features [30], and classifiers such as Support Vector Machines [3] generally show lower performance and robustness in complex railway environments compared to DL methods [23].

Significant challenges remain in railway image classification. Systems must cope with highly variable environmental factors like diverse illumination and adverse weather conditions, as well as motion blur from train speed [12]. Moreover, operational constraints like real-time processing demands and the need for extremely high reliability in safety-critical applications add further complexity. Addressing these challenges is crucial for the trustworthy deployment of AI in railways, highlighting the role of interpretability in validating model reasoning and ensuring reliance on valid features rather than dataset artefacts.

3   System Architecture and Experimental Design

This section details our approach to bridge the gap between model performance and transparency by employing ProtoPNet. ProtoPNet's architecture is inherently designed for interpretability, offering case-based explanations that resonate with human reasoning (*this part of the scene looks like this known example of an obstacle/pedestrian*). The aim is to develop a system that not only accurately identifies potential hazards but also provides transparent, verifiable justifications for its detections, thereby fostering trust and facilitating robust validation.

3.1   OSDaR23 Dataset

Open Sensor Data for Rail 2023 (OSDaR23) [28] is a multi-sensor dataset created to push automated railway operation research. Recorded in Hamburg, Germany, in September 2021, this dataset incorporates synchronised data from multiple sensors such as RGB cameras, infra-red cameras, LiDAR sensors, radar, and GNSS/IMU data. In addition, it includes manually labelled data of 20 different object classes relevant in railway environments, such as catenary poles, signals, vehicles, and pedestrians.

For the current work, we leverage the annotation within OSDaR23 dataset to extract information pertaining to individual objects. As the original images from this dataset can contain one or more annotated objects, a dedicated pre-processing pipeline was necessary to create a dataset consisting of individual object images. This process involves several steps: first, we extract the unique object identifiers linked to each annotation, then, we obtain precise positional information (bounding box coordinates) for each identified object in each image. Objects annotated with an occlusion level of 50% or higher are subsequently filtered out. Finally, each remaining object is extracted from the original image through cropping based on its spatial information. These crops are performed using two potential methods: either a square crop that includes the surrounding context around the object's bounding box, or a strict crop precisely defined by the bounding box, to which a black padding is added to ensure a square output image. The choice between these cropping methods is determined by the specific configuration of the model being trained, which will be further discussed in later sections.

Furthermore, during data capture, especially nearing a train station, train speed is decreased without reducing image capture frequency. This provokes an increase in the repetition of objects in subsequent images, which contain the same static objects, and similar poses for dynamic objects such as persons. To reduce this redundancy, we filter image crops of the same object with a process based on the Structural Similarity Index Measure to remove the image crops that exceed a similarity threshold of 0.3, ensuring significant differences between images. This process reduces the initial 67795 image dataset to a smaller set of 5922 relevant images.

After the filtering process is done, we separate the resulting dataset into sets of train, validation, and test, with a data percentage of %60, %25, and %15, respectively.

3.2 ProtoPNet Arquitecture and Configuration

ProtoPNet is a deep neural network based on the idea that, for image classification, an image can be decomposed into prototypical parts that can then be compared to similar parts of already known images. Its architecture (Fig. 1) includes three principal components: a classic CNN backbone, a prototypical layer and a dense layer.

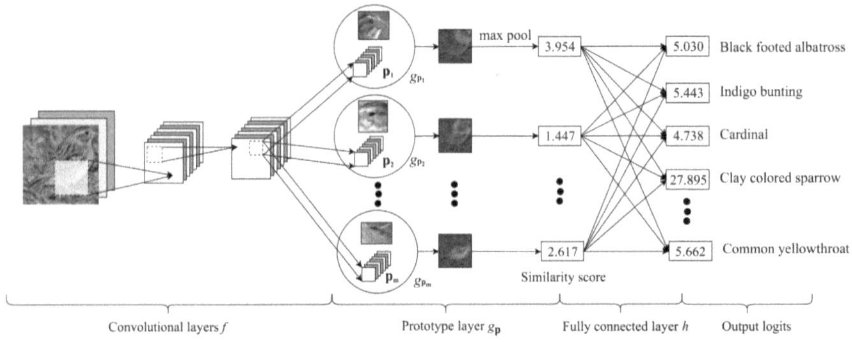

Fig. 1. Architecture of the ProtoPNet network [2].

- **Convolutional Neural Network:** a classical CNN such as VGG [27], or ResNet [9], is used as the backbone of the network to extract latent representations of features.
- **Prototypical Layer:** compares the extracted feature vectors within the latent space to a set of learned prototypes, calculating the euclidean distance between a feature vector and each prototype to get the most similar examples.
- **Dense Layer:** the last component of the architecture is a fully connected layer which produces class probabilities given the similarity scores calculated in the prototypical layer.

Since ProtoPNet's interpretability comes from comparing image features in the latent space produced by the backbone network, the selection of this backbone architecture is crucial for obtaining quality results in the final model. In addition, image preprocessing and augmentation settings, and hyperparameters such as the number of prototypes per class, and the number of prototype update operations play a fundamental role in the development of the model. Thus, in order to get an optimal system, extensive experimentation is required with different model configurations.

Table 1. Configuration of the eight initial ProtoPNet models.

Model name	Backbone	Prototype Update Freq.	Pre-processing
VGG-5P	VGG-19	5 Epochs	Padding
VGG-5N	VGG-19	5 Epochs	Cropping
VGG-10P	VGG-19	10 Epochs	Padding
VGG-10N	VGG-19	10 Epochs	Cropping
ResNet-5P	ResNet-50	5 Epochs	Padding
ResNet-5N	ResNet-50	5 Epochs	Cropping
ResNet-10P	ResNet-50	10 Epochs	Padding
ResNet-10N	ResNet-50	10 Epochs	Cropping

For this work, we train the ProtoPNet model with eight different configurations before choosing the setup for the training of the final model. These eight models have the same base configuration as proposed by the original paper [2] but differ in the backbone network setting, the prototype update operation frequency and the pre-processing done to the images before training. Details for each model are presented in Table 1.

4   Results and Final Model

This section presents the outcomes of our experimental investigations aimed at applying ProtoPNet to object classification in railway environments. We detail the performance metrics and qualitative analysis obtained from evaluating various model configurations and data processing methodologies. Based on these outcomes, we identify and present the final model configuration chosen for achieving robust and interpretable classification within the challenging railway environment.

Quantitative results of the eight initial models (Table 2) show that all of them achieve an *accuracy* over 0.9, with a maximum of 0.93 for the ResNet-5N model, measured in the validation set of the dataset. These metrics reveal that all tested configurations adapt well to the specifics of the railway domain. Nonetheless, there are subtle differences between them, for instance, models with ResNet backbone generally achieve better accuracy than those with VGG.

In a traditional deep learning development loop, we could stop our analysis here and choose the model with the better performance. However, since ProtoPNet also learns easily interpretable image prototypes, we can perform a qualitative evaluation of those to check if learned features and concepts align with human-perceived features of the classified labels.

Table 2. Quantitative metrics of the eight initial ProtoPNet models.

Model name	Accuracy	Precision	Recall	F1-Score
VGG-5P	0.9247	0.7515	0.6914	0.7012
VGG-5N	0.9139	0.7064	0.6264	0.6471
VGG-10P	0.9103	**0.7835**	0.6831	0.7135
VGG-10N	0.9031	0.7525	0.6989	0.7088
ResNet-5P	0.9291	0.7584	0.6934	0.6995
ResNet-5N	**0.9300**	0.7812	0.6982	**0.7278**
ResNet-10P	0.9202	0.6855	0.7079	0.6582
ResNet-10N	0.9175	0.7459	**0.7212**	0.7114

Fig. 2. Learned prototypes of the eight initial models for class *person*. Borders are colored green for prototypes that we consider present relevant information, orange for partial or low information and red for the ones that do not present information.

Qualitative exploration of the learned prototypes reveals that all eight configurations stand out negatively for the perceived quality of their prototypes (Fig. 2). All configurations suffer from the same effect that prototypes tend to converge into blurred, unfocused images or images with a great deal of padding. Looking into the model's implementation, especially the prototype selection process, we can infer that these blurry prototypes are probably chosen because the augmented dataset contains a fair amount of grey-like blur and sometimes black padding, which will probably end up being very close to each other in the learned latent space.

In addition, configurations with the ResNet backbone tend to learn the same prototype several times, which introduces a lack of intra-class diversity issue to the final model. Models with VGG as a backbone network seem to learn more visually robust prototypes, especially in the configurations without padding.

Our hypothesis is that the models are suffering from an effect called shortcut learning, a known issue with deep learning models which occurs when the models learn patterns on the training set that lead them to better benchmark results, thus receiving the name *shortcut*. However, this shortcut patterns or features do not align with real decisive features and result in models not performing well once they are deployed [8].

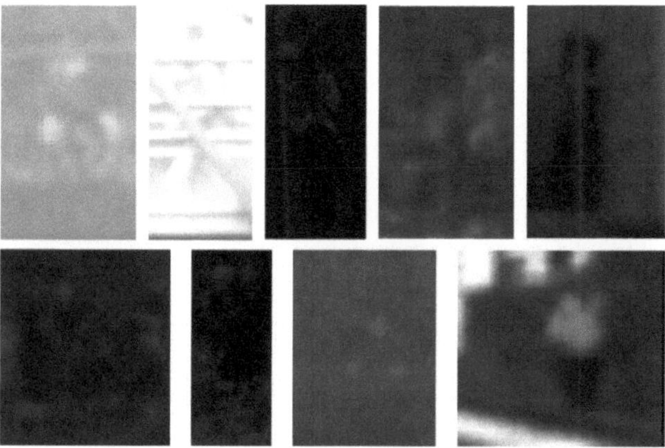

Fig. 3. Instances of the dataset from the *person* class with very small objects or no object at all.

Reflecting on these qualitative results and the possibility of shortcut learning, we take a step back and look into the issues the dataset could have introduced in our models:

- OSDaR23 is labelled in a very precise manner so even instances with very small bounding boxes ($< 15x15$) are taken into account as a valid instance (Fig. 3).
- The augmentation process proposed by the original authors of ProtoPNet [2] performs some zoom-in operations on images that may further degrade the information quality on the final dataset (Fig. 4).

Thus, a new filtering and augmentation process is needed for the OSDaR23 dataset. We perform a manual filtering that removes images that are too blurry or images with very low information of objects. This process reduces the 5922-image dataset to 5227 images. Additionally, we use a new data augmentation process that only uses horizontal flips, small rotations and controlled cut-out operations.

Given the results discussed above, there is a need to balance out quantitative and qualitative analysis when it comes to selecting a final model for training with the new approach for data processing. While *ResNet-5N* achieves the best metrics in terms of accuracy and F1-score, we consider that *VGG-5N* is the most appropriate model because it combines balanced quantitative metrics while obtaining one of the best sets of learned prototypes, achieving high performance and interpretability at the same time.

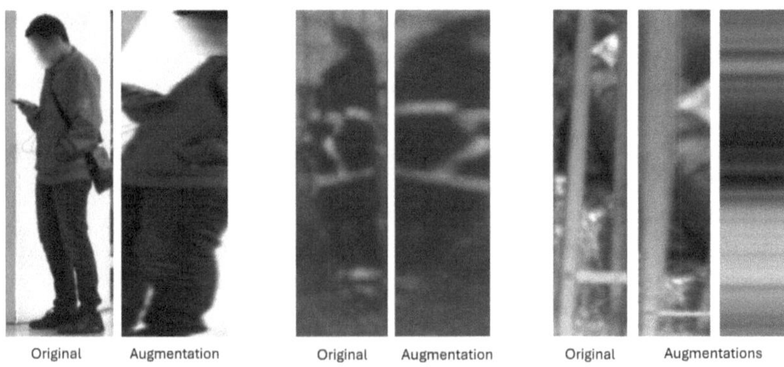

Fig. 4. Errors produced by the original data-augmentation process.

With this chosen model (*VGG-5N*), we perform a final training and validation of the model with the manually filtered OSDaR23 dataset. For data augmentation, we train 3 different configurations as seen in Table 3.

Table 3. Quantitative evaluation of the final ProtoPNet model.

Augmentation Type	Accuracy	Precision	Recall	F1-Score
No augmentation	0.8833	0.5618	0.6231	0.5598
ProtoPNet augmentation	**0.9054**	0.6028	0.6059	0.5994
Ours	0.9000	**0.7600**	**0.6678**	**0.6932**

Quantitative results of this final model reveal that the original data augmentation process enhances the model's performance in comparison to a model trained with no data augmentation. Our set of manually revised transformations proves to achieve a much higher performance in terms of *precision, recall* and *F1-score*.

Regarding qualitative results (Fig. 5), we observe an overall enhancement of proto-type quality in the non-augmented model and the model augmented with our manually

Fig. 5. Learned prototypes of the final models for class *person*. Borders are colored green for prototypes that we consider present relevant information, orange for partial or low information and red for the ones that do not present information.

defined augmentations, which further confirms that a careful selection of data filtering and augmentation is paramount.

5   Conclusions

This paper addressed the critical need for interpretable AI systems in the safety-critical railway environment, particularly for obstacle and pedestrian classification. Given the challenges posed by dynamic weather, illumination changes, cluttered backgrounds, and motion blur, relying on black-box models is problematic for verification and trust. To tackle this, we proposed and systematically adapted the Prototypical Part Network (ProtoPNet), an inherently interpretable model that provides case-based explanations, for this specific application. Utilising the multi-sensor OSDaR23 dataset, we detailed our system architecture and experimental design, exploring various ProtoPNet configurations based on different backbones, prototype update frequencies, and initial preprocessing methods. Our initial experiments showed that all eight tested configurations achieved promising quantitative metrics, with an accuracy over 0.9. However, a crucial qualitative analysis of the learned prototypes revealed significant limitations, with prototypes often appearing blurred, unfocused, or heavily padded. This qualitative issue was linked to characteristics of the OSDaR23 dataset, including very small objects, and potential problems introduced by the original data augmentation process. The rapid detection of issues on learned prototypes demonstrated the enhancement ProtoPNet presents over traditional *black-box* models for development and debugging.

Recognising the importance of prototype quality for interpretability, we refined our data filtering and augmentation process, implementing manual filtering for low-information images and a new augmentation strategy. Based on the balance between quantitative performance and prototype interpretability, the VGG-5N configuration was selected as the most appropriate model for training with this improved data handling. Our final experiments demonstrated that training with our manually revised augmentation significantly enhanced key performance indicators such as precision, recall, and F1-score compared to training without augmentation or with the original ProtoPNet augmentation. This indicates that the improved data processing is crucial for achieving both high performance and better prototype quality.

Future work could extend this research in several directions. Exploring alternative strategies for handling very small objects in the dataset and further refining data augmentation techniques specifically tailored for railway imagery could yield additional improvements. Investigating methods to quantify prototype clarity more objectively could also provide valuable insights. Furthermore, integrating ProtoPNet into a complete object detection framework to provide interpretable localisation could enhance its utility in real-world railway applications. Finally, examining the effectiveness of the prototype-based explanations with domain experts could validate their utility for certification and debugging processes.

Acknowledgments. This work has received funding from the European Union's Horizon Programme under Grant 101076754—AIthena Project.

Disclosure of Interests. The authors have no competing interests to declare that are relevant to the content of this article.

References

1. Chattopadhay, A., Sarkar, A., Howlader, P., Balasubramanian, V.N.: Grad-cam++: generalized gradient-based visual explanations for deep convolutional networks. In: 2018 IEEE Winter Conference on Applications of Computer Vision (WACV), pp. 839–847. IEEE (2018)
2. Chen, C., Li, O., Tao, D., Barnett, A., Rudin, C., Su, J.K.: This looks like that: deep learning for interpretable image recognition. Adv. Neural Inform. Process. Syst. **32** (2019)
3. Cortes, C., Vapnik, V.: Support-vector networks. Mach. Learn. **20**, 273–297 (1995)
4. Dalal, N., Triggs, B.: Histograms of oriented gradients for human detection. In: 2005 IEEE computer Society Conference on Computer Vision and Pattern Recognition (CVPR 2005), vol. 1, pp. 886–893. IEEE (2005)
5. Debole, N., Barbiero, P., Giannini, F., Passeggini, A., Teso, S., Marconato, E.: If concept bottlenecks are the question, are foundation models the answer? arXiv preprint arXiv:2504.19774 (2025)
6. Devireddy, K.: A comparative study of explainable ai methods: Model-agnostic vs. model-specific approaches. arXiv preprint arXiv:2504.04276 (2025)
7. Dosovitskiy, A., et al.: An image is worth 16x16 words: Transformers for image recognition at scale. arXiv preprint arXiv:2010.11929 (2020)
8. Geirhos, R., et al.: Shortcut learning in deep neural networks. Nat. Mach. Intell. **2**(11), 665–673 (2020)
9. He, K., Zhang, X., Ren, S., Sun, J.: Deep residual learning for image recognition. In: Proceedings of the IEEE Conference on Computer Vision and Pattern Recognition, pp. 770–778 (2016)
10. Hou, J., Liu, S., Bie, Y., Wang, H., Tan, A., Luo, L., Chen, H.: Self-explainable ai for medical image analysis: A survey and new outlooks. arXiv preprint arXiv:2410.02331 (2024)
11. Kamakshi, V., Krishnan, N.C.: Explainable image classification: the journey so far and the road ahead. AI **4**, 620–651 (2023)
12. Kapoor, R., Goel, R., Sharma, A.: An intelligent railway surveillance framework based on recognition of object and railway track using deep learning. Multimedia Tools Appli. **81**, 21083–21109 (2022)
13. Kim, B., Wattenberg, M., Gilmer, J., Cai, C., Wexler, J., Viegas, F., et al.: Interpretability beyond feature attribution: Quantitative testing with concept activation vectors (tcav). In: International Conference on Machine Learning, pp. 2668–2677. PMLR (2018)
14. Koh, P.W., et al.: Concept bottleneck models. In: International Conference on Machine Learning, pp. 5338–5348. PMLR (2020)
15. Liu, W., et al.: SSD: single shot multibox detector. In: Leibe, B., Matas, J., Sebe, N., Welling, M. (eds.) ECCV 2016. LNCS, vol. 9905, pp. 21–37. Springer, Cham (2016). https://doi.org/10.1007/978-3-319-46448-0_2
16. Liu, Z., et al.: Swin transformer: hierarchical vision transformer using shifted windows. In: Proceedings of the IEEE/CVF International Conference on Computer Vision, pp. 10012–10022 (2021)
17. Lundberg, S.M., Lee, S.I.: A unified approach to interpreting model predictions. In: Guyon, I., Luxburg, U.V., Bengio, S., Wallach, H., Fergus, R., Vishwanathan, S., Garnett, R. (eds.) Advances in Neural Information Processing Systems 30, pp. 4765–4774. Curran Associates, Inc. (2017)

18. Muhammad, M.B., Yeasin, M.: Eigen-cam: class activation map using principal components. In: 2020 International Joint Conference on Neural Networks (IJCNN), pp. 1–7. IEEE (2020)
19. Oh, S.C., Kim, G.D., Jeong, W.T., Park, Y.T.: Vision-based object detection for passenger's safety in railway platform. In: 2008 International Conference on Control, Automation and Systems, pp. 2134–2137. IEEE (2008)
20. Redmon, J., Divvala, S., Girshick, R., Farhadi, A.: You only look once: unified, real-time object detection. In: Proceedings of the IEEE Conference on Computer Vision and Pattern Recognition, pp. 779–788 (2016)
21. Ren, S., He, K., Girshick, R., Sun, J.: Faster r-cnn: towards real-time object detection with region proposal networks. Adv. Neural Inform. Process. Syst. **28** (2015)
22. Ribeiro, M.T., Singh, S., Guestrin, C.: why should i trust you? explaining the predictions of any classifier. In: Proceedings of the 22nd ACM SIGKDD International Conference on Knowledge Discovery and Data Mining, pp. 1135–1144 (2016)
23. Ristić-Durrant, D., Franke, M., Michels, K.: A review of vision-based on-board obstacle detection and distance estimation in railways. Sensors **21**(10), 3452 (2021)
24. Rudin, C.: Stop explaining black box machine learning models for high stakes decisions and use interpretable models instead. Nat. Mach. Intell. **1**, 206–215 (2019)
25. Sabnis, O.V., Lokeshkumar, R.: A novel object detection system for improving safety at unmanned railway crossings. In: 2019 Fifth International Conference on Science Technology Engineering and Mathematics (ICONSTEM), vol. 1, pp. 149–152. IEEE (2019)
26. Selvaraju, R.R., Cogswell, M., Das, A., Vedantam, R., Parikh, D., Batra, D.: Grad-cam: Visual explanations from deep networks via gradient-based localization. In: Proceedings of the IEEE International conference on Computer Vision, pp. 618–626 (2017)
27. Simonyan, K., Zisserman, A.: Very deep convolutional networks for large-scale image recognition. arXiv preprint arXiv:1409.1556 (2014)
28. Tagiew, R., et al.: Osdar23: open sensor data for rail 2023. In: 8th International Conference on Robotics and Automation Engineering (ICRAE), p. 270–276. IEEE (2023)
29. Tang, R., et al.: A literature review of artificial intelligence applications in railway systems. Trans. Res. Part C: Emerging Technol. **140**, 103679 (2022)
30. Viola, P., Jones, M.: Rapid object detection using a boosted cascade of simple features. In: Proceedings of the 2001 Ieee Computer Society Conference on Computer Vision and Pattern Recognition, CVPR 2001, vol. 1. IEEE (2001)
31. Winfield, A., Watson, E.N.: How to make autonomous systems more transparent and trustworthy (2022)

Efficient Construction of Interpretable Oblique Decision Trees

Vladimir Estivill-Castro[(✉)] and Nuru Nabuuso

Department of Engineering, Universitat Pompeu Fabra, Barcelona 08018, Spain
{vladimir.estivill,nuru.nabuuso}@upf.edu

Abstract. Human oversight of AI demands transparent models, especially in high-stakes contexts. While interpretable models like decision trees offer clarity, traditional axis-aligned trees struggle with complex features, and oblique trees, though more expressive, are hard to understand. This work introduces a human-centred framework using parallel coordinates (PCs) to visualise and construct oblique decision trees interactively. By enabling the visualisation of 3D hyperplanes and allowing users to manipulate them directly, the method enhances explainability without compromising accuracy. We present an efficient method for finding hyperplanes of up to 3 dimensions. We show experimentally by comparing several methods that our oblique trees do not sacrifice accuracy, but the explainability of the trees (due to their simplicity) is radically enhanced. This bridges the gap between model complexity and interpretability, supporting human oversight in AI decision-making.

Keywords: Decision Trees · Linear Regression · Parallel Coordinates · Human-Computer Interfaces Supporting XAI · Interpretable Machine Learning

1 Introduction

Explainable Artificial Intelligence (XAI) is crucial for human understanding, trust and oversight. These attributes are listed as fundamental values for the deployment of AI systems [15]. XAI has been considered essential for accountability (assign responsibility of a decision/recommendation), trust and adoption, error detection, regulatory compliance, ethical decision-making and human-AI collaboration [18,20,59,61]. In particular, *Interpretable AI* [38,46] focuses on making AI systems easier for humans to understand by designing models whose decision-making processes can be clearly examined. Such an approach is often called *transparent-by-design* [22] or explainablity from the beginning [50]. One achieves this by placing limits on the model's complexity, such as using simpler, human-friendly model types (like decision trees, linear models, case-based reasoning), keeping models sparse and compact (Binary ANNs [37]), or structuring how information is processed (e.g., disentangling neural networks) to improve human interpretability.

We offer a method for building oblique decision trees that use sparse [39] linear models with at most three attributes per decision boundary. We retain the accuracy of traditional oblique trees while being computationally faster, which is suitable for

F. Marcelloni et al. (Eds.): IJCCI 2025, CCIS 2829, pp. 98–115, 2026.
https://doi.org/10.1007/978-3-032-15638-9_7

human-in-the-loop learning. Sparse models are considered interpretable [16,57] and transparent [6]. We limit the number of attributes enhancing interpretability and transparency. We advocate the use of parallel coordinates [34] to visualise and interactively construct decision trees, allowing users to incorporate domain knowledge and better understand the data. Parallel coordinates [19,36] have been applied to enable the interactive construction of a decision tree following the general steps of the decision-tree building method.

Learning with decision trees is considered the most interpretable method. While deep learning continues to face challenges related to interpretability, hybrid approaches that integrate decision trees with reinforcement learning have been proposed as a way to maintain strong performance without sacrificing interpretability [14,23,27,58]. Previously, parallel coordinates were used successfully to build decision trees with human-in-the-loop, but this was limited to nodes that involve tests with at most two attributes [20]. So-called bi-variate splits make trees shallower (improving interpretability) while keeping the split human understandable [9]. However, algorithms such as CART (Classification and Regression Trees) [10,11] enable much more sophisticated boundaries at the nodes, typically using a linear hyperplane on all the attributes. These models are called oblique trees [12,25,44,49], and have been used to achieve remarkable systems (for instance, in combination with reinforcement learning [17] or clustering [25]). In many scenarios, the compact models are competitive in accuracy, but they are much more transparent and interpretable [37]. Humans can understand the decision-making process by inspecting the model. Moreover, as they usually require fewer parameters, they are less hungry for huge datasets.

In Sect. 2 we review decision tree construction. In particular, in Subsect. 2.1, we review oblique decision-trees. We define expressiveness as the capability of the model to fit boundary shapes; for instance, quadratic regression is more expressive than linear regression, axis-parallel boundaries are less expressive than oblique boundaries, and hyperplanes restricted to only k coefficients being different from zero are less expressive than hyperplanes without such restriction. Section 3 describes our new algorithm and Sect. 4 compares it against the original versions of oblique trees using the original code. The experiments across several datasets show our new algorithm produces shorter trees than the vanilla axis-parallel split and does not sacrifice accuracy. Thus, even with fully autonomous construction, the oblique trees constructed with our approach are expressive enough to reach high accuracy while remaining interpretable and explainable. Section 5 provides two reasons why our algorithm is suitable for interactive machine learning. First, in Subsect. 5.1 we demonstrate that our split is fast and thus responsive if it were to be used for proposing splits interactively with a human user. Subsection 5.2 shows that three-dimensional hyperplanes are also visualisable using parallel coordinates. Thus, linear decision boundaries involving one, two or three attributes can be presented to the human user.

2 Background on Decision Trees

Decision trees are highly regarded for their interpretability [24] and explainability [3,42] as their if-then rule structure closely mirrors human reasoning. This makes them especially attractive in domains requiring transparent decision support, such as healthcare, finance, and policy. The algorithmic construction of decision trees is founded on a recursive, greedy partitioning of the feature space, a strategy that seeks to maximise information gain or minimise impurity measures at each step. The earliest systematic approach to this problem is found in the Concept Learning System (CLS) developed by Hunt et al. [32,32], ID3 [47], and C4.5 [48]. These works provided the basis for the recursive partitioning method. Quinlan's ID3 [47] assumed categorical features but introduced information gain based on entropy as the criterion for preferring one split over another. C4.5 [48] added support for continuous features, pruning strategies, and better handling of missing values, significantly increasing its robustness. CART [10,11] further generalised the framework by supporting both classification and regression tasks through Gini impurity and variance reduction, respectively, and by enforcing binary splits, which simplify the tree structure.

More recent technical developments have explored how classical decision-tree construction can be improved to enhance interpretability without significantly sacrificing performance. One such line of work involves optimal decision tree learning, where algorithms attempt to construct globally optimal trees using combinatorial optimisation techniques (e.g., Mixed Integer Programming) rather than greedy heuristics [8]. Although computationally expensive, these methods produce smaller, more interpretable trees and have gained attention due to the availability of more computing power and the need for fairness and transparency in AI systems. In the context of explainable AI (XAI), decision trees also serve as surrogate models to explain blackbox models such as neural networks or ensembles [7,13]. In this role, a decision tree is trained to mimic the predictions of a more complex model, offering interpretable approximations that help users understand model behaviour on a local or global level. Research has shown that the structure of such surrogate trees can be regularised to maximise fidelity to the original model while maintaining comprehensibility [54].

We now introduce some necessary notation as we review that the primary objective of classification in supervised learning is to develop a classifier F from a labelled dataset T, which consists of pairs (x_i, y_i) for $i = 1, \ldots, n$, in a way that minimises the misclassification rate $MS = \mathbb{E}[F(x) \neq c(x)]$, where the expectation is over the distribution of unseen data. Here, x represents the input vector with d attributes, and T can be seen as a matrix with n rows and $d+1$ columns. A classifier $F : X \rightarrow Y$ assigns a label $F(x)$ to a previously unseen input x, choosing from a predefined set of categories $C = \{C_1, \ldots, C_k\} \subset Y$, with each $y_i \in C$. Supervised machine learning leverages the training data T to construct a model that predicts the output variable y. When $Y \subset \Re$, the challenge is called regression, and typically, minimisation of the squared error guides the selection of parameters for linear (or higher-order) fitting method [35]. We will focus our discussion on classification, and on continuous attributes ($x_i \in \Re$ and thus, $x \in \Re^d$) but it is simple to include nominal attributes or regression.

Since a classifier's interpretability is dependent on its length [41], and inspired by Occam's Razor guiding principle for induction, shallow trees are considered to have more generalisation power. However, the construction of the optimal decision tree is known to be computationally intractable [33]. As we mentioned before, the information gain [47] (although other options are possible [45]) is used as the criterion to choose a local improvement in the core step of the generic recursive greedy heuristic for decision-tree construction presented as Algorithm 1.

Algorithm 1. Construction of Decision Trees.

1: Initialise the root with T.
2: **while** there are nodes that have not been declared "pure"[11] **do**
3:     **if** T is not pure and the set T has all y_i values equal or satisfies a homogeneity criterion H **then**
4:         label T and its corresponding tree node as pure and terminate the processing of the node (a prediction using the decision procedure D and the data of the node will take place at execution).
5:     **else**
6:         Choose an informative criterion or question Q based on selected attributes, and split the set T horizontally into child (not pure) subsets $T_1, \ldots, T_t$ according to Q.
7:     **end if**
8: **end while**

Therefore, the tree-building process offers variants according to the specification of the three key components [11, Page 22]:

1. The type and method for selection of the split criteria Q. We already indicated that if a hyperplane in d dimensions (so called "variable combination" [11, Sect. 5.2]) is used to split the instances $T \subset X$, the trees are named oblique [12, 26, 30, 40, 44]).
2. The criteria H to declare that the dataset T is sufficiently homogeneous.
3. The method D used (at classification or regression) to declare the output at the corresponding terminal node.

2.1 Oblique Decision Trees

Oblique decision trees often result in more compact and accurate models, with standard decision trees being a special case of them. However, oblique trees tend to be harder to interpret [12], as their splits are defined by hyperplanes in a d-dimensional attribute space, making them less intuitive and more challenging to explain to humans.

So-called axis-parallel trees (the most popular version) define splits based on a test of a single attribute, such as $a_i < t_i$, $a_i \leq t_i$, $a_i > t_i$ or $a_i \geq t_i$ (for some constant t_i) correspond to an infinite interval defined by only one end-point t_i on the axis for A_i. Moreover, test with two bounds, for example $s_i \leq a_i < t_i$ correspond to bounded intervals with both end-points defined. We refer to these tests as one-dimensional splits, as they are equivalent to a boundary of the form $a_i/t_i = 0$. For nominal attributes, typically, a binary split is used by choosing a subset S of the possible values and labelling the branches with $a_i \in S$ versus $a_i \notin S$ (or branches that correspond to enumerating $a_i = t_j$ for all the values t_j of attribute A_i). The point being that the learning software is to compute the optimisation criteria (say information gain) for all d attributes, and then for all the possible points where to make the split. With a training set of size n, each continuous attribute will require testing $n - 1$ gaps (and will need pre-sorting the values for each dimension d).

This search for the best local split is even more complicated for oblique trees. A linear combination of the attributes defines the binary split

$$H : c_1 a_1 + c_2 a_2 + \ldots + c_d a_d + c_0 = 0. \tag{1}$$

The search for a split consists of finding the $d + 1$ coefficients c_j, ($j \in \{0, 1, \ldots, d\}$) so that the local optimality measure (assume information gain for now) is minimised. A naive exhaustive method would consist of exploring the 2^n partitions of the training data at the node, checking whether they are linearly separable, and if they are include the hyperplane as a candidate (with its information gain value) in a priority queue, so that when all possible hyperplanes have been considered, the one with optimal information gain is used to split the node further.

All methods for oblique decision trees use different types of heuristics to avoid such an expensive, exhaustive approach. Typically, CART [11] and OC1 [44] start a hill-climbing approach, seeding it with the best axis-parallel hyperplane H on one attribute found by the one-dimensional approach, and they perturb it to obtain one in d-dimensions. From there on, they chose one attribute a_i, and improve the hyperplane by the clever observation that considering the coefficient as a variable and the valuation of the hyperplane with all the data points, this reduces to finding a value of a split in a one-dimensional case. Different variants decide in different ways how to move across attributes, how to perturb or restart on local optima and how long to search for. In some cases, even more expensive, but potentially more exploratory forms, such as genetic algorithms [12] optimise for the hyper-planes. All evidence confirms that oblique trees achieve more accurate models (this is not surprising as they are at least as good as its axis-parallel counterparts). However, the hyperplane search is orders of magnitude more computationally costly than the axis-parallel version as each test of a new attribute (and an attribute may need to be revisited) is now a new sort against the normal vector of the candidate hyperplane. Moreover, we argue here that what may be won in accuracy may be lost in interpretability.

3   The New Algorithm

"It is often the case that some or many of the variables used in a multiple regression model are not associated with the response. By removing these variables— that is, by setting the corresponding coefficient estimates to zero—we can obtain a model that is more easily interpreted. " [35, Chapter 6].

The journal paper presenting OC1 [44] discusses the challenge of irrelevant attributes, noting that all methods struggle with them. It recommends using feature selection to reduce the number of attributes to only the meaningful ones before applying any method. However, integrating feature selection directly into the OC1 algorithm was left as future work. Naturally, we concur with the recommendation to globally eliminate irrelevant attributes at the start. But, to the best of our knowledge, our paper offers the first attempt to integrate the model selection at each step of the greedy expansion of a node in Algorithm 1. Note that one rapid alternative would be to apply Lasso regression [57] potentially seeded by the hyperplane found by the hill-climbing strategy of OC1 or CART. However, this is not computational efficient and interpretability remains uncertain. Instead, we favour a visualisation approach supporting human-in-the-loop learning is preferred, limiting split hyperplanes to three dimensions (i.e., involving no more than 3 attributes).

We design our approach inspired by *"forward stepwise selection"* [35]. Recall that we have the one-dimensional hyperplane found by the axis-parallel approach, and this identified the best attribute (by the splitting metric) among the d-attributes. We refer to this attribute by A_{i_1}. We now chose another attribute A_j with $j \neq i_1$ (that is $j \in \{1, 2, \ldots, d\} \setminus \{i_1\}$) and form the 2-dimensional candidates (A_{i_1}, A_j). We find the best hyperplane split (which is now 2-dimensional) among this $d - 1$ alternatives. If the best (A_{i_1}, A_j) improves the splitting metric over the one-dimensional (using only A_{i_1}), we adopt this 2-dimensional hyperplane as the new split at the node, we denote the two attributes now as (A_{i_1}, A_{i_2}) and move to three dimensions. If no j produces a better split, the hyper-plane selection exits with the one-dimensional hyperplane.

If a two-dimensional hyperplane was found as an improvement, then we attempt to add a third attribute by testing triples by adding from the remaining $d - 2$ attributes. Again, the best of this $d - 2$ triples, as evaluated by the splitting metric, is compared to our previous best—the two-dimensional hyperplane using (A_{i_1}, A_{i_2}). If it represents an improvement, then we adopt the 3D-hyperplane, otherwise we ese the earlier 2D of (A_{i_1}, A_{i_2})).

We summarise this approach in Algorithm 2 and refer to this decision-tree construction method as OP3.

Algorithm 2. OP3: Find low-dimensions hyperplane.

1: Let $p \leftarrow 1$, and adopt $H_p : c_{i_1} x_{i_i} + c_{1,0} = 0$ (the one-dimensional hyperplane found by an axis-parallel decision-tree algorithm optimising a metric m) as the decision boundary to return.

2: Let $D \leftarrow \{A_{i_1}\}$ be the current set of attributes used to define hyperplanes.

3: **for** $p \leftarrow 2$ **to** 3 **do**

4:     **for** $j \in \{1, 2, \ldots, d\} \setminus D$ **do**

5:         Find the best hyperplane H_p using the attributes in $D \cup \{A_j\}$.

6:     **end for**

7:     **if** H_p has a better m-value than H_{p-1} **then**

8:         Adopt H_p as the decision boundary to return

9:     **else**

10:         return H_{p-1} as the best decision boundary found.

11:     **end if**

12: **end for**

13: return H_3 as the best decision boundary found.

There are still some details to explain about our algorithm. Although we have a pair $(A_{i_1}, A_j$ or a triple $(A_{i_1}, A_{i_2}, A_j))$ of attributes, how do we efficiently find a high-quality splitting hyperplane? We also justify our approach by what we believe would be natural for humans participating in finding such a decision boundary. First, the split is binary, so the observable split (if we were to use any of the approaches to decision tree visualisation [5,20,55,56,60]) would be most noticeable in the two most populous classes (the two classes with more representative examples in the data being classified at this node). Also, we can not separate with a binary split into 3 or more classes. And the metrics of impurity or information gain do not improve significantly if the split separates only a few points. Therefore, we propose the following method. Find the two most populous classes (say C_i and C_j) and compute their centroids (estimated means). This is analogous to the means update-step of the well-known k-means clustering algorithm that finds the means of the k-classes, once each example has been labelled (but in our case $k = 2$). Finding the class counts and accumulating the membership counts of each class requires one pass over the data. Thus, it requires $O(nd)$ time. Note that k-means results in clusters whose decision boundaries are hyperplanes, and if the data originated from a mixture of multi-variate Gaussians, with similar covariance matrices, then the boundaries are hyperplanes.

Second, since we are focusing on the two populous classes, (and disregarding the others temporarily), the line connecting the two centroids is the best 1-dimensional projection direction, which is also the first PCA direction. In a sense, the direction explains the diversity. Thus, we take the line joining the centroids as the normal N of the hyperplane H_p we are looking for. We have found all the coefficients, and we are only missing the independent term c_0. Now, by projecting each data point into the line joining the two centroids, we have reduced the problem to a one-dimensional search for a gap in $n - 1$ places that optimises the metric for information gain (or similar).

The magnitude of the projection of a data point x_i into N is simply

$$\frac{x_i \cdot N}{N \cdot N}.$$

Thus, the complexity of the reduction to the one-dimensional case is just $O(dn)$ time. Because of sorting, the axis-parallel (one-dimensional) hyperplane finding takes $O(n \log n)$ find the constant c_0 Depending on the form of m, we do not need to evaluate it fully on each of the $n - 1$ gaps, as the previous evaluation varies in just one data point from the next that switches sides on the hyperplane. Note that this hyperplane finding algorithm applies to all dimensions, but because we use it within Algorithm 2, it only runs $d - 1$ times when $p = 2$ case and $d - 2$ times when $p = 3$.

Figure 1 provides a 2D illustration of our approach (although the method applies for any number of dimensions). In Fig. 1a, we see two classes, the computed centroids, and the parallel axis boundary in a thick line. The line between the centroids is drawn with a thin line as well as the orthogonal line through the mid-point of the centroids.

The axis-parallel boundary makes 4 classification errors, and the 2D-hyperplane (a line with negative slope) corrects for one error at the bottom, but makes 4 new errors for vertical y larger than 200. Figure 1bshows some dotted red lines which would be in the search for the independent term (the normal stays the same).

4   Experimental Comparison

We are fortunate that the C-implementation for version 3 of OC1 [44] is available publicly in at least two sites.

1. www.jair.org/index.php/jair/article/view/10121
2. github.com/Are52Tap/murthy-oc1.

We could patch errors and warnings for two C compiles (`gcc` under Ubuntu and `clang` under macOS). Using the available code, we reproduced the execution of the three variants presented in the original experiments [44, Table 1]. That is, an axis-parallel algorithm named OC1-AP, an implementation in adherence to the original of CART named CART-LC, and OC1. We implemented our algorithm reusing most of the code: we used the same data structures to represent tree nodes, and to allocate integer, float and double vectors. We reused the code for reading training and testing files, printing resulting trees and showing hyperplanes. Thus, our method is implemented by enabling a `-P` option in the command line invocation of the executable of the original `mktree` program.

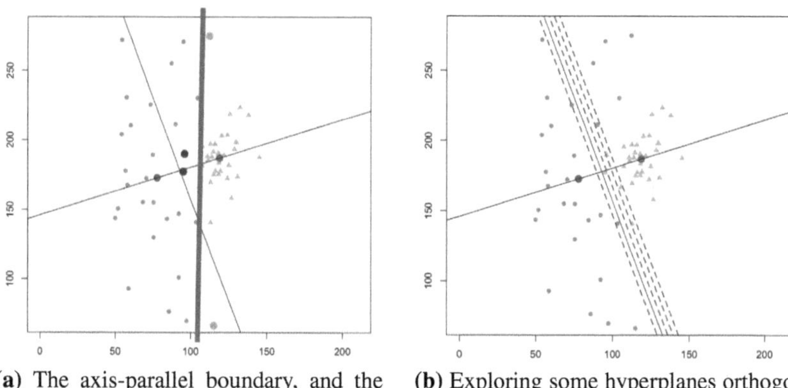

(a) The axis-parallel boundary, and the normal trough the centroids

(b) Exploring some hyperplanes orthogonal to the line trough the centroids.

Fig. 1. Two classes separated by an axis parallel boundary, and determination of a candidate normal for a 2D hyperplane as the line between the centroids of the two classes.

Thus, Table 1 follows the layout and reproduces the results [44, Table 1] for at least two data sets and the three algorithms already mentioned. The results show the standard deviation of 10 executions of 5-fold cross-validation, all algorithms using 10% for pruning. We ensured all parameters are set as per the original results. The two algorithms for oblique trees (OC1 and CART-LC) used their default parameters (20 restarts, and five jumps), and the same order of coefficient perturbation (sequential). All algorithms use the same pruning method (cost complexity), and the metric for node purity was also as per the original experiment for all algorithms, namely the "Twoing Rule". Not surprisingly, the results for OC1-AP, OC1 and CART-LC are essentially equivalent.

Table 1. Comparison of our algorithm OP3 with three variants of oblique decision trees, and the axis-parallel variant.

Accuracy as a percentage					
	Algorithm				
Dataset	OC1	CART-LC	OC1-AP	OP3	J48
Iris [44]	**94.8** ± 1.3	93.8 ± 1.4	93.8 ± 1.2	93.2 ± 1.4	92.1 ± 3.6
Housing [44]	**82.6** ± 1.1	82.0 ± 1.2	81.9 ± 1.2	82.4 ± 1.6	81.6 ± 3.1
Cancer [44]	**94.8** ± 0.3	93.3 ± 0.5	92.8 ± 0.5	93.3 ± 0.4	93.3
Diabetes [44]	68.8 ± 0.4	69.5 ± 1.3	70.0 ± 1.3	70.4 ± 0.9	**71.2** ± 4.0
Wine [19]	91.5 ± 1.9	90.8 ± 1.7	**91.8** ± 2.3	89.9 ± 1.9	89.3 ± 4.0
Waveform [11]	**79.4** ± 0.4	⋆	75.5 ± 0.5	79.3 ± 0.3	75.4 ± 1.2
Waveform noise [11]	78.8 ± 0.5	⋆	74.7 ± 0.5	**79.1** ± 1.3	75.8 ± 1.3
Segmentation [2]	94.8 ± 0.2	94.5 ± 1.8	95.7 ± 0.3	95.9 ± 0.4	**96.3** ± 0.8
Shuttle [1]	99.9 ± 0.0	⋆	99.9 ± 0.0	99.9 ± 0.0	99.9 ± 0.0
SAT [53]	85.6 ± 0.4	85.3 ± 0.4	85.2 ± 0.2	**85.9** ± 0.9	**85.9** ± 0.8
Letter recognition [52]	83.7 ± 0.8	84.7 ± 0.3	86.3 ± 0.1	85.7 ± 0.2	**87.8** ± 0.5
Depth of the tree					
Iris [44]	**3.4** ± 0.3	4.0 ± 0.3	4.2 ± 0.3	4.1 ± 0.3	4
Housing [44]	**7.0** ± 0.5	7.6 ± 0.4	9.0 ± 0.9	8.2 ± 0.7	**7**
Cancer [44]	**4.9** ± 0.4	5.7 ± 0.5	6.4 ± 0.3	6.3 ± 0.6	6
Diabetes [44]	10.3 ± 0.5	11.3 ± 0.6	11.9 ± 0.7	11.8 ± 0.5	**9**
Wine [19]	**3.7** ± 0.3	3.8 ± 0.2	4.0 ± 0.2	**3.7** ± 0.3	5
Waveform [11]	18.3 ± 0.8	⋆	16.8 ± 0.6	16.4 ± 1.0	**16**
Waveform noise [11]	18.5 ± 1.5	⋆	**17.7** ± 0.5	18.0 ± 1.1	19
Segmentation [2]	**9.5** ± 0.8	10.3 ± 0.6	12.5 ± 0.5	11.5 ± 0.8	12
Shuttle [1]	9.0 ± 0.3	⋆	8.5 ± 0.4	7.1 ± 0.4	**7**
SAT [53]	17.4 ± 0.8	17.4 ± 0.7	18.2 ± 0.9	**15.0** ± 0.7	23
Letter recognition [52]	**19.3** ± 0.4	20.6 ± 0.6	86.3 ± 0.1	19.4 ± 0.6	24
Number of leaves					
Iris [44]	**4.6** ± 0.3	5.6 ± 0.3	6.6 ± 0.6	5.6 ± 0.3	5
Housing [44]	23.6 ± 0.8	26.1 ± 2.1	35.6 ± 2.5	26.2 ± 2.1	**19**
Cancer [44]	**8.9** ± 0.7	12.2 ± 1.2	15.8 ± 0.3	14.0 ± 1.0	13
Diabetes [44]	55.9 ± 2.1	63.0 ± 2.0	82.1 ± 4.8	68.1 ± 4.5	**20**
Wine [19]	**6.0** ± 0.6	6.4 ± 0.4	6.9 ± 0.6	**6.0** ± 0.5	7
Waveform [11]	199.5 ± 3.0	⋆	333.4 ± 9.5	**187.1** ± 6.9	317
Waveform noise [11]	189.8 ± 6.4	⋆	307.9 ± 10.5	**162.2** ± 3.5	330
Segmentation [2]	39.6 ± 3.2	42.0 ± 1.9	57.0 ± 3.8	48.2 ± 3.6	**39**
Shuttle [1]	36.3 ± 1.2	⋆	32.0 ± 1.3	26.4 ± 1.0	**24**
SAT [53]	284.1 ± 5.9	308.1 ± 14.6	335.9 ± 16.5	**208.7** ± 6.2	323
Letter recognition [52]	1425.5 ± 62.7	1548.3 ± 55.0	1807.9 ± 20.4	1303.6 ± 37.8	**1227**

⋆ - Exceeded available time.

The column J48 corresponds to Weka's [29] implementation of C 4.5 [48] using the default settings (split is information gain and pruning is on) and 5-fold cross-validation running with its "Explorer tool" (version 3.8.6). The "Experimenter tool" reports the standard deviation for accuracy, but neither Weka tool reports the standard deviation for depth or the number of leaves of the decision tree.

The results in Table 1 are close to each other. A referee requested to highlight with boldface the best result for each row in Table 1 as OC1 is not always the top performer. Note that splitting metric can have an effect and even unexpectedly in some cases our method is the top performer. Our node-splitting does not sacrifice accuracy, and while not producing as small trees as OC1 or CART, it produces interpretable trees since the nodes of CART or OC1 can hold a boundary involving up to d (all of the) attributes. Thus, not all elements impacting complexity of the tree are reflected in Table 1.

5   Towards Human-in-the-Loop Learning

Human-in-the-loop learning (HILL) integrates human expertise into the model training process to improve accuracy, transparency, and explainability [43,62]. It is closely related to "interventional model training" [62] and "interactive machine learning" [4, 21,43]. In HILL, domain experts iteratively refine models, and through visualisation, they shape the learning process [51]. HILL emphasizes interpretable and trustworthy decision models aligned with human values and domain knowledge, enhancing their real-world relevance [21,28].

Therefore, we expose here why our methods are step in the direction of HILL. Early systems for human-participation while constructing AI systems were focused on capturing human expertise in the domain of the dataset [18,61]. Researchers noticed that the human user needed to visualise the data set [5,55,56]. Variants of Algorithm 1 where the human participated were introduced in Weka's *UserClassifier* [60] (still available in version 3.8.6). However, in this case, the splits were limited to 2 attributes and a Cartesian visualisation. Another variant of Algorithm 1 under the name of Nested-Cavities (NC) [36] enabled simultaneuous visualisation of the many attributes in the training set using parallel coordinates [34]. But the splits, as far as we can tell are all intersections of axis-parallel conditions. Using parallel coordinates for visualisation, Gilmore, Estivill-Castro and Hexel [19,20] enable decision-tree construction, and although they claim to use two-dimensional splits, the only demo we could find[1] uses only one-dimensional splits.

Thus, as far as we can see, the visualisation with parallel coordinates is a dominant approach that enables and assists the interaction by the user, but in defining a split Q at a node for Line 6 of Algorithm 1, anything above two-dimensional splits is not considered.

In this section we first show that our approach is efficient enough to respond rapidly in an interface and perhaps even suggest several alternatives and run with different criteria (twoing, information gain, or Gini index), and offering several split for the user to chose. Next we show that parallel coordinates can illustrate three dimensional splits.

---

[1] eugene-gilmore.com/hilltool.

5.1   Empirical Efficiency of Our Algorithm

Sect. 3 argued that our method to select a three-dimensional split is efficient, and in the experiments of Sect. 4 we noticed our methods performed particularly fast. The implementation our split was not optimized for efficiency or minimising function calls. It avoids some of the original code's legacy tricks. However, despite these straight forward approach, our current implementation still performs impressively fast.

To illustrate this point we generated datasets of 1,000 points and several dimensions $d \in \{5, 6, 7, 8, 11\}$. Eight datasets were generated as a mixture of 4 multivariate Gaussians with diagonal covariance matrices. The centres and the standard deviations on each coordinate are chosen randomly. The algorithms OC1, CART-LC, OC1-AP, and our own OP3 were run on each dataset under Ubuntu 22.04. We compiled the C-code with gcc and the option -pg to use gprof[2]. We compiled the C-code

Figure 2 shows a plot of the average CPU-time of each of the algorithms with 95% confidence intervals. We note that the use of gprof enables us to eliminate all the CPU time associated with reading the data, outputting the decision tree, or any other tasks that are not associated with computing the splits (this required all cases to add the CPU time of the function axis_parallel_split to the additional splits inspecting the output of gprof).

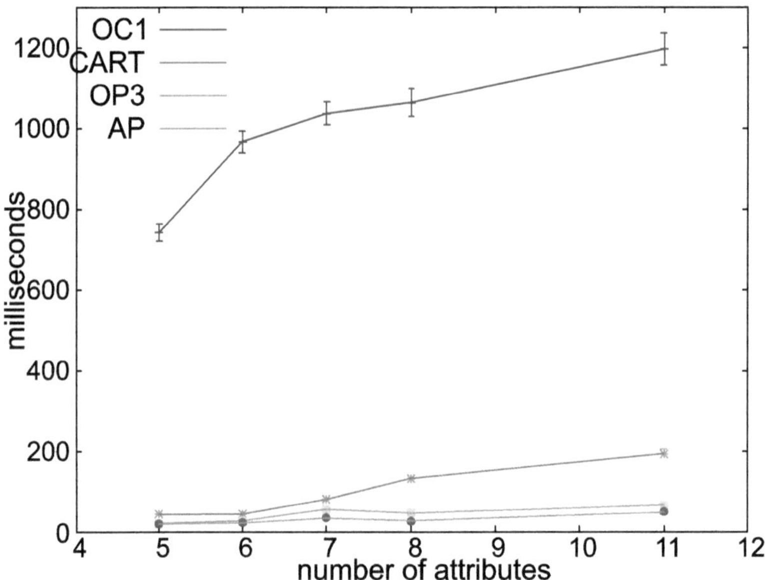

Fig. 2. Average CPU times (in ms) after each algorithm run on eight datasets for each dimension $d \in \{5, 6, 7, 8, 11\}$.

Also, gprof is a sampling approach and the results may seem rather noisy; however, Fig. 2 clearly shows that OC1 is much more costly and that both OC1 and CART

---

[2] ftp.gnu.org/pub/old-gnu/Manuals/gprof-2.9.1/html_mono/gprof.html.

have slopes that will grow rapidly with the number of dimensions. The time complexity of our method is not completely independent on the number of dimensions (neither is the axis-parallel approach as many attributes are explored as dimensions). However, our method required computational time stays remarkably close to the axis-parallel algorithm.

5.2 How 3D Hyperplanes Appear in Parallel Coordinates

Parallel coordinates are a visualisation method used to explore multi-dimensional data [31,34]. Each attribute A_i is shown as a vertical axis at $(0, i)$, and data points $x_j \in X$ are represented as poly-lines crossing these axes at $(i, x_{j,i})$. This technique allows all data instances to be viewed simultaneously, making it effective for detecting patterns such as correlations: parallel lines suggest positive correlation, while crossing lines may indicate negative correlation.

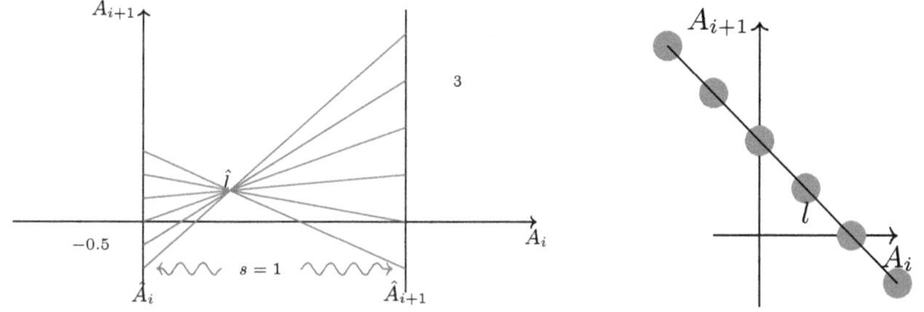

(a) The PC view with a separation $s = 1$, the point (-0.5,3) is a line segment. Other segments represent points in $\langle \hat{A}_i, \hat{A}_{i+1} \rangle$ that intersect at $\hat{l}$.

(b) In the orthogonal plane, the line l shows the points are co-linear.

Fig. 3. Illustration of the point-line duality of PC visualisation.

One crucial aspect of parallel coordinates is that there is a duality between the projective 2-dimensional plane and the d-dimensional space. Consider Fig. 3a that illustrates the parallel coordinates view (PC-view) for two attributes A_i and A_{i+1}. In the PC-view, we show a set $l \subset \Re^d$ of poly-lines that intersect at a point $\hat{l}$. The correspondence of $\hat{l}$ in the orthogonal (Cartesian) view is the line ls (refer to Fig. 3b).

In general, a line $l : a_{i+1} = ma_i + b$ in $\mathbb{R}^2$, maps to all the poly-lines in the parallel coordinates (PC) view that pass through the point

$$\hat{l} = \left(\frac{s}{1 - m}, \frac{b}{1 - m} \right), \tag{2}$$

where s is the distance between consecutive vertical axes, typically set to 1 for simplicity. This point $\hat{l}$ represents the common intersection of the segments corresponding to the line l in the PC view. This mapping illustrates the *duality* between the Cartesian view and the PC-view: a line corresponds to a point in the PC-view; conversely, a point in the Cartesian plane corresponds to a family of parallel polylines in the PC view.

However, when the slope $m = 1$, the denominator in Eq. 2 becomes zero, leading to a point at infinity (thus the use of the projective plane).

With the above background it becomes clear that splits based on a test of a single attribute, such as $a_i < t_i$, $a_i \leq t_i$, $a_i > t_i$ or $a_i \geq t_i$ (for some constant t_i) correspond to an infinite interval defined by only one end-point t_i on the axis for A_i. That is one-dimensional hyperplanes are easy to visualise. Moreover, test with two bounds, for example $s_i \leq a_i < t_i$ correspond to bounded intervals with both end-points defined. Alternatively, the intersection of two one-dimensional hyperplanes for the same attribute are also easy to visualise.

Thus, in HIIL [19,20], the human participant starts building a tree following Algorithm 1; that is, from a single root node. Repeatedly, the system suggest a node to split and displays the attributes with potential splits sorted down by information gain (or the metrics preferred by the user). The user can accept the default and go to another node, or modify the point t_i in the axis for A_i, and chose this as the split. Refer to Fig. 4, which shows the instances with a colour that corresponds to the class. The display will also shows the child nodes and the dominance of a colour indicates visualises the purity (which correlates with information gain).

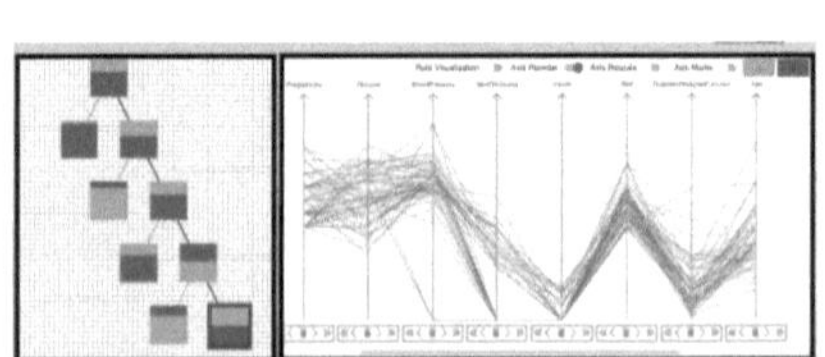

(a) System shows the decision tree and a suggested split.

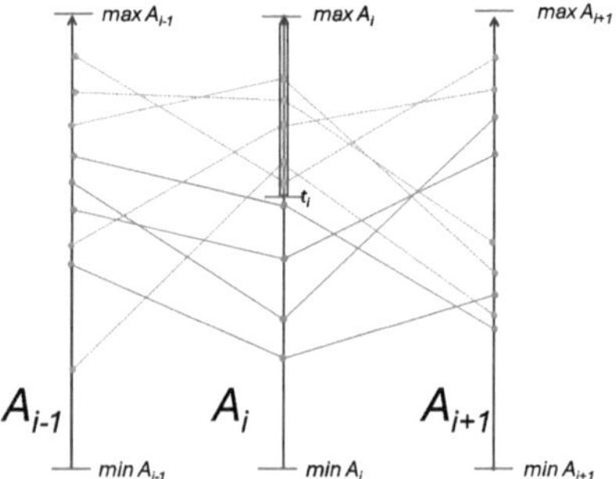

(b) In the PC-view, a point t_i in axis A_i defines a one-dimensional split $a_i > t_i$ separating two classes.

Fig. 4. Display of the interactive software where the user builds a decision tree selecting one-dimensional (axis parallel) splits.

Oblique tests using two attributes have the form $c_i a_i + c_{i+1} a + i + 1 < 0$ (or variants with $\leq$) when the two attributes are shown next to each other in the PC view. In the PC view, we know the line $l : c_i a_i + c_{i+1} a + i + 1 = 0$ corresponds to a point $\hat{l}$, and thus, these tests consist of a vertical unbounded interval (upwards or downwards) form the point $\hat{l}$.

In this paper we introduce oblique splits with three attributes; that is $c_i a_i + c_{i+1} a_{i+1} + c_{i+2} a_{i+2} + c_0 < 0$ (or variants with $\leq$). For this, we need to explain how is the hyperplane $c_i a_i + c_{i+1} a_{i+1} + c_{i+2} a_{i+2} + c_0 = 0$ visible in the PC view.

The first observation we make is that a plane in $\Re^3$ projects into a line for each pair of coordinates (by setting the third coordinate to zero). For example, consider the plane π^2 generated by the following three points in $A_i \times A_{i+1} \times A_{i+2} \subset \Re^3$:

$$\{(1,0,0),(0,2,0),(0,0,3)\}.$$

1. The plane π^2 intersects $a_{i+2} = 0$ and forms the line $l_{i,i+1} : a_{i+1} = -2a_i + 2$.
2. The plane π^2 intersects $a_{i+1} = 0$ and forms the line $l_{i,i+2} : a_{i+2} = -3a_i + 3$.
3. The plane π^2 intersects $a_i = 0$ and forms the line $l_{i+1,i+2} : a_{i+2} = -\frac{3}{2}a_{i+1} + 3$.

Recall that Eq. 2 enables us to find the point in the PC-view for each line $l_{i,i+1} : a_{i+1} = m_{i+1}, a_i + b_{i+1}$. Thus, in the PC view,

1. the line $l_{i,i+1}$ is represented by the point $\hat{l}_{i,i+1}$ on the PC-view, that is $(i+1/3, 2/3)$,
2. the line $l_{i+1,i+2}$ is represented by the point $\hat{l}_{i+1,i+2}$ on the PC-view, that is $((i+1) + 2/5, 6/5)$,
3. and the line $l_{i,i+2}$ is represented by the point $\hat{l}_{i,i+2}$ on the PC-view; here the separation $s = 2$, thus, this is the point (i+1/2,3/4).

The remarkable fact is that $\hat{l}_{i,i+1}$, $\hat{l}_{i+1,i+2}$, and $\hat{l}_{i,i+2}$ are co-linear passing through a line $\hat{L}$ on the PC-view (refer to Fig. 5a).

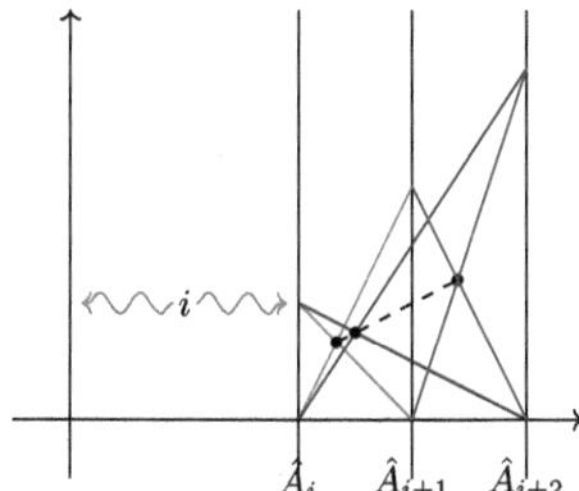

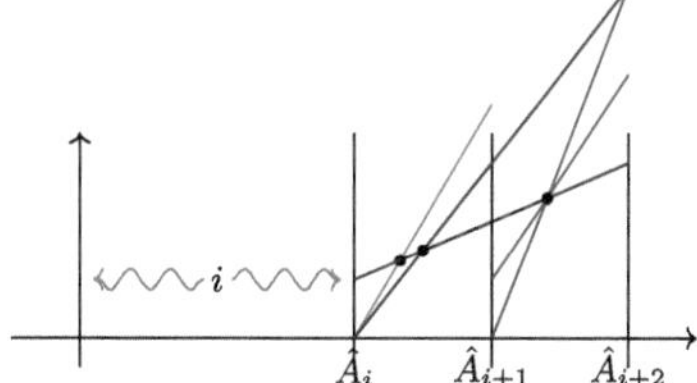

(a) Three lines embedded in a plane $\pi^2 \subset \Re^3$ are represented by co-linear points in the PC view.

(b) A second set of collinear points, from the projections $z = -3$, $y = -2$, and $x = -1$, we have 3 PC-view points (representing 3 lines in the plane $-3a_i - (3.2)a_{i+1} - a_{i+2} + 3 = 0$.

Fig. 5. A 2D plane of the PC view.

Moreover, this holds always. If we take three other lines l_i, l_j and l_k on the plane π^2, find their dual points $\hat{l}_{i,j}$, $\hat{l}_{j,k}$ and $\hat{l}_{i,k}$ on the PC-view, then $\hat{l}_{i,j}$, $\hat{l}_{j,k}$ are co-linear on a line $\hat{L}'$. More remarkable is that all lines $\hat{L}$ constructed this way meet at a point in the PC-view. This is the part of the plane π^2 in the PC view. In our running example, the plane is $-3a_i - \frac{3}{2}a_{i+1} - a_{i+2} + 3 = 0$. We can find three new lines. By setting $a_{i+1} = 3$, we have $\hat{l}_{i,i+1} : a_{i+1} = -2a_i$, which in the PC-view is the point $\hat{l}_{i,i+1} = (i + \frac{1}{3}, 0)$. By setting $a_{i+1} = 2$, we have $\hat{l}_{i,i+2} : a_{i+2} = -3a_i$, which in the PC-view is the point $\hat{l}_{i,i+2} = (i + 1 + \frac{1}{2}, 0)$. By setting $a_i = 1$, we have $\hat{l}_{i+1,i+2} : a_{i+2} = -\frac{3}{2}a_{i+1}$, which in the PC-view is the point $\hat{l}_{i+1,i+2} = (i + 1 + \frac{2}{5}, 0)$. These are clearly co-linear (they all have the second coordinate 0).

In general, given a plane π^2 embedded in $\Re^3$, we find a line $l \subset \pi^2$. We project l into three lines $l_{1,2}$, $l_{2,3}$, and $l_{1,3}$ (by making one out of the three axes set to zero). Then we obtain the points $\hat{l}_{1,2}$, $\hat{l}_{2,3}$, and $\hat{l}_{1,3}$ in the PC-view and the line $\hat{L}$ through them. Repeating this and obtaining two lines, $\hat{L}$ and $\hat{L}'$, they intersect at $(c_2 + 2c_3, c_0, c_1 + c_2 + c_3)$ in homogeneous coordinates. In the visualisation, this correspond to the point

$$\left(\frac{c_2 + 2c_3}{c_1 + c_2 + c_3}, \frac{c_0}{c_1 + c_2 + c_3} \right).$$

6 Final Remarks

We have presented a new version of oblique trees whose construction is remarkably efficient, and although the resulting models are less expressive than full multi-dimensional oblique trees, our evaluation shows the penalty in accuracy is minimal. We argue there are significant gains in transparency and interpretability, and we have argued for the possibility of enabling their construction interactively with a human user. Naturally, classification by our oblique trees are explainable. However, we leave for further work the evaluation with human participants of the merits of the approach. Nevertheless, even operating in fully autonomous mode, the construction of these oblique trees is an interesting trade-off in achieving transparency and explainability, and offering more expressive models than vanilla axis-parallel decision trees.

References

1. Statlog (Shuttle). UCI Machine Learning Repository. https://doi.org/10.24432/C5WS31
2. Image Segmentation. UCI Machine Learning Repository (1990). https://doi.org/10.24432/C5GP4N
3. Adadi, A., Berrada, M.: Peeking inside the black-box: a survey on explainable artificial intelligence (XAI). IEEE Access **VI**, 52138–52160 (2018)
4. Amershi, S., Cakmak, M., Knox, W.B., Kulesza, T.: Power to the people: The role of humans in interactive machine learning. AI Magazine **XXXV**(4), 105–120 (2014)
5. Ankerst, M., Elsen, C., Ester, M., Kriegel, H.P.: Visual classification: an interactive approach to decision tree construction. In: 5th ACM SIGKDD International Conference on Knowledge Discovery and Data Mining, pp. 392–396. ACM (1999)
6. Barredo Arrieta, A., et al.: Explainable artificial intelligence (XAI): concepts, taxonomies, opportunities and challenges toward responsible AI. Info. Fus. **58**, 82–115 (2020). https://doi.org/10.1016/j.inffus.2019.12.012
7. Bastani, O., Kim, C., Bastani, H.: Interpreting blackbox models via model extraction. In: Proceedings of the 2017 AAAI Workshop on Explainable AI (2017)
8. Bertsimas, D., Dunn, J.: Optimal classification trees. Mach. Learn. **106**(7), 1039–1082 (2017). https://doi.org/10.1007/s10994-017-5633-9
9. Bollwein, F., Westphal, S.: A branch & bound algorithm to determine optimal bivariate splits for oblique decision tree induction. Appl. Intell. **51**(10), 7552–7572 (2021). https://doi.org/10.1007/s10489-021-02281-x

10. Breiman, L., Friedman, J., Olshen, R.A., Stone, C.J.: Classification and Regression Trees. Routledge (2017)
11. Breiman, L., Friedman, J., Stone, C.J., Olshen, R.A.: Classification and Regression Trees. Wadsworth International Group, Monterrey, United States (1984)
12. Cantu-Paz, E., Kamath, C.: Inducing oblique decision trees with evolutionary algorithms. IEEE Trans. Evol. Comput. **7**(1), 54–68 (2003)
13. Craven, M., Shavlik, J.W.: Extracting tree-structured representations of trained networks. In: Advances in Neural Information Processing Systems, pp. 24–30 (1996)
14. Crespi, M., Ferigo, A., Custode, L.L., Iacca, G.: A population-based approach for multi-agent interpretable reinforcement learning. Appl. Soft Comput. **147**, 110758 (2023). https://doi.org/10.1016/j.asoc.2023.110758
15. Dignum, V.: AI value alignment: Guiding artificial intelligence towards shared human goals. White paper (2024)
16. Dwivedi, R., et al.: Explainable AI (XAI): core ideas, techniques, and solutions. ACM Comput. Surv. **55**(9) (2023). https://doi.org/10.1145/3561048
17. Engelhardt, R.C., Oedingen, M., Lange, M., Wiskott, L., Konen, W.: Iterative oblique decision trees deliver explainable RL models. Algorithms **16**(6) (2023). https://doi.org/10.3390/a16060282
18. Estivill-Castro, V.: Collaborative knowledge acquisition with a genetic algorithm. In: 9th International Conference on Tools with Artificial Intelligence, ICTAI '97, Newport Beach, CA, USA, November 3-8, 1997, pp. 270–277. IEEE Computer Society (1997). https://doi.org/10.1109/TAI.1997.632266
19. Estivill-Castro, V., Gilmore, E., Hexel, R.: Constructing Interpretable Decision Trees Using Parallel Coordinates. In: Rutkowski, L., Scherer, R., Korytkowski, M., Pedrycz, W., Tadeusiewicz, R., Zurada, J.M. (eds.) ICAISC 2020. LNCS (LNAI), vol. 12416, pp. 152–164. Springer, Cham (2020). https://doi.org/10.1007/978-3-030-61534-5_14
20. Estivill-Castro, V., Gilmore, E., Hexel, R.: Constructing explainable classifiers from the start - enabling human-in-the loop machine learning. Inf. **13**(10), 464 (2022). https://doi.org/10.3390/INFO13100464
21. Fails, J.A., Olsen, D.R.: Interactive machine learning. In: 8th International Conference on Intelligent User Interfaces, pp. 39–45. ACM (2003)
22. Felzmann, H., Fosch-Villaronga, E., Lutz, C., Tamò-Larrieux, A.: Towards Transparency by Design for Artificial Intelligence. Sci. Eng. Ethics (10), 1–29 (2020). https://doi.org/10.1007/s11948-020-00276-4
23. Ferigo, A., Custode, L.L., Iacca, G.: Quality diversity evolutionary learning of decision trees. In: Proceedings of the 38th ACM/SIGAPP Symposium on Applied Computing, pp. 425–432. SAC '23, Association for Computing Machinery, New York, NY, USA (2023). https://doi.org/10.1145/3555776.3577591
24. Freitas, A.A.: Comprehensible classification models: a position paper. SIGKDD Spec. Inter. Group on Knowl. Discov. Data Min. Explorat. Newsletter **15**(1), 1–10 (2014)
25. Gabidolla, M., Carreira-Perpiñán, M.Á.: Optimal interpretable clustering using oblique decision trees. In: Proceedings of the 28th ACM SIGKDD Conference on Knowledge Discovery and Data Mining, pp. 400–410 (2022)
26. Gaudart, J., Poudiougou, B., Ranque, S., Doumbo, O.: Oblique decision trees for spatial pattern detection: optimal algorithm and application to malaria risk. BMC Medical Res. Method. **5**(22) (2005)
27. Genetti, S., Longobardi, A., Iacca, G.: Evolutionary reinforcement learning for interpretable decision-making in supply chain management. In: García-Sánchez, P., Hart, E., Thomson, S.L. (eds.) Applications of Evolutionary Computation: 28th European Conference, EvoApplications 2025, Held as Part of EvoStar 2025, pp. 187–203. Springer-Verlag, Cham (2025). https://doi.org/10.1007/978-3-031-90062-4_12

28. Guidotti, R., Monreale, A., Ruggieri, S., Turini, F., Giannotti, F., Pedreschi, D.: A survey of methods for explaining black box models. ACM Comput. Surv. **LI**(5) (2018). Article 93

29. Hall, M., Frank, E., Holmes, G., Pfahringer, B., Reutemann, P., Witten, I.H.: The WEKA data mining software: an update. SIGKDD Explor. Newsl. **11**(1), 10–18 (11 2009).https://doi.org/10.1145/1656274.1656278

30. Heath, D. G. andJ Kasif, S., Salzberg, S.: Induction of oblique decision trees. In: Bajcsy, R. (ed.) Proceedings of the 13th International Joint Conference on Artificial Intelligence, pp. 1002–1007. Morgan Kaufmann, Chambéry, France (1993)

31. Heinrich, J., Weiskopf, D.: State of the art of parallel coordinates. In: Sbert, M., Szirmay-Kalos, L. (eds.) Eurographics 2013 - State of the Art Reports. The Eurographics Association (2013). https://doi.org/10.2312/conf/EG2013/stars/095-116

32. Hunt, E.B., Marin, J., Stone, P.J.: Experiments in Induction. Academic Press, Cambridge, United States (1966)

33. Hyafil, L., Rivest, R.L.: Constructing optimal binary decision trees is NP-complete. Inf. Process. Lett. **5**(1), 15–17 (1976). https://doi.org/10.1016/0020-0190(76)90095-8

34. Inselberg, A.: Parallel CoordinatesâĂŕ: Visual Multidimensional Geometry and its Applications. Springer, New York, United States (2009)

35. James, G., Witten, D., Hastie, T., Tibshirani, R.: An Introduction to Statistical Learning: With Applications in R. Springer Publishing Company, Incorporated (2014)

36. Lai, P.L., Liang, Y.J., Inselberg, A.: Geometric divide and conquer classification for high-dimensional data. In: 1st International Conference on Data Technologies and Applications, pp. 79–82. SciTePress (2012)

37. Leblanc, B., Germain, P.: Seeking interpretability and explainability in binary activated neural networks. In: Longo, L., Lapuschkin, S., Seifert, C. (eds.) Explainable Artificial Intelligence, pp. 3–20. Springer Nature Switzerland, Cham (2024)

38. Lipton, Z.C.: The mythos of model interpretability: in machine learning, the concept of interpretability is both important and slippery. Queue **16**(3), 31–57 (2018). https://doi.org/10.1145/3236386.3241340

39. Love, P.E., Fang, W., Matthews, J., Porter, S., Luo, H., Ding, L.: Explainable artificial intelligence (XAI): precepts, models, and opportunities for research in construction. Adv. Eng. Inform. **57**, 102024 (2023). https://doi.org/10.1016/j.aei.2023.102024

40. Manwani, N., Sastry, P.S.: A geometric algorithm for learning oblique decision trees. In: Chaudhury, S., Mitra, S., Murthy, C.A., Sastry, P.S., Pal, S.K. (eds.) Pattern Recognition and Machine Intelligence, pp. 25–31. Springer, Berlin Heidelberg, Berlin, Heidelberg (2009)

41. Mitchell, T.M.: Machine Learning. McGraw-hill, New York, United States (1997)

42. Moore, A., Murdock, V., Cai, Y., Jones, K.: Transparent tree ensembles. In: The 41st International ACM SIGIR Conference on Research & Development in Information Retrieval, pp. 1241–1244. SIGIR '18, Association for Computing Machinery, New York, NY, USA (2018). https://doi.org/10.1145/3209978.3210151

43. Mosqueira-Rey, E., Hernández-Pereira, E., Alonso-Ríos, D., Bobes-Bascarán, J., Fernández-Leal, Á.: Human-in-the-loop machine learning: a state of the art. Artif. Intell. Rev. **56**(4), 3005–3054 (2023). https://doi.org/10.1007/s10462-022-10246-w

44. Murthy, S.K., Kasif, S., Salzberg, S.: A system for induction of oblique decision trees. J. Artif. Intell. Res. **2**, 1–32 (1994)

45. Nowozin, S.: Improved information gain estimates for decision tree induction. In: Proceedings of the 29th International Conference on Machine Learning, pp. 571–578. ICML'12, Omnipress, Madison, WI, USA (2012)

46. Panigutti, C., et al.: The role of explainable AI in the context of the AI act. In: Proceedings of the 2023 ACM Conference on Fairness, Accountability, and Transparency, pp. 1139–1150. FAccT '23, Association for Computing Machinery, New York, NY, USA (2023). https://doi.org/10.1145/3593013.3594069

47. Quinlan, J.R.: Induction of decision trees. Mach. Learn. **I**(1), 81–106 (1986)
48. Quinlan, J.R.: C4.5: Programs for Machine Learning. Morgan Kaufmann, San Francisco, United States (1993)
49. Rivera-Lopez, R., Canul-Reich, J., Gámez, J.A., Puerta, J.M.: OC1-DE: a differential evolution based approach for inducing oblique decision trees. In: 16th International Conference on Artificial Intelligence and Soft Computing, pp. 427–438. Springer (2017)
50. Rudin, C.: Stop explaining black box machine learning models for high stakes decisions and use interpretable models instead. Nat. Mach. Intell. **1**(5), 206–215 (2019). https://doi.org/10.1038/s42256-019-0048-x
51. Sacha, D., Kraus, M., Keim, D.A., Chen, M.: VIS4ML: an ontology for visual analytics assisted machine learning. IEEE Trans. Visual Comput. Graphics **25**(1), 385–395 (2019). https://doi.org/10.1109/TVCG.2018.2864838
52. Slate, D.: Letter Recognition. UCI Machine Learning Repository (1991). https://doi.org/10.24432/C5ZP40
53. Srinivasan, A.: Statlog (Landsat Satellite). UCI Machine Learning Repository (1993). https://doi.org/10.24432/C55887
54. Tan, Y., Caruana, R., Hooker, G., Lou, Y.: Learning sparse decision trees for high-stakes decisions. Proc. Natl. Acad. Sci. **119**(1), e2100043119 (2022)
55. Teoh, S.T., Ma, K.L.: Paintingclass: interactive construction, visualization and exploration of decision trees. In: 9th ACM SIGKDD International Conference on Knowledge Discovery and Data Mining, pp. 667–672. ACM (2003)
56. Teoh, S.T., Ma, K.L.: StarClass: interactive visual classification using star coordinates. In: 3rd SIAM International Conference on Data Mining, pp. 178–185. SIAM (2003)
57. Tibshirani, R.: Regression shrinkage and selection via the Lasso. J. Royal Stat. Soc. Ser. B (Methodol.) **58**(1), 267–288 (1996)
58. Vasić, M., Petrović, A., Wang, K., Nikolić, M., Singh, R., Khurshid, S.: Moët: mixture of expert trees and its application to verifiable reinforcement learning. Neural Netw. **151**, 34–47 (2022)
59. Wang, Z., Huang, C., Yao, X.: A roadmap of explainable artificial intelligence: explain to whom, when, what and how? ACM Trans. Auton. Adapt. Syst. **19**(4) (2024). https://doi.org/10.1145/3702004
60. Ware, M., Frank, E., Holmes, G., A., H.M., Witten, I.H.: Interactive machine learning: letting users build classifiers. Int. J. Hum. Comput. Studi. **LV**(3), 281–292 (2001)
61. Webb, G.I.: Integrating machine learning with knowledge acquisition through direct interaction with domain experts. Knowl.-Based Syst. **9**(4), 253–266 (1996). https://doi.org/10.1016/0950-7051(96)01033-7
62. Wu, X., Xiao, L., Sun, Y., Zhang, J., Ma, T., He, L.: A survey of human-in-the-loop for machine learning. Futur. Gener. Comput. Syst. **135**, 364–381 (2022). https://doi.org/10.1016/j.future.2022.05.014

Extracting Deterministic Finite Automata from RNNs via Hyperplane Partitioning and Learning

Sandamali Yashodhara Wickramasinghe[1(✉)] [iD], Jacob M. Howe[1] [iD], and Laure Daviaud[2] [iD]

[1] City St George's, University of London, London, U.K.
{sandamali.wickramasinghe,j.m.howe}@citystgeorges.ac.uk
[2] University of East Anglia, Norwich, U.K.
L.Daviaud@uea.ac.uk

Abstract. Recurrent Neural Networks (RNNs) have achieved remarkable success in handling sequential data. However, they lack interpretability. Extracting Deterministic Finite Automata (DFAs) from black-box models can provide insight into their decision-making processes. This research focuses on extracting DFAs from RNNs trained on regular languages using an exact learning framework. The proposed approach employs the L^* algorithm to learn a DFA, and it demonstrates how a hyperplane-based method can be used to partition the RNN state space when evaluating equivalence queries.

Keywords: Recurrent Neural Networks · Finite State Automata · Explainable AI

1 Introduction

This research explores the extraction of automata from Recurrent Neural Networks (RNNs). Recurrent Neural Networks are known for successfully handling sequential data, making them widely used for language modelling tasks. Unlike feed-forward neural networks, RNNs' sequential processing architecture allows them to capture the patterns and dependencies of language data.

Regardless of their effectiveness, owing to their lack of interpretability, RNNs are considered to function as black boxes [2, 15, 22]. The internal decision-making process of RNNs is difficult to understand, and there is no straightforward way to interpret it or the network output at each step. This lack of transparency raises questions about the model training process and whether the model has captured the specifics of the problem it is supposed to address. This is not just a theoretical issue and has practical consequences, specifically in cases where understanding the decision-making process is important [4, 14].

To address this, researchers have been exploring methods to extract more interpretable models, such as automata, to approximate the functionality and behaviour of RNNs. Automata are symbolic, rule-based models that provide a clearer structure, making it easier to understand and verify the model's behaviour. While they can be used to

F. Marcelloni et al. (Eds.): IJCCI 2025, CCIS 2829, pp. 116–135, 2026.
https://doi.org/10.1007/978-3-032-15638-9_8

interpret the RNN's decision-making process, they can also serve as a substitute for the RNN when inferring.

One of the first attempts to extract an automaton from RNNs was presented in [7], where it was demonstrated that the structure of the state space of an RNN trained on a regular grammar could be closely approximated by a corresponding minimal finite-state automaton. The work in [11] further explored this connection by introducing a quantisation algorithm that divides the RNN state space into partitions for automaton extraction. Similarly, [23] showed that clustering the output space of an RNN is an effective method for extracting finite-state representations. Furthermore, [33] provided evidence that the hidden state space of an RNN forms cluster-like structures while learning a regular grammar. Also, [29] use k-means clustering to extract a Deterministic Finite Automata (DFA), but this requires predefining the number of DFA states as a hyperparameter [29].

The L^* *algorithm* [1] is a widely recognised *exact learning* method used to learn a minimal DFA that accepts the same language as an unknown target system. Exact learning is a framework where a learner interacts with an oracle (or teacher) to identify the target model. The goal of the learner is to construct a hypothesis, such as a DFA, that matches the behaviour of the target system, using a series of carefully chosen queries [12]. The L^* algorithm uses membership queries and equivalence queries to learn the DFA, where membership queries ask whether a specific string belongs to the target language, whilst equivalence queries check if a proposed automaton matches the target language and return a counterexample if it does not.

To apply L^* with RNNs as the oracle, answers to membership and equivalence queries are needed. Membership queries can be straightforwardly answered, but since RNNs are defined over high-dimensional continuous vector spaces, whereas DFAs have a finite set of discrete states, a notion of equivalence is not immediately available. Building on earlier work, [32] explores the extraction of DFAs from RNNs using the L^* algorithm. Their approach introduces an abstraction of the RNN state space for the equivalence query. This abstraction was constructed by partitioning the RNN hidden state space using support vector machine (SVM) classifiers trained on the hidden states, a potentially expensive operation. By iteratively partitioning the RNN state space only when necessary, they demonstrate that it is possible to extract a DFA from an RNN while mitigating state explosion.

In this work, an alternative DFA extraction method using the L^* algorithm is presented, where the equivalence queries are realised using a hyperplane-based partitioning of the RNN state space. This abstraction does not construct a DFA itself but serves as an intermediate representation that allows comparison of the behaviour of the RNN with the hypothesis DFA extracted by L^*. The main contributions include a hyperplane-based approach to partitioning the RNN state space, which is computationally attractive, and the way that this is integrated into a method for answering equivalence queries. In addition, using k-means clustering to start with an initial number of clusters, based on the natural clustering of RNN hidden states, in order to reduce the number of refinements needed in partitioning, is investigated. An experimental evaluation of the work on formal language problems is presented.

The remainder of this paper is organised as follows. Sections 2 and 3 review related work and provide background information and definitions relevant to the study.

Section 4 describes the proposed methodology, focusing on how the L^* algorithm is employed for automata extraction. Section 5 presents the experimental setup and results, followed by a discussion of the findings in Sect. 6 and conclusions in Sect. 7.

2   Related Work

Neural nets were first introduced in [20], with a model inspired by biological neurons. Foundational concepts involving the use of rational expressions and finite-state automata to describe the expressive power of such models were later established in [16]. Extending these foundational concepts, [27] explored the computational capabilities of first-order RNNs, where they argue that RNNs can simulate Turing machines under specific configurations, thereby establishing a theoretical framework that highlights the computational potential of neural networks. Building on this, recent research has focused on the extraction of automata from trained RNN models, which shows that there is a possibility to extract finite automata from RNNs trained for grammatical inferences [11, 15, 23].

The automata extraction methods can be divided into two main categories based on their underlying principles: compositional approaches and learning approaches [15]. Compositional approaches analyse and partition the RNN state space to identify representative state vectors and the transitions between them, typically using clustering to capture the discrete dynamics of the RNN. Learning approaches, such as exact learning, build an automaton by querying the target system's outputs and iteratively refining the model using counterexamples.

In the early days following the introduction of RNNs [9], various studies explored the possibility of obtaining finite automata by partitioning the state space of RNNs [7, 11, 23, 33], which laid the foundation for automata extraction. An early attempt to verify the structure of the RNN state space using clustering algorithms, specifically hierarchical clustering, was presented in [7]. This work aimed to closely approximate the corresponding minimal finite-state automaton by organising the states of the RNN into meaningful clusters. This approach provided an early demonstration of how unsupervised learning techniques could be applied to uncover the underlying state transitions of RNNs. Building on this work, further research in [11] explored the use of state space partitioning through clustering, highlighting its importance in understanding and modelling the behaviour of RNNs.

The study in [23] continued this line of inquiry by employing a simple clustering algorithm to partition the RNN state space, using input data generated from predefined DFAs. Their work used a quantisation method, which converts continuous internal states of RNNs into discrete representations. Here, an initial quantisation level has to be fixed before extraction. Their results indicated that a minimal DFA could be effectively extracted through the method by grouping like-valued state vectors together as a single cluster and treating a cluster as a state of a DFA. There are subsequent studies which have expanded and refined these approaches for automata extraction. For instance, one can use a clustering method such as k-means [29, 33], which does not need quantisation [29]. Quantisation can lead to a state explosion where the number of discrete states grows exponentially, making the extraction process more complicated.

The primary advantage of compositional approaches lies in their scalability, as these methods can process extensive datasets and derive automata representations without manual intervention. It makes them particularly suitable for applications where the RNN operates over a wide range of inputs and generates complex output patterns [15,29]. However, these methods also face significant challenges, such as ensuring the accuracy of the state space partitioning, as improper clustering can lead to the loss of important information and result in an incomplete or inaccurate automaton [8].

Learning approaches use exact learning algorithms, such as the L^* algorithm [1], to extract an automaton from a target system. Here, an automaton is extracted by asking queries to understand the underlying state transitions of a network. This approach has achieved wide success, specifically for regular languages. However, unlike compositional methods, such approaches have limitations in scalability [30].

A prominent example of work in this area is the research presented in [32], which employs the L^* algorithm originally developed in [1]. The L^* algorithm is a well-established exact learning method that uses an oracle to query a target network and incrementally construct an equivalent automaton. Variants of this algorithm have also been applied in other studies focused on automaton extraction [19].

The work in [32] adapted the L^* algorithm to extract a DFA from RNNs trained with perfect train and test accuracy. Their novel algorithm efficiently utilises the L^* algorithm and an abstraction of the state space of the RNN in extracting automata [32]. Later, this line of work was extended to extracting deterministic weighted automata using a variation of the L^* algorithm adapted to a probabilistic setting. However, this involved performing equivalence queries using the oracle itself [31]. Additionally, several studies have explored the extraction of weighted finite automata using variants of the L^* algorithm [10,21], originally proposed in [3].

3   Background

This section introduces the technical notation that will be used throughout the paper. See [26] for details.

3.1   Languages and Automata

Formal grammars are used to model and to study the properties, structures, and behaviours of computational systems from the Chomsky hierarchy [5]. Examples of common formal grammars include regular grammars, context-free grammars, and recursively enumerable grammars.

A finite alphabet Σ is a finite set of elements, called letters and the set of all finite words (a.k.a. strings, i.e. a finite sequence of letters) over Σ is denoted by Σ^*. The empty word of length 0 is denoted by ε. The length of a word $w \in \Sigma^*$ is denoted by $|w|$. Given two words u and v, their concatenation is the word $w = uv$, and u is said to be a prefix, and v a suffix of w, respectively.

Definition 1. *A Deterministic Finite Automata (DFA) A is a tuple $\langle \Sigma, Q, q_I, F, \delta \rangle$, where Σ is the finite alphabet, Q is the finite set of states, $F \subseteq Q$ are the accepting states, $q_I \in Q$ is an initial state and $\delta : Q \times \Sigma \to Q$ is the transition function.*

A word $w = \sigma_1\sigma_2\ldots\sigma_n \in \Sigma^*$, with $\sigma_i \in \Sigma$ for all i, is *accepted* by A, if there exists $q_0, q_1, q_2, \ldots, q_n \in Q$, such that $q_0 = q_I, q_n \in F$ and $\delta(q_{i-1}, \sigma_i) = q_i$. The regular language recognised by A, is denoted by L_A, which is defined by the set of words accepted by the DFA A. Two DFAs are called equivalent if they recognise the same language.

3.2   Recurrent Neural Networks

Recurrent Neural Networks (RNNs) are a class of neural networks that includes at least one feedback loop, allowing activations to circulate within the network. This recurrent structure makes RNNs particularly well-suited for modelling sequential data. By utilising their internal state, RNNs can retain memory of previous elements in a sequence, enabling them to consider the impact of past inputs when computing current outputs. Thus, the units in an RNN layer use not only the input from the previous layer but also the hidden state from the previous time step. The output at a given time step becomes part of the input for the next, facilitating the sequential mechanism within the RNN. Later advancements, such as Long Short-Term Memory (LSTM) [13] (and gated Recurrent Unit (GRU) [6]) have significantly improved RNNs' ability to handle long-range data dependencies, thus expanding their potential uses. In this work, the RNNs with LSTM and GRU architectures will be discussed.

An RNN that processes an input word $w \in \Sigma^*$ produces an output and can also provide access to the corresponding hidden state information. Formally [21,32], a *Recurrent Neural Network (RNN)* is a tuple $R = (h_0, f_R, g_R)$, where $h_0 \in \mathbb{R}^d$ is the initial state, where $d \in \mathbb{N}$ is the dimension of the RNN state space, $g_R : \mathbb{R}^d \times \Sigma \to \mathbb{R}^d$ is the transition function producing the next hidden state given the current hidden state and an input symbol, $f_R : \mathbb{R}^d \to \{0, 1\}$ is the output function. Here, the RNN R is a binary acceptor.

For any $w \in \Sigma^*$ the transition function g_R can be extended by defining the function $g_R^* : \mathbb{R}^d \times \Sigma^* \to \mathbb{R}^d$ as follows:

$$g_R^*(h, \varepsilon) = h, \quad g_R^*(h, w\sigma) = g_R(g_R^*(h, w), \sigma)$$

for $h \in \mathbb{R}^d$, $w \in \Sigma^*$, and $\sigma \in \Sigma$.
The hidden state of the RNN R after processing a word $w \in \Sigma^*$ is $g_R^*(h_0, w)$. The output of the RNN on input word w is then $f_R(g_R^*(h_0, w)) \in \{0, 1\}$. The set of all possible hidden states reachable from h_0 is denoted by $S \subseteq \mathbb{R}^d$.

3.3   *L** Algorithm for DFA Learning

Angluin's L^* algorithm [1] is an exact learning algorithm designed to learn a minimal DFA for regular languages through a series of *membership and equivalence queries*. It interacts with an oracle, known as a *minimally adequate teacher*, which is a theoretical construct capable of answering these two types of queries about an unknown regular language L over an alphabet Σ.

A membership query provides a word w from Σ^*, and the oracle classifies the word as either belonging to the language (`true`) or not (`false`). Based on these responses, the learner constructs a DFA A_L that hypothesises acceptance of the language L (see Definition 2). Equivalence queries are then used to verify whether the learned DFA A_L correctly accepts the language L. If A_L does not accept L, the oracle provides a counterexample. The learner uses this feedback to refine the DFA.

The algorithm iteratively refines the DFA through further membership queries until the oracle confirms that A_L accepts the language L. At this point, the algorithm terminates, and A_L is returned as the final DFA. In the L^* algorithm, an observation table is maintained to organise the results of membership queries. The table is incrementally filled by querying the oracle during the learning process.

The *observation table* T in the L^* algorithm consists of a matrix with rows indexed by a finite set of prefixes U and columns by a finite set of suffixes V. The entry at position (u, v) is 1 if $uv \in L$, and 0 otherwise, where $u \in U$ and $v \in V$. From an observation table (initially set with $U = V = \{\varepsilon\}$), the L^* algorithm performs membership queries to fill the table until it is *closed*.

The *closed condition* states that for every $u\sigma$, where u indexes a row and $\sigma \in \Sigma$, there exists a word u' indexing a row of T that is identical to what the row indexed by $u\sigma$ would be, i.e. the role of an extended word $u\sigma$, using $\sigma \in \Sigma$, is already covered by some existing row in T.

The membership queries are performed for all concatenations of prefixes in U with suffixes in V. When the table is not *closed*, U is updated by adding new prefixes that correspond to uncovered transitions, and when the table is not *consistent*, V is extended by adding distinguishing suffixes. The *consistency condition* states that for any two prefixes $s_1, s_2 \in U$, where U is the set of prefixes indexing the rows of the observation table, if the rows indexed by s_1 and s_2 are identical, for all $\sigma \in \Sigma$, rows indexed by $s_1\sigma$ and $s_2\sigma$ are equal.

When the observation table T is *closed* and *consistent*, the algorithm returns a DFA A_L, as defined in Definition 2, to consider for the equivalence queries. The equivalence queries verify whether the DFA A_L correctly recognises the target language. If it does not, a counterexample $w \in \Sigma^*$ is returned to the learner L^*. Then, there exist strings u and v such that $w = uv$, and by updating the observation table to include u in U (the set of prefixes) and v in V (the set of suffixes), the table remains consistent. Then the table T is extended by adding rows $u\sigma$ by using the membership queries for the missing entries where $\sigma \in \Sigma$.

Definition 2. *Let Σ be a finite alphabet. Given a closed and consistent observation table defined by two finite sets of words U and V and $T : U \times V \rightarrow \{0, 1\}$ (such that $T(u, v) = T(u', v')$ if $uv = u'v'$), the* hypothesis DFA *is defined by $A = (\Sigma, Q, q_I, F, \delta)$ where Q is the finite set of distinct rows indexed by words in U, $q_I \in Q$ is the row indexed by ε, $F \subseteq Q$ is the rows indexed by $u \in U$ such that $T(u, \varepsilon) = 1$, and $\delta : Q \times \Sigma \rightarrow Q$ is the transition function such that for a state q corresponding to a row indexed by a word u, and for a letter $\sigma \in \Sigma$, $\delta(q, \sigma)$ is the state corresponding to the row indexed by the prefix $u\sigma$. Note that δ is well defined, since the table is closed and consistent.*

4   Extracting a DFA from an RNN via Learning

This section describes the procedure for extracting a DFA from an RNN.

The DFA is extracted using the L^* algorithm, as explained in Sect. 3.3. Here, the oracle is the RNN R, and the learner L^* attempts to infer the language recognised by R. During the membership queries, words are queried through the RNN R, and the outputs are recorded in an observation table with $f_R(uv) \in \{0, 1\}$. Once the observation table T is closed, the algorithm returns a DFA to be evaluated through equivalence queries.

Membership queries pose no issue; the main challenge in using the L^* algorithm lies in answering equivalence queries. RNNs operate over high-dimensional continuous vector spaces, whereas DFAs consist of a finite set of discrete states. Due to the continuous and potentially infinite nature of RNN state representations, checking formal equivalence between an RNN and a DFA is undecidable in general [18]. One way to address this is by using an abstraction of the RNN's hidden state space – typically by clustering hidden states to induce a finite-state model [32]. This abstraction allows for the use of equivalence queries, in which the abstracted automaton is iteratively compared against the L^* DFA. Counterexamples identified through these queries are used to refine the DFA until convergence is reached or equivalence is reasonably approximated. In this work, this challenge is addressed using an alternative approach.

4.1   Partitioning the State Space of the RNN

The state space of an RNN can be partitioned based on the behaviour of its hidden states. During sequence processing, hidden states that exhibit similar dynamics tend to lie close to one another in the high-dimensional space. These hidden states can therefore be grouped into clusters, which reveal structural patterns in the RNN's internal representation. This abstraction can be used to perform equivalence queries and serve as the basis for constructing the L^* hypothesis DFA. Unlike in the other works, here a partitioning of the RNN state space is maintained *rather than* an exact DFA.

Given a recurrent neural network (RNN) R with a state space $S \subseteq \mathbb{R}^d$, a partitioning $P = \{p_i \subseteq \mathbb{R}^d | i, j \in \{0, \ldots, n\}, p_i \cap p_j = \emptyset, \cup_{i=0}^n p_i = \mathbb{R}^d\}$ divides space into non-overlapping regions. The abstraction A_P is the hypothesised abstraction of the RNN state space, represented as the collection of all partitions with each partition p_i mapped to 1 or 0, based on the classification of the first hidden state encountered within that partition.

To extract an abstract partitioning A_P, the state space of the RNN R can be explored according to the network's transition function g_R. This exploration begins with an initial partition $p(h_0)$, where h_0 represents the initial hidden state of the RNN. Initially, the entire state space is treated as a single partition. As exploration progresses through equivalence queries, this space is incrementally divided into smaller partitions. The classification of a partition is determined based on the first hidden state observed within it, during the processing of an input word. If the next hidden state, corresponding to a subsequent input symbol in the word, falls within the same partition, a loop transition is created. Otherwise, the transition leads to a different partition. The partitions are stored in a decision tree structure.

4.2 Identifying Conflicts

The equivalence between the L^* DFA A_L and the abstraction A_P should produce the same outputs for the same inputs, meaning they recognise the same language or accept the same set of input-output behaviours.

The equivalence query has two tasks: it should be able to return a counterexample in the event where the L^* DFA is incorrect, and it should be able to refine the partitioning P, if A_P is incorrect. As presented in [32], exploring the DFA and the partitioning in parallel allows for inconsistencies between them to be identified without fully exploring the abstraction partitioning A_P.

In equivalence queries, the extracted DFA A_L is compared against the current partitioning of the RNN state space, denoted by A_P. Each word $w \in \Sigma^*$, up to the maximum length of the training sequences, is submitted to both A_L and A_P. The states reached and their corresponding classifications are then compared to detect any discrepancies.

If the classifications of the reached states differ for a given input w, the input is passed to the RNN R to determine the source of the discrepancy. If A_L is incorrect according to R, i.e., $A_L(w) \neq f_R(w)$, the mismatch is referred to as a *conflict in the DFA*. Additionally, at each transition, the classification of the hidden states in R is compared with that of A_L to identify cases where both A_L and A_P may fail to represent the correct behaviour. If disagreement is found, these are also treated as *conflicts in the DFA*.

A *partition conflict* is identified when A_P behaves inconsistently with R, specifically when $A_P(w) \neq f_R(w)$. This indicates that a single partition in A_P maps to multiple states in the minimal DFA A_L. Such a one-to-many mapping is invalid, as each state in A_L must correspond to a unique behaviour.

Additionally, during the equivalence queries, each hidden state explored is checked against the current partition to ensure that no conflicting hidden states are grouped within the same partition. Specifically, this refers to a scenario where both a positively and a negatively classified hidden state (i.e., one leading to acceptance and one to rejection) reside within the same partition. Algorithm 1 details the procedure for performing the equivalence queries during the extraction.

4.3 Resolving Conflicts

There are two possible ways to resolve conflicts. If there are *conflicts in the DFA A_L*, the conflicting word w is returned to the learner L^* as a counterexample. The observation table T is then updated with the appropriate prefix and suffix of the counterexample, leading to a refined L^* DFA A_L.

If there is a *partition conflict*, it indicates an inconsistency in the partitioning of the RNN state space. Specifically, there exist two input sequences, w_1 and w_2, that follow distinct paths in A_L but are mapped to the same path in A_P. This means that two different states in A_L are mapped into a single partition in A_P, resulting in a conflict.

To resolve this, the conflicting partition is split using the hidden states corresponding to w_1 and w_2, leading to a partition refinement. These hidden states are selected by identifying sequences of equal length that behave differently in the L^* DFA but are assigned to the same abstract partition in A_P. This discrepancy reveals that the current partitioning fails to distinguish between sequences that should lead to different

---

Algorithm 1. Equivalence Queries.

Require: RNN $R = (h_0, f_R, g_R)$, learning DFA A_L, abstract partitioning A_P, $\texttt{visited} \leftarrow \emptyset$
Ensure: Counterexample, Refinement, or Equivalence
 1: **for** each input word $w \in \Sigma^*$ **do**
 2:     Add w to $\texttt{visited}$
 3:     Compute $h \leftarrow g_R^*(h_0, w)$
 4:     **if** $A_L(w) \neq A_P(w)$ and $A_L(w) \neq f_R(w)$ **then**
 5:         **return** w as a $\texttt{Counterexample}$
 6:     **else if** $A_L(w) \neq A_P(w)$ and $A_P(w) \neq f_R(w)$ **then**
 7:         Let $w_1 \leftarrow w$
 8:         **for** each $w_2 \in \texttt{visited} \setminus \{w_1\}$ **do**
 9:             **if** there exists a suffix s such that
                  $A_L(w_1 s) \neq A_L(w_2 s)$ and $A_P(w_1 s) = A_P(w_2 s)$ **then**
10:                 Let $h_1 \leftarrow g_R^*(h_0, w_1 s)$, $h_2 \leftarrow g_R^*(h_0, w_2 s)$
11:                 $\texttt{Refinement}(h_1, h_2)$
12:                 **return** refined A_P
13:             **end if**
14:         **end for**
15:     **else if** there exists a hidden state h' such that $f_R(h') \neq A_P(h')$ **then**
16:         conflicting h' in the same p_i, where $i \in \{0, 1, 2, ...\}$
17:         Let h_μ be the mean of all $h \in p_i \setminus \{h'\}$
18:         $\texttt{Refinement}(h', h_\mu)$
19:         **return** A_P
20:     **end if**
21: **end for**
22: **return** $\texttt{Equivalence}$

---

outcomes. By refining A_P—i.e., separating the conflicting hidden states into distinct partitions—the abstraction is updated so that w_1 and w_2 are no longer assigned to the same partitions.

Additionally, there can be partition conflicts due to conflicting hidden states that are grouped within the same partition. The refinement process is done by computing the mean of all the existing hidden states in the current partition and comparing it with the conflicting hidden state. These two representatives—the partition mean and the conflicting hidden state—are then passed to a refinement operation to split the conflicted partition, thereby creating more precise partitions that align with the actual behaviour of the RNN. Refinements are applied only when conflicts are detected within a partition.

Refinements were done using a hyperplane-based approach as described below.

Using a Hyperplane for the Partition Refinement. A hyperplane is a decision boundary that separates a high-dimensional space into two distinct regions. In the context of partition refinement, a hyperplane is used to divide a conflicted partition into two, ensuring that the points causing the conflict are assigned to distinct partitions, maximising the distance between them for better separation.

Suppose w_1 and w_2 are conflicting words identified under partition conflicts. Then, the hidden states $h_1 = g^*(h_0, w_1)$ and $h_2 = g^*(h_0, w_2)$, which exist in the conflicted partition, should be passed into a refinement operation, so that they are mapped into two separate partitions.

A hyperplane can be defined as $z.x + b = 0$, where $z \in \mathbb{R}^d$ is the normal vector to the hyperplane, $x \in \mathbb{R}^d$ is a point in the d-dimensional space, $b \in \mathbb{R}$ is the bias, determining the hyperplane's position relative to the origin. Hidden states $h_1, h_2 \in \mathbb{R}^d$ of the RNN are high-dimensional vectors. The hyperplane separating the two hidden states can be calculated as follows.

$$z = h_2 - h_1 \qquad m = \frac{h_1 + h_2}{2} \qquad b = -z \cdot m \qquad z \cdot x + b = 0$$

For a given hidden state h, the signed distance from the hyperplane is $d(h) = z \cdot h + b$. Thus, the partition refinement function is,

$$P(h) = \begin{cases} p_i, & \text{if } d(h) > 0, \\ p_{i+1}, & \text{if } d(h) \leq 0. \end{cases}$$

where, $i = 1, 2, 3...$ is the partition index.

The hyperplane passes through the midpoint of the two conflicting hidden states and is perpendicular to the line joining these vectors. This hyperplane divides the d-dimensional state space into two distinct partitions, ensuring h_1 and h_2 belong to separate partitions. This condition is checked before accepting the partition refinement.

During each iteration of the refinement process, a conflicted partition is divided into two, resulting in an incremental increase in the total number of partitions, i.e. abstract partitions in A_P. A decision tree structure was employed to manage and store these partitions. Each hyperplane that divides a partition is represented as a leaf node in the decision tree, with the resulting partitions as its child nodes.

This hierarchical structure allows the decision tree to dynamically expand with each new round of partition refinement, accommodating additional hyperplanes as they are introduced. The decision tree facilitates tracking transitions when a word w is fed into the abstract partitioning A_P. For each word w, the corresponding hidden states at each time step are obtained and then passed through the decision tree to determine their respective partitions.

The procedure is described in Algorithm 2, which elaborates the refinement operation mentioned in line 11 of Algorithm 1.

4.4 Initial Partitioning with k-Means Clustering

Initially, the RNN state space was treated as a single partition, and refinements were applied only when necessary. To potentially improve efficiency, an initial partitioning using k-means clustering is introduced. The hidden states of the RNN exhibit a natural clustering structure due to variations in their weight-based representations. This property was leveraged to divide the state space into multiple initial partitions, allowing the extraction process to begin with a more informative abstraction. As a result, fewer refinements were required later, reducing the overall effort needed for exploration and partition refinements.

k-means is a clustering algorithm that groups data points based on the Euclidean distance metric. The proposed method begins the extraction process with k clusters— that is, k abstract partitions. To determine an appropriate value for k, the elbow method was employed, which estimates the natural number of clusters present in the hidden state space [17].

Algorithm 2: Partition Refinement using Hyperplane.

Require: $R, A_P = \{p_i \mid i \in \{0, 1, 2, \ldots, n\}\}$, conflicted partition : p', decision tree : D, conflicted hidden states : h_1, h_2

Ensure: Refined A_P, updated : D

1: Define hyperplane: $z \leftarrow h_2 - h_1, b \leftarrow -z \cdot \frac{h_1 + h_2}{2}$

2: **for** each h in p' **do**

3:     Compute signed distance $d(h) \leftarrow z \cdot h + b$

4:     Assign h to p_i if $d(h) > 0$, otherwise to p_{i+1}

5: **end for**

6:     Store hyperplane $z \cdot x + b = 0$ at the conflicted node of D

7:     Add p_i and p_{i+1} as child nodes of D

8: **return** Refined A_P, updated : D

9: **Termination**: Proceed to the next iteration of Algorithm 1

4.5 Termination of the Equivalence Queries

The algorithm's termination strategy was adapted based on specific observations. In most cases, the extraction process completes within seconds. However, there were instances where the extraction algorithm required significantly more time due to having to evaluate the RNN model while doing the equivalence queries. To address such cases and prevent the algorithm from running indefinitely, terminating conditions were introduced.

Initially, a termination condition based on the length of a sequence was introduced. The counterexample search proceeds in lexicographical order. When a counterexample is identified, the equivalence check is repeated to verify whether the refined DFA A_L is equivalent to the RNN. If the equivalence is not satisfied, the process either provides a new counterexample or refines the partitioning A_P. If no counterexample is found and the sequence length exceeds the previous counterexample's length by more than five characters, the algorithm terminates.

The partition refinement process is performed to obtain an abstract partition A_P that can be explored in parallel to help identify counterexamples during the equivalence queries. However, an excessive number of refinements without yielding any counterexamples is inefficient and unproductive. Therefore, the algorithm is designed to terminate if more than 10 refinements are made without finding any counterexamples.

The algorithm is designed to operate until it can no longer identify any new counterexamples within the predefined word length limits. If the extraction process takes excessive time, a time limit is invoked. The set limit is 10 s. If no additional counterexamples are identified after exhausting all possibilities within that time, the algorithm terminates and outputs the final L^* DFA A_L, which is accepted as the final DFA.

5 Experiments and Results

5.1 Model Training

The implementation is in Python, using PyTorch for neural network construction and scikit-learn for additional machine learning functions [24, 25]. The target languages are infinite, so the training data is a sample from the language: for languages with an alphabet of 2 elements, sequences ranging from lengths 1 to 15 were used; if the alphabet contains more than 2 elements, the maximum sequence length was limited to 10. Each string was labelled as *true* or *false* depending on whether it belonged to the target language.

For each language, the sample data was initially divided into training (80%), validation (10%), and test sets (10%), with each selection reflecting the proportion of positive and negative samples in the whole dataset. Many of the languages are highly imbalanced; therefore, oversampling was used in the training set to ensure a balanced representation of labels, thereby mitigating imbalance and improving the model's ability to generalise.

The models were trained using one hidden layer with 16 hidden states, the Adam optimiser, and an initial learning rate of 0.0001. Each RNN was trained for up to 50 epochs on the training set, with an early stopping rule implemented, halting training if the validation accuracy did not improve for more than 5 consecutive epochs. If the desired accuracy was not achieved, training was extended up to 100 epochs. All models aimed to achieve 100% training and testing accuracy, with some achieving this benchmark, demonstrating perfect learning of the training data. However, the training process did not always result in perfect accuracy. The primary objective of this work is learning DFAs from RNNs, not training RNNs. Therefore, for some languages, the DFA extraction was performed on RNNs that have not completely captured the input language.

5.2 Datasets

Tomita Grammars. The Tomita grammars [28] are a set of seven regular grammars over the binary alphabet $\Sigma = \{0, 1\}$. These grammars are widely used as benchmarks for evaluating automaton learning, grammar inference, and explainable AI methods. Each grammar defines a distinct language with varying levels of complexity, making them valuable for testing the capabilities of models such as RNNs and automaton extraction techniques.

Further Regular Languages. Further sample regular languages were also used to test the algorithm. These languages primarily consisted of sequences over the binary alphabet $\Sigma = \{0, 1\}$ and alphabet of size 3: $\Sigma = \{a, b, c\}$. However, the algorithm is not limited to these and can be applied to RNNs trained on regular languages over any arbitrary alphabet. The 7 Tomita grammars and 20 regular languages are given below:

T1: 1^* T2: $(10)^*$ T3: no odd run of 1s follows an odd run of 0s T4: no substring 000 T5: even number of both 0s and 1s T6: $\#0 - \#1 = $ multiple of 3 T7: $0^*1^*0^*1^*$	RL1: string start with 1 RL2: string start with 0 RL3: string end with 0 RL4: string end with 1 RL5: string contains 101 RL6: string contains 0101 RL7: string start with 0, end 1 RL8: len$\geq$ 3, 3rd symbol 0 RL9: 0^*1^*01 RL10: string contains 00	RL11: string without 11 RL12: string contains 11 RL13: len $\geq$ 5, 5th from right 1 RL14: alternate 0, 1 RL15: every 0 follow by 1, end 1 RL16: a^*b or c^*b RL17: $(aa)^*(bbb)^*$ RL18: bc^*b or ac^*a RL19: 0^n1^m, $n \geq 3$, m even RL20: string with at most two 0s

5.3 Extracting DFA from LSTMs and GRUs

RNN models with LSTM and GRU architecture, for the regular language datasets, were trained as described in Sect. 5.1. The DFA extraction algorithm was then run on the resulting RNNs with a timeout of 10 s. Tables 1 and 2 collate the results. The dataset used for testing includes the full set of sequences generated up to the maximum sequence length, 15 when $|\Sigma| = 2$, and 10 when $|\Sigma| = 3$. Additionally, an experiment was conducted to compare the extraction time when SVMs were used instead of hyperplanes in the partitioning method.

The final set of experiments investigates the impact of initialising the DFA extraction algorithm with an initial partitioning determined by performing k-means clustering on the weight space of the trained RNN. Table 3 gives the result of repeating the DFA extraction from the LSTMs with this pre-processing step.

Table 1. Performance of automata extracted from LSTMs. Columns represents: target language L, performance of LSTM R compared to L, performance of the extracted DFA A_L compared to R, performance of A_L compared to L, number of partitions in A_P, number of states $|Q_A|$ in A_L, total time t for the extraction (seconds) using hyperplanes, and the time $ref.t$ of the final refinement/counterexample (seconds) using the hyperplanes, total time t for the extraction (seconds) using the SVM method, and the time $ref.t$ of the final refinement/counterexample (seconds) using the SVM method.

L	$R\,vs\,L$			$A_L\,vs\,R$			$A_L\,vs\,L$			$|P|$	$|Q_A|$	$Hyperplane$		SVM	
	$Acc.$	$Prec.$	$F1$	$Acc.$	$Prec.$	$F1$	$Acc.$	$Prec.$	$F1$			$t(s)$	$ref.t(s)$	$t(s)$	$ref.t(s)$
T1	1.00	1.00	1.00	1.00	1.00	1.00	1.00	1.00	1.00	2	2	0.23	0.08	2.18	0.13
T2	0.99	1.00	0.93	1.00	1.00	1.00	0.99	1.00	0.93	9	4	10.0	0.48	10.00	0.98
T3	1.00	1.00	1.00	1.00	1.00	1.00	1.00	1.00	1.00	5	5	3.39	1.42	8.09	1.56
T4	1.00	1.00	1.00	1.00	1.00	1.00	1.00	1.00	1.00	6	4	10.00	1.79	10.00	2.61
T5	0.83	0.50	0.67	1.00	1.00	1.00	0.83	0.50	0.67	7	4	10.00	1.31	10.00	3.5
T6	0.99	1.00	0.99	1.00	1.00	1.00	0.99	1.00	0.99	11	4	2.47	2.46	10.00	2.87
T7	1.00	1.00	1.00	1.00	1.00	1.00	1.00	1.00	1.00	12	5	10.00	4.24	10.00	4.07
RL1	1.00	1.00	1.00	1.00	1.00	1.00	1.00	1.00	1.00	3	3	0.63	0.15	10.00	0.19
RL2	1.00	1.00	1.00	1.00	1.00	1.00	1.00	1.00	1.00	3	3	0.64	0.20	10.00	0.53
RL3	1.00	1.00	1.00	1.00	1.00	1.00	1.00	1.00	1.00	3	2	1.10	0.10	2.01	0.17
RL4	1.00	1.00	1.00	1.00	1.00	1.00	1.00	1.00	1.00	2	2	0.22	0.10	1.59	0.16
RL5	1.00	1.00	1.00	1.00	1.00	1.00	1.00	1.00	1.00	12	4	10.00	1.74	10.00	6.67
RL6	1.00	1.00	1.00	1.00	1.00	1.00	1.00	1.00	1.00	11	5	10.00	6.64	10.00	9.73
RL7	1.00	1.00	1.00	1.00	1.00	1.00	1.00	1.00	1.00	7	4	10.00	0.81	10.00	0.90
RL8	1.00	1.00	1.00	1.00	1.00	1.00	1.00	1.00	1.00	8	5	10.00	5.85	10.00	0.44
RL9	1.00	1.00	1.00	1.00	1.00	1.00	1.00	1.00	1.00	10	7	10.00	0.79	10.00	1.74
RL10	1.00	1.00	1.00	1.00	1.00	1.00	1.00	1.00	1.00	7	3	10.00	0.78	10.00	1.57
RL11	1.00	1.00	1.00	1.00	1.00	1.00	1.00	1.00	1.00	3	3	0.66	0.16	10.00	0.21
RL12	1.00	1.00	1.00	1.00	1.00	1.00	1.00	1.00	1.00	6	3	10.00	0.74	10.00	2.00
RL13	0.99	0.99	0.99	0.68	0.75	0.64	0.68	0.75	0.64	17	28	10.00	9.98	10.00	9.99
RL14	1.00	1.00	1.00	1.00	1.00	1.00	1.00	1.00	1.00	5	4	3.08	0.43	6.92	0.52
RL15	1.00	1.00	1.00	1.00	1.00	1.00	1.00	1.00	1.00	4	3	10.00	0.41	10.00	0.69
RL16	1.00	1.00	1.00	1.00	1.00	1.00	1.00	1.00	1.00	8	5	10.00	3.47	10.00	8.82
RL17	0.99	1.00	0.98	1.00	1.00	1.00	0.99	1.00	0.98	10	7	10.00	0.98	10.00	2.17
RL18	0.99	0.80	0.89	1.00	1.00	1.00	0.99	0.80	0.89	16	11	10.00	8.86	10.00	6.29
RL19	0.93	0.01	0.02	0.94	0.94	0.19	0.99	0.09	0.16	14	13	10.00	9.79	10.00	9.98
RL20	1.00	1.00	1.00	1.00	1.00	1.00	1.00	1.00	1.00	9	4	10.00	4.25	10.00	2.62

Table 2. Performance of automata extracted from GRUs. Columns represents: target language L, performance of the GRU R compared to L, performance of the extracted DFA A_L compared to R, performance of A_L compared to L, number of partitions in A_P, number of states $|Q_A|$ in A_L, total time t for the extraction (seconds) using hyperplanes, the time $ref.t$ of the final refinement or counterexample (seconds) using the hyperplanes, total time t for the extraction (seconds) using the SVM method, and the time $ref.t$ of the final refinement or counterexample (seconds) using the SVM method. Each row represents languages for which the GRU models were trained.

L	$R\ vs\ L$			$A_L\ vs\ R$			$A_L\ vs\ L$			$|P|$	$|Q_A|$	$Hyperplane$		SVM	
	$Acc.$	$Prec.$	$F1$	$Acc.$	$Prec.$	$F1$	$Acc.$	$Prec.$	$F1$			$t(s)$	$ref.t(s)$	$t(s)$	$ref.t(s)$
T1	1.00	1.00	1.00	1.00	1.00	1.00	1.00	1.00	1.00	2	2	0.23	0.09	0.57	0.13
T2	0.99	1.00	0.93	1.00	1.00	1.00	0.99	1.00	0.93	8	4	10.00	0.57	10.00	0.98
T3	1.00	1.00	1.00	1.00	1.00	1.00	1.00	1.00	1.00	8	5	10.00	2.44	10.00	2.96
T4	1.00	1.00	1.00	1.00	1.00	1.00	1.00	1.00	1.00	8	4	10.00	1.74	10.00	3.57
T5	0.83	0.50	0.67	1.00	1.00	1.00	0.83	0.50	0.67	7	4	10.00	4.87	10.00	3.47
T6	1.00	1.00	1.00	1.00	1.00	1.00	1.00	1.00	1.00	5	3	10.00	0.17	10.00	0.25
T7	1.00	1.00	1.00	1.00	1.00	1.00	1.00	1.00	1.00	8	5	10.00	8.79	10.00	8.90
RL1	1.00	1.00	1.00	1.00	1.00	1.00	1.00	1.00	1.00	3	3	0.61	0.14	10.00	0.20
RL2	1.00	1.00	1.00	1.00	1.00	1.00	1.00	1.00	1.00	3	3	0.63	0.13	10.00	0.17
RL3	1.00	1.00	1.00	1.00	1.00	1.00	1.00	1.00	1.00	2	2	0.22	0.09	0.33	0.11
RL4	1.00	1.00	1.00	1.00	1.00	1.00	1.00	1.00	1.00	2	2	0.24	0.09	0.32	0.14
RL5	1.00	1.00	1.00	1.00	1.00	1.00	1.00	1.00	1.00	10	4	10.00	7.61	10.00	4.26
RL6	1.00	1.00	1.00	1.00	1.00	1.00	1.00	1.00	1.00	8	5	10.00	7.51	10.00	7.28
RL7	0.99	0.98	0.99	0.99	1.00	0.99	0.99	0.99	0.99	11	9	10.00	2.08	10.00	3.91
RL8	0.99	0.99	0.99	1.00	1.00	1.00	0.99	0.99	0.99	14	7	10.00	2.87	10.00	6.79
RL9	1.00	1.00	1.00	1.00	1.00	1.00	1.00	1.00	1.00	14	7	10.00	6.43	10.00	4.87
RL10	0.95	1.00	0.97	0.95	0.95	0.98	0.99	1.00	0.99	17	6	10.00	4.50	10.00	10.00
RL11	1.00	1.00	1.00	1.00	1.00	1.00	1.00	1.00	1.00	5	3	8.91	0.50	10.00	1.77
RL12	1.00	1.00	1.00	1.00	1.00	1.00	1.00	1.00	1.00	7	3	10.00	0.63	10.00	2.15
RL13	0.88	0.88	0.88	0.90	0.87	0.90	0.80	0.78	0.82	16	45	10.00	8.62	10.00	10.00
RL14	0.99	1.00	0.98	1.00	1.00	1.00	0.99	1.00	0.98	10	6	10.00	1.58	10.00	5.32
RL15	1.00	1.00	1.00	1.00	1.00	1.00	1.00	1.00	1.00	3	3	1.16	0.17	2.29	0.19
RL16	1.00	1.00	1.00	1.00	1.00	1.00	1.00	1.00	1.00	8	5	10.00	1.38	10.00	3.20
RL17	0.99	0.81	0.88	1.00	1.00	1.00	0.99	0.81	0.88	17	11	10.00	5.09	10.00	4.56
RL18	0.99	0.45	0.63	0.99	1.00	0.95	0.99	0.50	0.67	13	6	10.00	9.33	10.00	4.56
RL19	1.00	1.00	1.00	1.00	1.00	1.00	1.00	1.00	1.00	11	7	10.00	5.87	10.00	5.62
RL20	1.00	1.00	1.00	1.00	1.00	1.00	1.00	1.00	1.00	7	4	10.00	1.24	10.00	2.07

6 Evaluation and Discussion

The results show that the algorithm effectively extracts DFAs from both LSTM and GRU models. Among the 27 LSTMs, 20 achieved 100% training accuracy, and five reached 99%. However, among these, three models (T2, RL17, RL18) exhibited low F1-scores despite high accuracy, suggesting that they failed to capture the underlying language. Two models (T5, RL19) showed both low accuracy and F1-scores, with RL19 performing particularly poorly (F1-score 2%). DFA extraction was successful in 25 out of 27 cases, achieving 100% accuracy. The exceptions, RL13 and RL19, terminated due to time limits, producing DFAs with 68% and 94% accuracy, respectively, with the final counterexample or refinement occurring just before termination. In RL19, the low F1-score also indicated that the LSTM itself had not learned the language patterns.

For GRUs, 18 achieved 100% training accuracy, while two reached 99% accuracy and F1-score. Four models obtained 99% training but low F1-scores, again reflecting limited language generalisation, while three (T5, RL10, RL13) had both low accuracy and F1-scores. DFA extraction was fully successful for 23 of the 27 GRU models, each with 100% accuracy. The remaining four DFA extractions exceeded 90% accuracy and F1-scores, with reduced extraction accuracy correlating to weaker GRU training. Interestingly, in some cases, GRUs with less than perfect accuracy still produced 100% accurate DFAs.

The DFA extraction process was subject to a 10-second time limit. For LSTM models, most extractions continued until the limit, except for nine cases. Among these, 13 models completed their final refinement in under 1 s, and eight within 5 s, while only six required longer. A similar trend was observed for GRU models, where most extractions ran until the time limit, except for six cases. Notably, nine GRU models reached their final refinement in under 1 s and ten within 5 s. Only RL13 and RL18 extended close to the full 10-second limit. In both cases, the remaining time after the final refinement or counterexample was spent verifying equivalence against all provided sequences. This suggests that termination is influenced by the number of strings evaluated for equivalence.

Table 3. Performance of DFAs extracted from LSTMs using k-means clustering for initial partitioning. Columns represents: target language L, performance of the LSTM R compared to L, performance of the extracted DFA A_L compared to R, performance of A_L compared to L, number of initial partitions (clusters) k, total number of partitions in A_P, number of states $|Q_A|$ in A_L, total time t for the extraction (seconds), and the time $ref.t$ of the final refinement or counterexample (seconds). Each row represents languages for which the LSTM models were trained.

L	$R\ vs\ L$			$A_L\ vs\ R$			$A_L\ vs\ L$			k	$|P|$	$|Q_A|$	$t(s)$	$ref.t(s)$
	$Acc.$	$Prec.$	$F1$	$Acc.$	$Prec.$	$F1$	$Acc.$	$Prec.$	$F1$					
T1	1.00	1.00	1.00	1.00	1.00	1.00	1.00	1.00	1.00	2	3	2	1.86	0.90
T2	0.99	1.00	0.93	1.00	1.00	1.00	0.99	1.00	0.93	3	7	4	10.00	1.08
T3	1.00	1.00	1.00	1.00	1.00	1.00	1.00	1.00	1.00	3	5	5	4.08	2.26
T4	1.00	1.00	1.00	1.00	1.00	1.00	1.00	1.00	1.00	2	4	4	2.88	1.25
T5	0.83	0.50	0.67	1.00	1.00	1.00	0.83	0.50	0.67	5	7	4	10.00	1.48
T6	0.99	1.00	0.99	1.00	1.00	1.00	0.99	1.00	0.99	4	9	4	10.00	2.16
T7	1.00	1.00	1.00	1.00	1.00	1.00	1.00	1.00	1.00	2	9	5	10.00	2.11
RL1	1.00	1.00	1.00	1.00	1.00	1.00	1.00	1.00	1.00	2	3	3	1.42	0.85
RL2	1.00	1.00	1.00	1.00	1.00	1.00	1.00	1.00	1.00	2	3	3	1.28	1.03
RL3	1.00	1.00	1.00	1.00	1.00	1.00	1.00	1.00	1.00	2	3	2	1.83	0.97
RL4	1.00	1.00	1.00	1.00	1.00	1.00	1.00	1.00	1.00	2	3	2	1.83	0.99
RL5	1.00	1.00	1.00	1.00	1.00	1.00	1.00	1.00	1.00	2	12	4	10.00	2.89
RL6	1.00	1.00	1.00	1.00	1.00	1.00	1.00	1.00	1.00	2	11	5	10.00	6.04
RL7	1.00	1.00	1.00	1.00	1.00	1.00	1.00	1.00	1.00	3	7	4	10.00	1.13
RL8	1.00	1.00	1.00	1.00	1.00	1.00	1.00	1.00	1.00	3	9	5	10.00	1.23
RL9	1.00	1.00	1.00	1.00	1.00	1.00	1.00	1.00	1.00	3	9	7	10.00	1.39
RL10	1.00	1.00	1.00	1.00	1.00	1.00	1.00	1.00	1.00	2	7	3	10.00	1.49
RL11	1.00	1.00	1.00	1.00	1.00	1.00	1.00	1.00	1.00	2	3	3	1.40	0.91
RL12	1.00	1.00	1.00	1.00	1.00	1.00	1.00	1.00	1.00	2	6	3	10.00	1.45
RL13	0.99	0.99	0.99	0.68	0.75	0.64	0.68	0.75	0.64	4	11	19	10.00	9.73
RL14	1.00	1.00	1.00	1.00	1.00	1.00	1.00	1.00	1.00	2	5	4	3.50	1.18
RL15	1.00	1.00	1.00	1.00	1.00	1.00	1.00	1.00	1.00	3	6	3	10.00	2.96
RL16	1.00	1.00	1.00	1.00	1.00	1.00	1.00	1.00	1.00	3	8	5	10.00	4.93
RL17	0.99	1.00	0.98	1.00	1.00	1.00	0.99	1.00	0.98	3	9	7	10.00	1.66
RL18	0.99	0.80	0.89	1.00	1.00	1.00	0.99	0.80	0.89	3	13	11	10.00	7.26
RL19	0.94	0.01	0.02	0.94	0.94	0.19	0.99	0.09	0.16	2	14	13	10.00	9.99
RL20	1.00	1.00	1.00	1.00	1.00	1.00	1.00	1.00	1.00	2	8	4	10.00	4.26

In experiments where k-means clustering was used to partition the state space, the extracted DFAs achieved accuracy comparable to those obtained from LSTMs without initial partitioning [1,3]. Since the sample size was small (27 models) and normality assumptions were not met, the non-parametric Wilcoxon signed-rank test was applied to compare extraction times between the two settings. Contrary to expectations, no significant improvement was observed ($p = 0.97$). Although k-means clustering reduced the number of refinements, the cost of computing initial clusters offset these gains, compromising overall efficiency.

The proposed algorithm is designed to terminate once equivalence is established between the L^* DFA A_L and RNN R. In practice, additional termination strategies were introduced during experiments (Sect. 4.5), based on the length of the final counterexample and the number of excessive refinements. Crucially, these early termination conditions did not affect the correctness of the final DFAs, as the absence of counterexamples beyond these thresholds indicated a low likelihood of further discrepancies. Time constraints (10 s) also did not compromise correctness in most cases, with the exception of models RL13 and RL19.

Additionally, an experiment was conducted by substituting SVMs for hyperplane generation when partitioning the RNN state space, while keeping all other factors constant. In this setup, the SVMs separate the conflicted hidden states from others to form new partitions. The non-parametric Wilcoxon signed-rank test was then applied to both total extraction time and refinement/counterexample generation time for LSTM and GRU models.

For LSTMs, extraction with SVMs was significantly slower than with hyperplanes ($p = 0.004$ for both total extraction time and refinement/counterexample time). GRUs showed the same pattern, where extraction with SVMs was statistically slower than hyperplanes ($p = 0.046$ and $p = 0.042$, respectively), though the average time differences were smaller. Notably, total extraction time is not always a reliable measure, as many extractions ran until the time limit and termination was based on predefined factors.

When using hyperplanes, partitioning was refined upon detecting conflicts between hidden states, prioritising the separation of the most dissimilar states rather than isolating a single conflicted state. Subsequent refinements iteratively resolved remaining conflicts by grouping dissimilar states into distinct partitions, keeping the number of refinements minimal and ensuring computational efficiency.

7 Conclusion

This work presents an approach to extracting a DFA from an RNN to enhance the interpretability. The study employs the L^* algorithm for DFA extraction but introduces a novel approach to equivalence queries. The work in [32] proposed iteratively refining equivalence queries by comparing the L^*-learned DFA with an abstraction of the RNN state space generated by partitioning the RNN's state space using SVMs. Building on that, this work extends the methodology by using hyperplane-based state-space partitioning during equivalence checking. The main contribution of this work is the use of hyperplanes to partition the RNN's state space, ensuring the extracted DFA closely aligns with the RNN's decision-making process.

This method successfully extracts a sufficiently accurate DFA from a well-trained RNN. Similar to the work in [32], the algorithm needs only the language's alphabet and access to the RNN model; no additional details are required. The extraction process typically requires less than 10 s, as calculating hyperplanes to classify states in the abstract partitioning is computationally efficient.

If an RNN achieves a 100% training and testing accuracy, then the extracted DFA precisely represents the language on which the RNN was trained. Additionally, for some RNNs that did not achieve perfect training and testing accuracy, the extracted DFA still captured the RNN's behaviour with 100% accuracy. However, in such cases, the extracted DFA deviates from the target language, indicating that the RNN itself did not properly learn the intended language.

Although for some RNNs the extraction process continued until the time limit of 10 s, the final refinement or counterexample generation was typically completed well before that. The remaining time was spent confirming equivalence using all provided strings up to termination. If additional termination strategies or constraints had been introduced, DFAs could have been extracted more efficiently. In this study, no limitations were imposed on the set of strings used for equivalence queries; instead, all strings were explored in lexicographic order until the time limit or another termination condition (Sect. 4.5) was reached. Thus, the termination was not triggered earlier until all strings were inspected within the given time limit.

The accuracy of the extracted DFAs refers to their similarity in predictions compared to the original RNN. If the DFA can produce outputs that are exactly or closely aligned with the RNN's predictions (where the RNN acts as the oracle during extraction), it can be concluded that the DFA has successfully approximated the RNN's behaviour. However, this does not imply that it has learned the exact internal behaviour of the RNN. Instead, the DFA captures the decision-making logic of the RNN—that is, how the RNN processes input sequences to generate outputs.

In this work, counterexamples are not prioritised, and no limit is imposed on the number of strings evaluated to check equivalence. Future work aims to explore methods for managing the termination of the algorithm more effectively, without exhaustively evaluating all available strings within the given time limit.

The results of this work align with the findings reported in [32]. More importantly, the extraction method is simpler, as it exclusively uses hyperplanes to partition the state space. The accuracy of the extraction remains comparable, but the key contribution of this study lies in employing a simpler refinement approach for partitioning. The L^* algorithm inherently scales, increasing computational time as the number of states grows. By reducing the time required for equivalence queries, the proposed method improves the efficiency of DFA extraction from RNNs.

Acknowledgments. Work supported by the EPSRC grant EP/T018313/1.

References

1. Angluin, D.: Learning regular sets from queries and counterexamples. Inf. Comput. **75**(2), 87–106 (1987). https://doi.org/10.1016/0890-5401(87)90052-6
2. Ayache, S., Eyraud, R., Goudian, N.: Explaining black boxes on sequential data using weighted automata. In: Proceedings of the 14th International Conference on Grammatical

Inference. Proceedings of Machine Learning Research, vol. 93, pp. 81–103. PMLR (2018). http://proceedings.mlr.press/v93/ayache19a.html

3. Balle, B., Mohri, M.: Learning weighted automata. In: Maletti, A. (ed.) CAI 2015. LNCS, vol. 9270, pp. 1–21. Springer, Cham (2015). https://doi.org/10.1007/978-3-319-23021-4_1

4. Bojarski, M., et al.: End to end learning for self-driving cars (2016). http://arxiv.org/abs/1604.07316

5. Chomsky, N.: Three models for the description of language. IEEE Trans. Inf. Theory **2**(3), 113–124 (1956). https://doi.org/10.1109/TIT.1956.1056813

6. Chung, J., Gulcehre, C., Cho, K., Bengio, Y.: Empirical evaluation of gated recurrent neural networks on sequence modeling. In: NIPS 2014 Workshop on Deep Learning (2014). http://arxiv.org/abs/1412.3555

7. Cleeremans, A., Servan-Schreiber, D., McClelland, J.L.: Finite state automata and simple recurrent networks. Neural Comput. **1**(3), 372–381 (1989). https://doi.org/10.1162/neco.1989.1.3.372

8. Du, X., Xie, X., Li, Y., Ma, L., Liu, Y., Zhao, J.: DeepStellar: model-based quantitative analysis of stateful deep learning systems. In: Proceedings of the 2019 27th ACM Joint Meeting European Software Engineering Conference and Symposium on the Foundations of Software Engineering, pp. 477–487. ACM (2019). https://doi.org/10.1145/3338906.3338954

9. Elman, J.L.: Finding structure in time. Cogn. Sci. **14**(2), 179–211 (1990). https://doi.org/10.1207/s15516709cog1402_1

10. Eyraud, R., Ayache, S.: Distillation of weighted automata from recurrent neural networks using a spectral approach. Mach. Learn. **113**(5), 3233–3266 (2024). https://doi.org/10.1007/s10994-021-05948-1

11. Giles, C.L., Miller, C.B., Chen, D., Chen, H.H., Sun, G.Z., Lee, Y.C.: Learning and extracting finite state automata with second-order recurrent neural networks. Neural Comput. **4**(3), 393–405 (1992). https://doi.org/10.1162/neco.1992.4.3.393

12. Goldman, S., Kearns, M.: On the complexity of teaching. J. Comput. Syst. Sci. **50**(1), 20–31 (1995). https://doi.org/10.1006/jcss.1995.1003

13. Hochreiter, S., Schmidhuber, J.: Long short-term memory. Neural Comput. **9**(8), 1735–1780 (1997). https://doi.org/10.1162/neco.1997.9.8.1735

14. Holzinger, A., Biemann, C., Pattichis, C.S., Kell, D.B.: What do we need to build explainable AI systems for the medical domain? (2017). http://arxiv.org/abs/1712.09923

15. Jacobsson, H.: Rule extraction from recurrent neural networks: a taxonomy and review. Neural Comput. **17**(6), 1223–1263 (2005). https://doi.org/10.1162/0899766053630350

16. Kleene, S.C.: Representation of Events in Nerve Nets and Finite Automata, pp. 3–42. Princeton University Press (1956). https://doi.org/10.1515/9781400882618-002

17. Kodinariya, T.M., Makwana, P.R.: Review on determining number of cluster in k-means clustering. Int. J. Adv. Res. Comput. Sci. Manag. Stud. **1**(6), 90–95 (2013)

18. Marzouk, R., de la Higuera, C.: Distance and equivalence between finite state machines and recurrent neural networks: computational results. arXiv preprint arXiv:2004.00478 (2020). https://doi.org/10.48550/arXiv.2004.00478

19. Mayr, F., Yovine, S.: Regular inference on artificial neural networks. In: Holzinger, A., Kieseberg, P., Tjoa, A.M., Weippl, E. (eds.) CD-MAKE 2018. LNCS, vol. 11015, pp. 350–369. Springer, Cham (2018). https://doi.org/10.1007/978-3-319-99740-7_25

20. McCulloch, W.S., Pitts, W.: A logical calculus of the ideas immanent in nervous activity. Bull. Math. Biophys. **5**(4), 115–133 (1943). https://doi.org/10.1007/BF02478259

21. Okudono, T., Waga, M., Sekiyama, T., Hasuo, I.: Weighted automata extraction from recurrent neural networks via regression on state spaces. Proc. AAAI Conf. Artif. Intell. **34**(04), 5306–5314 (2020). https://doi.org/10.1609/aaai.v34i04.5977

22. Oliva, C., Lago-Fernández, L.F.: On the interpretation of recurrent neural networks as finite state machines. In: Tetko, I.V., Kůrková, V., Karpov, P., Theis, F. (eds.) ICANN 2019. LNCS, vol. 11727, pp. 312–323. Springer, Cham (2019). https://doi.org/10.1007/978-3-030-30487-4_25

23. Omlin, C.W., Giles, C.: Extraction of rules from discrete-time recurrent neural networks. Neural Netw. 9(1), 41–52 (1996). https://doi.org/10.1016/0893-6080(95)00086-0

24. Paszke, A., et al.: PyTorch: an imperative style, high-performance deep learning library. In: Proceedings of the 33rd International Conference on Neural Information Processing Systems. Curran Associates Inc. (2019)

25. Pedregosa, F., et al.: Scikit-learn: machine learning in python. J. Mach. Learn. Res. 12, 2825–2830 (2011)

26. Sakarovitch, J.: Elements of Automata Theory. Cambridge University Press (2009). https://doi.org/10.1017/CBO9781139195218

27. Siegelmann, H.T., Sontag, E.D.: On the computational power of neural nets. In: Proceedings of the Fifth Annual Workshop on Computational Learning Theory, pp. 440–449. ACM (1992). https://doi.org/10.1145/130385.130432

28. Tomita, M.: Learning of construction of finite automata from examples using hill-climbing RR: Regular set recognizer. Tech. Rep. CMU-CS-82-127, Carnegie-Mellon University (1982). http://shelf2.library.cmu.edu/Tech/8854663.pdf

29. Wang, Q., Zhang, K., II, A.G.O., Xing, X., Liu, X., Giles, C.L.: An empirical evaluation of rule extraction from recurrent neural networks. Neural Comput. 30(9), 2568–2591 (2018). https://doi.org/10.1162/neco_a_01111

30. Wei, Z., Zhang, X., Sun, M.: Extracting weighted finite automata from recurrent neural networks for natural languages. In: Formal Methods and Software Engineering. Lecture Notes in Computer Science, vol. 13478, pp. 370–385. Springer (2022). https://doi.org/10.1007/978-3-031-17244-1_22

31. Weiss, G., Goldberg, Y., Yahav, E.: Learning deterministic weighted automata with queries and counterexamples. In: Advances in Neural Information Processing Systems, vol. 32. Curran Associates, Inc. (2019). https://doi.org/10.48550/arXiv.1910.13895

32. Weiss, G., Goldberg, Y., Yahav, E.: Extracting automata from recurrent neural networks using queries and counterexamples (extended version). Mach. Learn. 113(5), 2877–2919 (2024). https://doi.org/10.1007/s10994-022-06163-2

33. Zeng, Z., Goodman, R.M., Smyth, P.: Learning finite state machines with self-clustering recurrent networks. Neural Comput. 5(6), 976–990 (1993). https://doi.org/10.1162/neco.1993.5.6.976

Profiling German Text Simplification
with Model-Fingerprints

Lars Klöser[(✉)], Mika Elias Beele, and Bodo Kraft

Aachen University of Applied Sciences, 52066 Aachen, Germany
`{kloeser,beele,kraft}@fh-aachen.de`

Abstract. While Large Language Models (LLMs) produce highly nuanced text simplifications, developers currently lack tools for a holistic, efficient, and reproducible diagnosis of their behavior. This paper introduces the Simplification Profiler, a diagnostic toolkit that generates a multidimensional, interpretable fingerprint of simplified texts. Multiple aggregated simplifications of a model result in a model's fingerprint. This novel evaluation paradigm is particularly vital for languages, where the data scarcity problem is magnified when creating flexible models for diverse target groups rather than a single, fixed simplification style. We propose that measuring a model's unique behavioral signature is more relevant in this context as an alternative to correlating metrics with human preferences. We operationalize this with a practical meta-evaluation of our fingerprints' descriptive power, which bypasses the need for large, human-rated datasets. This test measures if a simple linear classifier can reliably identify various model configurations by their created simplifications, confirming that our metrics are sensitive to a model's specific characteristics. The Profiler can distinguish high-level behavioral variations between prompting strategies and fine-grained changes from prompt engineering, including few-shot examples. Our complete feature set achieves classification F1-scores up to 71.9%, improving upon simple baselines by over 48% points. The Simplification Profiler thus offers developers a granular, actionable analysis to build more effective and truly adaptive text simplification systems.

Keywords: Automatic Text Simplification · Large Language Models · Evaluation · Explainable AI · German Text Simplification

1 Introduction

Automatic Text Simplification (ATS) endeavors to rephrase complex texts into more easily understandable versions to broaden information accessibility. Research in German ATS often focuses on replicating specific, fixed simplification styles inherent in reference texts [11,21,29]. However, simplification must be tailored to specific circumstances to enhance the practical relevance of ATS, which includes explicitly considering the overall context. This context encompasses both the target audience[1] and other situational requirements[2].

---

[1] For example, children, individuals with low literacy, or domain newcomers.
[2] such as desired output length, preservation of specific contents, or an appropriate tone.

© The Author(s), under exclusive license to Springer Nature Switzerland AG 2026
F. Marcelloni et al. (Eds.): IJCCI 2025, CCIS 2829, pp. 136–153, 2026.
https://doi.org/10.1007/978-3-032-15638-9_9

The rise of Large Language Models (LLMs) has made generating such context-aware simplifications more feasible. For example, through straightforward prompt variations. While this flexibility is a significant advancement, it introduces a critical challenge for developers: How do we move beyond simple *good/bad* scores to understand and steer the nuanced properties of these outputs? - To create solutions that are perfectly adjusted to the needs of different target groups, developers require tools that can characterize and diagnose model behavior.

Current approaches to ATS evaluation, however, do not always deliver nuanced insights in the model's behavior. The evaluation of ATS systems has long relied on reference-based metrics like BLEU [19] or SARI [31]. Reference-based metrics, by design, are inflexible. Evaluations with distinct language, domain, or simplification styles require creating novel reference simplifications to compare against. Reference-free approaches aim to avoid this need, notably learnable metrics such as REFeREE [9] or LENS [14].

These learnable metrics face further fundamental challenges. Firstly, there is the issue of *representational breadth*. As motivated, ATS should aim for versatility. Evaluation metrics should be flexible enough to manage this versatility reliably. However, existing human evaluation datasets, if used for training or validation, typically embody only one specific, often implicit, standard of *good* simplification [2, 14, 16, 25]. Creating the multitude of high-quality, comprehensive datasets needed to cover this broad application spectrum is likely prohibitively expensive and practically unfeasible. A challenge especially for ATS in non-English languages like German, which serves as the primary linguistic focus for the empirical validation in this paper [4, 17, 22].

Secondly, even when using the available data, there is the problem of *learning reliability*. We know from other natural language processing tasks like natural language inference (NLI), question answering, or relation extraction that AI models can learn *spurious patterns* – correlations present in limited data but irrelevant to the actual task [6, 7, 12, 24]. Given the limited scale of ATS data, learnable evaluation models are particularly susceptible to these pitfalls. These potentially faulty decision patterns create a high risk that these metrics will inadvertently embed biases and fail to generalize, especially when faced with diverse or out-of-domain simplifications.

Thirdly, many such learnable metrics operate as *black boxes*, presenting a significant challenge in explainability. When an evaluation metric does not provide clear insights into how it arrives at a score, it becomes exceedingly difficult to diagnose its failures, understand its sensitivities, or make targeted improvements to the evaluated ATS systems. This lack of transparency harms iterative LLM development, as it provides feedback that can hardly be used for model improvements. Low scores fail to inform developers what to fix, hindering the targeted improvements necessary for creating highly adaptive ATS systems.

These issues highlight the need for a paradigm shift from monolithic evaluation to granular analysis. This paper lays the conceptual and methodological foundations for the *Simplification Profiler*: a diagnostic toolkit designed to characterize ATS outputs across linguistically grounded properties. This approach provides a multi-dimensional, interpretable fingerprint of each simplification, making critical trade-offs (e.g., between simplicity and content preservation) transparent to the developer. These properties can be divided into universal quality criteria and context-dependent parameters:

- *Universal:* Linguistic Correctness (Grammar and Spelling), Factual Accuracy (Adequacy), Coherence.
- *Context-Dependent:* Linguistic Level (Complexity), Content Scope (Compression), Terminology Use, Text Length.

The key advantage of our approach is its compositional nature. Instead of building a single, monolithic, task-specific AI model for evaluation, we leverage robust, well-established tools for specific aspects, such as NLI models, grammar checkers, and readability indices. This compositional approach allows our Profiler to stand on the shoulders of significant existing expert knowledge and broad training data, potentially increasing the individual measurements' generalizability and reliability.

These observations lead to our central hypothesis: A multi-dimensional analysis of specific text properties provides an informative and nuanced *fingerprint* of ATS system performance, enabling targeted development. We test this by demonstrating that our toolkit is sensitive enough to detect the fingerprints of different development choices, using a linear classifier to distinguish outputs generated by various models and prompts. Our main technical contribution is the implementation of the toolkit and all code to reproduce the results of this paper in an open-source GitHub repository[3].

2   Related Work

Core Principles of ATS Evaluation. The evaluation of Automatic Text Simplification (ATS) targets qualities like simplicity, fluency, and adequacy [20,23], with recent LLM-focused work adding criteria like factuality and user engagement [18,30]. While human evaluation is the gold standard [1,28], its scalability issues necessitate automatic metrics [25].

Reference-Based and -Free Metrics. Automatic approaches include reference-based metrics like BLEU [19], despite its known correlation issues [26], the ATS-specific SARI [31], and semantic methods like BERTScore [32]. The dependency on high-quality references spurred reference-free approaches, from traditional readability formulas to linguistically-informed structural metrics like SAMSA [27].

Modern Model-Based Evaluation. A recent trend is the use of learnable metrics trained on human judgments (LENS [14], REFeREE [9]) and the direct use of *LLM-as-a-Judge* [8,13], though the latter faces reliability challenges. Research into controllable generation also implies evaluating specific output properties [15]. However, these approaches often still rely on datasets with a single, implicit standard of simplification quality [2,14,16,25].

---

[3] https://github.com/MSLars/German-Simplification-Profiling.

The Need for a German Diagnostic Toolkit. While comprehensive toolkits like EASSE [3] and its German version EASSE-DE [25] exist, they focus on aggregating overall scores. For German, property-focused analysis of novel systems often remains a manual process [4,5,11]. To our knowledge, a comprehensive *diagnostic toolkit* for German designed to generate an *interpretable fingerprint*—revealing the interplay of linguistic properties to guide development—remains an unexplored area.

3   Methodology for Generating Test Cases

To validate the Simplification Profiler's diagnostic capabilities, we needed to generate a wide spectrum of simplified texts with varied characteristics. Our goal was to create a testbed reflecting common LLM development scenarios, featuring both major and subtle differences in output style. This allows us to demonstrate that our toolkit can not only distinguish between fundamentally different approaches but is also sensitive enough for fine-grained analysis, such as evaluating the impact of prompt engineering. We selected German Wikipedia as our primary data source because its encyclopedic, fact-driven style provides a controlled and standardized domain. This allows us to more clearly isolate and measure the distinct effects of different models and prompting strategies, which is the core goal of our diagnostic validation, before extending the analysis to more heterogeneous text genres.

All simplifications were generated from 5-sentence German Wikipedia excerpts using the *Gemma* model family (1B, 4B, and 12B parameters), allowing us to analyze behavior at different model scales. We created our test cases using the following structured variations:

High-Level Behavioral Variations. To produce outputs with significant stylistic differences, we employed two core prompting strategies. Plain prompts provided minimal guidance and represent a common baseline approach: a generic version (*plain*) and one specifying a target audience (*target*). In contrast, property-oriented prompts gave explicit instructions to control specific output characteristics: one focused on maximizing content preservation (*coverage*) and another on adhering to specific linguistic rules (*correctness*, detailed in Appendix A). The combination of these distinct prompt strategies and model sizes creates a broad set of clearly differentiated behavioral profiles.

Fine-Grained Prompt Variations. To test the Profiler's sensitivity to more subtle changes, we created variants of the *coverage* and *correctness* prompts. For each of these, we generated outputs both with and without few-shot examples. This mimics a typical prompt engineering workflow and allows us to test whether our toolkit's *fingerprint* is sensitive enough to detect the nuanced impact of in-context learning.

This methodology provides a rich testbed of 18 distinct simplification configurations. In our analysis, we refer to these varied approaches by the labels introduced above—distinguishing between prompt strategies (*Plain, Target, Rules, Content*), model sizes (*1B, 4B, 12B*), and the use of few-shot samples—to investigate the unique fingerprint of each.

4   The Simplification Profiler Toolkit

At the heart of text simplification lies a fundamental tension of modifying the linguistic form of a text to increase simplicity while perfectly preserving its semantic content. Our Simplification Profiler is a diagnostic toolkit designed to deconstruct and quantify this tension. It provides a multidimensional fingerprint by measuring a set of well-motivated properties, each targeting an established dimension of text quality. The toolkit rests on the principle of *text-ground explainability*: every score is directly derived from detectable text segments, ensuring that the resulting profile is transparent and interpretable.

The Profiler's metrics are organized into three key areas of analysis:

4.1   Measuring Semantic Fidelity

The most critical aspect is whether the simplification preserves the original meaning. We assess this with two NLI-based metrics.

Content Correctness (COR). To ensure a simplification does not introduce factual errors
or distort the original meaning, we measure Content Correctness. For each sentence $(s_i^{T_O})$ in the original text (T_O), we use an NLI model to find the probability of a contradiction, $P(contradiction)$. We consider the complete simplification as hypothesis and test whether any $s_i^{T_O}$ as premise leads to a contradiction. The final score, S_{CCor} , reflects the aggregated absence of contradictions.

$$S_{CCor} = \left(\frac{1}{|T_O|} \sum_{s_i^{T_O} \in T_O} (1 - P(\text{contradiction})) \right) \cdot 100$$

Content Coverage (COV). To measure omissions, we calculate Content Coverage. The metric assesses each original meaning unit (the premise, p) against the simplification (the hypothesis, h). It combines the NLI model's entailment probability, $P(entailment)$, with a semantic similarity score, $sim(p, h)$, to reward outputs that retain the source content. This hybrid approach is robust because a high coverage score requires both strong logical entailment and high semantic similarity, making it less prone to errors than either metric would be individually.

$$Cov(s_i^{T_O}) = P(\text{entailment}) \cdot sim(p, h)$$

We compute the final S_{CCov} score similar to the Content Correctness.

4.2   Measuring Linguistic Quality

This analysis assesses the form and fluency of the simplified text.

Rule-Based Simplicity (SIM) and Correctness (LNG). To measure adherence to simplification conventions and basic grammar, we use a set of custom rules implemented in LanguageTool (see Appendix A). Violations of simplicity rules (e.g., passive voice) decrease the SIM score, while grammatical errors (e.g., typos) decrease the LNG score. The score (S) is the squared harmonic mean of the non-violating word (R_w) and sub-clause (R_{sc}) ratios:

$$ S = \left(\frac{2}{\frac{1}{R_w} + \frac{1}{R_{sc}}} \right)^2 $$

This formula uses both word and sub-clause ratios to ensure the score reflects both lexical and structural rule adherence. Squaring the result more heavily penalizes any violations, making the metric more sensitive to imperfections and better distinguishing nearly-perfect outputs from those with even a few errors.

Readability (FBR). We measure readability using a German readability index [10], which combines sentence and word length heuristics into a single score (FBR_{raw}). The detailed formula is in Appendix B. We normalize this score using a sigmoid function to a 0–1 scale, where higher values indicate better readability.

$$ FBR = \sigma(k \cdot (x_0 - FBR_{raw})) $$

The parameter x_0 centers the function, and we set it to 50 so that a raw FBR score of 50 corresponds to a normalized readability of 0.5.

Coherence (COH). To approximate local coherence, we calculate the mean cosine similarity between the sentence embeddings of adjacent sentences. Higher scores suggest smoother semantic transitions.

Simple Diagnostic Heuristics. Alongside our core metrics, we track three simple heuristics that serve as powerful, high-level indicators of simplification style. **Text Length (LEN)** measures the output's compression relative to the source text, while the **Named Entity (ENT)** ratio serves as a proxy for the retention of key factual information. Finally, we include the **Average Sentence Length (ASL)**. While a traditional feature, our experiments (detailed in Sect. 6) revealed it to have disproportionately high diagnostic power for identifying specific simplification strategies. Therefore, this expressive power empirically motivates its inclusion as a key component of the fingerprint.

4.3   Visualizing and Interpreting the Fingerprint

The true power of the Simplification Profiler lies not in any single score but in analyzing the complete, multidimensional fingerprint. To make these profiles tangible, we visualize them as spider diagrams, which excel at revealing the unique shape and trade-offs of a simplification profile. Therefore, the visualization via spider diagrams is not merely an illustration but a core component of our methodology. It is explicitly designed to manage the complexity of the multidimensional output, allowing developers to interpret the profile holistically by its overall shape and trade-offs at a glance rather than being overwhelmed by individual numeric scores.

We applied the Profiler to a complex German source text and three distinct, generated simplifications to provide a concrete example. The English translations below illustrate the characteristics of each text:

> *"Due to the extraordinary weather conditions and the resulting agricultural production losses, renowned economic institutes are forecasting a significant increase in inflation for Germany."*

S1: *"Because of bad weather, experts predict higher prices."*

S2: *"The weather was very unusual, which is why farmers could harvest less. Well-known economic experts therefore believe that everything will become significantly more expensive."*

S3: *"The weather was bad, the harvests may be broen. Experts say, the moey will worth less."*[4]

Applying our toolkit to the original German texts yields the distinct fingerprints visualized in Fig. 1. The diagram makes the trade-offs immediately apparent: a *Detailed* simplification (S2, dashed green) shows a high coverage, content is mostly preserved, but it uses some more sophisticated formulations. In contrast, the *Concise* version (S1, solid blue) reveals a clear deficit in Content Coverage (COV), while the *Erroneous* output (S3, dash-dot red) shows a dramatic failure in Linguistic Correctness (LNG).

5   Evaluation Data

Our analysis is reference-less, meaning our Profiler does not require pre-existing gold standard simplifications for evaluation. This section details the data-driven methodology we used to construct a compact yet representative evaluation subset from our primary data source, the German Wikipedia dataset[5]. The goal was to build a challenging testbed specifically designed to probe the fine-grained diagnostic capabilities of our toolkit.

5.1   Construction of the Evaluation Subset

A robust diagnostic toolkit requires a challenging and diverse test dataset. Our initial analysis of German Wikipedia revealed a highly skewed distribution of simplification phenomena, as detailed in Table 1. This analysis confirmed that random sampling would fail to capture many rare but important linguistic challenges. We employed a greedy

---

[4] The typos in this translation reflect grammatical errors present in the original German output from the Erroneous simulation.

[5] The German portion of the Wikipedia dataset is available at https://huggingface.co/datasets/wikimedia/wikipedia.

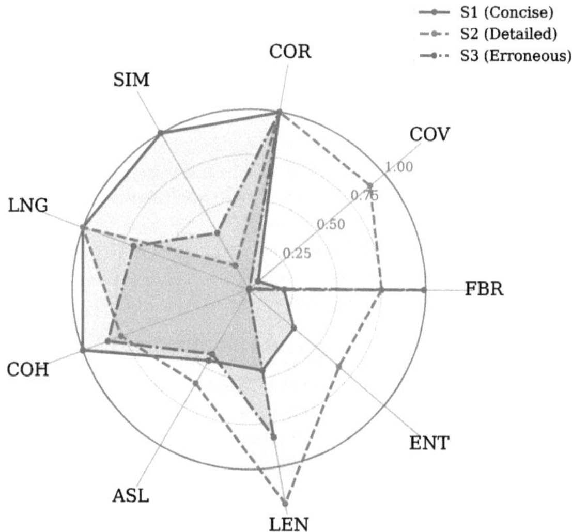

Fig. 1. The fingerprints of the three example simplification strategies (S1, S2, S3). Each distinct profile shape highlights the strengths, weaknesses, and trade-offs of the corresponding simplification, such as conciseness versus content coverage.

sampling algorithm to build a benchmark that explicitly contains these phenomena. This data-driven approach allowed us to construct a compact (1,000 examples) yet diverse dataset with a guaranteed high concentration of various simplification targets, making it ideal for fine-grained analysis. Each example consists of a five-sentence window, with sentences segmented using spaCy and filtered to ensure grammatical completeness (e.g., containing a verb and proper terminal punctuation).

6   Experimental Validation

To validate that the Simplification Profiler can generate meaningful and diagnostically useful fingerprints, we designed a series of classification experiments. The central hypothesis is that if the metric profiles are sufficiently sensitive, a simple linear classifier should be able to distinguish texts produced by different models and prompting strategies reliably. This section details this validation, moving from high-level distinctions to fine-grained analysis.

6.1   Experimental Setup

Data and Configurations. Our dataset consists of 14,868 simplified texts generated from 826 unique source excerpts using the 18 distinct configurations described in Sect. 3.

Table 1. Distribution of selected comprehensibility rule violations across 10,000 excerpts randomly sampled from German Wikipedia. Each excerpt consists of five consecutive sentences from a randomly selected article. The imbalance in rule frequency illustrates the need for targeted sampling to ensure coverage of rare but relevant linguistic phenomena.

Rule	Share (%)
Top 3 most frequent rule violations	
Sentence length	84.71%
Long words	72.99%
Abstract words	70.68%
Top 3 least frequent rule violations	
Anglicisms	12.57%
Negation	10.16%
Technical terms	3.37%

The Classification Task. For our main validation, we frame the problem as a series of binary classification tasks (e.g., *Was this simplification created with target prompt?* or *Was this text generated by the 1B model?*). Success in these tasks provides direct evidence of the Profiler's descriptive power.

Features and Baselines. The feature vector for each simplification comprises the 23 metrics from the Profiler toolkit. We compare the performance of a *LogisticRegression* classifier trained on this full feature set against two baselines: a 'Random' classifier and a *Simple Baseline* trained only on six basic length-related features. This allows us to quantify the added value of our comprehensive linguistic fingerprint.

6.2 Results: Diagnostic Sensitivity of the Fingerprints

A preliminary analysis reveals apparent differences in the average metric scores between strategies. For instance, *Target* prompts produce longer texts with higher FBR scores than *Plain* prompts, suggesting the existence of distinct profiles. Our classification experiments formally test this.

High-Level Distinguishability. Table 2 shows the main classification results. Our Profiler's complete feature set enables the classifier to outperform both baselines across nearly all categories significantly. The dramatic F1-score improvement for *Target* prompts (48.3% vs. 0.0% for the Simple Baseline) is particularly notable, confirming that the rich linguistic fingerprint captures nuances that simple length metrics miss entirely. The strong overall performance validates that the Profiler can reliably distinguish significant behavioral differences between models and prompt archetypes.

Fine-Grained Analysis. We conducted two further experiments to probe the Profiler's sensitivity. First, to investigate the poor performance for the 4b class in Table 2, we performed pairwise classification between model sizes (Table 3). The high success rate

Table 2. The diagnostic sensitivity of the Simplification Profiler. A linear classifier using the Profiler's rich linguistic metrics significantly outperforms simple baselines in identifying text origins, confirming the metrics provide a distinct and valuable fingerprint.

Target Label	Accuracy ($\pm$ std)	F1-Score ($\pm$ std)	Simple Baseline (Acc/F1 $\pm$ std)	Random Baseline (Acc/F1 $\pm$ std)
Plain	84.4 ± 0.7	39.1 ± 1.9	$(83.7 \pm 0.7 / 35.2 \pm 2.1)$	$(71.9 \pm 0.9 / 16.9 \pm 2.0)$
Target	86.6 ± 0.4	48.2 ± 1.7	$(83.2 \pm 0.0 / 0.0 \pm 0.0)$	$(71.8 \pm 0.2 / 16.4 \pm 1.4)$
Rules	78.9 ± 0.5	64.6 ± 1.1	$(78.0 \pm 0.8 / 62.4 \pm 1.6)$	$(55.6 \pm 0.5 / 33.6 \pm 0.5)$
Content	80.7 ± 0.5	67.1 ± 1.0	$(78.0 \pm 0.6 / 61.1 \pm 1.2)$	$(55.6 \pm 0.6 / 33.5 \pm 0.9)$
1b	81.7 ± 0.2	71.9 ± 0.3	$(77.7 \pm 0.5 / 65.7 \pm 0.7)$	$(54.8 \pm 0.6 / 32.4 \pm 0.9)$
4b	66.4 ± 0.2	3.4 ± 1.1	$(66.6 \pm 0.0 / 0.0 \pm 0.0)$	$(55.2 \pm 0.5 / 32.9 \pm 0.8)$
12b	80.1 ± 1.1	67.0 ± 1.8	$(78.6 \pm 1.7 / 63.5 \pm 3.0)$	$(55.9 \pm 0.6 / 33.9 \pm 1.0)$

in these head-to-head tests confirms the fingerprints are distinct and that the initial poor result was an artifact of the linear classifier's inability to handle the *middle-ground* problem, further highlighting the Profiler's diagnostic utility.

Second, as an ultimate sensitivity test, we checked if the Profiler could detect the subtle impact of few-shot prompting. As shown in Table 4, the classifier successfully distinguishes between the *FS* and *NoFS* variants for both *Rules* and *Content* prompts. The successful classification confirms the Profiler is sensitive enough for fine-grained prompt engineering analysis.

Table 3. Pairwise classification results between model sizes. The high head-to-head accuracy confirms each model size has a distinct fingerprint, indicating the *middle-ground* problem for the 4B model in the general classification task was a classifier artifact.

Target Labels	Accuracy ($\pm$ std)	F1-Score ($\pm$ std)	Simple Baseline (Acc/F1 $\pm$ std)	Random Baseline (Acc/F1 $\pm$ std)
1B vs. 4B	80.25 ± 0.3	79.77 ± 0.5	$(77.60 \pm 1.0 / 76.68 \pm 1.3)$	$(4997 \pm 0.4 / 4974 \pm 0.5)$
4B vs. 12B	68.96 ± 0.7	68.96 ± 0.7	$(65.94 \pm 1.1 / 66.00 \pm 1.1)$	$(49.97 \pm 0.4 / 49.74 \pm 0.5)$
1B vs. 12B	86.57 ± 0.0	86.04 ± 0.4	$(85.59 \pm 0.4 / 84.65 \pm 0.6)$	$(49.97 \pm 0.4 / 49.74 \pm 0.5)$

Table 4. The Profiler's sensitivity to fine-grained changes is confirmed, as its metric fingerprint allows a classifier to distinguish outputs generated with and without few-shot prompting.

Target Labels	Accuracy ($\pm$ std)	F1-Score ($\pm$ std)	Simple Baseline (Acc/F1 $\pm$ std)	Random Baseline (Acc/F1 $\pm$ std)
Rules	63.22 ± 1.5	61.77 ± 1.5	$(59.54 \pm 1.7 / 58.67 \pm 2.1)$	$(50.32 \pm 1.9 / 50.34 \pm 2.0)$
Content	77.30 ± 1.6	77.30 ± 1.5	$(76.49 \pm 1.1 / 76.39 \pm 1.1)$	$(50.32 \pm 1.9 / 50.34 \pm 2.0)$

6.3   Visualizing and Interpreting the Fingerprints

We can visualize the feature importance to understand why these fingerprints are so effective (Fig. 2). The heatmap clearly shows that different strategies have different

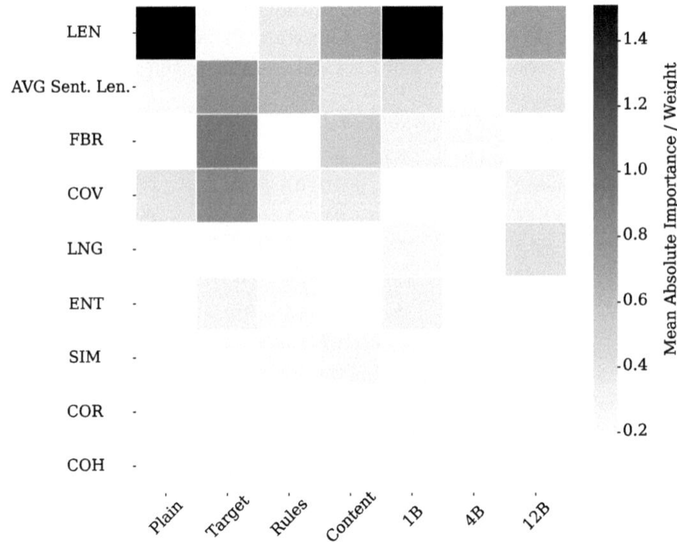

Fig. 2. Feature importance heatmap showing the distinct metric patterns for identifying each condition. For example, the *Target* prompt is strongly characterized by readability (FBR) and content coverage (COV), while the *1B* model is primarily identified by its length (LEN).

Table 5. Ablation study for Average Sentence Length. Removing this single feature significantly harms classifier performance for *Target* and *Rules* prompts, confirming its critical role in characterizing these simplification strategies.

Target Label	Full Feature Set (Acc/F1 $\pm$ std)	Without Avg. Sent. Len. (Acc/F1 $\pm$ std)
Plain	$84.4 \pm 0.6 / 39.1 \pm 1.7$	$84.2 \pm 0.6 / 38.2 \pm 1.9$
Target	$86.6 \pm 0.4 / \mathbf{48.3 \pm 1.7}$	$84.1 \pm 0.5 / \mathbf{24.9 \pm 2.1}$
Rules	$78.9 \pm 0.5 / \mathbf{64.6 \pm 1.1}$	$73.9 \pm 0.2 / \mathbf{52.1 \pm 0.4}$
Content	$80.7 \pm 0.5 / 67.1 \pm 1.0$	$80.8 \pm 0.6 / 67.4 \pm 1.1$
1B	$81.7 \pm 0.2 / 71.9 \pm 0.3$	$81.5 \pm 0.4 / 71.5 \pm 0.6$
4B	$66.4 \pm 0.2 / 3.4 \pm 1.0$	$66.6 \pm 0.1 / 0.6 \pm 0.4$
12B	$80.1 \pm 1.1 / 67.0 \pm 1.9$	$(8.8 \pm 0.8 / 64.7 \pm 1.3$

key identifiers. For instance, *Target* prompts are defined by FBR and COV, while LEN defines 1B model outputs. The ablation study in Table 5 quantifies this, showing that removing Average Sentence Length impedes the ability to detect Target and Rules prompts, confirming the critical and sometimes non-obvious role of simple heuristics within a complex fingerprint.

While the feature importance heatmap in Fig. 2 reveals which individual metrics are most discriminative, the overlay spider diagram in Fig. 3 provides a holistic, qualitative view of the archetypal fingerprint for each prompting strategy. The visual differences are striking. For example, the *Target* profile (blue, solid line) skews towards high Read-

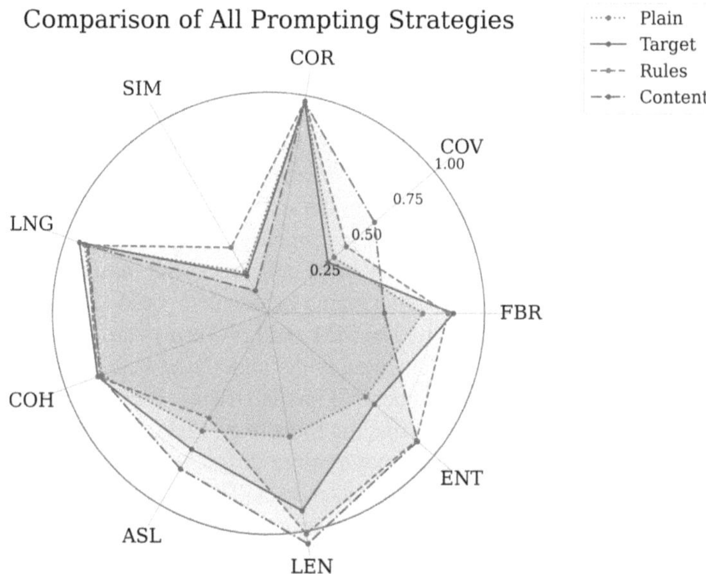

Fig. 3. Overlaid fingerprints of the four main prompting strategies. The distinct shape of each profile highlights the inherent trade-offs of each strategy, such as the focus on readability versus content coverage.

ability (FBR) but at the cost of lower Content Coverage (COV), while the *Content* profile (red, dash-dot line) shows the opposite trade-off. Furthermore, the plot reveals non-obvious relationships between metrics. One might assume that a lower Average Sentence Length (ASL) would directly lead to the highest FBR score, but the chart shows otherwise. The *Rules* prompt (green, dashed line) produces the shortest sentences, yet the *Target* prompt achieves superior readability, suggesting it simplifies via other means like vocabulary choice rather than sentence splitting. This direct visual comparison of strategic trade-offs is the core diagnostic power of the Simplification Profiler.

7   Limitations

Scope of Analysis: A Diagnostic Tool, Not a Human Proxy. The Simplification Profiler primarily serves as a *diagnostic toolkit for differential analysis*—that is, it helps developers understand the effects of their changes by comparing the fingerprints of System A versus System B. It does not substitute for holistic human quality judgment but is a complementary tool for the iterative development phase. Our goal is to empower developers to diagnose *how* their design choices (e.g., a new prompt) impact the output's characteristics *before* undertaking costly and time-consuming human evaluation.

Furthermore, a single human preference score becomes less meaningful as ATS systems become more adaptive. The goal is to produce varied outputs for different audiences (e.g., one optimized for conciseness, another for content preservation). Our toolkit allows developers to deliberately craft and verify these different profiles. Establishing the correlation of these multi-faceted fingerprints with human judgments for specific user groups is a crucial but distinct subsequent step in the development pipeline and a valuable area for future work.

Our toolkit's compositional design means its effectiveness is inherently tied to the performance of its constituent components. For instance, the Content Correctness (COR) and Content Coverage (COV) metrics rely entirely on an NLI model. Any biases, domain gaps, or inaccuracies in this NLI model will propagate directly into our evaluation scores. Similarly, the COH (Coherence) metric, based on the cosine similarity of sentence embeddings, is a heuristic for local coherence and may not capture more complex, discourse-level coherence phenomena or penalize valid simplifications that intentionally restructure sentence flow. The SIM and LNG scores are also limited by the scope and accuracy of the rules implemented in the underlying LanguageTool checker.

Our validation was limited to German Wikipedia articles. Porting the framework is a non-trivial effort due to its language-specific tools (e.g., FBR, German LanguageTool rules), and applying it to new domains may require different evaluation dimensions to account for their unique simplification characteristics. Therefore, the immediate generalizability of our findings and tool to other languages and domains is limited.

While our framework covers core aspects of simplification quality, such as linguistic correctness, adequacy, and complexity, the evaluated dimensions are not exhaustive. For specific applications, other attributes may be equally or more important. For example, our current metrics do not explicitly assess the appropriateness of the text's tone or register, nor do they capture aspects of user engagement or trust. A comprehensive evaluation might need to incorporate metrics for these more pragmatic qualities, depending on the specific simplification goal.

8   Conclusions

The increasing sophistication of Large Language Models in Automatic Text Simplification has exposed a critical gap: developers lack the tools for a holistic, efficient, and reproducible diagnosis of model behavior. This paper argued for a paradigm shift toward granular diagnostic analysis as an alternative to costly and often inflexible human preference studies. We introduced the Simplification Profiler, a diagnostic toolkit designed to provide a multi-dimensional, interpretable fingerprint of any simplified text. This approach is particularly vital to overcoming the problem of data scarcity in German ATS and for developing flexible models for diverse target groups where data scarcity makes traditional evaluation methods impractical.

Our empirical evaluation systematically tested our toolkit by generating simplifications across various models and prompting strategies. The results demonstrated that these different development choices leave distinct and measurable signatures on the output texts. We showed that the Simplification Profiler is sensitive enough to capture these differences. A simple linear classifier could reliably distinguish the originating experimental condition based solely on the generated fingerprint. These findings confirm our central hypothesis: a multi-dimensional analysis of specific text properties provides an informative and nuanced fingerprint of ATS system performance, enabling targeted development.

Our work's primary implication is enabling a more principled and efficient development and analysis for ATS. The Simplification Profiler allows moving beyond simple trial and error by making the trade-offs between properties like readability, content coverage, and linguistic correctness transparent. Researchers and practitioners diagnose the specific effects of optimizations, understand model behaviors, and make targeted adjustments to create ATS solutions that are truly adaptive to the varied needs of different target groups. Our main technical contribution, the open-source implementation of our toolkit, serves as a practical first step in this direction.

While this work establishes the Profiler's diagnostic utility, it also lays the groundwork for future research. The path forward involves addressing the current limitations by conducting human correlation studies, expanding to new languages, and enriching the fingerprint with more nuanced properties like tone and register. The evolution of such diagnostic toolkits is a key step towards creating an ecosystem where developers can craft text simplification systems with greater precision, responsibility, and awareness of their target users.

A   LanguageTool Rule Overview

Tables 6 and 7.

Table 6. Overview of LanguageTool community-edited rule categories. Grammar rules cover general spelling, punctuation, and usage issues, while style rules are often highly specific to particular phrases or terms.

Name	Description
Examples for grammar rules	
Typos	Detects potential misspellings and simple typing errors.
Confused words	Flags words that look or sound similar and are often mixed up (e.g. *plant* vs. *planet*).
Capitalization	Checks for incorrect uppercase/lowercase usage (e.g. nominalizations, sentence starts).
Correspondence	Validates greetings and sign-off formats in letters and emails.
Idioms	Ensures correct use of fixed expressions (e.g. *in droves* vs. *in streams*).
Examples for style rules	
Style suggestions	Broad category covering clarity, tone, redundancy, and formality.
Adjective→adverb	Recommends replacing rare adjective forms (e.g. *previous*) with adverbial forms (e.g. *previously*).
Preposition in phrase	Corrects fixed-phrase preposition use (e.g. *to note → for noting*).
Name consistency	Enforces standardized spelling of specific names and brands (e.g. *OpenStreetMap*).
Term standardization	Validates and corrects specialized terms (e.g. *JavaScript applet*).

Table 7. Overview of LanguageTool community-edited *understandability* rule categories. Sentence-level rules target clarity by avoiding complex structures, while word-level rules discourage difficult or technical vocabulary.

Name	Description
Sentence-level rules	
Sentence length	Warns when a sentence exceeds recommended length thresholds.
Perfect tense	Prefers present perfect over simple past for easier comprehension.
Negation	Advises removing negations to simplify statements.
Subordinate clauses	Suggests breaking up or removing subordinate clauses.
Relative clauses	Encourages avoiding relative clauses for direct phrasing.
Passive voice	Flags passive constructions; promotes active voice.
Genitive case	Recommends replacing genitive phrases with simpler constructions (e.g. "of the").
One idea per sentence	Ensures each sentence contains only one main idea.
Subjunctive mood	Discourages subjunctive/indirect speech for clarity.
Word-level rules	
Abstract words	Recommends concrete terms over abstract nouns.
Long words	Warns when individual words exceed recommended length.

B Details of the FBR Index

The FBR heuristic is calculated from six component indices, which are first normalized and then aggregated. The components are divided into two categories:

1. **S1: Average Sentence Length in Syllables**

$$S_1 = (x_{S1} - 12.37) \cdot \frac{100}{24.12 - 12.37}$$

   where x_{S1} is the average sentence length in syllables.

2. **S2: Sentences with > 6 Words**

$$S_2 = (x_{S2} - 41.77) \cdot \frac{100}{67.42 - 41.77}$$

   where x_{S2} is the percentage of sentences with more than 6 words.

3. **S3: Sentences with > 16 Words**

$$S_3 = (x_{S3} - 22.10) \cdot \frac{100}{52.59 - 22.10}$$

   where x_{S3} is the percentage of sentences with more than 16 words.

4. **S4: Sentences with > 20 Words**

$$S_4 = (x_{S4} - 21.90) \cdot \frac{100}{64.67 - 21.90}$$

   where x_{S4} is the percentage of sentences with more than 20 words.

5. **W1: Average Word Length in Syllables**

$$W_1 = (x_{W1} - 1.936) \cdot \frac{100}{2.339 - 1.936}$$

where x_{W1} is the average word length in syllables.

6. **W2: Words with > 3 Syllables**

$$W_2 = (x_{W2} - 10.75) \cdot \frac{100}{21.21 - 10.75}$$

where x_{W2} is the percentage of words with more than 3 syllables.

Final Score Calculation. The component indices are first averaged into a Sentence Complexity Score (K_S) and a Word Complexity Score (K_W):

$$K_S = \frac{S_1 + S_2 + S_3 + S_4}{4} \quad \text{and} \quad K_W = \frac{W_1 + W_2}{2}$$

The final index is the mean of these two scores: $FBR_{raw} = (K_S + K_W)/2$.

References

1. Alfear, N., Kazakov, D.L., Al-Khalifa, H.: Meta-evaluation of sentence simplification metrics (2024). https://aclanthology.org/2024.lrec-main.981.pdf
2. Alva-Manchego, F., Martin, L., Bordes, A., Scarton, C., Sagot, B., Specia, L.: ASSET: a dataset for tuning and evaluation of sentence simplification models with multiple rewriting transformations. In: Proceedings of the 58th Annual Meeting of the Association for Computational Linguistics, ACL (2020). https://doi.org/10.18653/v1/2020.acl-main.424, https://www.aclweb.org/anthology/2020.acl-main.424
3. Alva-Manchego, F., Martin, L., Scarton, C., Specia, L.: EASSE: easier automatic sentence simplification evaluation. In: Padó, S., Huang, R. (eds.) Proceedings of the 2019 Conference on Empirical Methods in Natural Language Processing and the 9th International Joint Conference on Natural Language Processing (EMNLP-IJCNLP): System Demonstrations, ACL (2019). https://doi.org/10.18653/v1/D19-3009, https://aclanthology.org/D19-3009/
4. Anschütz, M., Oehms, J., Wimmer, T., Jezierski, B., Groh, G.: Language models for german text simplification: Overcoming parallel data scarcity through style-specific pre-training. In: Findings of the Association for Computational Linguistics: ACL 2023 (2023). https://doi.org/10.18653/v1/2023.findings-acl.74, http://arxiv.org/abs/2305.12908
5. Carrer, L., Säuberli, A., Kappus, M., Ebling, S.: Towards holistic human evaluation of automatic text simplification (2024). https://aclanthology.org/2024.humeval-1.7.pdf
6. Chan, R., Amini, A., El-Assady, M.: Which spurious correlations impact reasoning in NLI models? a visual interactive diagnosis through data-constrained counterfactuals. In: Bollegala, D., Huang, R., Ritter, A. (eds.) Proceedings of the 61st Annual Meeting of the Association for Computational Linguistics, ACL (2023). https://doi.org/10.18653/v1/2023.acl-demo.44, https://aclanthology.org/2023.acl-demo.44/

7. Glockner, M., Shwartz, V., Goldberg, Y.: Breaking NLI systems with sentences that require simple lexical inferences (2018). https://doi.org/10.18653/v1/p18-2103, http://aclweb.org/anthology/P18-2103

8. Gu, J., et al.: A survey on LLM-as-a-judge. https://doi.org/10.48550/arXiv.2411.15594, http://arxiv.org/abs/2411.15594

9. Huang, Y., Kochmar, E.: REFeREE: a Reference-FREE model-based metric for text simplification. In: Calzolari, N., Kan, M.Y., Hoste, V., Lenci, A., Sakti, S., Xue, N. (eds.) Proceedings of the 2024 Joint International Conference on Computational Linguistics, Language Resources and Evaluation (LREC-COLING 2024), ELRA and ICCL (2024). https://aclanthology.org/2024.lrec-main.1200/

10. Kercher, J.: Verstehen und Verstandlichkeit von Politikersprache: Verbale Bedeutungsvermittlung zwischen Politikern und Burgern. Springer Fachmedien (2013). https://doi.org/10.1007/978-3-658-00191-9, https://link.springer.com/10.1007/978-3-658-00191-9

11. Klöser, L., Beele, M., Schagen, J.N., Kraft, B.: German text simplification: finetuning large language models with semi-synthetic data. In: Proceedings of the Fourth Workshop on Language Technology for Equality, Diversity, Inclusion, ACL (2024). https://doi.org/10.48550/ARXIV.2402.10675, https://arxiv.org/abs/2402.10675

12. Klöser, L., Büsgen, A., Kohl, P., Kraft, B., Zündorf, A.: Explaining relation classification models with semantic extents. In: Deep Learning Theory and Applications - 4th International Conference, DeLTA 2023, Rome, Italy, 13–14 July 2023, Proceedings, Springer (2023). https://doi.org/10.1007/978-3-031-39059-3_13, http://arxiv.org/abs/2308.02193

13. Liu, J., Nam, Y., Cui, X., Swayamdipta, S.: Evaluation under imperfect benchmarks and ratings: a case study in text simplification. https://doi.org/10.48550/arXiv.2504.09394, http://arxiv.org/abs/2504.09394

14. Maddela, M., Dou, Y., Heineman, D., Xu, W.: LENS: a learnable evaluation metric for text simplification. In: Proceedings of the 61st Annual Meeting of the Association for Computational Linguistics, ACL (2023). https://doi.org/10.18653/v1/2023.acl-long.905, https://aclanthology.org/2023.acl-long.905

15. Martin, L., et al.: Controllable sentence simplification. In: Calzolari, N., et al. (eds.) Proceedings of the Twelfth Language Resources and Evaluation Conference. European Language Resources Association (2020). https://aclanthology.org/2020.lrec-1.577/

16. Naderi, B., Mohtaj, S., Ensikat, K., Möller, S.: Subjective assessment of text complexity: a dataset for German language. https://doi.org/10.48550/arXiv.1904.07733, http://arxiv.org/abs/1904.07733

17. Nagai, Y., Oka, T., Komachi, M.: A document-level text simplification dataset for japanese. In: Calzolari, N., Kan, M.Y., Hoste, V., Lenci, A., Sakti, S., Xue, N. (eds.) Proceedings of the 2024 Joint International Conference on Computational Linguistics, Language Resources and Evaluation (LREC-COLING 2024). ELRA and ICCL (2024). https://aclanthology.org/2024.lrec-main.41/

18. OPSI: text simplification for citizens with GenAI, https://oecd-opsi.org/innovations/text-simplification-for-citizens-with-genai/

19. Papineni, K., Roukos, S., Ward, T., Zhu, W.J.: Bleu: a method for automatic evaluation of machine translation. In: Isabelle, P., Charniak, E., Lin, D. (eds.) Proceedings of the 40th Annual Meeting of the Association for Computational Linguistics, ACL (2002). https://doi.org/10.3115/1073083.1073135, https://aclanthology.org/P02-1040/

20. Qiang, J., Huang, M., Zhu, Y., Yuan, Y., Zhang, C., Yu, K.: Redefining simplicity: benchmarking large language models from lexical to document simplification. https://doi.org/10.48550/arXiv.2502.08281, http://arxiv.org/abs/2502.08281

21. Rios, A., et al.: A new dataset and efficient baselines for document-level text simplification in german. In: Carenini, G., Cheung, J.C.K., Dong, Y., Liu, F., Wang, L. (eds.) Proceedings

of the Third Workshop on New Frontiers in Summarization, ACL (2021). https://doi.org/10.18653/v1/2021.newsum-1.16, https://aclanthology.org/2021.newsum-1.16/

22. Ryan, M., Naous, T., Xu, W.: Revisiting non-english text simplification: a unified multilingual benchmark. In: Proceedings of the 61st Annual Meeting of the Association for Computational Linguistics, ACL (2023). https://doi.org/10.18653/v1/2023.acl-long.269, https://aclanthology.org/2023.acl-long.269

23. Shardlow, M.: A survey of automated text simplification (2014). https://doi.org/10.14569/SpecialIssue.2014.040109, http://thesai.org/Publications/ViewPaper?Volume=4&Issue=1&Code=SpecialIssue&SerialNo=9

24. Shinoda, K., Sugawara, S., Aizawa, A.: Which shortcut solution do question answering models prefer to learn? (2023). https://doi.org/10.1609/aaai.v37i11.26590, https://ojs.aaai.org/index.php/AAAI/article/view/26590

25. Stodden, R.: EASSE-DE & EASSE-multi: easier automatic sentence simplification evaluation for german & multiple languages. In: Proceedings of the Third Workshop on Text Simplification, Accessibility and Readability (TSAR 2024), ACL (2024). https://doi.org/10.18653/v1/2024.tsar-1.11, https://aclanthology.org/2024.tsar-1.11

26. Sulem, E., Abend, O., Rappoport, A.: BLEU is not suitable for the evaluation of text simplification. In: Proceedings of the 2018 Conference on Empirical Methods in Natural Language Processing, ACL (2018). https://doi.org/10.18653/v1/D18-1081, http://aclweb.org/anthology/D18-1081

27. Sulem, E., Abend, O., Rappoport, A.: Semantic structural evaluation for text simplification. In: Walker, M., Ji, H., Stent, A. (eds.) Proceedings of the 2018 Conference of the North American Chapter of the Association for Computational Linguistics: Human Language Technologies, ACL (2018). https://doi.org/10.18653/v1/N18-1063, https://aclanthology.org/N18-1063/

28. Säuberli, A., et al.: Digital comprehensibility assessment of simplified texts among persons with intellectual disabilities (2024). https://doi.org/10.1145/3613904.3642570, https://dl.acm.org/doi/10.1145/3613904.3642570

29. Toborek, V., Busch, M., Boßert, M., Bauckhage, C., Welke, P.: A new aligned simple german corpus. In: Rogers, A., Boyd-Graber, J., Okazaki, N. (eds.) Proceedings of the 61st Annual Meeting of the Association for Computational Linguistics, ACL (2023). https://doi.org/10.18653/v1/2023.acl-long.638, https://aclanthology.org/2023.acl-long.638/

30. Wu, X., Arase, Y.: An in-depth evaluation of large language models in sentence simplification with error-based human assessment (2025). https://doi.org/10.1145/3744744, https://dl.acm.org/doi/10.1145/3744744

31. Xu, W., Napoles, C., Pavlick, E., Chen, Q., Callison Burch, C.: Optimizing statistical machine translation for text simplification (2016). https://doi.org/10.1162/tacl_a_00107, https://aclanthology.org/Q16-1029/

32. Zhang, T., Kishore, V., Wu, F., Weinberger, K.Q., Artzi, Y.: BERTScore: evaluating text generation with BERT. https://www.semanticscholar.org/paper/BERTScore%3A-Evaluating-Text-Generation-with-BERT-Zhang-Kishore/295065d942abca0711300b2b4c39829551060578

Attention Maps in 3D Shape Classification for Dental Stage Estimation with Class Node Graph Attention Networks

Barkin Buyukcakir[1]([⊠]) [iD], Rocharles Cavalcante Fontenele[2,3] [iD], Reinhilde Jacobs[2,3] [iD], Jannick De Tobel[4] [iD], Patrick Thevissen[5] [iD], Dirk Vandermeulen[1] [iD], and Peter Claes[1,6] [iD]

[1] Department of Electrical Engineering (ESAT) - Processing Speech and Images (PSI), KU Leuven, 3000 Leuven, Belgium
barkin.buyukcakir@kuleuven.be
[2] Department of Imaging and Pathology, Oral and Maxillofacial Surgery - Imaging and Pathology (OMFS-IMPATH), KU Leuven, 3000 Leuven, Belgium
[3] Department of Oral and Maxillofacial Surgery, KU Leuven, University Hospitals Leuven, 3000 Leuven, Belgium
[4] Deparment of Diagnostic Sciences, Ghent University, 9000 Ghent, Belgium
[5] Department of Imaging and Pathology, Forensic Odontology, KU Leuven, 3000 Leuven, Belgium
[6] Department of Human Genetics, KU Leuven, 3000 Leuven, Belgium

Abstract. Deep learning offers a promising avenue for automating many recognition tasks in fields such as medicine and forensics. However, the "black box" nature of these models hinders their adoption in high-stakes applications where trust and accountability are required. For 3D shape recognition tasks in particular, this paper introduces the Class Node Graph Attention Network (CGAT) architecture to address this need. Applied to 3D meshes of third molars derived from CBCT images, for Demirjian stage allocation, CGAT utilizes graph attention convolutions and an inherent attention mechanism, visualized via attention rollout, to explain its decision-making process. We evaluated the local mean curvature and distance to centroid node features, both individually and in combination, as well as model depth, finding that models incorporating directed edges to a global CLS node produced more intuitive attention maps, while also yielding desirable classification performance. We analyzed the attention-based explanations of the models, and their predictive performances to propose optimal settings for the CGAT. The combination of local mean curvature and distance to centroid as node features yielded a slight performance increase with a 0.76 weighted F1 score, and more comprehensive attention visualizations. The CGAT architecture's ability to generate human-understandable attention maps can enhance trust and facilitate expert validation of model decisions. While demonstrated on dental data, CGAT is broadly applicable to graph-based classification and regression tasks, promoting wider adoption of transparent and competitive deep learning models in high-stakes environments.

Keywords: Class Node Graph Attention Networks · Explainable Artificial Intelligence · 3D Shape Analysis · Dental Stage Assessment

F. Marcelloni et al. (Eds.): IJCCI 2025, CCIS 2829, pp. 154–176, 2026.
https://doi.org/10.1007/978-3-032-15638-9_10

1 Introduction

The field of deep learning (DL) has witnessed remarkable advancements in recent years, leading to transformative capabilities across a multitude of domains, and becoming state-of-the-art in tasks such as image recognition, natural language processing, text generation, and increasingly, the analysis of 3D shapes. Despite these significant strides and a plethora of published studies showcasing their potential, the widespread adoption of DL methods in the practice of high-stakes applications, particularly in sectors like healthcare and forensics, remains surprisingly limited. A primary factor contributing to this cautious uptake is the inherent opaqueness of many cutting-edge DL architectures. The inability to fully understand or interpret the decision processes of these complex algorithms raises concern about the reliability, trustworthiness, and potential biases, especially when critical outcomes regarding human health and rights are of concern [10]. This underscores a pressing need for explainable artificial intelligence (XAI) methods that can provide transparency and insight into model behavior. With the recent global focus by organizations such as the U.S. Defense Advanced Research Projects Agency (DARPA) [28], and the European Union with its General Data Protection Regulation (GDPR) [35], on interpretable deep learning systems intensifying, explainability in deep learning has become more important.

DL models have been applied extensively in the relevant literature on a host of medical imaging modalities, for diagnostic purposes, such as disease detection and diagnosis, segmentation of biological structures, and image registration [29]. In the forensic domain, DL algorithms have found use in the evaluation of biometrics, for authentication and identification tasks on various data such as faces and fingerprints [26]. However, most of these applications have been based on two-dimensional (2D) data, where the primary means of explainability has been attention or saliency maps, where the most influential regions of the input images in model prediction are highlighted on the images as a form of decision explanation [21]. However, when 3D data is concerned, the direct applications of the tried and tested DL architectures were not immediately possible due to the inherent properties of 3D shape representations raising new requirements such as permutation invariance or incorporation of topology [6], and as such, the direct equivalent of attention maps for 3D data has not yet been proposed.

Geometric deep learning (GDL) subsequently emerged as a specialized subfield of DL, providing powerful frameworks for analyzing 3D structures. GDL models operate most commonly on one of four representations of 3D data, which are: (1) point clouds where the 3D shape is depicted as a collection of points in space, (2) meshes, which depict discrete surfaces defined by vertices and edges, highlighting the underlying topology of the shapes, and (3) graphs, which are abstract structures defined by a set of nodes with arbitrary features encoded in them, and the edges connecting these nodes together which represent some relation between them [5,14]. For point clouds, PointNet and PointNet++ have been the pioneering architectures due to their agnosticism to input permutation [30,31]. These models have been applied to human identification based on 3D teeth models [23], Alzheimer's disease diagnosis on hippocampal surfaces [42] and age prediction [41]. The main difficulty of point cloud representations is that due to the lack of geodesic connections, two points in close proximity can be valid neighbors even though they are topologically unassociated. This difficulty is alleviated by mesh and, more generally, graph representations, where connectivity is defined. Operating on meshes, methods such as MeshCNN [19] and SpiralNet++ [17] have been introduced, and applied in tasks such as syndrome classification [24] and age prediction based on cortical surface shape [39]. Although mesh-based approaches solve the topology problem in point clouds, many of them require mesh vertices to be in correspondence with each other across samples, thus requiring mesh registration as a preprocessing step. Graph neural networks (GNNs) do not have such requirements as they commonly feature a permutation-invariant convolution operation which operates solely on the features embedded in the nodes of input graphs. Some commonly used examples of GNN architectures are graph convolutional networks (GCNs) [22], Graph-SAGE [18] and graph attention networks (GATs) [38], which have been employed in tumor segmentation [27], age and sex prediction [2,36] based on anatomical scans, and facial landmarking [12].

Despite this extensive literature on 3D shape analysis using geometric deep learning, widespread clinical use of these methods has remained limited, due to the lack of an explainability mechanism that can provide the basis for model decisions, with perhaps the exception of the inherent attention mechanism in GAT [38], and the set of critical points in PointNet architectures [30]. The former, as originally proposed, however, produces attentions in between the nodes of the graphs, but does not offer explanations with regard to the prediction, and the latter is a simplification of the input point cloud, not indicative of the importance of the points to the model decision. In order to address this shortcoming, we propose the Class Node Graph Attention Network (CGAT), which can operate on graph representations of 3D shapes and can generate attention map-based explanations for model decisions. The CGAT architecture aims to provide competitive performance in graph-level tasks such as shape classification and regression, and we leverage the inherent learned attention of GAT convolution operation in order to depict the contribution of nodes to the final decision. To the best of our knowledge, this is the first study to propose an architecture capable of generating attention-based explanations for its predictions on 3D shapes. We demonstrate the CGAT architecture in the context of dental stage assessment of 3D third molar meshes, which is an important intermediate step in age estimation in forensics, in order to demonstrate the advantages of attention-based explanations in this high-stakes application. Dental stage assessment is the process of classifying individual teeth into developmental stages, which can then be utilized to predict the age of the subject [32]. This process is conventionally carried out manually by experts assessing 2D radiographs. Herein lie two major drawbacks: inter-observer variability between the experts and a loss of information due to the projection of 3D structures onto a 2D image.

To counter the first drawback, it is recommended that two observers assess the radiograph independently, and that they discuss in case of discrepancies to obtain a consensus [7]. Thus, the main role of DL methods for dental staging lies in eliminating inter-observer variability. Although promising results have been published, no explainability mechanisms have been reported in the literature for this task [25], while plausibility of the staging is essential in a forensic context. After all, forensic age estimation implies a high-stakes decision where the misclassification of a minor as an adult represents an ethically unacceptable error. Therefore, explainability mechanisms of automated staging can contribute to a well-founded age estimation, helping the end user in understanding the outcome. To counter the second drawback, 3D imaging modalities have been studied for dental stage allocation [13]. We chose to study CBCT because of its widespread use in dental practice, and thus an abundance of data [4]. In the current study we aimed to design the explainability mechanism of CGAT to support the plausibility of automated dental stage allocation on CBCT-derived meshes. The CGAT architecture belongs to the GNN family, and while GNNs in general, including the CGAT, are not limited to graph-level tasks, and can also perform node-level and edge-level operations, for a broader shape-informed decision making framework, we limit the application of CGAT to graph-level classification in this study.

2   Methodology

2.1   Data and Preprocessing

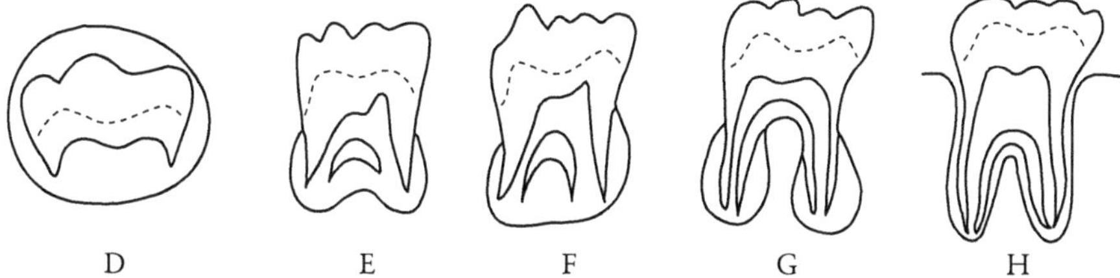

Fig. 1. A visualization of the criteria for stage allocation of stages D to H according to the Demirjian assessment system [8]. The staging criteria in this range depend mostly on root development. As the development progresses, the roots start to protrude from the crown and get longer, finally developing closed apices.

The dataset of focus for this study consists of triangular meshes of the four human third molars, corresponding to World Dental Federation (FDI) elements numbered 18, 28, 38 and 48 [20]. The original data collection was performed at five Belgian centers by Vranckx et al. in the form of CBCT images of prospective patients for third molar surgery [40]. Additional sample selection criteria were then applied on the 6010 patient records. This sample selection adhered to the criteria of patients aged 16 to 22 years old with at least two third molars, images containing no artifacts which affect third molars, such as metallic objects or presence of motion. Furthermore, third molars which were affected by pathological conditions were also excluded. The selection resulted in 138 patients in total, of which 67 were males and 71 were females, culminating in a total of 528 third molars. The third molars of the selected patients were then classified according to the developmental stage allocation system proposed by Demirjian et al. [8], by two dentists experienced in forensic dentistry and dentomaxillofacial radiology, respectively. The dentists carried out the stage assessment procedure on the corresponding panoramic radiographs for each CBCT recording [13]. In the cases where the assessors did not agree on the stage, the dispute was resolved by an oral and maxillofacial

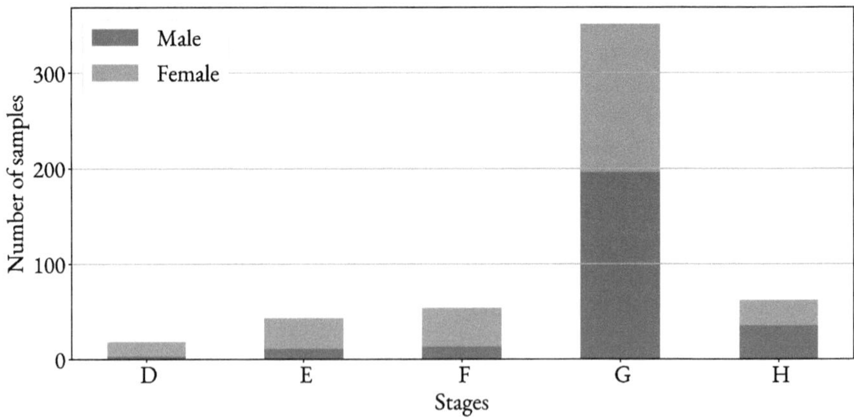

Fig. 2. Distribution of samples over the Demirjian developmental stages. Note that the stages contain high class imbalance, with the lowest stage making up approximately 3% of all data, and the stage containing most samples, G, accounting for 67% of the dataset.

radiologist with 8 years of experience. The third molars in the dataset ranged between Demirjian stages D to H, the criteria for which can be seen in Fig. 1.

These tooth samples contained high class imbalance, with the lowest representation belonging to the earliest stage, D, with only 18 samples, and the most well represented stage was G with 351 teeth. A distribution of samples over stages can be seen in Fig. 2. The CBCT images were then uploaded to the Relu® Creator software [33], with which all third molars in the images were segmented, and the triangular meshes were created for each of them.

In this study, we define all meshes as undirected graphs, and as such each sample $G(\mathcal{V}, \mathcal{E})$ is defined by a set of nodes $\mathcal{V} = \{v_1, v_2, ..., v_n\}$, and the edges $\mathcal{E} = \{e_1, e_2, ..., e_m\}$ between these nodes, where n and m are the respective number of nodes and edges of each sample. The Relu® Creator does not impose limits on the number of nodes and edges, and as such, the original outputs of the software had a varied number of these elements for each sample. More specifically, the output meshes consisted of 17205 ± 5389 nodes and 103225 ± 32336 edges on average. Thus, to remove the effect of this large variation in the number of graph elements, and to facilitate model training, these meshes were simplified using quadric mesh decimation to contain 751 ± 1 nodes, and 4497 ± 4 edges. Depictions of full resolution mesh samples from each category, along with their decimated counterparts can be seen in Fig. 3. It can be observed that the regions of interest for dental staging, e.g. structures such as the roots and the cusp are preserved in the decimated meshes, and shape information was not lost in these areas.

In order to use the decimated meshes for the model training, all of them were then scaled such that they all fit into a unit sphere in Euclidean space. The scaling was motivated by the desire to train the model on the shape information alone. If left in the original relative sizes, the teeth from earlier stages of formation would be smaller than those with fully formed roots. This would introduce bias into the learning process and could result in the model exploiting this information, which we consider an undesirable effect for the accurate depiction of the attention maps, relating to the shape of the teeth. This decision is further supported by the intrinsic variation in the size of third molars; some early stage third molars may turn out rather large, while some late stage third molars may turn out rather small. The changes in shape, on the other hand, are more consistent across individuals.

Upon preprocessing, 80% of the samples were assigned to be used in training, 5% was used in validation and 15% was reserved for testing, preserving the distribution of stage labels.

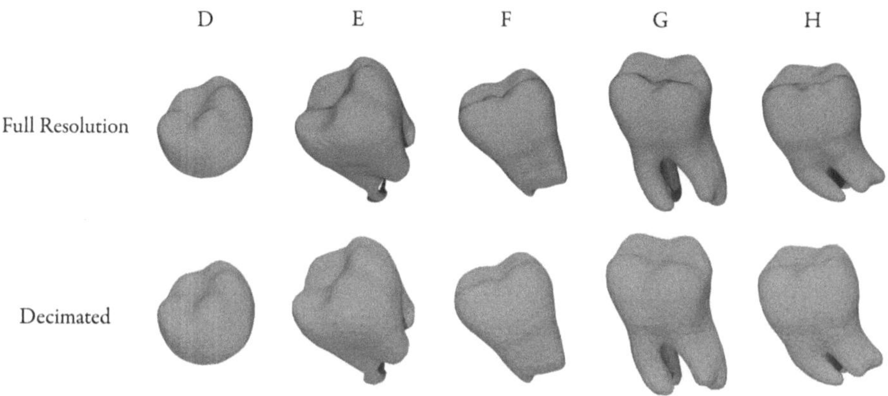

Fig. 3. Visualizations of sample meshes from each Demirjian stage in full resolution (a), and their decimated counterparts (b). The decimation allows for the elimination of the negative effects of the substantial variation in the number of nodes and edges, and makes model training and computational analyses feasible. It can be seen that the original outputs of the Relu® Creator software are overly dense, and the decimation step does not result in the loss of overall shape information.

2.2   Network Architecture

In order to facilitate the generation of attention maps on mesh classification, we employ the attention-weighted graph convolutions, first defined by Veličković et al. in [38]. The intrinsic attention mechanism allows for assigning weights to node features in graph convolutions through training. To generate attention-based explanations, we propose a modified architecture, which we refer to as the class-node graph attention network (CGAT).

Graph Attention Networks. Graph attention networks were introduced to rectify a common shortcoming in other GNN architectures, which is the aggregation of the node features in local neighborhoods with uniform weights. More specifically, the general framework of GNNs is a two step process of applying a (usually learnable) transformation to each node feature $h_i, i \in \mathcal{V}$, and aggregating the features around the neighborhood $\mathcal{N}_i = \{j \in \mathcal{V} \mid e_{j \to i} \in \mathcal{E}\}$. One layer in a graph convolution operation then allows each node to collect information from its 1-hop neighborhood, and via repeated convolutions this receptive field expands. Many common GNN architectures use uniform-weighted aggregation operations such as averaging node features, or max-pooling. The GAT convolution operation introduces a weighted aggregation operation as

$$h_i^{(l+1)} = \prod_{k=1}^{K} \sigma \left(\sum_{j:j \in \mathcal{N}_i} \alpha_{ij}^{(l,k)} \cdot W^{(l,k)} h_i^{(l)} \right) \tag{1}$$

where $h^{(l,k)}$ is the node feature vector, $\alpha_{ij}^{(l,k)}$ is the learnable attention weight and $W^{(l,k)}$ is the learnable transformation of attention head k in layer l. The GAT thus defines a multi-head attention mechanism to increase the robustness of the learned attention [37,38]. The learnable attention weights $\alpha_{ij}^{(l,k)}$ enable the higher consideration of some neighbors over others, and depend on the feature vectors of the neighboring nodes.

$$\alpha_{ij}^{(l,k)} = \text{softmax}(e(h_i, h_j))$$
$$e(h_i, h_j) = \text{LeakyReLU} \left(\alpha_{(l,k)}^{\top} \cdot \left[W^{(l,k)} h_i || W^{(l,k)} h_j \right] \right) \tag{2}$$

In Eq. 2, the $\alpha_{(l,k)}$ is the learnable attention parameter of layer l, from attention head k. However, this attention mechanism has diminished expressiveness, due to it being a form of static attention as defined by Brody et al. [3], where regardless of the target node, there is always a source node that will have the highest contribution in the aggregation operation. They propose a modification, dubbed GATv2, that reorders the application of the transformation $W^{(l,k)}$ and the attention parameter $\alpha_{(l,k)}$ as,

$$e(h_i, h_j) = \alpha_{(l,k)}^{\top} \text{LeakyReLU} \left(W^{(l,k)} \cdot [h_i || h_j] \right) \tag{3}$$

therefore rectifying the consecutive application of $W^{(l,k)}$ and $\alpha_{(l,k)}$, and using the attention parameter as an independent transform. In our experiments, we adopt the GATv2 convolution operation due to the improvements over the original proposal from [38] shown in Eq. 2.

Class Node Graph Attention Network. We propose the CGAT architecture in order to introduce explainability in the classification of the tooth meshes, in the form of attention maps. We leverage the intrinsic attention mechanism in the GATv2 convolution in order to do so, by the introduction of a virtual node, which we call the *CLS node* throughout this paper. The addition of a virtual node as a global information aggregator is an existing design philosophy. Among transformer models, the [CLS] token in BERT [9], Vision Transformer (ViT) [11] and the Graphormer [43] are notable examples implementing this design, the latter of which inspired by the usage of a *master node* by Gilmer et al. [15]. To the best of our knowledge, however, ours is the first study to use such a master node for the generation of attention maps on 3D shapes.

The core mechanic of the CGAT is the introduction of this virtual node as the first step of training to each input graph, individually. The CLS node is connected to each node in the graph, and treated no differently from the other nodes in the input graph

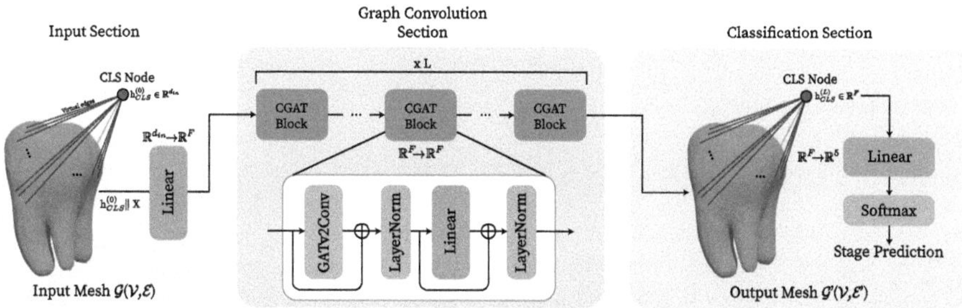

Fig. 4. Overview of the CGAT architecture. The overall architecture consists of three sections: (1) The input section is responsible for connecting the virtual CLS node to all nodes in the input graph, and applying a linear projection of all node features. (2) The graph convolution section of the model consists of stacked CGAT blocks, and is responsible for performing the attention-informed graph convolutions. (3) The classification section extracts the final embedding of the CLS node and applies one linear layer and a softmax function to produce the probability predictions of the stages. The architecture uses the attention scores of the CLS node to all other nodes to produce attention-based explanations.

throughout the graph-convolutional section of our model. The initial embedding of the CLS node h^0_{CLS} is a learnable parameter of the model. We experiment with connecting the CLS node to input graphs with directed edges $e_{i:i\in\mathcal{V}\rightarrow\mathrm{CLS}}$ which flow only towards the CLS node, and undirected edges $e_{i:i\in\mathcal{V}\leftrightarrow\mathrm{CLS}}$. This selection dictates whether the learned latent representations of the CLS node throughout the model can influence the latent embeddings of each node in the graph. We report on the effects of this choice in Sect. 3.

The CGAT consists of three functional sections; (1) the input section, where the input graphs are accepted, and the CLS node is appended to each of them, (2) the graph convolution section, where the CGAT blocks reside, and (3) the classification section, where the CLS node embedding is extracted and fed to a multilayer perceptron (MLP) head to reach the final classification output. An overview of this architecture can be seen in Fig. 4.

The input section of the model is only responsible for projecting all node features onto a latent space, and correctly appending one virtual node to each input graph in the batch. We note that regardless of the choice of using directed or undirected edges to connect the CLS node to the graph, all nodes *except for the CLS node* are also assigned a self-connection in order to ensure that the self-attention does not overshadow the attention to all other nodes, since the input degree of the CLS node $D^{\mathrm{CLS}}_{in} >> D^{v_i\in\mathcal{V}}_{in}$ is much larger than the average input degree of the original graph. We batch the input graphs, where each graph corresponds to a single tooth, by merging each sample to form an unconnected graph with a block-diagonal adjacency matrix. The input projection consists of one linear layer parametrized by $\boldsymbol{W}_{in} \in \mathbb{R}^{d_{in}\times F}$ where d is the unprojected dimensions of the node features $x \in \mathbb{R}^{d_{in}}$. The CLS node embedding $h^0_{CLS} \in \mathbb{R}^F$, as mentioned, is a learnable parameter of the model. We used the Glorot initialization for this parameter [16], as with all learnable parameters of the CGAT model. We allow the

---

Algorithm 1: CGAT Forward Pass.

Require: Graph $G(\mathcal{V}, \mathcal{E})$, node features $X \in \mathbb{R}^{|\mathcal{V}| \times d_{in}}$
Ensure: Class logits $\mathbf{y}_{\text{logits}} \in \mathbb{R}^{N_{\text{classes}}}$
 1: $\triangleright$ — Preprocessing & Input Section —
 2: Augment $\mathcal{E}$ with self-loops for all $v \in \mathcal{V}$
 3: $N \leftarrow |\mathcal{V}|$
 4: Augment $\mathcal{V}$ with a new CLS node: $\mathcal{V}' \leftarrow \mathcal{V} \cup \{v_{\text{CLS}}\}$
 5: Augment $\mathcal{E}$ to connect all nodes in $\mathcal{V}$ to v_{CLS}: $\mathcal{E}' \leftarrow \mathcal{E} \cup \{e_{i(\leftrightarrow,\rightarrow)\text{CLS}}\}_{i=1}^{N}$
 6: Initialize learnable CLS embedding: $\mathbf{h}_{\text{CLS}}^{(0)} \leftarrow \boldsymbol{\Theta}_{\text{CLS}} \in \mathbb{R}^{1 \times d_{in}}$
 7: Concatenate node features with CLS embedding: $X_{\text{aug}} \leftarrow \text{Concat}(X, \mathbf{h}_{\text{CLS}}^{(0)}) \in \mathbb{R}^{(N+1) \times d_{in}}$
 8: Project features into latent space: $\mathbf{H}^{(0)} \leftarrow \text{Linear}(X_{\text{aug}}) \in \mathbb{R}^{(N+1) \times F}$
 9: $\triangleright$ — Graph Convolution Section —
10: **for** $l = 0$ to $L - 1$ **do**
11:     $\mathbf{H}_{\text{conv}} \leftarrow \text{GATv2Conv}(\mathbf{H}^{(l)}, \mathcal{E}')$ $\triangleright$ Multi-head attention as in Eq. 4
12:     $\mathbf{H}_{\text{res1}} \leftarrow \text{LayerNorm}(\mathbf{H}^{(l)} + \mathbf{H}_{\text{conv}})$ $\triangleright$ First residual connection
13:     $\mathbf{H}_{\text{lin}} \leftarrow \text{Linear}(\mathbf{H}_{\text{res1}})$
14:     $\mathbf{H}^{(l+1)} \leftarrow \text{LayerNorm}(\mathbf{H}_{\text{res1}} + \mathbf{H}_{\text{lin}})$ $\triangleright$ Second residual connection
15: **end for**
16: $\triangleright$ — Classification Section —
17: Extract final CLS embedding: $\mathbf{h}_{\text{CLS}}^{(L)} \leftarrow \mathbf{H}^{(L)}[N + 1, :]$
18: Apply dropout: $\mathbf{h}_{\text{CLS}}^{(L)} \leftarrow \text{Dropout}(\mathbf{h}_{\text{CLS}}^{(L)})$
19: Compute logits: $\mathbf{y}_{\text{logits}} \leftarrow \text{MLP}(\mathbf{h}_{\text{CLS}}^{(L)})$ $\triangleright$ Classification head
20: **return** $\mathbf{y}_{\text{logits}}$

---

learning of h_{CLS}^{0} in order to find the best starting position in the input latent space via backpropagation.

The batched inputs with the CLS node appended and their node features projected are then fed into the graph convolution section of the CGAT. This section consists of sequentially stacked CGAT blocks, where each block is made up of one GATv2 convolution layer, and one linear layer (Fig. 4). It is worth noting that the term *block* for the CGAT architecture corresponds to the term *layer* in traditional GNN models. Similar to the GAT architecture, we employ multi-head attention in the GATv2 convolutions throughout the graph convolution section with $K = 8$ attention heads. Unlike the GAT architecture, however, where individual attention head outputs are concatenated, or averaged for the final layer, we merge attention heads with max-pooling, where only the maximum value of each node feature is kept.

$$h_i^{(l+1)} = \text{LayerNorm}\left(h_i'^{(l)} + \sigma_2 \boldsymbol{W}_2^{(l)} \cdot h_i'^{(l)} \right)$$

$$h_i'^{(l)} = \max(k_{1:K}) \left[\text{LayerNorm}\left(h_i^{(l)} + \sigma_1 \sum_{j:j \in \mathcal{N}_i} \alpha_{ij}^{(l,k)} \cdot \boldsymbol{W}_1^{(l,k)} h_i^{(l)} \right) \right] \tag{4}$$

Equation 4 describes a single CGAT block, where $\boldsymbol{W}_1^{(l,k)} \in \mathbb{R}^{F \times F}$ is the shared weight matrix from Eq. 3 of head k of block l, $\boldsymbol{W}_2^{(l)} \in \mathbb{R}^{F \times F}$ is the linear transform applied after the GATv2 convolution of block l, $h_i'^{(l)}$ is the output of the graph convolution, and σ_1, σ_2 are the nonlinearity functions, selected as the GeLU nonlinearity by default. We include residual connections around the graph convolutions and linear layers in order to rectify the vanishing gradient problem. The structure of the CGAT block is inspired by the transformer architecture from [37], and is designed to increase the expressive capability of the CGAT model by introducing additional parameters. In this sense, a CGAT block can be viewed as a transformer layer, using dynamic attention from [3] instead of the scaled dot product attention used by Vaswani et al., and without the positional encoding step. The graph convolution section of the CGAT architecture is then built by sequential application of block $l \in L$ to the output of the previous section, as detailed in lines 9–13 in Algorithm 1.

The classification section of the CGAT model, which follows the graph convolution section, takes the learned CLS node embedding $h_{\text{CLS}}^{(l+1)}$ after l blocks as *the only* input, therefore ensuring whatever information the final decision is based on flows through the CLS node, thus making the CLS node a global information aggregator. We employ an MLP head for classification parametrized by $W_{\text{MLP}} \in \mathbb{R}^{F \times 5}$, mapping $h_{\text{CLS}}^{(l+1)}$ to class logits to be used in the loss calculation. We further apply the softmax function to the logits to get class probabilities.

$$\tilde{y} = \text{softmax}(h_{\text{CLS}}^{(l+1)} \cdot W_{\text{MLP}}) \tag{5}$$

Throughout the three sections described, the model learns to produce CLS node embeddings that are informative of the developmental stage of the teeth, based on the input node features and the graph connectivity. The general aim of the CGAT architecture is to produce human-understandable representations of attention while preserving the predictive power of the GAT family of GNNs. In order to generate such representations, we leverage the attention weights.

Attention Map Generation. We design CGAT so that the information is funneled through the CLS node to the classification section. This design decision is the key factor in the ability of the CGAT to produce attention maps, where the learned attention scores with the target node as v_{CLS} can be utilized as a means of analyzing the contribution of each node to the model decision. Thanks to the inherent attention mechanism of the GATv2 convolutions, the visualization of attention maps to represent said contribution becomes possible. In order to bring about visual depictions of the regions of attention, we employ the attention rollout technique [1]. The attention rollout method was first proposed to display the attention of transformers with many layers, and has also been employed to depict the learned attention of the ViT models on 2D images in [11]. The attention rollout method can be formulated as,

$$\tilde{A}^{(l)} = (A^{(l)} + I)\tilde{A}^{(l-1)}$$
$$\tilde{A}^{(0)} = A^{(0)} + I \tag{6}$$

where $\tilde{A}^{(l)}$ is the rollout and $A^{(l)} \in \mathbb{R}^{n+1 \times n+1}$ is a sparse matrix containing the learned attention scores corresponding to each edge in the graph, of layer l, and I is the identity matrix, modeling the residual connections. With this recursive operation, the "flow" of the attention from the CLS node to each input node can be represented holistically by visualizing only the values $a_{\text{CLS}\rightarrow i} : e_{i\rightarrow\text{CLS}} \in \tilde{\mathcal{E}}$ on the corresponding nodes in the input meshes. Since our application is a graph-level classification task, we do not analyze the attentions between the nodes of the input graph $a_{j\rightarrow i} : e_{i\rightarrow j} \in \mathcal{E} \cap \tilde{\mathcal{E}}$. These values are only used in the calculation of the attention rollout.

In order to visualize the attention in an understandable manner, we performed an upper value clipping on the attention maps, where the values are clipped to the range of $[min(\tilde{A}), mean(\tilde{A}) + min(\tilde{A})]$. To determine these limits, we use the concatenated rollout scores from all the samples visualized in order to preserve the relative scales.

Model Training and Evaluation. In order to rectify the effects of large class imbalance, we employed weighted sampling in batch creation so that each batch represented a class-wise balanced selection of samples. The training optimized for the cross-entropy loss. The Adam optimizer was selected for all trainings with an initial learning rate of 0.001. A learning rate scheduler with a patience of 5 epochs was used to reduce the learning rate by a factor of 0.5 if no improvement was observed in the validation loss within the patience range. All models were trained for 150 epochs. A dropout operation with $p = 0.3$ probability was applied on the CLS node embedding after the graph convolutional section in order to increase the robustness of training against overfitting.

For the evaluation of the model performance, following a multiclass classification, the weighted F1 score was selected. This metric is an effective descriptor of classification capabilities in the presence of high class imbalance. The weighted F1 score is defined as,

$$\text{Weighted F1} = \sum_C w_c \text{F1}_c = \sum_C w_c \left(\frac{\text{TP}_c}{\text{TP}_c + 1/2(\text{FP}_c + \text{FN}_c)} \right)$$
$$w_c = \frac{\|c\|}{\|C\|} \tag{7}$$

where TP_c, FN_c and FP_c stand for one-versus-rest true positives, false negatives and false positives, w_c signifies the class weight, or ratio of the number of samples to the total number of samples for class c. The weighted F1 score is a useful metric which summarizes the precision and recall values, with regard to class imbalance. In addition to the metric evaluation, we perform qualitative evaluations of the attention maps generated by the model. We regard the conformity of these attention maps to human understanding to be of higher priority than the classification metrics, in the context of this paper. Although the metric performance of the model is important, if the models do not show attention patterns that overlap with expert knowledge to some extent, then the question of trust in deep models remains unanswered.

As for the software and hardware used in the implementation of CGAT and the experiments, all training and evaluation was done using the PyTorch and PyTorch Geometric frameworks with GPU accelerated computing with NVIDIA CUDA, and an NVIDIA A100 GPU with 80 GB of memory was utilized for computations.

2.3   Node Features

The selection of node features, while important for the classification performance, is more so for the visualization of attention, and the generation of human-understandable representation of the attention. Since the only available information to CGAT are the node features $X = \{x_1, ...x_n, ..., x_N\}$, and the presence of edges, the final attentions are learned *only* with respect to these elements. In order to approximate the human understanding of 3-dimensional shapes in the model attention, the selected node features must also reflect the qualities of the shapes that an expert would focus on. In dental stage allocation within the given range of stages in our dataset (D to H), the focus of experts is on the development of the root(s), and the closing of the root apices (Fig. 1). In order to represent the shape information of these areas in the local neighborhood, we encode each node in the input graphs with shape-relevant features, namely the mean curvature and the distance to mesh centroid. These features are selected due to their pose-invariant quality, since the samples in the dataset are not in alignment, pose-dependent node features such as coordinates would lead models to depend on the position of the meshes in space, and therefore disregard shape information. Due to the alignment of varied shapes such as teeth being non-trivial, we favor rotation and translation-invariant learning via the employment of these features instead of performing mesh alignment. Furthermore, after the calculation of all features we apply min-max scaling across all samples so that all features are in the $[-1, 1]$ range while preserving the relative differences between the samples.

Mean Curvature. The mean curvature of a point on a surface is a function of the principal curvatures k_1 and k_2, and when calculated locally around each node is indicative of the concavity of the neighborhood. Mean curvature, thus, is a highly desirable descriptor for teeth, as the third molars tend to have less peaks due to the lack of the root apices in the early development, and a high mean curvature value in a node can be indicative of the development of the roots. Moreover, as the teeth get more elongated through subsequent stages, they contain more flat surfaces with very low curvedness, which again can be captured by the mean curvature feature. The difference in the distributions of the mean curvature feature encoded in the nodes of each label is shown in Fig. 5. These distributions show that the features change significantly for different stages, and display the necessary discriminative qualities.

Distance to Centroid. The Euclidean distances of each node to the mesh centroid can be informative towards the overall roundness and elongation of the meshes. In the earlier stages, the teeth are more rounded due to the lack of roots, and as roots form during dental development, the nodes in the root area get further away from the centroid. As the centroid shifts due to the elongation throughout the stages, the distances of the nodes remain informative of their relative positions in the mesh. We calculate this distance on the meshes after the preprocessing is carried out, with all meshes simplified and scaled into a unit sphere. Figure 5 shows the distributions of the normalized features from all nodes per class, which, similarly to the distributions of the mean curvature, differ across the developmental stages.

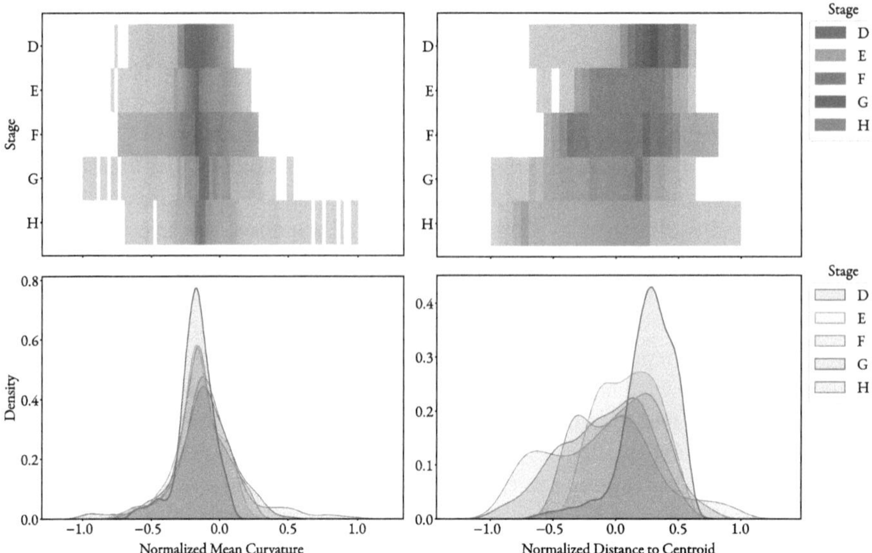

Fig. 5. Distributions of the normalized mean curvature feature for all nodes, separated by Demirjian stage. As dental development progresses from stage D to H, the distribution skews, showing an increase in nodes with high positive curvature. This corresponds to the formation of convex shapes like root apices. The clear separation of these distributions demonstrates that mean curvature is a highly discriminative feature, providing a strong signal for the classification task which can be leveraged by the CGAT.

Table 1. The metric results of the depth evaluation of CGAT variants. For each model family signifier, weighted F1 score is reported in the following row, where each columnar entry signifies the value of the metric corresponding to the model with the specified number of blocks. We evaluate models with 1 to 15 blocks, trained and tested using the mean curvature and the distance to centroid as node features, and another family of models using both these features. Each model is trained 10 times and the average value is reported. The standard deviations are not reported for the sake of brevity. The maximum value in each row, and values within the $\pm$ 0.01 range of it, are shown in bold for emphasis.

Model Family \ N.Layers	1	2	3	4	5	6	7	8	9	10	11	12	13	14	15
$^{1:15}\mathrm{CGAT}^{\leftrightarrow}_{curv}$	0.58	0.61	0.58	0.65	0.68	0.63	0.68	0.68	0.66	0.69	0.67	0.68	**0.71**	0.68	**0.72**
$^{1:15}\mathrm{CGAT}^{\leftrightarrow}_{dist}$	0.66	**0.70**	**0.71**	0.66	0.68	0.64	0.66	0.67	**0.70**	0.67	**0.70**	0.66	0.69	0.66	0.68
$^{1:15}\mathrm{CGAT}^{\leftrightarrow}_{both}$	0.68	0.69	0.68	0.69	0.70	0.71	0.73	0.71	0.74	**0.76**	0.70	**0.76**	0.71	0.70	0.68
$^{1:15}\mathrm{CGAT}^{\rightarrow}_{curv}$	0.62	0.60	0.62	0.62	0.66	0.68	0.65	0.72	0.69	0.69	**0.75**	0.71	0.64	0.60	0.66
$^{1:15}\mathrm{CGAT}^{\rightarrow}_{dist}$	0.67	**0.68**	**0.69**	**0.68**	0.67	**0.69**	0.67	0.66	0.67	0.63	0.66	**0.69**	**0.69**	**0.68**	0.65
$^{1:15}\mathrm{CGAT}^{\rightarrow}_{both}$	0.66	0.66	0.67	0.70	0.69	**0.76**	0.70	0.73	0.70	0.73	0.72	0.70	0.68	0.71	0.68

3 Experiments and Results

We devised several experiments in order to evaluate the attention maps and the predictive capability of the CGAT models, and demonstrate the effects of model depth and the direction of CLS edges. Throughout the rest of this paper, we refer to the models with the shorthand notation $^L\text{CGAT}_F^E$ for ease of discussion, where we denote the number of blocks $L \in \{1, 2, \cdots, 15\}$, the node features used $F \in \{curv, dist, both\}$ where *curv* stands for mean curvature, *dist* stands for the distance to centroid and *both* stands for a combination of both features and the type of the added edges that connect the CLS node to all edges $E \in \{\rightarrow, \leftrightarrow\}$ where $\rightarrow$ signifies directed edges from each node to the CLS node, and $\leftrightarrow$ indicates undirected edges, using superscripts and subscripts. As an example, $^3\text{CGAT}_{curv}^{\rightarrow}$ implies a model with 3 CGAT blocks, using mean curvatures as node features, connecting the CLS node to each node with directed edges $e_{i \rightarrow \text{CLS}}$.

For the experiments concerning the model depth and CLS edge direction, we compare the effects of the selected node features by training the models on mean curvature and the distance to centroid individually, and we also concatenate these features and train using the combined feature vectors. All the models, regardless of the other hyperparameters, have $K = 8$ attention heads. Figure 6 shows the weighted F1 score metrics for models with 1 to 15 blocks to demonstrate the effects of model depth. All models are trained using max-pooling to merge attention heads. Furthermore, we train all models with directed and undirected edges to the CLS node, and report on the resulting performance metrics. These results, displayed in Table 1 are discussed and the learned attention weights are depicted in the form of attention maps in the remainder of this section.

3.1 Model Depth and Edge Direction

GNNs are affected by the over-smoothing issue, where over multiple layers, the node features become more and more similar to each other due to continuous neighborhood aggregations. As a result, deep GNNs usually show a decaying performance as the number of layers increases [34]. It is therefore important to evaluate a number of CGAT models with varying number of layers. We train and compare the performances of models with the number of blocks increasing from 1 to 15 with unit step. We then demonstrate and discuss the effect of model depth on the generated attention maps, using a subset of this model pool to visualize learned attentions. Figure 6 shows the weighted F1 score metrics over the number of blocks. The direction of the edges used to connect the CLS node to all the other nodes in the input graphs has a direct impact on the learning process. If the edges are directed (from each node to the CLS node), the CLS node is allowed to attend to each node, and is only allowed to accumulate node features by the learned attention weights. Therefore, the embedding of the CLS node does not influence the node embeddings through aggregation. In this setting, the CLS node has an in-degree of $d_{\text{CLS}}^{in} = n$ and an out-degree of $d_{\text{CLS}}^{out} = 0$. However, if undirected edges are used, the learned embedding of the CLS node is also aggregated into each node along with the neighbors of the node, therefore, the node embeddings $h_i^{(l)}$ and $h_{\text{CLS}}^{(l)}$ both contribute to each other to learn an embedding that represents the stage information. Since the nodes in our input graphs have, on average, a degree of $\bar{d}_i^{in} = \bar{d}_i^{out} = 7$, the contribution of the CLS node in the aggregation step is significant compared to the contribution of the node embeddings to the CLS node, since $\bar{d}_i^{in} << \bar{d}_{\text{CLS}}^{in} = 751$. Furthermore, by using undirected CLS edges, all nodes in the graph are made to be in the 2-hop neighborhood of each other. This connectivity scheme, when 2 or more

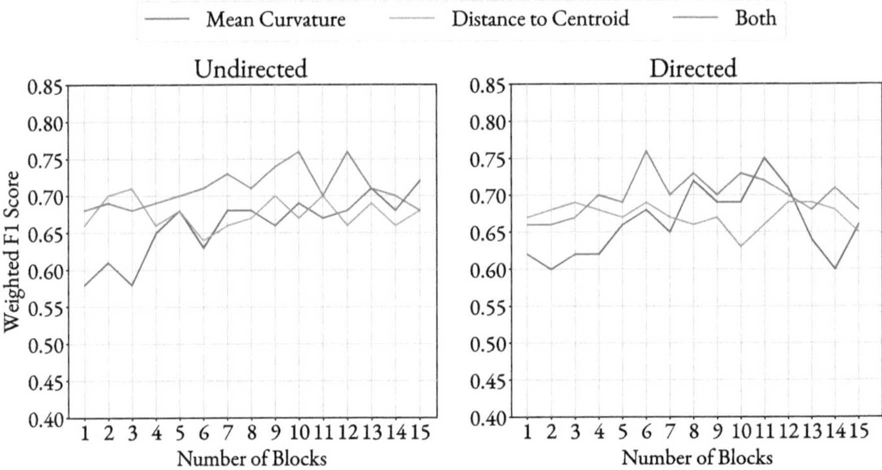

Fig. 6. Plots of the weighted F1 scores of the models with the number of blocks ranging from 1 to 15, trained using the mean curvature, the distance to centroid and both features combined. Overall, the CGAT model performance does not deteriorate significantly in the given range of blocks, and the effects of over-smoothing are not observed. There is a slight upward trend observed in the metrics with undirected edges, and with directed edges the performance metrics are mostly stable across differing number of blocks.

blocks are implemented, allows the information of each node to influence all other nodes, regardless of mesh topology and graph distance. Due to these fundamental differences brought on by the direction of the edges, it is important to analyze the effects on predictive performance and the attention maps.

It can be seen from the results that the CGAT shows a slight increase in performance as the model depth is increased from 1 to 15 when undirected CLS edges $e_{i \leftrightarrow CLS}$ are used. With the $^{1:15}\text{CGAT}^{\leftrightarrow}_{curv}$ family of models, the best mean weighted F1 scores are encountered with 13 and 15 blocks. A similar case is also seen with the models trained with both features, where $^{10,12}\text{CGAT}^{\leftrightarrow}_{both}$ show the best weighted F1 metrics, and F1 scores above 0.7 are seen with block numbers 4 through 14. As for the models with undirected edges using the distance to centroid as node features, a uniform performance profile can be observed where the variation of the mean metrics remains low across the number of blocks. These results indicate that the models with undirected edges did not suffer from performance loss due to oversmoothing in the given range of blocks. Contrastingly, $^{1:15}\text{CGAT}^{\leftrightarrow}_{curv}$ family of models consistently show a worse performance with $^{1:3}\text{CGAT}^{\leftrightarrow}_{curv}$ displaying the worst classification metrics, and are seen to benefit from an increased number of blocks, resulting in a larger receptive field for each node. Given all these observations, the best operating range for models with undirected edges can be said to be with 9 blocks and more.

When directed edges $e_{i \rightarrow CLS}$ are used to connect the CLS edges, a roughly stable trend in the metrics can be observed as the number of blocks is increased. The CGAT models trained with the mean curvature deviate from this observation the most, as the models $^{1:11}\text{CGAT}_{curv}^{\rightarrow}$ seem to benefit from the increased depth, but the performance starts deteriorating with more depth, with $^{14}\text{CGAT}_{curv}^{\rightarrow}$ showing the same mean weighted accuracy as $^{2}\text{CGAT}_{curv}^{\rightarrow}$, hinting at the initial symptoms of oversmoothing. The model family which uses both features, $^{1:15}\text{CGAT}_{both}^{\rightarrow}$ also mimics this trend, albeit less severely. The metrics increase for these models up until 6 blocks, after which a plateau in performance can be observed. As for the models trained with distance to centroid, we see a monotonic performance profile, with models yielding mean weighted F1 scores that are all in the $[0.63, 0.69]$ range, with the highest performances reported with shallower models as well as deeper ones. As per the results of the depth experiments, we suggest an operating range of 6 to 12 blocks, regardless of the node feature used. For both types of CLS edges, using both features together, instead of individually, shows better performance in the given operating ranges.

3.2 Attention Maps

Concerning the effects of number of blocks and the direction of the CLS edges on the attention maps, it is best to perform a qualitative comparison of the localization of attention for different model depths. For this comparison the models $^{1,7,14}\text{CGAT}_{F}^{\{\leftrightarrow,\rightarrow\}}$, $F \in \{curv, dist, both\}$ are selected, and the attention rollouts of these models for a sample tooth are visualized and directly compared in Fig. 7.

The attention maps for $^{1,7,14}\text{CGAT}^{\{\leftrightarrow,\rightarrow\}}$ models, displayed in Fig. 7, reveal the learned attention weights from each node to the CLS node. An immediate observation can be made on the difference of the attention maps of models with 1 block, and those with multiple blocks, with the former showing a patchy and noisy distribution, and examples of the latter being of smoother nature. This difference stems from neighborhood aggregations. With only one graph convolution operation, the attention weights of models with 1 layer are based on the initial embedding of the node features, and neighborhood aggregation does not affect the learned attention. As a result, due to the attention rollout of the first block being the attention weights of the block, the nodes show a more noisy contribution to the CLS embedding.

Another observation that can be drawn from Fig. 7 is that when the models get deeper, regardless of the feature used, the attention maps focus on the relevant anatomical regions such as the roots and root apices, the crown of the tooth, and the furcation point of the roots. This attention pattern generally aligns with the anatomical criteria shown in Fig. 1. One interesting aspect is that the attention seems to focus more on a single area in the deeper model with 14 blocks when undirected edges ($\leftrightarrow$) are used. This effect can be explained by the receptive field of each node. More specifically, with the deeper model, all nodes are informed by their 14-hop neighbors, leading to the gathering of the information in one area through weighted aggregation. This effect is much more pronounced for the $\text{CGAT}^{\leftrightarrow}$ than for $\text{CGAT}^{\rightarrow}$, where the attentions are spread out over several regions, due to each node being a 2-hop neighbor of every other node through the CLS node. When the CLS edges are directed, this effect is somewhat rectified, as the 14-hop neighborhood is locally limited to the mesh topology instead, and as a result the attention does not focus on one single area. It can be seen that the models using directed edges ($\rightarrow$) focus clearly on the aforementioned anatomical regions separately, even with the deepest model depicted in Fig. 7. This effect can reduce the

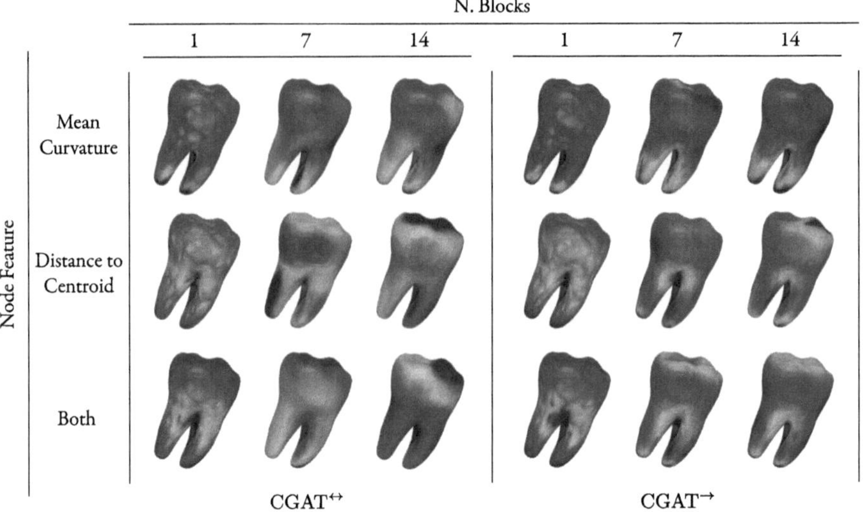

Fig. 7. The attention maps from the ^{1,7,14}CGAT$^{\{\leftrightarrow,\rightarrow\}}$ family of the tested CGAT models, visualized on a sample tooth mesh of stage H. Across both undirected ($\leftrightarrow$) and directed ($\rightarrow$) CLS edges, 1-block models exhibit noisy and mosaic attention, which becomes smoother and more focused in deeper models. A key difference emerges in deeper models: undirected edges tend to concentrate attention on a single anatomical region, while directed edges maintain a more distributed focus on multiple relevant areas like the root apices and furcation. This suggests that directed edges may yield more intuitive explanations, especially for deeper architectures.

transparency of the deeper CGAT$^{\leftrightarrow}$ models, as the focused attentions of the deep models seem to deviate from the human understanding faster than the attentions of their directed counterparts.

The attention maps directly reflect the effects of the chosen node features on the CLS node embedding. Because of this, the impact of the feature selection on the attention maps is highly non-trivial. This can be observed with the different regions of the sample tooth shown in Fig. 7. With the mean curvature, regardless of the CLS edge direction, the model places more emphasis on the root apices and the root furcation, which are areas of very high convexity and concavity. With the distance to centroid feature, these regions are still attended to, but the center region where the crown and root sections join as well as the top surface of the crown show higher attentions. As the distance to the centroid is much smaller in the middle section of the teeth with elongation, the effect of the feature in this region is greater compared to the mean curvature in the same area. When the two features are combined, it can be seen that the regions of focus of models with individual singular features are both represented in the final attention. This indicates that the CGAT indeed learned the effects of both features, and was able to leverage both features in the attention mechanisms, as opposed to relying on either one only. This is also reflected in the results depicted in Fig. 6 and Table 1, as the best classification performance was shown by the models using both features. This observation also corresponds with conventional clinical assessment, where both features are taken into account simultaneously by the human observer. Overall, the effects of the CLS edge direction seem to be more pronounced on the attention maps, compared to the effects of model depth. With a global 2-hop neighborhood enabled by undirected edges to the CLS node, in combination with the larger receptive field provided by deeper models, the attention maps seem to rely on singular regions rather than multiple relevant ones. For this, based on our observations we can state that using directed edges is more beneficial with regard to the interpretability and understandability of the attention maps, especially when the difference in the predictive performance of the two approaches is minimal. We also argue that the use of multiple features does not obscure the effects of the individual features in the final decision, and instead highlights the regions of interest covered by each feature, albeit to varying degrees. This aspect of the CGAT can help bring the explanations of the models by the way of attention maps closer to the attention patterns of human experts.

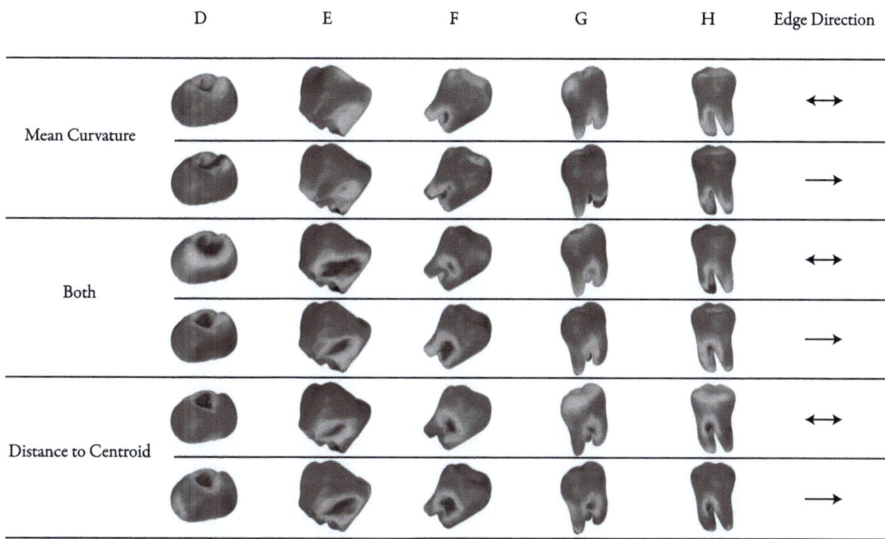

Fig. 8. The attention maps of ^{11}CGAT$^{\leftrightarrow,\rightarrow}$ for one sample tooth from all stages. At the earlier stages, the attentions are focused on the crown and the center regions of the teeth, and as the stages progress, the attention on the furcations and the roots also increase. This attention pattern is in line with the Demirjian criteria for stage assessment from Fig. 1.

It is also interesting to inspect how the attention evolves throughout the developmental stages. For this, we inspect the attention rollouts of ^{11}CGAT$^{\leftrightarrow}$ and ^{11}CGAT$^{\rightarrow}$ models, selected due to the metrics observed for these models in Fig. 6 indicating that this is the optimal depth for our task. An overview of the rollouts is depicted in Fig. 8. In this figure, we can see an overview of the attention rollouts of the selected models. An interesting pattern of attention can be seen across all features and edge directions; across different features and edge directions, the final localizations of attentions always have significant overlap. In stage D for example, the concavity of the occlusal surface is attended to in all models. Similarly, for the later stages, the furcation point of the roots is always considered. These results indicate that the CGAT models focus on the discriminative regions of the teeth regardless of the feature and edge direction used, since the local shape information of these regions describe the stage best. The effects of different features, then, are the amount of focus on these high-informative regions. For example, the mean curvature feature yields attentions to the crown, the furcation and the root apices in stages E and F, while the distance to centroid feature results in the attention mostly focusing to the root furcation. Similarly, for stages G and H, the mean curvature pronounces the root apices more heavily than the distance to centroid feature. We conclude that these differences stem from the distributions of the features across nodes, shown in Fig. 5. The magnitudes of these features end up being discriminative of the developmental stage in different ranges, and as such the CGAT models learn different levels of attentions relating to them. Once again we can observe the combination of the two features in training results in attention maps that are a combination of the individually learned attentions. This effect can clearly be seen in the example for stage F. The $^{11}\text{CGAT}_{curv}^{\rightarrow}$ highlights the roots of the tooth, while $^{11}\text{CGAT}_{dist}^{\rightarrow}$ implies the beginning of the roots to be informative, and both these areas are highlighted in the attention rollout of $^{11}\text{CGAT}_{both}^{\rightarrow}$. Interestingly, this effect seems to be more overtly displayed for the models using directed edges to the CLS node. This once again indicates using undirected edges can result in counter-intuitive attention maps, an outcome we attribute to the global 2-hop neighboring effect.

Overall, we conclude the CGAT architecture does indeed result in attention maps that align with the human understanding in the context of dental development, and consequently, of dental stage assessment, where the focus shifts from the crown, over the furcation, to the roots and finally the root apices.

3.3 Performance Comparison

To rigorously evaluate our proposed architecture, we compare its performance against several established baselines for 3D shape analysis, with the results detailed in Table 2. Our proposed CGAT models demonstrate superior performance, achieving a weighted F1-Score of 0.76 and a Mean Absolute Error (MAE) of 0.25, outperforming the strongest baseline, GAT (0.67 F1-Score, 0.34 MAE). This reduction in MAE is particularly significant for an ordinal task such as dental stage assessment, as it indicates that the model's prediction errors are of a lower magnitude. Furthermore, despite a higher parameter count, CGAT remains highly efficient with inference times (7–12 ms) comparable to GAT (9 ms) and significantly faster than the computationally intensive MeshCNN (275 ms) and PointNet++ (144 ms) models.

These performance differences are rooted in the models' distinct architectural philosophies. MeshCNN operates directly on mesh topology, using task-driven edge collapse for pooling, while PointNet++ builds a hierarchy on the metric space of an unstructured point cloud via farthest point sampling. Our CGAT, and the other graph-based approaches provide a balance, leveraging topological connectivity in an abstract form. Within the graph-based methods, the performance increase over GAT can be attributed to the architectural contribution of the CLS node, which acts as a global information aggregator to create a comprehensive graph-level embedding. Ultimately, while

Table 2. Performance comparison of CGAT against baseline models. We report the weighted Precision, Recall, and F1-Score, along with the Mean Absolute Error (MAE). For our proposed CGAT, we show the two best-performing variants for both directed ($\rightarrow$) and undirected ($\leftrightarrow$) CLS node edges.

Model	Precision	Recall	Weighted F1 Score	MAE	# of Parameters	Inference Time (ms)
Baseline Models						
PointNet++ [31]	0.60	0.63	0.60	0.48	1466950	144
MeshCNN [19]	0.57	0.67	0.61	0.44	981405	275
GCN [22]	0.66	0.65	0.64	0.39	215301	4
GAT [38]	0.69	0.65	0.67	0.34	1981765	9
Our Proposed Models (from the CGAT_{both} **family)**						
$^6\text{CGAT}_{both}^{\rightarrow}$	0.77	0.76	0.76	0.25	2499207	7
$^{10}\text{CGAT}_{both}^{\rightarrow}$	0.72	0.75	0.73	0.25	4153479	9
$^{10}\text{CGAT}_{both}^{\leftrightarrow}$	0.84	0.77	0.76	0.25	4153479	10
$^{12}\text{CGAT}_{both}^{\leftrightarrow}$	0.79	0.78	0.76	0.27	4980615	12

the predictive metrics are strong, the primary advantage of CGAT remains its unique synthesis of state-of-the-art performance, computational efficiency, and the inherent ability to produce intuitive, attention-based explanations—a capability the baseline models inherently lack.

4   Conclusion

The automation of dental stage assessment is desirable in order to eliminate the inter-observer variability, and to expedite the process. Deep learning models are the prime candidate in such applications due to their capabilities of learning complex decision patterns from the data. However, due to the high-stakes nature of dental staging in the forensic context, the opaque nature of deep learning models discourages their widespread use. In order to address this need, we introduced the CGAT architecture, which was applied to meshes of third molars derived from CBCT to determine the Demirjian stages. The CGAT model was based on the graph attention convolutions, and with its inherent attention mechanism can produce attention maps. These attention maps are produced via the attention rollout method, and are able to depict the magnitude of impact each node has on the final decision of the model, facilitated by the use of the CLS node as a global information aggregator. We evaluated the effects of different node features and the depth of the CGAT architecture on the classification performance, and the final visualization of attention maps. While all model variants were able to produce plausible attention maps, we argue that the models using directed edges to connect each node in the input meshes to the CLS node were clearer and more intuitive in their attention maps. The local mean curvature and the distance to centroid were both able to represent the local shape, from which the CGAT models were able to learn the shape structure. The combination of the two features in training not only slightly outperformed the models using individual features, but also incorporated elements from the attentions of all features into the final attention maps, which corresponds with the human expert's train of thought during conventional stage assessment. Therefore, we propose the use of $\text{CGAT}^{\rightarrow}_{both}$ as the best option to learn human-understandable attention maps.

Through the use of attention-based explanations produced by the CGAT, models can be judged with regard to their decision factors based on the attention patterns. This aspect of the CGAT architecture can increase the trust in the models and allow experts to accept or deny the model decisions. The CGAT architecture, although demonstrated on teeth meshes, is applicable to all graph data, and can perform any graph-level classification task, or regression task with small modifications to the classification section, and explain its decisions through generating attention maps. We conclude that the benefits of an architecture with increased decision transparency which can perform competitively will allow a more widespread adoption of such models with increased confidence in high-stakes settings.

5 Future Work

In order to further analyze the explanation capabilities of the CGAT architecture, we argue that the merging function of the attention heads is an important avenue to explore. Many permutation-invariant merging functions exist, such as the mean and concatenation, aside from the max-pooling that we employed. Since these functions directly influence the models during training, and with that the learned attention scores, the CGAT would benefit from the exploration of this design choice. Additionally, although we only demonstrate the CGAT on dental stage classification on meshes, it is not limited to such an application only, and can be used in any graph-level task. Further applications of the CGAT models on different mesh datasets where shape information is critical, and on different graph-definable data domains such as molecular interactions or image classification, can reveal more insight into the attention-based explanations generated by CGAT.

Disclosure of Interests. The authors declare no conflict of interest.

References

1. Abnar, S., Zuidema, W.: Quantifying attention flow in transformers. arXiv preprint arXiv:2005.00928 (2020)
2. Besson, P., Parrish, T., Katsaggelos, A.K., Bandt, S.K.: Geometric deep learning on brain shape predicts sex and age. Comput. Med. Imaging Graph. **91**, 101939 (2021)
3. Brody, S., Alon, U., Yahav, E.: How attentive are graph attention networks? arXiv preprint arXiv:2105.14491 (2021)
4. Brown, J., et al.: Basic training requirements for the use of dental cbct by dentists: a position paper prepared by the european academy of dentomaxillofacial radiology. Dentomaxillofacial Radiol. **43**(1), 20130291 (2014)
5. Cao, W., Yan, Z., He, Z., He, Z.: A comprehensive survey on geometric deep learning. IEEE Access **8**, 35929–35949 (2020)

6. Chen, S., Duan, C., Yang, Y., Li, D., Feng, C., Tian, D.: Deep unsupervised learning of 3d point clouds via graph topology inference and filtering. IEEE Trans. Image Process. **29**, 3183–3198 (2019)
7. De Tobel, J., et al.: Dental and skeletal imaging in forensic age estimation: disparities in current approaches and the continuing search for optimization. In: Seminars in Musculoskeletal Radiology, vol. 24, pp. 510–522. Thieme Medical Publishers (2020)
8. Demirjian, A., Goldstein, H., Tanner, J.M.: A new system of dental age assessment. Human biology, pp. 211–227 (1973)
9. Devlin, J., Chang, M.W., Lee, K., Toutanova, K.: Bert: pre-training of deep bidirectional transformers for language understanding. In: Proceedings of the 2019 Conference of the North American Chapter of the Association for Computational Linguistics: Human Language Technologies, Volume 1 (long and short papers), pp. 4171–4186 (2019)
10. Dhar, T., Dey, N., Borra, S., Sherratt, R.S.: Challenges of deep learning in medical image analysis–improving explainability and trust. IEEE Trans. Technol. Soc. **4**(1), 68–75 (2023)
11. Dosovitskiy, A., et al.: An image is worth 16x16 words: transformers for image recognition at scale. arXiv preprint arXiv:2010.11929 (2020)
12. Facchi, G.M., et al.: Graph neural networks for 3d facial morphology: assessing the effectiveness of anthropometric and automated landmark detection. Pattern Recogn. Lett. (2025)
13. Franco, A., Vetter, F., Coimbra, E.d.F., Fernandes, Â., Thevissen, P.: Comparing third molar root development staging in panoramic radiography, extracted teeth, and cone beam computed tomography. Int. J. Legal Med. **134**(1), 347–353 (2020)
14. Gezawa, A.S., Zhang, Y., Wang, Q., Yunqi, L.: A review on deep learning approaches for 3d data representations in retrieval and classifications. IEEE Access **8**, 57566–57593 (2020)
15. Gilmer, J., Schoenholz, S.S., Riley, P.F., Vinyals, O., Dahl, G.E.: Neural message passing for quantum chemistry. In: International conference on Machine Learning, pp. 1263–1272. PMLR (2017)
16. Glorot, X., Bengio, Y.: Understanding the difficulty of training deep feedforward neural networks. In: Teh, Y.W., Titterington, M. (eds.) Proceedings of the Thirteenth International Conference on Artificial Intelligence and Statistics. Proceedings of Machine Learning Research, vol. 9, pp. 249–256. PMLR, Chia Laguna Resort, Sardinia, Italy, 13–15 May 2010
17. Gong, S., Chen, L., Bronstein, M., Zafeiriou, S.: Spiralnet++: a fast and highly efficient mesh convolution operator. In: Proceedings of the IEEE/CVF International Conference on Computer Vision Workshops (2019)
18. Hamilton, W., Ying, Z., Leskovec, J.: Inductive representation learning on large graphs. In: Advances in Neural Information Processing Systems, vol. 30 (2017)
19. Hanocka, R., Hertz, A., Fish, N., Giryes, R., Fleishman, S., Cohen-Or, D.: Meshcnn: a network with an edge. ACM Trans. Graph. (ToG) **38**(4), 1–12 (2019)
20. Harris, E.F., et al.: Tooth-coding systems in the clinical dental setting. Dental Anthropol. J. **18**(2), 43–49 (2005)
21. Hassanin, M., Anwar, S., Radwan, I., Khan, F.S., Mian, A.: Visual attention methods in deep learning: an in-depth survey. Inf. Fusion **108**, 102417 (2024)
22. Kipf, T.N., Welling, M.: Semi-supervised classification with graph convolutional networks. arXiv preprint arXiv:1609.02907 (2016)
23. Liu, X., Yuan, L., Jiang, C., JiannanYu, Li, Y.: Human identification using tooth based on pointnet++. In: Chinese Conference on Biometric Recognition, pp. 129–139. Springer (2023)
24. Mahdi, S.S., et al.: Multi-scale part-based syndrome classification of 3d facial images. IEEE Access **10**, 23450–23462 (2022)
25. Matthijs, L., et al.: Artificial intelligence and dental age estimation: development and validation of an automated stage allocation technique on all mandibular tooth types in panoramic radiographs. Int. J. Legal Med. **138**(6), 2469–2479 (2024)

26. Minaee, S., Abdolrashidi, A., Su, H., Bennamoun, M., Zhang, D.: Biometrics recognition using deep learning: a survey. Artif. Intell. Rev. **56**(8), 8647–8695 (2023)
27. Mohammadi, S., Allali, M.: Advancing brain tumor segmentation with spectral-spatial graph neural networks. Appl. Sci. **14**(8), 3424 (2024)
28. National Science and Technology Council (US). Select Committee on Artificial Intelligence: The National Artificial Intelligence Research and Development Strategic Plan: 2023 Update. National Science and Technology Council (US), Select Committee on Artificial . . . (2019)
29. Piccialli, F., Di Somma, V., Giampaolo, F., Cuomo, S., Fortino, G.: A survey on deep learning in medicine: why, how and when? Inf. Fusion **66**, 111–137 (2021)
30. Qi, C.R., Su, H., Mo, K., Guibas, L.J.: Pointnet: deep learning on point sets for 3d classification and segmentation. In: Proceedings of the IEEE Conference on Computer Vision and Pattern Recognition, pp. 652–660 (2017)
31. Qi, C.R., Yi, L., Su, H., Guibas, L.J.: Pointnet++: deep hierarchical feature learning on point sets in a metric space. In: Advances in Neural Information Processing Systems, vol. 30 (2017)
32. Rahim, A.A., Davies, J., Liversidge, H.: Reliability and limitations of permanent tooth staging techniques. Forensic Sci. Int. **346**, 111654 (2023)
33. Relu BV: Relu Creator. https://www.relu.eu/creator (2022–2025)
34. Rusch, T.K., Bronstein, M.M., Mishra, S.: A survey on oversmoothing in graph neural networks. arXiv preprint arXiv:2303.10993 (2023)
35. Sartor, G., Lagioia, F.: The impact of the general data protection regulation (gdpr) on artificial intelligence. In: Panel for the Future of Science and Technology (2020)
36. Shehata, N., Bain, W., Glocker, B.: A comparative study of graph neural networks for shape classification in neuroimaging. In: Geometric Deep Learning in Medical Image Analysis, pp. 160–171. PMLR (2022)
37. Vaswani, A., et al.: Attention is all you need. In: Advances in Neural Information Processing Systems, vol. 30 (2017)
38. Velickovic, P., Cucurull, G., Casanova, A., Romero, A., Lio, P., Bengio, Y., et al.: Graph attention networks. Stat **1050**(20), 10–48550 (2017)
39. Vosylius, V., et al.: Geometric deep learning for post-menstrual age prediction based on the neonatal white matter cortical surface. In: Uncertainty for Safe Utilization of Machine Learning in Medical Imaging, and Graphs in Biomedical Image Analysis: Second International Workshop, UNSURE 2020, and Third International Workshop, GRAIL 2020, Held in Conjunction with MICCAI 2020, Lima, Peru, 8 October 2020, Proceedings 2, pp. 174–186. Springer (2020)
40. Vranckx, M., Fieuws, S., Jacobs, R., Politis, C.: Prophylactic vs. symptomatic third molar removal: effects on patient postoperative morbidity. J. Evidence Based Dental Pract. **21**(3), 101582 (2021)
41. Yang, X., Li, R., Yang, X., Zhou, Y., Liu, Y., Han, J.D.J.: Coordinate-wise monotonic transformations enable privacy-preserving age estimation with 3d face point cloud. Sci. China Life Sci. **67**(7), 1489–1501 (2024)
42. Yang, Z., et al.: Pre-training graph attention convolution for brain structural imaging biomarker analysis and its application to alzheimer's disease pathology identification. In: 2024 IEEE International Symposium on Biomedical Imaging (ISBI), pp. 1–5. IEEE (2024)
43. Ying, C., et al.: Do transformers really perform badly for graph representation? Adv. Neural. Inf. Process. Syst. **34**, 28877–28888 (2021)

Extensibility, Model Interpretability and Explainability, and Automation in ML.NET: A Comprehensive Analysis

Robin Nunkesser[(✉)]

Hamm-Lippstadt University of Applied Sciences, Marker Allee 76–78, 59063 Hamm, Germany
robin.nunkesser@hshl.de

Abstract. This paper presents an in-depth analysis of the extensibility, model interpretability and explainability, and automation capabilities of ML.NET, Microsoft's open-source machine learning framework for .NET developers. We identify and address the challenges faced by third-party developers, particularly due to ML.NET's restricted internal APIs and reliance on friend assemblies. We propose practical approaches for implementing custom estimators and evaluation metrics, enabling greater flexibility for external contributors. Furthermore, we examine the interpretability and explainability of models produced by ML.NET and its AutoML component, demonstrating both black-box and white-box strategies. Finally, we introduce an automated methodology for fair benchmarking and comparison of ML.NET's AutoML results with alternative frameworks, leveraging JSON-based configuration and reproducible evaluation pipelines. All proposed methods are supported by open-source code and validated through experiments on standard benchmark datasets. Our findings highlight both the strengths and current limitations of ML.NET, providing actionable guidance for practitioners and researchers seeking to extend and analyze machine learning workflows within the .NET ecosystem.

Keywords: Machine Learning · Classification · Regression · Interpretability · Explainability · ML.NET

1 Introduction

Microsoft's ML.NET framework [1] is a powerful tool for building machine learning applications. It provides a wide range of algorithms and tools for data preprocessing, model training, and evaluation. Although it is an open source project, a lot of functionality is deliberately only accessible to friend assemblies. This means that third-party developers can only use and extend the framework in a limited way. This limits the possibilities to extend the framework and to interpret or explain the results of the trained models.

Automated Machine Learning (AutoML) is a feature of ML.NET that automates the process of selecting the best machine learning algorithm and hyperparameters for a given dataset. It is designed to make it easier for developers to build machine learning applications without having to manually select and tune algorithms. However,

F. Marcelloni et al. (Eds.): IJCCI 2025, CCIS 2829, pp. 177–196, 2026.
https://doi.org/10.1007/978-3-032-15638-9_11

AutoML poses some additional challenges for interpretability, explainability and external automation.

This paper presents an approach to extend ML.NET in a way that allows third-party developers to implement custom algorithms and to use them in the framework. The approach is based on the `CustomMapping` functionality of ML.NET, which allows developers to implement custom data transformations. We show how to use this functionality to implement custom estimators.

This paper also discusses the challenges of interpreting and explaining the results of ML.NET and ML.NET's AutoML and additionally presents an approach to automate the comparison of AutoML results with other algorithms. The approach is based on the JSON configuration files used by AutoML, which can be generated and used to run AutoML in a reproducible way.

For each of the areas examined in this article, we have developed reusable code components that are suitable for integration into a library. The resulting library is available as open-source software[1] and can be accessed as a NuGet package[2].

The paper is structured as follows: Sect. 2 discusses related work. In Sect. 3, we give an overview of ML.NET and its features and limitations. Section 4 addresses the challenges associated with extending ML.NET and introduces our methodology for implementing custom estimators. Section 5 examines the interpretability and explainability of models produced by ML.NET's AutoML, outlining the specific obstacles encountered. Section 6 explores the analysis of ML.NET's AutoML results and details our approach for automating the comparison with alternative algorithms. Finally, in Sect. 7, we conclude the paper and discuss future work.

2   Related Work

Despite its relevance in the .NET ecosystem, ML.NET and its AutoML capabilities are largely absent from the academic literature. Comprehensive surveys and comparative studies of AutoML tools and frameworks, such as [9–11], do not include ML.NET in their analyses.

One possible explanation for this observed gap is that much of the research related to ML.NET appears to be conducted internally at Microsoft and Microsoft Research. For example, Psallidas et al. [19] note that their work was initially based on internal research at Microsoft and was only published after its "significant internal impact" became evident. Their study highlights the current dominance of Python in the machine learning community, while also emphasizing the importance of .NET in enterprise environments. The widespread adoption of .NET in the enterprise sector is a key reason why ML.NET remains a relevant framework for machine learning applications.

Ahmed et al. [1] give an overview of ML.NET and its features, but do not discuss the challenges of extending the framework or interpreting and explaining the results of the trained models.

---

[1] https://github.com/Italbytz/nuget-ml.
[2] https://www.nuget.org/packages/Italbytz.ML.

Papers using ML.NET for specific applications like Nematzadeh et al. [18] and Torkamanian-Afshar et al. [26] do not discuss the challenges of extending the framework or interpreting and explaining the results of the trained models either.

3   ML.NET

ML.NET is an open-source machine learning framework for .NET developers.

3.1   Overview

ML.NET provides a wide range of tools and libraries for building machine learning applications, including data preprocessing, model training, and evaluation. ML.NET supports various machine learning tasks, such as classification, regression, clustering, and anomaly detection. A typical ML.NET workflow consists of the following steps:

- *Data loading:* Load data from various sources.
- *Data preprocessing:* Clean and transform the data using various techniques.
- *Model training:* Train a model using one of the available algorithms.
- *Model evaluation:* Evaluate the performance of the trained model.
- *Model deployment:* Deploy the trained model to a production environment.

Microsoft's ML.NET uses the concept of *pipelines* and *trainers*. A trainer is a combination of a parameterized algorithm and a task and typically central part of a pipeline with additional pre- and postprocessing of the data.

3.2   Relevance

ML.NET, while not extensively covered in academic literature, plays a significant role within the .NET ecosystem. Although Microsoft's AutoML functionality is not broadly recognized in the research community, it provides robust capabilities for automating algorithm and hyperparameter selection, thereby streamlining the development of machine learning applications for practitioners. ML.NET is widely adopted in enterprise environments and benefits from an active community of developers and users. In June 2025, ML.NET had over 9.9M downloads of the NuGet package and 9.2k stars and 225 contributors on GitHub.

ML.NET also underpins several Microsoft products. In comparison, Microsoft's Azure Machine Learning focuses on cloud-based solutions and is not open source. Consequently, ML.NET represents a compelling option for .NET developers seeking to implement machine learning workflows without dependence on commercial services.

3.3   Source Code

Although ML.NET is open source[3], the extensibility is deliberately limited by strongly using the concept of friend assemblies, limiting the access to certain important internal APIs for third-party developers. Many Microsoft projects are Commercial Open

---

[3] https://github.com/dotnet/machinelearning.

Source [21], and while there is no apparent direct connection to commercial products in ML.NET, the source code contains strong hints that it was programmed for Commercial Open Source. Additionaly, there is an indirect connection to the commercial Azure Machine Learning that used NimbusML which in turn uses ML.NET. The use of the code by third-party developers is limited to the public API, which is often not sufficient to implement custom algorithms and to achieve full interpretability or explainability of the results.

It is possible to participate in the development of ML.NET by submitting issues and feature requests and by submitting pull requests. However, many developers would like to extend and fully use the framework without having to participate in the development process. Third-party developers and researchers can use ML.NET to build machine learning applications, but they are limited in their ability to extend the framework. Many of the underlying algorithms return interpretable or explainable models, but it is not possible to gain insight in a straightforward way.

This is why we will take a closer look at the extensibility of ML.NET and the interpretability and explainability of returned models in the following sections.

4 Extending ML.NET

Typical extensions third-party developers would like to make focus on preprocessing, training, and evaluation.

4.1 Extending Transformations

Extending preprocessing, postprocessing, and training in ML.NET presents significant challenges; however, these components share a unified extension mechanism. Specifically, each is implemented via classes adhering to the `IEstimator` interface, which

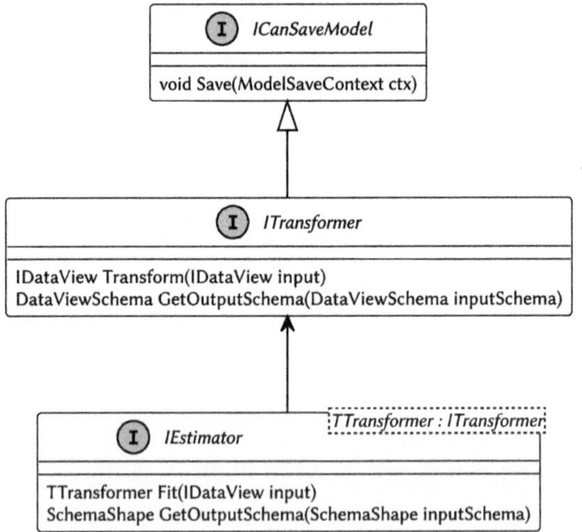

Fig. 1. Minimum interfaces to implement for a custom estimator in ML.NET.

are composed into an `EstimatorChain`—commonly referred to as a pipeline. The essential interfaces required for such extensions are depicted in Fig. 1.

However, as easy as this may seem at first glance, it is not. Some of the problems developers will encounter are the following:

- Model saving and loading is not easily possible for custom estimators. The class `ModelSaveContext` only provides necessary functionality for friend assemblies.
- Constructing correct output schemas from input schemas is not easily possible for similar reasons.
- Developing an implementation that operates robustly across arbitrary input data and diverse preprocessing steps within the pipeline presents significant challenges.

The only documented official way to write custom transformations without pull requests is to implement a `CustomMapping`[4]. However, this in itself does not solve the mentioned problems.

The best solutions for these problems with a balanced cost-benefit ratio from our point of view are described in the following subsections. The underlying idea is to implement a custom `IEstimator` and to use the `CustomMapping` functionality to implement the necessary functionality. This way, the implementation is not limited to friend assemblies and can be used in any ML.NET project. We also gain the necessary flexibility to implement the functionality we need in comparison to using the `CustomMapping` functionality directly. With regard to the problems mentioned above, we can solve them as follows:

Model Saving and Loading. It does not seem to be possible to save and load complex custom models with the official API with reasonable effort in ML.NET. It is therefore recommended to implement custom model saving and loading.

Constructing Output Schemas. The `IEstimator` interface requires the implementation of a `GetOutputSchema` method. This method is used to construct the output schema from the input schema. As we will use a `CustomMapping` as the core of our implementation, we will be able to use the `GetOutputSchema` method of the `CustomMapping` class for our custom estimator.

Implementing Generic Custom Estimators. The estimators provided by ML.NET rely on data preprocessing and postprocessing to work generically. They expect a feature column and a label column of a specific format to be present in the input data.

Custom estimators may therefore use the same approach and work with the same pipeline. There are only a few notable exceptions, which are discussed later.

The general approach of using ML.NET's `CustomMappingEstimator` to build custom trainers can be achieved with the code snippet shown in the following. The code snippet shows the implementation of a custom trainer that uses a custom mapping to

---

[4] https://github.com/dotnet/machinelearning/blob/main/docs/code/MlNetCookBook.md#how-can-i-define-my-own-transformation-of-data.

transform the input data to the output schema. The custom mapping is implemented in the `Map` method. The `PrepareForFit` method may be used to prepare the input data for fitting the model or for fitting a custom model to use in the `Map` method.

```
public abstract class
   CustomTrainer<TInput, TOutput> : IEstimator<
      ITransformer>
   where TOutput : class, new()
   where TInput : class, new()
{
   public SchemaShape GetOutputSchema(SchemaShape
      inputSchema)
   {
      return GetCustomMappingEstimator()
         .GetOutputSchema(inputSchema);
   }

   public ITransformer Fit(IDataView input)
   {
      PrepareForFit(input);
      return GetCustomMappingEstimator().Fit(input);
   }

   protected virtual void PrepareForFit(IDataView input)
      { }

   private CustomMappingEstimator<TInput, TOutput>
      GetCustomMappingEstimator()
   {
      return new MLContext().Transforms.CustomMapping(
            GetMapping(),null);
   }

   private Action<TInput, TOutput> GetMapping() => Map;

   protected abstract void Map(TInput input, TOutput
      output);
}
```

4.2 Extending Evaluation

Extending the evaluation capabilities is easy in comparison to other extensions. There are at least two ways to extend the evaluation capabilities of ML.NET: by implementing a custom evaluation method or by extending the built-in evaluation methods.

Custom Evaluation. The evaluation of a model is done by calling an `Evaluate` method on a data set and a trained model. Writing a custom evaluation is therefore straightforward.

Extending Built-in Evaluation. In many cases it will not be necessary to write a custom evaluation. ML.NET often does not provide the multitude of evaluation metrics that are available in other frameworks like scikit-learn[5], but in the classification case, many metrics can be computed from the confusion matrix. The confusion matrix is returned by the `Evaluate` method and can be used to compute additional metrics. The following code snippet shows how to compute the F1 macro score from the confusion matrix. The F1 macro score is a common evaluation metric for classification tasks and is defined as the harmonic mean of precision and recall. It is computed by averaging the F1 scores for each class, which is particularly useful in multiclass classification tasks.

```
public static double F1Macro(this ConfusionMatrix matrix)
{
    double f1Sum = 0;
    var classCount = matrix.NumberOfClasses;
    for (var i = 0; i < classCount; i++)
    {
        var dividend = matrix.PerClassPrecision[i] *
                    matrix.PerClassRecall[i];
        var divisor = matrix.PerClassPrecision[i] +
                matrix.PerClassRecall[i];
        if (divisor == 0) continue;
        f1Sum += 2 * (dividend / divisor);
    }
    return f1Sum / classCount;
}
```

The described approach is particularly advantageous for binary classification tasks, as AutoML frequently chooses multiclass classification with a `OneVersusAll` strategy for binary classification, which may result in a reduced set of reported metrics. By leveraging the confusion matrix, it is possible to compute additional evaluation metrics such as precision, recall, and F1 score, thereby enabling a more comprehensive assessment of model performance.

4.3 Examples

In this section, we focus on the primary machine learning tasks of binary classification, multiclass classification, and regression, which are most relevant for extending ML.NET. Accordingly, we provide illustrative examples for each of these tasks. As previously discussed, our approach leverages the `CustomMapping` functionality in conjunction with standard data preprocessing, which produces a feature column and a

---

[5] https://scikit-learn.org/stable/modules/model_evaluation.html.

label column in a predefined format. While custom mappings are applicable to arbitrary data, ML.NET enforces to explicitly define the required input columns and any additional output columns for each use case.

Input Schema. The modelling of the input columns for all tasks can be done in the same way:

```
public class CustomMappingInputSchema
{
    public float[]? Features { get; set; }
}
```

All features therefore need to be transformed to floating point vectors, classification labels need to be transformed to unsigned integer values. This is done automatically by ML.NET. Unfortunately, not all transformations may be easily used in custom pipelines. Table 1 shows the transformations ML.NET typically uses.

Table 1. Transformations typically used by AutoML.

Data	Transformation
Label	`ValueToKey`
Numerical	Conversion to `float`
Ordinal	Conversion to `float`
Categorical	`OneHotEncoding`
Textual	`FeaturizeText`

Custom mappings cannot use `ValueToKey`, because the mapping information is not accessible and therefore the reverse mapping cannot be done. Ordinal information is lost for official and custom algorithms. Both problems can be solved with ML.NET's `MapValue` which accepts custom value based mappings.

Output Schema. The recommended output differs between binary classification, multiclass classification, and regression. Figure 2 shows an approach that can be used to model the necessary classes.

Considered Examples. In the following, we consider examples of existing algorithms and their integration as custom trainers into ML.NET. We typically have to solve two problems: the first is to convert the generic feature column to the expected input of the existing algorithm, the second is to convert the output of the algorithm to the expected output schema of ML.NET.

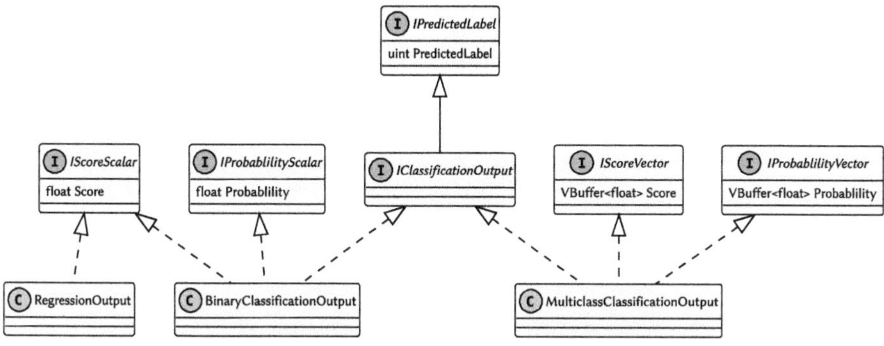

Fig. 2. Overview of custom output schema classification and regression.

Binary Classification. We start with `DecisionTreeLearner` from Russel and Norvig [22]. A C# implementation is available as open source code[6].

In the case of integrating `DecisionTreeLearner`, a key challenge arises from the standard featurization process, which removes the original categorical information from the input data. This issue can easily be addressed by identifying the unique values present in the transformed input and reconstructing the categorical variables accordingly. The transformation of the output is straightforward.

Multiclass Classification. `DecisionTreeLearner` can also be utilized for multiclass classification tasks. However, ML.NET's evaluation functions require the output score vector to have a statically defined size at compile time. Consequently, it is necessary to introduce helper classes with explicit vector size annotations, as illustrated below for the case of a quaternary classification task. This approach ensures that the output schema aligns with the expectations of ML.NET's evaluation functions.

```
public class QuaternaryClassificationOutput :
    MulticlassClassificationOutput
{
    [VectorType(4)] public override VBuffer<float> Score {
        get; set;}
}
```

While this is not an elegant solution, it is a practical workaround to ensure compatibility with ML.NET's evaluation functions. Furthermore, the necessary functionality only needs to be implemented once, as it can be reused for all multiclass classification tasks.

By leveraging the `CustomMapping` functionality as described, it becomes possible to utilize the standard evaluation functions provided by ML.NET. While these built-in functions offer a limited set of metrics, they do report the confusion matrix, which can serve as a basis for calculating additional evaluation measures. Furthermore,

---

[6] https://github.com/ltalbytz/nuget-adapters-algorithms-ai.

the confusion matrix can be extended with visualization capabilities. Figure 3 presents an example of an illustrated confusion matrix for an application to the National Poll on Healthy Aging (NPHA) data set [15].

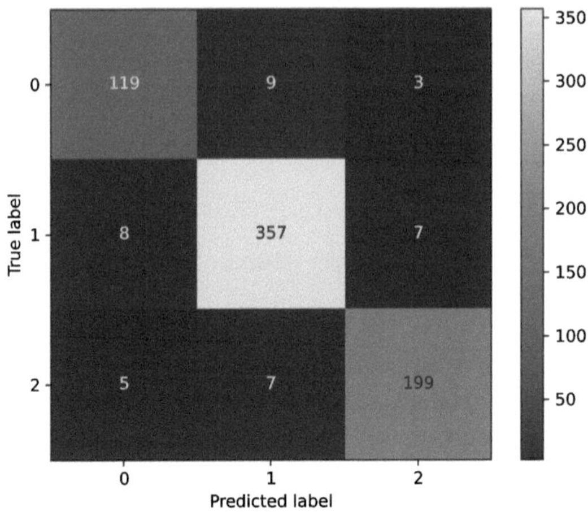

Fig. 3. Extended illustration of the confusion matrix.

Regression For regression tasks, `Fit.MultiDim` from the `MathNet.Numerics` library[7] serves as a representative example. Given that the ML.NET pipeline standardizes input data into a feature column, the implementation of a custom regression trainer is straightforward.

A detailed implementation of the considered examples is contained in the source code of the library accompanying this paper.

5 Interpreting or Explaining Models

ML.NET, and particularly its AutoML component—which returns models in binary format—does not provide a straightforward or documented mechanism for interpreting or explaining the results of trained models. This limitation poses a significant challenge for model transparency and trustworthiness. Although the framework's design intended to support parameter inspection to some extent[8], comprehensive interpretability remains insufficiently addressed.

The algorithms integrated within AutoML are predominantly established machine learning methods that inherently support interpretability and explainability. Specifically, AutoML incorporates the Stochastic Dual Coordinate Ascent (SDCA) method

---

⁷ https://www.nuget.org/packages/MathNet.Numerics.
⁸ https://github.com/dotnet/machinelearning/issues/1698.

presented in [24, 28], L-BFGS presented in [14], Gradient Boosting Decision Tree presented in [12], Fast Tree based on [20], and Fast Forest based on [16].

Models trained with ML.NET are returned as `ITransformer` objects or in a binary serialized format. While these formats facilitate efficient prediction on new data, they lack a documented and standardized mechanism for model interpretation or explanation. In particular, models produced by AutoML in binary format cannot be directly inspected for interpretability; they need to be loaded or retrained for more insight. Fortunately, AutoML generates the code necessary to retrain the model.

Interpretability and explainability of models trained with ML.NET can be addressed using two principal strategies: black-box and white-box approaches. The following subsections illustrate both methodologies, using ternary classification experiments on the Iris [27] dataset as representative examples.

5.1 Black Box Approaches

Black box approaches focus on interpreting or explaining the results of trained models without accessing the model parameters directly. These approaches are often used when the model is complex or when the model parameters are not accessible. We have at least three options to achieve this in ML.NET:

- Using Python.NET[9] to access the trained model from Python and use Python libraries like `dalex` [2] or InterpretML[10].
- Using built-in functionality of ML.NET.
- Implementing black box explanation methods in C#.

Using Python.NET is a viable option, but it requires additional dependencies and may not be suitable for all use cases. It is not considered here.

Built-In Functionality. To explain the results of the trained models, ML.NET offers some functionality out of the box:

- `ExplainabilityCatalog.CalculateFeatureContribution`
- `*Catalog.PermutationFeatureImportance` for binary classification, multiclass classification, regression, and ranking

These methods can be used to compute feature contributions and permutation feature importance (PFI) for trained models. However, the calculation of feature contributions is often not available for the models trained with AutoML.

The remaining PFI computation is easy to use and provides a good starting point for explaining the results of trained models. However, it is not sufficient for all use cases and more complex models.

Custom Black Box Explanation Methods. While widely used techniques such as partial dependence plots, SHAP values, and LIME explanations (see, e.g. [3] or [17]) are not natively supported in ML.NET, they can be implemented as black-box explanation methods. ML.NET offers the possibility to create a `PredictionEngine` from a trained model, which can be used to make predictions on new data. This allows the implementation of black-box explanation methods that use the trained model to compute explanations for new data.

Implementation. Similar to black box explanation methods in Python libraries like `dalex`, we can implement an `Explainer` class that takes a trained model and a data set as input and computes explanations for the data set. Figure 4 shows an overview of such a class.

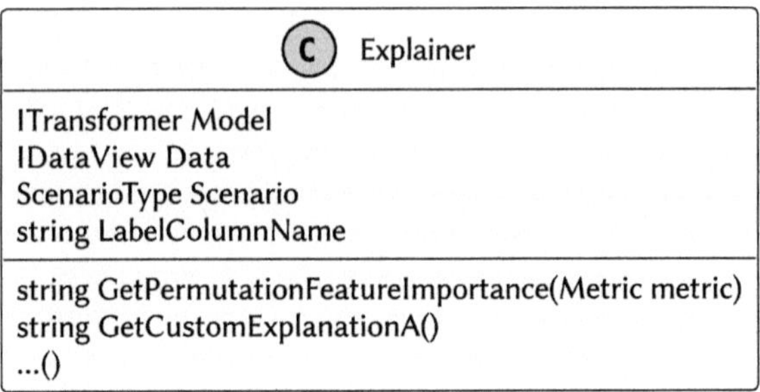

Fig. 4. Overview of a custom black box explanation class.

The `Explainer` class can be used to compute explanations for the data set by using the trained model and the data set. The class can be used and extended to implement different black box explanation methods.

Example As we are using black box explanation methods, we only need to provide the trained model in ML.NET's binary format and the data set to the `Explainer` class. The class can then be used to compute explanations for the data set.

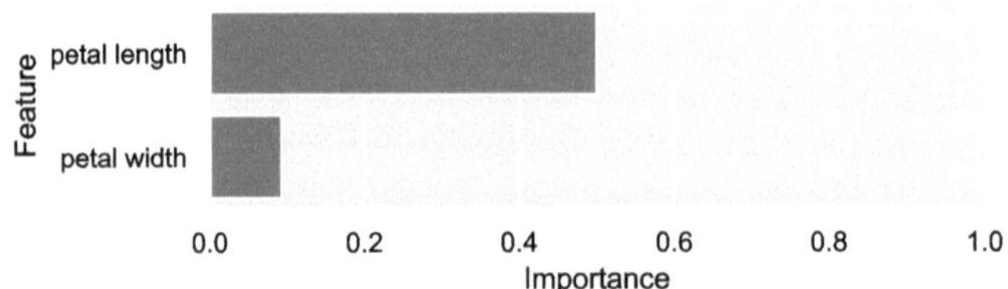

Fig. 5. Example of feature contributions computed with ML.NET's `Permutation-FeatureImportance` method.

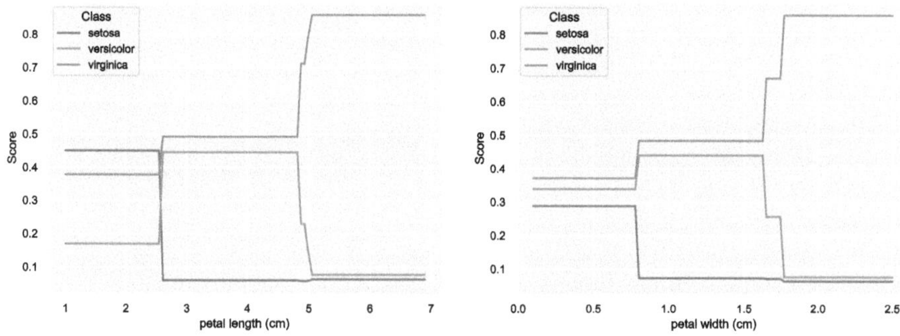

Fig. 6. Example of a Ceteris Paribus plot for the Iris data set.

Figure 5 shows an example result of the `GetPermutationFeature-Importance` method for a trained model on the Iris data set (with a custom visualization implementation). The feature contributions are computed for each feature and show how much each feature contributes to the prediction of the model.

As an example for custom black box explanation methods, we implemented a Ceteris Paribus explanation method in `Explainer`. The Ceteris Paribus plot is a common way to visualize feature contributions. It shows the predicted value of the model for each feature value, while keeping all other features constant. Figure 6 shows an example of a Ceteris Paribus plot for the Iris data set. The plot shows the predicted value of the model for each feature value, while keeping all other features constant. The plot can be used to understand how the model behaves for different feature values and how the features interact with each other.

Nevertheless, given that ML.NET is open source, white-box approaches are often feasible and, in many cases, preferable.

5.2 White Box Approaches

All trainers in ML.NET are implemented as `IPredictionTransformer` and contain model information as model parameters. The model parameters are stored in the `Model` property of the `IPredictionTransformer` interface (see Fig. 7). The model parameters can be used to interpret and explain the results of the trained models.

However, the design of ML.NET often does not allow to access the model parameters directly. For example, a one versus all classifier like `OneVersusAll` that is frequently used by AutoML does not publicly expose the model parameters of the underlying classifiers.

White box interpretions and explanations can therefore often only be achieved by e.g. forking the ML.NET repository and implementing the necessary functionality or by using a debugger and reflection. We recommend the latter approach, as it is easier to implement and does not require a deep understanding of the ML.NET code base.

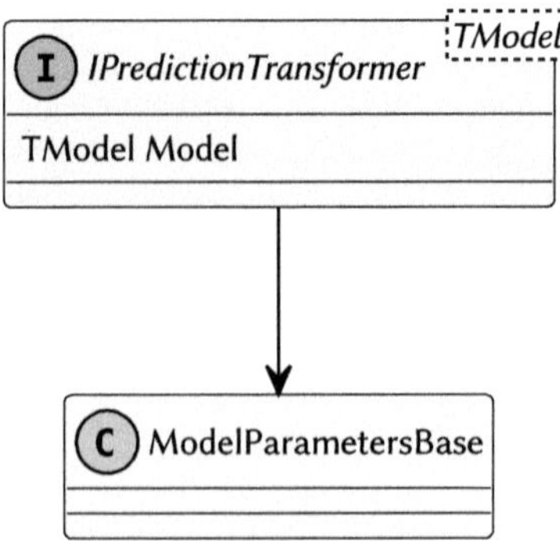

Fig. 7. Accessibility of model parameters.

Implementation. The central functionality of a white-box interpretation class is to systematically extract and reconstruct model parameters from trained models, thereby enabling detailed inspection and interpretation. This is typically achieved by leveraging reflection to access otherwise inaccessible internal fields of the model parameter objects. Once extracted, these parameters can be used to reconstruct the underlying model structure, facilitating comprehensive analysis and interpretation of the model's decision process. Figure 8 provides an overview of such a class.

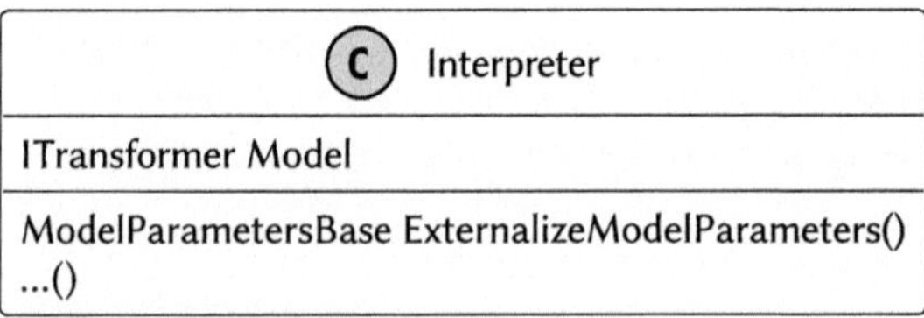

Fig. 8. Overview of a custom white box interpretation class.

Example. Similar to the black box explanation methods, we can use the `Interpreter` class to extract and reconstruct model parameters from trained models. Due to the use of reflection, the class can also be used on models in ML.NET's binary format.

For our example, we consider the higly interpretable Fast Tree trainers contained in the `Microsoft.ML.Trainers.FastTree` namespace, which are based on MART [20]. As they are binary classifiers and the data set has ternary output, AutoML adds a `OneVersusAll` classifier to the pipeline. In addition, Platt scaling is used to transform the output of the binary classifiers to probabilities. This decision from AutoML hides the model parameters of the underlying binary classifiers, which are not directly accessible. This happens quite frequently in ML.NET, and model parameters get hidden in sometimes complex hierarchies of classes with multiple levels of submodels.

In our case, the outermost model parameters of the `OneVersusAll` classifier can easily be accessed via the `Model` property of the `IPredictionTransformer` interface. However, the model parameters of the underlying binary classifiers are not directly accessible, but can be accessed via reflection. With reflection, we can access the inner `RegressionTree` trees and their parameters. Trees are reconstructable from the parameters, which allows us to interpret the results of the trained models. Figure 9 shows an example of such a reconstructed tree.

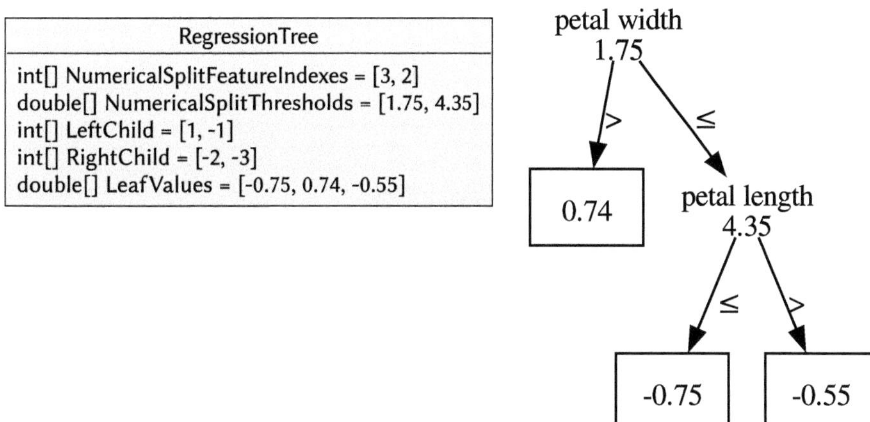

Fig. 9. Object obtained by reflection and corresponding regression tree.

This section has shown that it is possible to extend ML.NET to interpret and explain the results of trained models. The black box approaches are easy to implement and can be used for any trained model. The white box approaches may require more effort, but they allow for a deeper understanding of the trained models and their parameters. Based on the presented approaches, it is possible to implement custom interpreters and explainers for ML.NET that can be used to interpret and explain the results of trained models.

6 Analyzing ML.NET's AutoML

Microsoft's AutoML offers a convenient way to analyze data sets without having to manually choose the best algorithm and hyperparameter settings. Many practicioners will use AutoML's result without major modifications. However, at least one of the following problems may occur:

- A fair comparison of the results to alternative algorithms is not easily possible.
- Criteria like explainability and interpretability are not considered.

The runs of AutoML are configured via command line arguments, a JSON configuration file, or the Model Builder UI[11]. The most flexible and automatable way to configure AutoML is via JSON configuration files. The command line arguments do not offer all necessary configurations and the Model Builder UI is not automatable.

6.1 Approach

The JSON configuration files are not well documented and some reverse engineering is necessary to understand the possibilities. The JSON configuration files are used to configure the AutoML runs and can be used to specify the data sets, the algorithms, and the hyperparameters. Figure 10 shows an excerpt of the JSON configuration possibilities.

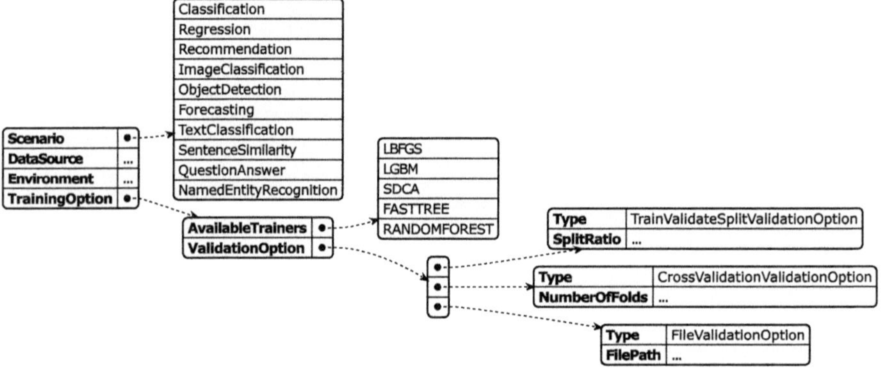

Fig. 10. Excerpt of JSON configuration possibilities for AutoML.

Interpretability and explainability can be achieved by restricting the algorithms used by AutoML to those that are inherently interpretable, such as decision trees or linear models or to models that can be better explained with the help of black box or white box approaches.

A problem for the fair comparison of the results to alternative algorithms is that AutoML does not support a train/validation/test split for the data. The command line parameter `-test-dataset` is not supported in newer versions of AutoML. A fair comparison of the results to alternative algorithms should therefore use the following approach:

- Split the data into train/validation/test sets.
- Generate the JSON configuration file with the train and validation data for AutoML.
- Run AutoML with the generated JSON configuration file.
- Use the returned model to evaluate the test data.

While it is not easy to automate this process, it is possible to do so with the help of C# code. The resulting code is available in the source code of the library accompanying this paper. We demonstrate the possibilities of this approach in the following section.

---

[11] https://dotnet.microsoft.com/en-us/apps/ai/ml-dotnet/model-builder.

6.2   Experimental Setup

In this section, we describe the experimental setup used to evaluate the proposed approaches and implementations. The main intention of this experiments is twofold: first, to demonstrate the feasibility of the proposed automation approach, and second, to provide a comparative analysis of the performance of ML.NET with the well known Python AutoML framework AutoGluon-Tabular [8].

The experiments were conducted using ML.NET 16.18.2 and AutoGluon 1.3.1. Table 2 provides an overview of the data sets used in the experiments. The data sets were selected to cover a range of machine learning tasks, including multiclass classification, binary classification, and regression. The data sets were split into training and test sets according to the specified ratios. The data sets are the most popular data sets from the UCI Machine Learning Repository[12] for the considered tasks and are widely used in machine learning research and practice.

Table 2. Overview of the used data sets.

Data Set	Task	Features	Observations	Split
Iris [27]	Multiclass (3)	4	150	80/20
Heart Disease [6]	Multiclass (5), Binary	13	303	80/20
Wine Quality [4]	Multiclass (7), Regression	11	4.898	90/10
Breast Cancer Wisconsin [25]	Binary	30	569	80/20
Adult Income [13]	Binary	14	48.842	95/5
Student Performance Final [5]	Regression	30	649	80/20
Automobile [23]	Regression	25	205	80/20

ML.NET and AutoGluon operated on the same 100 randomly selected train-test-splits of the data sets. Figure 11 shows the results of the experiments.

The experimental results indicate that ML.NET demonstrates a slight performance advantage over AutoGluon in binary classification tasks, whereas AutoGluon achieves superior results in multiclass classification and regression scenarios. The enhanced performance of AutoGluon in these tasks can be attributed primarily to its use of weighted ensembles incorporating algorithms not currently available in ML.NET, such as certain neural network architectures [8] and CatBoost gradient boosting trees [7]. It is noteworthy that the inclusion of neural networks, while beneficial for predictive accuracy, may reduce model interpretability.

Nevertheless, as outlined in this paper, the extensibility mechanisms described enable the integration of additional algorithms as custom trainers within ML.NET, which has the potential to further improve its performance in these domains. Ultimately, the choice between ML.NET and alternative Python-based AutoML frameworks should be guided by the software ecosystem and the specific requirements of the task at hand, including considerations of interpretability, extensibility, and performance.

---

[12] https://archive.ics.uci.edu/ml/index.php.

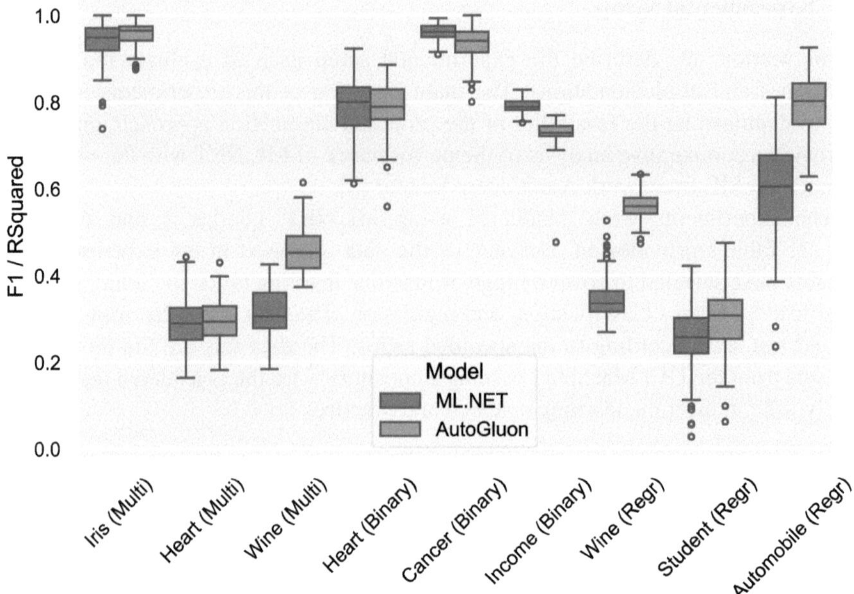

Fig. 11. Comparison of ML.NET and AutoGluon on various datasets.

7   Conclusion

This paper has presented a comprehensive analysis of the extensibility, interpretability, and automation capabilities of ML.NET, with a particular focus on its AutoML component. We have identified key challenges faced by third-party developers, notably the restricted access to internal APIs and the limitations imposed by friend assemblies. To address these issues, we proposed practical approaches for implementing custom estimators and evaluation metrics.

Our investigation into model interpretability and explainability revealed that, while ML.NET supports inherently interpretable algorithms, direct access to model parameters is often hindered by design choices. We demonstrated both black-box and white-box strategies for model explanation, including the implementation of custom explanation methods and the use of reflection to extract model parameters.

Furthermore, we introduced a methodology for benchmarking ML.NET's AutoML results against alternative frameworks, leveraging JSON-based configuration and reproducible evaluation pipelines. Our experimental results indicate that ML.NET is competitive with established AutoML frameworks such as AutoGluon in the considered classification tasks with few classes. Other tasks may require additional algorithms to be integrated into ML.NET to achieve similar performance.

Overall, our findings highlight both the strengths and current limitations of ML.NET. The framework offers robust capabilities for .NET developers but would benefit from improved extensibility and transparency. The open-source library accompanying this work provides practical tools to overcome many of these challenges. Future work will focus on extending these solutions to additional machine learning tasks and further enhancing the interpretability and automation features within the ML.NET ecosystem.

References

1. Ahmed, Z., et al.: Machine learning at microsoft with ML.NET. In: Proceedings of the 25th ACM SIGKDD International Conference on Knowledge Discovery & Data Mining, KDD 2019, pp. 2448–2458. ACM, New York, NY, USA (2019). https://doi.org/10.1145/3292500.3330667
2. Baniecki, H., Kretowicz, W., Piatyszek, P., Wisniewski, J., Biecek, P.: dalex: responsible machine learning with interactive explainability and fairness in python. J. Mach. Learn. Res. 22(214), 1–7 (2021). http://jmlr.org/papers/v22/20-1473.html
3. Biecek, P., Burzykowski, T.: Explanatory Model Analysis. Chapman and Hall/CRC, New York (2021). https://pbiecek.github.io/ema/
4. Cortez, P., Cerdeira, A.L., Almeida, F., Matos, T., Reis, J.: Modeling wine preferences by data mining from physicochemical properties. Decis. Support Syst. 47, 547–553 (2009). https://api.semanticscholar.org/CorpusID:2996254
5. Cortez, P., Silva, A.M.G.: Using data mining to predict secondary school student performance (2008). https://api.semanticscholar.org/CorpusID:16621299
6. Detrano, R.C., et al.: International application of a new probability algorithm for the diagnosis of coronary artery disease. Am. J. Cardiol. 64 5, 304–10 (1989). https://api.semanticscholar.org/CorpusID:23545303
7. Dorogush, A.V., Ershov, V., Gulin, A.: CatBoost: gradient boosting with categorical features support (2018). https://arxiv.org/abs/1810.11363
8. Erickson, N., et al.: AutoGluon-tabular: robust and accurate AutoML for structured data. arXiv preprint arXiv:2003.06505 (2020)
9. Ferreira, L., Pilastri, A., Martins, C.M., Pires, P.M., Cortez, P.: A comparison of AutoML tools for machine learning, deep learning and XGBoost. In: 2021 International Joint Conference on Neural Networks (IJCNN), pp. 1–8 (2021). https://doi.org/10.1109/IJCNN52387.2021.9534091
10. Gijsbers, P., et al.: AMLB: an AutoML benchmark. J. Mach. Learn. Res. 25(101), 1–65 (2024). http://jmlr.org/papers/v25/22-0493.html
11. Karmaker ("Santu"), S.K., Hassan, M.M., Smith, M.J., Xu, L., Zhai, C., Veeramachaneni, K.: AutoML to date and beyond: challenges and opportunities. ACM Comput. Surv. 54(8) (2021). https://doi.org/10.1145/3470918
12. Ke, G., et al.: LightGBM: a highly efficient gradient boosting decision tree. In: Proceedings of the 31st International Conference on Neural Information Processing Systems, NIPS 2017, pp. 3149–3157. Curran Associates Inc., Red Hook, NY, USA (2017)
13. Kohavi, R.: Scaling up the accuracy of naive-bayes classifiers: a decision-tree hybrid. In: Proceedings of the Second International Conference on Knowledge Discovery and Data Mining, KDD 1996, pp. 202–207. AAAI Press (1996)
14. Liu, D.C., Nocedal, J.: On the limited memory BFGS method for large scale optimization. Math. Program. 45(1), 503–528 (1989). https://doi.org/10.1007/BF01589116
15. Malani, P.N., Kullgren, J., Solway, E.: National Poll on Healthy Aging (NPHA), [United States], April 2017 (2019). https://doi.org/10.3886/ICPSR37305.v1
16. Meinshausen, N.: Quantile regression forests. J. Mach. Learn. Res. 7(35), 983–999 (2006). http://jmlr.org/papers/v7/meinshausen06a.html

17. Molnar, C.: Interpretable Machine Learning. 3 edn. (2025). https://christophm.github.io/interpretable-ml-book
18. Nematzadeh, S., Kiani, F., Torkamanian-Afshar, M., Aydin, N.: Tuning hyperparameters of machine learning algorithms and deep neural networks using metaheuristics: a bioinformatics study on biomedical and biological cases. Comput. Biol. Chem. **97**, 107619 (2022). https://doi.org/10.1016/j.compbiolchem.2021.107619
19. Psallidas, F., et al.: Data science through the looking glass: analysis of millions of GitHub notebooks and ML.NET pipelines. SIGMOD Rec. **51**(2), 30–37 (2022). https://doi.org/10.1145/3552490.3552496
20. Rashmi, K.V., Gilad-Bachrach, R.: DART: dropouts meet multiple additive regression trees (2015). https://arxiv.org/abs/1505.01866
21. Riehle, D.: The commercial open source business model. In: Nelson, M.L., Shaw, M.J., Strader, T.J. (eds.) Value Creation in E-Bus. Manage., pp. 18–30. Springer, Berlin Heidelberg, Berlin, Heidelberg (2009)
22. Russell, S., Norvig, P.: Artificial Intelligence: A Modern Approach, 4th edn. Pearson, Hoboken, NJ (2020)
23. Schlimmer, J.: Automobile. UCI Machine Learning Repository (1985). https://doi.org/10.24432/C5B01C
24. Shalev-Shwartz, S., Zhang, T.: Stochastic dual coordinate ascent methods for regularized loss. J. Mach. Learn. Res. **14**(1), 567–599 (2013)
25. Street, W.N., Wolberg, W.H., Mangasarian, O.L.: Nuclear feature extraction for breast tumor diagnosis. In: Electronic imaging (1993). https://api.semanticscholar.org/CorpusID:14922543
26. Torkamanian-Afshar, M., Nematzadeh, S., Tabarzad, M., Najafi, A., Lanjanian, H., Masoudi-Nejad, A.: In silico design of novel aptamers utilizing a hybrid method of machine learning and genetic algorithm. Mol. Diversity **25**(3), 1395–1407 (2021). https://doi.org/10.1007/s11030-021-10192-9
27. Unwin, A., Kleinman, K.: The iris data set: In search of the source of virginica. Significance **18** (2021). https://api.semanticscholar.org/CorpusID:244763032
28. Yu, H.F., Hsieh, C.J., Chang, K.W., Lin, C.J.: Large linear classification when data cannot fit in memory. ACM Trans. Knowl. Discov. Data **5**(4) (2012). https://doi.org/10.1145/2086737.2086743

SemantriX: An Explainable Hybrid Model for Aligning Vector Similarity and Semantic Relevance

Antony Seabra[✉], Claudio Cavalcante, Edward Hermann Hauesler, Daniel Schwabe, and Sergio Lifschitz

PUC-Rio - Departamento de Informática, Rio de Janeiro, Brazil
{amedeiros,cfraga,hermman,sergio}@inf.puc-rio.br

Abstract. Retrieval-Augmented Generation (RAG) systems combine large language models with retrieval mechanisms to generate contextually relevant answers. However, while vector similarity effectively retrieves geometrically close documents, it often lacks the semantic depth required to fully address user queries. This paper investigates the gap between vector similarity and semantic relevance through mathematical formulations and real-world examples. We propose SemantriX, an explainable hybrid retrieval strategy that integrates metadata enrichment and cross-encoder-based reranking. The model is evaluated within a question-answering system applied to the domain of contract management. Experimental results show significant improvements in precision, recall, and F1-score. By aligning retrieval with semantic relevance, our approach enhances the performance and explainability of RAG systems in real-world decision-support scenarios.

Keywords: Information Retrieval · Vector Similarity · Semantic Relevance · AI · Explainability · Contract Management

1 Introduction

Large Language Models (LLMs) have demonstrated remarkable capabilities in generating coherent and contextually relevant responses across various tasks. However, the informativeness and factual grounding of such responses are often constrained by the model's internal knowledge, which is limited to the data observed during pre-training. Retrieval-Augmented Generation (RAG) systems have emerged as a powerful solution to this limitation by integrating LLMs with external information retrieval mechanisms. These systems retrieve documents or passages from large corpora based on a query and incorporate them into the generation process, enabling LLMs to produce responses that are not only fluent but also grounded in external, factual evidence.

A key challenge in the design of RAG systems lies in determining which information should be retrieved. The dominant approach relies on vector similarity, in which both the query and candidate document chunks are represented in high-dimensional

F. Marcelloni et al. (Eds.): IJCCI 2025, CCIS 2829, pp. 197–212, 2026.
https://doi.org/10.1007/978-3-032-15638-9_12

embedding spaces. Ranking is typically based on measures such as cosine similarity or dot product. Although effective in capturing surface-level semantic associations, this approach often does not reflect the deeper semantic relevance required to fully satisfy the user's intent. A document may appear lexically or geometrically similar to a query but be semantically inadequate due to differences in context, purpose, or informational need.

This misalignment becomes critical in explainable systems, where LLMs rely on the retrieved content to produce outputs that must be traceable and justifiable. Vector similarity emphasizes geometric closeness in latent space, while semantic relevance considers how well a document satisfies a query with respect to task-specific constraints, context, and user intent. RAG systems based solely on vector similarity risk removing content that appears relevant but lacks true semantic alignment, undermining transparency and explainability in generated output.

In this paper, we investigate the divergence between vector similarity and semantic relevance in RAG-based question answering, grounded in formal mathematical definitions. We quantify vector similarity using cosine metrics and define semantic relevance within a probabilistic framework. Through concrete examples, we show that high vector similarity does not guarantee relevance, highlighting situations in which the retrieved chunks are geometrically proximate but informationally unhelpful.

To address this limitation, we propose SemantriX, an explainable hybrid retrieval model that aligns similarity and relevance through metadata enrichment and reranking with cross-encoders. These strategies support more interpretable and semantically faithful retrieval. Our approach is evaluated in a real-world question answering system focused on contract management, demonstrating significant improvements in both retrieval effectiveness and transparency of recommendations.

The remainder of the paper is organized as follows. Section 2 presents a technical background on RAG architectures, vector databases, and the concepts of vector similarity and semantic relevance. Section 3 discusses related work. Section 4 describes our proposed methodology and Sect. 5 presents the experimental evaluation. Finally, Sect. 6 concludes the paper and outlines directions for future work.

2   Background

2.1   Retrieval-Augmented Generation (RAG)

Retrieval-Augmented Generation (RAG) is a hybrid approach that combines the textual generation capabilities of large language models (LLMs) with external document retrieval mechanisms, enabling the model to ground its answers in up-to-date and relevant information. Unlike purely generative models, RAG systems rely on knowledge bases to retrieve relevant passages that serve as context for the final response [13].

A typical RAG system comprises two main components: the *Retriever*, responsible for locating documents similar to the query using vector similarity, and the *Generator*, which produces the final answer using both the original query and the retrieved documents. The *Retriever* transforms both queries and documents into a shared vector space, allowing efficient comparison using metrics such as cosine similarity. The *generator*, usually an LLM, is then conditioned on these data to generate more accurate and grounded responses. RAG models have proven effective in tasks that demand factual accuracy, such as open-domain question answering [12], informative dialogue generation [1], and retrieval-oriented pretraining [8].

2.2 Vector Storage

Vector stores are specialized databases designed to store and retrieve dense high-dimensional vector representations, widely used in information retrieval and machine learning tasks. In the context of RAG systems, they are essential for efficiently locating the documents that are semantically closest to a given query.

Documents and queries are converted into vector embeddings by models such as Sentence Transformers or LLM embedding layers. These vectors are stored and indexed in systems like Pinecone [20] or Milvus [26], which implement efficient Approximate Nearest Neighbor (ANN) search algorithms, such as HNSW graphs. The use of techniques such as metadata enrichment [29] has further improved performance, allowing the semantic and structural context of documents to be considered during retrieval.

2.3 Vector Similarity and Semantic Relevance

Vector similarity measures the geometric proximity between two vectors in a high-dimensional embedding space. In RAG systems, both queries and documents are encoded into dense vector representations by models such as *text-davinci-002* or other embedding techniques. Similarity measures like cosine similarity or dot product are then used to assess how "close" a document is to a given query.

Semantic relevance, on the other hand, reflects the degree to which a document satisfies the informational need expressed by a query. Relevance considers not only surface-level lexical similarities, but also deeper factors such as context, user intent, and task specificity. Thus, a document may have high vector similarity and yet be semantically irrelevant to the query - and vice versa.

Mathematical formalization provides a rigorous framework for quantifying the differences between these two concepts, providing insights into their strengths and limitations. By formalizing these ideas, researchers and practitioners can better understand why high vector similarity does not necessarily imply semantic relevance and how to design retrieval mechanisms more aligned with user intent. In addition, mathematical models help optimize the retrieval pipeline, guiding the development of hybrid systems that balance efficiency and accuracy.

Consider a query vector $\mathbf{q} \in \mathbb{R}^n$ and a document vector $\mathbf{d}_i \in \mathbb{R}^n$. The cosine similarity between the vectors $\mathbf{q}$ and $\mathbf{d}_i$ is defined as:

$$\text{cosine\_similarity}(\mathbf{q}, \mathbf{d}_i) = \frac{\mathbf{q} \cdot \mathbf{d}_i}{\|\mathbf{q}\| \|\mathbf{d}_i\|}$$

This metric measures the cosine of the angle between the vectors, producing a similarity score between -1 and 1.

The dot product between the vectors $\mathbf{q}$ and $\mathbf{d}_i$ is defined as:

$$\text{dot\_product}(\mathbf{q}, \mathbf{d}_i) = \mathbf{q} \cdot \mathbf{d}_i = \sum_{j=1}^{n} q_j d_{i,j}$$

The dot product is sensitive to vector magnitude and provides a non-normalized similarity score. Although these metrics are effective in capturing syntactic or surface-level similarities, they often overlook deeper contextual meaning. Semantic relevance, in contrast, can be modeled as a probability reflecting how well a document satisfies the user's information need given a query:

$$R(q, d_i) = P(d_i \mid q) \propto P(q \mid d_i) P(d_i)$$

Here, $P(q \mid d_i)$ measures how well the document explains or relates to the query, and $P(d_i)$ may represent prior knowledge, such as the popularity or authority of the document. Vector similarity emphasizes geometric alignment based on term frequency and vector magnitude, whereas semantic relevance is concerned with the likelihood that a document fulfills the user's intent, taking meaning and context into account.

2.4   RAG-Based Question Answering Systems

RAG-based question answering (QA) systems combine the precision of document retrieval with the generative flexibility of large language models. Unlike extractive approaches, which return literal excerpts from documents, or purely generative methods, which may lack factual grounding, RAG systems aim to synthesize accurate and context-aware answers using retrieved evidence [13].

In a standard pipeline, the retriever selects top-k relevant documents or passages from a large corpus using vector similarity methods such as dense embeddings [12], sparse lexical methods such as BM25 [22], or hybrid approaches [3]. These passages are then concatenated with the user query to form a prompt, which is processed by an LLM to generate an answer. This combination allows the model to respond with information grounded in external, up-to-date sources, which helps mitigate hallucination and improves factuality [8].

The effectiveness of RAG systems has been demonstrated in various domains, including legal assistance [9], biomedical QA [17], and chain-of-thought reasoning tasks [15]. These systems benefit from the flexibility of LLMs while maintaining a clear path to source attribution and content traceability, which are important for applications requiring transparency and user trust.

Advanced implementations also incorporate reranking mechanisms such as cross-encoders [18], structured document filtering [29], and prompt optimization techniques [16] to improve the selection of supporting evidence and refine response generation. In domains such as contract analysis, where information is often structured and distributed across multiple clauses, such refinements are critical to ensuring semantic alignment and completeness of answers [23].

As these systems continue to evolve, the integration of feedback loops, ontological constraints, and domain-specific embeddings is being explored to further improve adaptability, explainability, and performance in complex QA tasks [24].

2.5   Explainability in LLM-Based Question Answering

According to [19], explainability is a critical requirement for the adoption of LLM-based question answering (QA) systems, especially in high-stakes domains such as law, finance, and public administration. Users not only expect correct answers, but also need to understand the reasoning or evidence behind them [6]. In this context, the challenge lies in making the internal operations of large models and retrieval pipelines interpretable and traceable [21].

RAG systems introduce a natural interface for explainability by explicitly incorporating retrieved documents into the answer generation process [13]. However, explainability requires more than visibility into inputs: it involves articulating why specific documents were selected and how they influenced the final answer. Techniques such as metadata filtering, semantic highlighting of supporting text, and scoring justifications using cross-encoders contribute to this goal [18].

Hybrid retrieval methods like SemantriX offer additional avenues of explainability. By combining dense embeddings with structured filters and reranking, the system can not only improve answer quality but also expose a logical sequence of decision-making steps [29]. Moreover, interactive components such as user feedback loops and query clarification mechanisms can make the reasoning path transparent and adaptive [28].

From a technical point of view, explainable QA systems benefit from attribution of scores, justification of relevance, and traceable mappings between inputs (questions and context) and outputs (answers). These mechanisms are essential to build user trust and ensure accountability in AI-assisted decision making [10].

3   Related Work

Traditional information retrieval (IR) systems rely on sparse representations to match queries and documents [27]. Although effective for term-based matching, these approaches often fail to capture contextual relevance. The emergence of dense retrieval techniques based on Transformer models [25], such as BERT [5], has enabled more precise semantic search by encoding queries and documents into high-dimensional vector spaces [12]. These advances underpin modern RAG systems, which combine dense retrieval with generative models to produce more relevant and context sensitive responses.

RAG systems, introduced by [13], aim to improve the factual accuracy of the language model outputs. By retrieving documents from a knowledge base and conditioning the generation process on these documents, RAG systems overcome the limitations of purely generative models, such as hallucination and lack of factual grounding. Subsequent work has extended the use of RAG systems to various domains, including open-domain question answering [8] and dialogue generation [1]. These studies demonstrate the versatility of RAG systems while highlighting ongoing challenges related to retrieval relevance and generation fidelity.

To effectively apply RAG systems to specialized domains, additional task-specific adaptations are often required [7]. For example, previous research in the domain of contract management has explored the use of Transformer-based models for tasks such as clause classification, summarization, and information extraction [9]. However, these approaches typically depend on supervised fine-tuning and annotated datasets, which may not be readily available in all contexts [11]. Recent advances in few-shot and zero-shot learning techniques [15] have enabled the adaptation of pre-trained models to new domains with minimal labeled data, making them promising for contract management applications.

In terms of improving retrieval relevance, techniques such as dense passage retrieval [12], re-ranking with cross encoders [18], and hybrid models [29] - which combine sparse and dense methods - have been proposed to improve document selection accuracy. These approaches integrate both semantic understanding and lexical matching to ensure that the retrieved documents are relevant in terms of both content and language [3]. Moreover, enriching documents with metadata and employing structured representations have shown significant gains in retrieval tasks within specialized domains [29].

A growing body of work focuses on enhancing the explainability of QA and RAG systems. Miller [19] and Doshi-Velez and Kim [6] emphasize that explainable AI must be understandable to humans, especially when deployed in decision support contexts. In RAG architectures, transparency can be improved by exposing the retrieved evidence, scoring rationales, and input-output mappings [18]. These methods help mitigate the black-box nature of LLMs and build trust with end-users.

Recent research has also explored how pipeline components, such as retrievers, rerankers, and generators, can contribute to explainability. For example, cross-encoders not only re-rank but also provide interpretable relevance scores [29]. Metadata filtering helps justify retrieval through structured attributes, while prompt engineering techniques facilitate clearer associations between user intent and generated output [16]. Studies such as [10] formalize evaluation frameworks and criteria to assess the interpretability of outputs in QA systems.

In addition, explainable quality assurance systems increasingly incorporate interactive elements. Feedback mechanisms [24] and the generation of clarification questions [2] allow users to iteratively refine their queries and validate the reasoning steps behind the responses. This human-in-the-loop paradigm enhances transparency and adaptability while supporting dynamic knowledge exploration. As explainability becomes a standard requirement in critical applications, hybrid models like SemantriX offer promising solutions that align technical relevance with human interpretability.

Despite the significant progress in retrieval and generation techniques, the gap between vector similarity and semantic relevance remains a critical challenge. High-dimensional embeddings may retrieve geometrically close documents that are not truly informative or contextually appropriate. Therefore, methods that explicitly aim to align similarity measures with human-perceived relevance—such as hybrid retrieval strategies and semantic reranking—are still essential for improving the fidelity and explainability of question-answering systems.

4   Methodology

In this section, we present the methodology adopted to enhance relevance in RAG systems, focusing on improved semantic retrieval techniques, reranking strategies, and hybrid architectures. We begin by illustrating the mismatch between similarity and relevance, then introduce our proposed hybrid method.

4.1   High Similarity and Low Relevance

Consider the following query and two documents:

- *Query*: "Best hiking trails in Europe"
- *Document A*: "Top tourist destinations in Europe, including cities and cultural landmarks"
- *Document B*: "A guide to the best hiking trails in the Alps and Pyrenees"

Using a vector-based model, we represent the query and documents as vectors in a high-dimensional space. Let $\mathbf{q}$ be the query vector, $\mathbf{d}_A$ the vector for Document A, and $\mathbf{d}_B$ the vector for Document B. Cosine similarity is computed as:

$$\text{cosine\_similarity}(\mathbf{q}, \mathbf{d}_A) = \frac{\mathbf{q} \cdot \mathbf{d}_A}{\|\mathbf{q}\| \|\mathbf{d}_A\|} = 0.85$$

$$\text{cosine\_similarity}(\mathbf{q}, \mathbf{d}_B) = \frac{\mathbf{q} \cdot \mathbf{d}_B}{\|\mathbf{q}\| \|\mathbf{d}_B\|} = 0.92$$

Document A scores high in similarity due to overlapping terms like "Europe" and "destinations", but lacks direct relevance to hiking. Document B, although scoring similarly or higher, directly addresses the intent of the query. This example illustrates that high vector similarity does not guarantee semantic relevance.

4.2   SemantriX: A Hybrid Method for Aligning Semantic Relevance and Vector Similarity

The convergence between vector similarity and semantic relevance can be formalized as a limit expression. As the expressiveness of the model increases (parameters, layers, data), the learned similarity function $f_\theta(q, d)$ tends to approximate the ideal relevance distribution $P(d \mid q)$:

$$\lim_{\substack{\text{capacity} \to \infty \\ \text{data} \to \infty}} f_\theta(q, d) \approx P(d \mid q)$$

Alternatively, we may minimize a divergence function between the learned similarity and the true relevance:

$$\min_\theta \mathcal{D}(f_\theta(q, d), P(d \mid q))$$

where $\mathcal{D}$ denotes a divergence measure such as KL-divergence. This formalization guides the design of our architecture, which computationally aligns vector similarity with semantic relevance in RAG systems. Our hybrid method combines three components: (i) dense semantic embedding generation, (ii) metadata-based filtering, and (iii) reranking with cross-encoders.

Dense Semantic Embedding Generation. We encode queries and documents using Transformer-based models (e.g., BERT, Sentence-BERT) to obtain deep semantic embeddings:

$$\mathbf{q} = \text{Encoder}(q), \quad \mathbf{d}_i = \text{Encoder}(d_i)$$

The similarity is then computed using cosine similarity:

$$\text{sim}(\mathbf{q}, \mathbf{d}_i) = \frac{\mathbf{q} \cdot \mathbf{d}_i}{\|\mathbf{q}\| \|\mathbf{d}_i\|}$$

Top k most similar documents are selected:

$$\text{Top}_k(q) = \arg \underset{d_i \in \mathcal{D}}{\text{top-}k} \, \text{sim}(\mathbf{q}, \mathbf{d}_i)$$

In practical implementations, this step is performed by generating and storing dense embeddings for all documents offline in a vector database (e.g., FAISS, Pinecone, Milvus). At query time, the query is encoded and compared against these pre-computed vectors using approximate nearest neighbor (ANN) search algorithms like HNSW. This approach ensures low latency and high scalability in large collections. Embeddings are typically 384 to 768 dimensions, depending on the encoder model, and are normalized to unit vectors to improve cosine similarity comparisons. Additionally, domain-specific fine-tuning or prompt adaptation of the encoder can enhance the semantic alignment between queries and retrieved documents.

Metadata-Enriched Filtering. To reduce ambiguities and increase retrieval precision, we enrich each document vector in the vectorstore with metadata attributes. For instance, in the domain of contract management, each vector may be associated with structured fields such as contract identifier, clause type and subtype, validity period, contracting and/or contracted entity. These metadata are stored alongside the embedding vectors in vector databases that support filtering capabilities, such as Pinecone or Weaviate. During retrieval, before computing vector similarity, the system applies a metadata-based filter query to narrow down the search space. This filtered subset $\mathcal{D}_q$ is defined as:

$$\mathcal{D}_q = \{d \in \mathcal{D} \mid \text{match}(q, \text{metadata}(d)) = \text{true}\}$$

Then, vector similarity is applied over $\mathcal{D}_q$, improving the semantic relevance by constraining the search to the appropriate context.

Reranking with Cross-Encoders. Candidate documents are reranked via semantic analysis of each query-document pair using a cross-encoder:

$$\text{Relevance}(q, d_i) = \text{CrossEncoder}(q, d_i)$$

The cross-encoder jointly encodes the pair as a single sequence, enabling deep token-level interaction. The most relevant document is:

$$d^* = \underset{d_i \in \text{Top}_k(q)}{\arg\max} \ \text{CrossEncoder}(q, d_i)$$

In practical terms, this process takes each query-document pair and feeds it into a Transformer-based encoder, such as BERT or RoBERTa, with a special separator token (e.g., [SEP]) between the query and the document. This enables the model to compute a contextualized relevance score considering the entire sequence holistically. For instance, a query and a clause from a contract are concatenated as: "[CLS] query [SEP] clause [SEP]", allowing the model to capture fine-grained semantic alignments and dependencies between words in the query and terms in the clause. This architecture is computationally more expensive than dual-encoders but significantly enhances accuracy in domains that demand precision, such as legal or administrative contract management. The output score is then used to reorder the top-k documents retrieved by the vector similarity, ensuring that the final candidate is not only similar in vector space but also semantically the most relevant.

Workflow Summary. The proposed hybrid model integrates dense embeddings, structured metadata filtering, and cross-encoder reranking. This three-stage pipeline enhances semantic alignment while reducing noise. Figure 1 illustrates this flow and includes a feedback mechanism for user-guided refinement.

Unlike approaches that apply these techniques in isolation, SemantriX chains them as a coordinated refinement pipeline, aligning vector similarity with probabilistic relevance modeling. This sequential integration ensures that each component—embeddings, metadata filtering, and contextual reranking—contributes to the final relevance. The architecture also supports user feedback loops to dynamically refine inferences via explicit feedback or query reformulation.

Human-In-The-Loop. The final step of the SemantriX architecture, which incorporates user feedback, plays a crucial role in closing the loop between automated retrieval and human judgment. By allowing users to validate, reject, or refine retrieved content and generated answers, the system creates an adaptive layer capable of learning from domain-specific preferences and evolving informational needs. This feedback mechanism not only improves the precision of future responses but also enhances trust and transparency, particularly in high-stakes environments like legal or public-sector decision-making. Rather than treating the retrieval pipeline as static, this dynamic adjustment enables iterative alignment between model behavior and user expectations.

This feedback process is implicitly aligned with the Human-in-the-Loop (HITL) paradigm, wherein human oversight and interaction are essential to guide, correct, and contextualize the outputs of AI systems. In the case of SemantriX, the user does not merely consume the answer but actively participates in shaping it through corrective feedback. This design transforms the system from a passive retrieval engine into an interactive, explainable agent capable of learning from human input. Such integration of human agency is fundamental to the development of explainable and trustworthy AI, ensuring that the outputs remain aligned with institutional rules, contextual nuances, and evolving information needs.

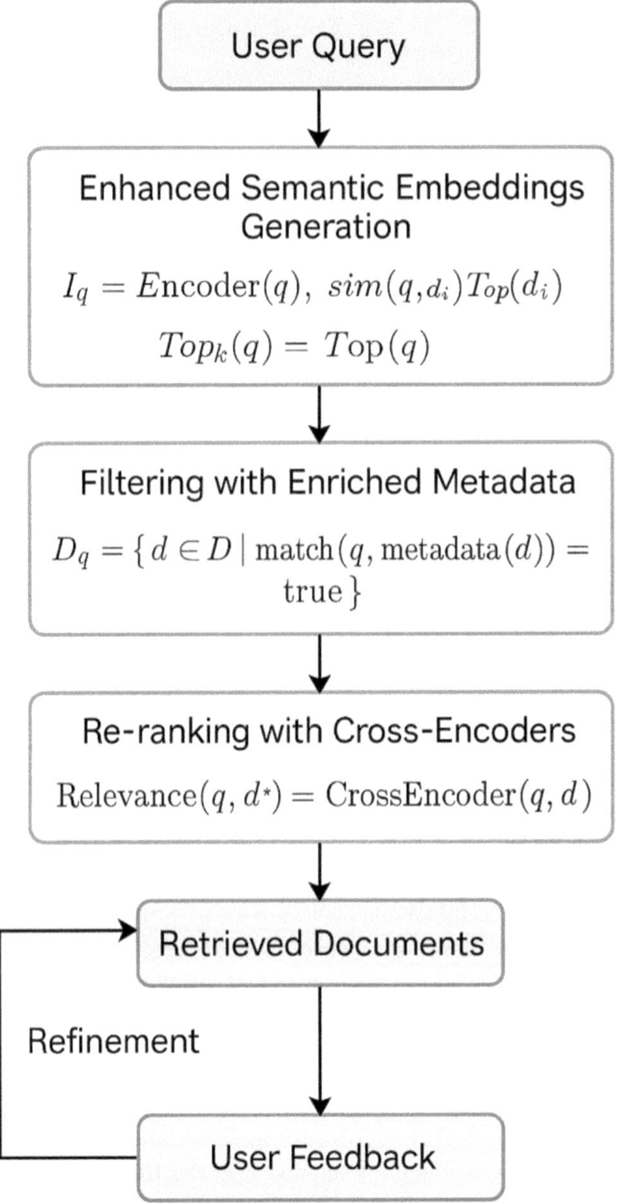

Fig. 1. SemantriX. Source: Authors.

Practical Implications. By incorporating semantic embeddings, enriched metadata, and cross-encoders, RAG systems become more capable of understanding user intent and retrieving more relevant content. Consider the following examples:

- *Query*: "Health benefits of apples"
- *Document*: "Nutritional advantages of fruits rich in fiber and antioxidants"

Here, even without the term "apple", semantic embeddings and cross-encoders enable accurate retrieval.

- *Query*: "Best hiking trails in Europe"
- *Retrieved Document*: "A guide to the best hiking trails in the Alps and Pyrenees"
- *Generated Answer*: "Some of the best hiking trails in Europe are located in the Alps and Pyrenees, offering breathtaking views and challenging routes."

Enriched metadata helps pre-filter relevant content, and cross-encoders rank the results accordingly. In ambiguous queries, such as:

- *Query*: "Apple"
- *Retrieved Documents*:
    - "Nutritional benefits of apples"
    - "Recent product launches by Apple Inc."
- *Generated Answer*: "Are you referring to the fruit or the company? If the fruit, apples are rich in fiber and antioxidants. If the company, Apple Inc. has recently released new products."

Unlike previous approaches discussed in the Related Work section, which often address retrieval, filtering, or reranking as isolated components, our methodology emphasizes a coordinated hybrid pipeline designed to progressively align vector similarity with semantic relevance. While prior work has demonstrated gains from metadata enrichment or cross-encoder reranking independently, SemantriX integrates these strategies within a unified architecture that emphasizes interpretability, domain adaptability, and explainable retrieval. This integration enables not only improved answer accuracy but also a clearer understanding of the reasoning behind each retrieval and generation step, which is particularly valuable in high-stakes structured domains.

5   Evaluation

The evaluation of SemantriX was conducted using real-world use cases from Contract Management in the Public Sector. By integrating semantic embeddings, metadata filtering, and cross-encoder reranking, the model effectively aligned vector similarity with contextual and semantic relevance, producing more accurate answers aligned with user intent.

5.1   Common Issues

SemantriX was evaluated in three typical recurring failure scenarios for RAG systems, demonstrating how its architecture efficiently addresses each one.

Ambiguous Queries. Vector similarity-based systems tend to retrieve semantically irrelevant documents when dealing with generic terms. SemantriX uses **metadata filtering** to restrict the search to the correct contract:

- *Query*: "What are the liability limits of the software vendor in the ABC-123 contract?"
- *RAG Answer:* "The vendor's responsibilities are detailed in the contract terms and may include service provisions."
- *SemantriX Answer*: "The liability limits of the vendor in contract ABC-123 are defined in Clause 7.2 and do not exceed the total contracted amount."

Vector-only systems often fail to resolve ambiguous terms, retrieving content based on lexical overlap rather than contextual appropriateness. In the case of the query regarding the liability limits of a software vendor, baseline RAG approaches might return documents from unrelated contracts containing similar terminology. In contrast, SemantriX employs metadata filtering to ensure that the retrieved content is specific to the referenced contract. This refinement significantly improves the accuracy of the answer by reducing semantic drift and ensures that only contract-specific obligations are presented.

Context Misalignment. By combining dense embeddings with semantic reranking, SemantriX can select the correct clause even within the right contract:

- *Query*: "What happens in case of a delivery delay according to contract DEF-789?"
- *RAG Answer:* "Delays may be subject to penalties as defined in the contract."
- *SemantriX Answer*: "According to clause 5.4 of contract DEF-789, delays exceeding 10 days will incur a 2% monthly penalty."

Contextual misalignment is a frequent issue in large documents where several clauses mention overlapping topics. Traditional models tend to retrieve the correct contract, but fail to isolate the clause most relevant to the specific user query. For instance, in the delay penalty example, previous models might select introductory or adjacent clauses lacking precise penalty conditions. SemantriX's use of cross-encoder reranking ensures that the clause explicitly governing delays is prioritized, providing an answer that is both legally accurate and directly responsive to the user's informational need.

Partial Information Retrieval. By synthesizing multiple clauses, SemantriX generates more complete and user-aligned answers:

- *Query*: "What intellectual property rights are granted in contract GHI-012?"
- *RAG Answer:* "The contract includes terms related to intellectual property."
- *SemantriX Answer*: "The licensee has non-exclusive usage rights for the software, and modifications made on demand become property of the contracting party."

In complex queries that require synthesizing dispersed information, standard RAG systems typically return a single, most similar chunk, which may only partially address the question. As seen in the intellectual property query, traditional methods might retrieve only one clause - omitting critical conditions regarding ownership or licensing. SemantriX improves over these models by integrating semantic embeddings with reranking logic that allows for the fusion of related clauses. This results in a more comprehensive and coherent response that reflects the full scope of contractual stipulations (Table 1).

Table 1. Performance comparison of SemantriX and traditional approaches.

Approach	Precision	Recall	F1 Score
Baseline (Cosine Similarity)	0.65	0.70	0.67
Metadata Enrichment	0.80	0.75	0.77
Hybrid Retrieval (BM25 + Dense)	0.87	0.83	0.85
Cross-Encoder Reranking	0.91	0.88	0.89
SemantriX (Full)	**0.95**	**0.92**	**0.93**

5.2   Results with SemantriX

The combination of SemantriX's three core components—semantic embeddings, metadata filtering, and cross-encoder reranking—enables it to overcome the limitations of traditional RAG retrieval. Below are examples of accurate responses generated by the system using the integrated workflow:

- *Query*: "Who is the manager of contract 123/2024?"
- *SemantriX Answer*: "The manager of contract 123/2024 is Jane Smith."

- *Query*: "What is the termination clause of contract 123/2024?"
- *SemantriX Answer*: "The termination clause of contract 123/2024 states that either party may terminate with 60 days' notice."

- *Query*: "What are the payment terms for contract 123/2024?"
- *SemantriX Answer*: "The payment terms for contract 123/2024 specify a due date of up to 30 days after invoice issuance. Late payments incur a 5% monthly fee and may be split into up to two installments."

5.3   Experimental Performance

A comparative experiment was conducted between SemantriX and traditional RAG approaches using a real dataset of administrative contracts and associated queries [23]. The baseline approach used only cosine similarity between embeddings generated by a general-purpose language model, without any customization or filtering. This configuration reflects the common setup in generic RAG applications.

The second evaluated strategy added metadata enrichment by incorporating contract identifiers and clause types into vectors, enabling refined retrieval. The third strategy tested hybrid retrieval, combining BM25 with dense embeddings to leverage both lexical and semantic strengths. BM25 (Best Matching 25) is a classic sparse vector retrieval algorithm based on term frequency (TF), inverse document frequency (IDF), and document length. It favors terms that are frequent in a document but rare in the collection and normalizes for document length. Finally, a cross-encoder reranking strategy was tested, applying models that jointly analyze the query and clause.

The results show that SemantriX consistently outperforms traditional methods in all metrics evaluated. While the baseline cosine-only approach achieved limited performance (F1 = 0.67), isolated strategies such as metadata enrichment (F1 = 0.77), hybrid retrieval (F1 = 0.85), and cross-encoder reranking (F1 = 0.89) yielded incremental gains. However, the coordinated integration of these techniques in the full SemantriX model achieved the highest performance, with an F1 score of 0.93, precision of 0.95, and recall of 0.92. These results demonstrate that the proposed architecture enables deeper semantic interpretation, precise contextualization, and more robust generation of complete, domain-specific answers.

While SemantriX demonstrates superior retrieval accuracy and answer quality across all evaluated scenarios, it also introduces trade-offs inherent to more sophisticated architectures. On the positive side, the hybrid combination of semantic embeddings, structured metadata, and cross-encoder reranking ensures high precision and relevance, particularly in domains requiring factual rigor and contextual fidelity. However, this comes at the cost of increased computational complexity and inference time, especially due to the reranking stage involving pairwise input processing. In practical deployments, these trade-offs must be balanced depending on the performance requirements and computational constraints of the target environment. Nonetheless, in decision-critical applications where explainability and precision are paramount, the benefits of SemantriX outweigh its costs.

6 Conclusions and Future Work

This work introduced SemantriX, a hybrid model that integrates dense embeddings, structured metadata filtering, and cross-encoder reranking to minimize the gap between vector similarity and semantic relevance in RAG systems. Through experiments in the domain of public contract management, the model demonstrated superior performance in retrieving specific clauses and generating complete, context-aware answers—even for ambiguous queries or those requiring multi-source synthesis. The proposed architecture yielded significant improvements in precision, recall, and F1 score, highlighting the importance of pipelines that combine deep semantic understanding with structured document attributes.

As future works, we propose expanding the application of SemantriX to other complex domains such as healthcare, education, and public finance, where documents exhibit distinct semantic and structural characteristics. We also intend to enhance the user feedback mechanisms, allowing for dynamic adjustment of relevance criteria based on explicit user corrections and preferences. Another promising avenue involves the development of an explainability layer to justify retrieval and generation decisions based on the activated components and selected evidence. This layer would provide users with transparent insights into why certain documents were retrieved, how they contributed to the final answer, and which semantic or structural cues influenced the model's decisions. By highlighting the role of embeddings, metadata filters, and cross-encoder scores, such a mechanism can enhance trust, support validation in critical domains, and enable iterative refinement through user feedback.

Finally, as a promising direction for future research, we propose extending SemantriX with formal semantic models and domain ontologies to further improve semantic accuracy and answer explainability. Such integration would allow the system to move beyond pattern-based retrieval and toward semantically grounded inference, increasing both transparency and reliability. Furthermore, by aligning the retrieval and generation steps with formally defined knowledge representations, it becomes possible to provide users with not only correct answers, but also justifiable explanations grounded in institutional logic, norms, and policies.

References

1. Cai, D., Wang, Y., Bi, W., Tu, Z., Liu, X., Shi, S.: Retrieval-guided dialogue response generation via a matching-to-generation framework. In: Proceedings of the 2019 Conference on Empirical Methods in Natural Language Processing and the 9th International Joint Conference on Natural Language Processing (EMNLP-IJCNLP), pp. 1866–1875 (2019)
2. Cao, S., Mirzaei, Y., Gao, J., Fan, Y., Zhang, Y., Chen, Y.: Asking clarifying questions in open-domain question answering. In: Proceedings of the 2021 Conference of the North American Chapter of the Association for Computational Linguistics: Human Language Technologies, pp. 1265–1277 (2021)
3. Chen, J., Lin, H., Han, X., Sun, L.: Benchmarking large language models in retrieval-augmented generation. arXiv preprint arXiv:2309.01431 (2023)
4. Brown, T., et al.: Language models are few-shot learners. arXiv preprint arXiv:2005.14165 (2020)
5. Devlin, J., Chang, M.W., Lee, K., Toutanova, K.: BERT: pre-training of deep bidirectional transformers for language understanding. arXiv preprint arXiv:1810.04805 (2018)
6. Doshi-Velez, F., Kim, B.: Towards a rigorous science of interpretable machine learning. arXiv preprint arXiv:1702.08608 (2017)
7. Feng, Z., Feng, X., Zhao, D., Yang, M., Qin, B.: Retrieval-generation synergy augmented large language models. In: ICASSP 2024 - IEEE International Conference on Acoustics, Speech and Signal Processing, pp. 11661–11665. IEEE (2024)
8. Guu, K., Lee, K., Tung, Z., Pasupat, P., Chang, M.W.: REALM: retrieval-augmented language model pre-training. arXiv preprint arXiv:2002.08909 (2020)
9. Hendrycks, D., Burns, C., Chen, A., Ball, S.: CUAD: an expert-annotated dataset for legal contract review. arXiv preprint arXiv:2103.06268 (2021)
10. Jacovi, A., Goldberg, Y.: Formalizing trust in explainable machine learning: a user-centered perspective. In: Proceedings of the CHI Conference on Human Factors in Computing Systems (CHI), pp. 1–12 (2021)
11. Jeong, C.: A study on the implementation of generative AI services using an enterprise database-based LLM application architecture. arXiv preprint arXiv:2309.01105 (2023)
12. Karpukhin, V., et al.: Dense passage retrieval for open-domain question answering. arXiv preprint arXiv:2004.04906 (2020)

13. Lewis, P., et al.: Retrieval-augmented generation for knowledge-intensive NLP tasks. Adv. Neural. Inf. Process. Syst. **33**, 9459–9474 (2020)
14. Lipton, Z.C.: The mythos of model interpretability. Commun. ACM **61**(10), 36–43 (2018)
15. Ling, Z., et al.: Deductive verification of chain-of-thought reasoning. In: Advances in Neural Information Processing Systems, vol. 36 (2024)
16. Liu, N.F., et al.: Lost in the middle: how language models use long contexts. In: Proceedings of the 2023 Conference on Empirical Methods in Natural Language Processing (EMNLP) (2023)
17. Mao, Y., et al.: Contextualized sparse representations for real-time open-domain question answering. arXiv preprint arXiv:2104.08767 (2021)
18. Mialon, G., et al.: Augmented retrieval for few-shot question answering. In: International Conference on Learning Representations (ICLR) (2023)
19. Miller, T.: Explanation in artificial intelligence: insights from the social sciences. Artif. Intell. **267**, 1–38 (2019)
20. Pinecone Systems, Inc.: Pinecone: vector database for vector search. https://www.pinecone.io/, Accessed 3 May 2025
21. Rajani, N.F., et al.: Explain yourself! Leveraging language models for commonsense reasoning. In: Proceedings of the 57th Annual Meeting of the Association for Computational Linguistics (ACL) (2019)
22. Robertson, S., Zaragoza, H.: The probabilistic relevance framework: BM25 and beyond. Found. Trends Inf. Retr. **3**(4), 333–389 (2009)
23. Seabra, A., Cavalcante, C., Nepomuceno, J., Lago, L., Ruberg, N., Lifschitz, S.: Contrato360 2.0: a document and database-driven question-answer system using large language models and agents. In: Proceedings of the 16th International Joint Conference on Knowledge Discovery, Knowledge Engineering and Knowledge Management (IC3K), vol. 1: KDIR, Porto, Portugal (2024)
24. Stumpf, S., et al.: Interacting meaningfully with machine learning systems: three experiments. ACM Trans. Comput.-Hum. Interact. (TOCHI) **17**(3), 1–28 (2009)
25. Vaswani, A., et al.: Attention is all you need. In: Advances in Neural Information Processing Systems (2017)
26. Wang, W., et al.: Milvus: a purpose-built vector data management system. Proc. VLDB Endow. **14**(12), 2536–2548 (2021)
27. White, J., et al.: A prompt pattern catalog to enhance prompt engineering with ChatGPT. arXiv preprint arXiv:2302.11382 (2023)
28. Wiegreffe, S., Marasović, A.: Teach me to explain: a review of datasets for explainable NLP. In: Proceedings of the 2021 Conference on Empirical Methods in Natural Language Processing (EMNLP), pp. 9365–9388 (2021)
29. Zhang, Y., Zhang, Z., Liu, Z., Sun, M.: Metadata-enhanced document retrieval with BERT-based models. arXiv preprint arXiv:2106.02276 (2021)

Explainable Knowledge Access: Recursive and Rerank-Based RAG for Interpretable QA

Melchiorre Cosseddu, Alessandro Giuliani, Marco Manolo Manca, Marco Pilloni[✉],
Alessandro Sebastian Podda, and Sandro Gabriele Tiddia

Department of Mathematics and Computer Science, University of Cagliari, Cagliari, Italy
{melchiorre.cosseddu,alessandro.giuliani,marcom.manca,
marco.pilloni,sebastianpodda,sandrog.tiddia}@unica.it

Abstract. In response to the crucial need for more transparent and reliable tools based on Large Language Models (LLMs), this work introduces a novel and explainable architectural enhancement to the Retrieval-Augmented Generation (RAG) pipeline. As organizations increasingly rely on LLMs to navigate and interpret vast, unstructured datasets, ensuring the factual accuracy and traceability of generated content has become essential. Our key contribution is designing a system that formally anchors all produced output in a formally verified knowledge base, thereby significantly alleviating challenges related to hallucination and verbosity. This focus on transparency is achieved by providing exact citations to the information used, enabling users to trace the source of generated content. Our work demonstrates significant improvements to chunk retrieval accuracy, achieved using a multi-granularity recursive chunking strategy and re-ranking with maximum information exposure to the LLM. We also provide extensive comparative analysis of various combinations of LLM and embedding models under this enhanced system. Our contributions lay the groundwork for developing more responsible and reliable AI solutions across knowledge-intensive domains, focusing on enhanced information retrieval and transparent response generation.

Keywords: Retrieval Augmented Generation (RAG) · Large Language Models (LLM) · Explainable Artificial Intelligence

1 Introduction

With the advent of the digital era, people have access to a vast amount of information, most of which appears voluminous and unorganized, making it challenging to handle effectively [3]. Enterprises and businesses must effectively manage the extensive amount of data. It concerns not merely organizational efficiency but rather represents a crucial factor in ensuring informed decision-making, maintaining a competitive edge, and fostering innovation. More broadly, issues of transparency, understandability, and interpretability remain open challenges that are increasingly relevant across many application domains where intelligent methods support human activity, such as medicine [29,34], finance [5,10,23], safety [28], data analysis [2], and industry [8,20,30,32]. In this scenario, finding valuable insights in such disorganized information remains

F. Marcelloni et al. (Eds.): IJCCI 2025, CCIS 2829, pp. 213–231, 2026.
https://doi.org/10.1007/978-3-032-15638-9_13

challenging, necessitating the development of new technologies and methods to tackle the task [35].

Indeed, while the recent rise of Large Language Models (LLMs) has represented a pivotal moment in providing more efficient and effective ways to process and interpret complex data [4,27], it is essential to acknowledge their limitations. These include the risk of hallucinations, producing overly verbose outputs and being unclear about their reasoning. To overcome the aforementioned drawbacks, the *Retrieval Augmented Generation* (RAG) pipeline has emerged as a promising solution [9]. RAG significantly enhances LLMs by incorporating an information retrieval feature that fetches relevant data before crafting a response. This approach effectively addresses issues like hallucination by connecting the LLM output to verified facts. Furthermore, the way RAG pipelines utilize chunks improves transparency, making it easier for users to trace the sources of the information in the generated text. However, maximizing RAG's performance and ensuring it operates reliably across various applications and document types remains hindered by several factors, such as low-quality or outdated data sources and ambiguous, complex, or incomplete queries [41].

This paper aims to address some of these ongoing issues by introducing a new architectural enhancement to the RAG pipeline. Our proposal focuses on enhancing the information retrieval stage through two key innovations: (i) a multi-granularity recursive chunking strategy and (ii) a re-ranking mechanism. The recursive chunking method enables us to capture and link information at various levels of detail, while the re-ranking mechanism utilizes an LLM to refine the order of the chunks presented, thereby maximizing their impact. This strategy ensures that the responses generated are firmly based on verified sources, increasing explainability, transparency and accuracy while significantly reducing the risk of hallucinations. A detailed comparative analysis of various LLMs within this improved framework will be conducted to highlight important concerns, evaluate improvement, and ultimately lay the basis for more reliable and efficient information usage.

The rest of the paper is structured as follows: after providing an overview of the state-of-the-art in Sect. 2, the proposed methodology is detailed in Sect. 3, while the experiments are described in Sect. 4. Section 5 reports the experimental results and their discussion. Section 6 ends the paper with the conclusions and future work.

2 Related Work

Retrieval-augmented generation (RAG) is shaping up to be an exciting approach for boosting the accuracy and relevance of answers from LLMs. This is especially true in areas such as open-domain question answering and tasks that require extensive knowledge. Recently, several studies have investigated methods for evaluating and benchmarking RAG systems across various large language model (LLM) architectures.

For example, ARES [33], an automated evaluation framework designed explicitly for RAG systems, breaks down performance into three essential dimensions: context relevance, response faithfulness, and response relevance. ARES is one-of-a-kind because it employs LLM-based evaluators to evaluate the quality of both retrieval and generation, eliminating the need for human annotations. It has been demonstrated to be more consistent and thorough compared to traditional metrics, such as BLEU or

ROUGE. It has been successfully used to benchmark a range of RAG pipelines across various datasets.

FRAMES [17] is a benchmark that has been human-annotated to evaluate the factuality, retrieval quality, and multi-hop reasoning abilities of RAG systems. The dataset is packed with complex queries that demand information synthesis from multiple sources, allowing for a nuanced performance evaluation from start to finish. The empirical findings indicate that LLMs without retrieval modules frequently encounter hallucinations and that RAG configurations significantly improve answer accuracy, underscoring the crucial role retrieval plays in generating grounded responses. Driven by this insight, we propose a novel and enhanced retrieval method, as described in the following sections.

A recent work investigated the intriguing relationship between RAG and long-context LLMs [18]. Analyzing models such as Gemini and GPT-4 across various datasets revealed that long-context LLMs can achieve remarkable accuracy when given access to entire input documents. RAG systems are notable for being a more computationally efficient choice while still providing similar performance levels. The authors also developed a creative hybrid routing mechanism that flexibly selects between retrieval-based and long-context strategies tailored to the input's features and the available resources.

Furthermore, recent work reveals that many LLMs give more weight to the information at the start and end of the context, leaving out crucial details that get *lost in the middle* [19]. The decision to introduce a re-ranker in our proposal was made in accordance with these findings. It is not merely an optimization but a focused approach to address a recognized problem with Transformer architectures.

An additional contribution is the review paper proposed by Gao et al. [9], which reports a survey of current RAG systems and the various components that can be involved. It is necessary for our work, as it outlines and describes some of the primary technologies we use.

Moreover, unlike earlier benchmarks that focus solely on single-hop queries (i.e., only one relevant chunk is retrieved), MultiHop-RAG [38] specifically addresses the challenge of multi-hop retrieval in RAG systems. The authors introduce a dataset containing over 2,500 complex questions, each requiring the integration of information from two to four different news article excerpts, with a gold-standard knowledge base with curated question-answer pairs and supporting evidence. To assess performance, they evaluate various embedding models and leading large language models, including GPT-4, PaLM, and LLaMA2-70B. Their findings emphasize the need for dedicated multi-hop benchmarks to effectively evaluate and enhance the retrieval and reasoning capabilities of RAG architectures. Our work, which falls into the same scenario, is compliant with this work, including reliance on a vast and heterogeneous corpus and comparisons of LLMs.

The reported works lay a strong foundation for grasping the trade-offs between retrieval-based and retrieval-free strategies in LLM applications. They emphasize the necessity of comprehensive, multidimensional assessments when evaluating RAG systems that utilize various language models.

3   Methodology

This section outlines the proposed methodology, starting with an introduction to the model architecture and then detailing its main steps.

3.1   Architecture Overview

The developed system was designed to improve and optimize the information processing workflow, enhancing the retrieval stage of the necessary context for users to address their needs. A crucial aspect is transparency, which is ensured by retrieving chunks that serve as the sole source of knowledge for the model. The goal is to provide validated and verified information, minimizing hallucinations. Figure 1 depicts the high-level architecture of the proposed strategy. The entire process consists of two main steps: *Data ingestion* and *Inference*, which are detailed in the following sections.

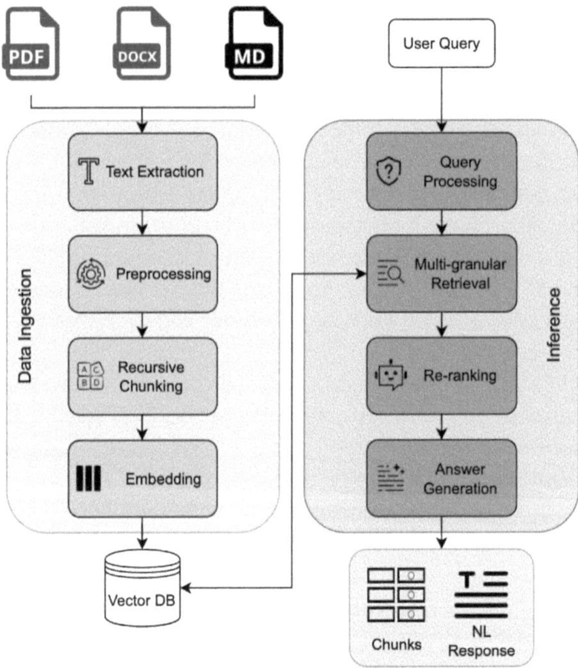

Fig. 1. RAG: High-level Architecture.

3.2   Data Ingestion

To ensure the RAG effectiveness, an initial knowledge base is created by preprocessing and storing the underlying document corpus for downstream reference. Given the architectural complexity of the system, its core phases can be outlined as follows.

Text Extraction. The architecture has been developed to ingest textual documents provided in various formats, including PDF, Markdown (MD), and Word documents. Preprocessing is essential for extracting and structuring as much useful information as possible. Specifically, the following activities are performed:

- Execution of OCR techniques in the case of images containing useful text.
- Extraction of main document structures to identify sections, lists, and tables.

Preprocessing. The preprocessing step involves cleaning and normalizing the extracted text to ensure consistency and facilitate downstream processing. This step includes tasks such as removing irrelevant characters, correcting encoding issues and properly formatting embedded data structures, which are often distorted during extraction. The result of the preprocessing is then formatted in *Markdown* form, as it proves to be highly effective in representing the extracted information and is efficiently processed by Large Language Models (LLMs) [7, 15].

Recursive Chunking. One of the most crucial phases of RAG involves splitting the document information into smaller atomic units, each containing as much valid information as possible. Chunking is necessary due to certain limitations of LLMs, such as their limited input context (which prevents them from considering an entire document) and their tendency to exhibit reduced performance with a large quantity of information, resulting in a high risk of overlooking practical details and diminishing context processing capabilities.

The proposed solution, which considers text excerpts (chunks) of varying granularity and detail, is particularly suitable for a highly heterogeneous corpus of documents and permits overcoming the aforementioned limitations. Specifically, recursive chunking has been adopted, starting with macro-chunks (of 1024 tokens) with character overlap, which are then recursively divided into smaller chunks (256 and 512 tokens). This approach allows for considering information at different levels of detail, ensuring high performance across a very general set of tasks (from targeted question-answering to information summarization).

Embedding. To correctly manage the various generated chunks, it is essential to represent textual information in an efficient format easily handled by different retrievers. For this reason, they are transformed into a vector format (*embeddings*) using precise models that will be discussed in Sect. 5. These embeddings are then stored in a single vector database (*VectorDB* in Fig. 1), which is capable of scaling and managing a large number of embeddings.

3.3 Inference

The inference phase focuses on a Question-Answering (QA) system capable of providing precise and detailed answers to specific user queries. Let us remark that the sole source of information (knowledge base) resides exclusively within the processed document base, and the system can provide precise references to the information used, thereby enhancing transparency.

Query Processing. First, the user's query is analyzed by an LLM, which will rephrase the query if necessary. Indeed, this step is crucial for managing poorly phrased, convoluted, or ill-formed questions, as the information retrieval phase primarily utilizes the concept of semantic similarity between the query and the chunks as the primary means of retrieving information.

Multi-granular Retrieval. The retrieval of chunks is a primary and fundamental aspect of RAG, and it is decisive for the entire system's performance. To fully leverage the hierarchical structure of the extracted chunks, a recursive retriever was chosen. Unlike a basic retriever, it utilizes semantic similarity to retrieve chunks most relevant to the query. In particular, *cosine similarity* is employed to compute the relevance between the query and the chunk embeddings, as it effectively measures the angular distance in high-dimensional vector spaces [36]. It also considers the parent-child relationships between chunks to retrieve information at different levels of granularity.

Re-ranking. Thereafter, a *re-ranking* process is also employed. Specifically, an LLM is used to reorder the chunks based on their relevance to the question and potential utility in answering it. Although this step may seem trivial, it proves to be fundamental and highly beneficial to the LLM [40]. By nature, LLMs tend to consider information provided at the beginning less important while giving more weight to information contained at the end. Through re-ranking, information is ordered precisely by exploiting this particular characteristic to achieve even higher performance. This re-ranking is performed through a prompt-based approach: the LLM is provided with a list of candidate chunks and explicitly instructed to reorder them based on their semantic relevance and informativeness concerning the query. This technique leverages the LLM's reasoning and comprehension capabilities to dynamically adjust chunk ordering, going beyond static similarity scores alone.

Answer Generation. The information extracted in the previous stages is leveraged to provide the model with a proper context. Let us remark that such information is the sole source of knowledge. This was made possible through a targeted prompting phase guided by several fundamental principles:

- Contextual Adherence: The system must rely solely on the information provided to it.
- Hallucination Reduction: Partially guaranteed by the previous point, the system must not make inferences that could lead to erroneous results and unreal information.
- Exhaustive and Clear Answers: The system must provide precise answers and offer additional contextual information for greater clarity.

In this final step, the system will return the generated response along with the correctly ordered text chunks. These are necessary to provide the user with a reference ground truth and ensure a more transparent and reliable system.

3.4   Inference Walkthrough

To illustrate the inference process, let us consider the example reported in Fig. 2. Given a user request (`User Query`), the system starts by rephrasing the question to make it more transparent and more precise (Reformulated Query). According to the refined question, the system then searches for the most relevant information in its knowledge base (Inferred Information). Employing this information, the system crafts the final answer (Generated Answer).

```
<User Query>:
What major event happened in Alaska on March 27,
1964 killing 133 people?

<Reformulated Query>:
What significant event occurred in Alaska on March 27,
1964, resulting in the death of 133 people?

<Inferred Information>:
On March 27, 1964, the massive Good Friday earthquake
killed 133 people and destroyed several villages...

<Generated Answer>:
On March 27, 1964, the massive Good Friday earthquake
occurred in Alaska, resulting in the deaths of 133 people.
The earthquake also destroyed several villages and
portions of large coastal communities, mainly due
to the resultant tsunamis and landslides. It was
the second most powerful earthquake in the recorded
history of the world, with a moment magnitude of 9.2.
```

Fig. 2. Example of answer generation.

This example showcases the entire RAG workflow: the user's question is first refined, then the most relevant information is retrieved, re-ranked, and finally used to create an accurate and well-grounded answer.

4   Experimental Setup

This section describes the experimental setup. In particular, the adopted dataset is first described, followed by a description of the evaluation metrics used.

4.1   Dataset Creation

The dataset employed for the comparative analysis across various systems and models is derived from SQuAD2.0 (The Stanford Question Answering Dataset) [31]. This comprehensive dataset comprises over 500 documents extracted from Wikipedia, along with more than 150,000 associated questions.

For our study, we created a dataset through a sampling process from SQuAD2.0. In detail, we selected 100 documents and 20 answerable questions for each of them. This methodology resulted in a final dataset containing 2000 questions, ensuring a perfectly balanced distribution of questions per document.

Each sample within our custom dataset is composed of the following items:

– *Document*: A reference to the source document.
– *Question*: The specific question posed.
– *Answer*: The ground truth answer to the question.
– *Chunk*: The precise portion of the document containing the information necessary to answer the question.

This dataset encompasses a highly heterogeneous corpus of topics, enabling a more accurate and generalizable analysis. All documents were grouped into broader macro-categories to facilitate further in-depth analysis, as illustrated in the graph depicted in Fig. 3.

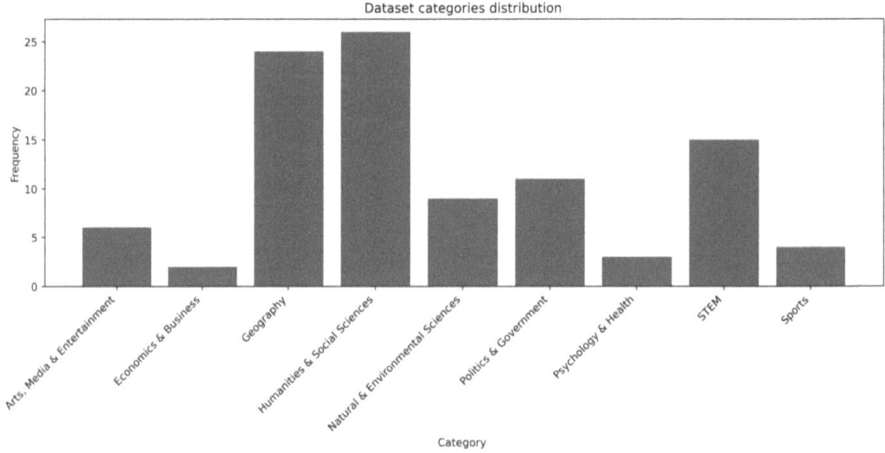

Fig. 3. Categories distribution in the dataset.

4.2   Evaluation Metrics

This work undertakes a comprehensive evaluation of the implemented RAG (Retrieval Augmented Generation) system, with a significant emphasis on a comparative analysis of the different embedding models and LLMs utilized. This aims to highlight performance differences across models and identify their respective strengths. We detail the structured tests and their execution below.

Retrieval Assessment. One of the essential aspects of RAG systems is the effective retrieval of relevant context to answer the given query. Performance in this area depends on the system's architecture and the embedding models' ability to represent information and perform semantic searches accurately.

For evaluation purposes, we employed two key metrics: *Precision@k* and *Normalized Discounted Cumulative Gain at k (nDCG@k)*.

The *Precision@k* metric [39] considers the number of correctly extracted chunks among the top k results. The Precision@k metric is defined as:

$$\text{Precision@k} = \frac{|\{r \in R_k \mid r \in G\}|}{|R_k|}$$

where:

- R_k is the set of the top k results returned by the system (the retrieved chunks).
- G is the set of relevant chunks (the gold set).
- $|\cdot|$ denotes the cardinality of the set.

Additionally, we utilized *nDCG@k* [1] to evaluate the ranking quality of the retrieved results, especially when relevance is graded. DCG@k (Discounted Cumulative Gain at k) is calculated as:

$$\text{DCG@k} = \sum_{i=1}^{k} \frac{2^{\text{rel}_i} - 1}{\log_2(i + 1)}$$

where:

- rel_i is the relevance score of the result at position i.

nDCG@k normalizes DCG@k by the Ideal DCG@k (IDCG@k), which is the DCG obtained by an ideal ranking where all relevant documents are ranked at the top:

$$\text{nDCG@k} = \frac{\text{DCG@k}}{\text{IDCG@k}}$$

This normalization ensures that nDCG@k values are between 0 and 1, where 1 represents a perfect ranking.

Answer Formulation Assessment. A further evaluation concerns the LLM's ability to generate an answer based on the previously extracted context. It is important to note that, unlike the assessment of the RAG architecture, where a baseline was established, a separate baseline for the LLMs in isolation is not required for this analysis. Our objective here is to conduct a direct comparison of how well different LLMs leverage the relevant information provided by our advanced retrieval pipeline to formulate responses.

To evaluate the capability of a given LLM to synthesize information from relevant retrieved content accurately, we developed a two-phase automated evaluation framework. This framework is designed to provide an objective, scalable, and robust assessment of LLM performance, leveraging a panel of several LLMs as evaluators. In the first phase, we first identified test samples for which the correct evidence chunks required to answer the question were successfully retrieved. These samples were then processed using several LLMs to generate responses. To ensure consistency and fairness in the evaluation, we employed an LLM-based majority voting system, in which each LLM adopted as an assessor was asked to judge the correctness of a response. In particular, a response was deemed *correct* if and only if it met all the following criteria:

- It accurately reflected the content of the ground truth.
- It did not use information external to the provided context.
- It did not generate hallucinations.
- It did not contradict the ground truth.

Although some of these criteria may intersect, their combination contributes to a more rigorous and reliable evaluation process. In the second phase of the framework, we scored models not only on correctness but also on the diversity and uniqueness of their correct solutions, further enriching our understanding of their synthesis capabilities. In the following, both phases are detailed.

Phase 1: Weighted Voting Correctness. Each answer produced by a model under evaluation is submitted to a panel of four distinct evaluator LLMs. The process employs a weighted-voting strategy, which assigns different levels of influence to evaluators based on their presumed reliability, as described in Sect. 5. This works as follows:

1. Each evaluator LLM assigns a score to the answer based on the following criteria:
   - **0 points:** for an answer deemed incorrect.
   - **0.8 points:** for a correct answer, as determined by a model from the Llama or Gemma family.
   - **1.2 points:** for a correct answer, as determined by a model from the GPT or Gemini family.
2. The scores from the four evaluators are summed. An answer is considered correct if, and only if, the total score is strictly greater than **2.0**.

This threshold ensures that an answer's validation requires a significant level of agreement, thereby mitigating the risk of errors from any single evaluator.

The choice of the weights (**0.8** and **1.2**) is the result of an empirical tuning of the subjective trustworthiness of the evaluator models. GPT and Gemini family models have generally demonstrated superior performance on various evaluation benchmarks, particularly in terms of alignment and factual consistency [14]. They are, therefore, accorded higher weight (**1.2** points) in the voting. Although the LLaMA and Gemma models exhibit competitive performance, they demonstrated slightly lower stability in evaluative consistency across diverse tasks. Consequently, they are assigned a moderate weighting factor (i.e., 0.8) in the scoring process.

The variable weighting is designed to balance the opinion diversity and resistance to over-reliance on any single evaluator. The response is assessed by a single evaluator from each of the model families: one GPT, one Gemini, one LLaMA, and one Gemma. The use of fixed weighting prevents any single model family from disproportionately influencing the outcome. By assigning higher weights to the more reliable evaluators (GPT and Gemini), the approach enhances the robustness of the scoring process, while still preserving a heterogeneous consensus as the basis for determining correctness. For instance, an answer labeled as correct by both GPT and Gemini evaluators ($1.2 + 1.2 = 2.4$) is accepted even if the other two evaluators label it as incorrect. Conversely, if only LLaMA and Gemma evaluators agree ($0.8 + 0.8 = 1.6$), the answer is not accepted, unless at least one of the stronger evaluators also agrees. The threshold of **2.0** for final answers has been derived from preliminary experiments on a validation set, which allowed us to infer the best trade-off between precision and recall in accepting correct answers. A lower threshold risked admitting borderline or wildly judged answers, while a higher one made the system over-cautious and lost correct answers. The strict "greater than"requirement (i.e., > 2.0) contributes to the robustness by demanding some consensus from at least one of the stronger judges.

Phase 2: Performance Scoring. Once an answer for a given test sample t has been validated as correct in Phase 1, a performance score σ_t is assigned to the model that generated it. This score is calculated using a shared credit mechanism designed to reward the ability to provide distinctive, correct solutions. In detail, σ_t is computed as follows:

$$\sigma_t = \frac{1}{k}$$

where k represents the total number of models that provided a correct answer for the sample t. The underlying criterion is to award **1 point** to a given LLM if it was the sole model to answer correctly. The value of σ_t decreases as the number of models producing the correct answer increases, reflecting a lower discriminative power of the test case. This scoring method effectively normalizes task difficulty, assigning greater value to correct answers that are also rare, thereby highlighting models with unique problem-solving capabilities.

Considering all the N test samples, we compute the cumulative score as follows:

$$\sigma = \sum_{t=1}^{N} \sigma_t$$

4.3   Models Selection

The experiments carried out involved four LLMs and four embedding models. In detail, the adopted LLMs are listed below:

- `Gemini 2.0 Flash`[13];
- `GPT-4o mini` [25];
- `LLaMA 3.1 8B`[21];
- `Gemma 2 9B`[11];

Furthermore, the selected embedding models are the following:

- `text-multilingual-002 (TME)` [12];
- `text-embedding-3-small (TES)` [26];
- `nomic-embed-text (NET)` [24];
- `snowflake-arctic-embed (SAE)` [37].

As summarized in Table 1, which reports the main characteristics of each model, the adopted LLMs differ in their weights, capabilities, and performance. In particular, the latter is reported on common and widely used benchmarks[1]: MMLU (Massive Multi-task Language Understanding) [16] and HumanEval [6].

Table 1. Comparative analysis of the selected Large Language Models.

Model	Parameters	Input/Output Size	MMLU (%)	HumanEval (%)
GEMINI 2.0 FLASH	Not disclosed	1M tokens	**83.4**	**90.0**
GPT-4O MINI	8B	128K tokens	82.0	87.2
LLAMA 3.1 8B	8B	128K tokens	69.4	73.0
GEMMA 2 9B	9B	8K tokens	71.3	40.2

As can be seen from Table 1, there is a significant performance gap between the first two models (GPT and Gemini) and the two open-source alternatives. Indeed, Gemini 2.0 Flash and GPT-4o mini demonstrate superior reasoning and language understanding capabilities. This insight motivated the decision to assign different weights in the voting strategy described in the previous section. Subsequent experiments will confirm and further validate these assumptions.

For the sake of completeness, the details of embedding models used in the experiments are listed in Table 2. Performance is reported as the average score on the Massive Text Embedding Benchmark (MTEB) [22], as documented in the model's documentation.

Table 2. Embedding models used in the experiments.

Model	Parameters	Max Context	MTEB avg (%)
TME	Not disclosed	2048 tokens	**62.16**
TES	Not disclosed	8191 tokens	54.26
NET	137M	8192 tokens	45.92
SAE	137M	~8K tokens	43.33

In this case, a clear benefit is observed for Google's *text-multilingual-embedding-002* model, followed by OpenAI's *text-embedding-3-small*.

To execute the tests, we decided to use the aforementioned LLMs paired with specific embedding models. Below, we propose the different pairs considered:

- GEMINI 2.0 FLASH and TME
- GPT-4O MINI and TES
- LLAMA 3.1 8B and NET
- GEMMA 2 9B and SAE

---

[1] For the sake of clarity, it is worth noting that performance on benchmarks is documented in the official model specifications and is not part of the empirical evaluation conducted in this study.

4.4 Baseline Model

To ensure a robust comparison, a baseline was designated using a simple RAG architecture. This allowed us to assess the improvement brought by the new architecture independently of the differences in the models used. The baseline architecture follows a structure similar to the more advanced one, always including a document pre-processing module. However, it features reduced complexity in the following modules:

- *Chunk Extraction*: Unlike the proposed solution, which employs recursive chunking for varying granularity, in the baseline, chunks are extracted based on a pre-defined and fixed size;
- *Query Processing*: The user's query is not reformulated by an LLM to manage poorly phrased or ill-formed questions;
- *Retrieval* and *Re-ranking*: No re-ranking process was employed after initial retrieval. Furthermore, unlike the recursive retriever used in the advanced architecture, the retrieval process does not leverage the chunk hierarchy (as it is not established during the ingestion phase).

5 Results

In this section, we report all the experimental results in accordance with the technological choices described in the previous section.

5.1 Chunk Retrieval

To evaluate the impact of the proposed architecture, we show the improvement obtained in chunk retrieval by the different embedding models. As previously explained, this is assessed by applying the Precision@k metric to both the baseline RAG architecture and the advanced one we developed. Table 3 reports the results in terms of Precision@5 and Precision@1. The percentage improvement ($\Delta\%$) achieved by our model, compared to the baseline, is also reported.

Table 3. Performance comparison between the baseline and our RAG model.

Model	Precision@5			Precision@1		
	Base	Adv	$\Delta\%$ P@5	Base	Adv	$\Delta\%$ P@1
GEMINI 2.0 FLASH + TME	0.403	**0.892**	**+121.4%**	0.568	**0.911**	**+60.4%**
GPT 4O-MINI + TES	**0.420**	0.747	+77.9%	**0.591**	0.843	+42.6%
LLAMA 3.1 8B + NET	0.411	0.732	+78.1%	0.586	0.741	+26.5%
GEMMA 2 9B + SAE	0.377	0.697	+84.9%	0.499	0.697	+39.7%

Significant improvements can be noted in all the tests performed, with an average improvement for Precision@5 exceeding **90%** and an average improvement for Precision@1 exceeding **40%**. Intuitively, it is evident that the proposed architecture has improved performance in retrieving relevant information across the different embedding models.

5.2 Models Comparison

The results presented in Table 3 also enable a comparative analysis of the models' capabilities within the context of our advanced RAG architecture. In detail, all models show important results, with the GEMINI + TME pair emerging as the absolute winner. An interesting finding concerns the open-source LLAMA + NET pair, which achieves results similar to GPT + TES in Precision@5.

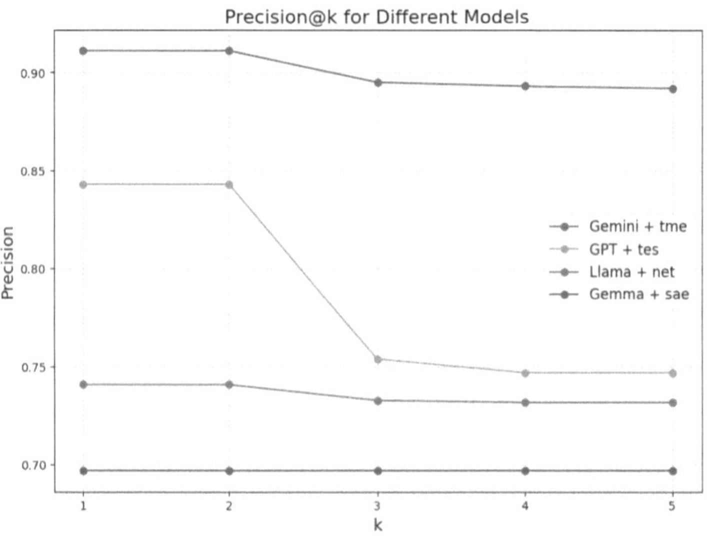

Fig. 4. Precision@k for different embedding models within the advanced RAG architecture.

To give an additional in-depth analysis, Fig. 4 shows the Precision@k for varying values of k within the advanced architecture. Consistent with the aggregate data in Table 3, the GEMINI + TME combination consistently maintains the highest precision across all k values, highlighting its superior capability in retrieving relevant information. The performance of GPT + TES shows an important drop after $k = 2$. In contrast, LLAMA + NET exhibits a more stable, albeit lower, precision across the range of k. The GEMMA + SAE pair consistently shows the lowest precision among the evaluated models. This detailed analysis reinforces the findings from the aggregate Precision@1 and Precision@5 values, providing a comprehensive understanding of the retrieval effectiveness of each model.

Besides looking into Precision@k, we also sought to understand the Normalized Discounted Cumulative Gain at k (nDCG@k) to assess further how well the sequences were ranked among the results retrieved, factoring in graded relevance.

Different models were compared regarding their nDCG@k values, which are shown in Fig. 5.

The figure highlights similar performances of the two settings GEMINI + TME and GPT + TES. Notably, both setups demonstrated excellent retrieval effectiveness, with

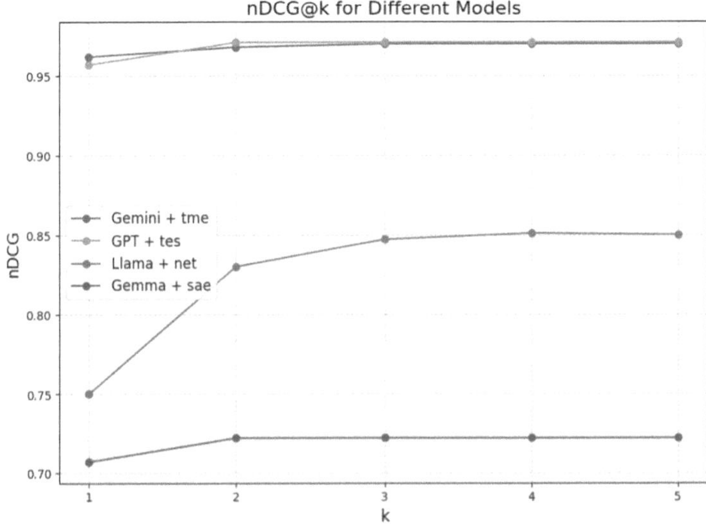

Fig. 5. nDCG@k for different embedding models within the advanced RAG architecture.

nDCG@k values exceeding 0.95 at every cutoff, implying a highly accurate ranking of relevant documents. Furthermore, Fig. 5 underlines how such models consistently achieve peak nDCG scores across all values of k, underscoring their dominance over the open models.

The most notable behavior was exhibited by combination LLAMA + NET, which displayed significant gains in its nDCG@k as k, starting around .75 at k = 1 and surpassing .85 by k = 5. This indicates that while perhaps less consistent in placing the most relevant documents at the highest positions compared to Gemini or GPT, NET retrieves a substantial number of relevant documents within the top k positions and ranks them reasonably well.

5.3   Large Language Model Comparison

As explained in the previous section, a further analysis focuses on LLMs and their ability to synthesize answers based on the retrieved context. Table 4 reports the cumulative performance score σ defined for assessing the answer formulation and described in Sect. 4.2.

Table 4. Evaluation scores for the answer generation capabilities of each LLM.

Models	σ
GEMINI 2.0 FLASH + TME	322.0
GPT 4O-MINI + TES	**322.2**
LLAMA 3.1 8B + NET	297.0
GEMMA 2 9B + SAE	273.8

From such results, we can observe nearly identical performance between the two top-performing models, GEMINI 2.0 FLASH and GPT-4O MINI. This agrees with the previously shown benchmarks, which described negligible differences. A clear gap exists between the two open-source models, which demonstrate lower performance than Gemini and GPT and differ from each other. With a difference of over **20** points, Llama appears more performant than Gemma. This partially agrees with the earlier results, where Llama performed better on one of the two considered metrics.

5.4 Strengths and Weaknesses

Strengths. The proposed system introduces several notable strengths. By grounding generated responses in a verified and structured knowledge base, the system improves both factual accuracy and interpretability, which are essential for high-stakes or domain-sensitive applications. Empirical evaluations demonstrate a significant enhancement in retrieval performance. The pairing of GEMINI 2.0 FLASH with TEXT-MULTILINGUAL-EMBEDDING-002 consistently achieved the best retrieval scores across test scenarios, validating the effectiveness of the embedding selection and retrieval pipeline.

The system also supports a modular evaluation of LLMs within the RAG setup, enabling systematic benchmarking. In this context, both GEMINI 2.0 FLASH and GPT-4O MINI emerged as top performers in response generation, producing high-quality, coherent, and contextually grounded outputs. Their comparable performance underscores the robustness and flexibility of the system across different model configurations.

Finally, a further strength lies in the inclusion and assessment of open-source models, such as LLAMA 3.1 8B and GEMMA 2 9B, within the same framework, contributing to a more inclusive and transparent evaluation landscape.

Weaknesses. During evaluation, a significant limitation noted was the system's occasional failure to retrieve relevant information, even when it was present in the indexed documents. This deficiency typically occurs during the retrieval phase due to embedding errors, retrieval threshold limits, or incompatibility between the query and document embeddings. Consistent with the previously noted advantage of the architecture, the system did not generate a false answer in such cases. Instead, it consistently provided a response stating that no relevant information was found. This behavior is valuable in domains where factual correctness is critical because it avoids generating misleading or fabricated content, a common problem in generation-first systems that are not tightly coupled with retrieval. In some systems, this behavior may seem less advanced due to the lack of an attempt to answer the query. However, such systems promote trust and transparency, ultimately benefiting the system.

6 Conclusion

This paper addresses the critical need for more reliable and transparent Large Language Model applications by introducing a novel architectural refinement to the Retrieval-Augmented Generation pipeline. Our primary contribution is the development of an explainable RAG system designed to mitigate pervasive limitations of LLMs, notably hallucinations and insufficient source transparency. This is achieved by rigorously grounding all generated responses in a verified and well-built knowledge base. A key demonstration of our work involves significant improvements in chunk retrieval performance, as demonstrated during the validation phase. Moreover, we present a comprehensive comparative analysis of various LLMs and embedding model pairings within this enhanced framework. Our experiments indicate that the GEMINI 2.0 FLASH + TEXT-MULTILINGUAL-EMBEDDING-002 combination demonstrated superior performance in retrieval tasks. Concurrently, GEMINI 2.0 FLASH and GPT-4O MINI exhibited comparable, high-tier performance in response generation, highlighting their advanced capabilities. We also contribute by evaluating the performance of well-known open-source LLMs, including LLAMA 3.1 8B and GEMMA 2 9B, within this improved RAG setup, providing valuable insights into their practical applicability. This research offers a concrete solution for constructing more robust and dependable LLM-powered information systems. Our contributions lay the groundwork for developing more trustworthy and responsible AI applications across knowledge-intensive domains by emphasizing enhanced information retrieval and transparent response generation.

Future developments will aim to improve both the accuracy and adaptability of the proposed architecture. For instance, one promising direction might be to fine-tune the model weights and threshold parameters used to evaluate LLM-generated responses, which would help the system better identify high-quality outputs. It would also be interesting to explore the integration of reasoning-oriented models, as they could enhance the system's ability to manage more complex queries. Furthermore, testing the pipeline on multilingual datasets could provide insights into its robustness across different languages and contexts. Finally, extending the system to support multimodal input, like adding images during the generation phase, could significantly enrich the responses in fields where visual information plays a crucial role.

Acknowledgments. We acknowledge financial support under the National Recovery and Resilience Plan (NRRP), Mission 4 Component 2 Investment 1.5 - Call for tender No.3277 published on December 30, 2021 by the Italian Ministry of University and Research (MUR) funded by the European Union - NextGenerationEU. Project Code ECS0000038 - Project Title eINS Ecosystem of Innovation for Next Generation Sardinia - CUP F53C22000430001 - Grant Assignment Decree No. 1056 adopted on June 23, 2022 by the Italian Ministry of University and Research (MUR).

We also acknowledge financial support under the National Recovery and Resilience Plan (NRRP), Mission 4 Component 2 Investment 1.5 - Call for tender No.328 published on March 27, 2023 by the Italian Ministry of University and Research (MUR), funded by the European Union - NextGenerationEU (Project Code ECS00000036 - Project NODES: -"Nord-Ovest Digitale E Sostenibile" - CUP J83B22000050001), through a cascade call (Project CAMELIA - "Creazione di conoscenzA per il turisMo culturalE tramite intelLIgenza Artificiale", CUP C29J24000260003).

We also acknowledge financial support under the National Recovery and Resilience Plan (NRRP), Mission 4 Component 2 Investment 1.3 - Call for tender No.341 published on March 15, 2022 by the Italian Ministry of University and Research (MUR), funded by the Euuropean Union - NextGenerationEU (Project Code PE0000013 - Project FAIR: Future Artificial Intelligent Research - CUP J53C22003010006), through a cascade call (Project ANSWER - innovAtive human-iN-the-loop-baSed knowledge undERstanding, CUP J23C24000080007).

References

1. Discounted cumulative gain. https://en.wikipedia.org/wiki/Discounted_cumulative_gain. Accessed 25 Jun 2025
2. Armano, G., Giuliani, A.: A two-tiered 2D visual tool for assessing classifier performance. Inf. Sci. **463–464**, 323–343 (2018)
3. Arnold, M., Goldschmitt, M., Rigotti, T.: Dealing with information overload: a comprehensive review. Front. Psychol. **14**, 1122200 (2023)
4. Carta, S., et al.: A zero-shot strategy for knowledge graph engineering using GPT-3.5. Procedia Comput. Sci. **246**, 2235–2243 (2024). https://doi.org/10.1016/j.procs.2024.09.573
5. Carta, S., Podda, A.S., Reforgiato Recupero, D., Stanciu, M.M.: Explainable AI for financial forecasting. In: Nicosia, G. (ed.) International Conference on Machine Learning, Optimization, and Data Science, pp. 51–69. Springer, Cham (2021). https://doi.org/10.1007/978-3-030-95470-3_5
6. Chen, M., et al.: Evaluating large language models trained on code. arXiv preprint arXiv:2107.03374 (2021)
7. Chen, Z., et al.: MDEval: evaluating and enhancing markdown awareness in large language models. In: Proceedings of the ACM on Web Conference, pp. 2981–2991 (2025)
8. Elia, N., et al.: Smart contracts for certified and sustainable safety-critical continuous monitoring applications. In: Chiusano, S., Cerquitelli, T., Wrembel, R. (eds) European Conference on Advances in Databases and Information Systems, vol. 13389, pp. 377–391. Springer, Cham (2022). https://doi.org/10.1007/978-3-031-15740-0_27
9. Gao, Y., et al.: Retrieval-augmented generation for large language models: a survey. arXiv preprint arXiv:2312.10997, **2**(1) (2023)
10. Giuliani, A., Savona, R., Carta, S., Addari, G., Podda, A.S.: Corporate risk stratification through an interpretable autoencoder-based model. Comput. Oper. Res. **174**, 106884 (2025)
11. Google: Gemma 2 (2025). https://ai.google.dev/gemma/docs/core/model_card_2
12. Google: text-multilingual-embedding-002 (2025). https://cloud.google.com/vertex-ai/generative-ai/docs/model-reference/text-embeddings-api
13. Google DeepMind: Gemini 2.0 flash (2025). https://developers.googleblog.com/en/gemini-2-family-expands/
14. Gu, J., et al.: A survey on LLM-as-a-judge. arXiv preprint arXiv:2411.15594 (2024)
15. He, J., Rungta, M., Koleczek, D., Sekhon, A., Wang, F.X., Hasan, S.: Does prompt formatting have any impact on LLM performance? arXiv preprint arXiv:2411.15594 (2024)
16. Hendrycks, D., et al.: Measuring massive multitask language understanding. arXiv preprint arXiv:2009.03300 (2020)
17. Krishna, S., et al.: Fact, fetch, and reason: a unified evaluation of retrieval-augmented generation. arXiv preprint arXiv:2409.12941 (2024)
18. Li, Z., Li, C., Zhang, M., Mei, Q., Bendersky, M.: Retrieval augmented generation or long-context LLMs? A comprehensive study and hybrid approach. In: Proceedings of the 2024 Conference on Empirical Methods in Natural Language Processing: Industry Track, pp. 881–893 (2024)
19. Liu, N.F., et al.: Lost in the middle: how language models use long contexts. arXiv preprint arXiv:2307.03172 (2023)
20. Mana, A., Allouhi, A., Hamrani, A., Rehman, S., El Jamaoui, I., Jayachandran, K.: Sustainable AI-based production agriculture: exploring AI applications and implications in agricultural practices. Smart Agri. Technol. **7**, 100416 (2024)
21. Meta AI: Llama 3.1 (2025). https://ai.meta.com/blog/meta-llama-3-1/
22. Muennighoff, N., Tazi, N., Magne, L., Reimers, N.: MTEB: massive text embedding benchmark. arXiv preprint arXiv:2210.07316 (2022)

23. Nallakaruppan, M., Balusamy, B., Shri, M.L., Malathi, V., Bhattacharyya, S.: An explainable AI framework for credit evaluation and analysis. Appl. Soft Comput. **153**, 111307 (2024)
24. Nomic AI: nomic-embed-text (2025). https://www.nomic.ai/blog/posts/nomic-embed-text-v1
25. OpenAI: GPT-4o mini (2025). https://openai.com/index/gpt-4o-mini-advancing-cost-efficient-intelligence/
26. OpenAI: text-embedding-3-small (2025). https://platform.openai.com/docs/models/text-embedding-3-small
27. Piano, L., Pisu, A., Tiddia, S.G., Carta, S., Giuliani, A., Pompianu, L.: LLIMONIIE: large language instructed model for open named Italian information extraction. J. Intell. Inf. Syst. (2025). https://doi.org/10.1007/s10844-025-00978-w
28. Pisu, A., Elia, N., Pompianu, L., Barchi, F., Acquaviva, A., Carta, S.: Enhancing workplace safety: a flexible approach for personal protective equipment monitoring. Exp. Syst. Appl. **238**, 122285 (2024)
29. Podda, A.S., Balia, R., Manca, M.M., Martellucci, J., Pompianu, L.: A deep learning strategy for the 3D segmentation of colorectal tumors from ultrasound imaging. Image Vis. Comput., 105668 (2025)
30. Podda, A.S., Balia, R., Pompianu, L., Carta, S., Fenu, G., Saia, R.: CARgram: CNN-based accident recognition from road sounds through intensity-projected spectrogram analysis. Digi. Sig. Proc. **147**, 104431 (2024)
31. Rajpurkar, P., Zhang, J., Lopyrev, K., Liang, P.: SQuAD: 100,000+ questions for machine comprehension of text. arXiv preprint arXiv:1606.05250 (2016)
32. Rong, Y., et al.: Du-bus: a realtime bus waiting time estimation system based on multi-source data. IEEE Trans. Intell. Transp. Syst. **23**(12), 24524–24539 (2022)
33. Saad-Falcon, J., Khattab, O., Potts, C., Zaharia, M.: ARES: an automated evaluation framework for retrieval-augmented generation systems. arXiv preprint arXiv:2311.09476 (2023)
34. Saia, R., Carta, S., Fenu, G., Pompianu, L.: Influencing brain waves by evoked potentials as biometric approach: taking stock of the last six years of research. Neural Comput. Appl. **35**(16), 11625–11651 (2023)
35. Shahid, N.U., Sheikh, N.J., et al.: Impact of big data on innovation, competitive advantage, productivity, and decision making: literature review. Open J. Bus. Manag. **9**(02), 586 (2021)
36. Singhal, A., et al.: Modern information retrieval: a brief overview. IEEE Data Eng. Bull. **24**(4), 35–43 (2001)
37. Snowflake: Snowflake arctic – open large language model (2025). https://www.snowflake.com/cn/blog/introducing-snowflake-arctic-embed-snowflakes-state-of-the-art-text-embedding-family-of-models/
38. Tang, Y., Yang, Y.: MultiHop-RAG: benchmarking retrieval-augmented generation for multi-hop queries. arXiv preprint arXiv:2401.15391 (2024)
39. Wikipedia Contributors: Evaluation measures (information retrieval) (2025). https://w.wiki/ATBY
40. Yu, Y., et al.: RankRAG: unifying context ranking with retrieval-augmented generation in LLMs. Adv. Neural. Inf. Process. Syst. **37**, 121156–121184 (2024)
41. Zhang, W., Zhang, J.: Hallucination mitigation for retrieval-augmented large language models: a review. Mathematics **13**(5), 856 (2025)

How Prompting Shapes Decisions: Analyzing LLM Behavior in XAI-Augmented Decision Support Systems

Finn Schwall[1,3]($\boxtimes$) (iD), Maximilian Becker[1,2] (iD), Anmol Ashri[1],
and Jürgen Beyerer[1,2]

[1] Fraunhofer IOSB, Karlsruhe, Germany
`https://www.iosb.fraunhofer.de/`
[2] Vision and Fusion Laboratory, Karlsruhe Institute of Technology,
Karlsruhe, Germany
`https://ies.iar.kit.edu/`
[3] Simula, Oslo, Norway
`finn@simula.no`
`https://www.simula.no`

Abstract. Large Language Models (LLMs) become increasingly prevalent in downstream tasks and user-facing applications. As explainable AI (XAI) strives to become more end-user-friendly, utilization of LLMs in XAI increases. However, the consequences of this are unclear. Do LLMs really improve user experience, or are there hidden problems that may limit their applicability? In this paper, we present results of experiments on the decision-making process of LLMs with the goal of evaluating their usefulness for such applications. By providing the LLM with different information and applying different metrics to evaluate their decisions, we present findings that should inform the applicability of LLMs. By analyzing nearly 300,000 prompts, we found that the LLM's decisions are only minimally influenced by XAI data. Secondly, the LLM's behavior can be changed significantly through prompting. This suggests that LLM behavior is more sensitive to presentation than to underlying model reliability, raising concerns about its role as a rational arbiter. If our results hold true, we advise caution when utilizing LLMs, especially when facing laypeople.

Keywords: XAI · LLMs · LLM Reasoning

1 Introduction

With the growing integration of LLMs in XAI applications, understanding their decision-making patterns becomes crucial. This paper presents an exploratory study into drivers for decision-making in a decision support system scenario. It is also a proof of concept for future research in this direction and highlights many open questions.

© The Author(s), under exclusive license to Springer Nature Switzerland AG 2026
F. Marcelloni et al. (Eds.): IJCCI 2025, CCIS 2829, pp. 232–248, 2026.
`https://doi.org/10.1007/978-3-032-15638-9_14`

This study emerged from an ongoing large-scale user study examining human behavior when solving tasks with a hybrid decision support system, which integrates XAI into an LLM-supported dashboard. We realized that the LLM in this and similar scenarios functions as an arbiter of information. However, key drivers, decision tendencies, biases, etc. of this arbiter have not been investigated yet.

In the user study, participants are asked to take on the role of a hotel manager and decide whether to accept an overbooking. To make an informed decision, an AI model has been trained and users are given access to various XAI methods through prompting of the LLM.

Here, we ask the LLM to make a decision, without the presence of a user, but with all of the information a user would have. By altering different aspects, such as the phrasing of prompts, the information available to the model, the presence of XAI methods and so forth, we aimed to discover, whether an LLM has decision tendencies and if so, what drives them. Our core research questions are:

1) Are LLMs consistent decision-makers?
2) What factors most strongly influence LLM decisions?
3) Are there any biases?
4) What implications does this have for system design?

Through systematic experimentation using nearly 300,000 prompts, we find that prompt phrasing is by far the strongest decision driver, while factors like model accuracy have minimal impact. These findings have important implications for designing LLM-based decision support systems.

The remainder of this paper is structured as follows. Section 2 reviews prior work on LLMs in the context of XAI. Our methodology and experimental setup are introduced in Sect. 3. Afterwards, Sect. 4 describes our findings and how we reached them. Section 5 concludes with key insights and limitations in the context of using LLMs in practical decision-support systems.

2   Background

Previous studies have investigated the ability of LLMs to accurately translate XAI outputs into natural language and simplify explanations for diverse audiences. However, few studies have investigated LLMs as decision-making tools, especially in relation to various "hyperparameters" such as prompt phrasing or explanation modes. This section reviews the relevant literature on LLMs in the context of XAI.

2.1   LLMs as Translators of XAI Explanations

In their study, Ichmoukhamedov et al. [6] systematically evaluated LLM-generated explanations using a set of quantitative metrics focused on faithfulness, human similarity, and plausibility of assumptions. They analyzed how well the LLM generated narratives aligned with underlying XAI outputs like SHAP. However, their work concentrated solely on narrative quality and did not involve any downstream decision-making tasks based on those explanations.

Susnjak et al. [8] proposed a system that combines explainable and prescriptive analytics to support at-risk students. This framework uses tools like SHAP and counterfactuals, as well as ChatGPT, to provide personalized feedback. The focus was on translating model outputs into clear, learner-friendly narratives. The evaluation was mostly qualitative, focusing on narrative clarity without verifying correctness or alignment with the ground truth. The study did not include a detailed examination of the impact of prompt variations or input changes on LLM behavior. It also acknowledged limitations such as reduced accuracy when using actionable features, high computing cost and limited automation possibilities.

Similarly, Burton et al. [3] addressed the complexity of making XAI outputs more accessible by fine-tuning BART and T5 language models to translate explainable output of XAI methods visualization plots into natural language narratives. They made a dataset of expert-written textual explanations with numerical XAI explanations, framing this as a data-to-text generation task. This approach achieved 91% accuracy on the question-answering tasks of explanation, demonstrating that LLMs can effectively convert the XAI output into fluent explanations. However, their work focused primarily on the translation quality and fluency of the generated narratives, without examining how variations in prompting strategies or input presentation might affect the translation process.

2.2 LLMs as Explanation Interfaces or Evaluation Agents

In a different direction, Mavrepis et al. [7] introduced an LLM-based interface whose goal is to simplify XAI by providing modified adaptive explanations for various levels of experts. This work improved the accessibility and comprehensibility of XAI details by translating complex technical outputs into natural language. However, it faced limitations such as the LLM occasionally attempting to explain the provided images rather than the XAI output. This introduced a potential ambiguity regarding the groundedness and precision of the explanation, which is not further analyzed in detail.

Expanding on the use of LLMs to improve explanation accessibility, Bokstaller et al. [2] proposed a reference architecture that integrates fine-tuned LLaMA-2 models into an interactive chatbot system designed to explain SHAP outputs for battery health predictions. Their approach follows a three-phase fine-tuning pipeline, including domain knowledge and explanation intent. It is evaluated using perplexity, model loss, expert interviews, and a user study with 61 participants. The system improved interpretability for users unfamiliar with XAI, particularly in real-time dialog settings. However, similar to previous work, this study focuses solely on how LLMs provide explanations, not how they reason or act upon them. The influence of prompt phrasing is not examined.

Lastly, De Bona et al. [4] investigated the potential of LLMs to replicate human participants in the evaluation of XAI tools. The study aimed to address the inherent cost, time, and scalability issues of traditional user studies. Multiple LLMs, including Llama 3, Qwen 2, and GPT-4o mini, were treated as participants, tasked with assessing the usefulness of explanations, predicting AI outputs and expressing confidence in their predictions. Their findings indicated that LLMs could replicate most conclusions from the original user study. The authors noticed that the LLM struggled with subjective and confidence-related questions. A notable limitation of this work was its focus on evaluating explanations presented to LLMs rather than examining explanations generated by the LLMs themselves.

3   Concept

This study complements an ongoing user study investigating user behavior in a decision support system that incorporates various XAI methods. Specifically, the accompanying study investigates the interaction of participants with an XAI-focused, LLM-based chat-dashboard hybrid. The users take on the role of hotel managers and need to decide whether to approve an overbooking in order to maximize the usage of hotel capacity. We trained a machine learning model (neural network) on a hotel booking dataset. Users see the model's prediction and have access to various XAI methods. The dataset contains various booking-related features, like the reserved room type, the tourist's country of origin, the number of adults and other booking-related data.

The study is based on RIXA [1]. RIXA translates user requests into code calls to generate various XAI methods and accompanying visualizations, namely: LIME, counterfactuals, global Shapley values, partial dependence plots and feature distribution histograms. Users can also ask for interpretations of the XAI plots to gain further understanding. RIXA therefore acts as a gatekeeper between the users and the various XAI methods. While users could ask directly for XAI methods, even specifying the hyperparameters, this is unlikely in practice as our sample consists of laypeople. Hence, RIXA usually takes a question and maps it onto an XAI method it deems fitting and then has to explain it. Using all of this information, users are asked to make decisions on up to 30 data points, either approving ("Check-Out") or rejecting ("Cancellation")[1].

A key point in this setup is that all information flows through RIXA and, more specifically, the LLM in the backend. It is therefore plausible that inherent biases, problems, preferences, etc. of the LLM may influence the user. Hence we evaluate whether we can determine whether such factors exist and, if so, what they are. It is important to stress that, given the novelty of LLMs in XAI decision-support systems, we did not have strong *a priori* hypotheses about what factors might drive LLM decisions or whether measurable patterns would emerge at all. Accordingly, this study is exploratory in nature.

---

[1] The choice of data points and the dataset is based on human-interaction-design considerations. This presents a limitation as will be discussed in Sect. 5.1.

To find key decision drivers, two approaches are used. In the first and primary approach, the LLM is prompted to make the same decision as the users. For this, the LLM is given the data point in tabular form, the model's prediction, its confidence level, and a textual representation of all XAI methods[2]. To determine the factors driving or influencing the decision, various parts of this were altered, e.g., by changing the model's confidence or excluding some XAI methods. It is important to note that all experimental conditions/manipulations are implemented in the prompting of the LLM; the underlying model remained unchanged.

In the second stage 138 chats[3] from the ongoing user study are used, to evaluate whether the results are confirmed in a realistic setting. Here, the LLM is also prompted to make a decision, but it is not explicitly given all XAI methods and only sees the data present in the chat i.e. what the user (implicitly) requested.

3.1 Experimental Setup

The LLM is prompted to make a decision on the same 30 data points as in the user study. We do not consider any reasoning of the LLM and measure how it decides using a binary scheme[4]. This is done in multiple runs, where each run is a different set of experimental conditions. All possible combinations in each set of experimental conditions are used for prompting. This process is illustrated by a short example:

If the first condition is this list of accuracies: 50.49%, 72.11%, 95% and the second condition is this list of confidences 49.69%, 78.96%, 98.36% the model would be prompted sequentially to make a decision on each data point for each combination i.e. first "accuracy = 50.49%, confidence = 49.69%", then again for each data point for "accuracy = 50.49%, confidence = 78.96%" and so forth. This in turn limits the number of categories or conditions that can be looked at in a single run, as the number of prompts scales exponentially with the conditions.

To evaluate a condition, it is "marginalized out". E.g., in the above example, the metrics (see Sect. 4.2) for the accuracy of 50.49% would be calculated by looking at the decisions for each of the possible confidence values across all data points.

There are no studies done to our knowledge that establish primary decision drivers for LLMs in an XAI context. Therefore, the choice of categories is unclear. We identified candidate categories that would plausibly drive LLM decisions under the assumption of rational behavior (e.g., XAI methods should be important) and then started the experiments in a staggered manner. We did smaller initial runs to see which categories appeared to influence decisions. Those categories were then used in larger runs, building on the previous results

---

[2] In theory, this gives the LLM access to the same information as the users. However, in practice, users most often did not choose to gather all available information.

[3] The chats contained on average 9 messages with a maximum of 24 messages per chat.

[4] The LLM may still provide reasoning, but we do not evaluate it.

and refining them. In total, the following categories have been altered in some combination in our runs:

- Model accuracy
- Model confidence
- Model prediction
- XAI method order
- XAI method inclusion or exclusion
- Bias to avoid false negatives or false positives
- Prompt type
  - Chain of thought vs. direct answer
  - Emotional vs. rational tone
  - AI trust vs. AI distrust

All experiments were carried out with GPT-4[5].

4   Results

This section presents the results of our experiments. We start by listing the most important findings. Afterwards we introduce the metrics used in the evaluation, followed by a detailed description of how we reached our findings.

4.1   Key Findings

1) Measurable Effects were Observed
A priori it was not clear that the chosen setup would yield anything other than noise. The experimental conditions could have been chosen so that they have insufficient impact on the LLM, resulting in no measurable differences. This could, for example, be due to the LLM being strongly influenced by small phrasing differences, or simply having no inherent tendencies. However, for multiple conditions we found a significant tension with the hypothesis of an unaffected LLM.

2) Prompt Phrasing is the Primary Decision Driver
The strongest tensions with the null hypothesis, with a considerable distance to all other conditions, have been achieved by changing the way the prompts are phrased. This is a consistent result across all experimental conditions including runs on real chats. However, while the data clearly shows the importance of phrasing, it gives no clear mechanistic understanding such as what follows from a more emotional tone.

3) XAI Methods are not a Primary Decision Driver
Looking at the exclusion of various individual, or even all, XAI methods yields inconclusive results. The presence or absence of XAI methods does not strongly correlate with the LLM's decisions.

---

[5] Specifically "gpt-4o-2024-08-06", the same as in the user study.

4) Confidence is Important
Manipulating the per-data point confidence of the model has a significant impact on the LLM's decision, especially on whether or not to agree with it. For low confidence values the LLM seems to actively decide against the model instead of deciding randomly.

5) Accuracy is not Important
Changing the model's test-set accuracy has no significant impact on the LLM's decisions.

6) Avoidance of False Negatives
The LLM seems to try to avoid false negatives (Chose "Cancellation" wrongly instead of "Check-Out").

4.2 Metrics

Here we present the metrics used to evaluate our experiments. Given the exploratory nature of this study, we interpret these metrics comparatively rather than against predetermined thresholds.

Krippendorff's Alpha (α): We adapt Krippendorff's Alpha α to measure decision consistency for a focal experimental condition. For a given condition we treat each combination of a data point and the other experimental condition as a separate rater. This means α measures the consistency of the LLM for a focal condition[6].

$\alpha = 1$ indicates decision consistency, suggesting the looked-at condition (strongly) influences the LLM's behavior. $\alpha = 0$ indicates random and/or inconsistent decisions. This implies the condition has no influence on the LLM or "confuses" it, leading to erratic behavior. α is a function of the other varied conditions. If the other conditions are more influential the looked at condition may be "drowned out", resulting in a low α. Hence α can only determine if a condition is meaningful in relation to others.

Avg. % Correct: This describes how often the LLM makes correct decisions measured against the ground truth in the dataset. This is calculated over all data points. The variance over all marginalized conditions is also given. The interpretation for this metric is difficult because a correct decision does not imply correct reasoning and vice versa. For example, the LLM could form a logical argument, but the data point is classified differently due to the aleatoric uncertainty.
Note: On the 30 data points the trained model is correct for exactly 70%. The points were chosen so to give a range of correct and false classifications and different confidences. On the test dataset the model has an accuracy of 84.38%.

Avg. % Agree: Measures how often the LLM agrees with the model's predictions. The agreement is calculated over all data points per condition. The variance from the mean over all marginalized conditions is also given.

---

[6] This is assuming the choice is non-trivial, e.g., the LLM could always choose to decline, which would lead to $\alpha = 1$. However, we did not encounter this behavior.

Avg. Corr. Correct: Pearson correlation coefficient between the ground truth and the LLM decision.

Avg. Corr. Agree: Pearson correlation coefficient between the model prediction and the LLM decision.

FP and FN Rate: False positive and false negative rates. We used the convention "Check-Out": 1, "Cancellation": 0 to calculate the values. The 30 data points of the study have the following distribution:

$$FP = 0.22$$
$$FN = 0.26$$

4.3   Detailed Results

Using the aforementioned marginalization (Sect. 3.1), summary results are calculated for each run. Each run summary contains a table for each of the experimental conditions where each row in a table is a value of this condition. Each row encompasses all valid metrics. As the tables are lengthy and difficult to read in a static form[7], they can be found in the accompanying repository with dynamic sort options. This repository also includes the raw data used to generate the tables as well as the code for table generation and the code used for the experiments.

The raw data from using the real chats is only partially available because the user data had to be omitted. The chats to reproduce the results will be made public when the evaluation of the user study is finished.

Given the exploratory nature, we interpret results comparatively, using variance as a proxy for uncertainty. The assumption being that the variance can be used to approximate the confidence interval. All values that fall within each other's combined variance are considered to be noise. It is important to note that this form of analysis, can only indicate trends but is insufficient to make definitive statements about effects and is not able to make out effect sizes.

Here we discuss results for each of the experimental conditions and general metrics in detail. Interpretations can be found in Sect. 5.

Correctness: Correctness was evaluated in every run, as it is not dependent on the varied conditions. We used 2 metrics to measure correctness: $r_{correct}$ and *avg. % correct*. Across most experimental conditions and runs both do not vary significantly. Most values are compatible within margins of error. The median is ≈70%. Values that deviate significantly from this show larger errors. The only exceptions are when varying data point confidence, global accuracy or changing the prediction. This is separately discussed in the below paragraphs.

It should be noted that the model's predictions are correct in 70 % of cases which is compatible with the *avg. % correct* values (except the aforementioned outliers). The LLM is on average as correct as the model. We investigate this further in the paragraph *"Model Decisions"*.

---

[7] In total 8 runs with 19 tables.

Confidence: Confidence was manipulated in Run 2 in combination with the original prompt types and accuracy. In the used model high confidence is an important predictor as can be seen in Fig. 1 which has been calculated on the test dataset.

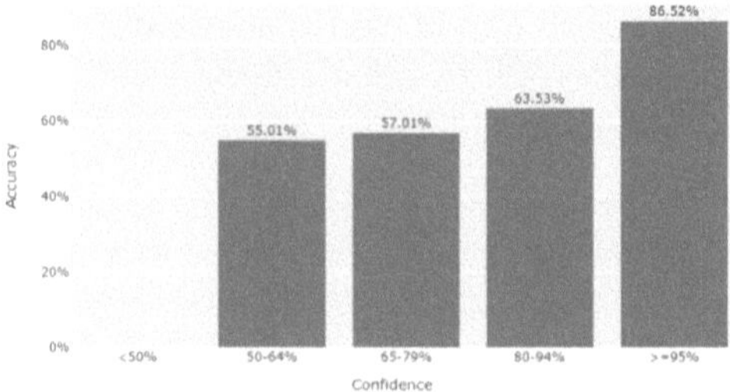

Fig. 1. Distribution of confidence and accuracy of the decision model used in the user study.

For lower values confidence becomes less and less relevant. So in an ideal case the LLM would put high emphasis on high confidence scores and gravitate more and more towards random decisions for lower scores. While the LLM is not aware of the model's probability calibration, a similar result can be expected generally. Table 1 shows the result of the LLM decision making.

Table 1. Run 2: Krippendorff's Alpha, Avg. % correct, Avg. % agree and correlation coefficient of agreement in dependence of a varied confidence score given to the LLM.

Confidence Score	α	Avg. % correct	Avg. % agree	r_{agree}
No mention	0.42	66.9 ± 7.0	80.9 ± 16.4	0.59 ± 0.37
49.69	0.41	42.9 ± 11.2	23.8 ± 18.7	-0.60 ± 0.30
60.10	0.18	60.8 ± 8.4	65.2 ± 16.8	0.22 ± 0.40
69.17	0.28	62.9 ± 7.7	73.8 ± 16.4	0.43 ± 0.38
78.96	0.42	66.1 ± 6.7	82.1 ± 17.4	0.61 ± 0.40
88.45	0.47	66.9 ± 6.4	83.5 ± 17.5	0.65 ± 0.39
98.36	0.63	67.4 ± 5.8	89.3 ± 14.4	0.78 ± 0.30

There is an upward trend between agreement and correctness with the model and the confidence score. For example for the *avg. % correct* and confidence score we can calculate $R^2 = 0.77$ which shows a correlation. This is a logical result as a lower confidence should lead an expert/informed decision maker to trust the system less and less and thus gravitate more towards random decisions. At 60% confidence the LLM starts to decide almost randomly as indicated by the low alpha value. However, we do see two unusual things:

1. At 50% we observe the only negative correlation coefficient in all experiments (correlation for both correctness and agreement), even when considering the error the values are still negative (for agree, we get a 2σ tension with 0).
2. At 50% we have the only true outlier for *avg. % correct*, a 2σ tension with the usual mean of about 70%.

The data suggests the LLM actively decides against the model's prediction when the confidence score falls below 60%. This explains both unusual findings. This constitutes illogical behavior. Low confidence does not give an indication to decide against the AI's decision, only that the AI's prediction is not meaningful for reaching a conclusion. However, it constitutes human-like behavior [9].

Accuracy: Accuracy was varied in the same run as confidence.

The results for varied accuracy are noisy across all metrics, and show a low α. As high variance and a low α indicate a tendency towards random choice, it is most likely that there is simply no unifying element in this condition. The data indicates that the given accuracy does not influence the LLMs decision.

However, it is possible that the accuracy gets "drowned out" since other experimental conditions are more impactful. We did not investigate this further as we are only looking for key factors.

XAI Methods: Three runs specifically looked at the impact of XAI methods:

1. Ordering of XAI methods (Run 3)
2. Exclusion of individual XAI methods and all possible exclusion combinations (Run 1)
3. Total inclusion/exclusion of all methods with improved prompts (Run 5)

1. showed that the ordering of XAI methods in the prompts has no measurable impact. All values are within the margins of error and have the same approximate median. Assuming applicability of the recency effect [5], this is a (weak) indication that XAI methods are not of high importance to the LLM.

2. and 3 showed that, within margins of error, the exclusion of individual or all XAI methods has no large impact. There are indications that the presence of some XAI methods has a form of influence on the decision. E.g. in Run 5 we find $\alpha_{\text{no SHAP}} = 0.56$ vs. $\alpha_{\text{SHAP}} = 0.68$ or in Run 1 we find something similar for the total exclusion of XAI methods ($\alpha_{\text{no xai}} = 0.65$ vs. $\alpha_{\text{xai}} = 0.53$). However other metrics give no real indication, so this may simply be noise.

Combining this we conclude that the presence of XAI methods is not a primary decision driver. Although it is possible that given more data, effects may be found.

Bias and General Tendency of FP and FN: In Run 6 the LLM was prompted to avoid false negatives (FN) (wrongly predicting "Cancellation" instead of "Check-Out") or false positives (FP) (wrongly predicting "Check-Out" instead of "Cancellation") using language that, e.g., emphasized how much underbooking could

Table 2. Run 6: Effect of prompting to avoid false positives/false negatives on Krippendorff's Alpha, mean false positive and mean false negative scores.

Bias category	α	$\overline{FP}$	$\overline{FN}$
Avoid FN	0.61	0.21 ± 0.06	0.10 ± 0.05
Avoid FP	0.43	0.19 ± 0.08	0.12 ± 0.08
None	0.62	0.19 ± 0.06	0.11 ± 0.06

hurt a hotel or how much a tourist may be displeased if experiencing an overbooking. The effect on the FP and FN rates can be seen in Table 2. All other metrics yield inconclusive results.

Neither FP nor FN seem to be affected significantly. If the LLM would always agree with the model, the expected rates would be:

$$FP = 0.22$$
$$FN = 0.26$$

For FP that is compatible. For FN we observe up to a 3σ tension. This is not an outlier but rather the rule. One of the largest runs, Run 5 gives on average:

$$FP = 0.20$$
$$FN = 0.13$$

Most runs show a similar result. So the LLM has a clear tendency to avoid false negatives, but no special tendency to avoid false positives. This could explain why adding a biased prompt has no strong effect as one is already present.

Model Decision: Two runs looked at the impact of changing the model's prediction. The decision gets displayed constantly as "Check-Out" or "Cancellation" or remains unchanged. This is done with the first iteration of prompt categories in Run 4 and then with the new categories in Run 7. Combined results from these runs using error propagation can be seen in Table 3.

Table 3. Combined Runs 4 & 7: Krippendorff's Alpha, Avg. % correct, Avg. % agree and correlation coefficient of correctness in dependence of a changed predicted target.

Predicted Target	α	Avg. % correct	Avg. % agree	r_{correct}
Check-Out	0.33	60.2 ± 4.4	83.1 ± 11.6	0.03 ± 0.12
Cancellation	0.20	50.1 ± 9.3	51.8 ± 20.0	0.01 ± 0.14
Unchanged	0.48	67.6 ± 4.9	82.7 ± 9.7	0.30 ± 0.09

Some facts need to be established before the interpretation. The data point distribution is:

$$N_{\text{Check-Out}} = 19$$
$$N_{\text{Cancellation}} = 11$$

So if the LLM would agree with the altered prediction one would expect a distribution of

$$\text{avg. } \% \text{ correct}_{\text{Check-Out}} \approx 63\%$$
$$\text{avg. } \% \text{ correct}_{\text{Cancellation}} \approx 37\%$$

Flipping only the model decision but leaving everything else the same should lead an ideal decider to gravitate towards random decisions. Because only flipping the decision causes visible discrepancies. E.g. by summing up the individual contributions in the LIME explanation and comparing it to the unchanged confidences.

This data can then be seen as further evidence for the LLM not being influenced by the XAI data. The correlation between being correct and the LLM decision is compatible with 0 for both alterations. There is also an avg. % agree of $83.06 \pm 11.58\%$ for "Check-Out". This is a significant tension with random decisions. Lastly, the error for *avg. % agree* for "Check-Out" is nearly the same as for the unchanged condition.

However, there is an *avg. % agree* of $50.05 \pm 9.26\%$ for "Check-Out", so there is a slight tension with random decisions. α is significantly lower than in the unchanged condition, especially for "Cancellation". This may be a hint that the LLM sees the discrepancies in the XAI methods and prediction.

There is something that is not explained by either hypothesis. There seems to be a meaningful difference for the LLM between the constant "Cancellation" and "Check-Out". The most likely explanation is that the LLM does not weigh making a false positive or false negative in the same way. The previous paragraph shows evidence for an inherent bias in the LLM that accounts for this. The evidence from altering decisions strengthens the hypothesis of the LLM ignoring XAI data. While the data here does not allow a definite conclusion, it points into the same direction as the data from *XAI Methods*.

Prompt Categories: The model needs to be prompted to make a final decision. The possibilities for prompts and how to categorize them are endless. The question what meaningful categories are is not trivial. As a first guess, we started with the following:

- Direct answer
- Chain of thought (CoT)
- Emotional tone
- AI distrust
- AI trust

To make sure all categories produce comparable results 10 different prompts per listed category have been generated using a mix of Claude, ChatGPT, Mistral, DeepSeek and handcrafting[8]. Due to resource constraints most runs used a smaller subset of the prompts (mostly 8 per category per run).

The prompts are a fundamental property since one always needs to ask the model for a decision. Thus, all of the above categories have been varied in Run 2, Run 3, Run 4 and a subset (no emotional tone) in Run 1.

To evaluate if the categories are meaningful/produce measurable results, the α values for all the mentioned runs are shown in Table 4.

Table 4. Krippendorff's Alpha for prompt categories (Direct Answer, CoT, Emotional Tone, AI Distrust, AI Trust) and Runs (Run 2: Confidence/Accuracy, Run 3: XAI Method Order, Run 4: AI Decision, Run 1: XAI Exclusion).

Condition	Confidence/Accuracy	XAI Method Order	AI Decision	XAI Exclusion
Direct Answers	0.42	0.89	0.50	0.78
CoT	0.16	0.48	0.16	0.55
Emotional Tone	0.08	0.65	0.26	–
AI Distrust	0.10	0.45	0.19	0.42
AI Trust	0.32	0.92	0.17	0.86

While the α values vary significantly, generally they are high compared to other varied conditions. We observe the two highest α values encountered in all conditions across all runs, 0.89 and 0.92. Other metrics support this as e.g. in the third run we find for AI trust an *avg. % agree* of 97.83 ± 3.07 and for direct answers of 95.25 ± 4.20. This means these conditions cause the LLM to be in almost perfect agreement with the AI model. E.g. for the Confidence/Accuracy run the α values and other metrics are less clear. However this is to be expected since it has already been elaborated in e.g. *Confidence* that varying confidence influences the LLM. In turn the effects or prompting become less pronounced. Even considering these runs other conditions do give substantially lower α values and show significantly less influence on other metrics. Hence it can be concluded that prompting has by far the most impact.

Given this result new runs have been done with refined prompt categories. Here 3 axes have been devised

- CoT vs. Direct answering
- Trust vs. Distrust in the AI model
- Rational vs. Emotional prompt phrasing

In this system each axis presents an own condition. The idea being that effects of emotional phrasing can be viewed independently of whether or not the model

---

[8] The prompts can be found under: https://github.com/finnschwall/explains2025/tree/main/rixeval/prompts.

has been asked to provide a chain of thought or if this is done with a trusting/distrusting tone. Whether this is valid, is discussed in the next paragraph.

These new prompts categories have been used in Run 5 and Run 7. In Run 7 values are noisy and difficult to interpret, most likely due to the strong influence of altering the models prediction. In Run 5 the results from the previous paragraph are recovered with high α values and low variances for other metrics. This is expected as this is essentially a repetition of the previous paragraph, with a much higher sample size.

As this resulted in the most significant results (measured by impact on the metrics) it was decided that this is the only condition that should be explored on real chat data. Since runs on user chats are resource intensive due to the number of chats, the number of messages (up to 24) per chat and correspondingly high token numbers, the number of explorable conditions is limited.

As can be seen in Run 8 the run on the real conversations again reproduces the results in terms of variances and α. We take this as evidence that the prompts are an important condition since both on artificial experiments in all different permutations and on real chats we see consistent results.

Hence, we conclude prompting is an influential factor on LLMs. In our data it is even the most important factor.

Prompt Axes: We observed that the way the LLM is prompted is the most important decision driver. Here we show what exactly the differences across each prompting category are.

- *CoT vs. Direct answer*: One can see as a general trend that the α values are close for CoT vs. direct answers but there is a consistent trend: α is higher for direct answers, i.e. the LLM is more consistent when directly answering. However, as mentioned in *Correctness* CoT does not make the model more "correct", but only less consistent. The effect on agreement with the AI model is inconclusive.
- *Emotional vs. Rational*: Some runs (e.g. Run 8) show no real difference but looking at Run 7, Run 5 or Run 3 one can see a pattern. Emotional phrasing makes the LLM less conclusive (lower α, higher variance for agreement). Actual correctness and agreement rates are similar. There are hints that rational prompting leads to higher agreement. However none of this is significant enough to draw definite conclusions.
- *Trust vs. Distrust*: Higher trust mostly leads to the LLM being more consistent and more agreeable with the model. The effect reverses for distrust. This is an expected result.

In conclusion our data indicates that these axis for devising prompts are at least partially meaningful (e.g. seen in *Trust vs. Distrust*) and do catch some form of correlation with the LLMs behavior. The data shows that categorizing prompts can result in measurable differences. However it also suggests that a better prompt categorization scheme can be found.

Human Impact on the LLM: As the user study is ongoing, no results from it can yet be published. We intend to do an in-depth analysis using the insights from this paper after the study has concluded.

That being said, the available data contains indications for a human impact on the LLM. Grouping the chats by users and then analyzing the r_{agree} and the *avg. % correct* we find stark differences between the LLM's behavior per user. The two most extreme users by r_{agree} can be seen in Table 5.

Table 5. Correlation coefficient of agreement, Avg. % agree and correlation coefficient to LLM decision of two people from the user study.

User	r_{agree}	Avg. % agree	r_{Human}
1	0.81	90.86%	0.45
2	0.14	54.09%	−0.52

r_{Human} is the correlation coefficient between the humans decision and the LLM decision. This data, among other already evaluated from the user study, hints towards the possibility, that prompting meaningfully influences the LLM, even in real applications.

5 Conclusion and Outlook

Several findings have been established in the previous sections. Among the most noteworthy is that LLM decision-making is systematically influenced by experimental conditions. *A priori* it was unclear whether the LLM would have decision tendencies, if our chosen categories would have enough influence, or if the chosen metrics are adequate to capture any findings. Hence, we conclude that this approach, while improvable, is valid.

The statistically most significant findings are the importance of prompts and their phrasing. Presentation seems to matter significantly. However, this study is unable to draw definitive conclusions as to how and why exactly presentation matters.

Our findings also indicate that the LLM is not just a rational actor. Evidence for this is found e.g. in modifying the confidence. Here, the LLM starts to actively decide against the model's prediction, instead of making random choices. This is behavior that humans display [9]. The importance of phrasing and impact of e.g. more emotionally loaded adjectives underscores this. Further, we established an inherent bias in the LLM towards avoiding false negatives. All of this points towards the possibility of the LLM being more in line with human behavior than that of a purely rational actor.

This is further underscored by the fact that we did not find any conclusive evidence of an impact of XAI methods on the LLM's decisions. While it is possible that no better decision can be made, it is unlikely that XAI methods would have no measurable impact. This questions the entire premise of LLM-mediated XAI systems as anything other than a translation layer.

However, a completely different take is also possible: As we established that presentation is important, so there may be a mismatch between the methods' presentation and the LLM. Maybe there are better ways to convey the information from the XAI methods to an LLM.

Given all of this and our key findings from Sect. 4.1 we think the following design considerations for an LLM-assisted XAI system are justified:

- A bias analysis before deployment is warranted. Especially since a bias might not be in line with actual requirements.
- The deploying party and corresponding users should be made aware that framing likely has an impact and that the LLM may seem more rational in its decisions and explanation than it is.
- We conclude that usage of such a system with laypeople should only be implemented with great caution. As laypeople are less capable of interpreting data correctly on their own, we see a danger of the LLM's tendencies or biases to impact their decision-making.

5.1 Outlook

While some findings have been made, the paper leaves many questions unanswered, either unintentionally or by design. A succession to this paper is warranted that is not just exploratory but defines a rigorous methodology, a proper statistical evaluation method and prior research questions. With such an updated approach, a comparison of multiple LLM architectures becomes possible. This is important to establish if these findings hold across architectures and to see what differences may be there. Finally, this update also needs to do an analysis with the full user study data.

Another possible research avenue is having a more detailed look at the decision tendencies which are hinted at in the data. Clearly, there are key parameters driving the used LLM's decision. But do these tendencies have an impact? Could we see different choices of presented XAI methods to a user or different lines of reasoning when prompting differently?

Thinking further, if such new studies find evidence that humans do impact both the LLM's decision and its apparent reasoning, presented XAI methods, etc. then a look at a possible human-LLM interaction dynamics is warranted. It is thinkable that an influenced LLM also influences the user. Maybe this even allows making a concrete connection between prompting strategies and user behavior. As the LLM is the arbiter of information, especially for laypeople, this is not an unreasonable assumption. It is even thinkable that feedback loops form.

There are limitations inherent to this study that can only be overcome by a wholly different design. As this study is accompanying a user study with laypeople, the breadth of XAI is inherently shallow. Other studies should address these and other limitations by not limiting the amount of data points, looking at other datasets or including more XAI methods.

A different and so far unaddressed research direction could explore a mechanistic understanding of these findings. For example, we know prompts matter, but not why. We also did not establish the exact impact of different prompt strategies. However, the search for a why and not only an if, requires completely different research strategies.

Finally, we see a clear lack of unified methodology in the area of LLM application in XAI tasks. There are no agreed upon strategies, metrics, values or similar. This study is not able to establish recommendations in this direction. It is also too complicated and costly to do on a regular basis. With future studies, one could come up with a more standardized test framework. Given previous insights, main decision drivers etc. it may be possible to reduce the complexity of such an analysis. We think an overview paper is best suited for such a task.

Acknowledgments. This work was supported by funding from the topic Engineering Secure Systems of the Helmholtz Association (HGF) and by KASTEL Security Research Labs.

References

1. Becker, M., Vishwesh, V., Birnstill, P., Schwall, F., Wu, S., Beyerer, J.: RIXA-explaining artificial intelligence in natural language. In: 2023 IEEE International Conference on Data Mining Workshops (ICDMW), pp. 875–884. IEEE (2023)
2. Bokstaller, J., et al.: Enhancing ml model interpretability: leveraging fine-tuned large language models for better understanding of AI. arXiv preprint arXiv:2505.02859 (2025)
3. Burton, J., Al Moubayed, N., Enshaei, A.: Natural language explanations for machine learning classification decisions. In: 2023 International Joint Conference on Neural Networks (IJCNN), pp. 1–9. IEEE (2023)
4. De Bona, F.B., Dominici, G., Miller, T., Langheinrich, M., Gjoreski, M.: Evaluating explanations through LLMs: beyond traditional user studies. arXiv preprint arXiv:2410.17781 (2024)
5. Guo, X., Vosoughi, S.: Serial position effects of large language models. arXiv preprint arXiv:2406.15981 (2024)
6. Ichmoukhamedov, T., Hinns, J., Martens, D.: How good is my story? Towards quantitative metrics for evaluating LLM-generated XAI narratives. arXiv preprint arXiv:2412.10220 (2024)
7. Mavrepis, P., Makridis, G., Fatouros, G., Koukos, V., Separdani, M.M., Kyriazis, D.: XAI for all: can large language models simplify explainable AI? arXiv preprint arXiv:2401.13110 (2024)
8. Susnjak, T.: Beyond predictive learning analytics modelling and onto explainable artificial intelligence with prescriptive analytics and ChatGPT. Int. J. Artif. Intell. Educ. **34**(2), 452–482 (2024)
9. Yadnakudige Subramanya, S., Watanabe, K., Dengel, A., Ishimaru, S.: Human-in-the-loop annotation for image-based engagement estimation: assessing the impact of model reliability on annotation accuracy. In: Kurosu, M., Hashizume, A. (eds.) International Conference on Human-Computer Interaction. LNCS, vol. 15770, pp. 169–186. Springer, Cham (2025). https://doi.org/10.1007/978-3-031-93864-1_12

Mechanistic Interpretability for Transformer-Based Time Series Classification

Matīss Kalnāre[(✉)] [ID], Sofoklis Kitharidis [ID], Thomas Bäck [ID], and Niki van Stein [ID]

Leiden Institute of Advanced Computer Science (LIACS), Leiden University, Einsteinweg 55, 2333 Leiden, CC, The Netherlands
{m.kalnare, s.kitharidis, T.H.W.Baeck, n.van.stein}@liacs.leidenuniv.nl

Abstract. Transformer-based models have become state-of-the-art tools in various machine learning tasks, including time series classification, yet their complexity makes understanding their internal decision-making challenging. Existing explainability methods often focus on input-output attributions, leaving the internal mechanisms largely opaque. This paper addresses this gap by adapting various Mechanistic Interpretability techniques; activation patching, attention saliency, and sparse autoencoders, from NLP to transformer architectures designed explicitly for time series classification. We systematically probe the internal causal roles of individual attention heads and timesteps, revealing causal structures within these models. Through experimentation on a benchmark time series dataset, we construct causal graphs illustrating how information propagates internally, highlighting key attention heads and temporal positions driving correct classifications. Additionally, we demonstrate the potential of sparse autoencoders for uncovering interpretable latent features. Our findings provide both methodological contributions to transformer interpretability and novel insights into the functional mechanics underlying transformer performance in time series classification tasks.

Keywords: Transformer Models · Mechanistic Interpretability · Activation Patching · Attention Saliency · Sparse Autoencoders · Time Series Classification · Explainable AI (XAI) · Causal Interpretability

1 Introduction

Deep learning models, ranging from convolutional networks to modern Transformer architectures, have achieved remarkable accuracy in domains as diverse as computer vision, natural language processing, and time series analysis. Yet this performance often comes at the cost of interpretability: practitioners and stakeholders lack insight into the internal decision-making processes of these "black-box" models, undermining trust in safety-critical applications such as healthcare, finance, and autonomous systems.

Explainable AI (XAI) has advanced numerous post-hoc and inherently interpretable techniques [11], such as feature attribution (e.g., SHAP, LIME) and self-explaining models, to clarify which inputs influence outputs. However, these approaches stop at the

F. Marcelloni et al. (Eds.): IJCCI 2025, CCIS 2829, pp. 249–270, 2026.
https://doi.org/10.1007/978-3-032-15638-9_15

model boundary and do not illuminate the causal mechanisms within complex architectures. Mechanistic Interpretability (MI) addresses this gap by treating internal components like layers, attention heads, and neurons as discrete causal units, probed through interventional and observational methods to reveal how information flows and decisions emerge.

Existing XAI work in TSC seldom moves beyond surface-level attributions [5]. Moreover, despite significant progress in MI for NLP transformers, Time Series Transformer models (TSTs) pose distinct challenges: the absence of discrete tokens, the fluid nature of temporal dependencies, and variable sequence lengths complicate direct application of language-based techniques. This raises the central question of our work:

What novel insights can Mechanistic Interpretability methods, specifically, activation patching, attention saliency, and sparse autoencoders, yield when adapted to Time Series Transformers?

To answer this question, we **contribute** the following[1]:

- We adapt *activation patching*, *attention saliency*, and *sparse autoencoders* to continuous, variable-length TSTs, enabling causal probing at the layer, head, and timestep levels.
- We build *directed causal graphs* that stitch these probes together, mapping information flow from input timesteps through attention heads to the class logits.
- We demonstrate that a single early-layer patch can recover up to **0.89** true-class probability in a misclassified instance, exposing discrete, manipulable circuits inside the model.
- We provide complementary sparse-autoencoder visualisations that surface class-specific temporal motifs, enriching the causal story with human-readable patterns.

Mechanistic Interpretability at this granularity is critical in safety-critical settings, such as ECG triage (the automated prioritization and classification of electrocardiogram signals to detect urgent cardiac arrhythmias), industrial fault detection, and algorithmic trading, because it converts opaque failures into actionable explanations that auditors can trust.

The remainder of the paper is structured as follows. Section 2 discusses background and related work. Section 3 details the proposed methodology, describing the transformer architecture and interpretability techniques. Section 4 outlines the experimental setup, including datasets and metrics. Section 5 presents our experimental results, followed by a discussion in Sect. 6. Finally, Sect. 7 concludes with key findings and future directions.

2 Background and Related Work

This work lies at the intersection of four research threads: time series classification, Transformer architectures, explainable AI, and particularly Mechanistic Interpretability.

---

[1] Source code and extended experimental results are available at our public GitHub repository [7].

Traditional approaches to Time Series Classification (TSC) have relied on distance-based methods (e.g., k-Nearest Neighbors with Dynamic Time Warping) and feature-based classifiers such as Catch22 [9] and ROCKET [3]. While computationally efficient, these methods depend heavily on handcrafted features and struggle to capture complex temporal dependencies. Deep learning models, particularly Convolutional Neural Networks (CNNs) and Recurrent Neural Networks (RNNs), have addressed some of these limitations but still face challenges in modeling long-range correlations and irregular sampling.

More recently, Transformer-based architectures have been adapted for TSC due to their self-attention mechanism's ability to model non-local interactions in sequence data [15,16]. Empirical studies report that TSTs consistently outperform CNNs and RNNs on multivariate benchmarks, achieving state-of-the-art accuracy with competitive efficiency [15]. Nevertheless, their superior performance comes at the cost of interpretability: the high-dimensional, distributed representations learned by attention layers and feed-forward blocks obscure how temporal features contribute to final decisions.

XAI has traditionally focused on input–output attribution, employing methods such as SHAP [10] and LIME [12] to assign importance scores to input features. In time series contexts, these techniques highlight which timesteps or channels influence predictions but do not reveal how internal computations interact to produce outcomes. Observational approaches like attention saliency further inspect self-attention weights, offering insight into routing patterns but lacking causal guarantees [14].

MI emerges as a complementary paradigm that probes model internals as causal systems. By treating layers, attention heads, or neurons as discrete intervention targets, MI methods such as activation patching and causal tracing allow researchers to test *necessity* and *sufficiency* of components via denoising and noising interventions [4]. Sparse autoencoders extend this viewpoint by disentangling latent activations into semantically meaningful codes under sparsity constraints [2], enabling targeted analysis and potential manipulation of high-level features.

Notably, this mechanistic approach is inspired by neuroscientific interventions and observations. For example, activation patching parallels transcranial magnetic stimulation (TMS) in neuroscience, by both selectively perturbing internal components to reveal causal effects, while observational saliency analyses are conceptually akin to EEG/MEG recordings that map normal and abnormal signal flows in the brain [1,13]. These analogies underscore the value of causal probing alongside attribution for truly understanding complex systems, whether biological or artificial.

3   Methodology

To effectively apply activation patching to our TST model, the first critical step is to carefully select source and target instances. We define **clean instances** as those from the test set where the model predicts the correct class with very high confidence ($P(y_{\text{true}}) > 0.95$). Conversely, **corrupt instances** are those misclassified or predicted with low confidence ($P(y_{\text{true}}) < 0.50$). This criterion ensures a meaningful contrast while maintaining inherent similarities, as both types originate from the same ground-truth class. Consequently, any observed differences primarily reflect the model's decision-making behavior rather than inherent class-level variance, making the results of subsequent interventions more interpretable.

To quantify the causal effect of interventions precisely, we use the change in the true-class prediction probability:

$$\Delta P = P_{\text{patched}}(y_{\text{true}}) - P_{\text{orig}}(y_{\text{true}}), \tag{1}$$

where P_{orig} is the baseline probability on the corrupt instance, P_{patched} is the probability after an interventional patch, and y_{true} is the true class of the instance.

3.1 Activation Patching

Activation patching is an interventional method in which activations from a clean instance are injected into a corrupt instance at specific model components. We implement this via PyTorch forward hooks, following a "denoising" protocol:

1. Run a forward pass on the clean instance, caching activations at the target component (layer, head, or head–timestep).
2. During the forward pass of the corrupt instance, replace the corresponding activations with cached values.
3. Compute ΔP to measure causal influence.

Notably, this technique can utilize logical structures such as *AND* logic (where multiple activations need simultaneous patching) or *OR* logic (where patching individual activations independently restores model accuracy). We apply this procedure at three levels of granularity:

- **Layer-Level:** replace all attention-head activations in a given encoder layer.
- **Head-Level:** replace activations of a single attention head.
- **Position-Level:** replace the per-timestep output of a specific head.

We systematically identify critical patches by exhaustively evaluating all layer, head, and position combinations. A patch is deemed *critical* when the resulting change ΔP surpasses a predefined threshold significantly favoring the correct classification. These identified critical patches form the basis for constructing detailed causal graphs, providing insight into the internal decision-making pathways of the model.

3.2 Attention Saliency

To complement activation patching, we utilize attention saliency as an observational interpretability tool. Attention saliency clarifies which timesteps each attention head focuses on during prediction. Specifically, we extract raw attention weights $A^{(h)} \in \mathbb{R}^{T \times T}$ for each head h and average them across all query positions:

$$S_t^{(h)} = \frac{1}{T} \sum_{i=1}^{T} A_{i,t}^{(h)}, \tag{2}$$

where $S_t^{(h)}$ represents the average attention score for each timestep t. While attention saliency does not imply causation, it highlights potential candidate timesteps for targeted causal experiments.

3.3 Sparse Autoencoders for Latent Feature Analysis

Direct neuron-level analysis in transformer encoder blocks is challenging due to high dimensionality and semantic entanglement among neuron activations. To overcome this, we introduce Sparse Autoencoders (SAEs) trained on internal activations from the first encoder block's MLP outputs. The SAE consists of an encoder network, converting high-dimensional activations into a sparse code, and a decoder network, reconstructing the original activations from this sparse code. SAE training uses a mean squared reconstruction loss combined with an L_1 sparsity penalty:

$$\mathcal{L}_{\text{SAE}} = ||x - \hat{x}||_2^2 + \lambda \sum_j |z_j|, \tag{3}$$

where $x \in \mathbb{R}^d$ is the original MLP activation vector at a single timestep, $\hat{x} \in \mathbb{R}^d$ is its reconstruction produced by the SAE decoder, $z \in \mathbb{R}^H$ is the sparse code produced by the SAE encoder, $j \in \{1, \ldots, H\}$ indexes SAE units, $\lambda > 0$ weights the ℓ_1 sparsity penalty, and $|| \cdot ||_2$ denotes the Euclidean norm. The sparsity term ensures a compact set of active neurons, aiding disentanglement and interpretability. After training, we analyze top-activating instances and timesteps per SAE neuron, forming hypotheses about semantically meaningful features in the model's internal representations.

4 Experimental Setup

Our classifier f_θ operates on inputs $X \in \mathbb{R}^{T \times C}$, where $T = 25$ is the sequence length and $C = 12$ is the number of channels. We use d to denote the model embedding (hidden) dimension used by the convolutional front end, positional embeddings, attention blocks, and MLPs; L the number of encoder layers; H the number of attention heads per layer; and K the number of target classes. The model consists of:

1. **Convolutional Front End:** Three 1D convolutional layers with kernel sizes $5, 3, 3$ and hidden-dimensional transitions $C \rightarrow d/4, d/4 \rightarrow d/2, d/2 \rightarrow d$, each followed by batch-normalization and ReLU.
2. **Positional Embedding:** A learnable embedding $P \in \mathbb{R}^{T \times d}$ added to the conv-net output.
3. **Transformer Encoder:** $L = 3$ layers of multi-head self-attention ($H = 8$ heads) and two-layer MLP blocks, each with residual connections, dropout, and layer-normalization. The Transformer's decoder portion is excluded, since its typical role in generating sequential outputs (e.g. next-word prediction in NLP or time-series forecasting) is unsuitable for fixed-length classification tasks. Omitting the decoder roughly halves the total number of parameters, significantly reducing computational cost without sacrificing classification performance.
4. **Classification Head:** A temporal max-pool over T to produce a d–dimensional vector, followed by a linear layer $\mathbb{R}^d \rightarrow \mathbb{R}^K$ and softmax to yield class probabilities.

We evaluate on the JapaneseVowels benchmark [8], which comprises LPC-derived 12-dimensional frames from nine speakers pronouncing Japanese vowels. To ensure uniform input shape, all sequences are truncated or zero-padded to $T = 25$ frames.

Table 1. Japanese Vowels dataset summary.

Dataset	Train	Test	Series Length	# Classes	Dimensionality	Type
Japanese Vowels	270	370	25	9	12	Audio

The dataset splits into 270 training and 370 test samples across $Y = 9$ classes. Table 1 summarizes its key statistics.

We train f_θ for 100 epochs using cross-entropy loss, with a batch size of 4. Optimization is performed with the Rectified Adam (RAdam) optimizer (learning rate 10^{-3}, weight decay 10^{-4}). These hyperparameters, particularly the small batch size, intensive weight decay, and RAdam, were chosen to ensure stable convergence given the small training set and to promote effective generalization while mitigating overfitting.

Models are implemented in PyTorch and trained on a single GPU. We report standard test accuracy for classification performance, and use the true-class probability change ΔP (see Sect. 3) as our primary metric for all interpretability experiments.

5   Results

Baseline Performance and Instance Selection. Our first objective is to confirm that the trained TST model achieves strong classification performance, ensuring that subsequent interpretability analyses probe meaningful representations rather than noise.

Figure 1 shows the model achieves **97.57%** accuracy, with errors confined to a few off-diagonal cells. Such strong performance ensures that any causal effects we measure reflect learned structure rather than noise. We therefore select a *clean* instance (ID 33, Class 2, confidence 100.00%) and a *corrupt* instance (ID 44, Class 2 misclassified as 3 with true class confidence of 10.72%) for our denoising-style patching pipeline (Table 2).

Table 2. Chosen instance pair for activation patching.

ID	Type	True Class	Pred. Class	$P(\text{true})$
33	Clean	2	2	100.00%
44	Corrupt	2	3	10.72%

The pair is ideal for denoising-style activation patching as both belong to the same true class, and the internal activations are aligned in semantic space. The chosen instances' raw time series are visualized in Fig. 16 in Appendix B. We report results on our primary clean–corrupt pair (IDs 33/44) and replicate every experiment from Sects. 5.1–5.5 on a second pair for robustness (see Appendix A).

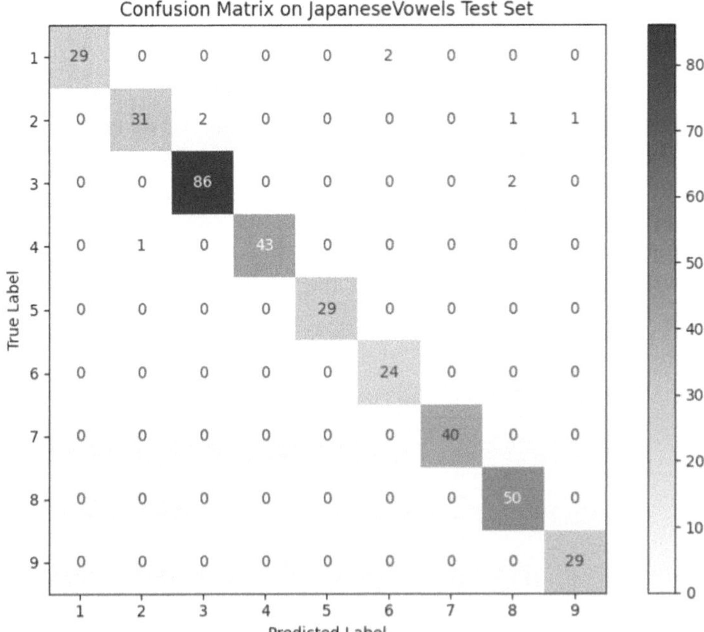

Fig. 1. Confusion matrix of the trained TST on the JapaneseVowels test set. Rows are true classes; columns are predicted classes.

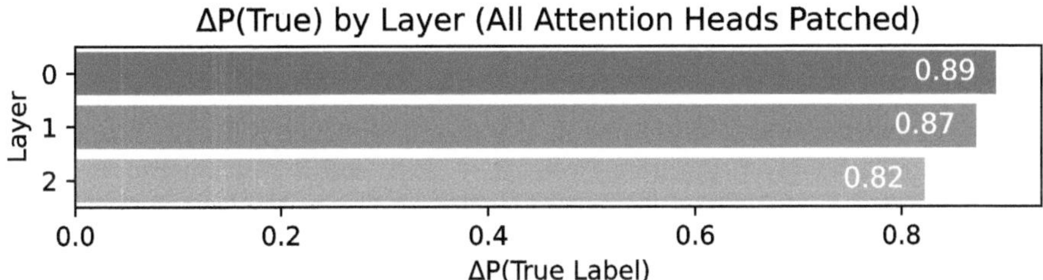

Fig. 2. Change in true-class probability (ΔP) when patching all heads of each encoder layer.

Layer-Level Causal Influence. We first ask: *Which encoder layer, if patched from the clean to the corrupt instance, most restores correct classification?* To answer this, we replace all attention-head activations in each layer of the corrupt instance with those from the clean instance, and measure the change in true-class probability ΔP.

Figure 2 reveals that patching **Layer 0** yields the largest increase ($\Delta P \approx 0.89$), with diminishing effects in Layers 1 and 2. This indicates that the earliest attention block carries the most critical causal signal, motivating a finer-grained head-level analysis.

The gradual decline in ΔP across deeper layers indicates a *distributed* representation: all layers encode useful information, but the causal signal is strongest at the model's entry point. This observation motivates our next step, drilling down to the head-level within Layer 0 to pinpoint the individual components driving the bulk of this effect.

Head-Level Causal Influence. To refine our causal analysis, we next inspect the influence of each individual attention head. Rather than patching an entire layer, we perform a sweep across all heads h in each layer: we inject its activations from the clean instance into the corrupt instance and measure the resulting true-class probability $P(y_{\text{true}})$.

The corrupt instance's baseline confidence is $P(y_{\text{true}}) = 0.1072$. Figure 3 presents a heatmap where each cell (ℓ, h) indicates the model's true-class probability change after patching head h in layer ℓ.[2]

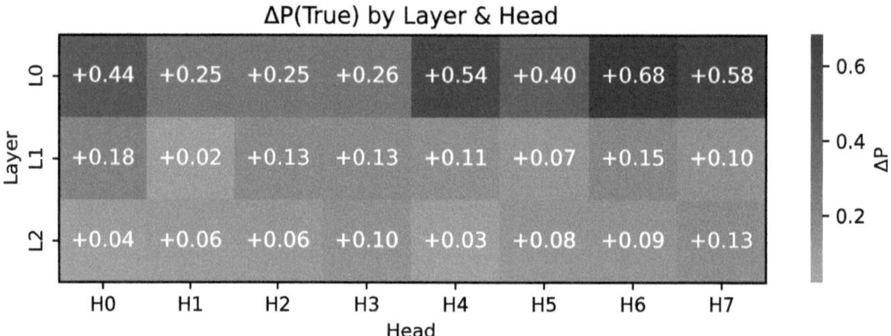

Fig. 3. Head-level activation patching: true-class probability change ΔP after patching each individual attention head.

A close examination of Fig. 3 reveals that causal influence is heavily concentrated in the first encoder layer: patching Heads 0, 4, 5, 6, and 7 individually raises true-class confidence to above 0.78. In particular, Head 6 boosts $P(y_{\text{true}})$ from 0.1072 to approximately 0.7872 ($\Delta P \approx 0.68$), echoing the strong Layer 0 effect observed earlier. In contrast, heads in Layers 1 and 2 each yield smaller gains (typically $\Delta P < 0.20$), with a more uniform distribution of influence and slightly weaker impact in Layer 2 than Layer 1, reinforcing the pattern of diminishing causal strength in deeper blocks. These results point to a sparse internal circuit, where only a handful of heads drive the bulk of the model's causal computation, guiding us to focus on these key heads for finer positional analyses.

Position-Level Causal Influence. Having pinpointed Head 6 in Layer 0 as the most influential, we now investigate its *timestep-specific* contributions: *which exact input positions does this head attend to in a causally meaningful way?* To answer this, we patch the per-timestep output of Head 6 at each position t (from 0 to $T - 1$) independently into the corrupt instance and record the resulting change ΔP_t in true-class probability.

---

[2] Layers and heads are zero-indexed; "Layer 0" denotes the first encoder block.

Figure 4 reveals a highly non-uniform distribution of causal impact across timesteps. Approximately half of the positions yield negligible effects ($\Delta P_t < 0.01$), and five additional positions barely exceed this threshold. Intriguingly, two positions (6 and 8) produce slightly negative ΔP_t, indicating that in isolation they introduce interference rather than denoising. In contrast, four positions: 1, 16, 21, and 23, stand out, each contributing $\Delta P_t > 0.05$. These peaks highlight the "when" of Head 6's causal intervention: specific segments of the input sequence are critical for steering the model back toward the correct classification.

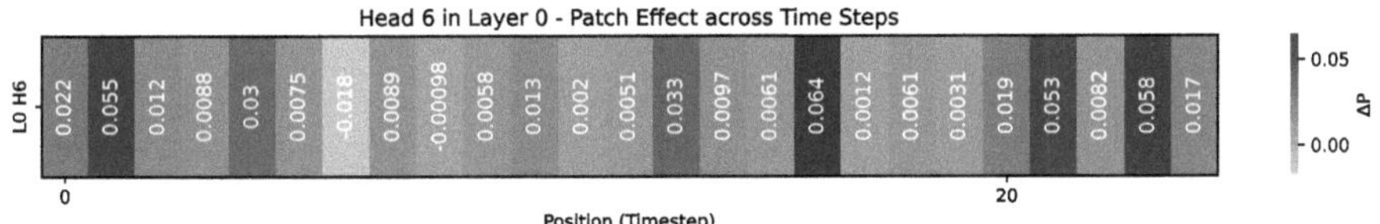

Fig. 4. True-class probability ΔP_t after patching each timestep of Layer 0 Head 6.

Importantly, summing these individual contributions

$$\sum_{t=0}^{T-1} \Delta P_t \approx 0.43$$

underestimates the full-head effect ($\Delta P_{\text{full head}} \approx 0.68$). This discrepancy underscores that Head 6's overall influence is *not* simply additive across timesteps. Instead, it arises from nonlinear synergies enabled by self-attention, residual connections, and layer normalization, components that integrate information across positions. Consequently, while individual patches reveal candidate "hotspots", only the simultaneous reinstatement of the entire head's activation fully recovers the lost confidence, as multiple positions interact constructively (or destructively) when combined.

Attention Saliency. To complement our interventional patching analyses, we apply an observational saliency method to the same head. Specifically, we extract raw self-attention weights $A_{i,t}^{(h)}$ from Head 6 in Layer 0 and average over all query positions i to compute a timestep saliency score $S_t^{(h)}$. We then overlay these scores on the raw input sequences of the clean and corrupt instances to visualize where the head attends.

Figure 5 shows that saliency is highly concentrated at a few timesteps, most notably some of the final positions. However, it is important to emphasize that attention saliency does *not* imply causation: attention weights can reflect routing or normalization behaviors rather than true feature importance, a limitation well documented in prior NLP studies [6]. Nonetheless, saliency remains a useful diagnostic, efficiently highlighting candidate positions for more targeted mechanistic interventions.

Accumulating Top-k Critical Patches. To evaluate how influential components interact when combined, we conduct a controlled multi-patching experiment. Starting from

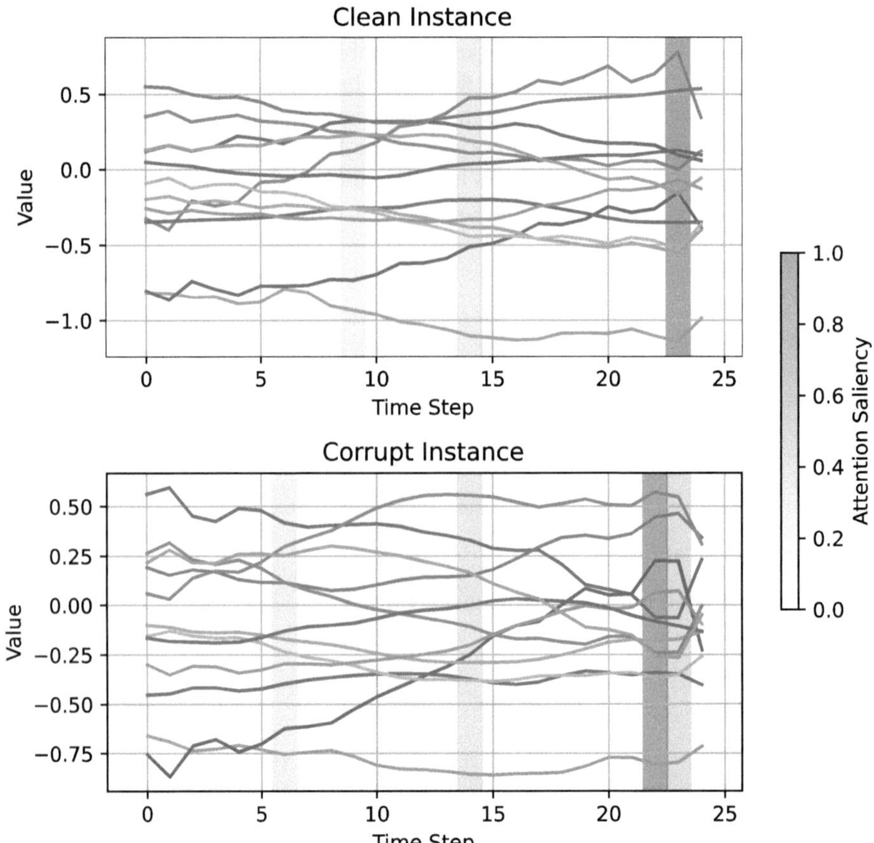

Fig. 5. Attention saliency scores of Layer 0 Head 6 overlaid on the raw input series for the clean (top) and corrupt (bottom) instances.

the most influential (layer, head, position) triplets, which were identified via individual ΔP scores, we sequentially apply patches from the clean instance into the corrupt instance.

For each $k \in \{1, \ldots, 10\}$, we apply the top-k critical patches and observe the resulting change in predicted probability for the true class. Table 3 summarizes the results.

The results show that the model's confidence increases steadily as more top-ranked patches are applied, but not monotonically. Notably, applying the 6th patch results in a **drop in confidence** compared to the 5-patch version, despite being ranked as a top contributor when evaluated in isolation.

This effect illustrates a key point already discussed previously: when multiple internal components are activated simultaneously, their influence may interfere, overlap, or even cancel out due to the model's nonlinear architecture. Components that are helpful alone may become redundant or conflicting when patched in tandem with others.

Table 3. Results of applying top-k critical patches. Shows cumulative ΔP and final predicted probability for the true class as more patches are applied.

Top-k Patches	ΔP(True)	P(True) Final
1	0.1270	0.2342
2	0.1716	0.2788
3	0.2058	0.3131
4	0.3055	0.4128
5	0.4324	0.5396
6	**0.3712**	**0.4784**
7	0.4502	0.5575
8	0.4835	0.5908
9	0.5772	0.6844
10	0.6375	0.7448

Table 4. Top-5 critical patches and their ΔP contributions.

Timestep	Head (LℓHh)	ΔP
T 21	L0H7	0.1270
T 21	L1H7	0.0870
T 22	L2H7	0.0810
T 16	L0H6	0.0640
T 17	L0H3	0.0620

Causal Graph Construction. To synthesize our interventional and observational findings into a unified representation, we build a directed *causal graph* that traces the flow of influence from input timesteps through internal attention heads to the final class prediction. We organize nodes into three tiers: (i) *Input nodes* T_t, $t \in \{0, \ldots, T - 1\}$, for each timestep in the time series instance (green), (ii) *Internal nodes* LℓHh for each attention head h in layer ℓ (blue) and (iii) *Output nodes* C_y for each class y (orange). Each patch corresponds to two directed edges: a timestep-to-head edge weighted by $\Delta P_{t \to (\ell,h)}$ and a head-to-class edge weighted by $\Delta P_{(\ell,h) \to \mathrm{class}}$.

Figure 6 shows the **top-5 patches**, those with the largest individual ΔP values, forming a minimal causal circuit that restores $P(y_{\mathrm{true}})$ to 0.5396 (above 0.5).

Notably, three of these patches involve Head 7 in Layers 0, 1, and 2, while the other two involve Heads 6 and 3 in Layer 0. Table 4 lists these patches in descending order of ΔP.

While this sparse view highlights the most pivotal pathways, it omits moderate but meaningful contributions. To capture a broader circuit, we construct a **thresholded graph** by including all head-to-class edges with $\Delta P \geq 0.10$ and all timestep-to-head edges with $\Delta P \geq 0.01$, retaining only heads that satisfy the former condition. Figure 7

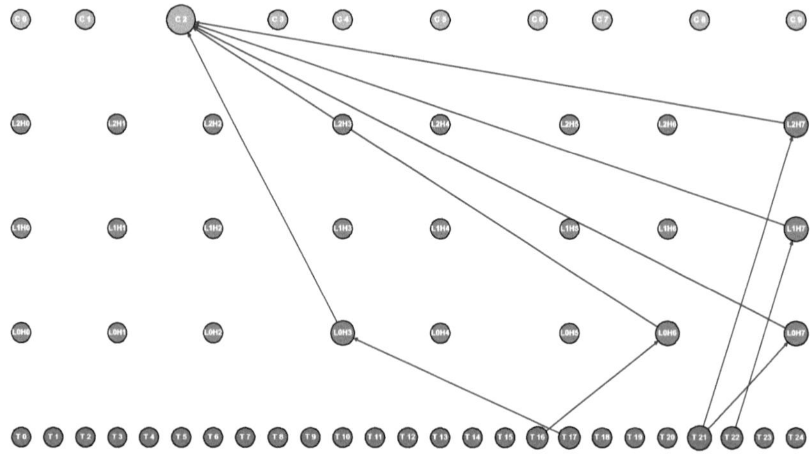

Fig. 6. Causal graph of the top-5 most influential patches (timesteps → heads → class).

shows this denser graph, revealing additional connections, particularly from timesteps near the end of the sequence, into key heads.

To quantify the prominence of each node in the thresholded graph, we compute degree centrality: the number of outgoing edges per timestep and the number of incoming edges per head (excluding head-to-class edges). Figure 8 plots these degrees.

On the input side, late timesteps (21–24) dominate outgoing connections, with an additional early peak at Time 1, indicating non-uniform reliance on sequence positions. On the internal side, eight of the top-ten most connected heads reside in Layer

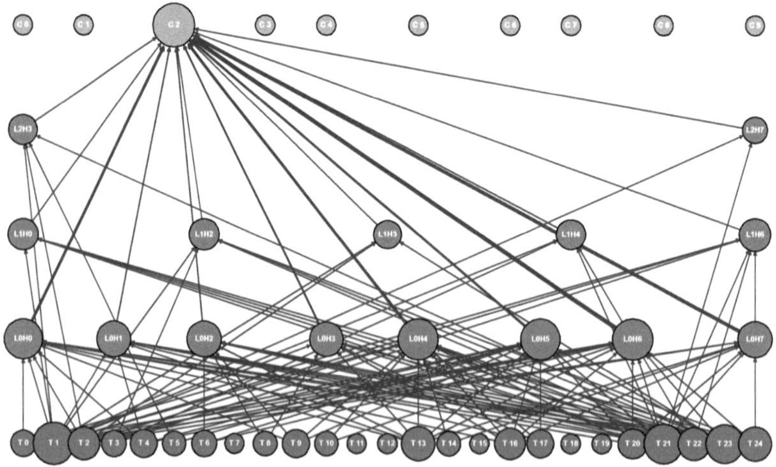

Fig. 7. Thresholded causal graph: head → class edges with $\Delta P \geq 0.10$, timestep → head edges with $\Delta P \geq 0.01$.

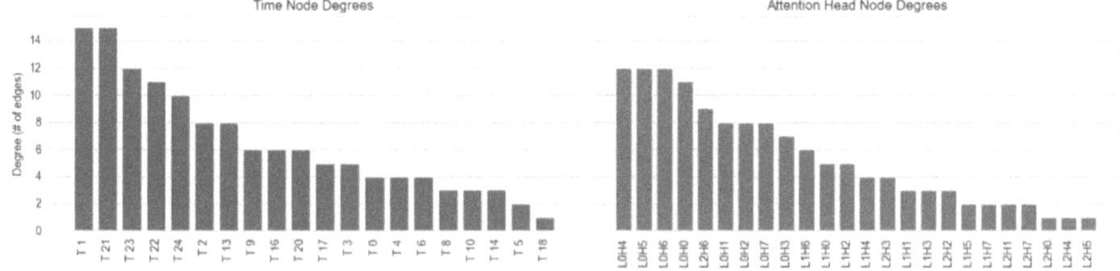

Fig. 8. Degree centrality in the thresholded causal graph. *Left:* outgoing degree of each timestep. *Right:* incoming degree of each head.

0, reinforcing our earlier observations that the first encoder layer concentrates causal influence.

5.1 Provisional Sparse Autoencoder Insights

To uncover whether the transformer's internal activations contain disentangled, semantically meaningful features, we train a SAE on the output of the MLP block (`linear2`) in the first encoder layer. The SAE encoder maps each activation vector $x \in \mathbb{R}^d$ at each timestep into a sparse code $z \in \mathbb{R}^H$, where H is the number of SAE neurons. A strong L_1 penalty on z encourages each code to activate only a few neurons, facilitating interpretability.

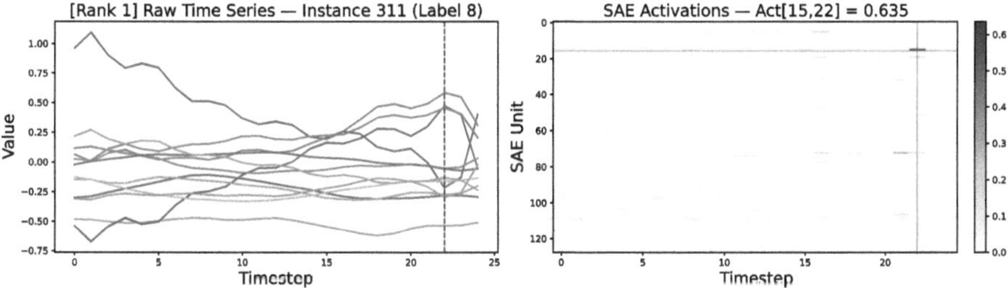

Fig. 9. Top-activating test instances for SAE neuron 15 at timestep 22. Each overlay shows the raw input time series, with neuron 15's activation highlighted. All top-activating instances belong to Class 8.

Figure 9 illustrates a representative top-activating instance for SAE neuron 15 peaking at timestep 22 (the remaining top-activating instances are shown in Appendix Fig. 17). All of these top-ranked instances share the ground-truth label (Class 8), and inspection of their raw inputs reveals a strikingly consistent waveform motif: a pronounced simultaneous peak in Channels 2, 5, and 9 accompanied by a dip in Channel 7 around that timestep. This recurring pattern across multiple speakers suggests that neuron 15 has specialized to detect this class-specific temporal feature, rather than idiosyncratic noise in any single instance.

To see how this sparse code behavior contrasts with our clean–corrupt pair, we plot full SAE activation heatmaps (neurons × timesteps) for Instance 33 (clean, Class 2) and Instance 44 (corrupt, misclassified) in Fig. 10.

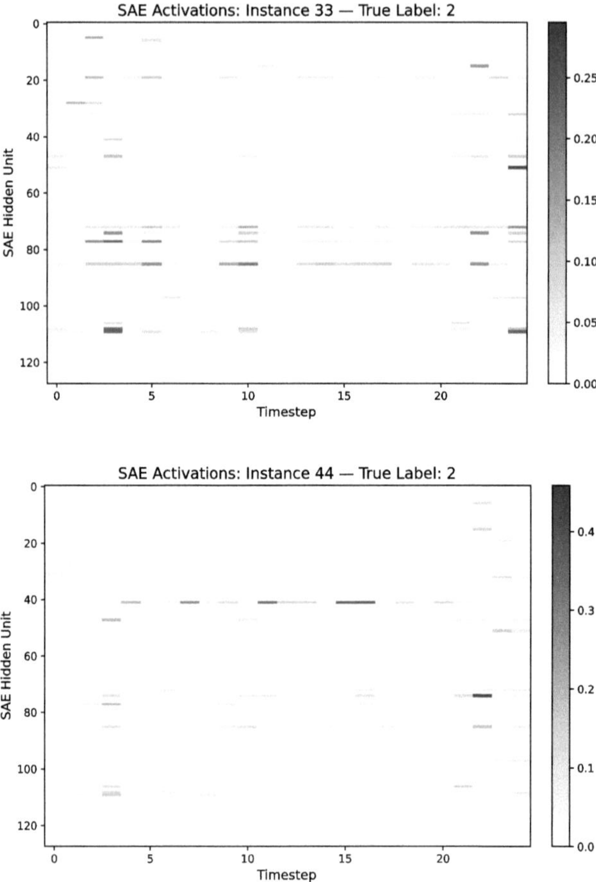

Fig. 10. SAE activation heatmaps (neurons × timesteps) for the clean and corrupt instances. Darker cells denote higher activation values.

Specifically, Fig. 10 shows that in the clean instance several neurons (notably 78 and 85) activate sharply at specific timesteps that align with class-discriminative patterns, whereas the corrupt instance exhibits a different activation profile, activating neuron 41 more broadly but failing to trigger the combination of neurons that uniquely identify Class 2. This divergence suggests that misclassification may arise when an instance erroneously engages features characteristic of another class. In Appendix B Fig. 18, we provide further visualizations showing the activation of these neurons across other Class 2 and Class 3 test samples.

Importantly, SAE neurons do not correspond one-to-one with the transformer's original neurons. Instead, each SAE neuron represents a learned direction in the transformer's latent space. To test the causal role of these sparse features, one could encode the transformer activations into z, manually amplify a chosen dimension z_j, decode back to $\hat{x}$ via the SAE decoder, and patch $\hat{x}$ into the transformer's MLP at the corresponding layer and timestep. Thus, measuring the resulting change in $P(y_{\mathrm{true}})$ would reveal the sufficiency of individual sparse features.

6 Discussion

Mechanistic Interpretability has emerged as a critical component for trustworthy AI, especially in safety-critical domains such as healthcare and finance where understanding internal decision pathways can build trust and facilitate risk mitigation. Our experiments demonstrate that MI techniques originally developed for NLP transformers: activation patching, attention saliency, and sparse autoencoders, can be effectively adapted to TSTs, yielding fine-grained insights into which layers, heads, timesteps, and latent features drive model decisions. The resulting causal graphs and sparse feature visualizations (Sect. 5.1) illustrate multilevel interpretability in TSTs and highlight both shared patterns across layers and instance-specific motifs.

Despite these promising results, several limitations constrain our current pipeline. First off, MI methods remain labor-intensive: selecting representative clean–corrupt pairs, choosing granularity levels, and interpreting interventional outcomes require substantial domain expertise. Second, our patching experiments focus solely on attention heads; other components like MLP blocks, residual streams, positional embeddings, and individual neurons likely encode additional causal signals that remain unexplored. Third, unlike NLP, where tokens often correspond to semantic concepts, time series features lack clear semantics, making interpretation of causal motifs (e.g., peaks, periodicities, anomalies) inherently more challenging. This raises questions of *universality*, whether the circuits we uncover generalize across datasets or reflect dataset-specific idiosyncrasies. Finally, our multi-patching results show non-additive effects; interventions can interfere or synergize due to the transformer's complex interplay of multi-head attention, residual connections, and layer normalization, complicating efforts to isolate independent causal factors.

To address these limitations, we envision several avenues for future research. First, expanding patching to include MLP activations, residual streams, and embeddings will provide a more comprehensive view of TST internals. Second, scaling analysis to multiple clean–corrupt pairs across classes and datasets, especially higher-dimensional or irregular real-world time series, will test the generality of discovered circuits. Third, integrating optimization-driven search (e.g., evolutionary algorithms) could automatically identify minimal, non-redundant patch sets, refining causal graph construction. Fourth, developing higher-level metrics beyond ΔP, as well as visualization and summarization techniques (e.g., clustering pathways, generating narrative descriptions), will enhance interpretability for non-expert stakeholders. Finally, leveraging MI for post-hoc error correction, dynamically patching critical components at inference time, offers a promising route to improve model robustness without retraining.

By systematically extending and automating these MI frameworks, we can move toward interpretable, reliable TSTs that not only perform well but also reveal their inner workings in a way that aligns with human understanding and ethical standards.

7   Conclusion

This work set out to determine whether Mechanistic Interpretability techniques, originally developed for NLP transformers, can be meaningfully adapted to TSTs. Through a series of structured experiments on a speaker-identification task using the JapaneseVowels dataset, we have shown that activation patching at the layer, head, and timestep levels reveals clear causal effects on model predictions. Patching individual attention heads and even specific timestep outputs can substantially restore correct classification in mispredicted instances, demonstrating that TST internals encode discrete, manipulable circuits akin to those observed in language models.

By synthesizing these interventional findings into causal graphs, we uncovered interpretable pathways from input timesteps through attention heads to class logits, and we observed nonlinear synergies and interference when combining multiple patches. Complementary observational methods, attention saliency, and sparse autoencoders, provided additional insights into where and when the model attends and what latent features it extracts. In particular, sparse autoencoders uncovered class-selective temporal motifs, highlighting the potential for disentangled feature discovery in time series domains.

While we stop short of a full reverse-engineering of TSTs, our results establish a robust proof-of-concept: MI tools can help in gaining understanding about the hidden logic of deep time series models. Ultimately, embedding Mechanistic Interpretability into TSTs can enhance transparency, accountability, and reliability of AI systems deployed in critical real-world applications.

A   Causal Influence via Activation Patching

To verify that our activation-patching pipeline generalizes beyond the primary example (Sect. 5), we repeat the full sequence of analyses on a second clean–corrupt pair. All panels below mirror Sects. 5.1–5.5.

A.1   Pair 2: Instance Selection and Overview

Table 5 lists the chosen pair. The "clean" instance 123 is classified correctly as class 3 with almost perfect confidence, while the "corrupt" instance 114 (also true label 3) is misclassified as class 8 with only 18.5% true-class probability.

Table 5. Second clean–and–corrupt instance pair for robustness testing.

ID	Type	True Class	Pred. Class	$P(\text{true})$
123	Clean	3	3	99.99%
114	Corrupt	3	8	18.49%

Figure 11 shows the raw 12-channel time series for both instances.

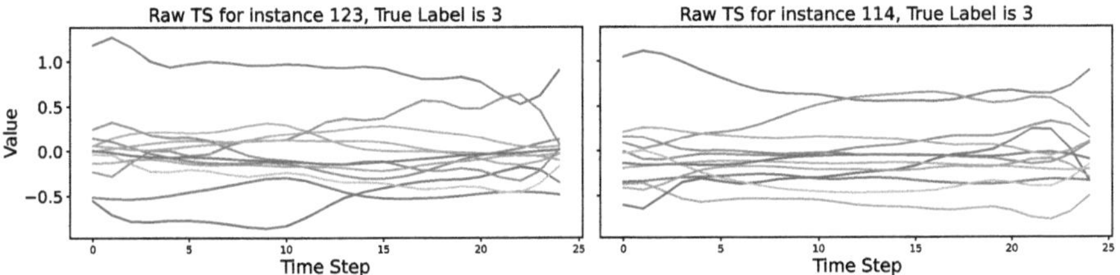

Fig. 11. Raw multivariate time series for instances 123 (clean) and 114 (corrupt), 12 channels over 25 timesteps.

A.2 Layer-Wise Patching

In Fig. 12, we patch all heads in each encoder layer of the clean into the corrupt instance and plot the resulting true-class probability. Layer 0 again exerts the strongest causal influence (increase from 0.185 to 0.855, $\Delta P_t = 0.67$), confirming early layers' dominance.

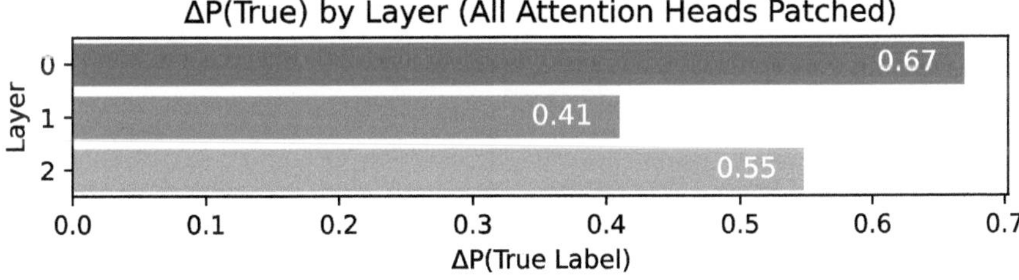

Fig. 12. True-class probability after patching all heads of each encoder layer.

A.3   Head-Wise Patching

Breaking down Layer 0 into individual heads (Fig. 13), head 3 contributes the largest improvement ($\Delta P_t \approx +0.19$), followed by heads 1 and 7. Other layers show far smaller or negligible effects.

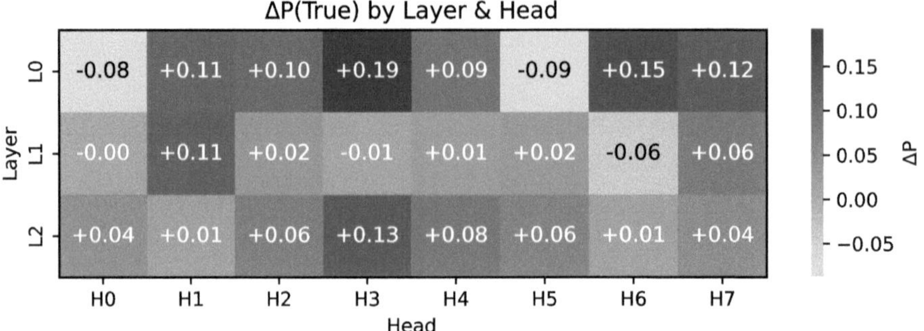

Fig. 13. Post-patching true-class probability for each attention head.

A.4   Position-Wise Patching

Focusing on Layer 0 Head 3, Fig. 14 plots true-class probability after patching each timestep in isolation. We again observe "hotspots" around timesteps 0, 3, 17, and 23, where a single-step patch raises $P(y_{\text{true}})$ by more than 0.02 (baseline $= 0.1849$).

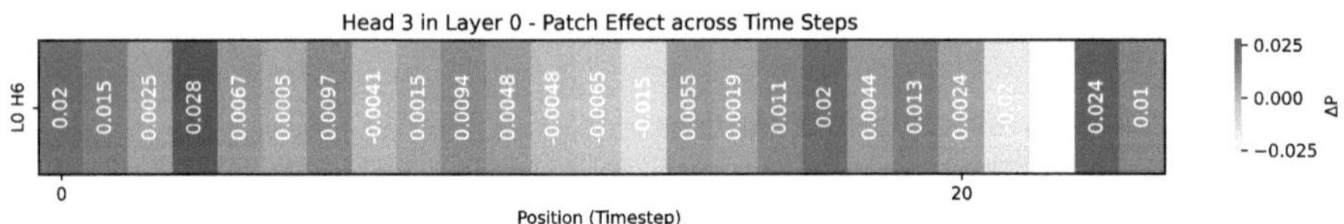

Fig. 14. True-class probability change ΔP_t after patching each timestep of Layer 0 Head 3.

A.5   Attention-Saliency Comparison

Finally, Fig. 15 overlays Layer 0 Head 3's observational attention-saliency scores on both inputs.

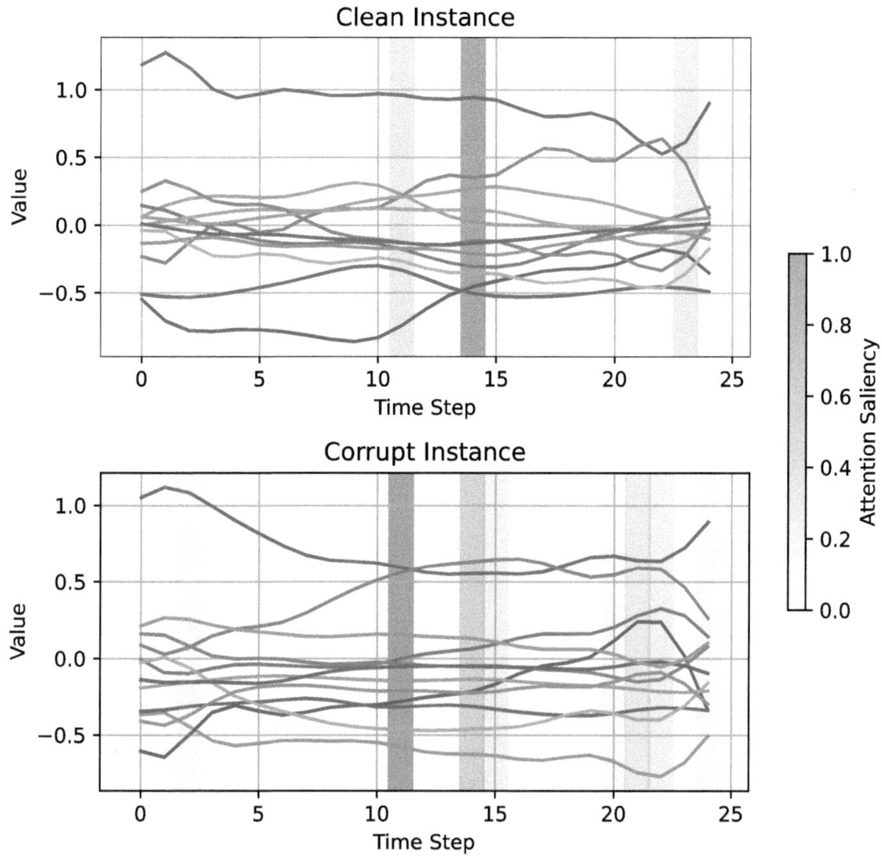

Fig. 15. Attention-saliency scores for Layer 0 Head 3 overlaid on raw input series. Top: clean instance; bottom: corrupt instance.

B   Additional Visualizations

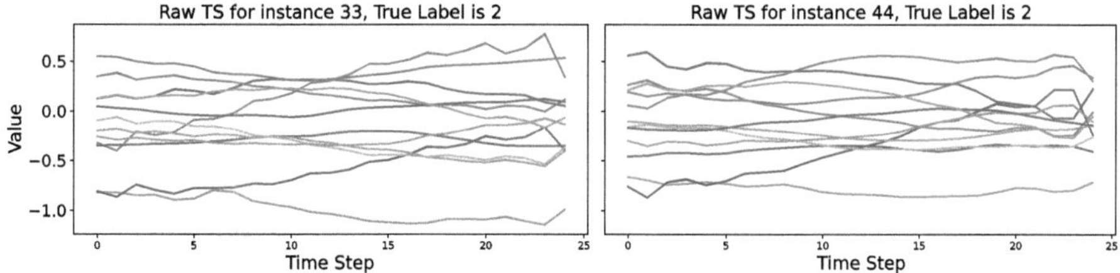

Fig. 16. Sample visualizations of multivariate time series instances 33 (clean) and 44 (corrupt) from the JapanseVowels dataset, showing 12-channel signal over 25 time steps.

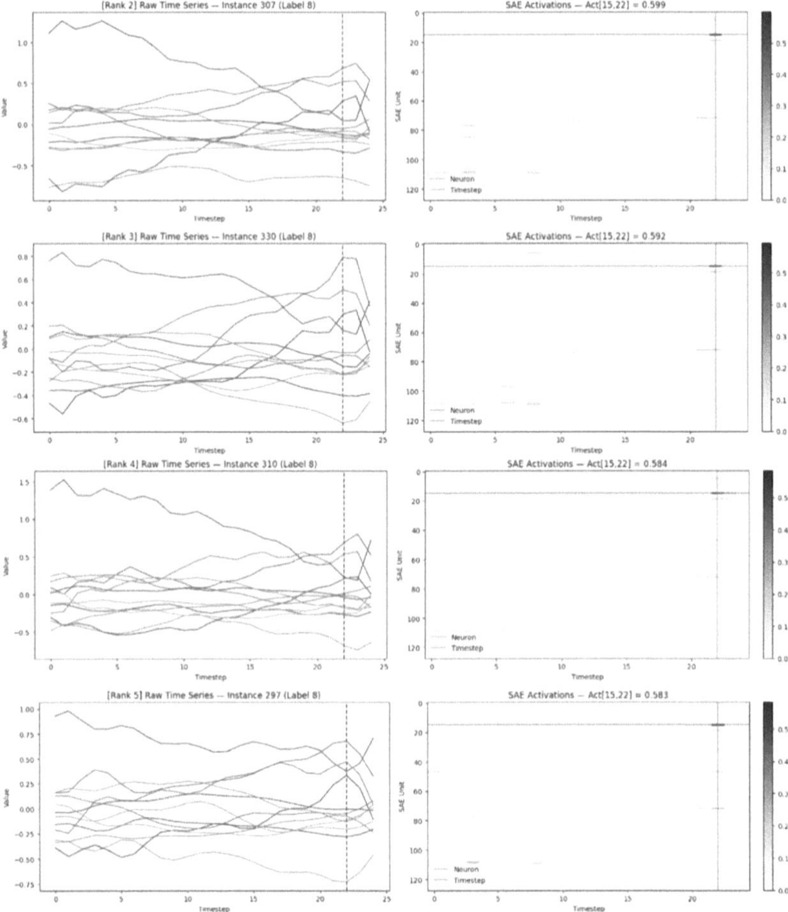

Fig. 17. Top-activating test instances for SAE neuron 15 at timestep 22. These instances all belong to Class 8 and exhibit similar temporal patterns, suggesting neuron 15 encodes a class-discriminative feature.

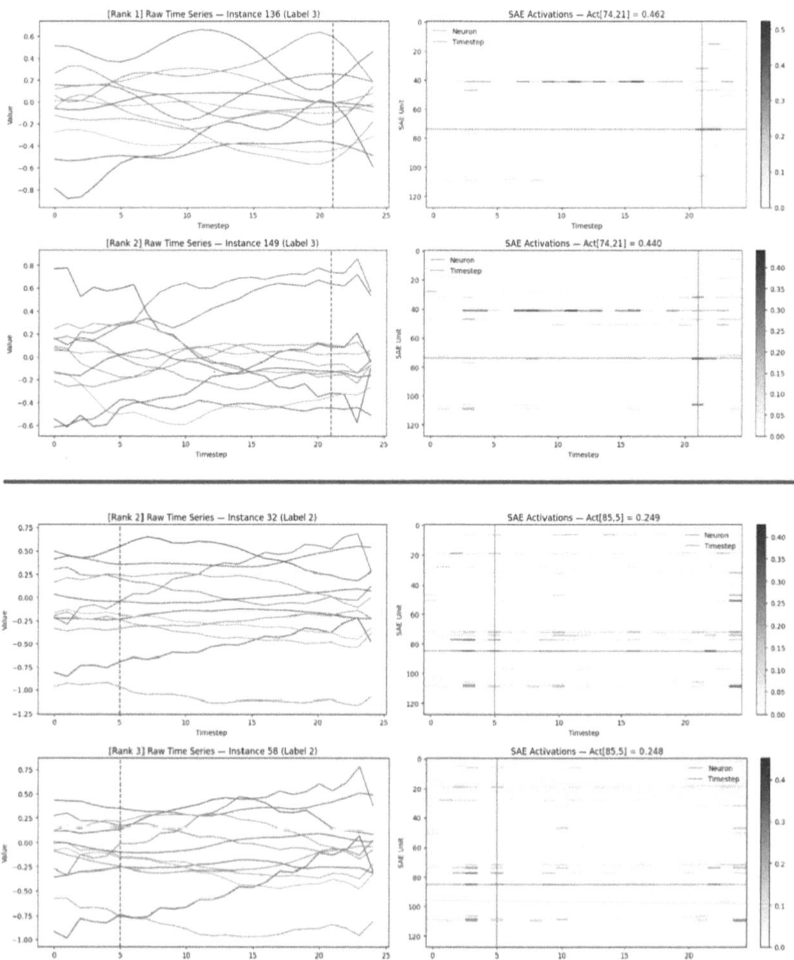

Fig. 18. Comparison of sparse activations in Class 2 vs Class 3 instances. Highlights class-selective neurons such as neuron 85 and 78 (Class 2) and neuron 41 (Class 3).

References

1. Akram, S., Presacco, A., Shamma, S., Babadi, B.: Robust decoding of selective auditory attention from meg in a competing-speaker environment via state-space modeling. NeuroImage **124** (2015). https://doi.org/10.1016/j.neuroimage.2015.09.048
2. Cunningham, H., Ewart, A., Riggs, L., Huben, R., Sharkey, L.: Sparse autoencoders find highly interpretable features in language models (2023). https://arxiv.org/abs/2309.08600
3. Dempster, A., Petitjean, F., Webb, G.I.: Rocket: exceptionally fast and accurate time series classification using random convolutional kernels. Data Min. Knowl. Discov. **34**(5), 1454–1495 (2020). https://doi.org/10.1007/s10618-020-00701-z
4. Heimersheim, S., Nanda, N.: How to use and interpret activation patching (2024). https://arxiv.org/abs/2404.15255
5. Islam, M.K., et al.: Interpreting time series transformer models and sensitivity analysis of population age groups to covid-19 infections (2024). https://arxiv.org/abs/2401.15119
6. Jain, S., Wallace, B.C.: Attention is not explanation (2019). https://arxiv.org/abs/1902.10186
7. Kalnāre, M.: TST-mechanistic-interpretability (2025). https://github.com/mathiisk/TST-Mechanistic-Interpretability, gitHub repository
8. Kudo, M., Toyama, J., Shimbo, M.: Japanese Vowels [dataset]. UCI Machine Learning Repository (1999). https://doi.org/10.24432/C5NS47
9. Lubba, C.H., Sethi, S.S., Knaute, P., Schultz, S.R., Fulcher, B.D., Jones, N.S.: catch22: canonical time-series characteristics (2019). https://arxiv.org/abs/1901.10200
10. Lundberg, S., Lee, S.I.: A unified approach to interpreting model predictions (2017). https://arxiv.org/abs/1705.07874
11. Meng, K., Bau, D., Andonian, A., Belinkov, Y.: Locating and editing factual associations in GPT (2023). https://arxiv.org/abs/2202.05262
12. Ribeiro, M.T., Singh, S., Guestrin, C.: "why should i trust you?": explaining the predictions of any classifier (2016). https://arxiv.org/abs/1602.04938
13. Romero, J., Ramirez, D., Aglio, L., Gugino, L.: Brain mapping using transcranial magnetic stimulation. Neurosurgery Clinics North Am. **22**, 141–52, vii (2011). https://doi.org/10.1016/j.nec.2010.11.002
14. Serrano, S., Smith, N.A.: Is attention interpretable? (2019). https://arxiv.org/abs/1906.03731
15. Wen, Q., et al.: Transformers in time series: a survey (2023). https://arxiv.org/abs/2202.07125
16. Zerveas, G., Jayaraman, S., Patel, D., Bhamidipaty, A., Eickhoff, C.: A transformer-based framework for multivariate time series representation learning. In: Proceedings of the 27th ACM SIGKDD Conference on Knowledge Discovery & Data Mining. KDD '21, New York, NY, USA, pp. 2114–2124. Association for Computing Machinery (2021). https://doi.org/10.1145/3447548.3467401

XAI-Driven Solutions to Enhance Safety for Limited-Mobility Road Users

Gianmarco Cherchi, Nicola Floris[✉], Alessandro Sebastian Podda, Livio Pompianu, Roberto Saia, and Riccardo Scateni

Department of Mathematics and Computer Science, University of Cagliari, Via Ospedale 72, Cagliari 09124, Italy
{gianmarco.cherchi,nicola.floris,sebastian.podda,livio.pompianu, roberto.saia,riccardo.scateni}@unica.it

Abstract. The large availability of urban surveillance video, combined with recent advances in deep learning, allows us to exploit them to support road safety by analyzing pedestrian behavior in real-time. This work proposes an XAI-driven approach for monitoring crosswalks with limited visibility, where the risk to Vulnerable Road Users (VRUs) is particularly high. We focus specifically on Limited-Mobility VRUs (LM-VRUs), including wheelchair users or adults with strollers, who may have a reduced ability to respond quickly to dangerous situations. The system processes live video streams from fixed surveillance cameras using deep learning-based computer vision to detect and track LM-VRUs in both crossing and waiting areas. Unlike data-intensive predictive models, our approach utilizes a lightweight, rule-based module that infers pedestrian crossing intent through human-understandable spatiotemporal heuristics. This explainable component ensures that the decision-making process remains transparent and auditable. Upon detecting a potentially dangerous crossing scenario, the system immediately activates acoustic and visual warnings for approaching drivers, improving safety, including for visually impaired pedestrians. Beyond its technical contribution, this work explores the social impact of AI technologies designed to protect mobility-impaired individuals in urban environments. Our goal is to reduce traffic-related accidents and contribute to more inclusive, intelligent city infrastructures.

Keywords: Vulnerable Road Users · XAI · Computer Vision · Object Detection · Object Tracking · YOLO · Urban Surveillance · Road Safety · Crossing Intention Prediction · Explainability

1 Introduction

The widespread deployment of urban video surveillance systems, driven by the need for enhanced safety and traffic monitoring, has significantly increased the availability of real-time visual data [6, 12]. Concurrently, recent advances in deep learning have enabled computer vision applications to autonomously extract semantic data from video

© The Author(s), under exclusive license to Springer Nature Switzerland AG 2026
F. Marcelloni et al. (Eds.): IJCCI 2025, CCIS 2829, pp. 271–289, 2026.
https://doi.org/10.1007/978-3-032-15638-9_16

streams [13,19], unlocking new opportunities for intelligent image analysis in road safety applications [4]. Through Internet of Things (IoT) technology, it is possible to create distributed systems that can process live surveillance video, autonomously make decisions, and interact with the physical world through actuators [26–28]. These systems can rapidly warn both pedestrians and drivers about potential dangers or alert human operators to unusual or risky events [3,7,23]. In the context of road safety [32], such technologies support a variety of applications, ranging from anomaly detection to forecasting pedestrian actions, such as crossing intent prediction.

Our contribution focuses on the development of an XAI-driven prototype system for real-time monitoring of pedestrian crossings with limited visibility, such as those located on multi-lane urban roads. These configurations pose an increased risk for *Vulnerable Road Users* (VRUs) with *limited mobility* (hereafter denoted as LM-VRUs), such as individuals in wheelchairs or those caring for small children in strollers, who may have limited capacity to react quickly in critical situations. The presence of large and obstructing vehicles (e.g., buses or trucks) may further reduce the mutual visibility between drivers and pedestrians, increasing the likelihood of accidents.

Specifically, the proposed system leverages real-time video analysis from strategically placed surveillance cameras that cover both the crosswalk and adjacent waiting areas, using deep learning-based computer vision models to identify and track VRUs of the category of interest. Then, a decision-making module is used to predict the crossing intent. Unlike existing approaches that rely on additional machine learning or deep learning models, our method is rule-based, utilizing a set of lightweight, interpretable, and transparent spatio-temporal conditions designed to ensure both effectiveness and explainability. When an LM-VRU pedestrian shows clear intent to cross, nearby drivers are promptly alerted via LED-based signaling and acoustic cues, increasing the safety of the crossing area, including visually impaired users.

Beyond the technical validation of the system, this project also aims to explore the *social impact* of deploying AI technologies to protect LM-VRUs, intending to contribute to the reduction of traffic-related accidents and assist in the development of safer and more inclusive urban environments.

The main scientific contributions of this work are:

- The design of an XAI-driven system for real-time monitoring of pedestrian crossings in low-visibility scenarios;
- A specific focus on Limited-Mobility Vulnerable Road Users (LM-VRUs), promoting inclusivity and safety in urban contexts;
- The development of a lightweight, rule-based reasoning module for interpretable and transparent intent prediction, fostering trust and accountability through human-understandable logic;
- The integration of multimodal alerts (LED and acoustic) to enhance drivers' situational awareness in potentially hazardous conditions;
- A critical reflection on the societal impact of deploying explainable AI technologies in the public space, with a focus on fairness and accessibility.

The remainder of this paper is organized as follows. Section 2 provides an overview of the literature relevant to our research scenario. Then, Sect. 3 formalizes and describes the proposed approach, while Sect. 4 presents the experimental environment. Section 5 discusses the obtained results, and, finally, Sect. 6 concludes the work and indicates our future research directions.

2   Related Work

Within this context, related research can be broadly divided into two main areas: (A) the detection of VRUs, especially LM-VRUs, and (B) pedestrian *Crossing Intention Prediction (CIP)*. This study makes contributions to both fields. In this section, we examine a range of significant works pertinent to each domain.

A. VRUs Detection. Several studies have explored pedestrian and generally VRU detection in traffic environments using deep learning–based object detectors. Notably, the *YOLO (You Only Look Once)* [24] family has become popular for real-time applications due to its balance of speed and accuracy [20].

In this work [30], the authors propose an enhanced lightweight detector designed for infrastructure-integrated systems, embedding dynamic detection heads into YOLOv7-tiny and introducing attention mechanisms like Dyhead and VoV-GSCSP to boost VRUs detection performance without increasing computational load.

Similarly, this work [22] investigates VRU detection—including strollers, motorbikes, and bicycles—using both one-stage and two-stage CNN architectures. They introduced a custom roadside dataset to address occlusion challenges and analyzed the trade-off between detection accuracy and inference speed.

Another work [18] demonstrates the versatility of visual pipelines by adapting YOLOv3 for indoor smart wheelchair mobility. The authors integrated depth estimation via Intel RealSense and tracking with SORT, showing the potential of vision-based detection and tracking in assistive mobility contexts.

Despite these advances, most state-of-the-art approaches focus primarily on detection and tracking. At the same time, the real-time interpretation of dangerous scenarios, such as low-visibility pedestrian crossings, and the implementation of driver alerting mechanisms remain underexplored. Our work addresses this gap by integrating real-time detection, explainable intent prediction, and proactive multimodal warnings for LM-VRUs.

B. Crossing Intention Prediction. A considerable body of research on CIP has been developed in the context of autonomous driving, where real-time perception is achieved using vehicle-mounted sensors or cameras [14,36]. These approaches are primarily designed to enable autonomous vehicles to anticipate pedestrian behavior and adjust their motion planning accordingly [29].

An alternative line of research, such as the one investigated in this article, examines CIP through fixed infrastructure, including roadside cameras, CCTV surveillance systems, and LiDAR units. This infrastructure-centric approach offers natural integration with smart city ecosystems, allowing for cooperative safety interventions that actively alert both pedestrians and vehicles interacting directly with the environment through actuators.

Several relevant approaches have been proposed in the CIP domain. A probabilistic model based on roadside LiDAR sensors is presented in [35], where a modified Naïve Bayes classifier is trained on annotated crossing and non-crossing sequences. In [17], the authors propose a camera-invariant pedestrian crossing direction prediction framework leveraging pose key points and trajectory data, using Transformer-based, GCN, and hybrid Transformer+GCN architectures.

VRU-CIPU [1] extends CIP to multiple VRUs (pedestrians, cyclists, and e-mobility users) through a two-stage YOLO pipeline for VRU detection and human pose estimation, combined with semantic scene segmentation and a GRU enhanced by Transformer self-attention. The system can proactively activate crossing signals and issue I2V warnings.

The VENUS system [34] focuses on non-motorized road users, including people with disabilities. It employs YOLOv4 and OpenPose for detection and pose estimation, followed by an additional neural network for mobility status classification. While explicitly addressing LM-VRUs, VENUS controls traffic lights based on direction estimation rather than probabilistic intention modeling.

Most of the aforementioned works do not explicitly address the detection and analysis of LM-VRUs. Additionally, these approaches rely entirely on end-to-end deep learning pipelines or apply additional deep learning at the decision-making stage. While such methods are powerful, they often suffer from limited transparency and interpretability.

Our proposal specifically targets the detection of LM-VRUs and introduces a rule-based crossing intention prediction module. Deep learning is employed solely within the perception layer, whereas the high-level decision-making process is fully rule-based. The outputs from the perception layer serve as structured inputs to a probabilistic reasoning algorithm, which infers crossing intent using transparent, hand-crafted spatio-temporal rules.

By decoupling perception and intention inference, it is possible to preserve the strengths of deep learning in visual recognition while also providing explanations that are comprehensible to humans, enabling the identification of both the moments and reasons for any failures in the decision-making process.

3 Methodology

This section describes our methodology. First, Sect. 3.1 provides an overview of our pipeline, which is primarily composed of the perception layer and the decision-making layer. Then, Sect. 3.2 to 3.4 details each pipeline module.

3.1 Pipeline Overview

The proposed system for predicting LM-VRU crossing intention is organized into two layers: the perception layer and the decision-making layer. The perception layer applies deep learning–based object detection and tracking to process real-time video streams from fixed surveillance cameras. The decision-making layer uses these outputs, together with the static geometric layout of the scene (e.g., waiting areas near crossings), to analyze trajectories over multiple frames. By applying predefined spatio-temporal rules, it assigns each entity a probability score of crossing intention. When the score surpasses the threshold, the system issues an alert. In the evaluation phase, this alert is simulated through on-screen notifications, whereas in real deployments, it is designed to interface with IoT devices and RESTful APIs to trigger visual or auditory warnings for both pedestrians and approaching drivers. A schematic representation of the proposed pipeline is provided in Fig. 1. The following sections will explore each component of the system in detail.

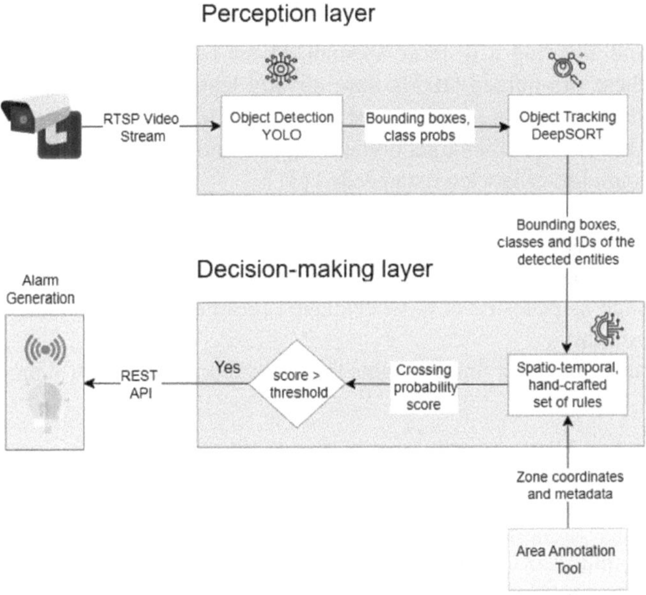

Fig. 1. System architecture of the proposed approach in real-world scenarios. Surveillance cameras stream video to the perception layer, which detects and tracks entities. The decision-making layer infers crossing intent using rule-based logic and scene geometry. If the alert threshold is exceeded, actuators are triggered to warn drivers.

3.2   Object Detection

The AI module of the proposed system combines object detection and object tracking. The detection component identifies and localizes entities of interest within each frame, while the tracking component preserves temporal consistency by assigning and maintaining object identities across consecutive frames.

For detection, YOLO was selected due to its balance between accuracy and real-time performance, making it well-suited for continuous video stream analysis. As a one-stage detector, YOLO performs classification and localization in a single pass over the image [2]. This design contrasts with two-stage approaches, such as the R-CNN family, where region proposal and classification are executed in separate phases, often at the expense of higher inference time [37].

Object Detection Models (ODMs) are typically data-driven and require vast amounts of labeled data to train effectively. Acquiring such large-scale annotated datasets is particularly challenging in practical applications, especially for underrepresented classes such as pedestrians with mobility impairments. Additionally, developing and training such models from scratch demands substantial computational resources and time.

To address these limitations, we adopted a *Transfer Learning (TL)* approach. TL helps to overcome data limitations by leveraging the general-purpose visual knowledge learned from models trained on large-scale datasets (such as COCO [21] or ImageNet [9]) and adapting it to more specialized but related domains [15]. The CNNs underlying these pre-trained ODMs have already learned key visual features such as edges, textures, and colors, and exhibit strong generalization capabilities. These learned features can be reused in the target domain, significantly reducing the required training time and the amount of labeled data needed [31].

We used pre-trained weights from the YOLOv5 model [16] trained on the COCO dataset, and applied transfer learning on a custom dataset sourced from the Roboflow computer vision platform [25]. The dataset includes a (non-exhaustive) set of VRUs categories, such as pedestrians in wheelchairs, pushing a stroller, using crutches, and regular pedestrians.

In particular, the pre-trained weights were based on the YOLOv5l architecture.

3.3 Object Tracking

Object Detection alone is capable of identifying entities within a single frame, but it lacks temporal awareness and does not ensure continuity across multiple frames. Object tracking algorithms are employed to maintain persistent identification of entities over time within video sequences.

We refer to tracking-by-detection methods as those that leverage the output of an ODM—such as YOLO—to perform the tracking task. These algorithms assign a unique identifier (ID) to each detected entity, enabling consistent recognition across successive frames, also estimating the trajectory of each entity to preserve temporal coherence, even in the presence of partial occlusions or short-term disappearances from the scene.

In our perception layer, we integrate the DeepSORT algorithm [33], which enhances the original SORT algorithm [5] by incorporating both motion and appearance cues. While SORT relies solely on motion estimation using Kalman filtering and the Hungarian algorithm for data association, DeepSORT adds a dedicated convolutional neural network (CNN) to extract appearance visual features, resulting in a more accurate identity matching [8].

In our implementation, DeepSORT communicates directly with YOLO, receiving its outputs in real time. Then, tracking-refined entities' information is forwarded to the decision-making layer.

3.4 Decision-Making

As discussed earlier, the perception layer provides essential input, but it is not sufficient on its own for the decision-making process to reliably infer pedestrian crossing intent. To this end, the system must first understand the spatial layout of the monitored scene. In particular, it requires prior knowledge of the areas adjacent to the curb—just before the crosswalk—where pedestrians are most likely to initiate a crossing.

To support this, we developed a custom annotation tool that allows users to define elliptical or semi-elliptical waiting zones directly on a reference video frame, as illustrated in Fig. 2. The tool supports rotation, scaling, and translation to adapt these zones to the geometry of each specific scene. Once defined, their spatial properties are imported into the processing pipeline (see Fig. 1).

Fig. 2. Example of area segmentation using the developed tool. Given a frame of the target scenario, elliptical shapes (highlighted in green) can be drawn to define the waiting areas, along with the intended crossing direction (indicated by a directional arrow). (Color figure online)

Each annotated zone is internally modeled as a set of concentric elliptical bands, as seen in Fig. 3. These subregions enable a nuanced assessment of crossing likelihood based on proximity to the curb and the crosswalk entry point.

Entities detected by the perception module are assigned a dynamic probability score representing the likelihood of an upcoming crossing. This score is computed by integrating three key components: *spatial proximity*, *temporal persistence*, and *directional intent*.

At each frame, the system evaluates which subregion the entity occupies. Higher weights are assigned to inner subregions closer to the curb, reflecting a greater probability of crossing. The score accumulates over time: the longer an entity remains within the area (especially in inner bands), the higher its estimated crossing probability. This time-weighted mechanism allows outer zones to trigger alerts only after sustained presence, reducing false positives from brief passersby.

When the accumulated probability exceeds a configurable threshold, the system issues a crossing alert. This approach introduces a flexible mechanism to balance sensitivity and specificity in detection.

Let p_t be the probability score at frame t, defined as:

$$p_t = c_i \quad \text{where} \quad c_i \in \{0.25, 0.50, 0.75, 1.0\}$$

where c_i is the contribution value associated with the elliptical subregion i in which the entity is located. The elliptical area is divided into 4 concentric bands, with contribution values increasing by 0.25 from the outermost to the innermost region.

Fig. 3. A visual representation of the waiting areas, segmented into four concentric bands. The innermost band (shown in red) corresponds to the region with the highest likelihood of crossing, while the outermost band (shown in green) contributes the least. The width of each band can be configured before execution. (Color figure online)

The final crossing probability score P is computed as:

$$P = \frac{1}{F} \sum_{t=1}^{F} p_t$$

where F is the *frame factor*, i.e., the number of frames over which the probability is accumulated (default: $F = 100$).

This formulation ensures that $P \in [0, 1]$, where a value of 1 corresponds to the entity being located in the innermost ring for the entire duration of the frame window.

An alert is triggered once the score P exceeds a predefined threshold $\tau \in [0, 1]$. By default, $\tau = 0.5$.

Figure 4 shows an example of how alert generation was simulated during initial testing: as soon as the crossing probability for a tracked LM-VRU exceeds the defined threshold, an on-screen warning is displayed. This allows for iterative validation of the signaling logic before integrating the system with external driver notification devices.

Fig. 4. Simulated warning display used to validate the signaling logic during testing in synthetic footage.

Directional Intention Estimation. Incorporating directional cues is crucial for distinguishing genuine crossing intentions from mere proximity to a crosswalk. Without this information, the system might misclassify pedestrians walking along the sidewalk near the crosswalk or pedestrians who have already crossed and reached the opposite side, resulting in duplicate alerts.

To address this, we include a *Movement Direction Detection (MDD)* module that estimates pedestrian orientation using:

- the short-term trajectory of the entity (i.e., centroid positions over recent frames);
- the spatial layout of the annotated area;
- the intended crossing direction, defined during configuration.

The annotation tool allows users to specify the intended crossing direction for each zone. This directional data, expressed in radians, is stored as part of the zone metadata.

The MDD module calculates a movement vector and its angular orientation by analyzing tracked positions over a configurable frame window. These values are then compared to the annotated crossing direction.

If the angular difference is within a defined threshold (default: $\pi/6$ radians), the entity is considered aligned with the intended path, and its crossing probability is positively reinforced. If the direction falls outside this range, no additional weight is applied.

To formally incorporate this directional constraint into the scoring mechanism, the original formula is updated as follows:

$$P = \frac{1}{F} \sum_{t=1}^{F} p_t \cdot w_t$$

where: $w_t \in \{0, 1\}$ is the direction weight at frame t:

$$w_t = \begin{cases} 1 & \text{if } \Delta\theta \leq \pi/6 \\ 0 & \text{otherwise} \end{cases}$$

This directional filtering mechanism significantly improves robustness, reducing false alerts and enhancing the system's ability to distinguish true crossing intent.

Minimizing Missed Alerts. Scenes captured by surveillance cameras and synthetic simulations are often affected by factors like limited visibility, occlusions, and poor lighting, which degrade the performance of object detection and tracking algorithms in the perception layer. These issues extend to the decision-making core, compromising reliable detection of LM-VRUs intending to cross. Missed alerts usually occur due to perception layer failures. Visual limitations cause models like YOLO to misidentify entities, so the system incorporates the DeepSORT tracking algorithm for maintaining identity tracking despite short-term detection gaps.

However, in simulated environments, during perceptual uncertainty when YOLO fails to provide consistent detections, DeepSORT may also lose tracking coherence, leading to missed associations or identity switches, culminating in missed alerts. To address this, a fault-tolerant mechanism was implemented, allowing configurable tolerance to short-term detection losses. The system stores the latest crossing contribution score for each entity. If an entity goes undetected for some frames, the system fills the gap by using the higher of the last known or current contribution value across missed frames. For instance, if a pedestrian in a high-contribution subregion becomes temporarily undetected and is then re-identified within the crosswalk, the system preserves alert conditions by multiplying the subregion's contribution by the number of missed frames.

Based on these considerations, we can define the final formulation for calculating the crossing probability score for a frame t:

$$\hat{p}_t = \sum_{j=1}^{M+1} \max(p_t, p_{\text{last}}) \tag{1}$$

where:

- $\hat{p}_t$: final contribution at frame t, incorporating fault tolerance for missing detections;
- p_t: score computed at the current frame based on the subregion occupied by the entity;
- p_{last}: last valid score available with frame $< t$;
- M: number of consecutive frames in which the entity was missing (missing frame factor).

This mechanism replicates the highest contribution between p_t and p_{last} for each missing frame, up to a total of $M + 1$ times, simulating continuous presence within the critical region. Adding one unit to M (i.e., $M + 1$) ensures that at least one contribution is counted when the number of missing frames M is zero. In this case, p_t coincides with p_{last}

Accordingly, the final probability score P is defined as:

$$P = \frac{1}{F} \sum_{t=1}^{F} \hat{p}_t \cdot w_t \tag{2}$$

4 Experiments

To assess the effectiveness of the proposed system, we adopt a two-level evaluation approach reflecting its modular architecture: (i) evaluation of the Perception Layer and (ii) evaluation of the Decision-Making Layer, responsible for inferring crossing intent.

4.1 Perception Layer

For the evaluation of the perception layer, we adopted a train/test/evaluation split on the custom LM-VRU dataset presented in Sect. 3.2, aiming to assess the effectiveness of transfer learning applied to the YOLOv5 model. The model was fine-tuned in terms of both YOLO-specific architecture and key hyperparameters to optimize performance.

Metrics. In object detection, model performance is typically assessed based on the accuracy of the predicted bounding boxes—rectangular regions that localize objects within an image—and their associated class labels. To evaluate the YOLO-based perception layer, we adopt standard object detection metrics derived from the COCO evaluation protocol [21]. These include:

- **Precision (P):** The ratio of correctly predicted bounding boxes to the total number of predicted boxes. It reflects how many of the detections made by the model are actually correct.
- **Recall (R):** The ratio of correctly predicted bounding boxes to the total number of ground truth boxes. It measures how well the model finds all relevant objects.
- **mAP@IoU=0.5:** Mean Average Precision computed at a fixed Intersection over Union (IoU) threshold of 0.5. A predicted bounding box is considered a correct detection if it overlaps with a ground truth box by at least 50% in terms of area. This metric evaluates how accurately the model detects and localizes objects.
- **mAP@[.5:.95]:** Mean Average Precision averaged across multiple IoU thresholds, from 0.5 to 0.95 with a step size of 0.05. This stricter metric rewards models that not only detect objects correctly but also localize them with high precision. It provides a more comprehensive assessment by considering both classification and bounding box accuracy.

Results will be discussed in Sect. 5.1.

4.2 Decision-Making Layer

To evaluate the effectiveness of the Decision-Making component, it was necessary to acquire a reliable source of data for testing purposes. An online search was conducted to identify publicly available video datasets or footage from public surveillance cameras. However, the availability of such video material proved to be very limited and insufficient to meet the requirements of this work. Regarding public surveillance camera recordings, although they represent a potentially valuable source, the likelihood of capturing relevant scenarios, such as LM-VRUs about to cross the street, is relatively low.

Given these limitations, the decision was made to rely on a virtual simulation environment to generate synthetic test footage tailored to system evaluation. This approach ensures full control over the scenarios and provides access to a wide range of heterogeneous situations that are difficult to capture in real-world settings.

We relied on the CARLA simulation engine [11], an open-source simulation platform widely adopted for autonomous driving research. This choice enables us to test the full perceptionâĂŞalerting pipeline under configurable urban conditions that closely resemble real-world traffic environments.

Although CARLA does not yet support the full spectrum of LM-VRUs included in the reference training dataset (e.g., pedestrians with strollers or crutches), it has recently introduced wheelchair users among its pedestrian actor models. Based on this, we designed synthetic test scenarios featuring:

- Multi-lane urban roads with limited visibility crosswalks
- A mix of vehicles and standard pedestrians
- A subset of pedestrians modeled as wheelchair users.

Figure 5 shows the main CARLA interface displaying one of the test scenes used during evaluation.

To simulate realistic surveillance conditions, static virtual cameras were strategically placed at typical lamp post height, ensuring a clear and complete view of both pedestrian crossings and waiting areas. The captured footage was recorded at 30 frames per second, providing sufficient temporal resolution for real-time detection and tracking tasks.

During video generation, some environmental and contextual variables were systematically varied, such as traffic density, number of pedestrians, percentage of LM-VRUs (wheelchair users), weather conditions, and lighting levels (day/night variations).

Metrics. During the evaluation phase, the generated videos were manually reviewed. From each observation, the following classification indicators, aligned with standard machine learning terminology, were extracted:

- **True Positive (TP):** Number of LM-VRU pedestrians correctly detected by the algorithm as intending to cross.
- **True Negative (TN):** Number of LM-VRU pedestrians correctly not flagged because they were not about to cross, or regular pedestrians correctly excluded.

Fig. 5. CARLA Simulation Environment.

- **False Positive (FP):** Incorrect alerts raised by the algorithm, including regular pedestrians and LM-VRUs who were not attempting to cross but passed nearby.
- **False Negative (FN):** LM-VRU pedestrians who did cross the street but were not detected by the algorithm.

It is important to note that no alerts are generated for pedestrians not classified as LM-VRUs. In some cases, pedestrian agents initially showed an intention to cross—e.g., by approaching the edge of the sidewalk—then changed their behavior. In such cases, the algorithm's response is still considered correct.

Based on these indicators, the following evaluation metrics were computed: *Accuracy, Precision, Recall,* and *F1-Score.* Results will be discussed in Sect. 5.2

5   Results and Discussion

Following the structural breakdown and the evaluation metrics introduced in Sect. 4, this section presents and discusses the results obtained during experimentation. Section 5.1 reports the performance of the YOLO-based perception layer. Section 5.2 focuses on the evaluation of the decision-making logic.

5.1   Perception Layer

This subsection presents the results obtained from the transfer learning-based training and evaluation of the YOLOv5 model, which constitutes the core of the perception layer. Experiments were carried out using the custom dataset described in Sect. 3.2. The following table summarizes the performance metrics achieved after hyperparameter and architecture fine-tuning (Table 1):

Table 1. Detection evaluation performance by LM-VRU class (YOLOv5).

Class	Precision	Recall	mAP@0.5	mAP@0.5:0.95
person-babywalker	0.891	0.866	0.866	0.619
person-crutches	0.860	0.936	0.944	0.694
person-wheelchair	**0.961**	**0.955**	**0.962**	**0.679**

Overall, the model demonstrates good precision and recall across all categories, confirming the effectiveness of the Transfer Learning approach. Among the classes, the *person-wheelchair* category achieved the highest performance, with a precision of 0.961, a recall of 0.955, and an mAP@0.5 of 0.962. The mAP@0.5:0.95 values, while lower, remain consistent and acceptable given the inherent difficulty of precise localization across variable poses and partial occlusions.

As previously mentioned, synthetic videos made on CARLA include only the person-wheelchair class among the LM-VRU categories. It is important to note, however, that both the perception and decision-making components have been designed to handle multiple LM-VRU categories. Although current experiments are limited to wheelchair users, future evaluations will aim to validate the entire system on additional vulnerable user types, such as individuals using crutches or baby walkers, for which the detection model has already shown promising results.

5.2   Decision-Making Layer

This subsection presents and discusses the results of the decision-making layer, which is responsible for the crossing intention prediction of LM-VRUs pedestrians —a task that represents the final objective of this work. The evaluation was conducted using synthetic video clips generated within the CARLA simulator. Results are summarized in the following table, based on the evaluation metrics introduced in Sect. 4.

The accuracy metric provides an overall measure of system performance by considering both positive and negative class samples. In this context, the positive class includes LM-VRUs who have exhibited an intention to cross the street, while the negative class comprises both regular pedestrians who cross and LM-VRUs who do not show crossing intent. Accuracy is a reliable metric when the two classes are well-balanced. In controlled scenarios, such as those generated by the simulator, it is possible to maintain this balance between pedestrians with and without mobility issues, thus making the accuracy metric meaningful. However, in future real-world tests, such a balance may not be preserved, and alternative evaluation metrics will be more appropriate.

To address this, precision and recall are introduced, as they provide more detailed insights into system errors for both positive and negative classes. In this context, precision refers to the proportion of correctly generated alerts out of the total number of alerts issued. It reflects the algorithm's ability to reduce false positives (FP), i.e., cases of unnecessary alerts. Conversely, recall measures the proportion of correctly detected alerts out of the total number of true events that should have triggered an alert. It reflects

Table 2. Performance metrics per method and scenario.

Method	Scenario	Accuracy	Precision	Recall	F1-Score
Base Method	Scenario 1	0.750	0.684	0.867	0.765
	Scenario 2	0.526	0.500	0.667	0.571
	Scenario 3	0.619	0.556	1.000	0.714
	Total	**0.645**	**0.582**	**0.886**	**0.703**
MDD	Scenario 1	0.938	1.000	0.867	0.929
	Scenario 2	0.842	1.000	0.667	0.800
	Scenario 3	0.952	0.909	1.000	0.952
	Total	**0.925**	**0.951**	**0.886**	**0.918**
MDD + FN Red.	Scenario 1	0.969	1.000	0.933	0.966
	Scenario 2	0.947	1.000	0.889	0.941
	Scenario 3	0.952	0.909	1.000	0.952
	Total	**0.958**	**0.955**	**0.955**	**0.955**

the system's ability to minimize false negatives (FN)—cases in which an LM-VRU actually crossed the street but was not detected by the system (missed alert).

Analyzing the results in Table 2, the introduction of the MDD module markedly reduced false positives, which mainly arose from: (i) redundant detections of pedestrians who had already crossed and reached the opposite side, and (ii) individuals walking along the roadside long enough to trigger an alert without any real intention to cross. By checking trajectory alignment with the predefined crossing direction, the MDD module effectively filters out these cases.

The false negative reduction mechanism also slightly improved recall by compensating for short-term detection gaps, although its effectiveness decreases under prolonged occlusions or poor visibility.

Future improvements will target both layers: at the perception layer, expanding the dataset with urban roadside samples and retraining with newer detectors (e.g., successors of YOLOv5); at the decision-making layer, refining outputs through entity-level cues such as bounding box shapes and motion speeds to reduce misclassifications, especially with pedestrians and cyclists.

Moreover, while pre-crossing zones are currently annotated manually once per camera, automatic scene segmentation represents a promising future direction.

Real-World Testing and Explainability Concerns. While the CARLA simulator has proven useful for creating controlled test scenarios, it presents limitations. Its LM-VRU actor set is restricted, and pedestrian behavior can be overly deterministic, for example, crossing without regard to traffic conditions. To compensate for this, the system's alert threshold was adjusted during simulation to better reflect real-world uncertainty. However, actual deployment and testing in real environments will be essential for comprehensive validation.

Real-world testing will involve embedding the system into surveillance cameras installed at lamp-post height to guarantee comprehensive coverage of pedestrian crossings, along with waiting areas. The system has been configured to operate in real time by accepting RTSP video streams from the cameras and interfacing with external actuators via REST APIs. Participants in the test will be tasked with mimicking LM-VRU behavior (such as individuals using wheelchairs, crutches, or pushing baby strollers) in controlled settings. The system has been designed to comply with privacy laws by not recording or storing any video data. Each frame is processed on-the-fly by the perception layer exclusively for extracting bounding boxes of identified entities. Once this process is complete, the frames are discarded immediately, ensuring no lasting visual data is kept.

As a proxy for explainability, future efforts will incorporate standard XAI evaluation benchmarks to assess the system's interpretability in a structured way. Given the rule-based architecture of the decision layer, we believe human-grounded evaluation methods are the most suitable. In particular, we plan to adopt *forward simulation* [10], in which participants are provided with an explanation and an input, and are asked to correctly simulate the system's output, independently of the actual decision made by the model.

To enable such testing, a set of video snippets taken from the same sequences used for intrinsic system evaluation will be presented to participants. Each clip will include overlaid visual elements representing the relevant input features available to the decision-making logic, such as:

- The segmented waiting areas, shown as concentric elliptical bands (see Fig. 3).
- The entity's estimated movement direction as computed by the decision layer.
- A pair of boundary angles ($\pm\pi/6$) from the expected crossing direction, used to determine directional intent.

These visual elements will constitute the *input* presented to the user. The *explanation*, on the other hand, will consist of a simplified presentation of the decision rules described in Sect. 3.4. Given these two elements, participants will be asked to determine the system's output (alert, no alert).

6   Conclusions and Future Work

The proposed system showed encouraging success in forecasting the crossing intentions of LM-VRUs, attaining an overall recall rate of 0.955. Recall is vitally important in safety-critical applications, as it indicates the system's efficiency in reducing missed alerts, which for us are more critical than false alerts. Remarkably, this level of performance was achieved without incorporating any extra deep learning elements into the decision-making core, thereby preserving transparency and human interpretability through rule-based reasoning. To formally assess the explainability of the system, we plan to adopt established XAI benchmarks—such as the simulatability metric—which evaluate whether humans can accurately anticipate the system's output given its decision logic and inputs.

The integration of a directional component via the MDD module significantly reduced false positives, improving decision reliability. Additionally, the false negative reduction strategy contributed to a higher recall by effectively recovering short-term detection gaps. However, its effectiveness is limited to a small number of consecutive missing frames; beyond that, the tracker loses the association with the original entity, resulting in a new ID assignment and ultimately a failure in the intention inference process.

The CARLA simulator provided a valuable environment for initial evaluation under controlled, repeatable conditions. However, the current experiments were limited to wheelchair users due to the lack of simulator assets representing other LM-VRU categories; also, such pedestrian agents do not accurately replicate real-world behaviours, such as pre-crossing hesitations or reactions to nearby vehicles. To address this limitation, further experiments are planned using real-world video sources and live urban testing, aiming to assess system robustness under natural behavioral variability and a wider set of users.

As part of future improvements, efforts will focus on enriching the training dataset and evaluating more advanced object detection models via transfer learning. Additionally, perception outputs may be refined through post-processing, such as the implemented false negative reduction mechanism, by incorporating entity-level cues like speed and typical shape to better guide class predictions and reduce misclassifications.

Acknowledgments. We acknowledge financial support from the project *FIATLUCS - Favorire l'Inclusione e l'Accessibilità a per i Trasferimenti Locali Urbani a Cagliari e Sobborghi*, funded by the PNRR RAISE Liguria, Spoke 01, CUP: F23C24000240006, and from the call for tender No.3277 published on December 30, 2021 by the Italian Ministry of University and Research (MUR) funded by the European Union âĂŞ NextGenerationEU. Project Code ECS0000038 âĂŞ title *eINS Ecosystem of Innovation for Next Generation Sardinia* âĂŞ CUP F53C22000430001- Grant Assignment Decree No. 1056 adopted on June 23, 2022 by the Italian Ministry of University and Research (MUR).

References

1. Abdelrahman, A.S., Abdel-Aty, M., Tran, Q.D.: Vru-cipi: Crossing intention prediction at intersections for improving vulnerable road users safety. In: Proceedings of the Computer Vision and Pattern Recognition Conference, pp. 5623–5632 (2025)
2. Al-E'mari, S., Sanjalawe, Y., Alqudah, H.: Integrating enhanced security protocols with moving object detection: a yolo-based approach for real-time surveillance. In: 2024 2nd International Conference on Cyber Resilience (ICCR), pp. 1–6 (2024)
3. Atzori, A., Barra, S., Carta, S., Fenu, G., Podda, A.S.: Heimdall: an ai-based infrastructure for traffic monitoring and anomalies detection. In: 2021 IEEE International Conference on Pervasive Computing and Communications Workshops and other Affiliated Events (PerCom Workshops), pp. 154–159. IEEE (2021)
4. Balia, R., Barra, S., Carta, S., Fenu, G., Podda, A.S., Sansoni, N.: A deep learning solution for integrated traffic control through automatic license plate recognition. In: International Conference on Computational Science and Its Applications, pp. 211–226. Springer (2021)
5. Bewley, A., Ge, Z., Ott, L., Ramos, F., Upcroft, B.: Simple online and realtime tracking. In: 2016 IEEE International Conference on Image Processing (ICIP), pp. 3464–3468 (2016)

6. Buch, N., Velastin, S.A., Orwell, J.: A review of computer vision techniques for the analysis of urban traffic. IEEE Trans. Intell. Transp. Syst. **12**(3), 920–939 (2011). https://doi.org/10.1109/TITS.2011.2119372

7. Carta, S., et al.: Roadsense3d: a framework for roadside monocular 3d object detection. In: Adjunct Proceedings of the 32nd ACM Conference on User Modeling, Adaptation and Personalization, pp. 452–459 (2024)

8. Chen, F., Wang, X., Zhao, Y., Lv, S., Niu, X.: Visual object tracking: a survey. Comput. Vis. Image Understanding **222**, 103508 (2022). https://doi.org/10.1016/j.cviu.2022.103508. https://www.sciencedirect.com/science/article/pii/S1077314222001011

9. Deng, J., Dong, W., Socher, R., Li, L.J., Li, K., Fei-Fei, L.: Imagenet: a large-scale hierarchical image database. In: Proceedings of the IEEE Conference on Computer Vision and Pattern Recognition (CVPR), pp. 248–255. IEEE (2009)

10. Doshi-Velez, F., Kim, B.: Towards a rigorous science of interpretable machine learning. arXiv preprint arXiv:1702.08608 (2017)

11. Dosovitskiy, A., Ros, G., Codevilla, F., Lopez, A., Koltun, V.: Carla: an open urban driving simulator. In: Conference on Robot Learning (CoRL), pp. 1–16. PMLR (2017)

12. Elia, N., Barchi, F., Parisi, E., Pompianu, L., Carta, S., Bartolini, A., Acquaviva, A.: Smart contracts for certified and sustainable safety-critical continuous monitoring applications. In: European Conference on Advances in Databases and Information Systems. pp. 377–391. Springer (2022)

13. Guo, Y., Liu, Y., Oerlemans, A., Lao, S., Wu, S., Lew, M.S.: Deep learning for visual understanding: a review. Neurocomputing **187**, 27–48 (2016). https://doi.org/10.1016/j.neucom.2015.09.116. https://www.sciencedirect.com/science/article/pii/S0925231215017634, recent Developments on Deep Big Vision

14. Ham, J.S., Kim, D.H., Jung, N., Moon, J.: Cipf: crossing intention prediction network based on feature fusion modules for improving pedestrian safety. In: Proceedings of the IEEE/CVF Conference on Computer Vision and Pattern Recognition (CVPR) Workshops, pp. 3666–3675, June 2023

15. Himeur, Y., et al.: Video surveillance using deep transfer learning and deep domain adaptation: Towards better generalization. Eng. Appl. Artif. Intell. **119**, 105698 (2023). https://doi.org/10.1016/j.engappai.2022.105698, https://www.sciencedirect.com/science/article/pii/S0952197622006881

16. Jocher, G., contributors: Yolov5 by ultralytics (2023). https://github.com/ultralytics/yolov5. Accessed 16 June 2025

17. Kim, Y., Abdel-Aty, M., Choi, K., Islam, Z., Wang, D., Zhai, S.: Pedestrian crossing direction prediction at intersections for pedestrian safety. IEEE Open J. Intell. Transp. Syst. **6**, 692–707 (2025). https://doi.org/10.1109/OJITS.2025.3574082

18. Lecrosnier, L., Khemmar, R., Ragot, N., Decoux, B., Rossi, R., Kefi, N., Ertaud, J.Y.: Deep learning-based object detection, localisation and tracking for smart wheelchair healthcare mobility. Int. J. Environ. Res. Public Health **18**(1), 91 (2021)

19. LeCun, Y., Bengio, Y., Hinton, G.: Deep learning. Nature **521**(7553), 436–444 (2015). https://doi.org/10.1038/nature14539

20. Liang, Z., et al.: Vehicle and pedestrian detection based on improved yolov7-tiny. Electronics **13**(20), 4010 (2024)

21. Lin, T.Y., Maire, M., Belongie, S., Hays, J., Perona, P., Ramanan, D., Dollár, P., Zitnick, C.L.: Microsoft coco: common objects in context. In: European Conference on Computer Vision, pp. 740–755 (2014)

22. Mammeri, A., Siddiqui, A.J., Zhao, Y.: Vru-net: convolutional neural networks-based detection of vulnerable road users. In: VEHITS, pp. 257–266 (2024)

23. Podda, A.S., Balia, R., Pompianu, L., Carta, S., Fenu, G., Saia, R.: Cargram: Cnn-based accident recognition from road sounds through intensity-projected spectrogram analysis. Digital Signal Processing **147**, 104431 (2024)
24. Redmon, J., Divvala, S., Girshick, R., Farhadi, A.: You only look once: unified, real-time object detection. In: 2016 IEEE Conference on Computer Vision and Pattern Recognition (CVPR), pp. 779–788 (2016).https://doi.org/10.1109/CVPR.2016.91
25. Roboflow: Roboflow: Organize, label, and deploy computer vision datasets (2023). https://roboflow.com. Accessed 16 June 2025
26. Saia, R., Balia, R., Podda, A.S., Pompianu, L., Carta, S., Pisu, A.: Eeg biometrics with gan integration for secure smart city data access. In: International Conference on Computer-Human Interaction Research and Applications, pp. 454–467. Springer (2024)
27. Saia, R., Carta, S., Fenu, G., Pompianu, L.: Influencing brain waves by evoked potentials as biometric approach: taking stock of the last six years of research. Neural Comput. Appl. **35**(16), 11625–11651 (2023)
28. Saia, R., Podda, A.S., Pompianu, L., Reforgiato Recupero, D., Fenu, G.: A blockchain-based distributed paradigm to secure localization services. Sensors **21**(20), 6814 (2021)
29. Sharma, N., Dhiman, C., Indu, S.: Pedestrian intention prediction for autonomous vehicles: a comprehensive survey. Neurocomputing **508**, 120–152 (2022)
30. Shi, J., Sun, D., Kieu, M., Guo, B., Gao, M.: An enhanced detector for vulnerable road users using infrastructure-sensors-enabled device. Sensors **24**(1), 59 (2023)
31. Situ, Z., Teng, S., Feng, W., Zhong, Q., Chen, G., Su, J., Zhou, Q.: A transfer learning-based yolo network for sewer defect detection in comparison to classic object detection methods. Developments Built Environ. **15**, 100191 (2023). https://doi.org/10.1016/j.dibe.2023.100191, https://www.sciencedirect.com/science/article/pii/S266616592300073X
32. Ukey, E., Varma, S.L.: Review on evolution of vehicle detection methods in computer vision for surveillance. In: 2024 2nd International Conference on Emerging Trends in Engineering and Medical Sciences (ICETEMS), pp. 1–6 (2024). https://doi.org/10.1109/ICETEMS64039.2024.10965053
33. Wojke, N., Bewley, A., Paulus, D.: Simple online and realtime tracking with a deep association metric. In: 2017 IEEE International Conference on Image Processing (ICIP), pp. 3645–3649. IEEE (2017). https://doi.org/10.1109/ICIP.2017.8296962
34. Yang, H.F., Ling, Y., Kopca, C., Ricord, S., Wang, Y.: Cooperative traffic signal assistance system for non-motorized users and disabilities empowered by computer vision and edge artificial intelligence. Transp. Res. Part C: Emerg. Technol. **145**, 103896 (2022). https://doi.org/10.1016/j.trc.2022.103896, https://www.sciencedirect.com/science/article/pii/S0968090X22003096
35. Zhao, J., Li, Y., Xu, H., Liu, H.: Probabilistic prediction of pedestrian crossing intention using roadside lidar data. IEEE Access **7**, 93781–93790 (2019). https://doi.org/10.1109/ACCESS.2019.2927889
36. Zhou, Y., Tan, G., Zhong, R., Li, Y., Gou, C.: Pit: progressive interaction transformer for pedestrian crossing intention prediction. IEEE Trans. Intell. Transp. Syst. **24**(12), 14213–14225 (2023)
37. Zou, Z., Chen, K., Shi, Z., Guo, Y., Ye, J.: Object detection in 20 years: a survey. Proc. IEEE **111**(3), 257–276 (2023)

User Fairness in Recommender Systems Using Beyond-Accuracy Basket Quality Metrics

Debarati Bhaumik[✉][iD], Diptish Dey[iD], Milan T. Bril[iD], and Vanessa Stoica[iD]

Amsterdam University of Applied Sciences, Amsterdam, The Netherlands
`{debarati.bhaumik,d.dey2,milan.t.bril,vanessa.stoica}@hva.nl`

Abstract. Recommender systems (RecSys) have conventionally assessed user fairness through accuracy-based metrics, such as Root Mean Squared Error and Hit Rate. These metrics are limited in their ability to evaluate fairness of user experiences, as they do not consider beyond-accuracy perspectives of recommendation basket quality. Key elements of beyond-accuracy perspectives - including novelty, relevancy, unexpectedness, coverage, serendipity, ranking, diversity, and recency - collectively enhance existing frameworks. They do so by enabling evaluation of user satisfaction and user fairness across different demographic groups.

Expanding upon established fairness metrics in the literature, this work introduces novel beyond-accuracy basket quality perspectives, including novelty-relevancy (NovRel) and recency, along with complementary metrics for computing relevancy and unexpectedness at individual and group levels. While leveraging existing accuracy-based fairness measures, our approach evaluates user fairness through the lens of beyond-accuracy basket quality measures and provides a comprehensive assessment of whether RecSys deliver comparable experiences to similar users and ensure equitable treatment across different sensitive groups.

Methods to calculate these metrics are demonstrated using common algorithms, SVD++, Q-SVD++, and MLP-SVD++, which are trained on two widely used benchmark datasets. The trade-off between accuracy metrics and beyond-accuracy fairness metrics is presented. Beyond-accuracy metrics provide valuable insights into how these algorithms differ in user fairness due to their inherent design considerations.

Keywords: Recommender System · User Fairness · Basket Quality

1 Introduction and Problem Statement

RecSys have become an integral part of modern digital platforms such as e-commerce, social media, news, and entertainment, impacting users' daily digital experiences worldwide. Based on user history, preferences, and interactions these systems aim to predict users' preferences to provide customized recommendations to enhance user experience and increase engagement and sales [3,51,52]. These systems also find applications in sensitive sectors such as healthcare, finance, and education, providing personalized recommendations for treatments, loans, insurance, and learning [21,53,56]. For trustworthy application and equitable access to services among users, it is of utmost

© The Author(s), under exclusive license to Springer Nature Switzerland AG 2026
F. Marcelloni et al. (Eds.): IJCCI 2025, CCIS 2829, pp. 290–308, 2026.
https://doi.org/10.1007/978-3-032-15638-9_17

importance to ensure fairness in these systems [16,30]. Afterall, biased recommendations may lead to discrimination against certain specific groups or individuals based on their preferences and demographics [25,44]. Driven by commercial incentives and the need for generating engagement in competitive markets like e-commerce and streaming services, researchers and developers predominantly prioritize building RecSys with enhanced accuracy and computational speed [4,7,34,65]. However, focusing solely on these measures can reinforce and amplify existing biases, resulting in unfair and discriminatory outcomes within these systems [33,67]. Biases that can lead to fairness issues in RecSys can be categorized into four key stages: (i) *data bias* from collection and preprocessing, (ii) *model bias* during algorithm design, (iii) *result bias* in the final recommendations presented to users, and (iv) *amplification bias*, where existing biases are reinforced over time [12]. Fairness issues in RecSys, often reflecting or amplifying such biases, arise at two levels: (i) users, which concerns whether similar users and users from different demographic groups (e.g., age, gender) receive equally relevant and satisfying recommendations, and (ii) items, which focuses on equitable exposure of items—ensuring that content from minority creators or niche categories is not systematically underrepresented [61].

User fairness in RecSys ensures that users with varying preferences, interaction patterns, or demographic characteristics receive recommendations that are comparably relevant, satisfying, and reflective of their individual interests. This ensures that algorithmic decisions do not systematically disadvantage specific sensitive user groups, thereby promoting consistent and fair allocation of user experience and utility across the user base [30,61,62]. Furthermore, user fairness is commonly evaluated at three distinct levels: (i) individual, (ii) group, and (iii) system. Individual fairness implies that users with similar preferences or characteristics should receive recommendations of comparable quality [19,29,37]. Group fairness ensures equitable treatment across different demographic groups or protected attributes, for example, to mitigate disparities in recommendation quality [28,30]. System-level fairness extends beyond individual and group fairness to consider the holistic impact of recommendations on the entire ecosystem [1,43].

Although the field is in its infancy [17], and of growing regulatory interest [13,14], contemporary research is uncovering the underlying mechanisms through which biases lead to disparate or unfair outcomes [5,12,16,26,61]. This expanding literature reflects a shift toward developing recommendation algorithms that are not only accurate, but also equitable and user-centric. However, such literature continues to evaluate fairness primarily through accuracy-based measures (e.g., RMSE, precision) or rank-based measures (e.g., NDCG) [5,15,61]. These measures, though useful for assessing predictive performance [10], provide a limited perspective on user experience and may obscure deeper fairness concerns. Specifically, they fail to capture how users perceive the quality, variety, or relevance of their recommendation baskets [18,32]. Although beyond-accuracy basket quality metrics, such as novelty, relevance, diversity, unexpectedness, serendipity, ranking, and coverage, have been adopted in literature to better assess recommendation basket quality, these dimensions have not been collectively integrated into fairness evaluations [40,48,60,66]. For instance, Li et al. [40] propose diversity-enhanced, accuracy-aware metrics but limit their scope to basket quality evaluation without addressing broader fairness concerns. Paparella et al. [48] introduce a multi-objective, user-centric evaluation framework that integrates only a subset of quality indicators—accuracy, diversity, and novelty. Similarly, Zhao et al. [66] investigate the relationship between fairness and diversity, focusing on user- and item-level fairness, but do not consider other basket quality dimensions.

To bridge this gap, this paper presents a collection of measures, a couple of which are novel, and proposes metrics and methods to evaluate such measures. For the sake of clarity, the terms *measures*, *metrics*, and *methods* are distinguished as follows: basket-quality measures refer to evaluation of a phenomenon or concept; these measures are quantified using various metrics; methods refer to the steps/process involved in computing the metrics. The presented measures, metrics, and methods collectively assess user fairness more extensively at both individual and group levels. This enables a user-centric understanding of equitable satisfaction in RecSys. Specifically, the two novel measures proposed for evaluating beyond-accuracy basket quality are: (1) NovRel, which simultaneously captures the novelty and relevance of recommended items, and (2) recency, which quantifies the temporal freshness of the recommendation basket. Additionally, existing measures such as relevancy and unexpectedness are evaluated using new methods, thereby collectively forming an integrated framework for fairness assessment grounded in user experience.

Sections 2 outlines the key elements of beyond-accuracy basket quality required for assessing user fairness. In Sect. 3, beyond-accuracy basket quality metrics and methods are proposed. In Sect. 4, the proposed metrics are demonstrated using two commonly used datasets in literature for a sensitive feature, gender. In Sect. 5, discussion, conclusions and future work are discussed.

2   Measures for Beyond-Accuracy Basket Quality

To gain a comprehensive understanding of beyond-accuracy basket quality, it is essential to examine its constituent elements. These elements collectively form the foundation for evaluating user satisfaction in RecSys. This section reviews established measures and their corresponding metrics used to assess beyond-accuracy basket quality. In addition, it introduces two novel measures—NovRel and Recency—aimed at further enhancing the evaluation of user experience in RecSys.

2.1   Existing Measures in Literature

Novelty. Item novelty in RecSys refers to the extent to which recommended items are unfamiliar to users [32]. As novelty does not imply relevance, unfamiliar items may still be not useful to the user. Novelty is commonly measured via (1) item popularity, which inversely correlates with novelty, or (2) assessing the dissimilarity between recommended items and a user's interaction history [33,54]. Popularity is typically based on the proportion of users who interacted with an item, while dissimilarity is calculated using metrics such as cosine similarity or Jaccard distance over item features [36,69].

Relevance. Item relevance in RecSys denotes how well a recommended item matches a user's interests or needs [36]. It is commonly evaluated using accuracy-based metrics such as precision, recall, F1-score, and absolute error [34,54]. More recent approaches emphasize ranking-based metrics—e.g., Mean Average Precision (MAP), Normalized Discounted Cumulative Gain (NDCG), Hit Rate (HR) and Mean Reciprocal Rank (MRR)—for offline evaluation [55]. Online performance of relevance is typically assessed via click-through rate (CTR) and retention rate [54]. These metrics are generally computed at the system level across users.

Unexpectedness. Unexpectedness measures how much recommended items deviate from a user's typical consumption patterns, enhancing user satisfaction through surprise [38]. Unlike novelty and relevance, it focuses on the element of surprise, which can apply even to familiar items [32,36]. Common metrics for unexpectedness include the negative Point-wise Mutual Information (PMI) for pairwise surprise [23,31,47,54], and distance-based measures comparing recommended items to the user's historical profile in item feature space [36,64]. Recent approaches also utilize Euclidean distance in latent space via autoencoder embeddings to capture unexpectedness more effectively [39].

Serendipity. Serendipity refers to the phenomenon of users discovering unexpectedly relevant or useful items that they might not have otherwise encountered. It combines elements of novelty, unexpectedness and relevance [36,69]. While recognized as key to enhancing user satisfaction, it remains difficult to quantify due to its subjective nature [24]. Serendipity is typically measured either holistically or via its components: novelty, relevance, and unexpectedness [18,69]. A simple metric computes the proportion of serendipitous items in a recommendation list [6]. For comprehensive overviews on serendipity quantification, see [36,54,69].

Coverage. Coverage in RecSys measures how broadly items are recommended across the catalog, impacting both user satisfaction and item exposure [51,54]. It is typically assessed at the system level and is categorized into three main types: item space coverage (proportion of items recommended to at least one user), user space coverage (proportion of users receiving recommendations), and genre space coverage (proportion of recommended items across genres) [6]. Key metrics include $COV(N)$—the ratio of top-N recommended items to the full item set, $CIL(N)$ for long-tail coverage [46], and distribution measures like Shannon's entropy [6] and the Gini coefficient [18]. Most existing metrics in literature focus on item space coverage.

Ranking. Item ranking in RecSys determines the order in which items are presented to users, influencing user engagement and satisfaction [9,51]. Key evaluation metrics include MAP, MRR, and NDCG, which assess whether more relevant items are ranked higher [6,63]. MAP averages (across users) precision at relevant item positions in recommendation list, MRR averages reciprocal ranks, and NDCG evaluates ranking quality by discounting relevance by position and normalizing against the ideal ranking [55]. Among these, NDCG is the most widely used for ranking-based relevance evaluation.

Diversity. Diversity in RecSys refers to the variety of items in a recommendation basket, promoting exposure to a broader range of genres or attributes, which helps prevent filter bubbles [18,49]. The most common diversity metric is intra-list similarity (ILS), which measures the aggregated similarity between items, with lower ILS indicating higher diversity [54,70]. Other metrics include Shannon's entropy, categorical entropy, and Hamming distance for intra- and inter-user diversity [6,32]. While related to coverage, diversity focuses on variety within an individual basket, whereas coverage assesses the system's overall inclusion.

2.2 Proposed Measures

NovRel. A new measure for evaluating utility of recommended items in RecSys is proposed- *NovRel*, which ensures a balanced integration of novelty and relevance. In RecSys, user satisfaction is influenced by the trade-off between these two elements. While novelty alone may result in the recommendation of items that are irrelevant to the user's preferences, and relevance can lead to the repetitive presentation of familiar items. NovRel optimizes the two elements by delivering items that are both novel and relevant. This dual focus mitigates the risk of recommending overly predictable items or irrelevant content, thereby enhancing the overall user experience. The NovRel measure is particularly advantageous in aligning with user expectations by maintaining diversity in the recommendation without compromising relevance. User satisfaction arises when recommendations introduce novel content that remains consistent with their preferences, avoiding the redundancy that often occurs when novelty or relevance is prioritized independently. Unlike serendipity, which introduces unexpected but non-essential items, NovRel guarantees that each recommendation is both useful and pertinent to the user's interests, thereby contributing to a more effective and satisfying recommendation process. Serendipity through its component of surprise can enhance long-term user experience, NovRel directly impacts immediate user satisfaction. It ensures that the RecSys consistently delivers relevant, novel, and personalized recommendations, fostering trust through a more tailored and meaningful recommendation experience [20].

Recency. Recency refers to the temporal relevance of items within a recommendation basket, recognizing that user preferences, item popularity, and availability are transient [22,45]. User preferences are not fixed; they evolve over time based on recent interactions, contextual shifts, and changing needs. In RecSys, it has been established that recent behavior is a stronger predictor of future preferences than historical interactions, making recency a critical factor in improving the system's responsiveness to dynamic user interests [50,59]. The temporal alignment enhances user engagement, as it demonstrates the system's responsiveness to evolving user interests, ultimately improving both the perceived and actual quality of the recommendation basket. Although predominantly utilized in the literature to enhance predictive accuracy, recency can also serve as a metric for evaluating recommendation basket quality. A recommendation basket dominated by outdated or stale items may not effectively capture user's current intent, even if those items remain relevant to their past preferences. Assessing basket quality through recency evaluates the system's responsiveness to user's evolving preferences.

Thus, recency acts as a proxy for measuring basket freshness, bridging the gap between system outputs and evolving user interests.

Beyond-accuracy basket quality is a multifaceted concept encompassing various elements discussed in this section, each playing a crucial role in user satisfaction and system effectiveness. These elements interact in complex ways, presenting challenges in balancing sometimes competing objectives to optimize overall basket quality while ensuring equitable outcomes for all users.

3   Methods to Compute Beyond-Accuracy Basket Quality Metrics

This section reviews existing methods and introduces several metrics and methods to quantify beyond-accuracy measures of recommendation basket quality, which can be used to assess fairness in RecSys at both the individual and group levels. Table 1 provides a list of notations frequently used in this section, along with their corresponding descriptions for better understanding.

Table 1. Notations frequently used in Sect. 3 along with their descriptions.

Notation	Explanation
u	A user in the RecSys
I_u	Recommendation basket (list) for user u
$|I_u|$	Size of recommendation basket I_u
i	An item in the Recsys
$\mathcal{G}$	User group set $\mathcal{G}$
$|\mathcal{G}|$	Number of users in set $\mathcal{G}$
$\mathbb{1}_{\{.\}}$	Indicator function returns 1 if the condition in the parenthesis is true, 0 otherwise.

3.1   Existing Methods and Metrics

Novelty Measure. As discussed in Sect. 2.1, novelty of an item indicates users' unfamiliarity with it. Item popularity is used as an indicator of the novelty of an item, recognizing that the two have an inverse relationship [6,11]. An item is considered popular if it has received a high number of user interactions, such as clicks, views, or ratings, indicating widespread engagement and preference within the user base [51]. Mathematically, popularity score of an item i, denoted as P_i, is calculated using the ratio of the number of users that have interacted with the item (N_i) to the total number of users in the system $|\mathcal{U}|$, where $\mathcal{U}$ is the set of all users. The popularity score is given by $P_i = \frac{N_i}{|\mathcal{U}|}$. Items with a high popularity score (e.g. top 10% to 20%) can be considered popular, however this threshold is use-case specific [33]. Given the inverse relationship of item popularity and novelty, an item i is considered novel to a user u if it is not popular, i.e.,

$$N_{i,u} := \mathbb{1}_{\{\text{if the item is not a popular item.}\}} \tag{1}$$

Novelty of a recommendation basket I_u for an individual user u and for a user group $\mathcal{G}$ are computed using:

$$N_{I_u,u} = \frac{1}{|I_u|} \sum_{i \in I_u} N_{i,u} \quad \text{and} \quad N_{\mathcal{G}} = \frac{1}{|\mathcal{G}|} \sum_{u \in \mathcal{G}} N_{I_u,u}, \tag{2}$$

respectively. The novelty metrics take values between $[0, 1]$. Higher values, approaching 1, indicate a greater degree of novelty for the item or the recommended basket.

Diversity Measure. As discussed in Sect. 2.1, a diverse recommendation list exposes users to a broader range of items, enhancing opportunities for novel and serendipitous item discoveries [11]. To evaluate the diversity of a recommended basket I_u for user u, we utilize the *Intra-List Distance* (ILD) metric, which computes the average pairwise distance among items within I_u [6,58]. Diversity of a recommended basket I_u for user u is computed using the ILD metric with cosine similarity as the distance measure [8]:

$$D_{I_u,u} = \frac{1}{|I_u|(|I_u| - 1)} \sum_{i \in I_u} \sum_{j \neq i \in I_u} \max(0, cos(i, j)), \tag{3}$$

where $\max(0, \cdot)$ enforces non-negativity to $D_{I_u,u} \in [0, 1]$. The diversity measure at a user group $\mathcal{G}$ level is computed using

$$D_{\mathcal{G}} = \frac{1}{|\mathcal{G}|} \sum_{u \in \mathcal{G}} D_{I_u,u}. \tag{4}$$

The diversity metrics range from 0 to 1, where 0 indicates perfect diversity (maximal dissimilarity between all recommended items) and 1 indicates perfect homogeneity (complete similarity among all items in the basket).

3.2 Proposed Novel Methods and Metrics

Relevancy Measure. Item relevance in recommendation systems refers to how well a suggested item matches a user's interests, preferences, or requirements [36]. Therefore, an item can be considered relevant for a user if it has high average feature-similarity with other items that are highly rated by the user (user profile). Hence, *relevancy* of an item i for user u is defined by the average cosine similarity between item i and other items that the user has highly rated. Mathematically, this is expressed as, $\mathrm{Rel}_{i,u} := \overline{\forall_{j \in HR^u \wedge j \neq i} \cos(j, i)}$, where HR^u is the set of items that the user has historically rated highly. An item i is considered relevant to user u if $\mathrm{Rel}_{i,u}$ exceeds a predefined high threshold, R_{Thres} i.e.,

$$R_{i,u} := \mathbb{1}_{\{\mathrm{Rel}_{i,u} > R_{\mathrm{Thres}}\}}. \tag{5}$$

Note that $R_{i,u}$ is a function of two thresholds: (i) high user rating $\mathrm{HR}^u_{\mathrm{Thres}}$ and (ii) high similarity threshold, R_{Thres}, both of which can be determined based on heuristics and are specific to use cases. The relevancy of a recommendation basket I_u for an individual user u and for a user group $\mathcal{G}$ are computed using:

$$R_{I_u,u} = \frac{1}{|I_u|} \sum_{i \in I_u} R_{i,u} \quad \text{and} \quad R_{\mathcal{G}} = \frac{1}{|\mathcal{G}|} \sum_{u \in \mathcal{G}} R_{I_u,u}, \tag{6}$$

respectively. The relevancy metrics take values between $[0, 1]$. Higher values, approaching 1, indicate a greater degree of relevancy of the item or the recommended basket. In the following sections, we will continue to use the same definition of item relevancy and the set of items that the user has rated highly in the past, denoted as HR^u, while introducing new methods to compute the metrics.

Unexpectedness Measure. An unexpected item in a recommendation list substantially diverges from a user's prior expectations or established consumption patterns, presenting recommendations that are atypical of their usual preferences [2,63]. Collaborative filtering-based recommendation algorithms such as SVD++ leverage users' past consumption behavior to enhance recommendation predictions, as reflected in the item composition of the recommended basket. Consequently, an unexpected item is defined as one that exhibits low feature similarity with the other items in the basket recommended to the user. Hence, we propose to measure the level of *unexpectedness* of an item i for user u using the distance of that item from the average feature values of the other items in the recommendation basket I_u, i.e.,

$$U_{i,u} := dist(i, \bar{I}_u), \tag{7}$$

where $dist(\cdot)$ can be any distance function such as cosine similarity and $\bar{I}_u$ is the average of the recommended basket item features. An item i is considered as unexpected if it is an outlier based on its distance from the basket average; specifically, if this distance exceeds a predefined threshold, U_{Thres}. Let $\tilde{U}_{I_u,u}$ denote the set of outliers (unexpected items) in the basket recommended to user u, identified using cosine similarity as the $dist(\cdot)$ function defined in Eq. (7). *Unexpectedness* of the recommended basket is defined using harmonic mean of two measures: (i) proportion of outlier items in the basket, given by $\frac{|\tilde{U}_{I_u,u}|}{|I_u|}$ and (ii) average distance of the outliers from the average of the basket item feature $\bar{I}_u$, i.e., $\bar{U}_{I_u,u} = \overline{\forall_{i \in |\tilde{U}_{I_u,u}|} cos(i, \bar{I}_u)}$. This method ensures that both metrics contribute significantly to the final unexpectedness measure, such that a low value in either metric leads to a proportionally lower overall score. Hence, unexpectedness of a recommendation basket I_u for an individual user u and for a user group $\mathcal{G}$ are computed using:

$$U_{I_u,u} = \frac{2 \cdot \frac{|\tilde{U}_{I_u,u}|}{|I_u|} \cdot \bar{U}_{I_u,u}}{\frac{|\tilde{U}_{I_u,u}|}{|I_u|} + \bar{U}_{I_u,u}} \quad \text{and} \quad U_{\mathcal{G}} = \frac{1}{|\mathcal{G}|} \sum_{u \in \mathcal{G}} U_{I_u,u}, \tag{8}$$

respectively. The unexpectedness metrics range from $-\infty$ to 0, where a more negative value indicates a higher degree of unexpectedness in the recommendation basket for an individual user or user group.

Serendipity Measure. An item is considered serendipitous if it is novel, relevant, and unexpected [69]. Explicitly, an item i is considered serendipitous to user u if,

$$S_{i,u} := \mathbb{1}_{\{\text{if the item is novel, relevant and unexpected.}\}} \tag{9}$$

Novelty, relevance, and unexpectedness of an item can be assessed using the methods outlined in Sects. 3.1, 3.2, and 3.2, respectively. *Serendipity* of a recommendation basket is quantified as the proportion of serendipitous items within the basket [6]. Hence, the serendipity measure of a recommendation basket I_u for an individual user u and for a user group $\mathcal{G}$ are computed using:

$$S_{I_u,u} = \frac{1}{|I_u|} \sum_{i \in I_u} S_{i,u} \quad \text{and} \quad S_{\mathcal{G}} = \frac{1}{|\mathcal{G}|} \sum_{u \in \mathcal{G}} S_{I_u,u}, \tag{10}$$

respectively. Serendipity metrics range from 0 to 1, with values closer to 1 indicating a higher degree of serendipity in the recommendation basket for an individual user or user group.

NovRel Measure. The *NovRel* measure for item i recommended to user u is computed using Eqs. (1) and (5),

$$NR_{i,u} := N_{i,u} \times R_{i,u}, \tag{11}$$

where $N_{i,u}$ and $R_{i,u}$ are the metrics to measure novelty and relevancy of the item, respectively. The NovRel measure of a recommendation basket I_u for an individual user u and for a user group $\mathcal{G}$ are computed using:

$$NR_{I_u,u} = \frac{1}{|I_u|} \sum_{i \in I_u} NR_{i,u} \quad \text{and} \quad NR_{\mathcal{G}} = \frac{1}{|\mathcal{G}|} \sum_{u \in \mathcal{G}} NR_{I_u,u}, \tag{12}$$

respectively. The NovRel metrics range from 0 (no novelty or relevancy) to 1 (maximal novelty and relevancy), quantifying how well recommendations simultaneously satisfy both criteria.

Recency Measure. The temporal relevance of an item in a recommendation basket is captured using an exponential time decay function [27,42]. *Recency* of an item i in a basket I_u recommended to user u is computed using:

$$\text{Rec}_{i,u} = e^{-\alpha \cdot (t - t_i)}, \tag{13}$$

where α is the decay rate which typically takes value between $0.01 \leq \alpha \leq 0.1$ [42], t is the current time, and t_i is the timestamp when item i was added, last updated or rated in the system. $\text{Rec}_{i,u}$ ranges from 1, indicating a perfectly fresh item, to 0, representing a completely stale item, with values decreasing exponentially as a function of item age and the tunable parameter α. The recency measure of a recommendation basket I_u for an individual user u and for a user group $\mathcal{G}$ are computed using:

$$\text{Rec}_{I_u,u} = \frac{1}{|I_u|} \sum_{i \in I_u} \text{Rec}_{i,u} \quad \text{and} \quad \text{Rec}_{\mathcal{G}} = \frac{1}{|\mathcal{G}|} \sum_{u \in \mathcal{G}} \text{Rec}_{I_u,u}, \tag{14}$$

respectively. The recency metrics range from 0 to 1, with higher values indicating a greater presence of fresh items in the recommendation basket.

3.3 Enhanced Methods and Metrics

The metrics presented in this section for measuring coverage and ranking are commonly used in the literature; however, the methods employed to compute them in this paper are based on the relevancy measure introduced in Sect. 3.2.

Coverage Measure. As mentioned in Sect. 2.1, coverage represents RecSys's ability to suggest items across the full catalog spectrum, guaranteeing a broad representation of available content [6,57]. To evaluate basket quality through the lens of coverage, we adapt genre-space coverage to a generalized category-space coverage. This metric quantifies a RecSys's capacity to proportionally represent each item category in user's recommendation basket, aligned with the user's historical category preferences. To measure *category-space coverage* of a basket I_u recommended to user u, we compute the similarity between the category distribution in I_u and the user's historical category preferences using Jensen-Shannon Divergence (JSD) [61],

$$C_{I_u,u} = 1 - \mathrm{JSD}(\mathcal{P}(I_u)\|\mathcal{P}(HR^u)), \tag{15}$$

where $\mathcal{P}(I_u)$ is the item category distribution in recommendation basket I_u, $\mathcal{P}(HR^u)$ is the item category distribution in the set of items that the user has historically rated highly HR^u (as discussed in Sect. 3.2), and $0 \leq \mathrm{JSD}, C_{I_u,u} \leq 1$. A JSD of 0 indicates perfect alignment between category distribution in recommendation basket and user preferences ($C_{I_u,u} = 1$, ideal coverage of preferred categories), while a JSD of 1 signifies complete mismatch ($C_{I_u,u} = 0$, no coverage of preferred categories). Similar to the other metrics, the group-level coverage measure is computed using the following metric:

$$C_{\mathcal{G}} = \frac{1}{|\mathcal{G}|} \sum_{u \in \mathcal{G}} C_{I_u,u}, \tag{16}$$

where $C_{\mathcal{G}} \in [0,1]$.

Ranking Measure. Optimal item ranking prioritizes the most relevant and engaging items in the recommendation basket, playing a key role in enhancing user satisfaction [9,58]. To evaluate the quality of item ranking in a recommendation basket I_u, the widely adopted NDCG metric is utilized, as detailed in Sect. 2.1. It measures how well the predicted ranking of items aligns with an ideal ranking by considering both the relevance of items and their positions in the list [6]. To compute NDCG, the Discounted Cumulative Gain (DCG) is calculated, which measures ranking quality by assigning higher importance to top-ranked items. DCG for a recommendation basket I_u is given by:

$$\mathrm{DCG} = \sum_{i=1}^{|I_u|} \frac{\mathrm{Rel}_{i,u}}{\log_2(i+1)}, \tag{17}$$

where $\mathrm{Rel}_{i,u}$ is the relevance score of the item at position i in the recommendation basket I_u presented to user u. The computation of $\mathrm{Rel}_{i,u}$ is detailed in Sect. 3.2. Then, the Ideal Discounted Cumulative Gain (IDCG) is computed, which is the DCG of the ideal ranking (i.e., items sorted by highest relevance). Finally, NDCG is computed by normalizing DCG with IDCG, serving as the *ranking* measure for the recommendation basket I_u provided to user u:

$$\text{Rank}_{I_u,u} = \text{NDCG} = \frac{\text{DCG}}{\text{IDCG}}. \tag{18}$$

NDCG (and consequently, $\text{Rank}_{I_u,u}$) values range from 0 to 1, with 1 indicating a perfect ranking alignment. The group-level ranking measure is computed using the following metric:

$$\text{Rank}_{\mathcal{G}} = \frac{1}{|\mathcal{G}|} \sum_{u \in \mathcal{G}} \text{Rank}_{I_u,u}, \tag{19}$$

where $\text{Rank}_{\mathcal{G}} \in [0, 1]$.

4   Demonstrations

The beyond-accuracy basket quality metrics introduced in Sect. 3 are demonstrated in this section. To evaluate user fairness, the variation of these metrics across user groups based on gender is examined, highlighting potential disparities in recommendation quality.

4.1   Datasets

In RecSys research, several datasets have become widely adopted as benchmarks, including the MovieLens 1M dataset, the Netflix dataset, the Amazon Product Datasets, and the Yelp dataset [16]. Given that this study assesses fairness between different user groups, datasets that include demographic information such as gender, ethnicity, or level of education are utilized. Both the MovieLens 1M[1] and Yelp[2] datasets contain sensitive demographic attributes, such as gender, making them suitable for this analysis. For the Yelp dataset, gender is inferred from user names using the Python package Gender-Guesser[3]. For pre-processing, users and items with fewer than 20 interactions were excluded to ensure adequate data density for the analysis in both datasets. Additionally, for the Yelp dataset, only restaurant businesses marked as open were used to train the recommendation algorithms and for subsequent user fairness analysis. Table 2 presents the distribution of users, items, and ratings across both datasets, obtained using stratified sampling to ensure both splits have equal number of users.

4.2   Algorithms

To compute the beyond-accuracy basket quality metrics, multiple collaborative filteringâĂŞbased recommendation algorithms were trained on the two datasets mentioned in the previous section. These algorithms include various variants of Singular Value

---

[1] https://grouplens.org/datasets/movielens/.
[2] https://www.yelp.com/dataset/.
[3] https://pypi.org/project/gender-guesser/.

Table 2. Statistics of users, items, and ratings in the complete, training, and test sets for each dataset.

Dataset	Split	Users	Items	Ratings
Movielens1M	Complete	6040	3706	1,000,209
	Train	6040	3677	800,167
	Test	6040	3487	200,042
Yelp	Complete	1146	2759	51,988
	Train	1146	2576	41,590
	Test	1146	2000	10,398

Decomposition (SVD)-based methods [35], specifically SVD++, Q-SVD++, and MLP-SVD++.

SVD++ is an extension of the SVD matrix factorization method that incorporates implicit feedback (e.g., user interactions like clicks, views) alongside explicit feedback (e.g., ratings) [35]. The hyperparameters of this model include: (i) the number of latent factors K, which governs the dimensionality of user and item embeddings, (ii) the learning rates α, which control the gradient descent step sizes, (iii) regularization terms λ, which prevent overfitting by penalizing large weights in user, item, and implicit feedback vectors, (iv) the feature weights, which adjust the influence of implicit interactions, and (v) the number of epochs E, which balances training duration and convergence.

The **Q-SVD++** algorithm, proposed by Zhou et al. [68], extends the SVD++ method by incorporating timestamped data and optimizing predictions through Q-learning. Unlike traditional SVD-based algorithms, which ignore the impact of timestamps, Q-SVD++ models user-item interactions as a Markov Decision Process (MDP), where state-action pairs are based on sequential item transitions, and rewards reflect rating differences. Q-learning is used to update Q-values, enhancing SVD++ predictions with feedback from the reward function. This results in more accurate recommendations by dynamically adjusting to users' evolving preferences, considering both historical interactions and temporal factors. In addition to the SVD++ hyperparameters, this algorithm includes two additional hyperparameters for the Q-learning updates in the Bellman equation.

Traditional SVD++ struggles with capturing nonlinear relationships data. Additionally, it cannot model complex patterns between users and items interactions. The **MLP-SVD++** algorithm addresses these limitations by first using an autoencoder to extract meaningful features from item data, such as genres, and then combining these features with the collaborative filtering power of SVD++. The autoencoder helps in learning nonlinear representations of item features, while the MLP further captures complex patterns. This hybrid approach enhances the model's ability to make more accurate and personalized predictions by integrating both explicit and implicit feedback, alongside rich item features, providing a significant improvement over traditional SVD++ methods. Similar combination of MLP and SVD has been used in the literature [41]. Along with the SVD++ hyperparameters, this algorithm includes additional hyperparameters for the MLP regressor model from Scikit-Learn[4]. Table 3 presents the hyperparameter settings used for the three algorithms, along with their corresponding performance accuracy metrics—HitRate@20 and RMSE—evaluated on the test dataset.

Table 3. Hyperparameters and accuracy of SVD++, Q-SVD++, and MLP-SVD++ models.

Model	SVD++	Q-SVD++	MLP-SVD++
Latent factors (K)	10	10	10
Learning rate (η_1)	0.007	0.007	0.007
Learning rate (η_2)	0.007	0.007	0.007
Regularization (λ_1)	0.005	0.005	0.005
Regularization (λ_2)	0.015	0.015	0.015
Feature weight (γ)	0.01	0.01	0.01
Q-learning rate (α)	N/A	0.01	N/A
Discount factor (γ_Q)	N/A	0.9	N/A
Hidden layers	N/A	N/A	(32,16,10,16,32)
Activation	N/A	N/A	ReLU
Optimizer	N/A	N/A	Adam
Epochs	20	20	500(20)
Random seed	N/A	N/A	42
HitRate@20 - Movielens1M	0.428	0.429	0.608
HitRate@20 - Yelp	0.086	0.093	0.177
RMSE - Movielens1M	0.89	0.86	0.85
RMSE - Yelp	0.99	0.97	0.96

4.3   Demonstration of Proposed Metrics on User Fairness

This section evaluates user fairness across the sensitive attribute of gender, specifically comparing male and female user groups through the beyond-accuracy basket quality metrics presented in Sect. 3. To quantify disparity across user groups, the relative absolute difference between the metric values for male and female users is computed and normalized by the overall metric value across all users. Table 4 presents the threshold and parameter values used for computing the beyond accuracy basket quality metrics discussed in Sect. 3 across the two datasets, Movielens1M and Yelp.

Figure 1 illustrates the relative difference in beyond-accuracy basket quality measures (computed for a basket size of 20) between males and females for the two datasets. The figure highlights how the measures perform across genders for the two datasets. The measures differ in their outcomes per algorithm, indicating their ability to discriminate

---

[4] https://scikit-learn.org/stable/modules/generated/sklearn.neural_network.MLPRegressor.html.

Table 4. Threshold and parameter values used for computing the metrics presented in Sect. 3 across different datasets.

Dataset	Popularity Score (%)	R_{Thres}	HR^u_{Thres}	U_{Thres}	α
Movielens1M	0.1	0.3	4.0	0.3	0.05
Yelp	0.1	0.3	4.0	0.5	0.05

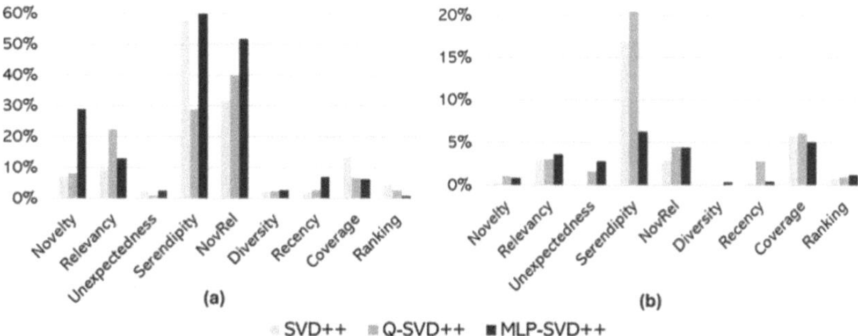

Fig. 1. Relative difference in beyond-accuracy basket quality measures (computed for a basket size of 20) between males and females for (a) Movielens1M and (b) Yelp datasets.

across algorithms. This is observed across most of the measures in the two datasets. Also, the outcomes of the measures differ substantially among themselves, indicating the mutual ability of the measures to potentially differentiate between themselves. This is observed more in the Movielens dataset than in the Yelp dataset. The novel measures, NovRel and Recency, have good differentiating power relative to the other measures. In the Movielens dataset, NovRel outcomes are high relative to the other measures except Serendipity. However, this is not true for the Yelp dataset. In addition, the two measures have good discriminatory power among the algorithms.

From Table 3, MLP-SVD++ achieves the best overall accuracy across both the datasets, showing the highest HitRate@20 and lowest RMSE. Q-SVD++ marginally performs better than SVD++ on both the datasets. However, as shown in Fig. 1, the relative differences in beyond-accuracy basket quality measures between males and females do not exhibit a clear preference for a specific algorithm. For example, with the Movielens dataset, whereas the disparities in the outcomes of Novelty and NovRel worsen as the algorithmic accuracy improves, the same is not observable for the outcomes of Serendipity or Coverage.

5   Discussion and Conclusion

This paper presents a comprehensive collection of measures, metrics, and methods for evaluating user fairness in RecSys. The collection includes established measures—such as novelty, relevance, diversity, unexpectedness, serendipity, ranking, and coverage—that are used to assess recommendation basket quality but not user fairness. Furthermore, it introduces two novel measures: novelty-relevancy (NovRel) and recency. Along with these measures, new methods for computing existing metrics for relevancy, unexpectedness, and serendipity are also proposed. The NovRel measure balances novelty and relevance in recommendations, ensuring items are both new and aligned with the user's preferences. Recency refers to the temporal relevance of items in a recommendation basket, reflecting how well a system adapts to users' evolving preferences and interests over time. Hence, the proposed recency measure assesses the freshness of the recommendation basket to ensure it aligns with users' evolving interests.

These nine measures are computed using various metrics and methods for three SVD-based algorithms, SVD++, Q-SVD++ and MLP-SVD++ on two benchmark datasets. To assess user fairness across the sensitive feature gender, the relative difference of these measures across males and females are computed. The results demonstrate that these beyond-accuracy basket quality measures exhibit discriminative power between algorithms and can potentially differentiate among each other. The results indicate that no clear trade-off exists between algorithmic accuracy and user fairness assessed using beyond-accuracy basket quality measures. Not all beyond-accuracy basket quality measures follow the same trade-off with accuracy. These findings suggest that to ensure equitable user satisfaction across user groups, it is important to collectively assess different beyond-accuracy basket quality measures, while also evaluating each measure individually to understand their unique impact on user experience and fairness.

Ongoing work involves linking these measures to the potential risks that RecSys pose to social justice principles in specific use cases, such as social media, online e-commerce, and others. Future research in this direction will provide a more comprehensive understanding of the societal impact of RecSys. Additionally, future work also involves evaluating these measures for more complex RecSys operating in dynamic environments.

References

1. Abdollahpouri, H.: Incorporating system-level objectives into recommender systems. In: Companion Proceedings of the 2019 World Wide Web Conference, pp. 2–6 (2019)
2. Adamopoulos, P., Tuzhilin, A.: On unexpectedness in recommender systems: Or how to better expect the unexpected. ACM Trans. Intell. Syst. Technol. (TIST) 5(4), 1–32 (2014)
3. Aggarwal, C.C., et al.: Recommender systems, vol. 1. Springer (2016)
4. Alamdari, P.M., Navimipour, N.J., Hosseinzadeh, M., Safaei, A.A., Darwesh, A.: A systematic study on the recommender systems in the e-commerce. IEEE Access 8, 115694–115716 (2020)
5. Alamgir Kheya, T., Reda Bouadjenek, M., Aryal, S.: Unmasking gender bias in recommendation systems and enhancing category-aware fairness. arXiv e-prints pp. arXiv–2502 (2025)
6. Alhijawi, B., Awajan, A., Fraihat, S.: Survey on the objectives of recommender systems: measures, solutions, evaluation methodology, and new perspectives. ACM Comput. Surv. 55(5), 1–38 (2022)
7. Almahmood, R.J.K., Tekerek, A.: Issues and solutions in deep learning-enabled recommendation systems within the e-commerce field. Appl. Sci. 12(21), 11256 (2022)

8. Avazpour, I., Pitakrat, T., Grunske, L., Grundy, J.: Dimensions and metrics for evaluating recommendation systems. Recommendation systems in software engineering, pp. 245–273 (2014)
9. Bae, H.K., Jang, H.R., Moon, Y.S., Kim, S.W.: Item-ranking promotion in recommender systems. In: Companion Proceedings of the ACM Web Conference 2024, pp. 505–508 (2024)
10. Bauer, C., Zangerle, E., Said, A.: Exploring the landscape of recommender systems evaluation: Practices and perspectives. ACM Trans. Recommender Syst. 2(1), 1–31 (2024)
11. Castells, P., Hurley, N., Vargas, S.: Novelty and diversity in recommender systems. In: Recommender Systems Handbook, pp. 603–646. Springer (2021)
12. Chen, J., Dong, H., Wang, X., Feng, F., Wang, M., He, X.: Bias and debias in recommender system: a survey and future directions. ACM Trans. Inf. Syst. 41(3), 1–39 (2023)
13. Commission, E.: Regulation (eu) 2022/2065 of the european parliament and of the council of 19 october 2022 on a single market for digital services (digital services act) and amending directive 2000/31/ec (2022). https://eur-lex.europa.eu/eli/reg/2022/2065/oj, accessed: 2024-06-28
14. Council, E.: Artificial intelligence act: council gives final green light to the first worldwide rules on ai (2024). https://encr.pw/EYR4W. Accessed 28 June 2024
15. Deldjoo, Y., Anelli, V.W., Zamani, H., Bellogin, A., Di Noia, T.: A flexible framework for evaluating user and item fairness in recommender systems. User Modeling and User-Adapted Interaction, pp. 1–55 (2021)
16. Deldjoo, Y., Jannach, D., Bellogin, A., Difonzo, A., Zanzonelli, D.: Fairness in recommender systems: research landscape and future directions. User Model. User-Adap. Inter. 34(1), 59–108 (2024)
17. Di Noia, T., Tintarev, N., Fatourou, P., Schedl, M.: Recommender systems under European ai regulations. Commun. ACM 65(4), 69–73 (2022)
18. Duricic, T., Kowald, D., Lacic, E., Lex, E.: Beyond-accuracy: a review on diversity, serendipity, and fairness in recommender systems based on graph neural networks. Front. Big Data 6, 1251072 (2023)
19. Dwork, C., Hardt, M., Pitassi, T., Reingold, O., Zemel, R.: Fairness through awareness. In: Proceedings of the 3rd Innovations in Theoretical Computer Science Conference, pp. 214–226 (2012)
20. Eirinaki, M., Louta, M.D., Varlamis, I.: A trust-aware system for personalized user recommendations in social networks. IEEE Trans. Syst. Man Cybernetics: Syst. 44(4), 409–421 (2013)
21. Etemadi, M., et al.: A systematic review of healthcare recommender systems: open issues, challenges, and techniques. Expert Syst. Appl. 213, 118823 (2023)
22. Faggioli, G., Polato, M., Aiolli, F.: Recency aware collaborative filtering for next basket recommendation. In: Proceedings of the 28th ACM Conference on User Modeling, Adaptation and Personalization, pp. 80–87 (2020)
23. Fu, Z., Niu, X.: Modeling users' curiosity in recommender systems. ACM Trans. Knowl. Discov. Data 18(1), 1–23 (2023)
24. Fu, Z., Niu, X., Maher, M.L.: Deep learning models for serendipity recommendations: a survey and new perspectives. ACM Comput. Surv. 56(1), 1–26 (2023)
25. Gómez, E., Shui Zhang, C., Boratto, L., Salamó, M., Marras, M.: The winner takes it all: geographic imbalance and provider (un) fairness in educational recommender systems. In: Proceedings of the 44th International ACM SIGIR Conference on Research and Development in Information Retrieval, pp. 1808–1812 (2021)
26. González, Á., Ortega, F., Pérez-López, D., Alonso, S.: Bias and unfairness of collaborative filtering based recommender systems in movielens dataset. IEEE Access 10, 68429–68439 (2022)

27. Guy, I., Zwerdling, N., Ronen, I., Carmel, D., Uziel, E.: Social media recommendation based on people and tags. In: Proceedings of the 33rd International ACM SIGIR Conference on Research and Development in Information Retrieval, pp. 194–201 (2010)
28. Hardt, M., Price, E., Srebro, N.: Equality of opportunity in supervised learning. Advances in neural information processing systems **29** (2016)
29. He, X., Liu, Q., Jung, S.: The impact of recommendation system on user satisfaction: a moderated mediation approach. J. Theor. Appl. Electron. Commer. Res. **19**(1), 448–466 (2024)
30. Jin, D., Wang, L., Zhang, H., Zheng, Y., Ding, W., Xia, F., Pan, S.: A survey on fairness-aware recommender systems. Inf. Fusion **100**, 101906 (2023)
31. Kaminskas, M., Bridge, D.: Measuring surprise in recommender systems. In: Proceedings of the Workshop on Recommender Systems Evaluation: Dimensions and Design (Workshop Programme of the 8th ACM Conference on Recommender Systems). Citeseer (2014)
32. Kaminskas, M., Bridge, D.: Diversity, serendipity, novelty, and coverage: a survey and empirical analysis of beyond-accuracy objectives in recommender systems. ACM Trans. Interact. Intell. Syst. (TiiS) **7**(1), 1–42 (2016)
33. Klimashevskaia, A., Jannach, D., Elahi, M., Trattner, C.: A survey on popularity bias in recommender systems. User Modeling and User-Adapted Interaction, pp. 1–58 (2024)
34. Ko, H., Lee, S., Park, Y., Choi, A.: A survey of recommendation systems: recommendation models, techniques, and application fields. Electronics **11**(1), 141 (2022)
35. Koren, Y., Bell, R., Volinsky, C.: Matrix factorization techniques for recommender systems. Computer **42**(8), 30–37 (2009)
36. Kotkov, D., Wang, S., Veijalainen, J.: A survey of serendipity in recommender systems. Knowl.-Based Syst. **111**, 180–192 (2016)
37. Leonhardt, J., Anand, A., Khosla, M.: User fairness in recommender systems. In: Companion Proceedings of the The Web Conference 2018, pp. 101–102 (2018)
38. Li, P., Que, M., Jiang, Z., Hu, Y., Tuzhilin, A.: Purs: personalized unexpected recommender system for improving user satisfaction. In: Proceedings of the 14th ACM Conference on Recommender Systems, pp. 279–288 (2020)
39. Li, P., Tuzhilin, A.: Latent unexpected recommendations. ACM Trans. Intell. Syst. Technol. (TIST) **11**(6), 1–25 (2020)
40. Li, X., Cong, G., Xiao, G., Xu, Y., Jiang, W., Li, K.: On evaluation metrics for diversity-enhanced recommendations. In: Proceedings of the 33rd ACM International Conference on Information and Knowledge Management, pp. 1286–1295 (2024)
41. Li, Z., Li, S., Liu, Y., Bu, Z.: Application of mlp based on joint similar groups in user interest expression and recommendation service. SN Comput. Sci. **3**(4), 309 (2022)
42. Liu, N.N., Zhao, M., Xiang, E., Yang, Q.: Online evolutionary collaborative filtering. In: Proceedings of the Fourth ACM Conference on Recommender Systems, pp. 95–102 (2010)
43. Mansoury, M.: Fairness-aware recommendation in multi-sided platforms. In: Proceedings of the 14th ACM International Conference on Web Search and Data Mining, pp. 1117–1118 (2021)
44. Mansoury, M., Abdollahpouri, H., Smith, J., Dehpanah, A., Pechenizkiy, M., Mobasher, B.: Investigating potential factors associated with gender discrimination in collaborative recommender systems. In: The Thirty-Third International Flairs Conference (2020)
45. Nitu, P., Coelho, J., Madiraju, P.: Improvising personalized travel recommendation system with recency effects. Big Data Mining Anal. **4**(3), 139–154 (2021)
46. Niu, K., Zhao, X., Li, F., Li, N., Peng, X., Chen, W.: Utsp: user-based two-step recommendation with popularity normalization towards diversity and novelty. IEEE Access **7**, 145426–145434 (2019)
47. Niu, X., Al-Doulat, A.: Luckyfind: leveraging surprise to improve user satisfaction and inspire curiosity in a recommender system. In: Proceedings of the 2021 Conference on Human Information Interaction and Retrieval, pp. 163–172 (2021)

48. Paparella, V., Di Palma, D., Anelli, V.W., Di Noia, T.: Broadening the scope: evaluating the potential of recommender systems beyond prioritizing accuracy. In: Proceedings of the 17th ACM Conference on Recommender Systems, pp. 1139–1145 (2023)
49. Ping, Y., Li, Y., Zhu, J.: Beyond accuracy measures: the effect of diversity, novelty and serendipity in recommender systems on user engagement. Electron. Commerce Res., 1–28 (2024)
50. Pradel, B., et al.: A case study in a recommender system based on purchase data. In: Proceedings of the 17th ACM SIGKDD International Conference on Knowledge Discovery and Data Mining, pp. 377–385 (2011)
51. Ricci, F., Rokach, L., Shapira, B.: Recommender systems: Techniques, applications, and challenges. Recommender Systems Handbook, pp. 1–35 (2021)
52. Roy, D., Dutta, M.: A systematic review and research perspective on recommender systems. J. Big Data 9(1), 59 (2022)
53. Sharaf, M., Hemdan, E.E.D., El-Sayed, A., El-Bahnasawy, N.A.: A survey on recommendation systems for financial services. Multimed. Tools Appl. 81(12), 16761–16781 (2022)
54. Silveira, T., Zhang, M., Lin, X., Liu, Y., Ma, S.: How good your recommender system is? a survey on evaluations in recommendation. Int. J. Mach. Learn. Cybern. 10, 813–831 (2019)
55. Sun, Z., Yu, D., Fang, H., Yang, J., Qu, X., Zhang, J., Geng, C.: Are we evaluating rigorously? benchmarking recommendation for reproducible evaluation and fair comparison. In: Proceedings of the 14th ACM Conference on Recommender Systems, pp. 23–32 (2020)
56. Urdaneta-Ponte, M.C., Mendez-Zorrilla, A., Oleagordia-Ruiz, I.: Recommendation systems for education: systematic review. Electronics 10(14), 1611 (2021)
57. Vargas, S., Baltrunas, L., Karatzoglou, A., Castells, P.: Coverage, redundancy and size-awareness in genre diversity for recommender systems. In: Proceedings of the 8th ACM Conference on Recommender Systems, pp. 209–216 (2014)
58. Vargas, S., Castells, P.: Rank and relevance in novelty and diversity metrics for recommender systems. In: Proceedings of the fifth ACM Conference on Recommender Systems, pp. 109–116 (2011)
59. Vinagre, J., Jorge, A.M., Gama, J.: Collaborative filtering with recency-based negative feedback. In: Proceedings of the 30th Annual ACM Symposium on Applied Computing, pp. 963–965 (2015)
60. Wang, N., Chen, L.: User bias in beyond-accuracy measurement of recommendation algorithms. In: Proceedings of the 15th ACM Conference on Recommender Systems, pp. 133–142 (2021)
61. Wang, Y., Ma, W., Zhang, M., Liu, Y., Ma, S.: A survey on the fairness of recommender systems. ACM Trans. Inf. Syst. 41(3), 1–43 (2023)
62. Wu, Y., Cao, J., Xu, G.: Fairness in recommender systems: evaluation approaches and assurance strategies. ACM Trans. Knowl. Discov. Data 18(1), 1–37 (2023)
63. Zangerle, E., Bauer, C.: Evaluating recommender systems: survey and framework. ACM Comput. Surv. 55(8), 1–38 (2022)
64. Zhang, M., Yang, Y., Abbas, R., Deng, K., Li, J., Zhang, B.: Snpr: A serendipity-oriented next poi recommendation model. In: Proceedings of the 30th ACM International Conference on Information & Knowledge Management, pp. 2568–2577 (2021)
65. Zhang, Y., Chen, X., et al.: Explainable recommendation: A survey and new perspectives. Foundations and Trends® in Information Retrieval 14(1), 1–101 (2020)
66. Zhao, Y., Wang, Y., Liu, Y., Cheng, X., Aggarwal, C.C., Derr, T.: Fairness and diversity in recommender systems: a survey. ACM Trans. Intell. Syst. Technol. 16(1), 1–28 (2025)
67. Zheng, Y., Wang, D.X.: A survey of recommender systems with multi-objective optimization. Neurocomputing 474, 141–153 (2022)

68. Zhou, Y., Zhang, X., Li, F., Liu, S., Jiao, J., Tian, D.: Svdpp recommendation algorithm optimized by q-learning algorithm. Comput. Eng. **47**(2), 46–51 (2021)
69. Ziarani, R.J., Ravanmehr, R.: Serendipity in recommender systems: a systematic literature review. J. Comput. Sci. Technol. **36**, 375–396 (2021)
70. Ziegler, C.N., McNee, S.M., Konstan, J.A., Lausen, G.: Improving recommendation lists through topic diversification. In: Proceedings of the 14th International Conference on World Wide Web, pp. 22–32 (2005)

Analyzing Accuracy and Consistency of GPT 4o Mini in Trivial Pursuit, and the Implications for Its Use in Professional Contexts

Keilen Smith[✉][ID] and Rich Maclin[ID]

University of Minnesota Duluth, Duluth, MN 55812, USA
`smi02720@d.umn.edu`

Abstract. Large language models such as OpenAI's GPT-4 have recently become a subject of interest for their use in education and information exchange. Such applications are hampered by the LLM's capacity to produce superficially plausible but ultimately incorrect information, called *hallucinations*. Using a bank of trivia questions sourced from the 1981 Genus Edition of Trivial Pursuit, we conducted a series of trials on GPT-4o Mini to measure the accuracy and consistency of the model and the characteristics of its hallucinations. The model demonstrated impressive accuracy across trials, averaging approximately 85%. Our results reinforced prior research, showing that the model's output is effectively non-deterministic and that its answers to identical questions can vary across sessions, meaning correct information cannot always be trusted to persist. We find evidence that GPT's demonstrated ability to improve incorrect answers when prompted is a consequence of random chance, that the model is more likely to break correct answers than fix incorrect ones, and that the strength of wording affects how likely this change is. We show that the model's factual *knowledge* is not discrete but probabilistic, with some questions being more prone to hallucination and the model's consistency being a reasonable metric of confidence in the information provided. Finally, we use the results to recommend best practices for future research examining the model's performance in professional and educational environments.

Keywords: Large Language Models · Hallucinations · OpenAI · Trivial Pursuit

1 Introduction

There is no doubt that ChatGPT, first unveiled by OpenAI in 2022, [23] has made major waves in the information age, but with that has come concerns regarding accuracy. The large language model provides an answer to virtually any question, yet makes face-value claims frequently lacking corroboration. Compounding this issue, ChatGPT's ability to cite sources, even when asked, is unreliable at best [38,39].

In previous research on both Versions 3 and 4 of the model, ChatGPT's responses were usually checked against a preexisting answer key (where appropriate) or verified with a panel of experts on the topic at hand. In such cases, the results can vary wildly based on discipline and context, or even simply by the particular test administered [10].

© The Author(s), under exclusive license to Springer Nature Switzerland AG 2026
F. Marcelloni et al. (Eds.): IJCCI 2025, CCIS 2829, pp. 309–327, 2026.
https://doi.org/10.1007/978-3-032-15638-9_18

Regardless of the exact format, virtually all of the research heretofore has concerned itself with a straightforward, objective measure of accuracy to demonstrate whether ChatGPT is or is not competent to answer queries within a particular discipline. Everything from programming education [4] to midwifery [5] has been used for examination. In this paper, we argue that ChatGPT and other LLMs like it do not store, handle, or express factual information in the same manner as human minds, and that this common quiz format is not sufficient to truly map their capabilities. We describe how a discrete accuracy score must be weighed in comparison with the model's general consistency, and how both metrics relate to the model's rate of hallucinations.

We describe a brief history of transformer-based LLMs and ChatGPT and the fundamental mechanics of their training, along with a deeper dive into previous research around the accuracy of ChatGPT and some of the shortcomings therein. We then identify a suitable question set (Trivial Pursuit) and design a system of repeating and varying trials using GPT-4o Mini's public-facing API, provided by OpenAI.

Delving into the results, we find that none of the trials scored significantly greater than any other, even when variability was introduced. The model scores very well across all trials, but exhibits certain behavior which demonstrates that its knowledge is not static. Hallucinations occur more commonly than a single raw score might suggest, and only a certain subset of information manifests consistently. The remainder is inconsistent and changes at seeming random, the results of which may be correct or a hallucination with no apparent pattern. The model also demonstrates itself to be poorly capable of self-reflection and correction, showing a random approach which can do as much harm as good.

In light of this information, we discuss its implications for past and future research delving into ChatGPT's accuracy. We make recommendations for future research in this topic, both for specific disciplines and for general information handling/hallucination research.

2   Background

We will start by discussing the basics of LLMs, their usage in creating tools like GPT, and prior research surrounding the GPT series of models.

2.1   History and Mechanics of LLMs

First formally described by Vaswani et al. in 2017 [37], transformer architecture has become the foundation for modern Large Language Models (LLMs), although the models themselves far predate it. An in-depth understanding of the accompanied logic is not necessary for this paper. It is sufficient to understand that transformer architecture utilizes artificial neural networks which are fed data quantitatively represented as *tokens*. Once trained, the logic of neural networks (and by extension transformers) is difficult or impossible to translate into human logic. The result is a *black box* which, when fed input, produces output. The underlying logical mechanisms created within the network are not understood, which makes manual adjustment impossible except by further training, called *fine-tuning*.

Currently, the best LLMs perform flawlessly in small-scale tasks, displaying impeccable grammar, language, and versatility on or above the level of average humans [8]. The models are capable of holding coherent, continuous conversations, which is learned implicitly as soft *rules* of grammar. In the typical datasets of books, scripts, and articles, the statement "Hello, how are you?" is typically followed by some variation of "I'm well. How are you?" This is a faux *rule* of grammar learned by the model, albeit one far less stringent than subject-verb pairings or sentence tense.

Furthering this, the question "Who was the first man on the moon?" is, in most textbooks and articles, followed by "Neil Armstrong." This, too, an LLM can learn, given sufficient training. As a result, LLMs have been widely observed to be capable of providing factual information in response to questions.

To this end, a particular model has risen to prominence today. ChatGPT, first released in 2022, is a sub-model built with the GPT series of models produced by OpenAI. GPT-1 and GPT-2 were created and tested, with the methods and results published publicly [28,33]. While successful in their own right, neither of these models reached mainstream renown. GPT-3 was released with an API for public and private access, at which point its capabilities began to draw attention. With GPT-3.5 came ChatGPT, a bot intended for conversation and linguistic aid [23] but which demonstrated remarkable accuracy in responding to common queries. Since then the models have only improved, with many users now treating GPT-4 as a substitute for internet search engines, making factual queries and acting on the results.

This functionality shows great promise for GPT-based models in a variety of fields, most notably education. Much research has been done into this, with greatly varying results.

2.2   Considerations and Previous Research

A brief sampling of existent literature finds numerous examples of GPT's capabilities in a variety of disciplines, with often eclectic and unpredictable results. It scores 59% on the Ophthalmic Knowledge Assessment Program [3], 60% on the United States Medical Licensing Exam [10], 67% in a bank of common questions posed by radiology professionals [38], 86% in a bank of common questions posed by bariatric surgeons [30], and anywhere from 57–79% accuracy correcting student code [4].

Dozens more such papers have been published, but this sampling is sufficient to show the variance. As the existing literature demonstrates, the model's accuracy varies significantly. The results of any particular study, seemingly, have little bearing on any other. The question of the model's viability for education and professional assistance remains very much in doubt, particularly for any field which does not already have studies. Nonetheless, certain trends can be identified.

First and most importantly, newer models in the GPT series generally perform better than their predecessors when directly compared [13,17,35], indicating that training data and techniques have a large impact on the model's success, unsurprisingly. As a byproduct of this rapid advancement, most research into GPT's accuracy quickly becomes obsolete as new GPT models release on a near-yearly basis and GPT is updated accordingly.

Secondly, the more easy or commonly known a question is, the more likely GPT is to get it correct. This inverse relation of difficulty and accuracy is often observed informally by subject matter experts, but papers which do use data classified by difficulty reflect this intuition [10,31].

Thirdly, GPT models are inconsistent, being greatly affected by surrounding text and highly susceptible to suggestion [40,43]. In the case of providing factual information, the same question phrased differently or placed in a different context will yield varying results, a trait exploited in techniques such as few-shot learning and persona suggestion [41,42]. This necessarily makes research difficult to replicate and further adds to the uncertainty. Trial-and-error mapping of a model's knowledge set will not be sufficient to vet it for a given purpose, as even the correct answers it gives are not certain to remain consistent.

Recognizing this, some research re-queries the model to try again on incorrect answers, which has been shown to increase the overall accuracy [26]. In a similar technique, a model can be prepared to give the correct answer using additional related text (priming), which has shown some success in improving the accuracy of models [1,41]. However, both of these techniques require foreknowledge, either to provide the model with information or to know that it must be corrected. Such conditions will present a problem for beginners employing a model to learn novel information and limit the applicability of any results.

By nature of their design, LLMs demonstrably do not *think* in the same manner as humans and are not capable of the same reasoning or problem solving, only of linguistically imitating the action [32]. Thus, when asked to reflect on its answers or reconsider them, these models can create a convincing imitation of critical thought which is likely to be correct in common cases but rapidly loses efficacy in more difficult cases [32].

These facts are overlooked in much of the research into the accuracy of GPT, particularly that conducted by subject matter experts in other fields (such as medicine). Questions are asked once and the answer is taken as final, even though the model demonstrably often varies its answers. The model's performance in a given trial, while interesting, has limited generalizability to other contexts of varying format or difficulty. Most relevantly, the results quickly become outdated as subsequent models release.

Causes of Misinformation. LLM models perform well, but almost never perfectly. *Hallucinations*, factually incorrect information which sounds sufficiently plausible to an uninformed user, is the primary cause. Due to the capabilities of LLMs, these hallucinations are often argued eloquently and supported with relevant secondary information due to the model's associative capabilities, making them particularly dangerous and difficult to identify for users seeking to learn.

Hallucinations can be *intrinsic*, a model-produced error which contradicts its training data, or *extrinsic*, a model-produced error making claims beyond the scope of the training data [14,36]. Despite having different causes, for the purposes of this paper they present the same issue. It is also possible that a functioning model may correctly repeat erroneous information present in the training data. In all cases, the effect on the user is the same. A user querying an LLM for novel information by definition does not have the tools to immediately prove or disprove its claims, and so can become victim to misinformation.

Despite this recognized risk of misinformation, LLMs are still frequently employed by learners across disciplines, education levels, and geography [2,9,22,34]. Attitudes in those surveyed are generally positive but frequently admit to reservations regarding the model's capabilities and trustworthiness.

The Link Between Accuracy and Consistency. Most factual queries have a single correct answer or at worst a finite answer set. The set of incorrect answers, on the other hand, is always infinite. In light of this, an LLM's consistency has been proposed as a means of estimating its accuracy and detecting hallucinations. If across multiple sessions a model gives varying answers for a question which can only have a single correct answer, then the model's knowledge of the topic becomes suspect as it is certainly hallucinating. On the other hand, a factual claim which the model knows well enough to repeat consistently is commonly present in the dataset and therefore likely true or, at worst, an error in the dataset itself.

Prior research pursuing this line of reasoning has explored using the probability tokens generated by an LLM's output as to judge how confident it is in an answer and how commonly it will repeat a claim [7,19,20]. While useful, this requires a user to have access to the model's tokens and enough technical savvy to interpret the results, which is impractical for most users. Similarly, black-box techniques such as RAG [16] or FLEEK [6] typically rely on outside tools to mitigate hallucinations.

A notable work, SelfCheckGPT [20], establishes a zero-resource technique using this reasoning which shows great potential. The fundamental technique involves querying the model repeatedly with an identical prompt and examining the results for consistency using various scoring methods, of which the best method asked the model to critically consider a given claim based on the context it generates across samples. It convincingly demonstrates that a model's consistency through token probability is a good way not only to identify hallucinations but also to judge the accuracy of its responses.

While the results are certainly impressive and a good blueprint for further research, they overlook several considerations. Most critically, it uses the WikiBio dataset, which is itself directly lifted from Wikipedia. It is known that Wikipedia is a substantial part of the training dataset of GPT (and most modern LLMs) [12], meaning that the model's performance could be heavily affected by simple data repetition rather than organic data synthesis, as should occur when users ask semi-novel questions. This research also limited itself to the largest 20% of articles to avoid obscure information. As stated previously, these models typically perform worse on rarer and higher-level information, so this restriction creates a best-case scenario for GPT.

Regardless of its shortcomings, the results convincingly demonstrate that consistency is a valuable metric to consider alongside accuracy, and so we will be measuring the consistency between trials in addition to their individual accuracy.

2.3 Summary

Large Language Models such as GPT are not designed to process or replicate *knowledge*, only to replicate *language*, and are prone to misinformation in the form of hallucinations. This creates unique challenges for its implementation in knowledge-dependent tasks such as medicine or education. Despite this, it is still under active consideration for such roles.

Significant research has been undertaken to evaluate the accuracy of information produced by GPT. This research has often been focused on single areas of knowledge, and also does not assess the repeatability of the responses. Those which do consider repeatability rarely perform enough trials to account for the model's inconsistency, and those which do consider its consistency ignore much of the real-world context of GPT's common use cases.

Given the existing research, we hypothesize that the standard knowledge tests typically applied to GPT are insufficient to capture its current behavior and do not accurately reflect how it would perform in common usage. Our goal in this research is to demonstrate and describe its true behavior.

3   Experimental Methods

In this section we will describe how we derived the material we used to evaluate GPT. Trivial Pursuit was used to provide a question bank and all questions were asked to GPT in a series of varying trials to better understand its behavior.

3.1   Question Bank and Qualifications

As previously shown, GPT's accuracy can vary significantly by subject field, so it was desirable to have a large bank of questions covering a wide range of topics and difficulty levels, and these questions needed to have short, definitive answers which could be reviewed and graded efficiently by a human reviewer. These questions also need to be phrased succinctly to minimize the token load and avoid overstepping the model's rate limit (see Sect. 3.3).

It was desirable that these queries not be directly available on the internet, if at all possible. The exact corpus used to train newer GPT models is not fully known, but models 1 and 2 were both heavily trained on parts of the internet judged suitably authoritative [12], and it is reasonable to assume that this has continued. If a question bank available on the internet happened to be included in GPT's training data, the model could provide correct answers through pure recitation. It is far more desirable to investigate the model's ability to synthesize information from its aggregate training data.

To cover all these needs, we employed a dataset of questions sourced from the original 1981 Genus Edition of Trivial Pursuit (owned by Hasbro). Trivial Pursuit's iconic six-category structure (geography, entertainment, history, arts & literature, science & nature, sports & leisure) provides a suitable breadth of categories to avoid narrowed focus and maximize applicability. To the best of our knowledge the questions and answers are not directly available on the internet, and the results bear this out. The questions and answers as provided in the game are brief and definitive, minimizing load on model and grader (Fig. 1).

Notably, given that this version of the game is decades old, some answers are almost certain to be incorrect. This is accounted for in the assessment structure, explained in Sect. 3.5.

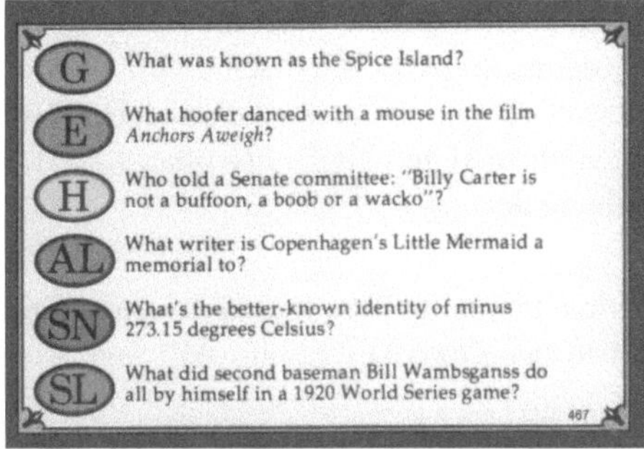

Fig. 1. A sample card from Trivial Pursuit.

A total of 80 cards were randomly chosen and scanned for a total of 480 questions and answers. All information was scanned using a commercial scanner and converted using Adobe Acrobat's OCR functionality, then manually reviewed by human researchers to clean up any remaining typos and make the data ready for submission to the model. A single typo which escaped this review only to be discovered during the trials was preserved in order to maintain consistency between trials.

3.2 Model Selection

LLM Choice. While this research focuses on GPT, the selection of this particular model requires some justifications. Many LLMs beside GPT are available and show results comparable to, or in some cases surpassing it. LLaMa, BERT, and Deepseek are some popular contemporaries of GPT, showing results competitive with the model and sometimes surpassing it [15,18,21,29,31]. In light of this, the selection of GPT is to a certain degree arbitrary, but motivated by several factors.

At the time of writing this, GPT remains one of the most popular and is frequently spoken of and used by both laymen and professionals outside computational academia. Given that the results of this paper are most novel and relevant to those looking to employ LLMs in disciplines discrete from computer science, we selected the most common line of models for the purpose.

OpenAI's openness and support of independent research around their models was also attractive, as with their support we were able to obtain a research account for using the model free of charge. Additionally, the company is continually refining and improving their models and anticipated to remain competitive for years to come, and while specific models may change, for the time being the GPT series continues functioning on the same architecture and thus these results should remain generalizable even as GPT 4o becomes outdated.

We do not claim or believe that GPT is the most universally powerful model and did not select it expecting that it would be the best scorer. As previously stated, models quickly become outdated, and as such any specific accuracy score is near-meaningless. Our objective is to describe how the model behaves and how users can maximize confidence in the answers provided by it or other models on the same architecture.

In light of this, we opted not to include other models as competitors. The individual accuracy scores mean little in comparison to each other, and our other primary metric, consistency, is an irrelevant measurement across models. While the methods of this research could easily be replicated and applied to other LLMs, we focused our efforts and discussion to a single model.

GPT Version. Rather than manually pass hundreds of questions through ChatGPT's public portal, we opted to use OpenAI's public API and a custom python script to feed all questions to the model and record results. Notably, this API talks to the GPT model itself, not ChatGPT. ChatGPT is built upon GPT using further fine-tuning [23] so while specific results of accuracy and response structure may vary, general trends should nonetheless apply across the models.

In addition to the logistical convenience, we chose to use GPT over the more popular ChatGPT due to the comparative usage policies. Per ChatGPT's usage policy, data is collected and may be used for future training, which presents a concern for researchers looking to ask the same questions multiple times (as we are). If the model were to learn from our questions as we ask them, this would naturally invalidate future results. GPT's API, in contrast, has been explicitly stated by OpenAI to not use provided data for training, making it the better option for our purposes [24].

Using GPT does raise one additional concern which must be accounted for. GPT's API does not by default *remember* previous queries, even from the same user, a conversation features for which ChatGPT is famous. Including the contextual features is desirable to better understand the model's behavior in real-world applications, something which OpenAI accounts for. In the API, previous conversation can be included in queries as additional inputs and will be taken into account when generating output. See Sect. 3.3 for further details.

For the duration of the research we used GPT-4o Mini, a version of 4o, that, per OpenAI, is optimized to run much faster with a small penalty to performance. Much like the choice to use to use GPT over ChatGPT, this means that exact output should not be expected to repeat across different models. Also, in all likelihood the main model would achieve slightly higher accuracy scores, and based on previous research it can be reasonably expected that later models will score higher still.

A free research account graciously provided by OpenAI was used for these queries, which provided free access in exchange for daily rate limits. GPT-4o mini was last updated in August of 2024 [25], and did not undergo any known changes over the course of these trials.

3.3   Query Structure

After prepocessing, all collected trivia questions from Trivial Pursuit were stored in an xlsx (Microsoft Excel) file as plaintext with minimal formatting, with each question and answer stored in a matching pair of cells and tagged by the category and an arbitrary numerical ID (Fig. 2).

The trial script imported this file using the Python pandas [27] library and incremented through the list to import all questions and answers as strings and generate queries to send them to the API.

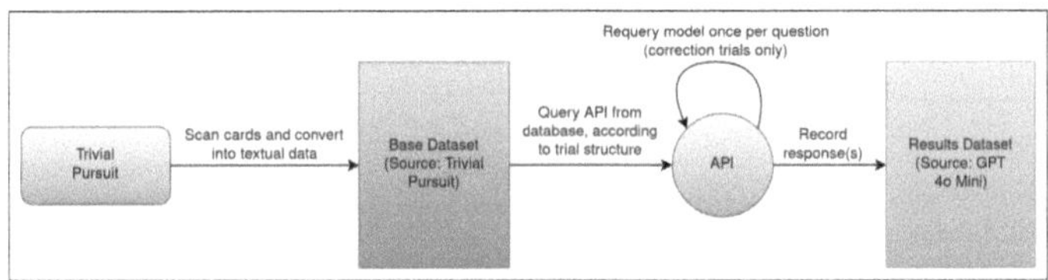

Fig. 2. Workflow for data collection. The novel dataset was copied from randomly chosen Trivial Pursuit cards. The completed dataset was then used to quiz GPT.

GPT's API provides robust options to customize its behavior such as token limits, assigned personas, and response temperature. As the purpose of this research is exploratory rather than competitive, these were left at their default values for all trials.

For all trials, the previous two trivia questions and model-sourced answers were included in the input so that the model effectively *remembered* them when generating its next answer. For most trials this meant two additional inclusions each of user and model input, for the double-check and second-guess trials, this included both the initial answers and the corrections. The first two questions of each trial batch had suitably reduced or absent input rather than carrying over conversation from previous trials.

GPT's API institutes rate limits, so each trial was spread over multiple days with the final few results of each day provided as context at the start of the following day to mimic a continuous conversation.

All results obtained from the model were collected and recorded in a separate file, then later manually graded by a human reviewer (see Sect. 3.5). The model was never apprised of its performance and never shown the official Trivial Pursuit answers in any way.

3.4   Trial Structure

Rather than limiting ourselves to a single pass-through of the model, we opted for a series of trials varying certain characteristics. Every trial used all 480 questions verbatim and was manually graded. A list of trials, as well as their rationale, is provided here.

Control Trial: An initial control test presenting all 480 questions in the default order (card by card, making for a perfect cycle of categories).

Repeat Control Trial: A second test following the exact format of the control trial. At a fundamental level all LLMs are deterministic, but it is quite common to artificially introduce variation, which OpenAI is known to include in its model as *temperature*. This trial, then, is to demonstrate if the model will merely vary superficial wording or if factual information is variable as well, and to serve as control when comparing the agreement between more varied trials.

Random Order Trial: Functionally similar to previous trials, this randomizes the order of questions. Previous context is known to impact future output, so if the first two trials show high parity, this may differ more strongly.

Grouping Trial: All questions are grouped by category and asked sequentially. All geography questions are asked in sequence, then all history questions, and so on. Similar to few-shot priming, this will attempt to use nominally similar information to get the model in a "state of mind" to better answer questions.

Appeal to Authority Trial: Default ordering. All questions have the attached prefix, "According to Wikipedia," prompting the model to retrieve specific information. Wikipedia is known to be in the training dataset of GPT [12], so this trial will measure how well the model can refer to specific sources, if at all.

Soft Correction Trial: Default ordering. All questions are met with the automatic response "That may not be correct, are you sure?" regardless of its accuracy, prompting the model to reconsider its output. Both the first and second responses to each question are recorded. Previous research has shown that the model can improve its overall score and self-correct when prompted to fix incorrect answers [26]. This trial will measure if it is any more likely to change factually incorrect answers than correct ones.

Hard Correction Trial: Default ordering. Functionally identical to the previous trial and conducted for the same reason, but using a much stronger prompt "That is wrong. Please correct your answer." By including this as well, we may additionally measure if the strength of the user's response is more persuasive to the model.

3.5   Assessment

Once all trials were run and recorded, they were manually reviewed and graded by a human. Trivial Pursuit was treated as the primary authority; all GPT-sourced answers which agreed with the Trivial Pursuit answer were immediately graded as true without additional research. Any ancillary information provided by the model was not considered or included as part of the grade regardless of its accuracy (Fig. 3).

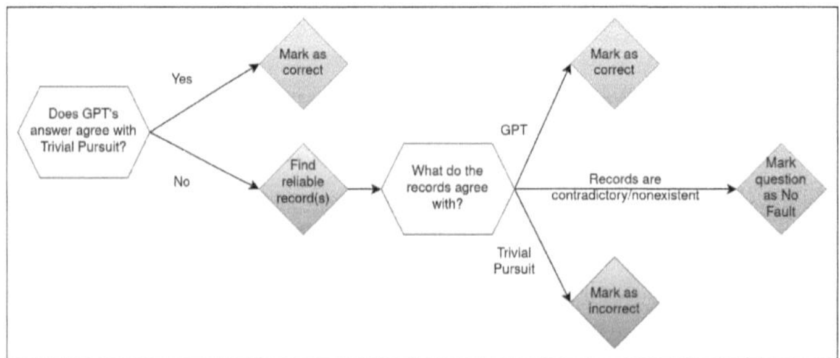

Fig. 3. Workflow for human review of GPT's answers in each trial. All questions which are marked as "no fault" are given this rating across all trials.

For answers which were incorrect, the reviewer conducted a brief, impromptu research to determine if Trivial Pursuit's answer is outdated or inaccurate. If not, the

GPT-provided answer was marked as false. Due to the wide spread of trivia and relatively niche nature of many questions, no unifying authority or knowledge source could be identified, and questions had to be handled on a case-by-case basis.

Questions in which Trivial Pursuit's answer is incorrect *and* GPT provided the correct answer according to a reputable source are marked as *no fault* and separated from the valid questions for purposes of reporting. Additionally, several questions were later found to be sufficiently open or vague as to have multiple correct answers, such as "What's 10 in. tall, weighs seven pounds and is gold-plated?" In such cases, the question will likewise be marked as *no fault*.

4   Results

In this section we provide the experimental results of our work. Accuracy remained relatively steady across all trials, although the model showed high inconsistency and would give varying answers to identical questions. Correct answers would not necessarily persist between trials, but consistency showed positive correlation with the model's average accuracy.

4.1   Exclusions

Ideally, all questions used in this study ought to still be valid today, but manually confirming this prior to testing would be too time intensive. However, it could reasonably be assumed that any questions in which Trivial Pursuit (released in 1981) and GPT-4o (trained in 2023/2024) wholly agreed were still valid (Table 1).

Table 1. No Fault rulings by category.

Arts & Literature	6/80
Entertainment	3/80
Geography	7/80
History	3/80
Science & Nature	11/80
Sports & Leisure	6/80
Total	36/480

Certain questions were ruled as *No Fault* due to having a large number of correct or disputed answers. For example "What is the fastest-growing thing in the plant and animal kingdoms?" can have many answers, depending on whether this growth is referring to length, mass, or volume. "Where is Nelson's Monument?" is likewise open when at least five monuments are named and erected in honor of men named Nelson.

While questions such as "Who was the oldest U.S. President inaugurated?" always have a single correct answer, (Reagan in 1981, Biden at the time of data collection) for the sake of simplicity these questions were excluded from all scoring for being partially outdated. Other outdated questions, such as "What bullfighter earned more than $3 million a year at the height of his career?" now have gained more correct answers and thus would make direct comparison between answers difficult.

During grading and evaluation, a total of 36 questions were ruled *No Fault* and excluded from all further scoring and data analysis.

4.2   Trial Accuracy

Based on the results summarized in Table 2, no technique can be definitively stated to improve the model's score at this time. The highest scorers are still closer to the first control trial than the two control trials are to each other, so this difference cannot be meaningfully attributed to the trial formats. The differing scores could easily be conjectured as resulting from random chance, though further research and larger datasets would be necessary to solidify this claim.

Table 2. Total correct answers in all trials.

Trial	Raw Score	Percentage
Control	385/444	86.71%
Repeat Control	382/444	86.04%
Random Order	384/444	86.26%
Grouping	388/444	87.39%
Appeal to Authority	378/444	85.14%
Soft Correction (initial)	383/444	86.26%
Soft Correction (corrected)	382/444	86.04%
Hard Correction (initial)	387/444	87.16%
Hard Correction (corrected)	343/444	77.25%

Control Trials. When directly compared, both control trials usually agreed on answers but nonetheless had some disagreement. In examining the answers, the model generally agreed with itself on the main points of matching answers but would use subtle differences in wording and sometimes disagreed on ancillary information (which was not checked or graded for this paper). The context-based nature of LLM generation could conceivably have contributed to a *butterfly effect* skewing future answers based on minor differences in prior conversation. However, even in the first question, which was free of context and therefore immune to this effect, the model's answers for each control trial agreed but were worded differently. This indicates that GPT's final output is not fully deterministic and can vary even in identical preconditions.

Correction Trials. Interestingly, neither correction trial saw ChatGPT improve at all when prompted to correct its answers. Both trials had a number of answers corrected after prompting, regardless of whether they were initially correct or not.

In the soft correction trial, 54 answers were adjusted, of which 15 were initially correct and changed to incorrect answers by the model. Of the 39 answers which were initially incorrect, only 14 were successfully corrected. The remainder were adjusted but remained incorrect. The model was more likely to adjust incorrect answers than correct ones, but the odds of a new answer being correct (regardless of the starting state) were significantly lower than 50%.

Hard correction showed even worse results. The stronger wording in our prompt resulted in the model being far more likely to change answers, with 111 answers being adjusted. 60 correct answers were adjusted to become incorrect. Of the 51 initially incorrect answers, only 16 were successfully corrected. When change was demanded rather than suggested, the model would more readily discard correct answers and achieved a substantially lower score after correction. Hard correction proved far more persuasive, much to the model's detriment, dropping the score by almost ten percentage points.

Authority Trial. Prompting the model to reference a source of authority did not result in any improvement to the score, and may even have hindered it. For all questions the model incorrectly answered in this trial, a reasonable effort was made to search related Wikipedia articles for the information. If Wikipedia indicated the same incorrect answer, then it could indicate that the model's hallucinations were extrinsic rather than intrinsic. In all cases, the information was not found or indicated an answer which disagreed with GPT and agreed with Trivial Pursuit.

This disconnect between GPT and a subset of its training data is ultimately unsurprising, as the methods used to train LLMs rarely require the model to distinguish where a particular data sample comes from. Interestingly, the prefix"According to Wikipedia," served primarily to dictate how the model formatted its response. All answers were likewise formally prefixed "According to Wikipedia..." and were notably shorter than all other output, giving direct answers with little ancillary information.

Grouping Trial. The grouping trial showed the highest overall score, perhaps as a soft form of priming, but this is difficult to say with certainty, particularly when the hard correction trial's initial answers come in second, and the distance from each of them to the first control is no greater than that between the two control trials.

4.3   Inter-trial Consistency

For the purpose here, consistency between two trials is a measure of how many answers match in their primary claim, whether that is correct or incorrect. Consistency results reported in Table 3 are limited to the first five trials, as both correction trials showed a reduction in the overall score and were specifically instructed to change their answers, thereby introducing artificial variation.

Table 3. Answer consistency between trials.

	Control 2	Random	Grouping	Authority
Control 1	88.29%	87.39%	87.16%	86.71%
Control 2	.	87.61%	87.16%	85.59%
Random	.	.	88.06%	86.49%
Grouping	.	.	.	86.49%

The authority trial showed the lowest overall consistency with the other trials, perhaps as a byproduct of also having the lowest accuracy score. The control trials, notably, did not correlate much more strongly with each other than they did with the random and grouping trials, which featured identical questions in a different ordering. The random and grouping trials also show agreement which is shockingly close to the control trials, further suggesting that these results are largely a product of chance rather than correlation.

For the first five trials, the model had total agreement in its answers for 353 questions, of which 346 were correct and, therefore, correct in every trial. Thus, for any question in which the model demonstrates complete agreement, its odds of being correct jump to an impressive 98.02%. Of the remaining 91 questions, 65 had a majority agreement, but of those only 44 saw the majority being correct and the other 21 had the correct answer present but not in the majority.

In total, across the first five trials GPT had perfect accuracy on 346 questions and perfect inaccuracy on 38 questions. When also including the correction trials, GPT was perfectly accurate on 300 questions and perfectly inaccurate on 30 questions. It should be noted that when all trials were considered together, not a single perfectly inaccurate question also had perfect agreement. Put another way, any answer which had perfect agreement across all seven trials was also guaranteed to be correct, at least within our test data.

5 Discussion

GPT 4o is, unsurprisingly, highly capable at basic trivia, scoring an average of 84.98% across all trials. This is particularly relevant as the question bank was chosen for its obscurity on the internet. While Trivial Pursuit itself is a household name, to the best of our knowledge there is no question bank, official or otherwise, which is available online, let alone present in the model's training dataset. Therefore, the model's answers are synthesized from its aggregate rather than directly mimicking particular question-answer pairings.

5.1 Results by Trial

Although GPT scored well, none of the techniques demonstrated convincing improvement in the model's score.

While previous research has demonstrated that prompting GPT and similar models to correct erroneous information raises the overall accuracy, these results indicate that it is more so an outcome of simple probability rather than a true method of improvement. Any test taker, human or machine, if prompted to fix truly incorrect answers alone, cannot possibly decrease their score, and even lucky guesses on a low probability will inevitably raise the score.

In contrast, when applied to a claim of unknown accuracy, GPT's"corrections" appear to be guesses with a success rate less than 50%, making their use inadvisable. The results imply that a gentle suggestion is less likely to change answers, but that regardless of the wording the model is more likely to undo its good work than correct its bad.

Based on the consistency results, if a question is asked in a context that is effectively random, the model's output will see little overall difference. Grouping the questions by category, original order, or randomly does little to affect the overall result. This is encouraging, as it means that random noise in a conversation is unlikely to affect the overall accuracy so long as the actual question is unchanged. On the other hand, it also means that the model's correct answers cannot be relied upon to consistently repeat even if the question is phrased identically, let alone varied in wording.

When asked to reference an informational authority (Wikipedia), the model's answers showed no correlation with the authority itself. GPT had clearly learned that the question "According to Wikipedia, what is X?" ought to be answered by "According to Wikipedia, X is..." but had no ability to reference the actual information, instead placing arbitrarily chosen information in and claiming its origin to be Wikipedia. This correlates with previous research asking GPT to write a scientific paper with references, where it showed ability to write a convincing-looking reference which usually turned out to be made up or completely unrelated [38,39].

5.2   Informal Observations

Several trends can be observed anecdotally but not measured empirically, and are described here for informational purposes.

When giving answers to "unknown" questions, GPT-4 had a tendency to produce "informed bullshit" [11] which would sound sufficiently credible to an uninformed reader. For example, when questioned "Which Presidential ticket was dubbed'Bozo and the Pineapple'?" GPT never provided an accurate response, yet every hallucination featured real presidential candidates (Gerald Ford and Bob Dole earned the nickname in the 1976 election). Likewise, questions such as "Whose acceptance speech of more than 30 min prompted a time limit on Academy Award thank-yous?" will see GPT always answering with actors (or other film professionals). Whether the knowledge is implicit or explicit, GPT has sufficient associations to recognize key terms in the question context and narrow its list of answers accordingly. In all cases except the Authority trial (noted to be uncharacteristically short), it would provide related claims and supporting evidence which might or might not be true, yet lends credence to the main claim.

Likewise, and more curiously, the model would sometimes pick answers which sound similar to the correct answer, as if partially recalling the correct information. For example, the Trivial Pursuit answer to "Who lives upstairs from Felix Unger and

Oscar Madison in Neil Simon's *The Odd Couple*?" is "the Pigeon Sisters." GPT-4 was never able to give this complete answer, yet guessed "Mrs. Muriel Pigeon" (an incorrect name) and "Miss Pigeon" independently.

5.3   Probabilistic Knowledge

The model's answers do not show perfect consistency, even in identical preconditions. While this is concerning for anyone wishing to use the model in a situation where inconsistency is unacceptable, it may also present opportunity.

As previously stated, if the model had perfect agreement on an answer, it was 98% accurate, a rate far higher than the overall accuracy. As such, anyone seeking to test the model's true knowledge should consider repeatedly asking questions. If the model's answer is perfectly consistent, then we may be very confident in the answer, and may presume that this answer is likely to persist. If the model is *uncertain* about a given fact, this becomes evident in repeat questioning. Higher agreement can generally mean higher confidence, although anything less than perfect agreement is suspect and warrants further investigation.

5.4   Implications for Present and Future Research

As our data seems to broadly indicate, in a standard trial one pass is enough to establish a ballpark for the average accuracy of the model in a given knowledge set, but not sufficient to indicate its consistency on any given question. Therefore, research which finds a certain accuracy on a given test or in a given field is basically valid but does not guarantee that those specific questions will always be answered correctly.

Nor is simply asking the model to self-correct sufficient. When asked to self-correct, the model is more likely to sabotage itself than to improve its score. While this will still result in improvement if the model is only asked to correct hallucinations, it demonstrates that the model's abilities of self-reflection are more a product of chance than any cognition or reasoning. Researchers should reconsider using this method without additional measures of control.

GPT cannot be trusted to reference or isolate sources in its training data and should not be trusted with this task. For researchers looking to employ it in information-critical contexts such as finding citations should reconsider or reorient their approach. The model does not possess any innate ability to search the current internet and likewise cannot *recall* particular instances in its training data.

The consistency results bear several important conclusions. First, they indicate that the model's answer to questions can vary even when the question is identically phrased. Thus, any experiments varying the query content are advised to make multiple trials of each variety before drawing conclusions.

Second, arbitrary conversation context (that which does not contain specific instructions or deliberate priming) may have an impact upon the final answer GPT generates, but it will not significantly affect the overall score of a given trial.

Third, the majority of GPT's correct knowledge does manifest consistently across all trials, but inconsistent knowledge remains and must be accounted for. The questions which this model is capable of answering correctly are fewer in number than those it can consistently answer. Those looking to employ GPT (or similar LLMs) for information generation and retrieval should consider asking questions multiple times. Strongly held information is likely to persist across multiple instances of questioning, while weakly held or missing information will become apparent as hallucinations in subsequent questioning. Those looking to test GPT's true knowledge would be best served by asking questions repeatedly in discrete instances (to avoid contamination by context), then aggregating the results to identify probable fact and probable hallucination.

In all cases, for any researchers intending repeat trials, it is necessary to carefully investigate whether the model used will remember/learn from prior input and plan accordingly to ensure that early trials do not corrupt future ones.

References

1. Agnes, C.K., Rahman, M.R., Maass, W.: Semantic priming via knowledge graphs to analyze and treat language model's honest lies (2024)
2. Ajlouni, A.O., Wahba, F.A.A., Almahaireh, A.S.: Students' attitudes towards using ChatGPT as a learning tool: the case of the university of Jordan. Int. J. Interact. Mob. Technol. **17**(18), 99–117 (2023)
3. Antaki, F., Touma, S., Milad, D., El-Khoury, J., Duval, R.: Evaluating the performance of chatgpt in ophthalmology: an analysis of its successes and shortcomings. Ophthalmol. Sci. **3**(4), 100324 (2023). https://doi.org/10.1016/j.xops.2023.100324, https://www.sciencedirect.com/science/article/pii/S2666914523000568
4. Balse, R., Valaboju, B., Singhal, S., Warriem, J.M., Prasad, P.: Investigating the potential of GPT-3 in providing feedback for programming assessments, p. 292–298 (2023). https://doi.org/10.1145/3587102.3588852
5. Barbounaki, S., Sarantaki, A., Lykeridou, A., Gourounti, K.: Can chat-GPT be an effective tool for midwifery education? Eur. J. Midwifery **7**(Supplement 1) (2023). https://doi.org/10.18332/ejm/172493
6. Bayat, F.F., et al.: Fleek: factual error detection and correction with evidence retrieved from external knowledge (2023). https://arxiv.org/abs/2310.17119
7. Chatterjee, A., Goel, Y., Chakraborty, T.: Hide and seek: detecting hallucinations in language models via decoupled representations (2025). https://arxiv.org/abs/2506.17748
8. Echavarria, R.: ChatGPT-4 in the turing test. Minds Mach **35** (2025). https://doi.org/10.1007/s11023-025-09711-6
9. Farhi, F., Jeljeli, R., Aburezeq, I., Dweikat, F.F., Al-shami, S.A., Slamene, R.: Analyzing the students' views, concerns, and perceived ethics about chat GPT usage. Comput. Educ. Artif. Intell. **5**, 100180 (2023). https://doi.org/10.1016/j.caeai.2023.100180, https://www.sciencedirect.com/science/article/pii/S2666920X23000590
10. Gilson, A., et al.: How does ChatGPT perform on the united states medical licensing examination? The implications of large language models for medical education and knowledge assessment. JMIR Med. Educ. **9**, e45312 (2023). https://doi.org/10.2196/45312, https://mededu.jmir.org/2023/1/e45312
11. Hicks, M., Humphries, J., Slater, J.: ChatGPT is bullshit. Ethics Inf. Technol. **26**, 1–10 (2024). https://doi.org/10.1007/s10676-024-09775-5
12. Hua, S., Jin, S., Jiang, S.: The limitations and ethical considerations of ChatGPT. Data Intell. **6**(1), 201–239 (2024)
13. Hussain, W., Grundy, J.: Advice for diabetes self-management by ChatGPT models: challenges and recommendations (2025). https://arxiv.org/abs/2501.07931

14. Ji, Z., et al.: Survey of hallucination in natural language generation. ACM Comput. Surv. **55**(12) (2023). https://doi.org/10.1145/3571730
15. Jiang, Q., Gao, Z., Karniadakis, G.E.: Deepseek vs. ChatGPT vs. claude: a comparative study for scientific computing and scientific machine learning tasks (2025). https://arxiv.org/abs/2502.17764
16. Lewis, P., et al.: Retrieval-augmented generation for knowledge-intensive NLP tasks. In: Larochelle, H., Ranzato, M., Hadsell, R., Balcan, M., Lin, H. (eds.) Advances in Neural Information Processing Systems, vol. 33, pp. 9459–9474. Curran Associates, Inc. (2020). https://proceedings.neurips.cc/paper_files/paper/2020/file/6b493230205f780e1bc26945df7481e5-Paper.pdf
17. Liu, J., et al.: The diagnostic ability of GPT-3.5 and GPT-4.0 in surgery: comparative analysis. J. Med. Internet Res. **26**, e54985 (2024). https://doi.org/10.2196/54985
18. Liu, R., Liu, J., Yang, J., Sun, Z., Yan, H.: Comparative analysis of ChatGPT-4o mini, ChatGPT-4o and Gemini advanced in the treatment of postmenopausal osteoporosis. BMC Musculoskeletal Disord. **26** (2025). https://doi.org/10.1186/s12891-025-08601-3
19. Liu, T., et al.: A token-level reference-free hallucination detection benchmark for free-form text generation. In: Muresan, S., Nakov, P., Villavicencio, A. (eds.) Proceedings of the 60th Annual Meeting of the Association for Computational Linguistics (Volume 1: Long Papers), pp. 6723–6737. Association for Computational Linguistics, Dublin, Ireland (2022). https://doi.org/10.18653/v1/2022.acl-long.464, https://aclanthology.org/2022.acl-long.464/
20. Manakul, P., Liusie, A., Gales, M.J.F.: SelfcheckGPT: zero-resource black-box hallucination detection for generative large language models (2023). https://arxiv.org/abs/2303.08896
21. Maniaci, A., Hoch, C., Sogalow, L., Schmidl, B., Lechien, J.: AI in clinical decision-making: ChatGPT-4 vs. llama2 for otolaryngology cases. Eur. Arch. Oto-Rhino-Laryngol. 1–10 (2025). https://doi.org/10.1007/s00405-025-09371-3
22. Ngo, T.T.A.: The perception by university students of the use of ChatGPT in education. Int. J. Emerg. Technol. Learn. (iJET) **18**(17), pp. 4–19 (2023). https://doi.org/10.3991/ijet.v18i17.39019, https://online-journals.org/index.php/i-jet/article/view/39019
23. OpenAI: Introducing ChatGPT (2022). https://openai.com/index/chatgpt/
24. OpenAI: How your data is used to improve model performance (2025). https://help.openai.com/en/articles/5722486-how-your-data-is-used-to-improve-model-performance
25. OpenAI: Models (2025). https://platform.openai.com/docs/models
26. Pan, L., Saxon, M., Xu, W., Nathani, D., Wang, X., Wang, W.Y.: Automatically correcting large language models: surveying the landscape of diverse self-correction strategies (2023). https://arxiv.org/abs/2308.03188
27. pandas: Pandas Python library (2025). https://pandas.pydata.org/
28. Radford, A., Narasimhan, K., Salimans, T., Sutskever, I.: Improving language understanding by generative pre-training (2018). https://api.semanticscholar.org/CorpusID:49313245
29. Ramchandani, R., et al.: Comparison of ChatGPT-4, copilot, bard and Gemini ultra on an otolaryngology question bank. Clin. Otolaryngol. 1–8 (2025). https://doi.org/10.1111/coa.14302
30. Samaan, J., et al.: Assessing the accuracy of responses by the language model ChatGPT to questions regarding bariatric surgery. Obesity Surgery **33**, 1–7 (2023). https://doi.org/10.1007/s11695-023-06603-5
31. Shakya, R., Vadiee, F., Khalil, M.: A showdown of ChatGPT vs deepseek in solving programming tasks (2025). https://arxiv.org/abs/2503.13549
32. Shojaee, P., Mirzadeh, I., Alizadeh, K., Horton, M., Bengio, S., Farajtabar, M.: The illusion of thinking: understanding the strengths and limitations of reasoning models via the lens of problem complexity (2025). https://ml-site.cdn-apple.com/papers/the-illusion-of-thinking.pdf

33. Solaiman, I., et al.: Release strategies and the social impacts of language models (2019). https://arxiv.org/abs/1908.09203
34. Strzelecki, A.: To use or not to use ChatGPT in higher education? A study of students' acceptance and use of technology. Interact. Learn. Environ. **32**, 5142–5155 (2024). https://doi.org/10.1080/10494820.2023.2209881
35. Taloni, A., et al.: Comparative performance of humans versus GPT-4.0 and GPT-3.5 in the self-assessment program of American academy of ophthalmology. Sci. Rep. **13** (2023). https://doi.org/10.1038/s41598-023-45837-2
36. Tonmoy, S.M.T.I., et al.: A comprehensive survey of hallucination mitigation techniques in large language models (2024). https://arxiv.org/abs/2401.01313
37. Vaswani, A., et al.: Attention is all you need (2017). https://arxiv.org/abs/1706.03762
38. Wagner, M.W., Ertl-Wagner, B.B.: Accuracy of information and references using ChatGPT-3 for retrieval of clinical radiological information. Canadian Assoc. Radiol. J. 08465371231171125 (2023). https://doi.org/10.1177/08465371231171125
39. Walters, W., Wilder, E.: Fabrication and errors in the bibliographic citations generated by ChatGPT. Sci. Rep. **13**, 14045 (2023). https://doi.org/10.1038/s41598-023-41032-5
40. Wang, B., Yue, X., Sun, H.: Can ChatGPT defend its belief in truth? evaluating LLM reasoning via debate. In: Bouamor, H., Pino, J., Bali, K. (eds.) Findings of the Association for Computational Linguistics: EMNLP 2023, pp. 11865–11881. Association for Computational Linguistics, Singapore (2023). https://doi.org/10.18653/v1/2023.findings-emnlp.795, https://aclanthology.org/2023.findings-emnlp.795/
41. Wang, Y., Yao, Q., Kwok, J.T., Ni, L.M.: Generalizing from a few examples: a survey on few-shot learning, **53**(3) (2020). https://doi.org/10.1145/3386252
42. White, J., et al.: A prompt pattern catalog to enhance prompt engineering with ChatGPT (2023). https://arxiv.org/abs/2302.11382
43. Zuccon, G., Koopman, B.: Dr ChatGPT, tell me what i want to hear: how prompt knowledge impacts health answer correctness (2023). https://arxiv.org/abs/2302.13793

Interpretable Explainable AI: Comparing Bayesian Structural Equation Modelling with Other Algorithms

Corina Ilinca[1,2]([✉]) [iD]

[1] Faculty of Mathematics and Computer Science and Faculty of Sociology and Social Work, University of Bucharest, 050663 Bucharest, Romania
corina.ilinca@unibuc.ro
[2] Softbinator Technologies, 050663 Bucharest, Romania

Abstract. This research study compares various machine learning algorithms for predicting self-assessments of cognitive functioning, specifically memory, using data from the Health and Retirement Study involving individuals over 50 years old. The study reveals that while different algorithms yield only minor differences in predictive performance, the theoretical and social context of the data plays a significant role in model construction. Study Objective: The aim is to identify which machine learning algorithm best predicts memory assessments using a dataset from the Health and Retirement Study. Algorithms Compared: The study examines several algorithms including ordinal regression, linear regression, neural networks, and Bayesian approaches, evaluating their effectiveness in predicting memory outcomes. Predictors Used: Key predictors in the models include self-rated hearing, age, education, work status, gender, marital status, and life satisfaction, derived from a sample of 15,408 respondents. Results Summary: The results indicate moderate levels of variance explained by the models, with significant predictors identified across all algorithms, particularly highlighting the importance of hearing and education. Model Performance: The goodness of fit measures show that the models provide similar levels of predictive accuracy, indicating modest explainability. Conclusions: The study concludes that while predictive performance varies slightly among algorithms, understanding the underlying theoretical connections is crucial for developing explainable machine learning models in social sciences.

Keywords: Bayesian Structural Equation Modelling · Neural Networks · Bayesian Linear Regression · Structural Equation Modelling · Ordinal Regression · Linear Regression · Comparison Between Algorithms · Memory Prediction · Cognitive Self-Assessment · Health and Retirement Study

1 Introduction

Comparing between different algorithms that can be used to predict ordinal features is useful in testing which algorithm would work better. Are neural networks better than ordinal regression, linear regression, or structural equation modelling? If we consider a

© The Author(s), under exclusive license to Springer Nature Switzerland AG 2026
F. Marcelloni et al. (Eds.): IJCCI 2025, CCIS 2829, pp. 328–338, 2026.
https://doi.org/10.1007/978-3-032-15638-9_19

probabilistic approach, would Bayesian linear regression or Bayesian structural equation modelling be better than neural networks? What kind of aspects can be considered when choosing between these algorithms? Are the coefficients showing similar connections with cognitive assessments across all the models tested and what kind of differences can be noticed? These questions can be particularly relevant for data at individual level in the social and health sciences, where the number of cases is variable depending on the magnitude of the study, but with limited data available on particular issues that need prediction, but with moderate number of cases. For picturing a use case, a model for predicting self-assessments of cognitive functioning is proposed for each algorithm trained.

Models for prediction purposes, on one hand aim at creating new outputs based on the variation that may exist in other samples, as well as are more and more accurate and reliable as the training data is better at predicting the outcome, as well as the existing theoretical connections between the variables, to have a story that creates a prediction model with connected features, especially relevant when the samples have limited number of cases. In social and health sciences, the available data for research purposes may have samples that vary from hundreds to thousands cases, with theoretical models that can address meaningful research questions (see [1]). Data mining for prediction purposes is important to exist across different fields of study in order to create systems that learn automatically from data for performance purposes, for example in educational systems [2].

2 Theoretical Assumptions

A hypothesis can be tested using different algorithms depending on available choices, given different assumptions related to the algorithms used, as well as exposure to different methodologies that researchers may have. Given the growth of developed algorithms, the available data and the software and programming languages available for creating statistics and estimating outcomes based on data, a comparison between analysis outputs and documentations on which algorithm would be better for different research purposes is interesting.

Ordinal regression is created to account for the ordinal level of measurement of the outcome feature, the dependent variable in the model, which, in comparison with the linear regression that is created for interval and numeric variables as outcomes, is considered to do a better work in prediction and model creation [3]. Structural equation modelling adds to the regression approaches possibilities to account for concepts that are not directly measurable, latent features, such as intersectional disadvantage [4], as well as the path analysis between the features included [5–7].

Bayesian structural equation model (SEM) is an alternative to the frequentist approach toward structural equation modelling [8, 9] where an estimation method is a based on maximum likelihood estimation with means estimated instead of distributions of probabilities for the features included in the analyses. Bayesian linear regression [10] adds to the frequentist linear regression the possibility to account for distributions of probabilities, without the capabilities of structural equation modelling for path trajectories and factorial modelling. The variables included in the model to predict memory assessment were documented in the past to be part of the theoretical model [11–13].

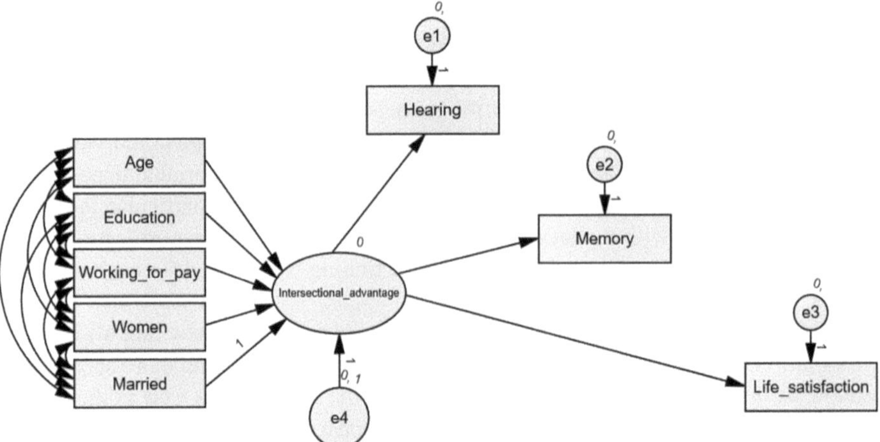

Fig. 1. Bayesian structural equation model (SEM) predicting memory using a formative factor for accounting for intersectional disadvantage (ID).

3   Methods

Using data from the Health and Retirement Study, from 2020–2021, the model for predicting memory assessments included as predictors self-rated hearing, age, years of education, if respondents are working for pay, gender with women compared to men, married respondents compared with other types of marital statuses, and perceived life satisfaction (see Figs. 1, 2 and 3). The total number of cases is 15408 people from the United States of America, above 50 years of age, within a nationally representative sample, using IBM SPSS Statistics v. 29 for the regressions and neural network created and IBM AMOS v. 29 for the structural equation models. The algorithms used for prediction purposes and their goodness of fit reported statistics [2] are ordinal regression (Cox and Snell R-Square), Linear regression (R-Square), Neural networks (Percent Incorrect Predictions - Testing), Bayesian linear regression (R-square), Structural equation modelling (R-square Memory), and Bayesian structural equation modelling (Posterior predictive p).

4   Results

Descriptive statistics show that 4 in 10 respondents have good levels of memory, whereas only 1 in 20 respondents reported an excellent memory assessment. Completely or very satisfied with their life are 7 in 10 respondents. The sample is composed of people of 50 years of age and above, with a mean of 69 years of age. 1 in 3 respondents in the sample are working for pay. On average, people in the sample have approximately 13 years of education completed. 6 in 10 respondents are women. Slightly over half of the sample is composed of people married or living with a partner. 7 in 10 persons are declaring an excellent hearing capability (see Table 1).

The coefficients for goodness of fit for each model show moderate levels of variance explained by the models or percentages of incorrect predictions, but with coefficients

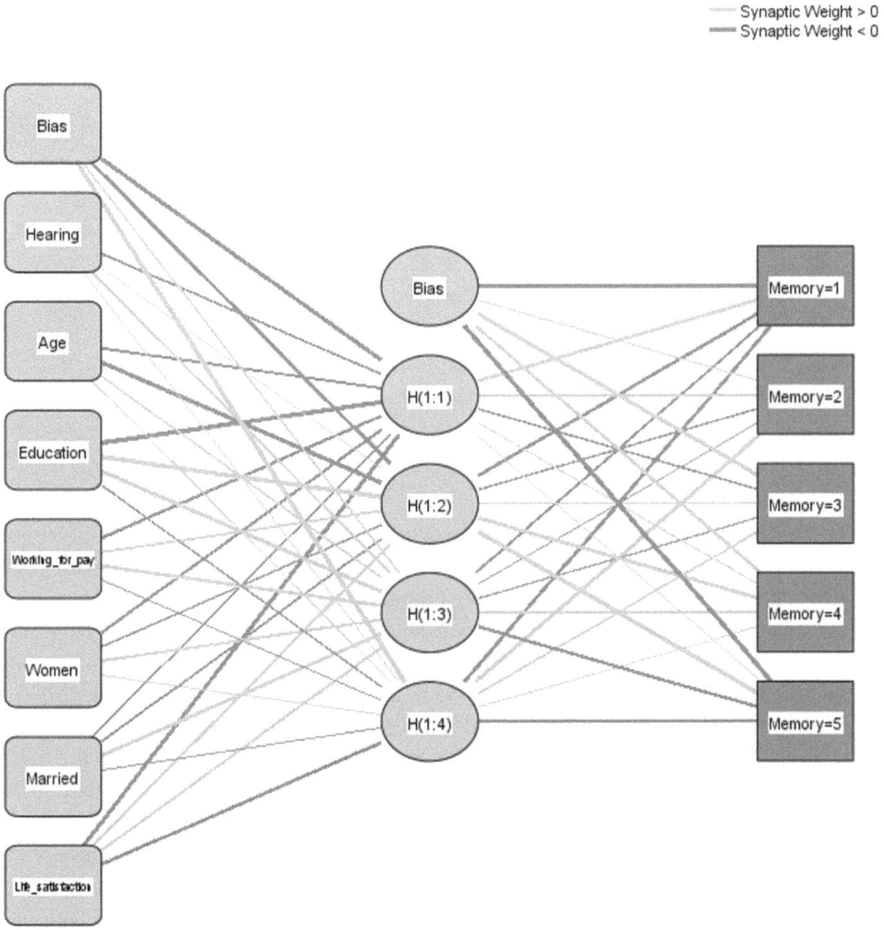

Fig. 2. Model for the neural network predicting memory.

showing the connection between variables similar across models, e.g. hearing or education being connected with memory, with statistically significant coefficient ($p < 0.001$). Ordinal regression obtained a Cox and Snell R-Square of 0.108, Linear regression an R-Square coefficient of 0.102, Neural networks had a 56.1% Incorrect Predictions in testing, Bayesian linear regression had an R-square of 0.102, the structural equation model had an R-square for memory of 0.087, whereas the Bayesian structural equation model had a Posterior predictive p of 0.5 at an acceptance rate of 0.289 (see Table 2).

Memory is predicted by hearing, age, education, age, work status, marital status and life satisfaction at a 95% level of confidence (see Table 3).

Except for the connection with gender, memory assessment is predicted by all the variables in a similar way in the linear regression model (see Table 4).

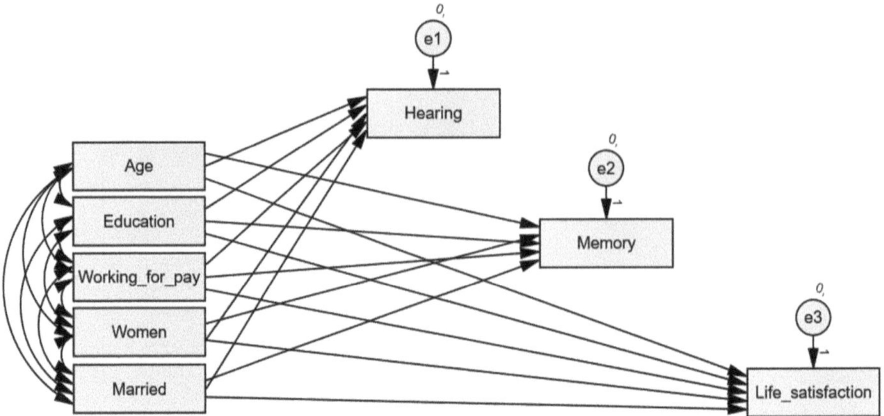

Fig. 3. Structural equation model predicting memory.

Evaluating the importance that each variable has for cognition in the analysed sample, it can be noticed that the lowest importance is from gender as well, but with lower importance for marital status and working categories as well (see Table 5).

Except gender, as well, all the other variables have a credible interval at 95% level that do not cross 0 and we can consider that our variables are connected to memory either positively, or negatively, in a similar way as in the previous analyses (see Table 6).

Structural equation modelling provides the possibility to create a path diagram that takes into account a causal relationship between variables, as allowing for having multiple outcomes and at the same time predicting each of them. It can be noticed that, in this extended model, the importance of marital status for memory assessment is no longer statistically significant, above and beyond the lack of connection between gender and memory (see Table 7).

Creating a factor that considers some of the predictors as part of a concept measured as intersectional disadvantage (ID) allows for measuring the connections between the concept and the outcome, whereas estimating with a Bayesian approach allows for creating distributions of probabilities instead of estimating only means for the connections included. It can be noticed that We can noticed that intersectional disadvantage is connected to memory assessments, as well as being composed of the other metrics included (see Table 8).

Table 1. Descriptive statistics for the analysed data.

Feature	% (Mean)
Memory	
1. Poor	4.4%
2. Fair	24.3%
3. Good	42.2%
4. Very good	24.2%
5. Excellent	4.9%
(N)	14667
Life satisfaction	
1. Not at all satisfied	1.0%
2. Not very satisfied	3.8%
3. Somewhat satisfied	27.9%
4. Very satisfied	45.2%
5. Completely satisfied	22.2%
(N)	14487
Age	69
(N)	15408
Working for pay	
0. No	64.5%
1. Yes	35.5%
(N)	15303
Years of education	13
(N)	15179
Gender	
0. Men	41%
1. Women	59%
(N)	15404
Married	
0. Not married	47.1%
1. Married	52.9%
(N)	15408
Hearing	
1. Poor	0.0%

(*continued*)

Table 1. (continued)

Feature	% (Mean)
2. Fair	1.2%
3. Good	4.4%
4. Very good	20.9%
5. Excellent	73.5%
(N)	13374

Table 2. Comparison between algorithms.

Algorithm and goodness of fit measure	Value of the goodness of fit measure
Ordinal regression (Cox and Snell R-Square)	0.108
Linear regression (R-Square)	0.102
Neural networks (Percent Incorrect Predictions - Testing)	56.1%
Bayesian linear regression (R-square)	0.102
Structural equation modeling (R-square Memory)	0.087
Bayesian structural equation modeling (Posterior predictive p; Acceptance rate)	0.5; 0.289

Table 3. Coefficients for the ordinal regression model.

	Estimate	Sig
[Memory = 1]	−0.051	0.817
[Memory = 2]	2.266	<.001
[Memory = 3]	4.236	<.001
[Memory = 4]	6.352	<.001
Hearing	0.164	<.001
Age	−0.013	<.001
Education	0.13	<.001
Working_for_pay	0.482	<.001
Women	−0.002	0.95
Married	−0.072	0.039
Life_satisfaction	0.372	<.001

Table 4. Coefficients for the linear regression model.

	B	Sig
(Constant)	1.61	<.001
Hearing	0.073	<.001
Age	−0.006	<.001
Education	0.056	<.001
Working_for_pay	0.228	<.001
Women	−0.006	0.722
Married	−0.043	0.01
Life_satisfaction	0.175	<.001

Table 5. Coefficients for the Neural network model.

	Normalized Importance
Hearing	45.0%
Age	61.9%
Education	100.0%
Working_for_pay	26.2%
Women	17.6%
Married	18.0%
Life_satisfaction	62.6%

Table 6. Coefficients for the Bayesian regression model.

		95% Credible Interval	
	Mean	Lower Bound	Upper Bound
(Intercept)	1.61	1.408	1.812
Hearing	0.073	0.047	0.099
Age	−0.006	−0.008	−0.005
Education	0.056	0.051	0.061
Working_for_pay	0.228	0.191	0.265
Women	−0.006	−0.038	0.027
Married	−0.043	−0.075	−0.01
Life_satisfaction	0.175	0.156	0.193

Table 7. Coefficients for the structural equation model.

			Estimate	Sig
Hearing	<	Age	−0.009	<.001
Hearing	<	Education	0.011	<.001
Hearing	<	Working_for_pay	0.04	0.002
Hearing	<	Women	0.041	<.001
Hearing	<	Married	0.044	<.001
Memory	<	Age	−0.005	<.001
Memory	<	Education	0.061	<.001
Memory	<	Working_for_pay	0.242	<.001
Memory	<	Women	0	0.997
Memory	<	Married	0.013	0.382
Memory	<	Hearing	0.078	<.001
Life_satisfaction	<	Age	0.01	<.001
Life_satisfaction	<	Education	−0.009	<.001
Life_satisfaction	<	Working_for_pay	0.082	<.001
Life_satisfaction	<	Women	0.021	0.136
Life_satisfaction	<	Married	0.304	<.001
Life_satisfaction	<	Memory	0.158	<.001

Table 8. Coefficients for the Bayesian SEM.

	Mean	95% Lower bound	95% Upper bound
ID <− Age	−0.057	−0.104	−0.032
ID <− Education	0.381	0.257	0.614
ID <− Working_for_pay	1.662	1.133	2.602
ID <− Women	0.258	0.069	0.483
Hearing <− ID	0.053	0.033	0.073
Memory <− ID	0.141	0.087	0.192
Life_satisfaction <− ID	0.035	0.011	0.064

5   Conclusions

With a large dataset for health and social sciences, the current analysed sample showed moderate explainability of the predicted outcome independently of the algorithm used, but with documented connected variables as predictors at a high level of statistical significance. Theoretical documentation of the features included in the model is also relevant given in the construction of the models, as well as the causal path considered. Available samples for data on social and health issues may be restricted to a limited short list of predictors and limited data availability especially when analysing in deeper detail features at individual level. Accepting for analysis lower levels of explained variance in the outcome feature and taking into account the magnitude of coefficients showing connections between variables are aspects that need to be taken into account for creating artificial intelligence models using data from social and health sciences. A collaboration between structural equation modelling and neural networks in analysing data can be also considered for future research [14].

Considering structural equation modelling as a framework, in line with other machine learning algorithm or adding it to the other algorithms may allow for a better understanding of the models that are trained and for creating explainable machine learning algorithms [15]. Above and beyond the data and the aims for prediction and data mining and analysis, the stories that lay behind the data may be important in choosing how to model it and what kind of algorithms to test.

To predict an ordinal outcome, i.e. memory assessments based on data from the United States Health and Retirement Study, a nationally representative sample with people over 50 years of age, several algorithms were tested and compared assuming that each of them is better for either accounting for the level of measurement of the data or for the theoretical relationships considered. To what extent do the results show improvements in the models as the algorithms are considered better and more sophisticated from different perspectives? Results showed that there are only small differences between models, whereas the story behind the data and both social and health theoretical aspects and technical aspects related to algorithms construction, either frequentist or Bayesian aspects, may result important in the model construction. The usage of a better description of the connections between the data through structural equation modelling and neural networks assessments may be a step forward toward explainable machine learning.

6 Disclosure of Interests.

No competing interests.

Acknowledgments. Data used for processing was provided by the Health and Retirement Study, HRS - RAND Study within the University of Michigan, grant numbers NIA U01AG009740 and NIA R01AG073289 [16]. The data can be downloaded from the HRS website after registering.

References

1. Ilinca, C.: Digital Skills Nowadays. Lambert Academic Publishing, Beau Bassin (2020)
2. Pop, I.-D.: Prediction in pre-university education system using machine learning methods. In: Proceedings of 16th International Conference Agents on Artificial Intelligent (ICAART 2024), vol. 3, pp. 430–437 (2024)
3. IBM: Ordinal Regression. https://www.ibm.com/docs/en/spss-statistics/31.0.0?topic= features-ordinal-regression#idh_plum. Accessed 12 Sept 2025

4. Dubrow, J.K., Ilinca, C.: Quantitative approaches to intersectionality research: new methodological directions and implications for policy analysis. In: Hankivsky, O., Jordan-Zachary, J. (eds.) The Palgrave Handbook of Intersectionality in Public Policy. The Politics of Intersectionality. Palgrave, Cham (2019)
5. Tufiş, P.A.: Status Attainment: Predictable Patterns or Trendless Fluctuation? Iaşi: Institutul European (2012)
6. Byrne, B.M.: The AMOS approach to analysis of categorical variables. In: Structural Equation Modeling with AMOS: Basic Concepts, Applications, and Programming, pp. 151–160. Routledge (Tayler & Francis Group), New York and London (2010)
7. Hair, J.F., Hult, G.T.M., Ringle, C.M., Sarstedt, M., Danks, N.P., Ray, S.: An introduction to structural equation modeling. In: Partial Least Squares Structural Equation Modeling (PLS-SEM) Using R, pp. 1–29 (2021)
8. Lee, S.-Y.: Structural Equation Modeling: A Bayesian Approach. Wiley, Padstow (2007)
9. Palomo, J., Dunson, D.B., Bollen, K.: Bayesian structural equation modeling. Handb. Latent Var. Relat. Model. 163–188 (2007)
10. Koehrsen, W.: Introduction to Bayesian Linear Regression: An explanation of the Bayesian approach to linear modeling. https://towardsdatascience.com/introduction-to-bayesian-linear-regression-e66e60791ea7/. Accessed 12 Sept 2025
11. Cutler, S.J., Ilinca, C.: On the relationship between hearing and cognitive limitations: evidence from 51 countries. J. Aging Health 32(10), 1309–1315 (2020)
12. Ilinca, C., Cutler, S.J.: On the relationship between self-rated hearing and memory in Romania. Soc. Work Rev. 19(2), 1–5 (2020)
13. Ilinca, C., Cutler, S.J.: Romanian disability statistics: the relationship between self-rated hearing and memory. In: The Social Work International Conference, Bucharest (2019)
14. Albahri, A.S., Alnoor, A., Zaidan, A.A., et al.: Hybrid artificial neural network and structural equation modelling techniques: a survey. Complex Intell. Syst. 8, 1781–1801 (2022)
15. Li, J., Sawaragi, T., Horiguchi, Y.: Introduce structural equation modelling to machine learning problems for building an explainable and persuasive model. SICE J. Control Meas. Syst. Integrat. 14(2), 67–79 (2021)
16. RAND HRS Longitudinal File 2020 (V2). Produced by the RAND Center for the Study of Aging, with funding from the National Institute on Aging and the Social Security Administration. Santa Monica, CA (May) (2024)

Unsupervised Hierarchical Growing Neural Architecture for Sensorimotor Map Learning

Abu Eyo Abu$^{(\boxtimes)}$ and Andrew Starkey

University of Aberdeen, Aberdeen, Scotland, UK
`a.abu.20@abdn.ac.uk`

Abstract. The ability to build cognitive maps of unknown environments in a continuous, unsupervised manner is an important capability for autonomous agents. Deep-reinforcement neural networks, despite demonstrating impressive capabilities across diverse domains, fail to rival mammalian proficiency in this critical navigation task due to their lack of explainability, sample inefficiency, and limited capacity to generalize to new environments. This paper presents a Modular and Incremental Network with Enhanced Representation and Vertical Abstraction (MINERVA), a bioinspired and explainable architecture for sensorimotor map learning that aims to extend the capabilities of growing neural architectures by incorporating principles observed in mammalian spatial cognition, including distributed and hierarchical processing of inputs and sparse coding mechanisms. The algorithm is compared with the Temporospatial Merge Grow When Required (TMGWR) network, which was previously demonstrated in a maze navigation context to be superior to algorithms such as growing neural gas (GNG), Grow When Required (GWR) and time GNG (TGNG) in terms of disambiguation performance, sensorial representation accuracy, and sensorimotor-link error. From the experiments conducted, MINERVA demonstrated more robust performance in these metrics with better multi-sensorial processing capabilities, which can be leveraged in solving more complex challenges in more difficult sensorial environments.

Keywords: Autonomous AI · Hierarchical Processing · Bioinspired AI · Binary Encoding · Sensorimotor Mapping · Navigation · Explainability · Cognitive Maps

1 Introduction

Deep learning architectures, despite their success in various aspects of robotics, face significant limitations in their application in autonomous navigation systems, including excessive data requirements [1], lack of explainability, and poor online adaptation and generalization capabilities [2]. This "black-box" nature is particularly problematic in safety-critical domains, where understanding the

F. Marcelloni et al. (Eds.): IJCCI 2025, CCIS 2829, pp. 339–355, 2026.
https://doi.org/10.1007/978-3-032-15638-9_20

rationale behind decisions is essential. In contrast, self-organising map-based approaches like TMGWR have been shown to provide inherent explainability while maintaining competitive performance [3]. Furthermore, Reinforcement Learning (RL) architectures, which are typically combined with deep learning in navigation tasks, conventionally assume Markovian environments; however, partial observability or sensor inacurracies frequently violate this assumption in the real world, leading to inaccuracies in prediction and model performance [4].

The field of autonomous robotics has seen significant growth in recent years, driven by an increasing demand for systems capable of independent operation in complex environments. This increased demand for autonomy has been in response to scenarios that require quick and data-rich decision making, such as navigation of hazardous environments, self-driving [5], disaster relief [6], and planetary exploration [7].

Although maintaining an internal model of the environment is crucial for intelligent behavior in agents [8], current Simultaneous Localization and Mapping (SLAM) algorithms, despite their advancement, still face limitations in achieving true autonomy, such as real-time processing limitations and challenges with environmental dynamics such as environmental variability. The biological solution found in the mammalian brain, specifically the hippocampal-entorhinal system (H/E-S), offers an alternative through cognitive mapping. In particular, the sparse coding observed in this system, where information is encoded by the activity of a small subset of neurons, offers some insights for the development of truly autonomous navigation systems [9].

Early approaches to the development of cognitive maps, such as [2,10,11] based on the Kohenen self-organizing map [12] focused on the addressing the need to decide *a priori*, the number of nodes required to map an input distribution, which was the key challenge with the Kohenen network, especially when trying to apply it in scenarios where the distribution of observations was unknown. Researchers have also developed several unsupervised learning techniques based on progress from previous work to address challenges, such as the coupling of sensory prototypes with motor signals and the disambiguation of aliased states in these cognitive maps [13,14]. However, TMGWR and similar monolithic architectures suffer from two critical limitations: (1) they couple all sensory dimensions during processing, leaving room for the possibility of noise in one dimension affecting all others, and (2) they struggle with high-dimensional sensory inputs, as demonstrated by experiments presented in later sections of this work.

The technique proposed in this paper follows a different approach from conventional cognitive maps. Just as the H/E-S employs distinct neural populations for different aspects of spatial processing [15], this technique separates inputs into specialized streams for processing before integration, giving it a hierarchical structure. More importantly, it adopts the principle of sparse coding by keeping track of binary firing patterns, where only a small subset of nodes are active for any given input, mirroring the efficient coding strategies observed in biological neural circuits [16–18]. Recent work has further demonstrated how these sparse representations support flexible cognition and navigation beyond purely spatial tasks [19, 20].

Building upon Tolman's argument that cognitive maps are necessary for complex inferences from sparse observations [12], our approach decouples raw sensory processing from higher-level cognitive learning. This separation mirrors the biological organization where sensory inputs undergo multiple stages of processing before integration into cognitive maps [7], potentially allowing stability even when some sensory channels are temporarily unavailable or degraded.

The main contribution of this paper is a hierarchical architecture for building explainable representations. The algorithm described in this work is a multilevel architecture, where each level's decisions can be interpreted and traced, while maintaining a dichotomy between raw sensory processing and cognitive map building. In addition, by processing sensory dimensions independently before integration, this approach builds in noise robustness through dimensional separation and provides for scalability to multisensorial input scenarios.

In the context of autonomous navigation systems, explainability refers to an algorithm's ability to elucidate its internal decision-making mechanisms, including explanations of what changes occurred in the environment, how these changes were detected and processed, and the resulting navigation decisions [3]. This capacity goes beyond mere performance metrics to include real-time articulation of decision-making processes, enabling robots to communicate their reasoning to human operators for validation and trust building. MINERVA advances this explainability paradigm by providing multilayer transparency: at the sensory layer through inherently interpretable SOM prototypes, at the cognitive layer through binary firing patterns that serve as symbolic addresses to the sensory input, and at the decision layer through traceable connections between sensorimotor states and actions. This inherently transparent learning model has been adopted as opposed to post-hoc processing which adds computational overhead to the process of training and deploying a learning algorithm.

2   Related Work

Growing neural architectures address a fundamental challenge in autonomous spatial learning, which is the need to adapt structure based on environmental complexity without predefined constraints. As highlighted by Ezenkwu and Starkey [14], traditional SOMs have significant limitations in their applicability in the development of autonomous agents, particularly their fixed network size that requires prior knowledge of the input distribution. In response, numerous authors have developed variants of the self-organising map, such as Growing Neural Gas (GNG) [9], Grow When Required (GWR) [10], and Grow Cell Structure (GCS) [11] that can dynamically adjust their structure and size based on input data. Parisi et al. [20] demonstrated that GWR networks can effectively support lifelong learning while avoiding catastrophic forgetting, a critical feature for systems operating in dynamic environments. The TMGWR (Temporospatial Merge Grow When Required) algorithm [14] extends these capabilities by combining the structural flexibility of GWR with temporal context processing, enabling more robust operation in noisy and partially observable environments.

Hierarchical processing has emerged as a critical approach for efficient spatial representation. Hawkins' Thousand Brains Theory [22] proposes that the neocortex builds multiple models of objects and concepts through parallel processing pathways rather than constructing a single unified model. Kreiser et al. [23] implemented a neuromorphic architecture that models the role of hippocampal formation in spatial mapping, demonstrating how specialized neural circuits can efficiently represent space. Their work draws direct inspiration from mammalian neural organization to create computationally efficient navigational systems.

A key distinction among spatial representation algorithms lies in their approach to temporal context. In addition, they differ in how they incorporate sensorimotor relationships. Toussaint [12] pioneered the concept of sensorimotor maps that couple sensory prototypes with motor signals, while Butz et al. [13] introduced Time Growing Neural Gas (TGNG), which creates directed links between nodes based on sensorimotor proximity rather than just sensory proximity and TMGWR represents an advancement in this area by incorporating both temporal context and sensorimotor relationships.

Adam et al. [15] proposed Generalized Simultaneous Localization and Mapping (G-SLAM) as a unification framework connecting natural and artificial intelligence approaches to spatial mapping. Their work draws parallels between computational SLAM algorithms and the neural mechanisms observed in the hippocampal and entorhinal cortex. This connection between engineering approaches and biological implementations has provided valuable insights for our own approach, particularly regarding the integration of separate spatial encoding networks. The principle of maintaining multiple partial models that can operate independently when faced with sensory degradation aligns closely with the mammalian brain's strategy for robust spatial navigation.

The architecture proposed in this paper builds on these foundations while making additional contributions. First, it implements a fully hierarchical processing approach in which spatial dimensions are encoded separately before integration, allowing greater robustness when faced with partial information. Second, it employs binary firing patterns that create natural discretization of space, providing a form of implicit filtering mechanism. The results presented in this paper align with and extend previous observations about the benefits of incorporating temporal context and adaptive structure in spatial representation algorithms [14].

Previous research [14] compared TMGWR with GNG, TGNG and GWR, and demonstrated that TMGWR is superior in terms of sensorial and sensorimotor representation in noisy environments, maintaining consistent performance with increasing sensor noise. For this reason, in this work, we compared MINERVA with the TMGWR benchmark.

3 Modular and Incremental Network with Enhanced Representation and Vertical Abstraction (MINERVA)

MINERVA extends traditional growing neural architectures by introducing a hierarchical processing structure inspired by biological neural systems. The algorithm consists of three major components, which can be expanded at individual hierarchical levels or incremented by layers: lower-level networks specialized for processing individual sensory inputs or dimensions of sensory inputs; higher-level integration network(s) that learn from the lower-level neurons instead of the raw sensory inputs and mechanisms for bidirectional information flow between different levels of the hierarchy, allowing for the mapping of higher-level neurons to actual sensory input resulting in explainability. The central idea of MINERVA is the decoupling of raw sensory inputs from the sensorimotor map by first encoding the input space via specialised lower-level neurons and in turn learning from those as opposed to learning directly from the input space.

In the context of MINERVA, explainability refers to the ability to trace any node in the sensorimotor map back to its original sensory inputs through the binary firing patterns. Unlike TMGWR, where the nodes directly encode continuous sensory values that become entangled during learning, MINERVA maintains a clear separation between sensory processing and map building. Each high-level node stores a unique binary pattern (fx, fy) that explicitly identifies which lower-level neurons were fired for that state. This binary encoding acts as an interpretable address that can be decoded to retrieve the original continuous sensory values. The explainability emerges from two properties: (1) reconstruction accuracy - the ability to recover original sensory input from binary patterns, (2) traceability - the hierarchical levels that can be traversed bidirectionally without information loss.

3.1 Lower-Level Processing Networks

The foundation of MINERVA consists of specialised Grow When Required Self-Organizing Maps (GWRSOM), each dedicated to processing a single sensor input. In the simplest case, these inputs can be spatial dimensions (x- and y-coordinate information), resulting in two sets of GWRSOM for each coordinate. Each specialised GWRSOM network maintains its own set of weight vectors that effectively encode the raw input from the environment as shown in Fig. 1. For each input dimension $d \in \{x, y\}$, we define:

1. Node set $N_d = \{n_1, n_2, ..., n_k\}$ where $k \leq N_{max}$
2. Weight vectors $W_d = \{w_1, w_2, ..., w_k\}$ where $\boldsymbol{w}_i \in \mathbb{R}^1$
3. Connection matrix $C_d \in \{0, 1\}^{k \times k}$

The activation function for a node i given input x is defined as:

$$a_d(i, x) = \exp(-\|x - \boldsymbol{w}_i\|^2) \tag{1}$$

Node Creation Criterion in GWRSOM:

$$\text{if } \min_{i}(a_d(i, x)) < \alpha \wedge |N_d| < N_{max} : \begin{cases} N_d \leftarrow N_d \cup \{n_{new}\} \\ \boldsymbol{w}_{new} = x \end{cases} \tag{2}$$

3.2 Bidirectional Information Flow

Having established lower-level processing structure, we now describe how information flows between levels. MINERVA implements bidirectional information flow between levels as follows.

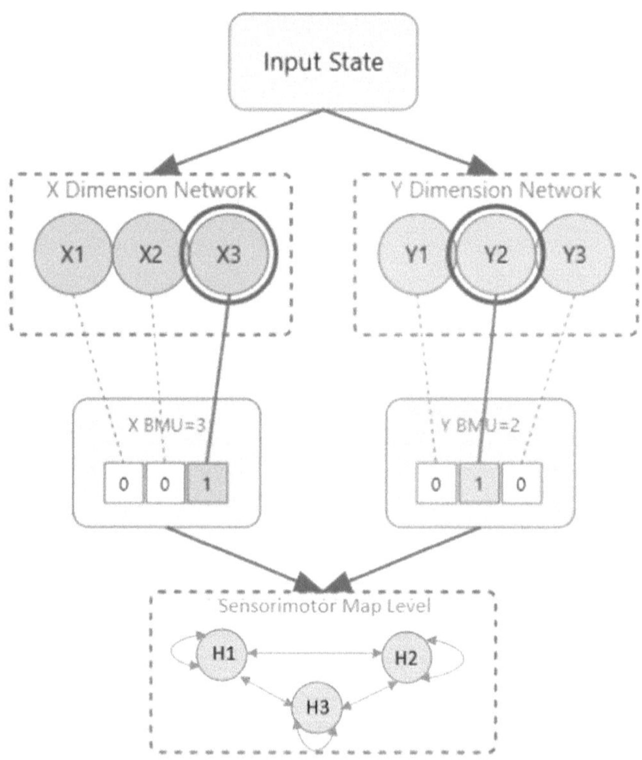

Fig. 1. MINERVA network organization.

Bottom-up encoding for input state $s = (x, y)$:

1. Lower-level networks identify BMU for each coordinate:

$$b_x = \arg\min_{i \in N_x} \|x - \boldsymbol{w}_i\| \tag{3}$$

$$b_y = \arg\min_{i \in N_y} \|y - \boldsymbol{w}_i\| \tag{4}$$

2. The algorithm creates binary firing patterns where only the BMU is active:

$$\mathbf{f}_x[i] = \begin{cases} 1 & \text{if } i = b_x \\ 0 & \text{otherwise} \end{cases} \tag{5}$$

$$\mathbf{f}_y[i] = \begin{cases} 1 & \text{if } i = b_y \\ 0 & \text{otherwise} \end{cases} \tag{6}$$

3. MINERVA combines the firing patterns:

$$\mathbf{f}(s) = (\mathbf{f}_x, \mathbf{f}_y) \tag{7}$$

Input information encoded in BMU firing patterns can be retrieved at any point, making the process explainable.

3.3 Higher-Level Integration Networks

The higher-level network integrates information from the lower-level networks. It uses a firing pattern recognition mechanism to achieve this integration. When a state is processed:

1. Each lower-level network identifies its BMU for that state
2. The combination of BMUs creates a unique firing pattern.
3. Nodes are created in the higher-level network for each unique pattern formed.
4. If a non-unique firing pattern is observed, no node is created.
5. The previous node is connected to the current node, and the connection is encoded as the action that led to that transition.

The higher-level network or sensorimotor map maintains a set of nodes representing states in the agent's world and connections representing temporal relationships in the form of actions between the nodes. Figure 1 shows the organisation of the MINERVA network. Active nodes (solid lines) correspond to 1s in binary representation. Inactive nodes (broken lines) correspond to 0s in binary representation. BMUs highlighted in red generate the binary pattern. The binary pattern feeds into the higher-level network, which is the sensorimotor map layer. Colour-coding indicates the dimensional separation of inputs. The MINERVA algorithm is shown in Algorithm 1.

4 Methods

Both TMGWR and MINERVA were evaluated on the basis of their ability to learn spatial representations of the agent's environment. The setup was similar to that used in an earlier work that compared TMGWR with other growing neural architectures [14].

Although the 2D maze environment setup may appear simplistic compared to the rather complicated scenarios agents are exposed to in the real-world, it remains a useful way of demonstrating some of the gaps of current approaches that need to be closed such as computational scalability, memory management for lifelong learning, and robustness to sensory noise.

4.1 Experimental Setup

The two maze environments used for the experiments included the following: Maze A, a 7×7 grid with minimal obstacles as shown in Fig. 2(A) and Maze B, a 15×15 grid with multiple corridors and dead ends as shown in Fig. 2(B).

In the first experiment, the objective was to demonstrate that MINERVA can produce sensorimotor maps identical to those produced using TMGWR. The inputs were x and y coordinate values or 2D inputs, similar to those used in previous work [14] and the maze environment used for this experiment was Maze B (the more complicated maze). The state space consists of (x, y) coordinates. Four discrete actions are available: up, down, left, and right. If there is no wall in the way of the way of the agent, each action will move the agent by 1 unit in the desired direction. The agent has no prior knowledge of the environment, including wall locations, and does not perform wall detection prior to action selection. When the agent attempts to move into a wall, the collision is handled by the maze environment. The environment keeps the agent in its current position. This allows implicit learning of obstacle locations through failed movement attempts. The algorithm only encodes observations when an action produces a change in the environment.

In the second experiment, the objective was to evaluate the agent's ability to create accurate sensorimotor maps in a higher-dimensional input space. Four beacons were added at the corners of the simulation environment of Fig. 2A, emitting pulsating signals which the agents receive as continuous distance readings [d1, d2, d3, d4]. TMGWR algorithm was modified to handle this 4D sensor input instead of 2D, while additional GWRSOMs were added to MINERVA to make four input streams.

Exploration Strategy. Exploration of the environment was performed with a random walk, allowing the agent to learn sensorimotor maps of the environment by creating state representations in hierarchical networks and developing environmental knowledge by building connections between states based on actions. The random walk step count was kept at 2000 time steps, during which time the agent learned the topology of the maze online.

Noise Modelling. To evaluate the robustness of the algorithms to noise, Gaussian perturbations were used at the sensory inputs $\epsilon \sim \mathcal{N}(0, \sigma^2)$. Gaussian noise represents a reasonable approximation for electronic sensor noise. This approach is widely adopted in robotics literature for mathematical traceability [24, 24]. However, we acknowledge that real-world localisation systems may exhibit more complex error patterns including outliers, systemic biases, and temporal correlations. For the chosen noise variance $\sigma^2 = 1/6$, there is a Signal-to-Noise Ratio (SNR) of 12.6 dB (power) and 23.3 dB (amplitude), which represents moderate sensor uncertainty typical of indoor localisation systems using low-cost IMU sensors. To prevent out-of-bounds measurements, noisy coordinates are clipped to environment boundaries.

Algorithm 1: MINERVA Sensorimotor Map Building Algorithm.

---

procedure PRETRAIN($\mathcal{D}_{maze}$)
    $X \leftarrow \{x : (x, y) \in \mathcal{D}_{maze}\}$ $\triangleright$ Extract all x-coordinates from maze data
    $Y \leftarrow \{y : (x, y) \in \mathcal{D}_{maze}\}$ $\triangleright$ Extract all y-coordinates from maze data
    GWRSOM-Train($X, epochs$), GWRSOM-Train($Y, epochs$) $\triangleright$ Train lower-level
networks using growing topology
end procedure
procedure GETFIRINGPATTERN($\mathbf{s} = (x, y)$)
    $b_x \leftarrow \text{FindBMU}_x(x)$ $\triangleright$ Find best x-coordinate unit
    $b_y \leftarrow \text{FindBMU}_y(y)$ $\triangleright$ Find best y-coordinate unit
    $\mathbf{f}_x \leftarrow \mathbf{0}_{|W_x|}, \mathbf{f}_y \leftarrow \mathbf{0}_{|W_y|}$ $\triangleright$ Create zero binary vectors
    $\mathbf{f}_x[b_x] \leftarrow 1, \mathbf{f}_y[b_y] \leftarrow 1$ $\triangleright$ Activate winning units only
    return $(\mathbf{f}_x, \mathbf{f}_y)$ $\triangleright$ Return binary firing pattern
end procedure
procedure UPDATEMODEL($\mathbf{s}_{next}, a$)
    $\mathbf{f} \leftarrow \text{GetFiringPattern}(\mathbf{s}_{next})$ $\triangleright$ Convert position to binary pattern
    $n \leftarrow \text{FindNodeIndex}(\mathbf{f})$ $\triangleright$ Search for existing pattern node
    if $n = \text{null}$ **then** $\triangleright$ Pattern never seen before
        $\mathcal{N} \leftarrow \mathcal{N} \cup \{(\mathbf{f}, \mathbf{s}_{next})\}$ $\triangleright$ Create new node
        $n \leftarrow |\mathcal{N}| - 1$ $\triangleright$ Get new node's index
        $\mathbf{C} \leftarrow \text{ExpandMatrix}(\mathbf{C})$ $\triangleright$ Expand high-level connection matrix
        $\mathbf{T} \leftarrow \text{ExpandMatrix}(\mathbf{T})$ $\triangleright$ Expand high-level age matrix
    else
        $\mathbf{p}_{old} \leftarrow \mathcal{N}[n].\text{position}$ $\triangleright$ Get stored continuous position
        $\mathcal{N}[n].\text{position} \leftarrow 0.9 \cdot \mathbf{p}_{old} + 0.1 \cdot \mathbf{s}_{next}$ $\triangleright$ Update with moving average
    end if
    if $n_{prev} \neq \text{null}$ **then** $\triangleright$ Previous node exists
        $\mathbf{C}[n_{prev}, n] \leftarrow 1$ $\triangleright$ Create temporal sequence link
        $\mathbf{T}[n_{prev}, n] \leftarrow 0$ $\triangleright$ Set connection age to zero
        $\mathcal{M}[(n_{prev}, n)] \leftarrow a$ $\triangleright$ Store action that caused transition
        $\mathbf{T}[n_{prev}, :] \leftarrow \mathbf{T}[n_{prev}, :] + 1$ $\triangleright$ Age all other connections
        $\mathbf{C}[\mathbf{T} > T_{max}] \leftarrow 0$ $\triangleright$ Remove connections exceeding max age
    end if
    $n_{prev} \leftarrow n$ $\triangleright$ Update previous node for next iteration
end procedure
procedure SELECTACTION($\mathbf{s}$)
    return $a \sim \text{Uniform}(\{0, 1, 2, 3\})$ $\triangleright$ Random exploration for map building
end procedure

Parameters. Both algorithms were configured with comparable parameters, as shown in Table 1:

4.2 Evaluation Metrics

We evaluated the performance of MINERVA using performance metrics that were used in previous experiments between TMGWR and other algorithms and that had previously demonstrated the superiority of TMGWR over other approaches [14].

Table 1. Parameter settings for TMGWR and MINERVA.

Parameter	TMGWR	MINERVA
Learning rates	$\epsilon_b = 0.35$	$\epsilon_b = 0.35$
	$\epsilon_n = 0.15$	$\epsilon_n = 0.15$
Maximum nodes	$N_{max} = 200$	$N_{max} = 200$
Activity threshold	$\delta = 0.79$	$\delta = 0.79$
Maximum edge age	$T_{max} = 20$	Not required
Context factor	$\beta = 0.7$	Not required
Temporal importance	$\alpha = 0.5$	Not required
Temporal parameters	$\tau_B = 0.3$	Not required
	$\kappa = 1.05$	Not required

Fig. 2. Agent Environments: **Maze A**: 7×7 maze **Maze B**: 15×15 maze.

Network Growth. Number of nodes created by each algorithm.

Sensorial Representation Metric (Purity). A measure of the ability of each node in the sensorimotor map to represent a known state in the agent's world.

$$\text{purity} = \frac{\sum_{n \in \mathcal{N}} \max_{s \in \mathcal{S}} |n \cap s|}{M} X 100\% \tag{8}$$

$\mathcal{N}$ is the set of nodes that make up the sensorimotor map, $\mathcal{S}$ is the set of world states in the agent's environment, while M is the total number of observations made by the agent while learning in the environment.

Sensorimotor Representation Metric. The Sensorimotor-Link Error (SE), which is the ratio of the number of impossible transitions to the total number of transitions in the map is calculated as follows:

$$\text{SE} = \frac{\sum_{i=1}^{|E|} \mathcal{I}\left\{E^i\left[\boldsymbol{w}^{\{1\}}{}_{t-1}, \boldsymbol{w}^{\{1\}}{}_t\right] \notin \mathcal{H}\right\} + \tau}{|E| + \tau} \tag{9}$$

where E is a set of connections learned by the map; $|E|$ is the number of transitions in E, $\mathcal{I}\{.\}$ is an indicator function that returns 1 when the argument evaluates true or 0 otherwise, $\mathcal{H}$ is a set of all possible transitions, while τ (set at 0.0001) is a sufficiently small positive number that prevents division by zero if no transition has been learned.

4.3   Results

Sensorimotor Map Building. In the simple maze environment, MINERVA, like TMGWR, was able to create a sensorimotor map that represents the environment with precision, as shown in Appendix A.

Quantifying the Goodness of the Map from 2D Inputs. The goodness of the sensorimotor maps due to TMGWR and MINERVA using 2D input in Maze B (the more complicated maze) was evaluated based on the purity and sensorimotor link error metrics discussed in Sect. 5.2. For each noise level, the agent builds a map, using both algorithms, while exploring the maze. The sensory and sensorimotor properties of the maps are evaluated. Figure 3 presents the percentage accuracy of the observations identified by the agent in the form of nodes.

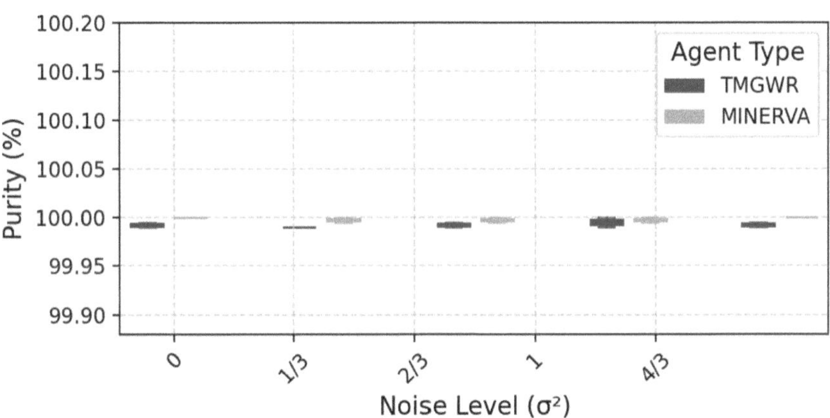

Fig. 3. Percentage accuracies of the algorithms at different noise levels in the environment of Fig. 2(b).

The sensorimotor link error, which is a measure of the number of impossible transitions to the total number of transitions learned by the map, was also tracked, and MINERVA showed better performance compared to TMGWR, as shown in Fig. 4.

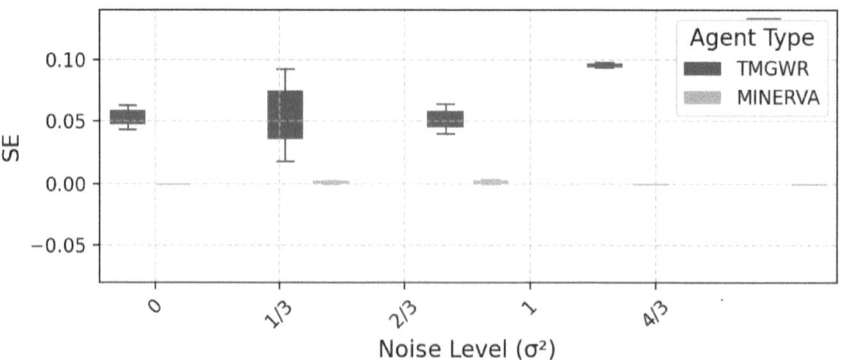

Fig. 4. SEs of the algorithms at different noise levels in the environment of Fig. 2(b).

Significance Test. To validate the statistical significance of these improvements, Tukey's Honestly Significant Difference (HSD) analysis shown in Table 2 confirmed MINERVA'smmm superiority over the benchmark on both sensorimotor error ($0.008537 > 0.005435$ threshold) and purity $0.1487\% > 0.1363\%$ threshold), establishing statistical evidence for algorithm performance differences. The temporal nature of TMGWR which explicitly gives it a memory of transitions between observations, ensured that incorrect connections are removed from the network; however, this pruning happens too late sometimes as TMGWR prunes incorrect connections only after they exceed T_{max} age. Wrong connections can hence influence learning and navigation for many timesteps before being removed, during which time the agent can learn suboptimal policies that rely on these wrong connections, creating patterns that persist even after pruning. MINERVA on the other hand had even lower errors because of its immediate correctness imposed by the binary pattern matching and uniqueness check before nodes are added to the sensorimotor map, ensuring that connections that are included in the map between these nodes are topologically correct from the onset.

Table 2. Tukey HSD Results: Algorithm Performance Comparison.

Metric	TMGWR	MINERVA	Mean	HSD	Result
	Mean	Mean	Difference	Threshold	
Sensorimotor Error	0.009513	0.000975	0.008537	0.005435	**Significant**
Purity (%)	99.8375	99.9862	0.1487	0.1363	**Significant**

Sample Size: n = 40 per algorithm (80 total observations)
Test: Tukey HSD with $k = 2\,groups, \alpha = 0.05, q = 2.83$
Performance Winner: MINERVA significantly superior on both metrics
Effect Direction: Lower SE and higher purity indicate better performance

Quantifying the Goodness of the Map from Multisensorial 4D Inputs.
The goodness of the sensorimotor maps due to TMGWR and MINERVA using
4D input in Maze A was evaluated based on purity and sensorimotor link error
metrics. At each noise level, the purity of the map due to each of the algorithms
is calculated and reported in Fig. 5. The results show that, while MINERVA
consistently maps the environment correctly over different noise levels, TMGWR
only gets the right number of nodes in the noiseless case and the node count
explodes when noise is added to the multisensorial input as the algorithm ends
up creating duplicate nodes as it categorises slightly different sensor readings as
completely new states.

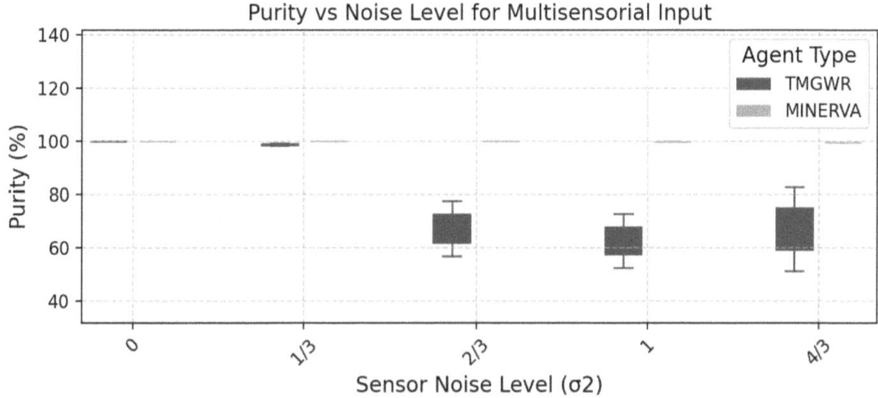

Fig. 5. Purity vs noise level for multisensorial input in the environment of Fig. 2(a).

The sensorimotor link error is also tracked and reported in Fig. 6. The results
show that TMGWR fails catastrophically with noise in the higher sensor space
as noise is added. It is important to note that the formation of spurious nodes by
TMGWR could be handled by adjusting the growth parameter, whereas MIN-
ERVA by design, is better suited to handle multisensorial input as the algorithm
clearly delineates between sensor processing in the lower layers and sensorimotor
map building in the higher layer of the network.

Significance Test. To validate the statistical significance of these improvements, Tukey's HSD analysis was conducted. Tukey's HSD analysis in Table 3
confirmed MINERVA's significant superiority over TMGWR in the 4D sensor
space. The analysis compared TMGWR and MINERVA performance across two
metrics using conservative multiple comparison testing ($k = 2$ groups, $\alpha = 0.05$,
$q = 2.83$). Tukey HSD analysis established MINERVA's superior performance
over TMGWR across all performance dimensions (SE: $0.288800 > 0.091888$
threshold, Purity: $20.7267\% > 12.7513\%$ threshold), demonstrating algorithm
robustness to sensor ambiguity and multisensorial input challenges.

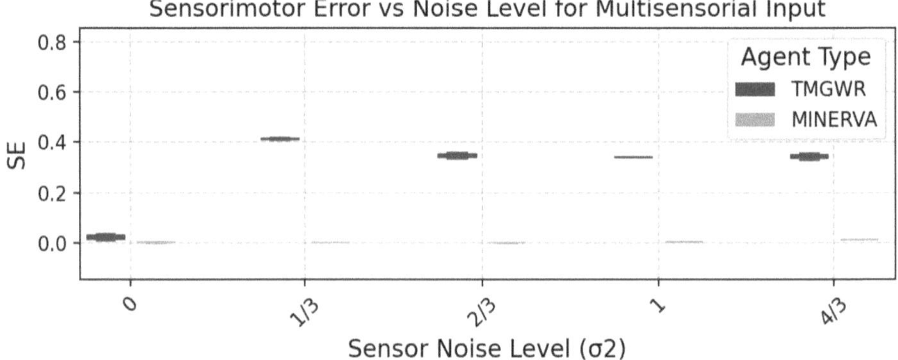

Fig. 6. Sensorimotor Link Error vs noise level for multisensorial input in the environment of Fig. 2(a).

Table 3. Comprehensive Tukey HSD Results: Multisensory (4D input) Algorithm Performance Comparison.

Metric	TMGWR Mean	MINERVA Mean	Mean Difference	HSD Threshold	Result
Sensorimotor Error	0.293400	0.004600	0.288800	0.091888	**Significant**
Purity (%)	79.1333	99.8600	20.7267	12.7513	**Significant**

Sample Size: n = 40 per algorithm (80 total observations)
Test: Tukey HSD with $k = 2$ groups, $\alpha = 0.05$, $q = 2.83$
Performance Winner: MINERVA significantly superior on all metrics
Environment: Multi-sensor range-based 4D input
Effect Direction: Lower SE and higher purity indicate better performance

4.4   Conclusion

This paper presents MINERVA, a hierarchical and explainable framework that advances beyond traditional monolithic architectures for sensorimotor map learning of unknown, noisy, or partially observable environments. MINERVA's hierarchical structure enables traceability from high-level decisions to sensory input through binary firing patterns. The algorithm has been benchmarked with TMGWR on the basis of sensorial and sensorimotor representation accuracies, as well as handling of multi-sensory inputs and explainability through reconstruction accuracy. Our findings demonstrate that MINERVA's bioinspired hierarchical processing can give more robust sensorimotor maps than those formed by monolithic networks like TMGWR. MINERVA's separation of input processing and sensorimotor map building also provides inherent filtering of noise.

Further statistical testing is required to demonstrate the individual contributions of components in the hierarchical structure. The algorithm is yet to be tested in a real-world setting and current implementation can only handle static environments with noisy observations, but this could be enhanced for real-world environments with dynamic elements or changing topologies. Promising research directions include exploiting the multisensorial capabilities of MINERVA in feature selection based on task relevance.

Acknowledgments. This study was funded by the Petroleum Technology Development Fund (PTDF) (Grant Number: 19PHD164).

A   Sensorimotor Maps

See Fig. 7.

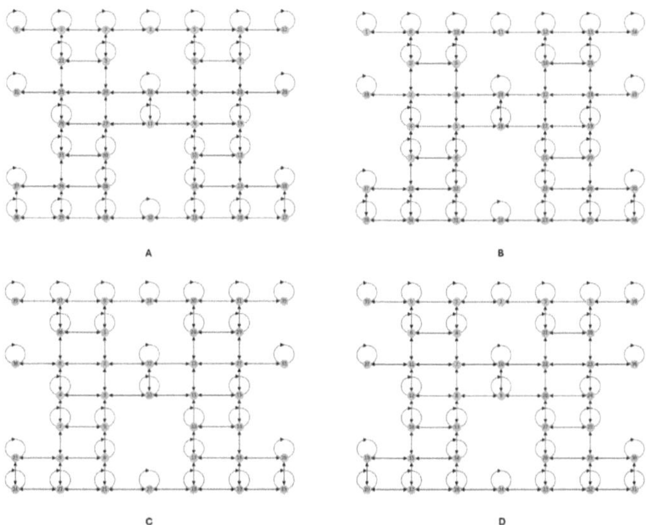

Fig. 7. Sensorimotor maps: (A) MINERVA at $\sigma^2 = 0$: Number of nodes $= 40$ (B) MINERVA at $\sigma^2 = \frac{1}{6}$: Number of nodes $= 40$ (C) TMGWR at $\sigma^2 = 0$: Number of nodes $= 40$ (D) TMGWR at $\sigma^2 = \frac{1}{6}$: Number of nodes $- 40$

References

1. Zhuang, Y.: Progress and challenges in applying deep reinforcement learning to intelligent navigation. Int. J. Comput. Sci. Inf. Technol. **3**(3), 173–185 (2024). https://doi.org/10.62051/ijcsit.v3n3.17
2. Fritzke, B.: Growing cell structures–a self-organizing network for unsupervised and supervised learning. Neural Netw. **7**(9), 1441–1460 (1994)
3. Starkey, A., Ezenkwu, C.P.: Towards autonomous developmental artificial intelligence: case study for explainable AI. In: IFIP International Conference on Artificial Intelligence Applications and Innovations, Switzerland, pp. 94–105 (2023)

4. Wisniewski, M., Chatzithanos, P., Guo, W., Tsourdos, A.: Benchmarking nDeep Reinforcement Learning for Navigation in Denied Sensor Environments, October 2024. http://arxiv.org/abs/2410.14616arXiv:2410.14616 [cs]
5. Ma, Y., Wang, Z., Yang, H., Yang, L.: Artificial intelligence applications in the development of autonomous vehicles: a survey. IEEE/CAA J. Automatica Sinica **7**(2), 315–329 (2020)
6. Ebadi, K., Palieri, M., Wood, S., Padgett, C., Agha-mohammadi, A.-A.: DARE-SLAM: degeneracy-aware and resilient loop closing in perceptually-degraded environments. J. Intell. Robot. Syst. **102**(1), 2 (2021)
7. Agha, A., Mitchell, K.L., Boston, P.J.: Robotic Exploration of Planetary Subsurface Voids in Search for Life, P41C-3463, vol. 2019, December 2019
8. George, D., Rikhye, R.V., Gothoskar, N., Guntupalli, J.S., Dedieu, A., Lázaro-Gredilla, M.: Clone-structured graph representations enable flexible learning and vicarious evaluation of cognitive maps. Nat. Commun. Nature Publishing Group **12**(1), 2392 (2021)
9. Moser, E.I., Moser, M.B., McNaughton, B.L.: Spatial representation in the hippocampal formation: a history. Nat. Neurosci. **17**(11), 1448–1464 (2017)
10. Marsland, S., Shapiro, J., Nehmzow, U.: A self-organising network that grows when required. Neural Netw. **15**(8), 1041–1058 (2002)
11. Fritzke, B.: Growing cell structures—a self-organizing network for unsupervised and supervised learning. Neural Netw. **7**(9), 1441–1460 (1994)
12. Kohonen, T.: The self-organizing map. Proc. IEEE **78**(9), 1464–1480 (1990)
13. Carvalho, L., Starkey, A.: Unsupervised neural architecture for sensorimotor mapping in perceptually aliased environments. In: 2024 IEEE International Conference on Omni-layer Intelligent Systems (COINS), London, United Kingdom, pp. 1–6 (2024). https://doi.org/10.1109/COINS61597.2024.10622113
14. Ezenkwu, C.P., Starkey, A.: Unsupervised temporospatial neural architecture for sensorimotor map learning. IEEE Trans. Cogn. Dev. Syst. **13**(1), 223–230 (2021). https://doi.org/10.1109/TCDS.2019.2934643
15. Moser, E.I., Kropff, E., Moser, M.B.: Place cells, grid cells, and the brain's spatial representation system. Annu. Rev. Neurosci. **31**, 69–89 (2008)
16. Barlow, H.B.: Possible principles underlying the transformation of sensory messages. Sens. Commun. **1**, 217–234 (1961)
17. Olshausen, B.A., Field, D.J.: Sparse coding of sensory inputs. Curr. Opin. Neurobiol. **14**(4), 481–487 (2004)
18. Richards, B.A., et al.: A deep learning framework for neuroscience. Nat. Neurosci. **22**(11), 1761–1770 (2019)
19. Bellmund, J.L., Gärdenfors, P., Moser, E.I., Doeller, C.F.: Navigating cognition: spatial codes for human thinking. Science **362**(6415), eaat6766 (2018)
20. Grieves, R.M., Jeffery, K.J.: The representation of space in the brain. Behav. Proc. **135**, 113–131 (2017)
21. Parisi, G.I., Tani, J., Weber, C., Wermter, S.: Lifelong learning of spatiotemporal representations with dual-memory recurrent self-organization. Front. Neurorobot. **12** (2018)
22. Hole, K.J., Ahmad, S.: A thousand brains: toward biologically constrained AI. SN Appl. Sci. **3**(8), 1–14 (2021). https://doi.org/10.1007/s42452-021-04715-0
23. Kreiser, A., et al.: A neuromorphic approach to path integration: a head-direction spiking neural network with vision-driven reset. In: 2019 IEEE/RSJ International Conference on Intelligent Robots and Systems (2019)

24. Wisniewski, M., et al.: Benchmarking deep reinforcement learning for navigation in denied sensor environments, 18 October 2024. arxiv.org/abs/2410.14616. Accessed 3 July 2025
25. Mysore, N.: Quantifying First-Order Markov Violations in Noisy Reinforcement Learning: A Causal Discovery Approach. https://arxiv.org/pdf/2503.00206. Accessed 3 July 2025

Rule Extraction from Fake News Classifiers

Fatima Iqbal and Jacob M. Howe[(✉)]

City St George's, University of London, London, UK
`{fatima.iqbal,j.m.howe}@citystgeorges.ac.uk`

Abstract. This study explores the decision-making processes of machine learning models for text classification problems. Using a fake news dataset as a test case, the study compares neural networks and machine learning approaches to fake news detection. This text-based problem has a feature space of 12,569 dimensions, and the necessity of the full dimensionality of the data is explored. In addition to developing effective classifiers, this study aims to investigate neural network interpretability by applying an explainable AI framework to extract human-understandable rules from trained models. The rule extraction process, taking a pedagogical approach, investigates the decision making of models. A Boolean function based model was developed, and the extent to which this rule-based system over a reduced feature set is successful is evaluated.

Keywords: Fake News Detection · Machine Learning · Explainable AI · Natural Language Processing · Rule Extraction

1 Introduction

The increasing application of machine learning across a variety of domains has highlighted the lack of transparency and interpretability of many models. Models, in particular neural networks, can achieve high predictive accuracy, but their black-box nature makes it difficult to understand their decision making and this lack of transparency forms a critical challenge in terms of establishing trust in their predictions [1].

The deliberate spread of misinformation online on social media platforms has been of some concern, particularly since 2016 US presidential elections [2]. Such misinformation can be quickly re-shared, and even if quickly recognised can be hard to mitigate against. This has led to interest across several disciplines in identifying fake news and tackling its spread [22]. Using machine learning to identify social media posts constituting fake news is of current interest and application of techniques is ongoing.

Rule extraction from neural networks is the process of generating clear and logical rules that represent the patterns learned by a trained neural network [3]. According to [8] rule extraction is when a trained neural network's predictions

F. Marcelloni et al. (Eds.): IJCCI 2025, CCIS 2829, pp. 356–376, 2026.
https://doi.org/10.1007/978-3-032-15638-9_21

are made easy to understand whilst still close to the model's actual performance. Rule extraction can help experts cross-check or verify neural network systems [14]. In [18] neural networks providing binary classifiers for a computer security problem, detecting cross-site scripting, were developed where the features were all Boolean valued, hence the neural network described a Boolean function. This makes the neural network an attractive target for rule extraction. The behaviour of this neural network was approximated using a pedagogical approach (querying the neural network, but not examining its internal structure) to learn the Boolean function it represented.

This research studies the decision-making of machine learning models for fake news detection, exploring interpretable techniques to explain black-box models in order to improve transparency. Multiple machine learning models trained on a fake news dataset demonstrate strong performance in discriminating fake from real news, including a Boolean featured classification model. Given that the dataset is small relative to the high-dimensionality of the features (10,700 data points, 12,569 features), this raises the question of what patterns are actually being learned and which elements from the dataset are impacting the model. To answer this, this paper analyses what features are necessary to train classifiers and applies rule extraction methods to trained neural networks. The feature importances in the models and the dataset are reviewed, and the impact of restricting the number of features is investigated. [18] gave an approach to rule extraction from a neural network trained on 62 Boolean features. Inspired by that work, this paper investigates how well the approach to rule extraction works when the feature space is in the thousands.

This paper makes the following contributions:

- Develops machine learning models, in particular neural networks, to provide classifiers for fake news detection on a dataset relating to COVID-19. This includes models using Boolean features. The models are compared.
- Investigates how the feature space impacts the models' decision making.
- Develops a rule extraction framework using the Boolean featured neural network model to generate interpretable rules.

The rest of this paper is structured as follows, Sect. 2 surveys related work on fake news detection and rule extraction. Section 3 covers training of classifiers, then Sect. 4 further considers the dataset and how curtailing the number of features used in training impacts the performance of classifiers. Section 5 gives a methodology for rule extraction and its results. Section 6 concludes.

2   Background and Related Work

This section surveys related work on fake news and its detection, explainable AI and rule extraction, and the overlap between the two.

2.1 Fake News and Fake News Detection

The term "fake news" is complicated and lacks a clear definition, for example [32] gives seven categories of fake news ranging from parody, through misdirection to demonstrably false content. This work follows [2] in that, "A news item or a part of a news item will be considered fake if it can be verified that its content is false." The majority of approaches to fakes news detection focus on identifying fake news as above, many using neural network models. Rodriguez and Iglesias [28] applied neural network models to classify fake news using text features. Their BERT model achieved accuracy of 97%. Umer et al. [31], developed a hybrid model of Convolutional Neural Networks (CNN) and Long Short-Term Memory (LSTM) networks. They evaluated their model on a dataset of news articles related to the 2016 US election and demonstrated that it outperformed baseline models. Patwa et al. [23] curated the COVID-19 Fake News Dataset used in this research. Their work used four machine learning models with Support Vector Machine (SVM) the best performing in terms of accuracy. Whitehouse et al. [33] investigated several state-of-the-art approaches for knowledge integration. They conducted experiments on the COVID-19 dataset used in this study and LIAR, a politics dataset. They study how incorporating knowledge bases into pre-trained language models can improve the detection of fake news. The findings reveal that the effectiveness of this integration depends on the choice of suitable knowledge bases and the overall data quality. The knowledge-enhanced models significantly improve fake news detection within the LIAR dataset when the knowledge base is relevant and up-to-date. However, for the COVID-19 dataset, results were mixed, highlighting the reliance on stylistic features and the importance of domain specific and current knowledge bases. Das et al. [9] applied a number of pre-trained language models as backbone models for text classification on the COVID-19 dataset. They applied an ensemble prediction method consisting of 'soft voting' and 'hard voting' along with heuristic post processing where they take into account the effect of username handles and URL domains present in the news posts. Their best performing model achieved accuracy of 98.31%.

2.2 Explainable AI and Rule Extraction

There are two key terminologies for understanding Neural Networks – Interpretability and Explainability. Interpretability allows developers to analyse the model's decision making process, offering insights into where the results were derived from, helping with understanding the system functioning. Explainability focuses on producing clear, user-friendly explanations of the model to the end user, producing accurate and unbiased decisions which the end user can trust [1]. The focus here is on the latter, Explainable Artificial Intelligence (XAI).

The concept of transparency is discussed in [20]. It refers to the extent to which the internal workings of a machine learning model are accessible and interpretable. Miller [20] emphasises the importance of clear and valuable insights, but that revealing too much detail of the internals does not aid explanations.

There are three approaches to explanation in XAI. The pedagogical approach treats the neural network as a black-box, ignoring its internal workings. Input-output pairs are generated to approximate how the network makes decisions [14]. It allows the production of interpretable rules which are human-readable, allowing people and organisations to leverage their capabilities. TREPAN [8] is a classic pedagogical algorithm where a Decision Tree (DT) is trained to mimic a neural network's predictions. TREPAN has recently been revisited to incorporate domain knowledge in the rule extraction process [6]. Decompositional approaches look at the neural network as a white-box, investigating the internal working of the model to extract rules. Logic is derived from analysing weights, activation functions, neurons etc. This can be a computationally expensive and challenging approach [14]. Hybrids of pedagogical and decompositional approaches are often referred to as eclectic approaches. They aim to balance the internal workings of a model with externally learned rules [14].

Explainable AI seeks to develop models which are interpretable whilst maintaining high predictive accuracy. It also aims to help users understand, trust and implement AI systems more effectively [13] in sensitive domains like healthcare, finance and fake news detection where it is crucial to understand and trust the process behind how the models make their predictions. Initial approaches to XAI focused on DTs and rule-based systems which offer transparency. There is some scepticism of extracted rule-based system as most ML models focus on accuracy over interpretability and in order to enhance transparency predictive performance can be lost which is not seen as a worthwhile trade off [15]. There is also no clear definition of what interpretability is [15], which could also be cause for inconsistency in evaluating XAI methods. However, more regulation and greater focus beyond just having high performance models hints at a shift towards more transparent models for ethical implementation of AI.

As well as extracting rules, there is work considering how features impact on decision making processes, discussed in [4] in the context of challenges in the field and the balance between explainability and adaptability of XAI models. For example, SHAP [16] and LIME [27] are commonly used techniques to help understand what a model has learned. They do not directly produce rules to explain model decisions but can be helpful in identifying features which influence a model's decision making process that can then be used to construct rules.

Two of the big challenges in current work are producing Boolean functions as output, and handling high dimensional datasets in novel rule extraction algorithms. In [29] three different techniques to extract rules from neural networks are presented. The first technique focuses on extracting rules from any trained neural network called Binarised Input-Output Rule Extraction, an approach which extracts binary rules from neural network trained on binary inputs. The other two focus on partial rule extraction and logical rule-extraction which works with networks trained on normal, or binary inputs, with no restrictions on the input values. A neuro-fuzzy framework called AdaTSK for rule-extraction from a high-dimensional dataset (7,000 features) was presented in [34]. A novel gate function was developed to identify useful features and rules by combining feature selection and rule extraction. Effectiveness of their approach was demonstrated through 19 different datasets. [30] presents a decompositional method, which looks at the internal working of the neural network to extract rules. It is applied to a neural network with a monotone output function, where the units of the neural network are approximated using Boolean functions. [10] proposed a novel pedagogical rule extraction technique based on active learning. They created artificial data points near uncertain predictions and used a black-box model to label them. The approach performed well on 25 different datasets. They concluded that rules generated explain the black-box models well and help improve model accuracy. The motivation for this work comes from [18,19], where rule extraction from neural networks trained to detect cross-site scripting attacks was investigated. It is noted that when a neural network uses Boolean features, it represents a Boolean function mapping inputs to either malicious or benign (that is, a Boolean value). For a low-dimensional feature space, it is possible to list every possible input combination and create a truth table. However, as the number of features grows, this becomes impractical. Therefore, for larger feature spaces, a sampling method is proposed where rule-based approximations to the behaviour of the neural network over varying granularity are built, thereby enabling explainability [18]. This approach can be extended to other problems with Boolean feature sets, such as the one in this research. The fake news dataset consists of real news posts and fake news posts so the output is again Boolean featured. The challenge here is the size of the feature set, which is considerably larger compared to [17]. The fake news dataset is a text-based dataset where each word is a token which is then used as a feature to train the model.

2.3   Rule Extraction and Fake News Detection

The nature of fake news detection in itself presents a challenge for explainable AI. Given that it is a text classification problem using a content based approach, the feature space is massive, which poses various challenges. There are studies focusing on extracting and analysing linguistic and semantic features of fake news [25,35]. These studies, however, do not focus on extracting rules from trained ML models, and there is a limited amount of research focusing on extracting explainable rules from natural language problems.

Bhattarai et al. [5] utilise a Tsetlin Machine for interpretable fake news detection. It focuses on capturing semantic properties of fake news posts through conjunctive clauses which allows extraction of human readable rules. Fitzpatrick et al. [11], study a rule-fact network to assess truthfulness of news stories based on factors such as emotion, the speaker's political affiliation, status and job. Gorai and Shaw [12] introduced an approach which focuses on analysing semantic discrepancies between news article headings and their contents. [21] proposed a novel framework for interpreting LSTM networks by extracting human-interpretable decision rules. They decompose the LSTM's prediction into additive contributions from individual input words using forest gate dynamics. This decomposition helps identify key phrases influencing the model's output helping construct a rule based classifier mimicking the original LSTM. This is applied to sentiment analysis and question-answering with minimal reduction to the model's predictive accuracy, which supports further research into model interpretability in NLP. DEXiRE [7] is focused on creating interpretable explanation of ML models which are reliable and stable. Their white-box approach produced distributed explanations across various different parts of the system, for example across the different layers of the neural network.

3   Classifiers for Fake News Detection

This section utilises machine learning to identify patterns in a fake news dataset. Firstly, the dataset is described and explored. Secondly, a number of classifiers to distinguish fake news posts from real are trained. These classifiers include some novel treatments of the feature space and model architectures.

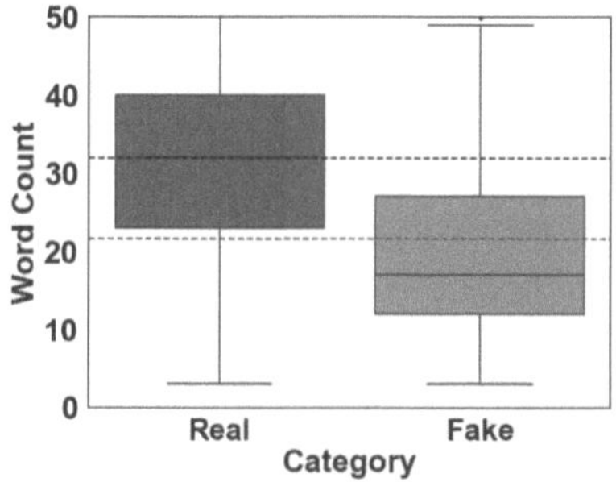

Fig. 1. Boxplot comparing lengths of Real and Fake news posts.

3.1   Dataset Description and Exploration

An existing fake news dataset [23] was used in this study. It consists of fake and real news posts in English related to COVID-19. The dataset was hand curated, with each news post manually checked to increase the authenticity and accuracy of the dataset. Fake news posts about COVID-19 were collected from social media websites. They were fact checked using fact-checking websites such as PolitiFact, Snopes, Boomlive. Real news data is a collection of tweets from verified twitter accounts. The dataset is class-wise balanced. In total the dataset contains 10,700 news posts, 5600 of the consisting of real news and 5100 of fake.

Preprocessing removed punctuation and special characters, then all characters were converted to lower case. The data is tokenised giving 42,436 tokens, most of which result from hyperlinks being tokenised into several tokens. Removing the hyperlinks leaves 15,514 tokens. Each hyperlink is unique and the presence of a link might be meaningful in itself. All links were converted into a single placeholder token 'https' indicating some website link in a post. 5323 real news posts contained URL links compared to 2062 in fake news posts. This is unlike [23] where they removed all the links during preprocessing. Their trained DT model achieved 85% accuracy, less than the accuracy of the DT in this study. Stopword removal and stemming is performed for some models, after which 12,569 unique tokens remain in the corpus.

The mean length of fake posts was 21.65 tokens, and for real post 31.96 tokens. This is illustrated in Fig. 1, where the boxes represent the Interquartile Range and the dotted lines show the mean length of posts. That is, real news posts typically contain more words than fake news posts. In addition, real news posts more regularly include hyperlinks as a means of referencing sources.

3.2 Model Composition

Two different approaches to encoding the text data were explored. Firstly, words were passed into a pre-trained word embedding before vectors of embedded words are trained on a convolutional neural network. Secondly, each word in the dataset was converted into a Boolean feature, where 1 indicates presence of a word from the corpus in the news post and 0 indicates absence of the word. Then each post is converted to a binary word occurrence vector that can be passed to a variety of machine learning models. The two approaches were implemented (along with a naïve baseline) to compare and contrast their performance.

Baseline Classifier. This naïve baseline approach assumes news posts exhibit systematic differences in length. The classifier categorises data points on length thresholds, since fake news articles tend be shorter in length. Whilst the model is primitive, this classifier serves as a benchmark, providing insights into whether simple textual patterns impact accuracy of fake news detection models.

Word Embedding and CNN. This approach converts the data points into a list of words before passing them through a pre-trained word embedding (GloVe[1] [24]), followed by training and testing on a One-Dimensional Convolutional Neural Network (1D-CNN).

After tokenisation, stopword removal was performed. The embedding used maps each word into a vector in a 100 dimensional space, and a data point (social media post) is then a tensor of these embedded words, in order, with padding by 0 valued vectors to deal with out-of-vocabulary words and to give a consistent sized input to the neural network. With the GloVe embedding being pre-trained, the corpus to which it applies is defined along with it. Words not part of the GloVe embedding corpus must be eliminated, and this was done as part of preprocessing. This was a cause of concern as frequently occurring words in the dataset include 'Covid' and 'Coronavirus' which were newly coined during the pandemic and are not included in the GloVe corpus which predates this. The

---

[1] Code for GloVe can be downloaded from http://nlp.stanford.edu/projects/glove/.

order of words is preserved in the embedding and the goal of the embedding is to capture textual meaning.

A 1D-CNN was used to train a classifier. 1D-CNNs are a natural choice of architecture for text, since text has a one dimensional structure. The 1D-CNN was constructed with two one-dimensional convolutional layers. The filters would slide over the vectors extracting features from them, passing them through to the fully connected layers making predictions of whether posts are fake or real.

The embedding layers form an input layer into the 1D-CNN. The architecture of the 1D-CNN is as follows.

- First Conv1D layer: Number of Filters: 128, Kernel Size:1
- Pooling layer: MaxPooling1D
- Second Conv1D layer: Number of Filters: 128, Kernel Size:3
- Pooling layer: GlobalMaxPooling1D
- One fully connected layer with 128
- Two output nodes (with a softmax to give a class)
- Activation function: ReLu
- Loss Function: Binary Cross-entropy
- Optimiser: Adam
- Number of Epochs: 10

Boolean Valued Features. This approach considers for each word in the corpus whether or not it occurs in a post. These Booleans lead to binary word occurrence vectors that can be used to train classifiers with a variety of models.

After tokenising the data points, stop word removal and stemming was applied. Stemming was performed using Porters Stemmer, a suffix stemming algorithm [26]. Stemming groups words of similar meaning together and reduces features, thereby aiding the extraction of meaningful information from the dataset. Below are examples of news posts from the dataset before and after preprocessing, stop word removal and stemming. Firstly, the original posts:

0 Florida Governor Ron DeSantis Botches COVID-19 Response - By banning Corona beer in order to flatten pandemic curve.
1 WHO warns against consuming cabbage to prevent COVID-19

Secondly the posts with special characters removed, conversion to lower case, stop word removal and stemming applied.

0 florida governor ron desanti botch covid respons ban corona beer order flatten pandem curv
1 warn consum cabbag prevent covid

After processing as described above, the dataset is converted to a binary sparse matrix, where each value in the matrix is either 0 or 1, with a 1 corresponding to a word appearing in a post. Only the presence or absence of a word is taken into account. That is, each post becomes a binary word occurrence vector

364   F. Iqbal and J. M. Howe

with length equal to that of the corpus. This was implemented using CountVectorizer with the binary parameter set to True. Continuing the example, below is an example of the two news posts in their vectorised format (with the many thousands of additional 0 values omitted):

florida	pandem	consum	botch	beer	ron	covid	respons	ban	governor	corona	order	prevent	flatten	curv	desanti	warn	cabbag
1	1	0	1	1	1	1	1	1	1	1	1	0	1	1	1	0	0
0	0	1	0	0	0	1	0	0	0	0	0	1	0	0	0	1	1

Order of words is lost here. An encoded post represents Boolean values whether or not a word occurs in a post but does not contain any relational information between those words. In addition, stopword removal and stemming whilst reducing the number of words in the corpus can lead to information loss. For example, the acronym "WHO" is interpreted it as a pronoun and stopword removal omits it. Additionally, stemming can sometimes merge different words into the same root form, which can potentially lead to inaccurate interpretation.

This feature set was used to train feed-forward neural network, random forest and decision tree classifiers.

Random Forest. A Random Forest classifier was trained on the Boolean feature set. The model was configured with the following hyper-parameters: 128 estimators, entropy as the splitting criterion.

Table 1. Accuracy, Precision, and Recall for Fake News Classifiers.

ML Model	Accuracy	Precision	Recall
1D-CNN + GloVe	92.14±0.24	92.49±0.69	91.09±0.57
Random Forest	93.07±0.26	92.58±0.52	92.84±0.41
Neural Network	93.64±0.21	93.70±0.21	93.59±0.22
Decision Tree	89.62	89.52	89.51
Naïve Classifier	70.32	73.34	66.90

Decision Tree. A DT classifier was trained on the Boolean features. The pre-processed data was used for training, with entropy as the splitting criterion.

Neural Network. A feed-forward neural network was trained on the Boolean feature set. The hyper-parameters values used for the network are given below:

- Single hidden layer
- Number of nodes: 150
- Two output nodes (with a softmax to give a class)
- Activation: ReLu

- Loss Function: Binary Cross-entropy
- Optimiser: Adam
- Number of Epochs: 10

3.3 Results

The implementation was developed in Python utilising TensorFlow and Keras for neural networks. Scikit-learn was used for the Random Forest and Decision Tree classifiers. Additional libraries include NumPy, Pandas and Matplotlib. The data is labelled with 1 for Fake news and 0 for Real news. The dataset was split 80% for training and validation, whilst 20% was retained as a holdout test set. Table 1 gives results for the classifiers, including 95% confidence intervals (noting that the decision trees and the naïve classifier are deterministic).

Figure 2 gives a confusion matrix for the naïve classifier, and Fig. 3 gives sample confusion matrices for single runs of the machine learning classifiers.

The naïve classifier gives a baseline result to contextualise the machine learning models, and it is noted that the pattern this captures gives results well above chance. The Decision Tree model over Boolean features works well, but less well that the other three machine learning models. In terms of accuracy, the Neural Network trained with the Boolean featured model performs marginally better than the Random Forest, which in turn performs marginally better than the 1D-CNN model. Whilst the differences are small, they are statistically significant. The Neural Network with Boolean features likewise has marginally better precision and recall than the other machine learning models. The performance of the Neural Network with Boolean features is encouraging, and motivates further study of what it has learned.

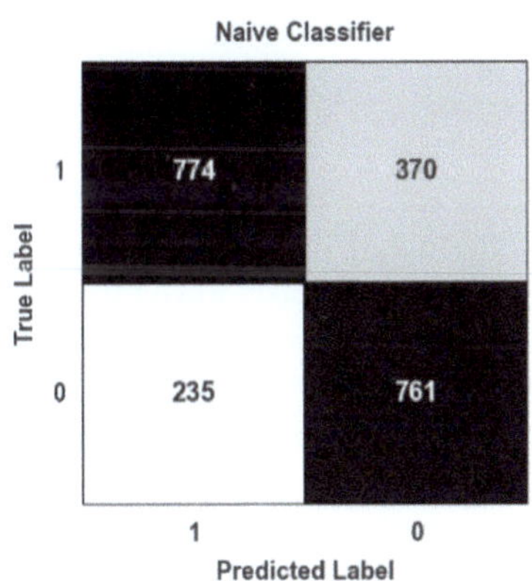

Fig. 2. Confusion matrix for the naïve classifier.

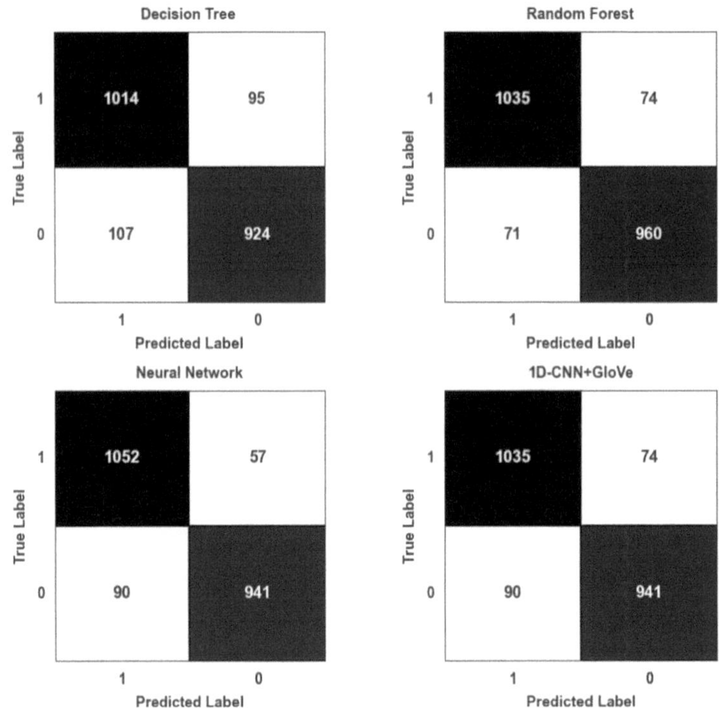

Fig. 3. Confusion matrices for the ML classifiers.

4   Feature Selection and Classification

Post preprocessing the corpus consisted of a high dimensional feature space of 12,569 words. This raises the question whether all of these features make a meaningful contribution to the classifier's performance or if there are redundancies. This section covers the feature selection and finalisation process for training the rule extraction model. First, a comprehensive analysis on the occurrences of words (features) was conducted on both the real and fake datasets to gain insight into any linguistic patterns. Words were ranked based on how many times they appear in each dataset. This process provides a way of selecting features by taking the top ranked features whilst filtering out rare or anomalous terms that might introduce noise into the analysis. Figure 4 gives the top 20 most occurring words in the real and fake posts.

However, high occurrence features do not necessarily produce the best classifiers. Feature Importances from the trained random forest classifier were considered as a ranking. This can provide insights into which features strongly influence the model's decisions. However, random forest feature importances can be unreliable with large feature sets, and this limitation was observed, where the resulting ranking were inadequate for effective feature selection.

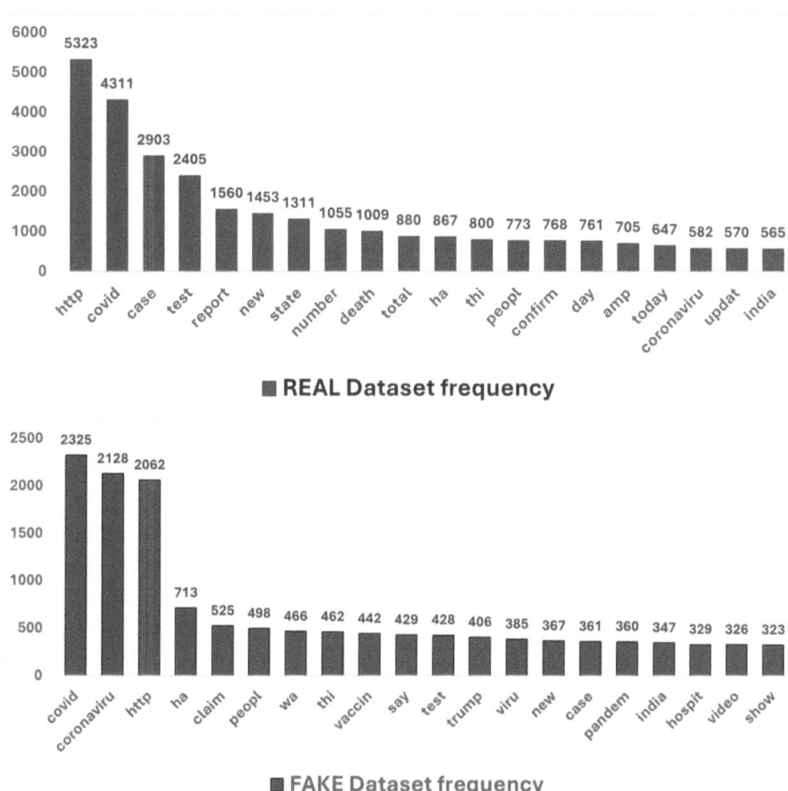

Fig. 4. Real and Fake Dataset Word Counts post pre-processing.

Therefore, a third option was considered. An upfront static order was desired, so entropy was used to calculate the information gain for each feature at the top level of decision making, and features with the highest information gain values were ranked highest. The top ranked features can be seen in Table 2 and this is the ranking used in the rest of this work. Information gain was calculated using the following standard formula:

Table 2. Features with the highest information gain value.

Feature	Information Gain	Feature	Information Gain
https	0.102279104	test	0.046422567
case	0.079177549	indiafightscorona	0.046089615
coronavirus	0.079136309	claim	0.042695365
report	0.06521603	number	0.042245714
total	0.047700305		

$$\text{Information Gain}(A) = \text{Entropy}(S) - \sum_{v \in \text{Values}(A)} \frac{|S_v|}{|S|} \text{Entropy}(S_v)$$

After establishing the feature ranking, the impact of reducing the size of the feature set was investigated. The binary word occurrence vectors were sliced by the n top ranked features for a range of values of n. For each n, a neural network was trained with the same parameter settings as in Sect. 3.2. The accuracy on testing data of each of these has been plotted in Fig. 5.

Considering Fig. 5, the initial experiment was with a single feature (https), yielding predictive accuracy of 67.75%, representing a significant result demonstrating the neural network's capability to achieve significant predictive accuracy using a single feature. As expected, performance accuracy improves with increasing number of features. There is a steady increase until reaching a plateau at 90% after 400 features. A similar pattern was observed when the model was trained using features given by word occurrences. This baseline provides a reference point and helps better understand the effectiveness of the Boolean extraction approach. Compared to [18], this study requires a substantially larger number of features to achieve close approximation to the neural network. The above findings also emphasise the inherently complex nature of text classification problems, particularly in the case of fake news detection owing to subtle relationships between words and their interactions. Section 5 builds on these observations to develop a rule extraction methodology from the Boolean feature model.

A SHAP plot (Fig. 6) was constructed on the top 20 ranked features. It illustrates the impact of key features on the model's decision making process. Red indicates 1 values (fake) and blue indicates 0 (real). It highlights the following.

- https is the most influential feature. This aligns with the occurrences ranking and information gain feature ranking. 'https' ranks highest on both lists. Its presence strongly influences model predictions, increasing the likelihood of classifying data points as real.
- several features show negative SHAP values (pushing towards real 0). 'test', 'case', 'report' can indicate real content

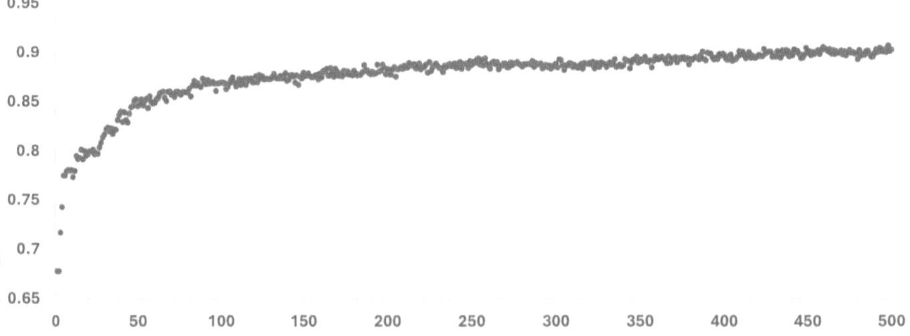

Fig. 5. Neural network performance on features selected by information gain.

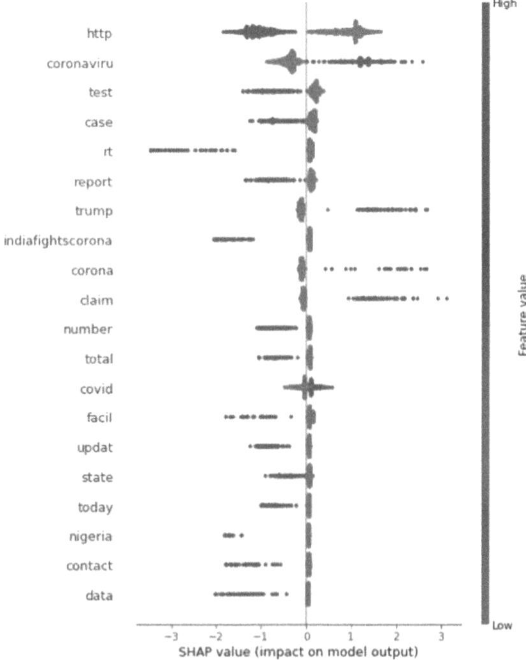

Fig. 6. SHAP analysis plot.

5   Rule Extraction

One of the main research questions addressed in this work is the extent to which a Boolean function can be extracted from the Boolean featured neural network (Sect. 3.2) such that it accurately represents the behaviour of that network. Such rule extraction can allow a better understanding of feature interaction and their contribution to the neural network's output. A pedagogical approach is taken. The neural network over the full Boolean feature set is queried repeatedly and the results of these queries are used to build a Boolean function. The points to be considered are what binary word occurrence vectors should be used as inputs to the neural network and how to collate the answers to synthesise a Boolean function. Given that the neural network has been trained on 12,579 features, one point investigated is how this might be reduced for an extracted Boolean function, a question addressed by the sampling method proposed by [18] and built upon here, although in that case the dataset had only 69 features.

The overall approach is to build rules (in the first instance the rows of a truth table over a restricted number of Boolean features) by building synthetic data points based on the variables involved in the rule and the training dataset. These are used to query the neural network. The results of these sample queries are then distilled into the output of the Boolean function, and the final learned

Boolean function that approximates the behaviour of the neural network is used as a classifier. This is detailed below.

Data Points. In order to reduce dimensionality, features were ranked by their top-level information gain (as in Table 2). The goal is to produce a Boolean function that generalises the behaviour of the neural network beyond the training data. This is done by fixing the features of concern and then populating the rest of the features. Simply doing so at random gives data points that look little like the dataset itself, so hybrid data points are built in two ways. Some take data points from the dataset and substitute in fixed values for the selected features. Others substitute these fixed values into all 0 vectors. The number of these samples and the balance between the two are hyper-parameters to the method.

For example, suppose the goal is to build a Boolean function over just the top two ranked features (http and coronaviru, stemmed in this case). Then a data point would be formed by selecting a random point from the dataset (or a 0 vector) and substituting in the 00 value (then 01, 10, 11). This gives a data point with the full 12,569 features which can be used as input to the neural network. Below, suppose that the top row was a data point from the dataset, then for the 00 input to the Boolean function, these features (http and coronaviru) would be fixed to 00, as in the second row.

$$1\ 1\ 0\ 1\ 0\ 1\ 0\ 1\ ...$$
$$0\ 0\ 0\ 1\ 0\ 1\ 0\ 1\ ...$$

Generate Predictions. For each Boolean input combination, 37 samples were created to query the neural network, 27 as hybrid points with an element of the training data, 10 with 0 vectors (since as shown in Fig. 6, 1 values count more heavily than 0 to classification).

Construct Truth Table. The predictions are then compiled into a table recording the predicted classification of the data points, with the output of the Boolean function decided by the majority prediction, hence the table becomes a truth table. Where n features are considered, the table has 2^n rows, and 4 columns. For example, if just 3 features were considered, the table would look something like Table 3. This describes a Boolean function, with some decisions fairly clear from the sampling, but others less so, for example the last row.

Table 3. Three feature table output.

Index	Fake	Real	Label	Index	Fake	Real	Label
(0,0,0)	25	12	1	(1,0,0)	15	22	0
(0,0,1)	22	15	1	(1,0,1)	13	24	0
(0,1,0)	22	15	1	(1,1,0)	24	13	1
(0,1,1)	24	13	1	(1,1,1)	18	19	0

Table as a Classifier. The table then forms a classifier. For a data point, and the features to be considered, the values of an input data point for those features are used to determine the relevant row of the table (and the rest discarded) and the prediction looked up. The results on the test data of building a truth table as above for between 2 and 16 top ranked features were calculated. Figure 7 plots the accuracy for the sampling table alongside the results for neural network and DT classifiers trained on the same features. Table 4 tabulates these values.

5.1   Results

As also seen in Fig. 5, a small number of features achieve considerable classification accuracy, demonstrating effective dimensionality reduction. However, the number of features is not enough to fully capture the neural network's behaviour.

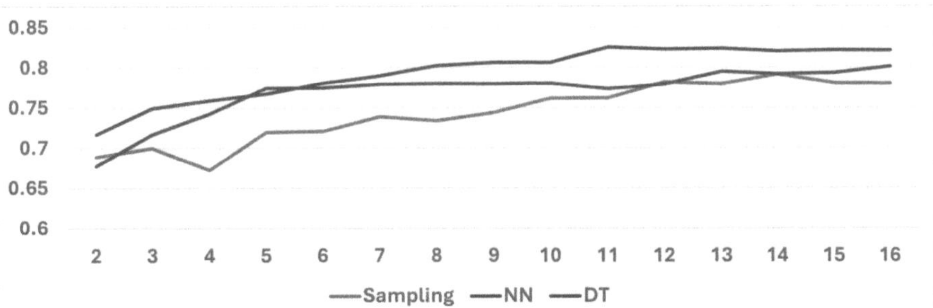

Fig. 7. Sampling Results vs NN and DT results trained on individual features.

Since the size of the truth table grows exponentially in the number of features, building the full truth table for all 12,569 features is infeasible. However, the Boolean function might be extended to additional features by adding a limited number of additional rules. To determine which rules to add, error analysis was performed on the training data and the 16 feature sampling Boolean function, which achieved 78% accuracy. The top problem cases are detailed below.

Data point	Misclassification/Occurrences
0000000000000000	356/1579
1000000000000000	416/1198
1010000000000000	200/584
0000010000000000	146/242

Table 4. Comparison of Sampling, NN, and DT by Number of Features.

Features	Sampling	NN	DT	Features	Sampling	NN	DT
				9	74.39%	77.99%	80.61%
2	68.89%	67.75%	71.68%	10	76.16%	78.04%	80.61%
3	70.37%	71.68%	74.95%	11	76.21%	77.34%	82.52%
4	67.24%	74.25%	75.89%	12	78.17%	77.94%	82.29%
5	71.82%	77.48%	76.73%	13	77.99%	79.49%	82.34%
6	72.05%	77.48%	78.04%	14	79.15%	79.21%	82.01%
7	73.78%	77.94%	78.97%	15	78.01%	79.30%	82.15%
8	73.36%	78.04%	80.23%	16	78.03%	80.14%	82.10%

In order to reduce the misclassification, features were augmented up to 500 features, but the full truth table was not built. Instead, a number of special cases were added with values given by the neural network. It is referred to as the branching approach as it adds long branches beneath the problem cases. The additional cases considered extend the problematic cases listed above with the first 1 encountered. For example, as below, the branch below the 0000000000000000 case has a linear number of additional rules, one for each of the possible first matches against a 1. If the input consists of 500 0s, then the 0000000000000000 result is given. Similarly for the other four cases.

```
17 00000000000000001
18 000000000000000001
19 0000000000000000001
 . . .
500 0000000000000000...01
```

The misclassified data points when applying the branching approach are:

Data point	Misclassification/Occurrences
0000000000000000	153/1579
1000000000000000	337/1198
1010000000000000	153/584
0000010000000000	30/242

After implementing the branching approach the misclassification rate on the training data was reduced, as can be seen above. This represents the final model for rule extraction. Figure 8 gives two confusions matrices. The first gives the results of the final Boolean function against the ground truth labels of the testing dataset. The second gives results of the final Boolean function against the neural network, as a measure of how well the Boolean function has captured the behaviour of the neural network. There was improvement in testing accuracy from 78.03% to 81.72%. The branches were considered incrementally, one at a time, beginning with the branch below

0000000000000000 then combining it with the second branch below 1000000000000000 followed by combining in the third and fourth data points. The final accuracy after branching across the four incremental cases is presented in Table 5. The branching approach validates well on both the training and testing data.

Table 5. Branching of the misclassified data points.

Branching data points	Accuracy
0000000000000000	80.67%
1000000000000000	81.61%
1010000000000000	82.17%
0000010000000000	83.59%

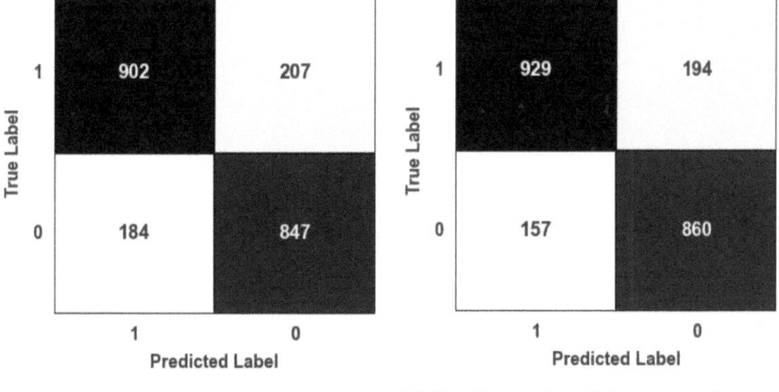

(a) Testing data labels vs sampling predictions   (b) Neural network predictions vs sampling predictions

Fig. 8. Comparison of neural network and sampling predictions.

This analysis identifies a key issue, Fig. 6 highlights that 'https' might have a dominant influence on the classifications. This can be seen in the error analysis of the 1000000000000000 case, as it is one of the most misclassified data points. The branching technique has little impact on improving this misclassification, since it adds only single additional feature which often is not weighted strongly enough to counterbalance the top ranked feature.

6 Conclusion

Models for fake news detection with novel architectures and features were developed to establish performance benchmarks. Two distinct approaches were taken, both yielding good results, as expected given previous research findings. However, during pre-processing of the text features, it was noted that the 1D-CNN combined with GloVe model failed to capture keywords from the dataset, 'COVID' and 'Coronavirus'. Despite this limitation, the model performed comparatively well. Given that the binary word occurrence vector model performed the best of the models, and that this model essentially describes a Boolean function, this motivated rule extraction from the model, with the aim of extracting interpretable patterns that would help explain the models high predictive accuracy and uncover potentially simpler underlying patterns.

Given the inherently complex nature of the dataset, with thousands of dimensions, there are challenges in producing rules. Appropriate pre-processing techniques to reduce the corpus size were applied. The accuracy of the neural network with a restricted number of Boolean features was plotted (Fig. 5), and it was observed that it flattens out after 400 features. This was followed by extracting meaningful features to develop a rule-based methodological framework, with a variety of feature rankings explored. Given that this is a text classification task and, as highlighted in Fig. 6, words in the dataset interact with each other to create semantic meaning, simply selecting words based on occurrences proved insufficient. However, the Boolean feature model based on solely feature occurrences yielded high predictive accuracy. This suggests some words present a strong discriminative signal for real versus fake news regardless of context.

The predictive power of a small number of features indicates potential for extracting human-readable rules using these features and motivated developing a interpretable XAI framework. Extracting rules in the form of Boolean functions was explored, with the best performing model giving 83.5% accuracy. Whilst this is not comparable with the target neural network, the rules are clear and interpretable and represent a reasonable approximation of the full network, given the size of the feature space.

This research focused on COVID-19 related content. There is a curiosity in applying rule extraction and the interpretable XAI framework across different domains. Future work will evaluate how the methodology for feature identification and rule formation generalises on other fake news datasets, as well as for other text-based problem domains.

References

1. Ali, S., et al.: Explainable Artificial Intelligence (XAI): what we know and what is left to attain trustworthy artificial intelligence. Inform. Fusion **99**, 101805 (2023)
2. Allcott, H., Gentzkow, M.: Social media and fake news in the 2016 election. J. Econ. Perspect. **31**(2), 211–36 (2017)
3. Andrews, R., Diederich, J., Tickle, A.B.: Survey and critique of techniques for extracting rules from trained artificial neural networks. Knowl.-Based Syst. **8**, 373–389 (1995)
4. Barredo Arrieta, A., et al.: Explainable Artificial Intelligence (XAI): concepts, taxonomies, opportunities and challenges toward responsible AI. Inform. Fusion **58**, 82–115 (2020)

5. Bhattarai, B., Granmo, O.C., Jiao, L.: Explainable Tsetlin machine framework for fake news detection with credibility score assessment. In: Proceedings of the Thirteenth Language Resources and Evaluation Conference, pp. 4894–4903. European Language Resources Association (2022)
6. Confalonieri, R., Weyde, T., Besold, T.R., del Prado Martín, F.M.: TREPAN reloaded: a knowledge-driven approach to explaining black-box models. In: European Conference on Artificial Intelligence. Frontiers in Artificial Intelligence and Applications, vol. 325, pp. 2457–2464. IOS Press (2020)
7. Contreras, V., et al.: A DEXiRE for Extracting propositional rules from neural networks via binarization. Electronics $11(24)$ (2022)
8. Craven, M.W.: Extracting comprehensible models from trained neural networks. Ph.D. thesis, The University of Wisconsin-Madison (1996)
9. Das, S.D., Basak, A., Dutta, S.: A heuristic-driven ensemble framework for COVID-19 fake news detection. In: Chakraborty, T., Shu, K., Bernard, H.R., Liu, H., Akhtar, M.S. (eds.) CONSTRAINT 2021. CCIS, vol. 1402, pp. 164–176. Springer, Cham (2021). https://doi.org/10.1007/978-3-030-73696-5_16
10. De Fortuny, E.J., Martens, D.: Active learning-based pedagogical rule extraction. IEEE Trans. Neural Netw. Learn. Syst. $26(11)$, 2664–2677 (2015)
11. Fitzpatrick, B., Liang, X.S., Straub, J.: Fake News and Phishing Detection Using a Machine Learning Trained Expert System. arXiv e-prints (2021), arXiv:2108.08264v1
12. Gorai, J., Shaw, D.K.: Semantic difference-based feature extraction technique for fake news detection. J. Supercomput. $80(15)$, 22631–22653 (2024)
13. Gunning, D., Stefik, M., Choi, J., Miller, T., Stumpf, S., Yang, G.Z.: XAI – explainable artificial intelligence. Sci. Robot. $4(37)$ (2019)
14. Hailesilassie, T.: Rule extraction algorithm for deep neural networks: a review. Inter. J. Comput. Sci. Inform. Sec. (2016)
15. Lipton, Z.C.: The mythos of model interpretability: in machine learning, the concept of interpretability is both important and slippery. Queue $16(3)$, 31–57 (2018)
16. Lundberg, S.M., Lee, S.I.: A unified approach to interpreting model predictions. In: Advances in Neural Information Processing Systems, pp. 4765–4774. Curran Associates, Inc. (2017)
17. Mereani, F., Howe, J.M.: Preventing cross-site scripting attacks by combining classifiers. In: International Joint Conference on Computational Intelligence, pp. 135–143. SCITEPRESS (2018)
18. Mereani, F., Howe, J.M.: Exact and approximate rule extraction from neural networks with Boolean features. In: International Joint Conference on Computational Intelligence, vol. 1, pp. 424–433. SCITEPRESS (2019)
19. Mereani, F.A., Howe, J.M.: Rule extraction from neural networks and other classifiers applied to XSS detection. In: Merelo, J.J., Garibaldi, J., Linares-Barranco, A., Warwick, K., Madani, K. (eds.) IJCCI 2019. SCI, vol. 922, pp. 359–386. Springer, Cham (2021). https://doi.org/10.1007/978-3-030-70594-7_15
20. Miller, T.: Explanation in artificial intelligence: insights from the social sciences. Artif. Intell. 267, 1–38 (2019)
21. Murdoch, W.J., Szlam, A.: Automatic rule extraction from long short term memory networks. In: International Conference on Learning Representations (2017)
22. Yafooz, W.M.S., Emara, A.-H.M., Lahby, M.: Detecting fake news on COVID-19 vaccine from youtube videos using advanced machine learning approaches. In: Lahby, M., Pathan, A.-S.K., Maleh, Y., Yafooz, W.M.S. (eds.) Combating Fake News with Computational Intelligence Techniques. SCI, vol. 1001, pp. 421–435. Springer, Cham (2022). https://doi.org/10.1007/978-3-030-90087-8_21

23. Patwa, P., et al.: Fighting an infodemic: COVID-19 fake news dataset. In: Chakraborty, T., Shu, K., Bernard, H.R., Liu, H., Akhtar, M.S. (eds.) CONSTRAINT 2021. CCIS, vol. 1402, pp. 21–29. Springer, Cham (2021). https://doi.org/10.1007/978-3-030-73696-5_3

24. Pennington, J., Socher, R., Manning, C.D.: GloVe: global vectors for word representation. In: Empirical Methods in Natural Language Processing, pp. 1532–1543. Association for Computational Linguistics (2014)

25. Pérez-Rosas, V., Kleinberg, B., Lefevre, A., Mihalcea, R.: Automatic detection of fake news. In: International Conference on Computational Linguistics, pp. 3391–3401. Association for Computational Linguistics (2018)

26. Porter, M.F.: Snowball: A language for stemming algorithms. Tech. rep. (2001), http://snowball.tartarus.org/texts/introduction

27. Ribeiro, M.T., Singh, S., Guestrin, C.: Why Should I Trust You?: explaining the Predictions of Any Classifier. In: International Conference on Knowledge Discovery and Data Mining, pp. 1135–1144. ACM Press (2016)

28. Rodriguez, Á.I., Iglesias, L.L.: Fake news detection using deep learning. arXiv preprint arXiv:1910.03496 (2019)

29. Taha, I.A., Ghosh, J.: Symbolic interpretation of artificial neural networks. IEEE Trans. Knowl. Data Eng. **11**(3), 448–463 (1999)

30. Tsukimoto, H.: Extracting rules from trained neural networks. IEEE Trans. Neural Netw. **11**(2), 377–389 (2000)

31. Umer, M., Imtiaz, Z., Ullah, S., Mehmood, A., Choi, G.S., On, B.W.: Fake news stance detection using deep learning architecture (CNN-LSTM). IEEE Access **8**, 156695–156706 (2020)

32. Wardle, C., Derakhshan, H.: Information disorder: toward an interdisciplinary framework for research and policy making. Tech. rep., Council of Europe (2017). https://edoc.coe.int/en/media/7495-information-disorder-toward-an-interdisciplinary-framework-for-research-and-policy-making.html

33. Whitehouse, C., Weyde, T., Madhyastha, P., Komninos, N.: Evaluation of fake news detection with knowledge-enhanced language models. In: AAAI Conference on Web and Social Media, vol. 16, pp. 1425–1429 (2022)

34. Xue, G., Chang, Q., Wang, J., Zhang, K., Pal, N.R.: An adaptive neuro-fuzzy system with integrated feature selection and rule extraction for high-dimensional classification problems. IEEE Trans. Fuzzy Syst. **31**(7), 2167–2181 (2022)

35. Zhang, W., Paudel, B., Zhang, W., Bernstein, A., Chen, H.: Interaction Embeddings for prediction and explanation in knowledge graphs. In: Conference on Web Search and Data Mining, pp. 96–104. ACM (2019)

Contrasting Human and Emergent Concepts in Image Classifiers

Tamara Bíla[✉][iD] and Igor Farkaš[iD]

Department of Applied Informatics, Comenius University Bratislava, Bratislava, Slovakia
{tamara.bila,igor.farkas}@fmph.uniba.sk

Abstract. In the age of AI becoming an everyday partner and support tool in both professional and private domains, users and stakeholders are increasingly confronted with the question of interpretability. This work contributes to the search for answers by exploring hidden meanings in the internal layers of convolutional neural networks trained on image classification tasks. Combining supervised concept-based and unsupervised learning paradigms, the goal is to discover semantically meaningful representations that can be contrasted with human-defined concepts (defined once per dataset). More specifically, we extracted layer-wise network-inherent clusters using hierarchical agglomerative clustering. To evaluate their semantic fidelity, we trained auxiliary classifiers on concepts as well as cluster memberships, and evaluated across all layers. Our experiments reveal a higher classification accuracy for clusters extracted from each layer compared to human-defined concepts, indicating better separability and indication that clusters may capture patterns beyond human labels. Additionally, the classification accuracy increases for both clusters and concepts toward the output layer. Beyond quantitative evaluation, we provide qualitative insights using visualization techniques such as UMAP projections and Concept Localization Maps. Our findings highlight the potential of hybrid approaches for post hoc explainability and point to promising directions in uncovering emergent structures within deep neural networks. The source code and results are provided at https://github.com/rikkathee/CE-HM.git.

Keywords: Neural Network Interpretability · Concept-Based Explanations · Unsupervised Clustering · Latent Space Analysis

1 Introduction

The rapid and continual growth of neural networks in their size and scope of application directly raises questions regarding their safety, trustworthiness, and thereto related applicability in sensitive and high-stakes domains. The research field of eXplainable AI (XAI) aims to address these and other questions by introspecting the working mechanisms and internal reasoning of the target models. In addition, understanding the model decision making process enables failure detection, debugging, diagnostics, and even reverse engineering solutions (referred to as "microscoping" AI) [17]. Explainable artificial intelligence encompasses different categories such as *data explainability*, using

F. Marcelloni et al. (Eds.): IJCCI 2025, CCIS 2829, pp. 377–397, 2026.
https://doi.org/10.1007/978-3-032-15638-9_22

statistical methods and visualization to detect biases and issues, *model explainability*, creating ad hoc interpretable models, and *post hoc explainability*, providing explanations after the inference process, often employing a separate machine learning (ML) technique [1].

The demand for high performance drives stakeholders to choose models with increasing complexity, which is usually inversely correlated with interpretability. For example, models with transformer architecture outperform DNNs in natural language processing (NLP), computer vision, and even multimodal tasks; however, from the stance of XAI they can be considered black-boxes. On the other hand simple ML algorithms like clustering, regression analysis or decision trees are basically self-explanatory. However, the relationship between model complexity and interpretability is perhaps not strictly negatively correlated, as a better understanding of the model structure could help improve its performance. In addition, traditional convolutional neural networks (CNNs) still achieve higher accuracy on small datasets, as transformers easily overfit, and in real-time computation on constrained devices.

In this work, we focus on understanding the internal structure of a few simple CNNs trained for the image classification task. However, the presented method can in principle be applied to more complex architectures with access to their hidden layers as well. It is considered a form of *post hoc explainability*, since it involves training an auxiliary model on the internal representations of a pre-trained target network.

In terms of learning paradigms, our scheme combines two primary methods, *supervised* concept activations regions (CARs) [7] and *unsupervised* agglomerative clustering. Both methods fall into the *concept-based* interpretability paradigm, which focuses on extracting learned representations of high-level concepts from the embedding space [3]. In particular, our approach localizes human-defined concepts and compares them with emergent, i.e., network-inherent, clusters of the same latent layer space. The joint evaluation of the two concept methods points towards varying semantic levels of concepts throughout network layer depth.

The main objective of our approach is to enable a quantitative comparison between a fixed set of human-defined concepts, defined once per dataset and kept constant across all models and layers, and data-driven clusters, which are extracted separately at each layer and vary with both model architecture and layer depth. Our expectation is that network-inherent clusters facilitate a more faithful representation of the *intrinsic structure* and will hence have higher accuracy, when measured via an auxiliary classifier.

More specifically, we employ the support vector classifier (SVC) to distinguish both concept positive from concept negative samples, as well as cluster positive from cluster negative ones. SVC were used in [7] named Concept Activation Regions (CARs) to extract representations of human-defined concepts and determine their importance with respect to the model's classification task. CARs build on assumption of *Concept Smoothness*, which may loosely speaking be translated as separability of the concept-positive and the -negative regions. Hence, we interpret the classification accuracy of the SVCs as a measure of how well these regions reflect the *intrinsic structure* of the high-dimensional latent space manifold.

The rest of the paper is organised as follows: Sect. 2 describes related work. In Sect. 3 we introduce the computational pipeline for concept and cluster extraction and the evaluation method for quantitative comparison. In Sect. 4, we provide visual support for the intermediate steps of the experiment to illustrate similarities and differences between cluster and concept representations, and to discuss the final results.

2 Related Work

The task of concept extraction can be framed as either a supervised or unsupervised learning problem. Supervised methods seem more popular in the literature, probably because of their reliance on predefined, semantically meaningful concepts, which facilitate human interpretability. On the contrary, unsupervised approaches often unveil latent structures that may not directly correspond to human-understandable abstractions, but can still yield valuable insights into the network's internal organization.

Among supervised ad hoc approaches, which intervene directly in the network architecture to engrave concept structure the following methods can be found: Concept Bottleneck Models (CBM) [10], Incremental Residual Concept Bottleneck Models [19], and Concept Whitening [5], all of which replace or modify an intermediate (often penultimative) layer to align latent directions with a predefined set of concepts. On the other hand, supervised post hoc methods analyze learned representations without altering the target model. One prominent example is the Concept Activation Vector (CAV) method [9] using a binary linear classifier to assign a latent vector to the sought for concept and, in addition to that, introducing a conceptual sensitivity measure of the target prediction using directional derivative with respect to the latent representation, called Testing with Concept Activation Vectors (TCAV). In [18] two explanation techniques were combined to produce local (sample-specific) as well as global (class-wide) explanations based on Integrated Gradients (IG) [20] and TCAVs.

Despite the universality of the model-agnostic CAV method, the underlying assumption of linear separability of concept positive from negative instances is not guaranteed to hold in all situations. In particular, the dimension of the latent space must be much higher than the size of the concept set in order to ensure separability by a linear surface [6]. To address this, follow-up work proposed an alternative approach employing Support Vector Classifier (SVC) with radial kernel termed Concept Activation Regions (CARs) achieving higher overall concept accuracy [7]. In [14], hypotheses regarding three properties of CAVs, namely: consistency across layers, entanglement between each other, and spatial dependence (with respect to input space), were postulated and tested on a configurable synthetic dataset. The findings point to challenges in ensuring robustness and interpretability across the network of CAVs.

A generalized Concept Gradient (CG) formula to facilitate concept-importance explanations, applicable to differentiable, also non-linear, concept classifier was proposed in [2].

In the unsupervised setting, the Automatic Concept-based Explanations (ACE) framework [8] introduced a pipeline that segments input images, embeds the segments in the latent space, clusters the resulting representations, removes outliers, and evaluates their importance for final classification. Similarly, hierarchical agglomerative clustering was applied to disentangle the 100 strongest activations of a polysemantic neurons, and thereby revealing hidden subclasses (clusters) within a single latent direction [15].

Building on this body of work, our prior research *reference will be added later* introduced a hybrid approach that combines supervised concept-based methods with unsupervised cluster analysis, applied to the MNIST dataset. The results suggested that latent clusters may offer a more faithful account of the internal manifold structure than human-annotated concepts, opening new pathways for introspective analysis of learned representations.

3 Method

Firstly, let us briefly discuss what we mean by the term *concept* and how it is represented in the network's activations. A concept is a user defined, in general human recognizable, generalization or abstraction of a portion of the underlying data distribution that captures one or a combination of attributes of the corresponding samples.

Hence, we can identify the subset of the training dataset $\mathcal{D}_{train}$, which includes only instances of the concept under examination, denoted by $\mathcal{P}^c = \{x^c\}$, and its complement $\mathcal{N}^c = \{x^{\neg c}\}$, which, in contrast, contains only samples where the concept is missing. Furthermore, we assume that the target network learns to map these concepts onto distributions in their latent space $f(\mathcal{P}^c)$, whereas the complete network can be decomposed as $h = g \circ f : \mathcal{X} \to \mathcal{Y}$ and $f : \mathcal{X} \to \mathcal{L}$ maps features to latent representations. In addition, $\mathcal{X} \subseteq \mathbb{R}^d$ is the input space, $\mathcal{Y} \subseteq \mathbb{R}^k$ the target classification output space, and $\mathcal{L} \subseteq \mathbb{R}^l$ the latent space.

Fig. 1. Concepts used in the CIFAR10 experiments, with an example of each present class.

Therefore, as a first step, a set of seven CNNs with depths ranging from three to nine layers was prepared and trained on two datasets: CIFAR-10 and CIFAR-100 [11]. Then, concept discovery tools can be applied to each hidden layer of all target models to analyze the development of learned representations.

In order to extract the concepts, an auxiliary concept classifier is trained to distinguish between $\mathcal{P}^c$ and $\mathcal{N}^c$ and will be denoted by $l_c : \mathcal{L} \to \mathcal{C}$, where $\mathcal{C} = [0,1]$ is the unit interval indicating concept membership. In particular, starting with a predefined set of m concepts $\{c_1, ..., c_m\}$ for a given $\mathcal{D}_{train}$, for any choice of concept c an l_c can be found. In the following, the support vector classifier (SVC) will be used as all l_c and trained on a randomly sampled subset $\mathcal{D}' \subset \mathcal{D}_{train}$ of latent representations that preserves the original data distribution.

The set of predefined concepts for CIFAR-10 is illustrated in Fig. 1. For each concept c a list of classes containing the concept was created, all instances of which were labeled as *concept positive* and instances of the remaining classes as *concept negative*. In case of CIFAR-100 the set of 20 superclasses (coarse labels) was adopted and used as concept set. Table 1 displays the concepts (superclasses) with the corresponding classes (fine labels).

Table 1. CIFAR100 concepts and classes.

Superclass	Labels
Aquatic mammals	beaver, dolphin, otter, seal, whale
Fish	aquarium fish, flatfish, ray, shark, trout
Flowers	orchids, poppies, roses, sunflowers, tulips
Food containers	bottles, bowls, cans, cups, plates
Fruit and vegetables	apples, mushrooms, oranges, pears, sweet peppers
Household electrical devices	clock, computer keyboard, lamp, telephone, television
Household furniture	bed, chair, couch, table, wardrobe
Insects	bee, beetle, butterfly, caterpillar, cockroach
Large carnivores	bear, leopard, lion, tiger, wolf
Large man-made outdoor things	bridge, castle, house, road, skyscraper
Large natural outdoor scenes	cloud, forest, mountain, plain, sea
Large omnivores and herbivores	camel, cattle, chimpanzee, elephant, kangaroo
Medium-sized mammals	fox, porcupine, possum, raccoon, skunk
Non-insect invertebrates	crab, lobster, snail, spider, worm
People	baby, boy, girl, man, woman
Reptiles	crocodile, dinosaur, lizard, snake, turtle
Small mammals	hamster, mouse, rabbit, shrew, squirrel
Trees	maple, oak, palm, pine, willow
Vehicles 1	bicycle, bus, motorcycle, pickup truck, train
Vehicles 2	lawn-mower, rocket, streetcar, tank, tractor

The next step consists of unsupervised (automatic) concept extraction via agglomerative clustering, starting from the same latent layer embeddings $\mathcal{L}$. The approach is inspired and adapted from [15], where, however, clustering was used to disentangle the strongest activations of a single neuron. On the contrary, in this work an agglomerative clustering algorithm was applied to latent embeddings $f(\mathcal{D}')$ mimicking the distribution of the underlying dataset. Note that agglomerative clustering does not require a fixed number of clusters. Hence, a dynamic distance threshold was implemented to ensure that the number of discovered clusters remains within a range between two and twice the size of the concept set. Additionally, the *euclidean* distance metric and the *ward* linkage criterion were used.

The clustering algorithm was applied to each hidden layer of the target model to investigate whether different levels of semantic abstraction emerge as the network depth increases. Another motivation for using unsupervised learning is the potential to uncover network-inherent abstractions that may complement human-defined concepts.

A set of clusters $\{cl_1^\ell, ..., cl_n^\ell\}$ was obtained for each layer ℓ of the target model. This enables a comparative study by training an auxiliary classifier l_{cl}^ℓ for each cluster cl and l_c^ℓ for each concept c in layer ℓ. As mentioned above, we employ a binary Support Vector Classifier (SVC) with a radial basis function (RBF) kernel, which preserves isometries in the latent space. Unlike the original implementation by [7], where the positive and negative sample sets were balanced to improve SVC training, we adopt a sampling strategy that draws examples uniformly from the original dataset distribution to better reflect its natural structure. A concise overview of the pipeline is provided in Algorithm 1.

Algorithm 1. Concept Search and Accuracy Evaluation

Input: Dataset $\mathcal{D}_{\text{train}}$, subset size n, initial distance threshold d
Output: Test accuracies of SVC classifiers trained on concept and cluster sets across
network layers

Create subsets $\mathcal{D}'_{\text{train}}, \mathcal{D}'_{\text{test}}$ of size n;
foreach *layer ℓ* **do**
    Extract latent representations $\mathcal{R}_{\text{train}} = f(\mathcal{D}'_{\text{train}})$, $\mathcal{R}_{\text{test}} = f(\mathcal{D}'_{\text{test}})$;
    Perform hierarchical clustering on $\mathcal{R}_{\text{train}}$ starting with d;
    Adjust d if necessary ;
    Assign $\mathcal{R}_{\text{test}}$ to nearest cluster centroids;
    Save discovered *cluster set* $= \{cl_1^\ell, ..., cl_n^\ell\}$;
    foreach *concept c_i* **do**
        Assign ground truth *concept membership* $\in \{0, 1\}$ to $\mathcal{R}_{\text{train}}$ and $\mathcal{R}_{\text{test}}$;
        Train SVC classifier for *concept membership*;
        Save training and testing *accuracies*
    foreach *cluster cl_j* **do**
        Assign ground truth *cluster membership* $\in \{0, 1\}$ to $\mathcal{R}_{\text{train}}$ and $\mathcal{R}_{\text{test}}$;
        Train SVC for the *cluster membership*;
        Save training and testing *accuracies*

Aggregate and save results;

After training a separate SVC $\sim l_{cl}^\ell \; \forall \; cl \in \{cl_1^\ell, ..., cl_n^\ell\}$ for all clusters in every layer ℓ and $\sim l_c^\ell \; \forall \; c \in \{c_1, ..., c_n\}$ for all concepts, their test accuracy was evaluated on a separate hold-out set and was averaged over the respective set in each layer ℓ. The test accuracies were documented and compared for each of the target models. Since, all components of our combined approach are *post hoc* and *model agnostic*, they can be applied to any target model with white-box access to its latent activations. However, we do assume the dataset is labeled with respect to concept membership.

4   Experiments

Initially, a set of seven target CNNs with the number of layers varying from 3 to 9 was trained on the CIFAR-10 and CIFAR-100 datasets. The total number of trainable parameters for each network is provided in Table 2.

Table 2. Number of trainable parameters for target networks trained on CIFAR-100.

Layers	3	4	5	6	7	8	9
Params	209,988	2,205,764	599,684	705,412	274,564	276,420	439,140

Provided the trained target models, the latent activations of the sampled subset $\mathcal{R}_{\text{train}}$ were clustered with a dynamic distance threshold. Since the concept membership of each sample is also known we can compare the sample counts in a cluster–concept co-occurrence matrix to inspect potential overlap of the groups. A heatmap representation of the co-occurrence matrix computed in the fourth layer of the 8-layer deep target model can be seen in Fig. 2. In this setting, two clusters were discovered, one of which has a substantial overlap with the concept of *animal*.

According to Algorithm 1, auxiliary SVCs $l_{cl,\ell}$ for each cluster cl and $l_{c,\ell}$ for each concept c, were trained. These classifiers partition the high-dimensional latent space of the target layer into concept (or cluster) positive regions $\mathcal{H}^c$ and their complements, the negative regions $\mathcal{H}^{\neg c}$.

To visualize this partitioning, we applied UMAP [13], a manifold learning algorithm that projects high-dimensional data into two dimensions while preserving local and global distances as faithfully as possible. As a result, cluster-like structures often remain visible in the projected space, making it easier to qualitatively assess the separation achieved by the SVCs.

Figure 3 shows the UMAP projection of latent representations from the fourth convolutional layer of an 8-layer deep CNN trained on CIFAR-10. The black line between the contoured areas represents the decision boundary of the trained SVC, i.e. the concept (cluster) region $\mathcal{H}^{c/cl}$. The blue points represent concept (cluster) - negative and the orange ones concept (cluster) - positive activations with respect to the predefined concept *animal* and clusters 0 and 1.

The dimensionally reduced projections qualitatively support the observations from Fig. 2. In particular, Fig. 3a shows the decision region associated with the concept *animal*, which closely matches the shape and extent of cluster 1 in Fig. 3b. In contrast, cluster 0 (Fig. 3c) shows minimal or no overlap with the *animal* concept.

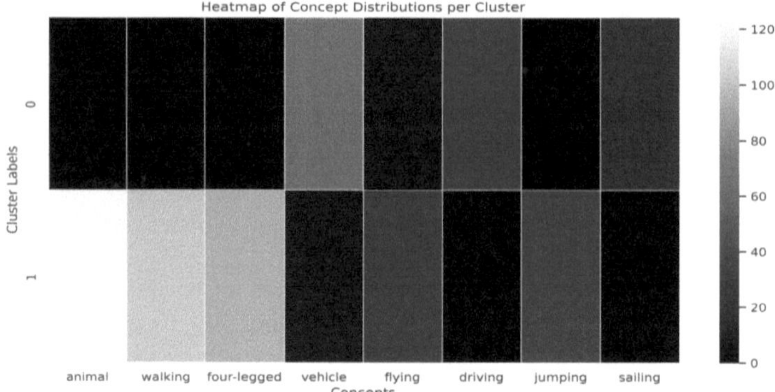

Fig. 2. A heatmap comparing cluster - concept co-occurrences in the fourth layer of target CNN trained on CIFAR-10. The highest overlap arises between 1st cluster and the concept *animal*.

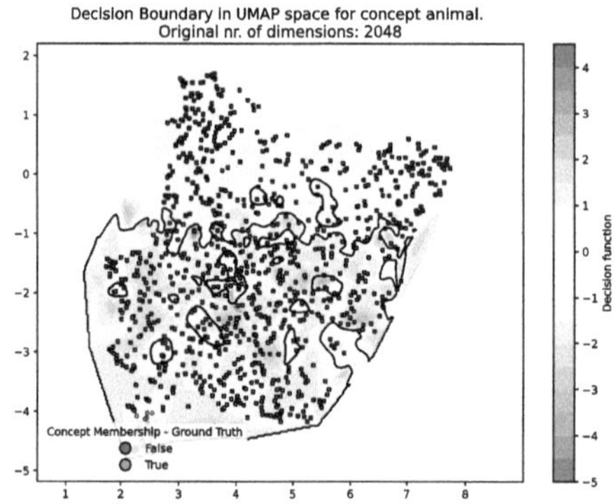

(a) Animal concept 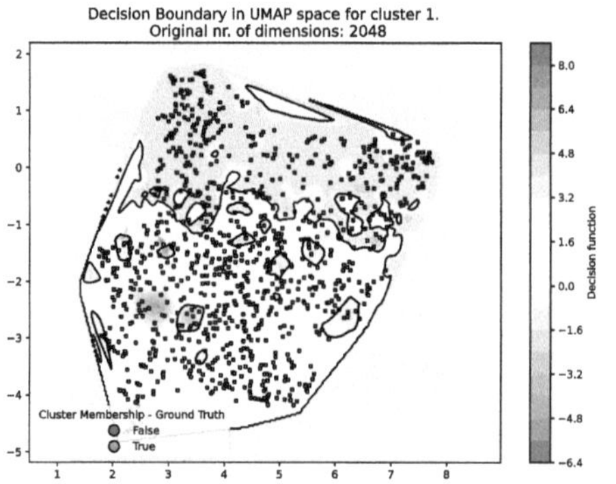   (b) Cluster 1

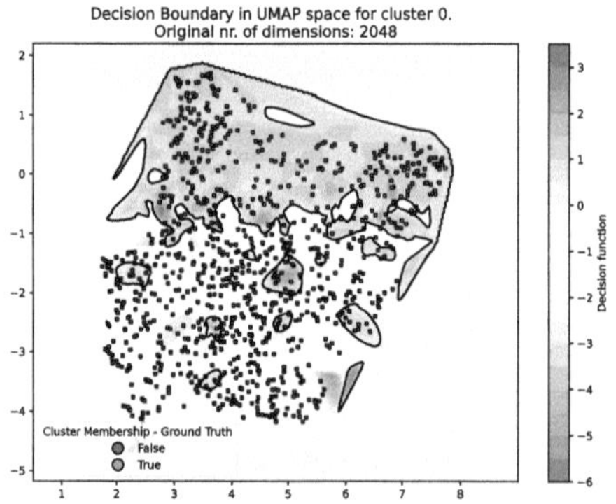

(c) Cluster 0 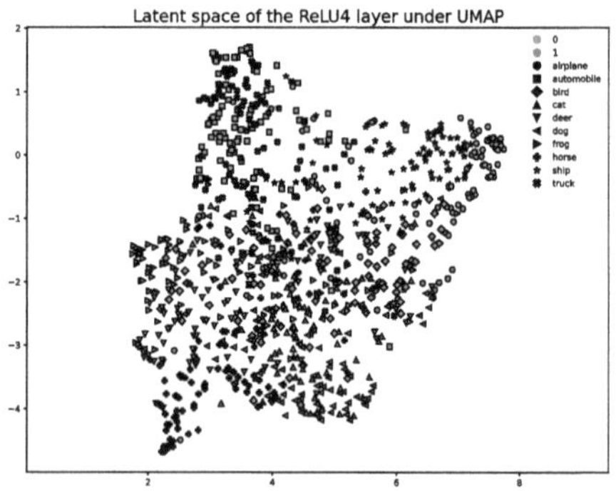   (d) Clusters vs. target labels

Fig. 3. UMAP projections of $\mathcal{R}'_{\text{train}}$ for CIFAR-10 onto a two-dimensional space, showing positive activations $f(\mathcal{P}^c)$ (orange) and negative ones $f(\mathcal{N}^c)$ (blue), along with SVC decision boundaries (black line) for concept *animal* (a), and discovered clusters 0, 1 (b), (c). (d) shows UMAP reduction of the latent distribution. Marker color indicates ground truth cluster identity and marker shape denotes original target class. (Color figure online)

To further examine the internal composition of the clusters, Fig. 3d encodes target class membership using marker shape and cluster identity using marker color. The distribution suggests that while cluster 1 primarily consists of samples labeled as *animals*, cluster 0 is dominated by vehicle-related classes such as *airplane, automobile, ship, truck*, again confirming the co-occurrence matrix counts from Fig. 2.

5   Results

Having examined the latent distributions and trained the corresponding concept and cluster classifiers, we proceed to evaluate their classification performance on held-out test samples. For each layer of the target networks, we computed the SVC test accuracies, i.e. accuracies of the trained concept and cluster classifiers, separately across all layers of the target CNN. Figure 4 presents the results for a 9-layer convolutional neural network trained on CIFAR-10, while Fig. 5 shows the results for a 9-layer network trained on CIFAR-100. For results from target networks with 4, 6, and 8 layers, see Appendix A.

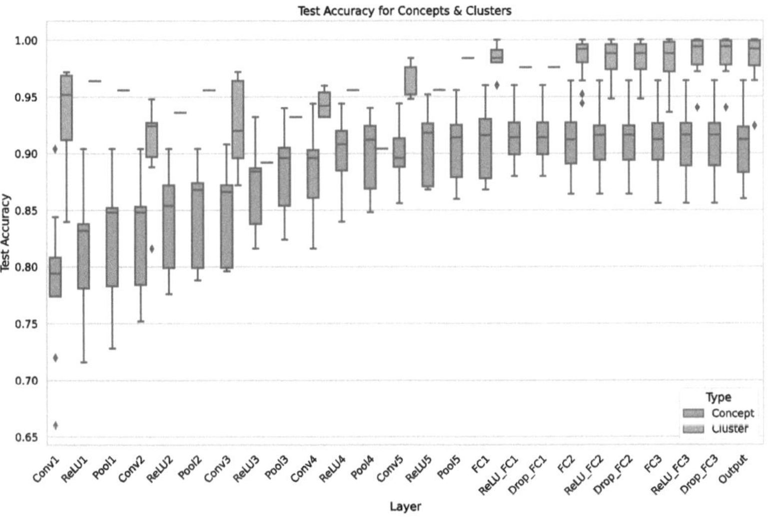

Fig. 4. SVC accuracy on a 9-layer CNN trained on CIFAR-10 dataset.

Across almost all layers, the SVCs trained to predict cluster membership consistently achieve higher test accuracy compared to those trained for concept membership. Additionally, for both classifier types, accuracy increases towards the output layer.

These findings are consistent with established expectations regarding internal representations in deep networks. The superior performance of cluster classifiers is attributable to the inherent geometric coherence of clusters. By construction, clusters consist of spatially proximate samples in latent space, making them more separable.

In contrast, concept instances may be distributed across multiple regions of the latent space without a consistent spatial pattern, meaning their separability is not guaranteed and must be empirically validated.

Furthermore, the increase in accuracy with layer depth supports the manifold disentanglement hypothesis, originally proposed by [4] and later empirically tested by [16].

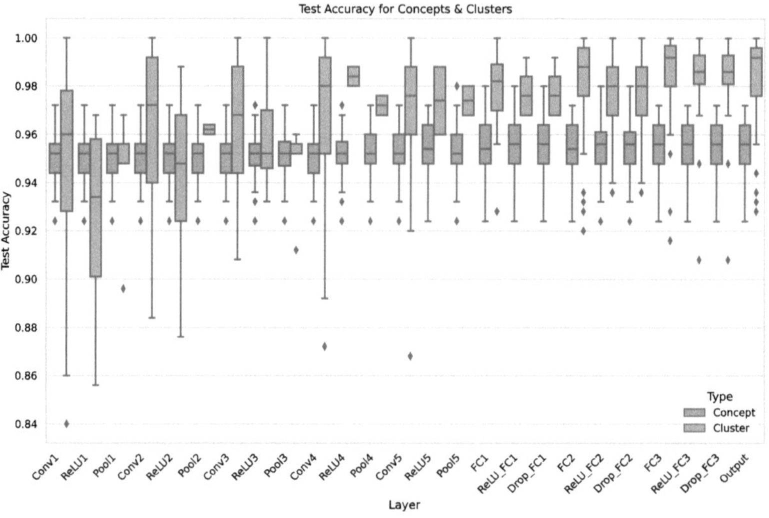

Fig. 5. SVC accuracy on a 9-layer CNN trained on CIFAR-100 dataset.

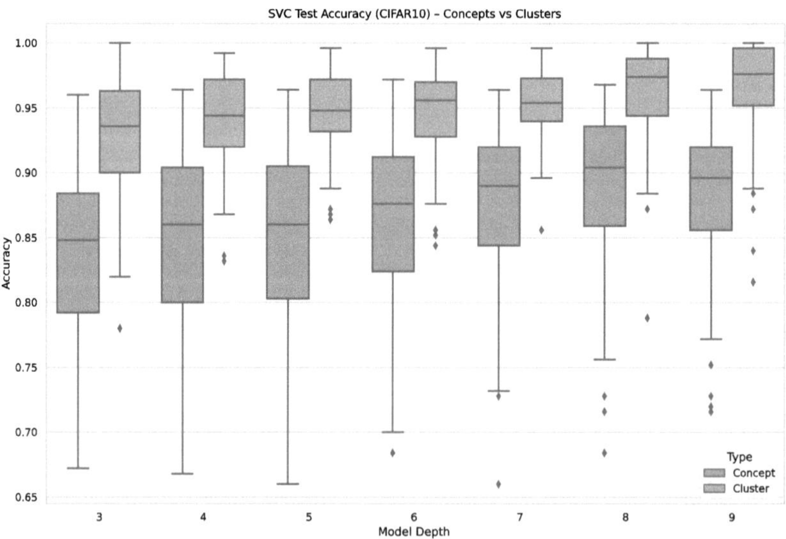

Fig. 6. SVC accuracy across all CNNs trained on CIFAR-10 dataset averaged over available layers.

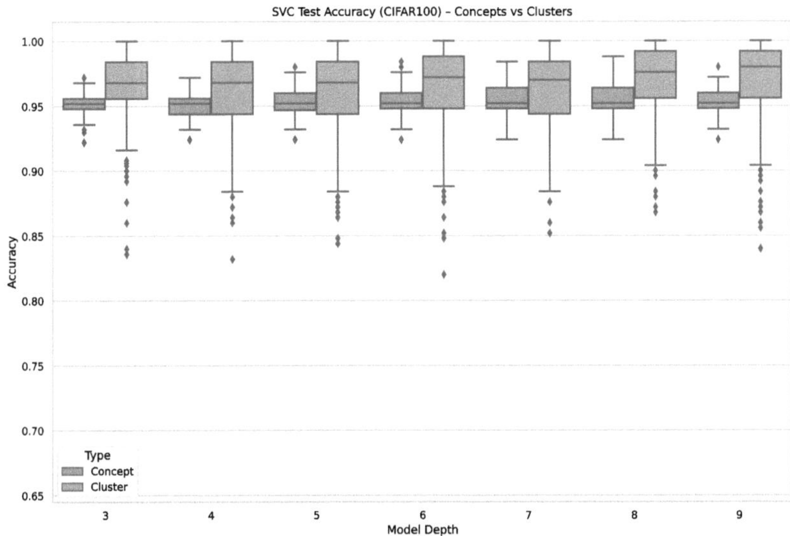

Fig. 7. SVC accuracy across all CNNs trained on CIFAR-100 dataset averaged over available layers.

This hypothesis posits that as the data propagate through the network, the underlying class manifolds gradually become more linearly separable. Since concepts are often associated with specific classes or combinations thereof, their representations are likewise expected to become more coherent and distinguishable in deeper layers.

An overall evaluation across all trained networks yields similar results to those above, demonstrating higher accuracy of auxiliary classifiers recognising clusters compared to predefined concepts.

In order to compare the SVCs performance in different networks against each other, the accuracy was averaged over the available layers in each CNN and plotted over the total network depth. Figures 6, 11, 12 and 7 corroborate the observation from Fig. 4 and 5 that cluster classifiers perform significantly better than concept classifiers, regardless of the network depth.

However, a new trend seems to appear, especially in the CIFAR-10 plot, suggesting an increase in mean performance with growing network depth. This is most arguably due to the fact that later layers achieve higher accuracy in general, hence the deeper the network, the more layers which allow higher performing SVCs are included. In other words, deeper CNNs enable the latent manifold to become less entangled with respect to the target classes, which as a secondary effect separates both cluster and concept regions spatially as well.

In Fig. 8 the differences between the determined cluster and concept SVC accuracies are displayed. The distinction seems to be more significant for networks trained on CIFAR-10 dataset as opposed to CIFAR-100. Nevertheless, all the distribution medians lie above the zero line. For the given sample of accuracy differences the p-values were evaluated via a statistical t-test for the null hypothesis that the expected value of these

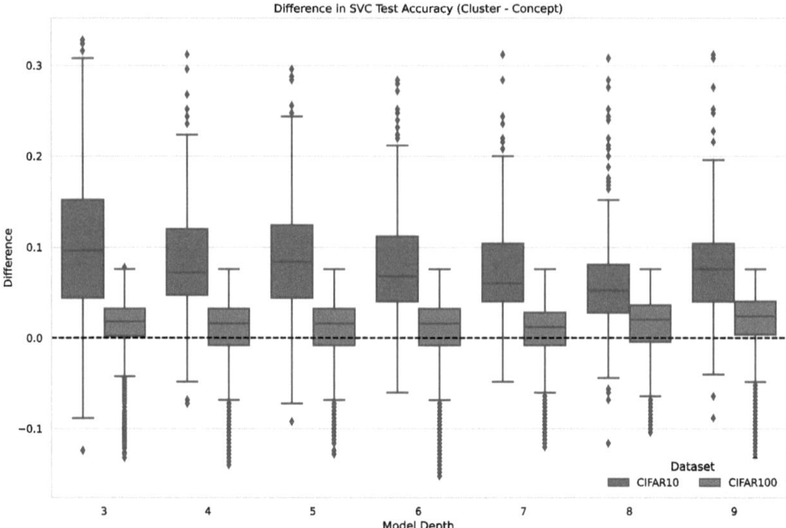

Fig. 8. Difference between cluster and concept SVC accuracies for all considered CNNs and both training sets.

observations is equal to zero. The p-value results are below the order of magnitude of 10^{-65} for all target models and datasets.

6 Cluster Visualization

To gain insight into the semantic content and meaning of the clusters, two visualization techniques were introduced and applied to every layer of the target model. Before visualization, only the five largest clusters were selected and the number of data points in each discovered cluster was reduced to five samples closest to the centroid vectors of the cluster $\mathbf{v}^{cl}$ (distance measured by the Euclidean norm), to maintain manageable size and provide clarity and readability.

The aforementioned UMAP projection tool was adapted so that, instead of data points, the input images x_i^{cl} are displayed at the position of their projected activation vector $UMAP(f(x_i^{cl}))$ for each of the selected clusters cl.

Moreover, we implemented a method called perturbation-based *Concept Localization Maps* (CLM) proposed in [21] (see also [12] for application) to highlight the most important input regions with respect to the cluster membership score. To define the score, the dot (scalar) product of the sample activation vector with the centroid of the cluster was calculated and taken as the membership score $\mathcal{S}_{cl,i}$ with cl being the cluster under consideration and i the index of the sample. Formally, the score can be expressed as

$$\mathcal{S}_{cl}(x_i) = \langle f(x_i), \mathbf{v}^{cl} \rangle = \sum_{k=1}^{l} f_k(x_i) \cdot v_k^{cl} \tag{1}$$

where l is the dimension of the latent space $\mathcal{L}$.

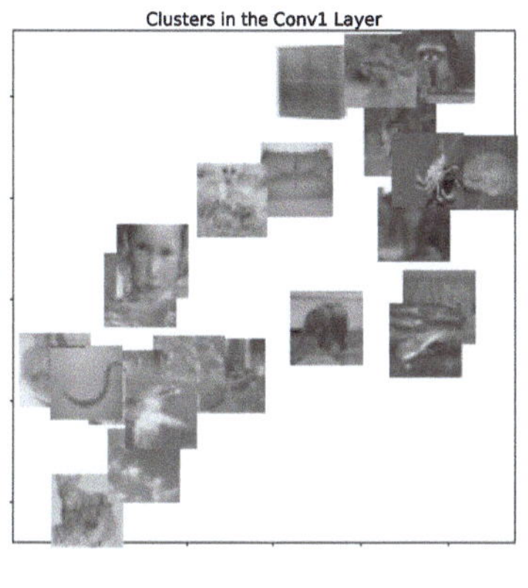

(a) 1st convolutional layer

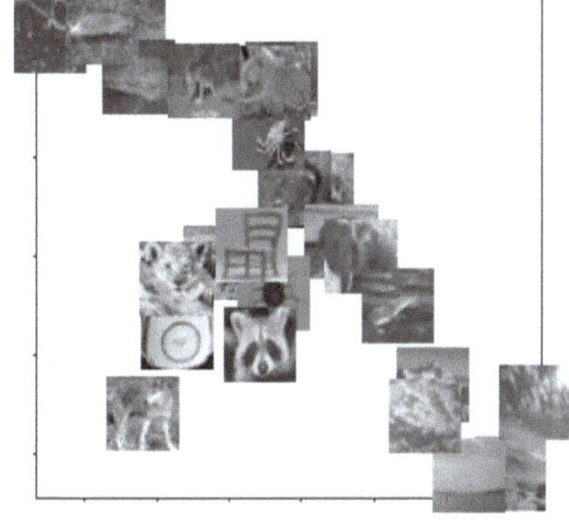
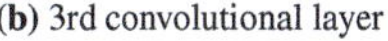

(b) 3rd convolutional layer

(c) 5th convolutional layer

(d) 2nd fully connected layer

Fig. 9. UMAP projections of five largest clusters with their five closest samples.

In order to obtain a saliency map for an input x_i with respect to the prediction of cluster membership, parts of the input were iteratively occluded by black patches with size 4×4 and stride 2. The membership score $\mathcal{S}_{cl,i}$ was calculated anew at each iteration and the values were stored and averaged for each pixel.

We applied the above techniques to the five most populous clusters extracted from four layers { Conv 1, Conv 3, Conv 5, FC 2 }, where Conv stands for a convolutional layer and FC for fully connected, of the nine layer CNN, i.e. the same target network as in Fig. 5.

Figure 9 displays sample activation distributions in four out of nine layers, specifically, every other layer starting from the first hidden one. The first noticeable characteristic is that in the first convolutional layer, images are primarily distributed according to their dominant color. In the third convolutional layer, the distribution is mainly influenced by the background's type and color. By the fifth convolutional layer, similar objects begin group together based on patterns and shapes. In the final displayed layer (the second fully connected layer), the clusters are almost completely separated: vehicles occupy one corner, while humans, still slightly overlapping with animals, are located in the opposite corner of the projected space. Between these extremes lie mixed clusters, comprising utility objects and animals in natural environments.

Here we take a closer look at the input features most responsible for the cluster membership of the same set of samples as in the previous UMAP visualization. Figures 10 to 13 display the five closest samples for each of the five clusters in a "double column", where the original images are on the left side and the same images overlaid with a p-CLM saliency map are on the right.

Fig. 10. P-CLM for clusters extracted from the 1st convolutional layer.

As for the first set of clusters in Fig. 10, the regions playing the key role are very diverse. The first cluster from the left seems to be interested in the middle part of the image, particularly searching for a bright spot. Second and third clusters notice corner and smaller middle regions, whereas fourth and fifth concentrate on the background close to the border of the image. Nevertheless, it is worth noting that the color scheme of the images seems to play a significant role in the cluster assignment.

Fig. 11. P-CLM for clusters extracted from the 3rd convolutional layer.

In 3rd convolutional layer almost all clusters focus on the background pixels with few exceptions of sharp contrast areas such as the eyes of the racoon in the second cluster from the left, and the dark bug in the fifth cluster.

Fig. 12. P-CLM for clusters found in the 5th convolutional layer.

The situation changes progressively as we investigate deeper into the network. In Fig. 12 central areas of the images are assigned the highest importance with respect to cluster membership. These include specific body parts of animals and humans, defining structures of human constructions and edges between objects and their surrounding.

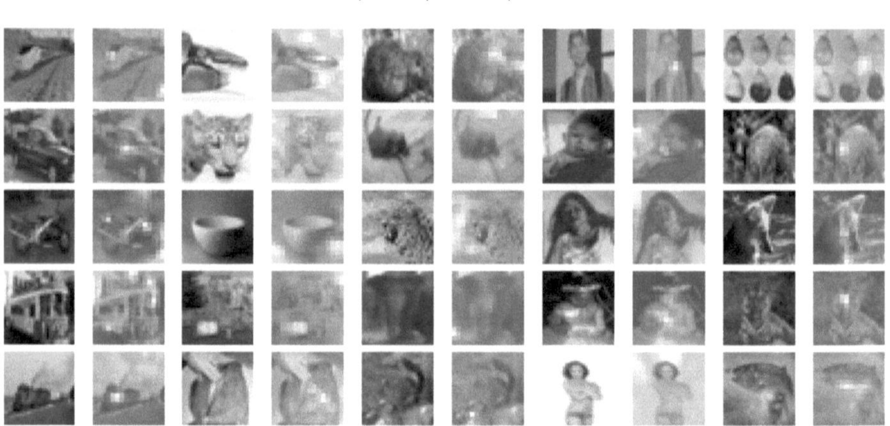

Fig. 13. P-CLM for clusters found in the 2nd fully connected layer.

In the last examined layer, the 2nd fully connected one a combination of 3 semantically pure and 2 mixed clusters was extracted. The first cluster contains instances of vehicles only, with parts such as front windows, doors or locomotive front carrying the highest score. The third cluster consists of three small animals and two large ones and the fourth includes only images of humans, where the salient features cover neck, chin, hand, torso and head. The second cluster is a little bit harder to interpret, since there are man-made objects as well as animals among the samples with areas from the background being highlighted. Except for one image of pears the fifth cluster reveals animal images with head neck and belly regions being attributed highest importance.

7   Conclusion and Future Work

The present work introduces a new paradigm in concept-based explainability, which aims to contrast human-defined (i.e. familiar) semantic meanings and feature abstractions with the emergent network-inherent data structures. The prospect of this endeavor is building a solid framework for understanding neural networks in a more faithful manner. This is admittedly not a simple task, since the process of reading out the network-inherent structures is equally subject to human interpretation and hence bias. However, a first step in this direction was taken by computing clusters of latent representations throughout the network layers and then quantitatively comparing the accuracy of an auxiliary classifier trained to distinguish human-defined concepts as opposed to the discovered clusters. The experiments with CIFAR-10 and CIFAR-100 datasets confirm our presumption that cluster-based regions are a more fitting representation of the latent manifolds than human-defined concepts.

Furthermore, the evolution of the semantic content of the discovered clusters over the network depth was observed by utilizing two techniques. The five cluster-closest activations of the input images were plotted via UMAP projection, in order to gain insight into their spatial distribution. Secondly, perturbation-based Concept Localization Maps (p-CLMs) were calculated for the same five inputs with respect to cluster membership score. Our findings demonstrate varying semantic depth as data flows through the network, with simpler meanings such as color or texture being more relevant in the earlier layers and shapes, body parts and complex structures more prominent in the latter layers.

Finally, there are still open research questions worth exploring, such as whether more hidden patterns emerge in the mixed clusters that were not detectable with the present visualization techniques. Another interesting pursuit would be investigating whether similar results can be obtained by applying the presented approach to deeper and more complex target model architectures and higher-resolution datasets.

Acknowledgments. This research was supported by Horizon Europe MSCA project TRAIL, no. 101072488 and by Slovak Grant Agency for Science (VEGA), project 1/0373/23.

Disclosure of Interests. The authors have no competing interests to declare which are relevant to the content of this article.

Appendix

A   Evaluation of CNNs Across Varying Depth

Here, additional results from the comparison of SVC accuracy between the predefined concepts and extracted clusters are presented. Figure 14 shows the 4-layer CNNs for the two training sets. Generally, the similar trends as in Figs. 4 and 5 are observed here: the accuracy increases toward the output layer, and cluster evaluation yields higher values than concepts.

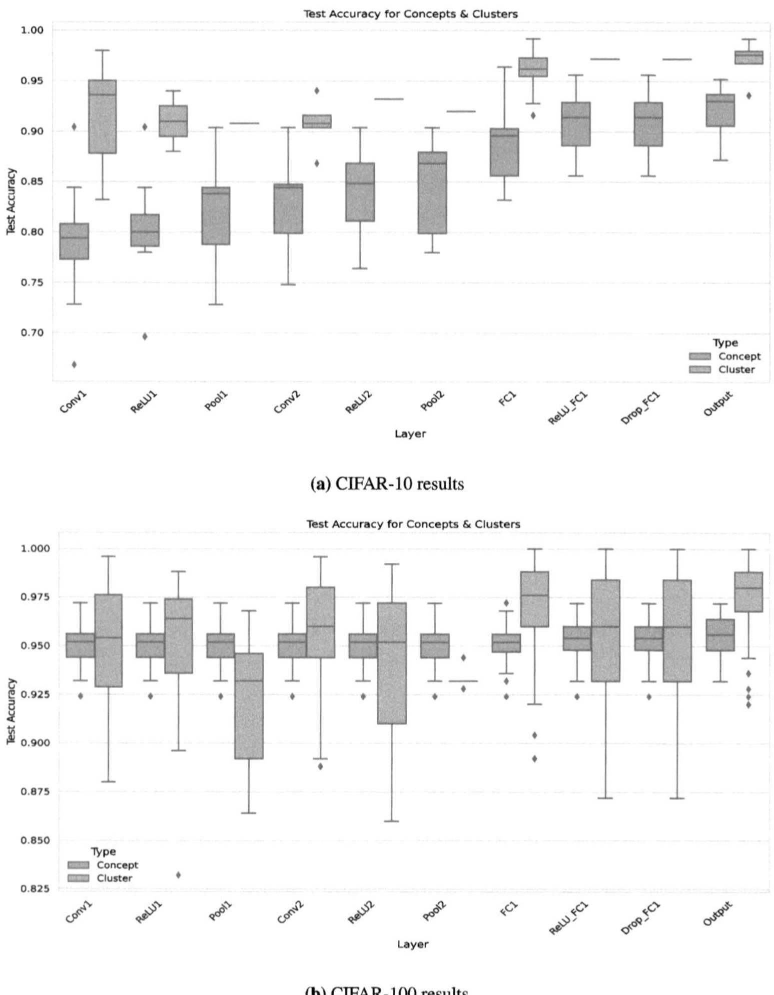

(a) CIFAR-10 results

(b) CIFAR-100 results

Fig. 14. SVC accuracies evaluated for 4-layer deep CNNs on CIFAR-10 and CIFAR-100.

The same behavior persists for the other two network depths shown in Figs. 15 and 16. However, it should be noted that there are several differences between SVC accuracy when trained on the smaller CIFAR-10 in contrast to the larger CIFAR-100, the former of which contains 10 classes and the latter 100. In addition, there are 10 predefined concepts in CIFAR-10 that substantially overlap, whereas 20 concepts in CIFAR-100 that do not overlap at all, see Fig. 1 and Table 1.

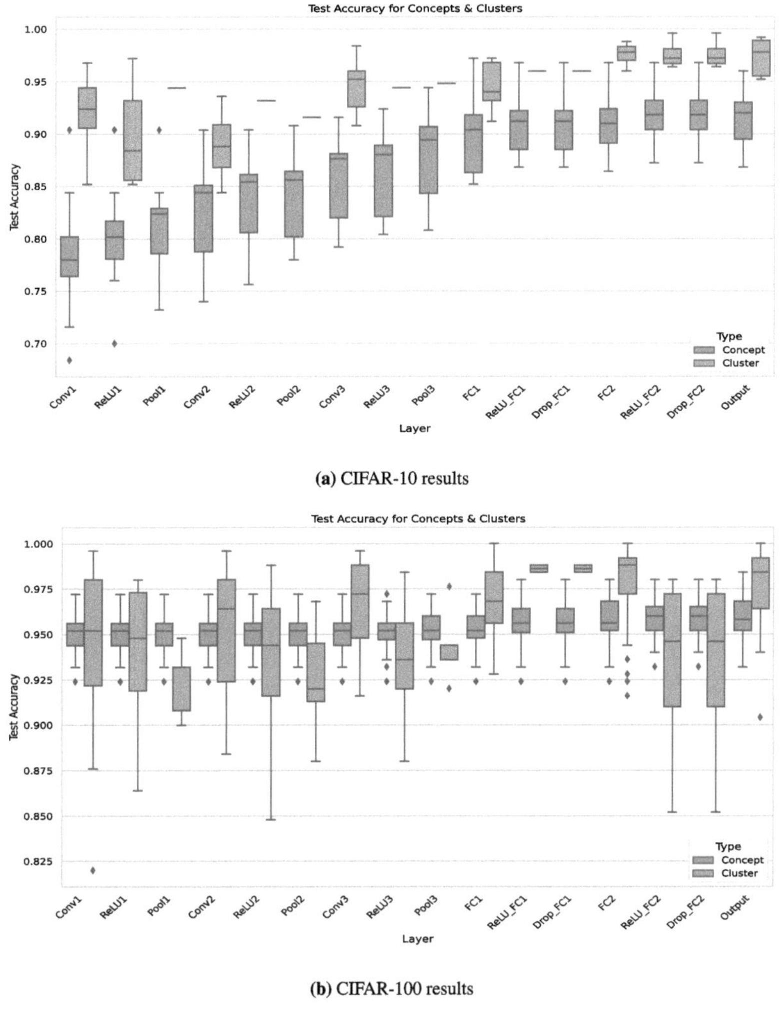

(a) CIFAR-10 results

(b) CIFAR-100 results

Fig. 15. SVC accuracies evaluated for 6-layer deep CNNs on CIFAR-10 and CIFAR-100.

For CIFAR-10 both effects can be observed consistently: across all networks accuracy rises with depth of the layer, and in each layer clusters outperform concepts. Although in the case of CIFAR-100 the results exhibit substantial variability, the overall trend remains consistent. For an summary of the results, see Fig. 8.

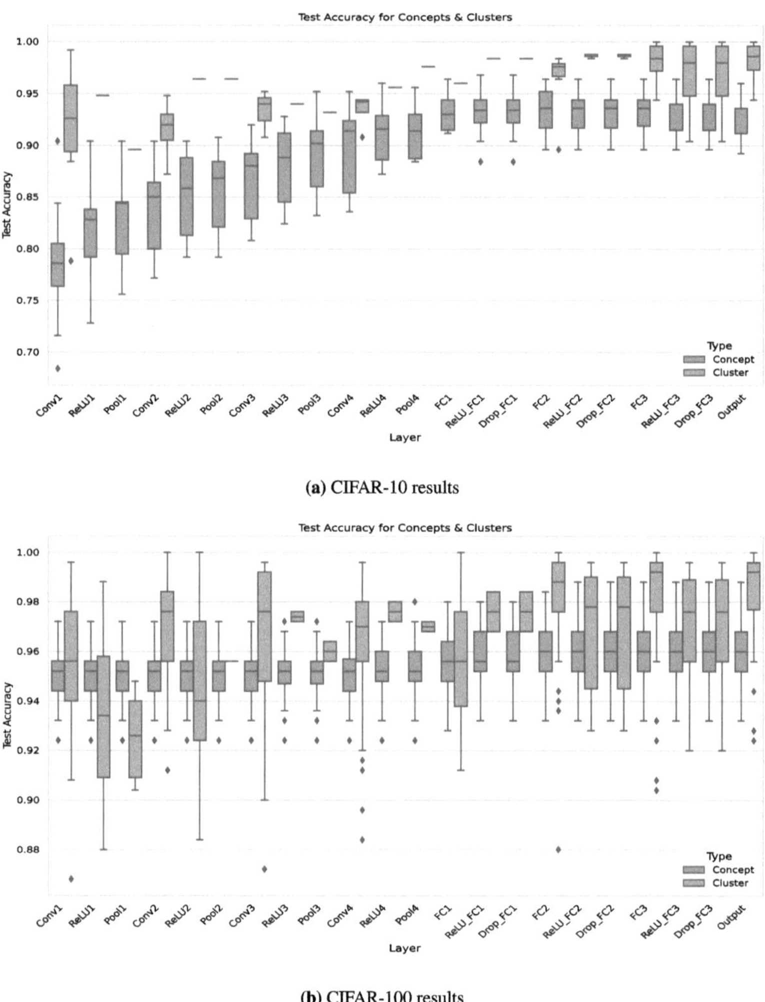

(a) CIFAR-10 results

(b) CIFAR-100 results

Fig. 16. SVC accuracies evaluated for 8-layer deep CNNs on CIFAR-10 and CIFAR-100.

References

1. Ali, S., et al.: Explainable artificial intelligence (XAI): what we know and what is left to attain trustworthy artificial intelligence. Inf. Fusion **99** (2023)
2. Bai, A., Yeh, C.K., Lin, N.Y., Ravikumar, P.K., Hsieh, C.J.: Concept gradient: concept-based interpretation without linear assumption. In: International Conference on Learning Representations (2023)
3. Bereska, L., Gavves, S.: Mechanistic interpretability for AI safety - a review. Trans. Mach. Learn. Res. (2024). https://openreview.net/forum?id=ePUVetPKu6
4. Brahma, P., Wu, D., She, Y.: Why deep learning works: a manifold disentanglement perspective. IEEE Trans. Neural Netw. Learn. Syst. **27**, 1–12 (2015)

5. Chen, Z., Bei, Y., Rudin, C.: Concept whitening for interpretable image recognition. Nat. Mach. Intell. **2**, 1–11 (2020)
6. Cover, T.M.: Geometrical and statistical properties of systems of linear inequalities with applications in pattern recognition. IEEE Trans. Electron. Comput. **14**(3), 326–334 (1965). https://doi.org/10.1109/PGEC.1965.264137
7. Crabbé, J., van der Schaar, M.: Concept activation regions: a generalized framework for concept-based explanations. In: Neural Information Processing Systems (2022)
8. Ghorbani, A., Wexler, J., Zou, J., Kim, B.: Towards automatic concept-based explanations. In: Neural Information Processing Systems (2019)
9. Kim, B., et al.: Interpretability beyond feature attribution: quantitative testing with concept activation vectors (TCAV). In: International Conference on Machine Learning (2018)
10. Koh, P.W., et al.: Concept bottleneck models. In: International Conference on Machine Learning, vol. 119, pp. 5338–5348 (2020)
11. Krizhevsky, A.: Learning multiple layers of features from tiny images. Technical report, University of Toronto (2009). https://www.cs.toronto.edu/~kriz/learning-features-2009-TR.pdf
12. Lucieri, A., Bajwa, M.N., Dengel, A., Ahmed, S.: Explaining AI-based decision support systems using concept localization maps. In: Yang, H., Pasupa, K., Leung, A.C.-S., Kwok, J.T., Chan, J.H., King, I. (eds.) ICONIP 2020. CCIS, vol. 1332, pp. 185–193. Springer, Cham (2020). https://doi.org/10.1007/978-3-030-63820-7_21
13. McInnes, L., Healy, J., Melville, J.: UMAP: uniform manifold approximation and projection for dimension reduction (2020). arXiv:1802.03426
14. Nicolson, A., Schut, L., Noble, J.A., Gal, Y.: Explaining explainability: recommendations for effective use of concept activation vectors (2025). arXiv:2404.03713
15. O'Mahony, L., Andrearczyk, V., Müller, H., Graziani, M.: Disentangling neuron representations with concept vectors. In: Conference on Computer Vision and Pattern Recognition, pp. 3770–3775. IEEE (2023)
16. Pócoš, S., Bečková, I., Kuzma, T., Farkaš, I.: Assessment of manifold unfolding in trained deep neural network classifiers. In: Trustworthy AI - Integrating Learning, Optimization and Reasoning, pp. 93–103. Springer Nature (2021)
17. Räuker, T., Ho, A., Casper, S., Hadfield-Menell, D.: Toward transparent AI: a survey on interpreting the inner structures of deep neural networks. In: Conference on Secure and Trustworthy Machine Learning (SaTML), pp. 464–483. IEEE (2023)
18. Schrouff, J., et al.: Best of both worlds: local and global explanations with human-understandable concepts (2022). arXiv:2106.08641
19. Shang, C., Zhou, S., Zhang, H., Ni, X., Yang, Y., Wang, Y.: Incremental residual concept bottleneck models (2024). arXiv:2404.08978
20. Sundararajan, M., Taly, A., Yan, Q.: Axiomatic attribution for deep networks. In: International Conference on Machine Learning, vol. 70, pp. 3319–3328 (2017)
21. Zeiler, M.D., Fergus, R.: Visualizing and understanding convolutional networks. In: European Conference on Computer Vision, pp. 818–833. Springer (2014)

An Explainable Multi-domain Document Summarization Framework Using Domain-Aware Fine-Tuned Large Language Models

Donggyu Kim, Kunyoung Kim, and Mye Sohn(✉)

Department of Industrial Engineering, Sungkyunkwan University, Suwon 16419, Korea
{dkpotter,kimkun0,myesohn}@skku.edu

Abstract. Large Language Model (LLM)-based text summarization has significantly advanced the generation of coherent and semantically rich summaries. However, these models still face critical limitations in real-world applications, particularly in multi-domain scenarios. Moreover, lack of explainability hinders the trust and reliability of generated summaries. To address these issues, we propose a novel eXplainable Multi-domAin document Summarization (X-MAS) framework that enhances both the performance and explainability of multi-domain text summarization. X-MAS utilizes semantic clustering of documents using BERT embeddings and HDBSCAN to discover the domains of each document in the corpus. Based on the domains, X-MAS utilizes domain-specific fine-tuned LLMs to generate the summary. Finally, X-MAS utilizes keyword matching and BERT to map summary content back to its source documents. We evaluate X-MAS on a real-world multi-domain document dataset and demonstrate that it outperforms existing methods in both summary quality and explainability.

Keywords: Document Summarization · Explainable AI · Large Language Model · Natural Language Processing

1 Introduction

With the explosive growth of textual data on the internet, individuals and organizations suffer from information overload. This challenge has led to the development of automatic text summarization techniques, especially for multi-document settings [1]. Early research predominantly focused on extractive summarization, which constructs summaries by directly selecting sentences or phrases from the source documents [2]. While this approach preserves fidelity to the original content, it often relies on shallow heuristics, such as sentence position or word frequency, resulting in summaries that may lack coherence and semantic richness.

Recent advances in Natural Language Processing (NLP), particularly the emergence of large language models (LLMs), have shifted the focus toward abstractive summarization [3, 4], which generates summaries that mimic human writing by capturing the

© The Author(s), under exclusive license to Springer Nature Switzerland AG 2026
F. Marcelloni et al. (Eds.): IJCCI 2025, CCIS 2829, pp. 398–413, 2026.
https://doi.org/10.1007/978-3-032-15638-9_23

underlying meaning of the source documents. However, abstractive summarization still faces a major limitation which is lack of explainability. Users often cannot trace how or why certain content appears in a generated summary, making it difficult to verify the accuracy or trustworthiness of the output.

To overcome this issue, several studies have proposed explainable summarization methods for LLM-based abstractive models [5]. Some approaches identify relevant topics [6] or highlight portions of source documents associated with the summary [7]. However, these methods continue to struggle in real-world multi-domain document scenarios, where documents vary widely in format, content, and writing style. Problems such as factual inconsistency, hallucination, and domain misalignment become more pronounced when models are applied to documents beyond their training distribution. Furthermore, explainability is often compromised in such settings, as existing methods typically overlook the underlying contextual background specific to each domain.

To address these challenges, we propose a novel eXplainable Multi-doMain document Summarization (X-MAS) framework that is designed to enhance both performance and explainability in multi-domain summarization tasks. X-MAS integrates semantic clustering of documents, domain-aware fine-tuning of LLMs, and traceable alignment between summary content and source materials.

The framework operates in three main stages. First, rather than relying on predefined domains, X-MAS automatically discovers domain clusters from the full document set using BERT-based embeddings [8] and HDBSCAN clustering [9]. For each domain, representative keywords are identified using TF-IDF [10], and domain-specific fine-tuning is performed using GPT-based models [11].

Second, for any given document set to be summarized, X-MAS uses named entity recognition (NER) to determine the most relevant domains by matching terms with the identified domain keywords. The corresponding domain-specific LLMs are then used to generate individual summaries. These are concatenated and refined using a general GPT-based summarization model to reduce redundancy and produce a coherent final summary.

Finally, to provide traceability, each sentence in the final summary is analyzed through keyword matching to associate it with its domain. BERT embeddings are then used to identify the most similar document and sentences within that domain. Relevant sections are highlighted to explain how the summary content was derived, without needing to exhaustively examine the entire corpus.

Our key contributions can be summarized as follows:

- We propose a domain-sensitive architecture that dynamically adapts LLMs to semantically clustered document groups, enabling more context-aware summarization.
- We introduce an explanation mechanism that maps summary content to its corresponding domains, documents, and sentences, thereby ensuring transparency and accountability.

The rest of this paper is organized as follows. Section 2 reviews related works. Section 3 presents the X-MAS framework in detail. Section 4 reports experimental results that demonstrate the effectiveness of our approach. Finally, Sect. 5 concludes the paper and outlines future directions.

2 Related Works

Methods for multi-document summarization are generally categorized into extractive and abstractive approaches. Extractive summarization involves selecting key sentences directly from the source documents to construct a summary. Representative extractive techniques include ontology-based, term-based, Rhetorical Structure Theory (RST)-based, and graph-based methods [12]. However, each approach has inherent limitations: ontology-based methods require extensive effort in constructing and maintaining ontologies; term-based methods often lack semantic understanding due to their reliance on word frequency; RST-based methods necessitate a deep understanding of complex discourse structures; and graph-based methods may generate summaries with grammatically unnatural connections.

In contrast, abstractive summarization generates novel sentences to convey the core information across multiple documents [13]. This can be further divided into three main categories: conventional methods, deep learning-based methods, and large language model-based methods. Among conventional methods, [14] decomposes source documents into phrases and selects those with high importance scores. These are then combined through Integer Linear Programming (ILP) to satisfy multiple constraints, though this often leads to grammatical errors. Another approach, [15], utilizes Semantic Role Labeling (SRL) to transform sentences into predicate-argument structures. It then identifies important semantic frames and generates summaries based on predefined templates. While effective, the template-based generation inherently limits the diversity of linguistic expressions.

DL-based approaches have primarily aimed to overcome the input length limitations of earlier deep learning models in multi-document summarization. For example, [16] filters important paragraphs using TF-IDF before applying a Transformer decoder for summarization. Similarly, [17] employs an LSTM-based extractor to identify key paragraphs prior to generating summaries. Although such preprocessing improves efficiency, it risks omitting essential content. To address this, [18] models inter-paragraph semantic relationships using graph structures and incorporates these into a GRU-based decoder to enhance semantic coherence. However, the graph structures are statically defined, limiting their adaptability across varying document configurations.

Finally, LLM-based methods have emerged as state-of-the-art in multi-document summarization. In [19], a QA-prompting technique is used, wherein domain-specific questions are predefined and posed to an LLM to generate answers and corresponding summaries. While effective, this approach relies heavily on human expertise to design domain-relevant questions. Meanwhile, [20] enhances summarization performance by identifying main event-centric sentences and their contextual complements across multiple documents. These are input to a fine-tuned LLM for summary generation. However, this method may underperform in non-news domains where information is not structured around central events.

3   Proposed Framework

The proposed X-MAS framework is designed to address the limitations of general LLMs in multi-domain environments by introducing semantic document clustering, domain-specific summarization, and semantic similarity-based explanation into a unified pipeline. The architecture is composed of three main stages: (1) domain assignment and domain-specific LLM fine-tuning, (2) domain-aware summarization and (3) semantic similarity-based explanation. Figure 1 provides an overview of this architecture.

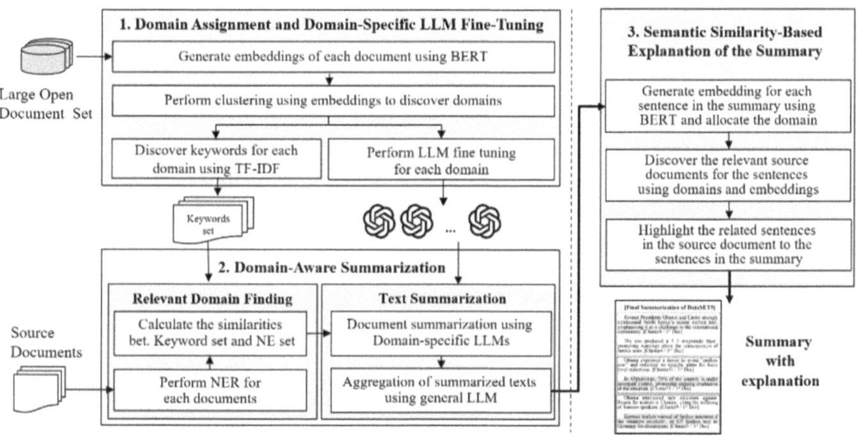

Fig. 1. Systematic procedure of the X-MAS framework.

3.1   Domain Assignment and Domain Specific LLM Fine-Tuning

This stage identifies latent domains within the document corpus through semantic clustering and fine-tunes LLMs for each discovered domain. The goal is to deal with the limitations of general LLMs by ensuring that each model is trained on a semantically coherent subset of documents, thereby improving relevance and minimizing hallucination. The procedure of this module is described in Fig. 2.

Document Clustering and Domain Assignment. Before performing fine-tuning, it is essential to group documents into semantically coherent domains. This step enables each LLM to be adapted to a specific domain. To achieve this, we apply unsupervised semantic clustering over a large document collection, as described below. From the large open document set, the corpus of documents (D) can be represented as follows.

$$D = \{d_1, d_2, \ldots, d_n, \ldots, d_N\} \qquad (1)$$

where d_n is the nth document in the corpus, and N is the total number of documents. Each document is semantically embedded using BERT as represented as follows.

$$e_{d_n} = BD(d_n) \qquad (2)$$

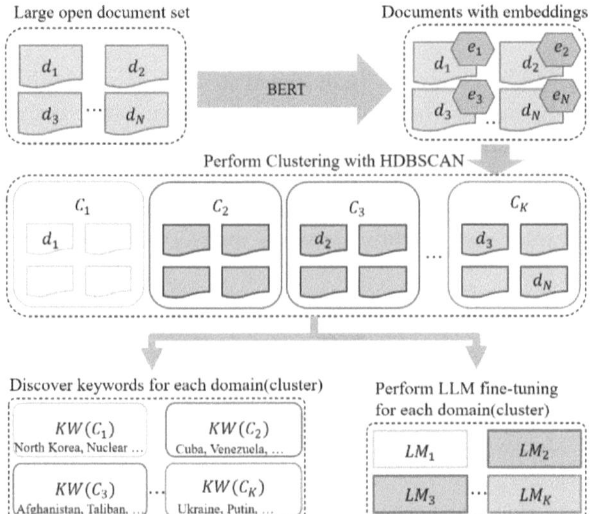

Fig. 2. Domain assignment and domain-specific LLM fine-tuning procedure

where e_n is the embedding of the document d_n and $BD(d_n)$ represents BERT-based document embedding of d_n.

The embedding set $\{e_{d_1}, e_{d_2}, \ldots, e_{d_N}\}$ is utilized for clustering documents using HDBSCAN. By doing so, the set of clusters C can be obtained as described as follows.

$$C = \{C_1, C_2, \ldots, C_k, \ldots, C_K\} \tag{3}$$

where K is the total number of clusters, and C_k is the kth cluster in C, which is a subset of D ($\bigcup_{k=1}^{K} C_k = D$).

By organizing documents in this way, we can uncover hidden domain structures and enhance domain adaptation for subsequent model training.

Discover Keywords For Each Domain Using TF-IDF. To identify salient keywords that characterize each domain (cluster), we adapt the classic TF-IDF scheme from the document level to the cluster level. For a word w in cluster C_k, the term frequency (TF) is the average of the document-level relative frequencies inside the cluster as depicted as follows.

$$TF(w, C_k) = \frac{1}{|C_k|} \sum_{d_n \in C_k} \frac{CNT(w, d_n)}{\sum_{w' \in d_n} CNT(w', d_n)} \tag{4}$$

where $TF(w, C_k)$ refers to TF of the term w on the cluster C_k, and $CNT(w, d_n)$ represents the number of times that the word w is included in the document d_n.

Simply summing counts would bias the score toward clusters that contain unusually long documents. Averaging the document-normalized counts gives each document equal weight.

The inverse document frequency (IDF) measures how unevenly a term is distributed across clusters and it can be calculated as represented as follows.

$$IDF(w) = \log K - \log\left(\sum\nolimits_{k=1}^{K} \frac{|\{d_n | d_n \in C_k \text{ and } w \in d_n\}|}{|C_k|}\right) \tag{5}$$

If w appears in every cluster at a similar rate, the second log term becomes large and $IDF(w)$ becomes small, down-weighting uninformative words. Standard IDF uses raw document counts. We instead use within-cluster proportions, so a term common in one cluster but scarce in the others receives a higher weight, even if a few very large clusters dominate the corpus.

Finally, the TF-IDF can be calculated by multiplying TF and IDF.

$$TFIDF(w, C_k) = TF(w, C_k) \cdot IDF(w) \tag{6}$$

A high $TFIDF(w, C_k)$ indicates that w is frequent within C_k, but relatively rare in other clusters, making it a strong candidate keyword for the cluster C_k. Finally, the keyword set TC_k for each cluster C_k is obtained by selecting the terms with the highest TF-IDF scores.

Domain-Specific LLM Fine-Tuning. In this step, we fine-tune an LLM for each document cluster, allowing the summarization process to adapt to the specific characteristics of each domain. Unlike using a single general-purpose LLM, this approach ensures that domain-specific nuances are effectively captured in the generated summaries.

We fine-tune a domain-specific summarization model LM_k for the cluster C_k from a general LLM LM_{gen} as represented as follows.

$$LM_k = FineTune\left(LM_{gen}, C_k\right) \tag{7}$$

where $FineTune\left(LM_{gen}, C_k\right)$ represents performing fine-tuning the LLM model LM_{gen} using the document set C_k.

Through this process, each LM_k acquires the ability to generate summaries that reflect the semantic characteristics of its domain. After domain specific summaries are generated, a general-purpose LLM is employed in the second stage to integrate these outputs into a single, coherent summary. This model operates as a unifying abstracter, ensuring that the final summary maintains global consistency and fluency across different domain-specific contents while preserving their semantic contributions.

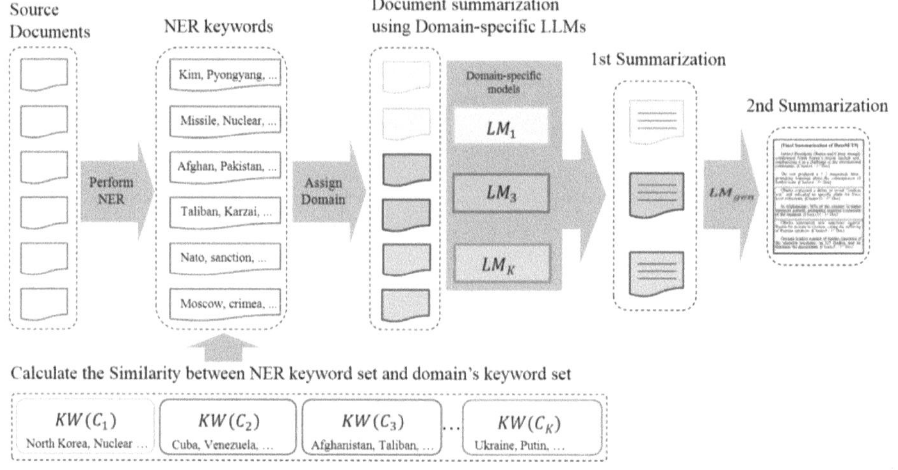

Fig. 3. Domain-aware text summarization procedure

3.2 Domain-Aware Summarization

In the second stage, we perform domain-aware summarization for a given set of source documents. The procedure of the summarization is depicted in Fig. 3. The goal is to assign each document to its most relevant domain (i.e., cluster) and generate a coherent summary using the corresponding domain-specific language model.

Domain Assignment via NER and TF-IDF Similarity. Let D^* denote the set of source documents to be summarized into a single summary. For each document $d \in D^*$, we apply Named Entity Recognition (NER) to extract a set of salient terms $NE(d)$ that represent d. To assign d to a domain, we compare its entity set $NE(d)$ to each cluster's keyword set TC_k, previously derived using TF-IDF, using cosine similarity between BERT-based word embeddings. The similarity between a document and a cluster is computed as follows.

$$SIM(NE(d), TC_k) = \max_{w \in NE(d), w' \in TC_k} Cosim\big(BW(w), BW(w')\big) \tag{8}$$

where $SIM(NE(d), TC_k)$ represents the similarity between the term sets $NE(d)$ and TC_k, $BW(w)$ represents the BERT-based word embedding of w, and $Cosim$ represents the cosine similarity between embeddings.

Each document is then assigned to the cluster with the highest similarity score as depicted as follows.

$$Cl(d) = arg \max_{C_k} SIM(NE(d), TC_k) \tag{9}$$

where $Cl(d)$ represents the assigned cluster of d.

This process yields domain-to-document mapping that informs the selection of a corresponding domain-specific summarization model. By tailoring the summarization process to the assigned domain, we ensure that the generated summary better reflects the specific terminology, context, and characteristics of that domain.

Domain-Aware Summarization. Once each source document $d \in D^*$ is assigned to its corresponding domain C_k, we perform a two-phase summarization process. In the first phase, each document is routed to its respective domain-specific summarization model LM_k, which has been fine-tuned as described in Sect. 3.1. The model generates a domain-aware preliminary summary as described as follows.

$$s_d^{(k)} = M_k(d) \tag{10}$$

where $s_d^{(k)}$ the summary of d when $Cl(d) = C_k$.

This yields a set of domain-aware summaries $(S_{D^*} = \left\{ s_d^{(k)} | d \in D^* \right\})$ of the documents belong to D^*. In the second phase, to reduce redundancy and enhance global coherence, the set S_{D^*} is passed to a general-purpose summarization model M_{gen}, which produces the final output summary ($sfin_{D^*}$) of D^* as depicted as follows.

$$sfin_{D^*} = M_{gen}(S_{D^*}) \tag{11}$$

This two-phase summarization approach preserves domain-specific nuances in individual summaries while ensuring that the final output is coherent, concise, and informative across the entire set of input documents.

3.3 Semantic Similarity-Based Explanation

To enhance transparency and interpretability, we incorporate an explainability module that traces the origin of each sentence in the final summary. This module identifies the domain, source document, and sentence-level provenance of each generated sentence. By doing so, it provides a structured semantic trace that improves user trust and enables detailed analysis of the model's behavior. The explainability process operates hierarchically in three stages: domain attribution, document alignment, and sentence highlight, as described below.

Domain Attribution. For each sentence $s \in sfin_{D^*}$, we first determine its corresponding domain cluster to narrow the search space for provenance tracking. The average embedding of each domain cluster C_k is computed as follows.

$$e_{C_k} = \frac{1}{|C_k|} \sum_{d \in C_k} e_d \tag{12}$$

where e_d is the embedding of document d, and e_{C_k} represents the centroid embedding of cluster C_k.

The embedding e_s of each sentence s in $sfin_{D^*}$ is computed using a BERT-based sentence encoder. The sentence is then assigned to the most semantically similar cluster using cosine similarity as represented as follows.

$$SCl(s) = \underset{C_k}{\operatorname{argmax}}\, cosim(e_s, e_{C_k}) \tag{13}$$

where $SCl(s)$ represents the allocated cluster for $s \in sfin_{D*}$ and $cosim$ represents cosine similarity of the two embeddings as depicted as follows.

$$cosim(e_a, e_b) = \frac{e_a \cdot e_b}{\|e_a\| \|e_b\|} \tag{14}$$

Document Alignment. Given the assigned domain $SCl(s) = C_k$ for each sentence $s \in sfin_{D*}$, we identify the most relevant source document within that cluster. This is done by comparing the sentence embedding e_s to each document embedding e_d as represented as follows.

$$SD_{C_k}(s) = \underset{d \in C_k}{\operatorname{argmax}} \, cosim(e_s, e_d) \tag{15}$$

where $SD_{C_k}(s)$ denotes the source document in cluster C_k that is most semantically aligned with sentence s.

Sentence Highlight. Finally, within the identified source document, we locate the specific sentences that most closely align with the summary sentence $s \in sfin_{D*}$. This is achieved through a combination of keyword matching and sentence-level cosine similarity between theses and sentence embeddings in the source document.

By integrating domain-level classification, document-level alignment, and sentence-level similarity analysis, this hierarchical explanation framework ensures that every sentence in the summary can be semantically traced to its origin. This not only enhances transparency and user trust but also facilitates validation and feedback in multi-domain summarization systems.

4 Performance Evaluation

To prove the superiority of the X-MAS framework, we performed experiments using real-world datasets. In the domain clustering phase, we employed the labeled AG News dataset to evaluate the performance of various clustering algorithms. Subsequently, we adopted HDBSCAN to cluster 10,180 CNN/DailyMail articles related to former U.S. President Barack Obama into 20 domains. For each cluster, we fine-tuned an LLM using the corresponding domain-specific data. We then constructed experimental datasets consisting of source documents related to multiple domains and compared the summarization results generated by our method against those produced by GPT-4, one of the most advanced language models. Finally, in the explanation stage, we presented the provenance-tracing process, which attributes each generated summary sentence to its originating domain, source document, and specific sentence. All experiments were conducted on a machine equipped with an Intel i5-13600K CPU, 96 GB RAM, and an NVIDIA GeForce RTX 3080 GPU (10 GB VRAM).

4.1 Domain Clustering

As a preprocessing step for domain-aware summarization, we performed semantic clustering of multi-domain documents using sentence-level embeddings. Specifically, we sampled 10,000 instances from the AG News dataset and embedded each document using the Sentence-BERT model (all-MiniLM-L6-v2). Based on these embeddings, we applied five representative clustering algorithms and evaluated their ability to recover the original category labels (World, Sports, Business, Sci/Tech).

We employed Adjusted Rand Index (ARI) and Normalized Mutual Information (NMI) as evaluation metrics. ARI measures the degree of agreement between the predicted clusters and ground truth labels, while NMI quantifies the mutual information between them, accounting for label permutations.

Table 1. ARI and NMI results by clustering methods.

Methods	ARI	NMI
K-Means	**0.516**	**0.522**
DBSCAN	0.066	0.409
HDBSCAN	0.295	0.467

In Table 1, we can find that K-Means yielded the highest performance in both ARI and NMI, it inherently requires the number of clusters k to be predefined. This constraint is incompatible with our research goal of adaptively discovering domain clusters, where the number of domains may not be known in advance. In contrast, HDBSCAN automatically determines the number of clusters, accommodates varying density structures, and is robust to noise. While its ARI is lower than that of K-Means, the NMI score (0.467) suggests that the semantic grouping of documents is reasonably well preserved.

In summary, HDBSCAN is more aligned with our experimental setting, which prioritizes semantic coherence and dynamic domain discovery. Therefore, we selected HDBSCAN as the default clustering engine in our domain assignment pipeline.

4.2 General LLM vs Fine-Tuned LLM Summarization

To quantitatively evaluate the effectiveness of domain-specific fine-tuning, we compare the performance of a general-purpose Large Language Model (GPT-4) with domain-specific GPT-3.5 models. However, our summarization framework operates in open domain environments where reference summaries are not available. This makes ROUGE less appropriate for evaluation.

Instead, we adopt BERTScore as the primary evaluation metric, which measures semantic similarity between generated summaries and source documents. BERTScore computes Precision, Recall, and F1 scores using contextual embeddings and is well-suited for abstractive summarization tasks where exact wording may vary but semantic equivalence should be preserved.

Dataset Construction. We use the "Obama" related CNN/DailyMail dataset. The original CNN/DailyMail dataset comprises a wide range of news articles and their associated abstractive highlights. We extracted 42,556 articles explicitly mentioning "Obama" in the headline or article body. This filtered corpus was subjected to extensive preprocessing, including normalization, deduplication, and noise removal, resulting in a final dataset of 10,180 articles suitable for clustering and summarization tasks.

Domain Assignment via Clustering. To automatically induce domain labels without relying on human-annotated categories, we applied HDBSCAN (Hierarchical Density-Based Spatial Clustering of Applications with Noise) on sentence embeddings obtained via Sentence-BERT. HDBSCAN was chosen for its ability to discover an optimal number of clusters in a data-driven fashion, without requiring a priori specification of the number of domains. This clustering process yielded 20 semantically coherent clusters, each representing a distinct latent domain. Clusters were verified by inspecting high TF-IDF keywords and representative documents per cluster, as described in Sect. 4.1. Documents identified as noise by HDBSCAN were discarded from further summarization experiments to maintain domain clarity.

Table 2 presents five representative clusters selected from the 20 discovered, showing the number of documents assigned to each and the top TF-IDF keywords extracted per cluster. These keywords were used to semantically label the clusters.

Table 2. Representative 5 Clusters and Their Keywords.

Cluster	Documents	Keywords
3	215	['cuba', 'cuban', 'president', 'castro', 'gross', 'chavez', 'government', 'venezuela', 'united']
4	202	['korean', 'nuclear', 'south', 'kim', 'united', 'pyongyang', 'missile', 'launch', 'test']
5	421	['ukraine', 'putin', 'ukrainian', 'sanctions', 'crimea', 'nato', 'military', 'moscow', 'united']
6	625	['iran', 'israel', 'nuclear', 'israeli', 'netanyahu', 'palestinian', 'iranian', 'gaza', 'hamas']
11	523	['afghanistan', 'pakistan', 'afghan', 'taliban', 'troops', 'karzai', 'pakistani', 'president', 'forces']

Evaluation Procedure. Each document in the Obama-specific subset was assigned to one of the 20 clusters based on HDBSCAN results. For each domain cluster, we fine-tuned a separate LLM using the associated documents and highlights. The fine-tuning process was conducted on GPT-3.5-turbo-1106 models ($\approx$175B parameters, not publicly disclosed in detail), leveraging domain-specific knowledge to enhance contextual relevance and coherence. The model was trained for 3 epochs with a batch size of 1 and a learning rate multiplier of 2, and the random seed was fixed to 777 to ensure reproducibility. The training dataset was prepared in chat format, consisting of article–summary pairs, and comprised approximately 677k training tokens.

To evaluate the effectiveness of our summarization framework across different domain distributions, we constructed six source datasets (SET1 to SET6) with varying domain compositions (Table 3). Each dataset comprises a small number of news articles, selectively drawn from three geopolitical domains through web scraping: North Korea, Taliban, and Ukraine respectively related to cluster number 4, 5 and 11.

Table 3. Source Datasets for Evaluation.

Datasets	Composition
SET1	5 articles on North Korea
SET2	3 articles on North Korea and 2 on the Taliban
SET3	2 articles each on North Korea and the Taliban, and 1 article on Ukraine
SET4	1 article from each of the three domains (North Korea, Taliban, Ukraine)
SET5	3 articles from each of the three domains (North Korea, Taliban, Ukraine)
SET6	5 articles from each of the three domains (North Korea, Taliban, Ukraine)

This progressive expansion of domain diversity from mono-domain to tri-domain settings with increasing scale allows us to examine how the model's performance is influenced by domain complexity and content heterogeneity in the summarization input. We compared the summarization output of (1) the latest general-purpose GPT-4 model used in zero-shot summarization and (2) two-phase summarization model (Domain-specific GPT-3.5 models fine-tuned on the clustered source datasets and general-purpose GPT-4 model). Generated summaries were compared with their source documents using BERTScore F1, computed via the all-MiniLM-L6-v2 Sentence-BERT model. This enabled a direct evaluation of semantic alignment without requiring reference summaries.

Evaluation Result. Across all six datasets, our summarization framework demonstrated comparable or superior BERTScore performance relative to the general-purpose GPT-4 model as depicted in Table 4. The improvements were particularly noticeable in multi-domain documents where semantic alignment and domain-aware summarization are crucial.

Table 4. Summarization Performance Comparison.

Data	Model	BERTScore	Data	Model	BERTScore
SET1	**GPT-4**	**0.8361**	SET4	GPT-4	0.8237
	Ours	0.8350		**Ours**	**0.8410**
SET2	GPT-4	0.8123	SET5	**GPT-4**	**0.8180**
	Ours	**0.8176**		Ours	0.8180
SET3	GPT-4	0.8251	SET6	GPT-4	0.8211
	Ours	**0.8262**		**Ours**	**0.8215**

Across all six datasets, our summarization framework demonstrated comparable or superior BERTScore performance relative to the general-purpose GPT-4 model. The improvements were particularly noticeable in multi-domain documents where semantic alignment and domain-aware summarization are crucial. In SET1, GPT-4 slightly outperformed our model (0.8361 vs. 0.8350), though the margin was minimal. This suggests that GPT-4 is already well pre-trained in domain related to North Korea. In contrast, SET2 and SET3, which included content from two or three domains, showcased our model's superior performance. In SET2, our model achieved a higher BERTScore (0.8176 vs. 0.8123), indicating improved semantic preservation despite the heterogeneous input. Similarly, in SET3, we slightly surpassed GPT-4 (0.8262 vs. 0.8251), suggesting that our domain-adaptive approach maintains coherence even when domain complexity increases.

A particularly significant improvement was observed in SET4, a balanced tri-domain setting with minimal document input (one article per domain). Our model recorded a BERTScore of 0.8410, outperforming GPT-4's 0.8237 by a substantial margin. This indicates that our framework effectively integrates sparse yet diverse information by leveraging fine-tuned domain-specific representations, whereas GPT-4's generality may dilute its contextual accuracy. For SET5 and SET6, both models exhibited almost identical performance. In SET5, the scores were matched (0.8180 vs. 0.8180), while in SET6, our model slightly exceeded GPT-4 (0.8215 vs. 0.8211). These datasets represent large-scale, high-domain-mix conditions, where domain-specific fine-tuning has limited influence unless further abstraction fidelity mechanisms are applied.

4.3 Explanation

To enhance the interpretability of the generated summaries, our framework incorporates a multi-level explanation module that traces the provenance of each sentence in the final summary. As illustrated in Fig. 4, we provide domain-level, document-level, and sentence-level attribution for each summary sentence. This hierarchical explanation not only supports transparency but also offers valuable insights into the model's decision-making process.

Domain Attribution. Each sentence in the summary is semantically compared with the centroids of each domain cluster (e.g., North Korea, Taliban, Ukraine), computed from the sentence embeddings of the source documents. The domain whose centroid exhibits the highest cosine similarity to the summary sentence is assigned as its origin. For instance, the sentence "Former Presidents Obama and Carter strongly condemned North Korea's recent nuclear test…" is most similar to the North Korea domain cluster, thereby assigned accordingly.

Document Attribution. After determining the most likely domain, we identify the specific document within that domain that most influenced the summary sentence. This is achieved by comparing the sentence embedding with the average embedding of each document. In the example above, the sentence corresponds to a source article in Cluster 4, which discusses official U.S. reactions to North Korea's nuclear tests.

Sentence Attribution. Finally, we trace the most semantically aligned sentence at the source sentence level. Using cosine similarity, the summary sentence is matched to its closest sentence from the source documents. In the highlighted case, the summary sentence directly corresponds to the sentence: "President Barack Obama and Defense Secretary Ash Carter today condemned North Korea's latest nuclear test in the strongest possible terms…".

This multi-stage attribution process is repeated for each sentence in the summary. For example, the sentence "In Afghanistan, 20% of the country is under insurgent control…" is attributed to Cluster 11, corresponding to a document discussing Taliban territorial gains. Also, the final sentence "The German leaders warned of further sanctions…" is traced to a Ukraine-related document in Cluster 5, which includes direct remarks from German Chancellor Merkel.

Error Analysis. While the proposed framework consistently improved both summarization quality and explainability, several failure cases were observed. Particularly, sentences describing events relevant to multiple domains were occasionally misattributed to the dominant domain cluster. Furthermore, when two source documents exhibited highly overlapping content, certain summary sentences were incorrectly traced to the less relevant source. These errors underscore the intrinsic challenges of handling ambiguous or multi-domain content. Although such mismatches were relatively rare, they highlight the need for more robust disambiguation mechanisms and finer-grained attribution methods, which we identify as promising directions for future research.

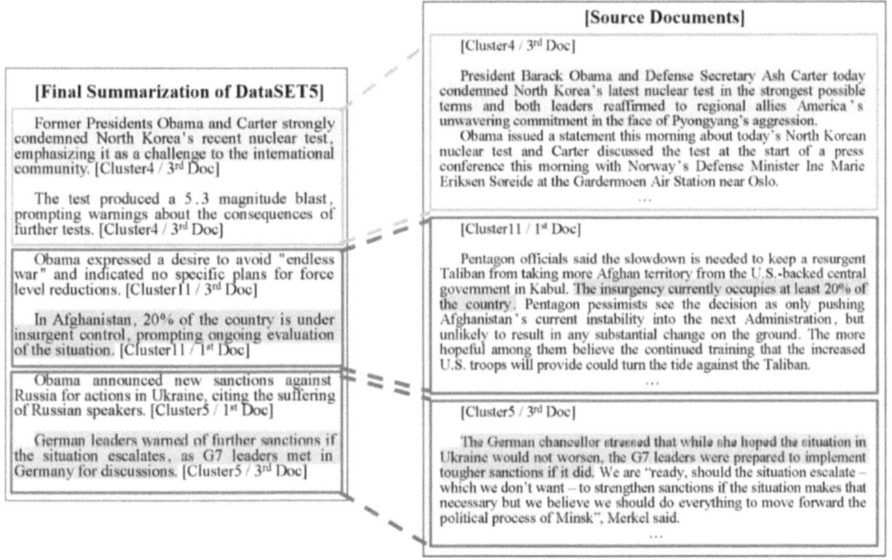

Fig. 4. Example of Explanation.

5   Conclusions

This paper presented a novel multi-document summarization framework that integrates domain-specific fine-tuning with hierarchical provenance-based explainability. Addressing the limitations of general-purpose LLMs in handling semantically heterogeneous document sets, we proposed a pipeline that combines document clustering, domain-adaptive modeling, and multi-level explanation tracing to improve both summarization quality and transparency.

Our contributions are summarized as follows. First, we utilize semantic embeddings to cluster documents into latent topical domains. These clusters are used to fine-tune dedicated summarization models, each tailored to a specific domain. Second, our framework provides domain-, document-, and sentence-level attribution for each summary sentence, allowing users to trace its semantic origin. This interpretability enhances trustworthiness and accountability, which are critical in high-stakes domains. Empirical results across six datasets with varying domain compositions demonstrate that our method consistently outperforms or competes with GPT-4 in both BERTScore-based metrics and interpretability.

Nevertheless, several limitations remain. Ambiguous or mixed-topic documents pose challenges for domain clustering, potentially leading to misclassification and inconsistent summaries. Also, as each domain requires a separate fine-tuned model, the framework faces scalability and computational efficiency issues when deployed over large and diverse domain sets. In this study, we deliberately employed full fine-tuning for each domain-specific model to establish an upper bound on the effectiveness of domain adaptation and to clearly isolate the contribution of domain clustering. While this approach allowed us to maximize domain alignment, we acknowledge that parameter-efficient tuning strategies could provide a more scalable alternative.

To address these issues, we will conduct research on employing soft clustering or mixture-of-experts mechanisms to account for documents with multiple latent domain affinities. This could improve summary coherence for mixed-topic inputs and reduce misclassification-induced noise. Also, to improve scalability, we will adopt parameter-efficient adaptation strategies such as LoRA, Adapter modules, or Prompt Tuning, reducing memory and computation costs when maintaining multiple domain-specific models.

References

1. Widyassari, A.P., et al.: Review of automatic text summarization techniques & methods. J. King Saud University – Comput. Inf. Sci. **34**(4), 1029–1046 (2022)
2. Moratanch, N., Chitrakala, S.: A survey on extractive text summarization. In: 2017 IEEE International Conference on Computer, Communication, and Signal Processing (ICCCSP), Chennai, India, pp. 1–6. IEEE (2017)
3. Shakil, H., Farooq, A., Kalita, J.: Abstractive text summarization: state of the art, challenges, and improvements. Neurocomputing **603**, 128255 (2024)
4. Zhang, H., Yu, P., Zhang, J.: A systematic survey of text summarization: from statistical methods to large language models. ACM Comput. Surv. **57**(11), 1–41 (2025)
5. Dhaini, M., Erdogan, E., Bakshi, S., Kasneci, G.: Explainability meets text summarization: a survey. In: 17th International Natural Language Generation Conference, pp 631–645. Association for Computational Linguistics, Tokyo, Japan (2024)
6. Xie, Q., Tiwari, P., Ananiadou, S.: Knowledge-enhanced graph topic transformer for explainable biomedical text summarization. IEEE J. Biomed. Health Inform. **28**(4), 1836–1847 (2024)

7. Norkute, M., Herger, N., Michalak, L., Mulder, A., Gao, S.: Towards explainable AI: assessing the usefulness and impact of added explainability features in legal document summarization. In: CHI Conference on Human Factors in Computing Systems Extended Abstracts, pp. 1–7, ACM, Yokohama, Japan (2021)
8. Devlin, J., Chang, M., Lee, K., Toutanova, K.: BERT: pre-training of deep bidirectional transformers for language understanding. In: 17th Annual Conference of the North American Chapter of the Association for Computational Linguistics, pp. 4171–4186. Association for Computational Linguistics, Minneapolis, Minnesota (2019)
9. Malzer, C., Baum, M.: A hybrid approach to hierarchical density-based cluster selection. In: 2020 IEEE International Conference on Multisensor Fusion and Integration for Intelligent Systems, pp. 223–228. IEEE, Karlsruhe, Germany (2020)
10. Jones, K.S.: A statistical interpretation of term specificity and its application in retrieval. J. Documentation **28**, 11–21 (1972)
11. Achiam, J., et al.: GPT-4 technical report. arXiv preprint arXiv:2303.08774 (2023)
12. Jalil, Z., Nasir, J.A., Nasir, M.: Extractive multi-document summarization: a review of progress in the last decade. IEEE Access **9**, 130928–130946 (2021)
13. Ma, C., Zhang, W.E., Guo, M., Wang, H., Sheng, Q.Z.: Multi-document summarization via deep learning techniques: a survey. ACM Comput. Surv. **55**(5), 1–37 (2022)
14. Bing, L., Li, P., Liao, Y., Lam, W., Guo, W., Passonneau, R.J.: Abstractive multi-document summarization via phrase selection and merging. arXiv preprint arXiv:1506.01597 (2015)
15. Khan, A., Salim, N., Kumar, Y.J.: A framework for multi-document abstractive summarization based on semantic role labelling. Appl. Soft Comput. **30**, 737–747 (2015)
16. Liu, P.J., et al.: Generating Wikipedia by summarizing long sequences. arXiv preprint arXiv: 1801.10198 (2018)
17. Liu, Y., Lapata, M.: Hierarchical transformers for multi-document summarization. arXiv preprint arXiv:1905.13164 (2019)
18. Li, W., Xiao, X., Liu, J., Wu, H., Wang, H., Du, J.: Leveraging graph to improve abstractive multi-document summarization. arXiv preprint arXiv:2005.10043 (2020)
19. Sinha, N.: QA-prompting: improving summarization with large language models using question-answering. arXiv preprint arXiv:2505.14347 (2025)
20. Kurisinkel, L.J., Chen, N.F.: LLM based multi-document summarization exploiting main-event biased monotone submodular content extraction. arXiv preprint arXiv:2310.03414 (2023)

SPAX: A Shapley-Based Point Attribution eXplanation for Interpreting 3D Point Cloud Classification

Marc F. Harinck[1], Muhammad Shoaib Sarwar[1(✉)], Bram Ton[2], and Faizan Ahmed[1,2]

[1] Department of Computer Science, University of Twente, Drienerlolaan 5,
Enschede, The Netherlands
`m.f.harinck@student.utwente.nl`,
`{shoaib.sarwar,faizan.ahmed}@utwente.nl`
[2] Ambient Intelligence Group, Saxion University of Applied Sciences,
Van Galenstraat 19, Enschede, The Netherlands
`b.t.ton@saxion.nl`

Abstract. In this paper, we propose an Explainable Artificial Intelligence methodology to interpret point cloud classification models. Our approach uses Shapley values from cooperative game theory which assigns contribution values to individual points towards a classification. Point cloud data contains rich spatial information, therefore it requires sophisticated deep learning models to make accurate predictions. However, these models often lack interpretability. Our methodology, Shapley-based Point Attribution eXplanations (SPAX), addresses this issue by quantifying and visualising the contribution of each point in a point cloud towards a prediction. To make it computationally feasible, a Monte Carlo approximation is used to estimate the Shapley values. To evaluate the method, a PointNet classifier trained on the ModelNet10 dataset is used to determine the Shapley values. These values are then colored according to their magnitude and used to visually interpret their impact on the classification. The results show how spatial features become more interpretable because they identify key components to determine classification results. This study creates foundational principles to enable better point cloud interpretation, which produces opportunities to boost real-world deployments of transparent algorithmic systems.

Keywords: Explainable AI · PointNet · Point Cloud · Shapley Values · Interpretability · Classification · Monte Carlo Estimation

1 Introduction

Artificial Intelligence (AI) has accelerated technological advancements [26] and greatly impacted the influence of technology on society [40]. Despite the many benefits of AI, the adoption of AI technology remains a multifaceted challenge. For instance, Sun and Medaglia describe the challenges from stakeholders associated with the adoption of AI in healthcare [34]. A limiting factor of widespread AI adaptation is its lack of transparency as AI models are considered as a "black box" [32]. This is especially true for

---

The original version of the chapter has been revised. The Chapter 24 author's name has been corrected. A correction to this chapter can be found at
https://doi.org/10.1007/978-3-032-15638-9_46

© The Author(s), under exclusive license to Springer Nature Switzerland AG 2026, corrected publication 2026
F. Marcelloni et al. (Eds.): IJCCI 2025, CCIS 2829, pp. 414–433, 2026.
https://doi.org/10.1007/978-3-032-15638-9_24

Deep Learning models leading to scepticism for critical applications such as healthcare [34] and autonomous driving [7]. Consequently, the field of Explainable Artificial Intelligence (XAI) was developed, designed for human interpretability as part of the model design [4]. This leads to machine learning models that are more interpretable and increase the trustworthiness for users [4,7,8,23].

Interpretability can be achieved by understanding the effect of individual components towards an outcome. The Shapley value, introduced by Lloyd Shapley in 1953, is a fundamental concept in cooperative game theory that provides a fair allocation of a total payoff among players based on their individual contributions [30]. In an n-player game, the Shapley value assigned to each player, reflects their average marginal contribution across all possible coalitions, computed as a weighted sum over all possible subsets of players [16,19,30] making the Shapley value viable for interpreting the outcome of a process.

The Shapley value theory has been extended to machine learning by Lundberg and Lee [16] and uses the Shapley value as the theoretical foundation for their Shapley Addative exPlanations (SHAP) [16,19]. SHAP applies Shapley value theory to machine learning, but differs in that it provides an optimised approximation for practical use in AI models. This framework has been successfully applied to lower-dimensional data problems such as textual sentiment analysis, text translation [16], image classification [11,15], and for medical use cases [3,22]. The adaptation of SHAP for higher-dimensional data such as point clouds has not been extensively studied yet [20].

Point cloud data plays an important role in autonomous vehicles [10] and is used to digitize road [12,42] and rail environments [27,38]. These datasets pose unique challenges due to their high dimensionality, unordered structure, and the large number of points per sample, necessitating more complex models. As model complexity increases, and given the critical nature of its application areas such as autonomous vehicles and infrastructure like rail and road systems, the demand for Explainable AI (XAI) frameworks [14,17] becomes more pressing.

This study addresses the demand for better explainability by developing a model-agnostic methodology that adapts Shapley values to point cloud classifiers. We propose a Monte Carlo approximation framework [33] adapted to the task of point cloud classification, which we dubbed Shapley-based Point Attribution eXplanations (SPAX). SPAX estimates the Shapley value of individual points within a point cloud and thus provides insights into the contribution of individual points towards a certain classification.

To validate our proposed framework, we visualize the contribution (Shapley value) of individual points by coloring them accordingly to their influence on the classification. These visualizations provide qualitative feedback on the contributions of individual points towards a certain outcome of the classifier. The scientific contribution is a model-agnostic XAI framework developed to interpret point cloud classifiers. The code and experiments are publicly available[1].

The remainder of this paper is structured as follows. Section 2 presents a focused review of the literature on modern XAI methods in the context of point cloud analysis, followed by an introduction of our proposed SPAX framework in Sect. 3 Sect. 4 provides empirical findings, as well as analyses of the findings with regard to the theoretical and practical implications, in several experimental scenarios. Finally, Sect. 5

---

[1] https://github.com/Mavisis/SHAP-for-Point-Cloud.

summarises the key findings, limitations, and research directions in interpretable point cloud models of the article.

2   Related Work

XAI is a rapidly growing field, with numerous approaches developed to interpret deep neural network models (DNNs) [20]. Some of these methods include creating saliency maps to highlight crucial input regions [39] or constructing proxy models that mimic the original network but are easier to understand [9]. While XAI techniques have been extensively explored for 2D data, the investigation into the explainability of 3D DNNs, particularly those processing point clouds, remains relatively limited [20].

Saranti et al. [28] deliver an extensive state-of-the-art review about LiDAR-based 3D point-cloud data acquisition methods that lead to applications of explainable geometric deep learning. The authors present research on how point cloud data can be converted to understandable formats, together with their analysis of XAI techniques in 3D data environments and future predictions of industry challenges.

Feature-based interpretability is a computationally efficient XAI method that evaluates the contribution of individual points in a point cloud to classification tasks [14]. It computes the L^1 norm of features for each point before the network's bottleneck layer, ranking points by their semantic significance [14]. This simple approach provides insights into which points are critical for classification, enabling real-time feedback during inference.

BubblEX is a multimodal framework designed to improve explainability in 3D point cloud classification [17]. It combines visualisation and interpretability modules to highlight salient features influencing model decisions [17]. The visualisation module uses techniques like t-SNE and UMAP to project high-dimensional features into interpretable clusters, while the interpretability module, inspired by Grad-CAM [29], identifies neighbouring points critical to feature extraction [17]. BubblEX generates visual explanations without requiring architectural changes, making it adaptable to various DNN-based models and is a model-agnostic approach.

Matrone et al. [18] propose BubblEX, a two-part explainable model comprised of a visual and interpretability aspect. BubblEX adapts the salience mapping technique for 2D images to handle semantic segmentation of point clouds. A multi-dataset assessment using S3DIS, SynthCity, Semantic3D, and KITTI enables the analysis of Point-Net, DGCNN, and BAAF-Net network designs. Their research examines how various factors shape data acquisition in addition to network configuration within the geomatics domain based on dataset type and scene type analysis.

The study by Arnold et al. [1] introduces an explainable and transparent point cloud classifier. The main contribution of their work is the introduction of compound prototype clouds as a distinct visual representation that enables explanation mappings in 3D space. Their proposed classification network achieves better results in accuracy compared to PointNet and PointNet++.

PointMask is a model-agnostic method, it uses regularisation that maximises information gain between masked points and class labels [35]. PointMask removes (masks) irrelevant points by identifying which points are highly correlated between the point cloud and the predicted class and removing non-correlated points [35].

Tan and Kotthaus [36, 37] developed two distinct point cloud explainability methods. First, they proposed explainability techniques using local surrogate models for neural networks based on point clouds while establishing method-independent verification through positive and negative point flips to evaluate existing point cloud explainability techniques in their initial work [36]. Second, they introduced generative Attribution Maps (AMs) as a way to visualize global interpretations of point cloud DNNs in their study [37]. These AMs were shown to work effectively with different common point cloud networks such as PointNet, PointNet++, and DGCNN.

The study conducted by Atik et al. [2] delved into XAI through machine learning-based photogrammetric point cloud classification. This research examined how SHAP analysis performed when combined with standard filter-based feature selection methods such as Information Gain(IG) [13] and ReliefF [25] to select features before applying an ensemble-based machine learning approach. Their research focused on interpreting SHAP values on the feature level. In contrast, our work focuses on interpreting SHAP values on the individual point level.

Chen et al. [6] developed an XAI framework that utilizes SHAP and Random Forest to rectify urban Ground Digital Elevation Models. The method specializes in the management of prediction elements that have spatial variability alongside their optimization measures. Their research achieves model optimization by enhancing feature elements as well as establishing enhanced regional machine learning models for risk assessment and urban planning.

The work of Shen et al. [31] also uses SHAP values for adding interpretability to point cloud models. In their work, regions of points are used instead of individual points. These regions are determined by first sampling the n farthest points from the point cloud as seed points. The remaining points are then assigned to each of these n nearest seed points. A benefit of this approach is that it becomes computationally easier, but as a downside, the defined regions do not have any semantic meaning. Therefore, our approach focuses on the individual points within a point cloud.

In contrast to the other methods described above, our method does not need any information about or access to the model itself. The model being analysed can remain a truly "black box".

3 SPAX Framework

In this section, the model-agnostic methodology to add interpretability to existing point cloud classification models is presented. This section starts by presenting the theory behind the Shapley value. How this method is applied to point cloud data, and the approximation of the Shapley value is presented thereafter.

3.1 Shapley Value Computation

The Shapley value considers the marginal contribution of players by forming a coalition. The contribution is computed by defining a valuation function that evaluates the contribution of the player. In our setting, each point j is considered a player and the contribution is computed using the valuation function val_j. The contribution to a coalition S is computed by the following formula.

$$\phi_j(val_x) = \sum_{S \subseteq \{1,\dots,N\} \setminus \{j\}} \frac{|S|! \, (N - |S| - 1)!}{N!} \left(val_x(S \cup \{j\}) - val_x(S) \right). \quad (1)$$

Here, S is a subset of the points belonging to a coalition of j excluding j from this coalition. $S \cup \{j\}$ is a subset of the points belonging to a coalition of j including j. The vector x of coalition values of the instance to be explained. N is the total number of points. The function $val_x(S)$ represents the prediction for points values in set S, marginalised over the features not included in S. To quantify the contribution of a feature subset S to the classification model's prediction, we use the following,

$$val_x(S) = \int \hat{f}(x_1, \ldots, x_N) \, dP_{x \notin S} - \mathbb{E}_X \left[\hat{f}(X) \right].$$ (2)

In Eq. 2 the first term, $\int \hat{f}(x_1, \ldots, x_p) \, dP_{x \notin S}$, computes the expected prediction when only the features in S are known, integrating over the distribution of the remaining features. The second term, $\mathbb{E}_X \left(\hat{f}(X) \right)$, represents the model's average prediction over all inputs. Their difference, $val_x(S)$, measures how much knowing S shifts the prediction. There are $2^N - 1$ possible coalitions, meaning the computational complexity grows exponentially with N. Therefore, we use Monte Carlo estimation to manage this complexity efficiently.

3.2 Weighted Marginal Point Contribution

The evaluation of the point cloud is conducted on a per-point basis, with each point denoted as the *target point*. Given a point cloud comprised of N points, the evaluation process should theoretically consider all possible coalitions (subsets) of each target point in N. In theory, the total number of such coalitions would be $2^N - 1$. As an example, consider a point cloud consisting of 1024 points, this would lead to a total number of possible coalitions of $2^{1024} - 1$, or approximately 1.8×10^{308}. A number vastly exceeding computational feasibility. Consequently, an exact enumeration of all coalitions is unrealistic. Therefore, an approximation such as Monte Carlo sampling is used in practice.

The methodological steps are combined into a pipeline which is depicted in Fig. 1. The pipeline uses the PointNet [24] network architecture, although it can be replaced by other deep learning architectures developed for point clouds.

3.3 Monte Carlo Approximation

Given the exponential complexity associated with evaluating all coalitions, a Monte Carlo approximation is employed [33] to estimate the value function. The objective is to discretise the expectation integral (Eq. 2) over M sampled coalitions:

$$val_x(S) \approx \frac{1}{M} \sum_{m=1}^{M} \hat{f}(x_S, x_{\sim S}^{(m)}) - \mathbb{E}_x[\hat{f}(x)].$$ (3)

The size of M is dependent on the statistical relevance needed for generalization. Larger M results in a prediction more in line with the ground truth at the cost of an increase in computation costs. Each point has S coalitions. M coalitions are sampled from S of varying sizes and are chosen at random. Pseudo code is included in the

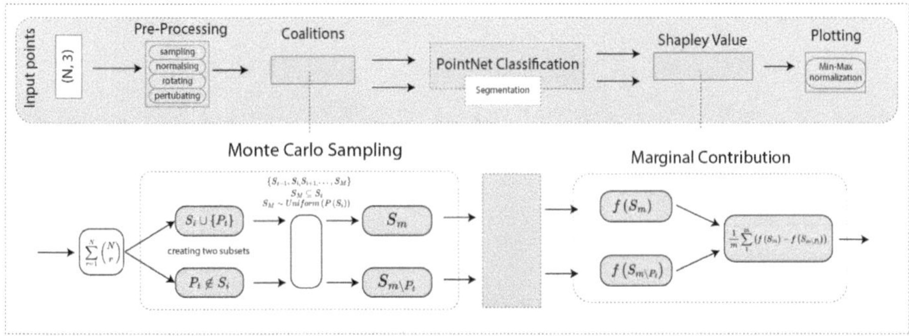

Fig. 1. The SPAX pipeline that integrates Shapley value attribution into PointNet. The key steps include preprocessing, creating two sets (S_m and $S_m \setminus P_t$), enumerating through each set and classifying point coalitions, calculating marginal contributions of points, mapping Shapley values to colors, and finally visualizing the results.

Appendix and clarifies how to calculate the Shapley value of a point in Algorithms A.1, A.2, A.3.

The method provides contribution values that indicate how individual input points influence the model's output, but these values do not reflect classification confidence or correctness. In other words, the method explains how input features contribute to the model's decision, regardless of whether that decision is actually correct. It does not verify whether the predicted class matches the true label. While it offers insight into the internal reasoning of the model, such as which features push the output toward a particular class, it does not confirm whether the final classification is accurate. To assess correctness, additional validation or analysis is needed.

4   Experiments

First, the dataset and the pre-processing steps are presented. Thereafter, the effect of different Monte Carlo sample sizes is explored. In addition, an analysis is done by comparing the Shapley values with the highest magnitude to the values with the lowest magnitude.

4.1   Data Processing

To ensure compatibility with existing point cloud classifiers and to improve model generalization, the preprocessing steps follow those proposed by Qi et al. [5]. The dataset used is the Princeton ModelNet10 [41], which includes ten classes of 3D CAD meshes: bathtub, bed, chair, desk, dresser, monitor, nightstand, sofa, table, and toilet. Each mesh is provided as an Object File Format (OFF) file, containing vertices, edges, and faces. Each instance in the dataset is sampled 1024 times to create the corresponding point cloud with exactly 1024 points.

The processing steps for instance 888 (class chair) are shown in Fig. 2. **Sampling:** Points are uniformly sampled on faces of the mesh, the sampling density for each face is

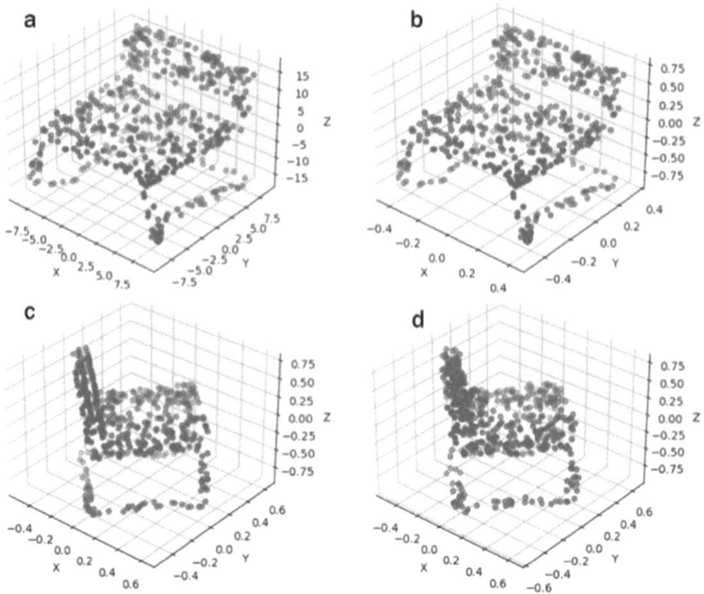

Fig. 2. Four processing steps applied to the 3D mesh. (a) Uniform sampling across surfaces. (b) Normalization into a unit sphere. (c) Rotation θ around the z-axis. (d) Perturbation with Gaussian noise.

proportional to the surface area. This ensures that the sampled point cloud has a uniform density (Fig. 2a). **Normalizing:** Identical to Qi et al. [5], the point cloud is translated and scaled to fit within a unit sphere for size invariance across classes (Fig. 2b) based on the min-max point coordinates in each cardinal direction. **Rotating:** The point cloud is rotated around the z-axis (up-axis) by a random angle θ (Fig. 2c). **Perturbation:** Additive Gaussian noise ($\mu = 0, \sigma = 0.002$) is applied to each point. (Fig. 2d) [5]. These four preprocessing steps ensure that the input data closely aligns with the conditions under which the model was trained, resulting in reliably classifying the performance and enabling a clearer assessment of the model's behaviour through XAI as the input data is controlled.

4.2   Monte Carlo Sample Size Analysis

The effect of different Monte Carlo sample sizes on the Shapley values is analysed for four different classes: monitor, sofa, table, and toilet. These classes were selected due to their distinct geometric characteristics, which are particularly well-suited for analysing spatial features and their influence on the model's predictions. A representative instance within each of these classes is sampled at varying Monte Carlo sample sizes. The following sample sizes are considered: 256, 1024, 4096, and 8192. This progression should allow for an examination of how the sampling rate influences the stability and precision of the Shapley values. A subset of the visual results is provided in Fig. 3, while the complete set is presented in Fig. 5. Figure 3 comprises eight subplots, corresponding to all combinations of class and Monte Carlo sample size. For each subplot, the same instance is depicted from two viewpoints, with individual point Shapley values encoded by color. Table 1 summarizes the associated descriptive statistics, including the mean, standard deviation, median, minimum, and maximum of the example instances.

Table 1. Shapley value statistics summary table of classes monitor, sofa, table, and toilet at Monte Carlo sample size 256, 1024, 4069, and 8192.

Class	MC Size	Mean	Std	Median	Min	Max
Monitor	256	0.0189	0.0069	0.0186	0.0029	0.0437
	1024	0.0191	0.0036	0.0190	0.0089	0.0309
	4096	0.0193	0.0017	0.0192	0.0126	0.0254
	8192	0.0191	0.0013	0.0191	0.0151	0.0233
Sofa	256	0.0512	0.0106	0.0510	0.0191	0.0907
	1024	0.0510	0.0052	0.0512	0.0336	0.0680
	4096	0.0509	0.0026	0.0509	0.0433	0.0600
	8192	0.0509	0.0019	0.0509	0.0439	0.0594
Table	256	0.0098	0.0062	0.0093	−0.0044	0.0341
	1024	0.0098	0.0032	0.0098	0.0006	0.0226
	4096	0.0100	0.0015	0.0100	0.0048	0.0163
	8192	0.0100	0.0011	0.0100	0.0067	0.0133
Toilet	256	0.0145	0.0056	0.0139	0.0006	0.0335
	1024	0.0145	0.0029	0.0145	0.0063	0.0254
	4096	0.0144	0.0014	0.0143	0.0102	0.0202
	8192	0.0144	0.0010	0.0144	0.0109	0.0183

In Table 1, the mean Shapley value is comparable across varying Monte Carlo sample sizes for each respective class and does not drift significantly as the Monte Carlo sample size increases. Furthermore, the standard deviation decreases across all four classes as the sample size for Monte Carlo increases. This suggests an inverse correlation between Monte Carlo sampling size and the standard deviation. This could signify that the confidence and stability of the prediction increase as the Monte Carlo sample size increases. This trend is observed in Table 1. The observed trend is inline with the Monte Carlo method, which states that if the number of samples is increased, a better approximation of the ground truth is achieved. Another observation shows that the min and max converge towards the mean, which is a direct result of the standard deviation decreasing. The median, just as the mean, does not seem to drift significantly as the Monte Carlo sample size increases.

4.3 Point Contributions Across Classes

With the Monte Carlo sample size fixed at 4096, one instance from each class is analysed. The corresponding descriptive statistics are reported in Table 2. A subset of the

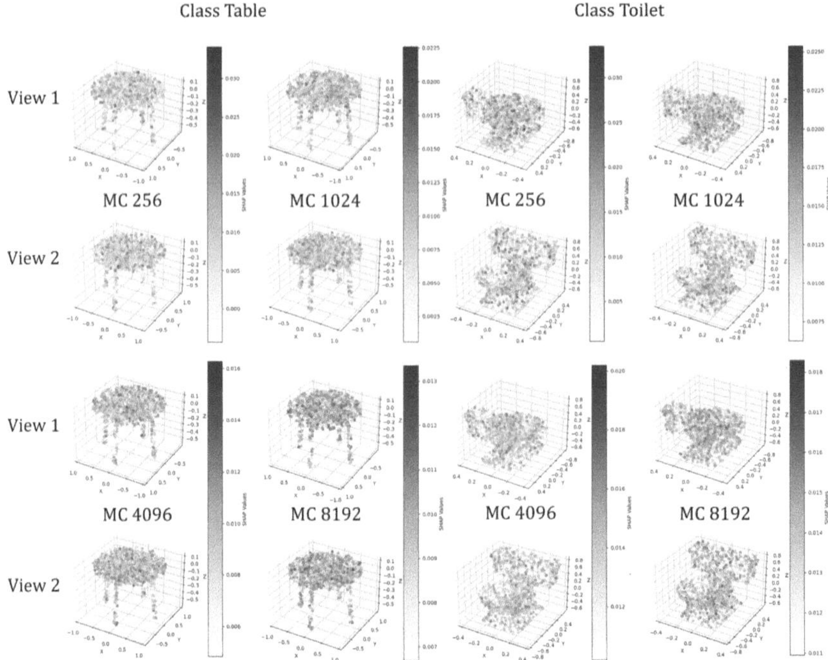

Fig. 3. Subplots for instances of class table and toilet at incrementally increasing Monte Carlo sample sizes.

visual results is shown in Fig. 4, while the complete visual analysis is presented in Fig. 6. An analysis of noteworthy classes is conducted, and the analysis is done in combination with the figures and the table. Figure 6 displays ten subplots, with each subplot containing two views of the same data instance. A Monte Carlo sample size of 4096 is used here. Within each subplot in Fig. 6, the x, y, and z coordinates of the original point cloud are displayed alongside each point's Shapley value represented by color. A Shapley value equal to zero signifies that point does not influence the prediction of the model, whereas higher Shapley values indicate significant contributions towards the prediction. Conversely, negative Shapley values suggest that these points contribute negatively to a prediction. A statistical summary for each subplot is provided in Table 2.

The statistical summary alongside the visualisation of Shapley values for points across ten classes reveals insights into the model's decision-making patterns. The bed class ($\mu = 0.0958$, $\sigma = 0.0039$) demonstrates the highest mean Shapley value of all classes. This may be a result of large flat surfaces being predicted as impactful or relatively large volumetric objects' points being perceived as impactful. Similarly, the dresser class ($\mu = 0.0845$, $\sigma = 0.004$) exhibits comparably high influence of point contribution also attributed to the potential over-reliance of local patterns of flat surfaces and box-like point structures, which both classes share. A causal link to geometric properties remains unproven without spatial attribution maps.

The desk instance ($\mu = -0.0620$, $\sigma = 0.0028$) presents consistently negative Shapley values. In general, a negative Shapley value for individual points does not inherently

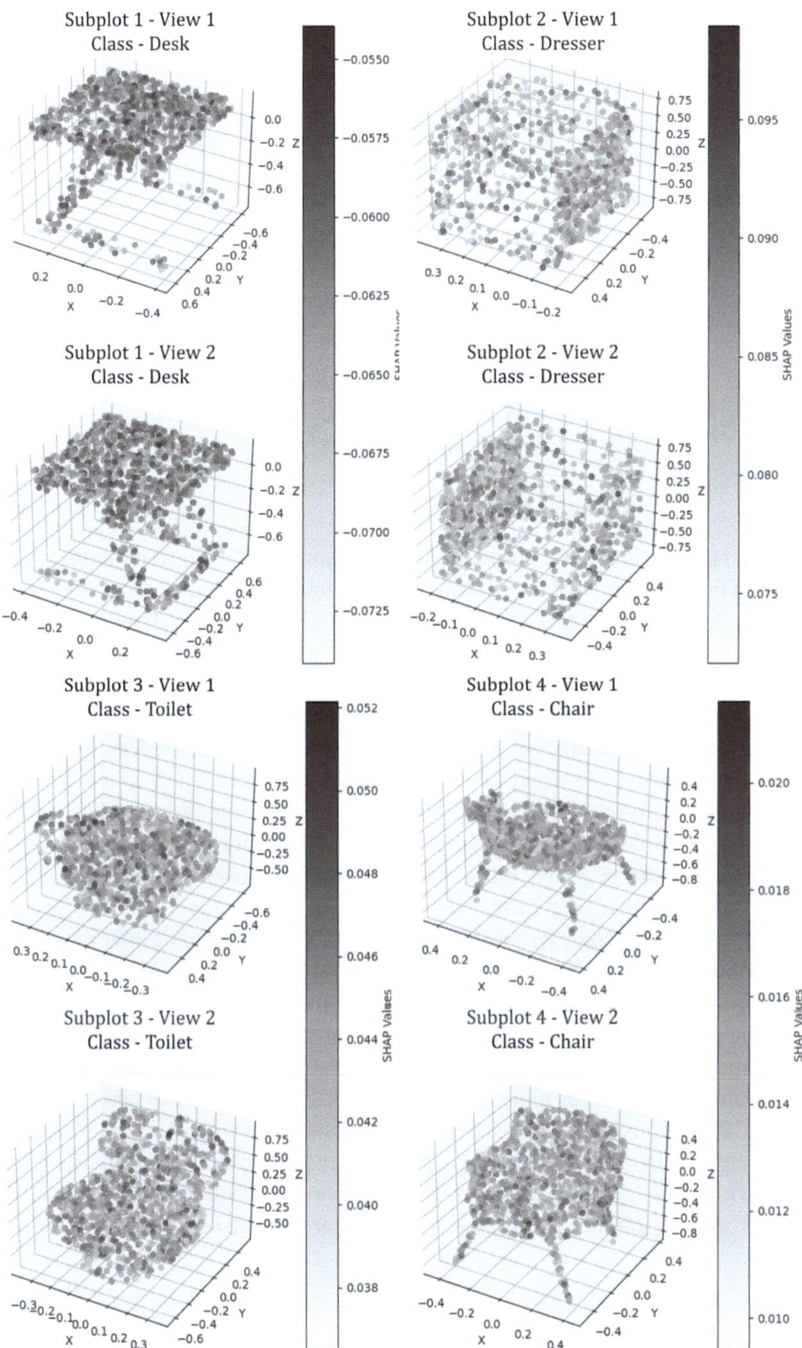

Fig. 4. Per point Shapley value of classes desk, dresser, toilet and chair at a Monte Carlo sample size of 4096.

Table 2. Shapley value statistics summary table for all classes present in Princeton ModelNet10 dataset at fixed Monte Carlo sampling rate of 4096 steps.

Class	Mean	Std	Median	Min	Max
Bathtub	0.0439	0.0030	0.0438	0.0342	0.0544
Bed	0.0958	0.0039	0.0957	0.0837	0.1098
Chair	0.0154	0.0016	0.0155	0.0094	0.0215
Desk	-0.0620	0.0028	-0.0619	−0.0742	−0.0540
Dresser	0.0845	0.0043	0.0846	0.0721	0.0990
Monitor	0.0359	0.0025	0.0359	0.0291	0.0440
Nightstand	0.0099	0.0032	0.0099	−0.0005	0.0236
Sofa	0.0293	0.0021	0.0293	0.0230	0.0375
Table	0.0372	0.0019	0.0371	0.0318	0.0439
Toilet	0.0444	0.0023	0.0443	0.0365	0.0522

imply misclassification of the entire point cloud as negative values in multi-class Shapley frameworks may represent features that are associated with other classes as well resulting in a negative Shapley value. However, a mean negative Shapley value would imply that for this specific instance, the model does misclassify the instance, and the mean of all points can only be negative if its majority of subsequent individual points also reflect a negative Shapley value. A visual analysis between subplots four and subplot seven in Fig. 6 visually portrays how similar these two instances look. A likely explanation is that subplot four, labelled as a table, was classified as a different class (through inference perhaps the model predicted the point cloud as a table) and shows significantly more dark red points in comparison to the other subplots, alongside all the Shapley values being negative. Further investigation into per-class error rates is essential to increase explainability further for this specific instance.

The instance of a chair ($\mu = 0.0154$, $\sigma = 0.0016$) exhibits a low mean Shapley value attributable to the chair class being overrepresented in the training dataset (by three times more in comparison to other classes) or through its distinct spatial relationship between points (e.g., legs, backrests, armrest combination). These two aspects might contribute to the class chair having the lowest mean Shapley value for all classes. Intermediate classes such as monitor ($\mu = 0.0359$, $\sigma = 0.0025$) and table ($\mu = 0.0372$, $\sigma = 0.0019$) demonstrate moderate point importance, possibly tied to including a flat surface paired with an elevation (e.g., flat surfaces for table, rectangular screens for monitor with table legs and a monitor base). Bathtub and toilet show nearly identical contributions, possibly reflecting shared curvilinear geometries that the model predicts similarly.

4.4   Shapley Value Quantile Analysis

All points are sorted in ascending order based on their Shapley value magnitudes. Subsequently, 75% of the points are removed, leaving only the 128 most and least impactful points in the point cloud. The points with the highest and lowest absolute Shapley values. This reduced point cloud is then evaluated for point contribution using the same methodology as in the previous experiment. Notably, the reduced point cloud instance is not regenerated, and the spatial positions of the retained points remain identical to those in the original instance from Table 2. The aim of this experiment is to remove reliance on specific clusters of points that may disproportionately influence Shapley value attribution.

The results of this experiment are presented in Table 3, where it can be observed that, for all classes, the top 128 points exhibit higher-magnitude Shapley values compared to the bottom 128 points from the original point cloud. Notably, the Nightstand class shows the smallest absolute difference between the high and low Shapley values. This may be attributed to the geometric simplicity of the selected Nightstand instance, which closely resembles a cube. As a result, the spatial properties are more evenly distributed, leading to a minimal distinction in the importance of individual points as captured by their Shapley values.

Table 3. Shapley value statistics summary table of partial point clouds for each class.

Class	Mean Shapley value of top 128 points	Mean Shapley value of bottom 128 points	Absolute Shapley Value Difference
Bed	−1.7045	−1.1675	0.5370
Chair	−0.4291	−0.1807	0.2484
Desk	0.2361	0.4210	0.1849
Dresser	0.8961	1.4815	0.5854
Monitor	−1.1987	−0.9312	0.2675
Nightstand	0.0846	0.1064	0.0218
Sofa	−1.6100	−1.1450	0.4650
Table	−0.7229	0.0558	0.6671
Toilet	−0.8414	0.4083	0.4331

The observed negative signage may reflect misclassification by the model, as the reduced point cloud (256 points per instance) could introduce ambiguity or alter the decision boundaries. This is especially true as this point cloud is not resampled, but rather reduced and reused. A potential median Shapley sampling statistic could highlight the skewedness of the points distribution, as the spatial mean of this reduced point cloud is not at the origin. All this combined might result in a prediction as the non-target classification label.

5 Conclusion

This paper introduced SPAX, an XAI model-agnostic framework for interpreting 3D point cloud classification using Shapley values. As it would be computationally infeasible to calculate Shapley values directly, a Monte Carlo approximation is used. With the aid of Shapley values a point cloud classification becomes interpretable as the contribution of individual points is captured.

The research presents visualizations that illustrate the influence of individual points within their respective coalitions, offering insights into their role in the model's predictions. These visualizations enhance the interpretability by highlighting spatial features relevant to classification.

Future work should focus on refining and evaluating this methodology, especially for practical scenarios, such as digitising railway infrastructure [38]. Beyond classification, this approach could be extended to segmentation tasks, expanding its applicability. Additionally, while this study focused on the ModelNet10 dataset, future research should explore how these findings translate to real-world data. Further investigation into a quantitative evaluation method, for instance the method proposed by Nauta et al. which provides a more structured framework for evaluating interpretability [21] is an interesting future direction.

Declarations

Author Contributions. Marc F. Harinck: Conceptualization, Technical Work, Original Draft, Editing and Reviewing. Muhammad Shoaib Sarwar: Conceptualization, Technical Work, Original Draft, Editing and Reviewing. Bram Ton: Conceptualization, Editing and Reviewing. Faizan Ahmed: Conceptualization, Editing and Reviewing.

Data Availability. In this study, a public dataset is used. The public dataset is available at https://modelnet.cs.princeton.edu. Since the data is publicly accessible, no ethical approval or informed consent was required.

Code Availability. Our code is available at https://github.com/Mavisis/SHAP-for-Point-Cloud.

Conflict of Interests. The authors declare that they have no conflict of interest.

Ethical and Informed Consent for Data Used. The data used in this study was a publicly available dataset on the Internet. No animals or humans were victims.

Appendix A

A.1   Algorithm 1: Generate Unique Coalitions per Target

- Let $indices \leftarrow \{0, 1, ..., num\_points - 1\}$
- Initialize $coalition\_dict \leftarrow \emptyset$
- For each $target \in indices$:
  - $other\_indices \leftarrow indices \setminus \{target\}$
  - $max\_possible \leftarrow 2^{|other\_indices|}$
  - $effective\_M \leftarrow \min(M, max\_possible)$
  - Initialize $unique\_subsets \leftarrow \emptyset$, $attempts \leftarrow 0$
  - $max\_attempts \leftarrow 10 \cdot effective\_M$
  - While $|unique\_subsets| < effective\_M$ and $attempts < max\_attempts$:
    * If $stratified$:

$\qquad \cdot$ Sample $k \in \{0, ..., |other\_indices|\}$ uniformly at random

$\qquad \cdot$ If $k > 0$: choose subset $S \subset other\_indices$ with $|S| = k$

$\qquad \cdot$ Else: $S \leftarrow \emptyset$

$\quad *$ Else:

$\qquad \cdot$ For each $i \in other\_indices$, include i in S with probability 0.5

$\quad *$ Let $T \leftarrow$ sorted tuple of S

$\quad *$ If $T \notin unique\_subsets$, add T to $unique\_subsets$

$\quad *$ Increment $attempts$

- Let $coalitions\_without \leftarrow$ list of elements in $unique\_subsets$
- Shuffle $coalitions\_without$
- Let $coalitions\_with \leftarrow \{S \cup \{target\} \mid S \in coalitions\_without\}$
- Set $coalition\_dict[target] \leftarrow \{$"without" $: coalitions\_without,$ "with" $: coalitions\_with\}$

– Return $coalition\_dict$

A.2   Algorithm 2: Evaluate Target Impact

– Inputs: model, $points\_tensor, coalition\_dict, target\_idx$

– Let $coalitions\_without \leftarrow coalition\_dict[target\_idx]["without"]$

– Let $coalitions\_with \leftarrow coalition\_dict[target\_idx]["with"]$

– Initialize $scores\_without \leftarrow [], scores\_with \leftarrow []$

– For each $C \in coalitions\_without$:

- Let $score \leftarrow model(points\_tensor$ restricted to $C)$
- Append $score$ to $scores\_without$

– For each $C \in coalitions\_with$:

- Let $score \leftarrow model(points\_tensor$ restricted to $C)$
- Append $score$ to $scores\_with$

– Let $impact\_scores \leftarrow [w - wo \mid (w, wo) \in zip(scores\_with, scores\_without)]$

– Return $impact\_scores$

A.3   Algorithm 3: Evaluate All Targets Impact

– Inputs: model, $points\_tensor, coalition\_dict$

– Initialize $all\_impact\_scores \leftarrow \emptyset$

– For each $target\_idx \in$ keys($coalition\_dict$):

- Let $impact\_scores \leftarrow$ EvaluateTargetImpact($model, points\_tensor, coalition\_dict, target\_idx$)
- Let $avg \leftarrow$ mean($impact\_scores$)
- Set $all\_impact\_scores[target\_idx] \leftarrow \{$"impact_scores" $: impact\_scores,$ "average_impact" $: avg\}$

– Return all_impact_scores

A.4   Comparison

Table 4 analyses 50 classifications, 5 instances of each class. In contrast to Table 3 who's results are specific to a unique instance, the Intersection over Union (IoU) table is a comparison between PointNet's Critical point sets described in Sect. 5.3 by Qi et al. [5] and the Shapley value agnostic method presented in this study. Each of the 50 experiments have instances sampled at a size of 1024 and a Monte Carlo sampling size of 4096.

Table 4. IoU statistics model comparison.

Metric	$K = 64$	$K = 256$	$K = 1024$
Mean IoU	0.031	0.073	0.173
Max IoU	0.094	0.117	0.212
Min IoU	0.0	0.043	0.114

There is an overall trend that as K increases, there is greater overlap, which is not surprising, as with increased K, more Shapley values are selected for comparison with the critical points from the PointNet. Analysing $K = 256$, which corresponds to 25% of the points in the entire point cloud, the mean Intersection over Union (IoU) score is 0.073, or an overlap of 7.3%. This indicates that for all the critical points identified by Qi et al. [5], only approximately 7% of the points are also included in the top 256 Shapley values deemed important by both methods, equating to 18 points on average where the model and Shapley values agree on critical importance. The results for $K = 256$ show a notable improvement over $K = 64$, with the mean IoU increasing to 0.073 and no classes exhibiting an IoU of zero, suggesting better class separation at this intermediate value of K.

These findings highlight the anecdotal aspect of the paper, as data-based results show limitations of this method as SPAX's reliance on coalition selection and size is great. As a result of approximating the prediction function of the model using Monte Carlo approximation and selecting a small sampling size relative to the number of points in the point cloud, this method sees a reduction in validity.

Appendix B

B.1   Variable MC

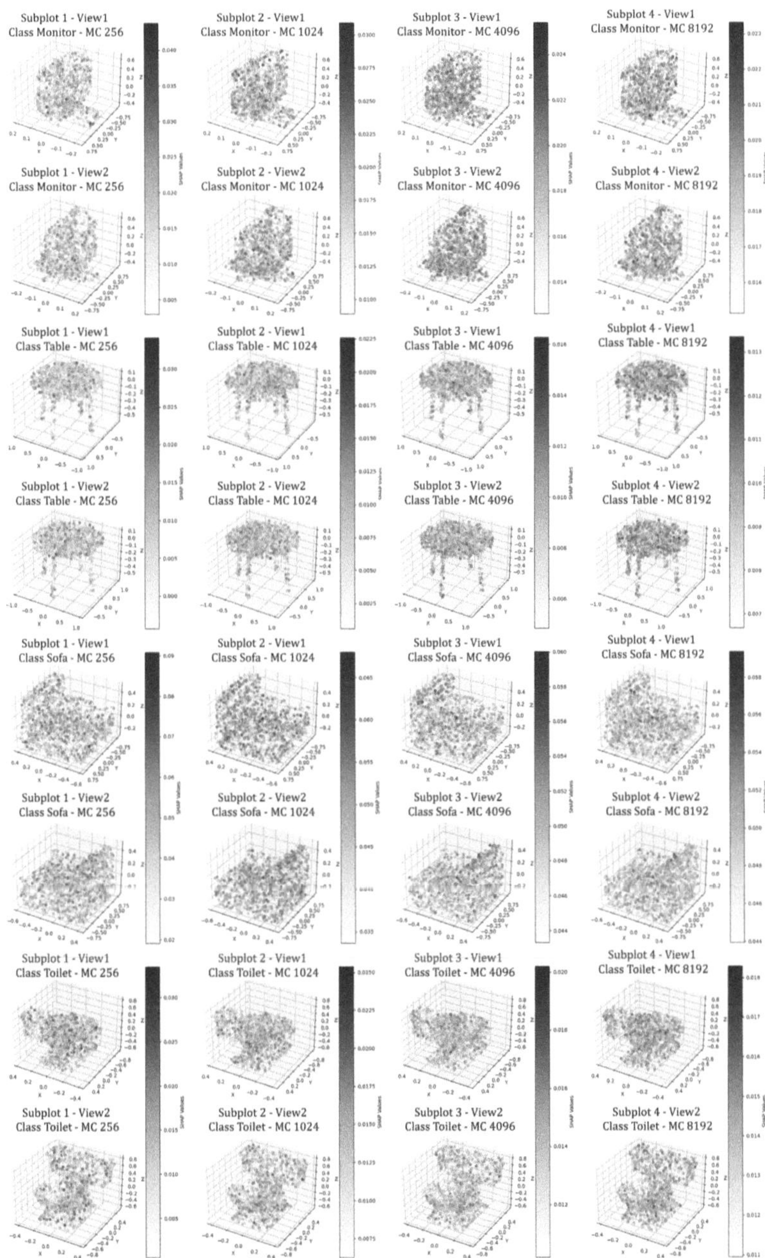

Fig. 5. Subplots representing various data instances at incrementally increasing Monte Carlo sample sizes. From left to right, the number of samples increases. From top to bottom in descending order, the classes monitor, sofa, table, and toilet are shown.

B.2   All Classes

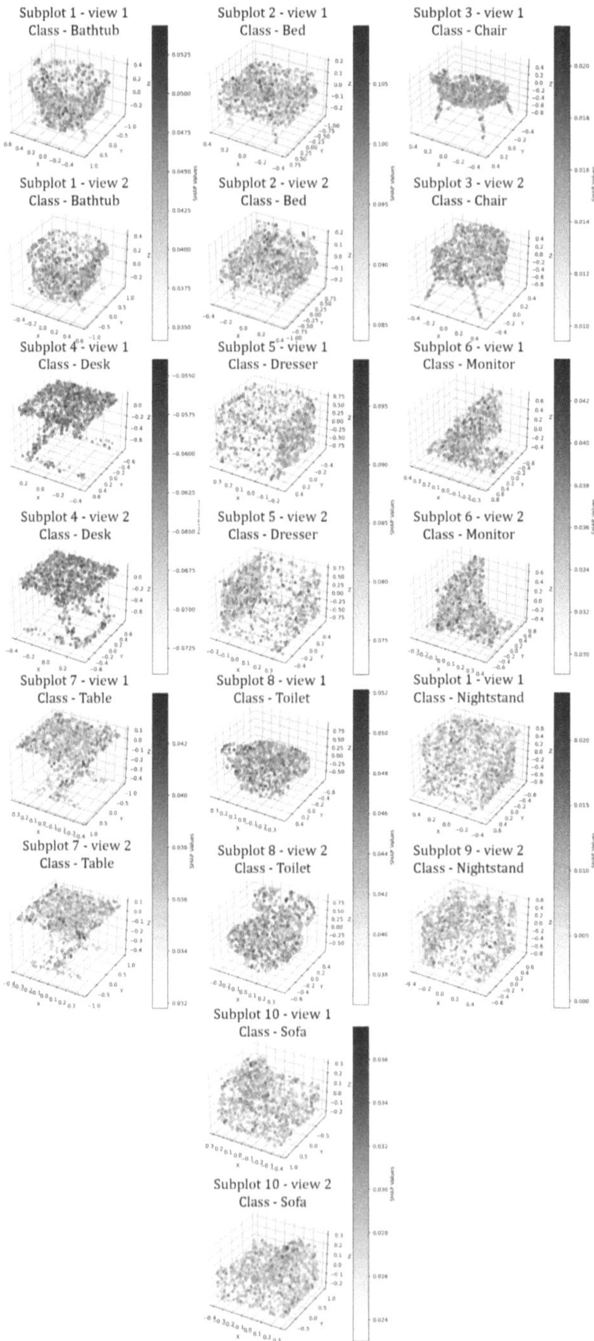

Fig. 6. Per point Shapley value of classes bathtub, bed, chair, desk, dresser, monitor, table, toilet, nightstand, and sofa at a Monte Carlo sample size of 4096.

References

1. Arnold, N.I., Angelov, P., Atkinson, P.M.: An improved explainable point cloud classifier (XPCC). IEEE Trans. Artifi. Intell. **3**(6), 785–797 (2022). https://doi.org/10.1109/TAI.2022.3150647
2. Atik, M.E., Duran, Z., Seker, D.Z.: Explainable artificial intelligence for machine learning-based photogrammetric point cloud classification. IEEE J. Selected Topics Appli. Earth Observat. Remote Sensing **17**, 5834–5846 (2024). https://doi.org/10.1109/JSTARS.2024.3370159
3. Ballester, P.L., et al.: Gray matter volume drives the brain age gap in schizophrenia: a SHAP study. Schizophrenia **9**(1), 3 (2023)
4. Barredo Arrieta, A., et al.: Explainable Artificial Intelligence (XAI): concepts, taxonomies, opportunities and challenges toward responsible AI. Information Fusion **58**, 82–115 (2020). https://doi.org/10.1016/j.inffus.2019.12.012, https://www.sciencedirect.com/science/article/pii/S1566253519308103
5. Charles, R.Q., Su, H., Kaichun, M., Guibas, L.J.: PointNet: deep learning on point sets for 3D classification and segmentation. In: 2017 IEEE Conference on Computer Vision and Pattern Recognition (CVPR), pp. 77–85 (2017). https://doi.org/10.1109/CVPR.2017.16
6. Chen, C., Liu, Y., Li, Y., Chen, D.: An explainable artificial intelligence (XAI)-based framework for urban GDEM correction considering spatial heterogeneity and factor optimization. Int. J. Appl. Earth Obs. Geoinf. **129**, 103843 (2024). https://doi.org/10.1016/j.jag.2024.103843
7. Goodman, B., Flaxman, S.: European Union regulations on algorithmic decision-making and a "right to explanation." AI Mag. **38**(3), 50–57 (2017)
8. Gunning, D., Stefik, M., Choi, J., Miller, T., Stumpf, S., Yang, G.Z.: XAI–explainable artificial intelligence. Sci. Rob. **4**(37), eaay7120 (2019)
9. Gupta, A., Watson, S., Yin, H.: 3D point cloud feature explanations using gradient-based methods. In: Proceeding of International Joint Conference on Neural Networks, pp. 1–8 (2020)
10. Han, F., Liang, T., Ren, J., Li, Y.: Automated extraction of rail point clouds by multi-scale dimensional features from MLS data. IEEE Access **11**, 32427–32436 (2023)
11. Kochhar, A., Arya, A.K., Thapliyal, A., Kumar, D., Khan, F.: Animal species classification using deep learning. In: Doctoral Symposium on Computational Intelligence, pp. 175–190. Springer (2024). https://doi.org/10.1007/978-981-97-6726-7_14
12. Kumar, P., Lewis, P., McElhinney, C.P.: Parametric analysis for automated extractions of road edges from mobile laser scanning data. ISPRS Annals of the Photogrammetry, Remote Sensing Spatial Inform. Sci. **2**, 215–221 (2015)
13. Lei, S.: A feature selection method based on information gain and genetic algorithm. In: International Conference on Computer Science and Electronics Engineering, vol. 2, pp. 355–358 (2012). https://doi.org/10.1109/ICCSEE.2012.97
14. Levi, M.Y., Gilboa, G.: Fast and simple explainability for point cloud networks (2024). https://arxiv.org/abs/2403.07706
15. Li, R., et al.: Machine learning-based interpretation and visualization of nonlinear interactions in prostate cancer survival. JCO Clinical Cancer Inform. **4**, 637–646 (2020)
16. Lundberg, S.M., Lee, S.I.: A unified approach to interpreting model predictions. In: Guyon, I., et al. (eds.) Advances in Neural Information Processing Systems, vol. 30, pp. 4765–4774. Curran Associates Inc, California (2017)
17. Matrone, F., Paolanti, M., Felicetti, A., Martini, M., Pierdicca, R.: BubblEX: an explainable deep learning framework for point-cloud classification. IEEE J. Selected Topics Appli. Earth Observations and Remote Sensing **15**, 1–18 (2022). https://doi.org/10.1109/JSTARS.2022.3195200

18. Matrone, F., Paolanti, M., Frontoni, E., Pierdicca, R.: Enhancing explainability of deep learning models for point cloud analysis: a focus on semantic segmentation. Inter. J. Digital Earth **17**(1), 2390457 (2024). https://doi.org/10.1080/17538947.2024.2390457
19. Molnar, C.: Interpretable Machine Learning A Guide for Making Black Box Models Explainable. Independently published, Digital (2022)
20. Mulawade, R.N., Garth, C., Wiebel, A.: Explainable artificial intelligence (XAI) for methods working on point cloud data: a survey. IEEE Access **12**, 146830–146851 (2024). https://doi.org/10.1109/ACCESS.2024.3472872
21. Nauta, M., et al.: From anecdotal evidence to quantitative evaluation methods: a systematic review on evaluating explainable AI. ACM Comput. Surv. **55**(13s) (Dec 2023). https://doi.org/10.1145/3583558
22. Oh, M.Y., Kim, H.S., Jung, Y.M., Lee, H., Lee, S.B., Lee, S.M.: Explainable automated non-linear computation scoring system for health (EACH) score: a machine learning based explainable automated nonlinear computation scoring system for health and an application for prediction of perioperative stroke. J. Med. Internet Res. **27**, 58021 (2025). https://doi.org/10.2196/58021
23. Preece, A., Harborne, D., Braines, D., Tomsett, R., Chakraborty, S.: Stakeholders in explainable AI. arXiv preprint arXiv:1810.00184 (2018)
24. Qiao, Y., et al.: Point clouds segmentation of rapeseed siliques based on sparse-dense point clouds mapping. Front. Plant Sci. **14**, 1188286 (2023)
25. Robnik-Šikonja, M., Kononenko, I.: Theoretical and empirical analysis of relieff and rrelieff. Machine Learning **53** (2003). https://doi.org/10.1023/A:1025667309714
26. Russell, S., Norvig, P.: Artificial Intelligence: A Modern Approach, 3rd edn. Prentice Hall, New Jersey (2010)
27. Sahebdivani, S., Arefi, H., Maboudi, M.: Rail track detection and projection-based 3D modeling from UAV point cloud. Sensors **20**(18), 5220 (2020)
28. Saranti, A., Pfeifer, B., Gollob, C., Stampfer, K., Holzinger, A.: From 3D point-cloud data to explainable geometric deep learning: State-of-the-art and future challenges. WIREs Data Min. Knowl. Discovery **14**(6), e1554 (2024). https://doi.org/10.1002/widm.1554
29. Selvaraju, R.R., Cogswell, M., Das, A., Vedantam, R., Parikh, D., Batra, D.: Grad-CAM: visual explanations from deep networks via gradient-based localization. In: 2017 IEEE International Conference on Computer Vision (ICCV), pp. 618–626 (2017). https://doi.org/10.1109/ICCV.2017.74
30. Shapley, L.S.: A value for n-person games. In: Kuhn, H.W., Tucker, A.W. (eds.) Contributions to the Theory of Games II, pp. 307–317. Princeton University Press, California (1953). https://doi.org/10.1515/9781400881970-018
31. Shen, W., Ren, Q., Liu, D., Zhang, Q.: Interpreting representation quality of DNNs for 3D point cloud processing. In: Proceedings of the 35th International Conference on Neural Information Processing Systems, NIPS 2021, Curran Associates Inc., Red Hook, NY, USA (2021). https://doi.org/10.5555/3540261.3540939
32. Shrikumar, A., Greenside, P., Shcherbina, A., Kundaje, A.: Not just a black box: learning important features through propagating activation differences. CoRR abs/ arXiv: 1605.01713 (2016)
33. Štrumbelj, E., Kononenko, I.: Explaining prediction models and individual predictions with feature contributions. Knowl. Inf. Syst. **41**, 647–665 (2014)
34. Sun, T.Q., Medaglia, R.: Mapping the challenges of artificial intelligence in the public sector: Evidence from public healthcare. Gov. Inf. Q. **36**(2), 368–383 (2019). https://doi.org/10.1016/j.giq.2018.09.008
35. Taghanaki, S.A., Hassani, K., Jayaraman, P.K., Khasahmadi, A.H., Custis, T.: PointMask: towards interpretable and bias-resilient point cloud processing. arXiv preprint arXiv:2007.04525 (2020)

36. Tan, H., Kotthaus, H.: Surrogate model-based explainability methods for point cloud NNs. In: 2022 IEEE/CVF Winter Conference on Applications of Computer Vision (WACV), pp. 2239–2248 (2022). https://doi.org/10.1109/WACV51458.2022.00233

37. Tan, H., Kotthaus, H.: Visualizing global explanations of point cloud DNNs. In: 2023 IEEE/CVF Winter Conference on Applications of Computer Vision (WACV), pp. 2269–2278 (2023). https://doi.org/10.1109/WACV56678.2023.00226

38. Ton, B., Ahmed, F., Linssen, J.: Semantic segmentation of terrestrial laser scans of railway catenary arches: A use case perspective. Sensors **23**(1), 222 (2022). https://doi.org/10.3390/s23010222

39. Wang, Y., Sun, Y., Liu, Z., Sarma, S.E., Bronstein, M.M., Solomon, J.M.: Dynamic graph CNN for learning on point clouds. ACM Trans. Graph. **38**(5) (2019). https://doi.org/10.1145/3326362

40. West, D.M.: The future of work: Robots, AI, and automation. Brookings Institution Press, Washington DC (2018)

41. Wu, Z., et al.: 3D ShapeNets: a deep representation for volumetric shapes (2015). https://modelnet.cs.princeton.edu/

42. Zeybek, M.: Extraction of road lane markings from mobile lidar data. Transp. Res. Rec. **2675**(5), 30–47 (2021)

A Privacy-Preserving and Explainable Approach for Anomaly Detection in Substation Networks

Paul Tavolato[1], Oliver Eigner[2], Philipp Kreimel-Haindl[3], Patrizia Agnello[4],
Marta Petyx[4], Antonella Santone[5], Fabio Martinelli[6], and Francesco Mercaldo[1,5(✉)]

[1] Faculty of Computer Science, University of Vienna, Vienna, Austria
paul.tavolato@univie.ac.at
[2] Department of Computer Science and Security, St. Pölten University of Applied Sciences,
Sankt Pölten, Austria
oliver.eigner@fhstp.ac.at
[3] OMV, Vienna, Austria
Philipp.KreimelHaindl@omv.com
[4] Istituto Nazionale per l'Assicurazione contro gli Infortuni sul Lavoro (INAIL), Roma, Italy
{p.agnello,m.petyx}@inail.it
[5] Department of Medicine and Health Sciences "V. Tiberio", University of Molise,
Campobasso, Italy
{antonella.santone, francesco.mercaldo}@unimol.it
[6] Institute of High Performance Computing and Networking, National Research Council of
Italy (CNR), Rende, Italy
fabio.martinelli@icar.cnr.it

Abstract. Electrical substations play a crucial role in managing electrical energy, making them critical targets cyber-attacks on these systems could severely impact the general population, hospitals, and both critical and non-critical infrastructure. Several methods, from both academic than industrial world, propose anomaly detection in substation networks but currently there is a lacking of explainability on the reasons why an anomaly is detected. Moreover, privacy is also an issue, as a matter of fact current methods require that network logs are sent to a centralized server for model training, exposing sensitive and confidential network traces outside the electrical substation network infrastructure. To overcome both of these limitations, in this paper, we present an explainable and privacy-preserving approach for detecting possible anomalies within electrical substations. The proposed method analyzes network logs to identify potential anomalies in substation networks. We convert a network trace into an image and we consider a Vision Transformer model for anomaly detection. We also consider prediction explainability by highlighting specific areas in the image generated from the network trace that the classifier identifies as indicative of an anomaly, with the aim to visually show which area of the image is symptomatic of a possible anomaly.

Keywords: Substation · SCADA · Modbus · Anomaly · Vision Transformer

© The Author(s), under exclusive license to Springer Nature Switzerland AG 2026
F. Marcelloni et al. (Eds.): IJCCI 2025, CCIS 2829, pp. 434–450, 2026.
https://doi.org/10.1007/978-3-032-15638-9_25

1 Introduction and Related Work

Critical infrastructure is increasingly under threat, with cyber-attacks surging by 30% in just one year [7]. Among the most frequently targeted sectors are energy, transportation, and telecommunications. This trend is not surprising, especially in developed nations, where these sectors are becoming progressively reliant on digital technologies—exposing them to new cyber-vulnerabilities. According to a report published in August 2024 by KnowBe4 security experts[1], such attacks may lead to catastrophic consequences for national stability. Consequently, geopolitical adversaries increasingly exploit these weaknesses, positioning cyber-attacks as powerful tools in their digital arsenals. Widespread service disruptions, including those affecting healthcare, emergency services, and governmental agencies, can result. Hospitals, for example, may lose access to essential data, while surgeries and medical appointments may face delays. Cybercriminals frequently capitalize on such chaos by registering phishing domains and impersonating support personnel.

A continuous and escalating wave of cyber-attacks on critical infrastructure is being observed globally, posing a profound threat with the potential for far, reaching social and economic disruption [2].

One of the most concerning scenarios involves cyber-attacks on the energy sector, encompassing systems such as power generation, water treatment, and electricity distribution [12, 17]. An attack on these interconnected systems could plunge communities into chaos. For example, during wartime, a targeted power outage could severely impede hospitals, emergency responders, and military operations [15].

In fact, as of November 2023, the average weekly number of cyberattacks against utilities had more than doubled between 2020 and 2022. In 2024, the North American Electric Reliability Corporation reported that the number of vulnerable points within U.S. power grids is growing at an alarming rate of approximately 60 per day. The total number of susceptible points increased from 21,000 in 2022 to an estimated 23,000–24,000 in 2024.

Between January 2023 and January 2024, over 420 million attacks targeting critical infrastructure were recorded, according to Forescout Research—Vedere Labs[2]. This corresponds to approximately 13 attacks per second, reflecting a 30% rise from 2022. These attacks impacted 163 countries, with the United States being the primary target, followed by the United Kingdom, Germany, India, and Japan. Additionally, the report notes that China hosts the highest concentration of threat actors targeting critical infrastructure, followed by Russia and Iran.

As power systems have evolved, substations have grown in size and complexity, increasing their exposure to cyber threats. The rising demand for transformer and switchyard substations, compared to generator stations, has driven the shift from manual to remote monitoring and control [10, 11]. Ensuring continuous, high-availability operation of substations is a key priority for utilities, as frequent faults can disrupt service and lead to revenue losses [1].

---

[1] https://industrialcyber.co/critical-infrastructure/critical-infrastructure-faces-30-percent-surge-in-cyber-attacks-knowbe4-report-highlights/.

[2] https://www.forescout.com/research-labs/.

To support this need, advanced supervisory systems have been developed, such as the Supervisory Control and Data Acquisition (SCADA) system. SCADA collects data from Intelligent Electronic Devices (IEDs) and Remote Terminal Units (RTUs) using standard communication protocols (e.g., IEC 60870-5-104, IEC 61850, or Modbus). These systems facilitate device monitoring and control via sophisticated visual interfaces and can automate tasks using predefined logic and algorithms, greatly improving operational efficiency and decision-making [2].

A Human-Machine Interface (HMI) is deployed in each substation to enable local monitoring and control. These interfaces are essential during commissioning, configuration, or maintenance operations, ensuring safe and efficient on-site management.

During the early digital era, manufacturers implemented proprietary standards for intelligent devices, resulting in poor interoperability and vendor lock-in. To address this, the Modbus protocol was introduced—initially for use with Programmable Logic Controllers (PLCs), and has since become a widely adopted communication standard across various industrial devices.

The purpose of the Modbus protocol is to facilitate interaction between IEDs and SCADA HMIs. An IED periodically changes voltage values either randomly or in response to SCADA HMI requests. Conversely, the SCADA HMI adjusts tap-changes or opens/closes circuits in response to voltage anomalies [6].

Anomaly detection in substation networks is critical to the safety and reliability of electrical grids. In recent years, Deep Learning (DL) has demonstrated strong potential for detecting such anomalies. Specifically, Convolutional Neural Networks (CNNs) have been widely adopted for pattern recognition, failure prediction, and anomaly assessment [4,5].

For instance, Kreimel *et al.* propose a hybrid deep learning method using CNNs for substation anomaly detection. While the approach achieves high detection accuracy, it lacks interpretability, which limits its usability for operators.

Similarly, in [15], a DL-based Intrusion Detection System (IDS) is introduced to detect malicious Generic Object-Oriented Substation Event (GOOSE) communications over process and station bus networks. However, the study does not address the model's explainability.

Valdes *et al.* [18] employ unsupervised learning to infer physical invariants for anomaly detection in substations. Although innovative, the approach does not provide mechanisms to interpret model decisions.

Most *et al.* [14] use non-negative tensor decomposition to detect anomalies in SCADA systems. While effective for statistical behavior modeling, this approach does not prioritize explainability.

Marino *et al.* [8] propose a DL model leveraging graph structures in communication networks. Their method improves interpretability by offering hierarchical features and supports self-supervised training without labeled data.

Despite these advances, DL models face major limitations in explainability. In particular, there is no explicit rationale for predictions from the model's internal perspective.

Furthermore, centralized training of DL models requires extensive datasets, raising privacy concerns due to the need for data sharing among multiple institutions.

To address these limitations, this paper proposes a method for explainable anomaly detection in substation networks. We introduce a Federated Machine Learning (FML) approach using a Vision Transformer (ViT) model to classify network traces as either anomalous or legitimate. FML enables collaborative training across distributed nodes or devices, sending only model updates (thus, not raw data) to a central server.

The data for ViT training and testing comprises images generated from network traces stored in PCAP files. As shown in literature, DL models achieve strong performance when trained on image-based datasets [4,5,9,16], this is the reason why we convert PCAP files into images.

To provide explainability behind the decision of the model, we identify regions within each image (representing a network trace) that are most influential in the model's classification decision. Thus, when a trace is labeled anomalous, we visually highlight the areas responsible for the prediction. This transparency improves operator trust and enhances model acceptance in practical substation management.

The remaining of this paper is structured as follows: Sect. 2 presents the proposed explainable and privacy-preserving method for anomaly detection in substation networks; Sect. 3 describes the experimental analysis aimed to evaluate the effectiveness of the proposed approach and; finally, Sect. 4 draws the conclusions and outlines future research directions.

2　The Method

In this section, we introduce the designed and developed privacy-preserving and explainable method for anomaly detection in electrical substation, by exploiting FML. The proposed method enables collaborative training across multiple decentralized electrical substation (i.e., the clients) without transferring or sharing raw network data. Each client retains its local dataset and participates in a federated training process by sending model updates to a central server, thereby ensuring data confidentiality.

The overall architecture of the proposed method is shown in Fig. 1. The proposed method is designed to achieve three primary objectives:

- *Data privacy*: ensured through decentralized training without any raw data exchange.
- *Model generalization*: achieved by aggregating knowledge from diverse electrical substations.
- *Prediction Explainability*: enabled by incorporating an attention-based visualization strategy suitable for models.

As shown from Fig. 1, the proposed method is organized into the following components: *Network logs Acquisition and Image Generation, Local Training with Vision Transformer, Federated Aggregation*, and *Explainability with Attention Rollout*, described in detail below.

2.1　Network Log Acquisition and Image Generation

Each client i.e., electrical substation, store network traces in PCAP (Packet Capture) format, that are transformed into image representations with the aim to exploit the capabilities of computer vision-based models for anomaly detection. To obtain this, a script

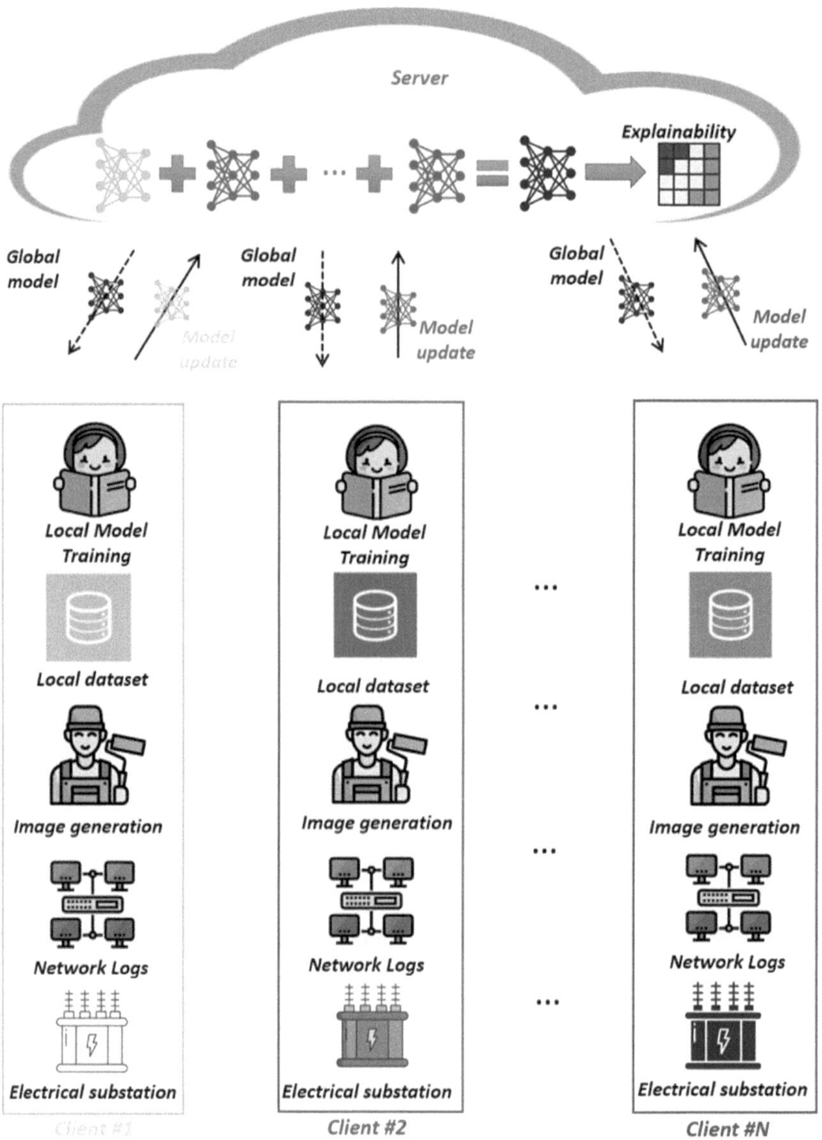

Fig. 1. The workflow of the proposed method for privacy-preserving anomaly detection in substation networks by exploiting FML.

in the Python programming language was developed by the authors. This script reads the raw binary content of each PCAP file and converts it into a numerical array using the NumPy library. The resulting array is then reshaped to match a fixed image resolution of 100×100 pixels.

To ensure consistency across all samples, if the binary data is smaller than the target size, zero-padding is applied; conversely, if the data exceeds the desired dimensions, it is truncated. Each final image is composed of three channels (i.e., in RGB format), enabling compatibility with standard deep learning architectures. The processed images are subsequently saved in PNG format and used as input for the anomaly detection model.

The proposed method consider a binary classification task where each network log is labeled as either:

- *Attack:* related to the presence of an anomalous traffic in the electrical substation under analysis;
- *Benign:* related the legitimate traffic in the electrical substation under analysis.

2.2   Local Training with Vision Transformer

Each client performs local training using a Vision Transformer (ViT) model, an architecture for image classification. ViT divides input images into fixed-size patches and processes them as token sequences via a transformer encoder. This enables the model to learn both local and global dependencies in the images we obtained from PCAP files.

In particular, we employ a pretrained ViT backbone[3] obtained from the Hugging Face transformers library. Its parameters are frozen to preserve general visual features learned from large-scale image datasets. On top of this, a custom classification head is appended and trained locally by each client, as shown from Python code snippet in Listing 1.1.

```python
# Load the pretrained "WinKawaks/vit-tiny-patch16-224" ViT
    model, a lightweight ViT with 16$\,\times\,$16 input
    patches and 224$\,\times\,$224 image resolution.
vit_base = ViTModel.from_pretrained("WinKawaks/vit-tiny-
    patch16-224").to(device)

# Freeze all layers of the pretrained ViT to use it as a fixed
    feature extractor.
for param in vit_base.parameters():
    param.requires_grad = False

# Define a trainable classifier head on top of the frozen ViT
    encoder.
classifier = nn.Sequential(
    nn.Linear(vit_base.config.hidden_size, 128),    # Project
    ViT output to 128-dim space
    nn.ReLU(),                                      # Add non-
    linearity
    nn.Dropout(0.3),                                # Apply
    dropout for regularization
```

---

[3] https://huggingface.co/WinKawaks/vit-tiny-patch16-224.

```
13    nn.Linear(128, number_classes)                      # Final
      classification layer (2 classes)
14  ).to(device)
```

Listing 1.1. Python code snippet for ViT model initialization and custom classifier.

The key components of the ViT model shown in Listing 1.1 are the following:

- *Feature Extraction with ViT:* The ViT model is pretrained and frozen, acting as a robust feature extractor. This form of transfer learning is especially effective when local training data is limited.
- *Custom Classification Head:* A fully connected (FC) classifier is attached to the output of ViT. It includes:
    1. A linear layer mapping the ViT output to a 128-dimensional feature space.
    2. A ReLU activation for introducing non-linearity.
    3. A dropout layer (30%) for mitigating overfitting.
    4. A final linear layer that maps to the binary output space (seizure vs. normal).

2.3   Federated Aggregation with FedAvg and FedProx

After local training, clients send their updated model weights (not raw data) to a central server. The server computes a global model via federated aggregation. Two aggregation strategies are considered by the proposed method:

Federated Averaging. Federated Averaging (FedAvg) computes the global model by averaging client models proportionally to the number of local data samples:

$$\theta_{t+1} = \sum_{k=1}^{K} \frac{n_k}{n} \theta_t^k \tag{1}$$

where θ_t^k is the local model of client k, n_k is the number of data samples on client k, and $n = \sum_{k=1}^{K} n_k$ is the total number of samples across all clients.

Federated Proximal. Federated Proximal (FedProx) extends FedAvg by adding a proximal term to the local loss function, which improves stability in heterogeneous environments:

$$\min_{\theta} \; f_k(\theta) + \frac{\mu}{2} \|\theta - \theta_t\|^2 \tag{2}$$

where $f_k(\theta)$ is the local training loss, θ_t is the global model received from the server, and μ is a regularization constant controlling the strength of the proximal term.

2.4   Explainability with Attention Rollout

To promote transparency in electrical substation anomaly detection decision-making, we adopt a ViT-specific explainability technique i.e., the Attention Rollout. This method exploits the self-attention mechanisms within transformer layers to identify which regions of the input image influence the model's predictions [13].

Unlike Grad-CAM, which is tailored to CNNs, Attention Rollout computes a joint attention map by recursively aggregating the attention matrices across all transformer layers. This includes residual connections, ensuring that key spatial dependencies are preserved across the entire network depth. The result is an attention heatmap indicating how much each image patch contributes to the final classification decision, specifically by analyzing its influence on the *[CLS]* token.

The final heatmap is overlaid on the original network trace-derived image, highlighting the regions that most influenced the model's output. This enables cybersecurity analyst to visually inspect the decision process and determine whether the model focused on meaningful signal patterns consistent with anomalous (or legitimate) activity.

3 Experimental Analysis

To demonstrate the effectiveness of the proposed method, real-world network logs from electrical substations are gathered from the CIC Modbus 2023 dataset [3], which is publicly accessible for research purposes[4]. This dataset comprises network traffic captures and associated attack logs generated within a simulated substation environment. The data is categorized into two primary groups: benign and malicious traces.

The malicious network logs simulate a diverse range of Modbus protocol attacks in a substation setting. These include reconnaissance, query flooding, payload injection, delayed responses, length field manipulation, false data injection, Modbus frame stacking, brute-force write operations, and baseline replay attacks. The attack scenarios are modeled based on techniques described in the MITRE ICS ATT&CK framework[5]. On the other side, the benign subset consists of standard Modbus communications representing legitimate substation operations. For manageability, all capture files are split into 100MB segments.

In Table 1 we report the set of experiments aimed to evaluate the proposed method, for each experiment we report the hyper-parameters we considered.

Table 1 presents the set of 48 experiments conducted to evaluate the performance of the proposed method with different configurations. The experimental setup systematically varies several hyper-parameters: aggregation strategy, number of local epochs, number of clients, and learning rate, while keeping the number of communication rounds fixed at 10 and the batch size constant at 32 across all experiments. Two aggregation algorithms i.e., FedAvg and FedProx, are considered in the experiments. The experiments are organized by progressively increasing the number of clients, starting from 2 and extending up to 10, in order to assess how client participation affects the learning process. Each configuration is tested with two values for the number of local training epochs (50 and 100), and three different learning rates (0.0001, 0.0005, and 0.001). This results in a total of 48 experimental runs, equally distributed between the FedAvg and FedProx strategies, with the aim to provide a comparison of the proposed method when these parameter are varying.

The results of the experiments in Table 1 are shown in Table 2, where we show the performances related, for each experiment, to the centralized model (i.e., the server).

---

[4] https://www.unb.ca/cic/datasets/modbus-2023.html.

[5] https://attack.mitre.org/techniques/ics/.

Table 1. The hyper-parameters for each experiment.

Exp	Epochs	Rounds	Aggregation	Clients	Learning Rate	Batch Size
1	50	10	FedAvg	2	0.0001	32
2	50	10	FedProx	2	0.0001	32
3	50	10	FedAvg	2	0.0005	32
4	50	10	FedProx	2	0.0005	32
5	50	10	FedAvg	2	0.001	32
6	50	10	FedProx	2	0.001	32
7	100	10	FedAvg	2	0.0001	32
8	100	10	FedProx	2	0.0001	32
9	100	10	FedAvg	2	0.0005	32
10	100	10	FedProx	2	0.0005	32
11	100	10	FedAvg	2	0.001	32
12	100	10	FedProx	2	0.001	32
13	50	10	FedAvg	3	0.0001	32
14	50	10	FedProx	3	0.0001	32
15	50	10	FedAvg	3	0.0005	32
16	50	10	FedProx	3	0.0005	32
17	50	10	FedAvg	3	0.001	32
18	50	10	FedProx	3	0.001	32
19	100	10	FedAvg	3	0.0001	32
20	100	10	FedProx	3	0.0001	32
21	100	10	FedAvg	3	0.0005	32
22	100	10	FedProx	3	0.0005	32
23	100	10	FedAvg	3	0.001	32
24	100	10	FedProx	3	0.001	32
25	50	10	FedAvg	5	0.0001	32
26	50	10	FedProx	5	0.0001	32
27	50	10	FedAvg	5	0.0005	32
28	50	10	FedProx	5	0.0005	32
29	50	10	FedAvg	5	0.001	32
30	50	10	FedProx	5	0.001	32
31	100	10	FedAvg	5	0.0001	32
32	100	10	FedProx	5	0.0001	32
33	100	10	FedAvg	5	0.0005	32
34	100	10	FedProx	5	0.0005	32
35	100	10	FedAvg	5	0.001	32
36	100	10	FedProx	5	0.001	32
37	50	10	FedAvg	10	0.0001	32
38	50	10	FedProx	10	0.0001	32
39	50	10	FedAvg	10	0.0005	32
40	50	10	FedProx	10	0.0005	32
41	50	10	FedAvg	10	0.001	32
42	50	10	FedProx	10	0.001	32
43	100	10	FedAvg	10	0.0001	32
44	100	10	FedProx	10	0.0001	32
45	100	10	FedAvg	10	0.0005	32
46	100	10	FedProx	10	0.0005	32
47	100	10	FedAvg	10	0.001	32
48	100	10	FedProx	10	0.001	32

Table 2. Experimental results related to the server model.

Exp	Loss	Accuracy	Precision	Recall	F-measure	AUC
1	0.5590	0.7500	0.7550	0.7500	0.7488	0.8040
2	0.5733	0.8000	0.8125	0.8000	0.7980	0.7884
3	0.4536	0.7700	0.7701	0.7700	0.7700	0.8796
4	0.5378	0.7600	0.8378	0.7600	0.7453	0.8608
5	0.4220	0.7900	0.7997	0.7900	0.7883	0.8852
6	0.4000	0.8200	0.8200	0.8200	0.8200	0.9076
7	0.5842	0.7600	0.7617	0.7600	0.7596	0.7736
8	0.5662	0.7300	0.7323	0.7300	0.7293	0.8076
9	0.4846	0.7400	0.7435	0.7400	0.7391	0.8592
10	0.5080	0.7300	0.7527	0.7300	0.7238	0.8596
11	0.4353	0.8300	0.8731	0.8300	0.8249	0.8912
12	0.4047	0.8400	0.8400	0.8400	0.8400	0.9008
13	0.6184	0.7400	0.7400	0.7400	0.7400	0.7556
14	0.5960	0.7500	0.7509	0.7500	0.7498	0.7896
15	0.5192	0.7700	0.8053	0.7700	0.7632	0.8284
16	0.5031	0.7900	0.8047	0.7900	0.7874	0.8464
17	0.5164	0.7700	0.8278	0.7700	0.7594	0.8828
18	0.5317	0.7100	0.7800	0.7100	0.6907	0.8672
19	0.6132	0.7300	0.7301	0.7300	0.7300	0.7660
20	0.6129	0.7500	0.7627	0.7500	0.7469	0.7444
21	0.4972	0.8400	0.8450	0.8400	0.8394	0.8624
22	0.5082	0.8100	0.8162	0.8100	0.8091	0.8412
23	0.4698	0.7700	0.7727	0.7700	0.7694	0.8500
24	0.5523	0.6700	0.7562	0.6700	0.6397	0.8828
25	0.6418	0.6700	0.6922	0.6700	0.6602	0.7568
26	0.6127	0.7200	0.7204	0.7200	0.7199	0.7712
27	0.5702	0.6700	0.6987	0.6700	0.6576	0.8244
28	0.5558	0.7500	0.7501	0.7500	0.7500	0.8056
29	0.5141	0.8000	0.8447	0.8000	0.7933	0.8472
30	0.5243	0.7800	0.8038	0.7800	0.7756	0.8332
31	0.6138	0.7800	0.7841	0.7800	0.7792	0.7908
32	0.6441	0.7800	0.7874	0.7800	0.7786	0.7380
33	0.5564	0.7700	0.7790	0.7700	0.7681	0.8092
34	0.5424	0.7200	0.7204	0.7200	0.7199	0.8240
35	0.5738	0.6400	0.7040	0.6400	0.6094	0.8440
36	0.4989	0.7600	0.7600	0.7600	0.7600	0.8400
37	0.6358	0.7500	0.7681	0.7500	0.7457	0.7744
38	0.6418	0.6300	0.6578	0.6300	0.6129	0.7372
39	0.6221	0.6600	0.6603	0.6600	0.6599	0.7508
40	0.5723	0.7100	0.7121	0.7100	0.7093	0.8068
41	0.6260	0.6000	0.6694	0.6000	0.5544	0.7960
42	0.6161	0.6000	0.6457	0.6000	0.5660	0.7908
43	0.6767	0.6800	0.6953	0.6800	0.6736	0.6568
44	0.6565	0.7000	0.7003	0.7000	0.6999	0.7356
45	0.5785	0.7100	0.7142	0.7100	0.7086	0.8008
46	0.5809	0.7000	0.7053	0.7000	0.6981	0.7964
47	0.5708	0.7000	0.7122	0.7000	0.6956	0.7936
48	0.6287	0.6500	0.7941	0.6500	0.6011	0.8188

Table 2 presents the experimental results (of the centralized model) obtained from 48 experiments that vary in terms of aggregation strategy, number of clients, number of local epochs, and learning rate. Each row corresponds to a distinct experiment and reports the loss, accuracy, precision, recall, F-measure, and area under the ROC curve (AUC).

From the results shown in Table 2 we note that comparisons between FedAvg and FedProx under identical settings consistently show that FedProx often yields slightly better generalization: for instance, comparing Experiments 1 and 2 (both with 2 clients, 50 epochs, and a learning rate of 0.0001), FedProx (Exp 2) achieves higher accuracy (0.80 vs. 0.75), precision (0.8125 vs. 0.7550), and F-measure (0.7980 vs. 0.7488), although with a slightly higher loss.

Moreover, as the learning rate increases from 0.0001 to 0.001, there is an overall trend toward improved classification performance, with particular regard to precision and AUC. For instance, within the 2-client setup with 50 epochs, Experiment 5 (FedAvg, learning rate 0.001) and Experiment 6 (FedProx, learning rate 0.001) report AUC values of 0.8852 and 0.9076, respectively, compared to lower AUCs in their lower learning rate counterparts (Experiments 1–4). However, this benefit is not linear; excessively high learning rates in more complex configurations (e.g., 10 clients with high heterogeneity) can lead to instability or reduced precision, as seen in Experiments 41–44.

The number of local training epochs also impacts model performance: as a matter of fact, by increasing the number of epochs from 50 to 100 often enhances metrics across the board, likely due to better local convergence before aggregation. For example, Experiment 11 (FedAvg, 100 epochs, 2 clients, 0.001 learning rate) outperforms its 50-epoch counterpart (Experiment 5) in accuracy (0.83 vs. 0.79), precision (0.8731 vs. 0.7997), and F-measure (0.8249 vs. 0.7883).

Moreover, as the number of clients increases from 2 to 10, a general decline in performance is observed, particularly when using FedAvg. For example, Experiment 23 (FedAvg, 3 clients) achieves an AUC of 0.85, whereas Experiment 47 (FedAvg, 10 clients, same learning rate and epochs) drops to 0.7936. FedProx shows slightly better resilience to increased client counts, though its performance also suffers at higher client numbers.

Additionally, experiments such as 24, 35, 41, and 48 illustrate cases with low recall and F-measure despite moderate precision. These suggest that while the model may be confident in positive predictions, it fails to identify a substantial portion of true positives, resulting in imbalanced classification performance. For example, Experiment 48 (FedProx, 10 clients, 100 epochs, 0.001 learning rate) reaches a precision of 0.7941 but only 0.6500 in recall, yielding a relatively low F-measure of 0.6011.

FedProx generally outperforms FedAvg, particularly with higher learning rates and fewer clients.

Once discussed in details the experimental results, obtained from the server model, shown in Table 2, in the follow we focus in detail on two experiments (with the aim to discuss also the performances on the clients): the first experiment is the one that obtained the best accuracy among all 48 experiments (i.e., Experiment 12), while the second one is the experiment with 10 clients that obtained the highest accuracy (we consider this experiment in more detail since, being with 10 clients, it represents a more realistic environment for using the proposed method) i.e., Experiment 37.

We show the performances across all the clients respectively involved in Experiments 12 and 37, with the aim to understand how the proposed method is able to detect anomalies in each client. Thus in the follow we focus on the performances obtained from the clients.

Table 3 shows the performances from each client obtained in Experiment 12.

Table 3. Experimental results related to the two clients involved in the Experiment 12.

Client	Loss	Accuracy	Precision	Recall	F-measure	AUC
1	0.5364	0.7700	0.8425	0.7700	0.7572	0.8664
2	0.4047	0.8400	0.8400	0.8400	0.8400	0.9008

Between the two clients, Client 2 is able to reach better performance if compared to Client 1. Specifically, Client 2 shows lower loss (0.4047 vs. 0.5364), higher accuracy (0.8400 vs. 0.7700), and better F-measure (0.8400 vs. 0.7572). Although Client 1 has a slightly higher precision (0.8425 vs. 0.8400), this comes at the cost of lower recall (0.7700 vs. 0.8400), indicating a more conservative model. Additionally, Client 2's AUC (0.9008) is higher than Client 1's (0.8664), confirming better overall anomaly detection performances.

Table 4 shows the performances from each client obtained in Experiment 37.

Table 4. Experimental results related to the eight clients involved in the Experiment 37.

Client	Loss	Accuracy	Precision	Recall	F-measure	AUC
1	0.6369	0.7200	0.7292	0.7200	0.7172	0.7548
2	0.6378	0.5900	0.6052	0.5900	0.5746	0.7492
3	0.6407	0.7300	0.7527	0.7300	0.7238	0.7628
4	0.6381	0.6100	0.6395	0.6100	0.5882	0.7584
5	0.6349	0.6500	0.6609	0.6500	0.6440	0.7624
6	0.6426	0.5700	0.6055	0.5700	0.5305	0.7548
7	0.6444	0.5600	0.5938	0.5600	0.5165	0.7376
8	0.6377	0.6000	0.6240	0.6000	0.5797	0.7592
9	0.6382	0.6100	0.6395	0.6100	0.5882	0.7584
10	0.6358	0.7500	0.7681	0.7500	0.7457	0.7744

The performance across the 10 clients in Experiment 37 shows variability: client 10 achieved the best overall performance, with the highest accuracy (75%), F-measure (0.7457), and AUC (0.7744). Client 3 also performed well, reaching 73% accuracy and an F-measure of 0.7238. In contrast, Clients 6 and 7 showed the weakest performance, with accuracies of 57% and 56%, and notably lower F-measures of 0.5305 and 0.5165, respectively, despite having similar precision levels.

Client 1 showed moderate performance (72% accuracy, 0.7172 F-measure), while Clients 2, 4, 5, 8, and 9 clustered around 59–65% accuracy, with F-measures ranging from 0.5746 to 0.6440. AUC values were relatively stable across clients, mostly in the range of 0.74 to 0.76.

Figures 2 and 3 show several examples of explainability, respectively obtained with the client 2 in Experiment 12 and with the client 8 in Experiment 37.

Each examples of explainability provided by the proposed method includes three images: the first is the input image, extracted from the network log saved in PCAP format; the second is the attention rollout heatmap generated by the model; and the third is a composite image created by overlaying the heatmap onto the input image. This overlay helps visualize which regions of the image are most relevant from the anomaly detection classification perspective, for the identification of parts associated with anomalies or benign network traces. In the attention rollout heatmaps, colors reflect the relative importance of pixels or regions in the model's decision-making process. Warm colors (such as red, orange, and yellow) highlight areas of high attention where the model focuses most, while cool colors (blue, green, purple) indicate regions with low attention and minimal influence on the prediction. The color scale generally progresses from blue to red (Blue $\rightarrow$ Green $\rightarrow$ Yellow $\rightarrow$ Red), with red representing the highest relevance.

In detail, in Figs. 2 and 3 we show four different examples of images obtained from network traces, the first two related attack logs, while the remaining two related to benign logs. We show the same input images in both Figs. 2 and 3 with the aim to show that the proposed method, regardless of the number of clients used in training, it is able to provide the same level of explainability i.e., to show similar areas in the input image regardless of the number of clients used in training.

Figures 2 is related to Client 2 in Experiment 12, which is the best-performing configuration in the entire set of 48 experiments. This experiment achieved an accuracy of 84%, an F-measure of 0.84, and an AUC of 0.9008. In contrast, Figure Figs. 3 depicts explainability outcomes for Client 8 in Experiment 37, which is particularly relevant because it involves 10 clients, representing a more realistic deployment scenario. Despite the increased heterogeneity of data in this larger federated environment, Client 8 maintained reasonable performance, with an accuracy of 60%, F-measure of 0.5797, and AUC of 0.7592, providing a valid case for comparison.

Figures 2 and 3 presents four input images derived from PCAP network traces: the first two correspond to attack samples, while the remaining two are benign. For each case, three visual elements are shown: the original input image created by converting the PCAP network trace into a 100×100 RGB image; the attention rollout heatmap, which aggregates attention weights across the layers of the Vision Transformer to identify influential regions; and the composite image where the heatmap is overlaid onto the input image. This last visualization highlights the spatial regions that were most decisive in the model's classification.

The attention rollout mechanism assigns relative importance to different image patches by analyzing how much each contributes to the [CLS] token's final representa-

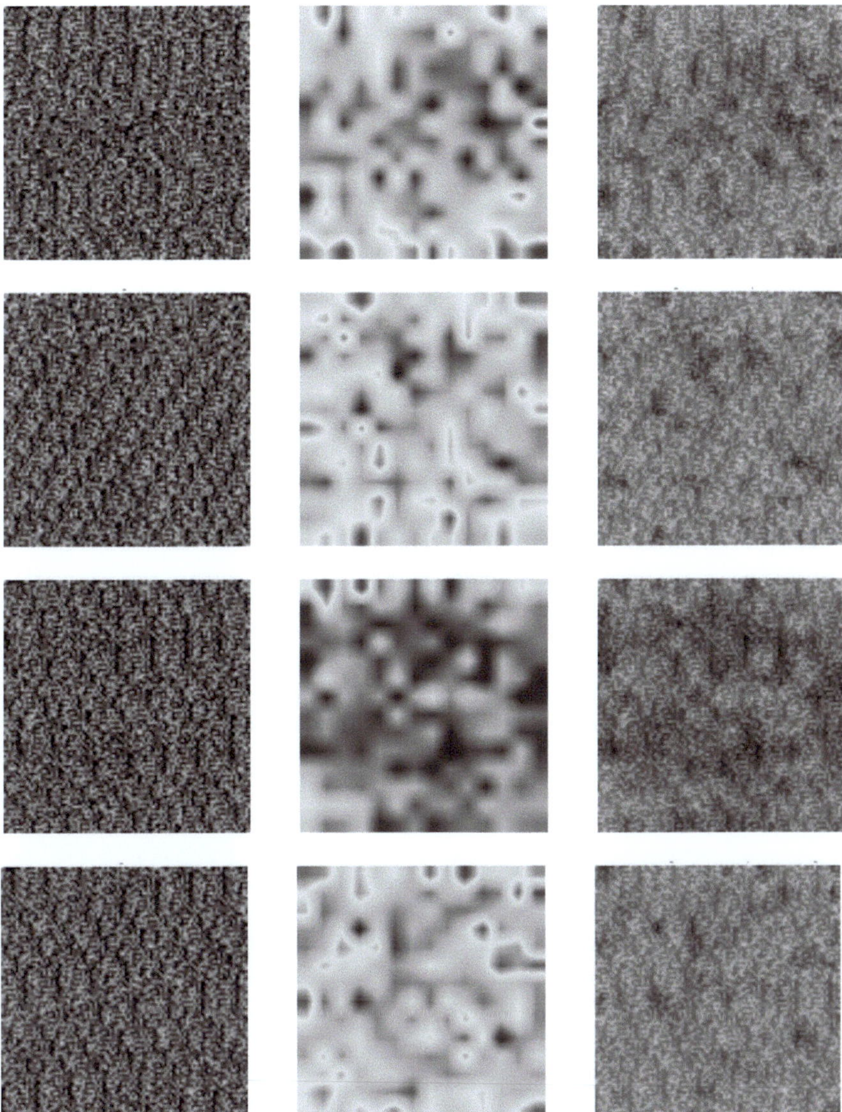

Fig. 2. Examples of explainability obtained with the client 2 in the experiment 12.

tion. Warm colors, such as red, yellow, and orange, indicate regions of high attention, suggesting that the model identified these areas as indicative of anomalous activity. Conversely, cool colors like blue and green correspond to regions with low attention and minimal influence on the prediction.

We consider the same input images in both figures, which enables a direct visual comparison of model behavior between Experiment 12 (2 clients) and Experiment 37

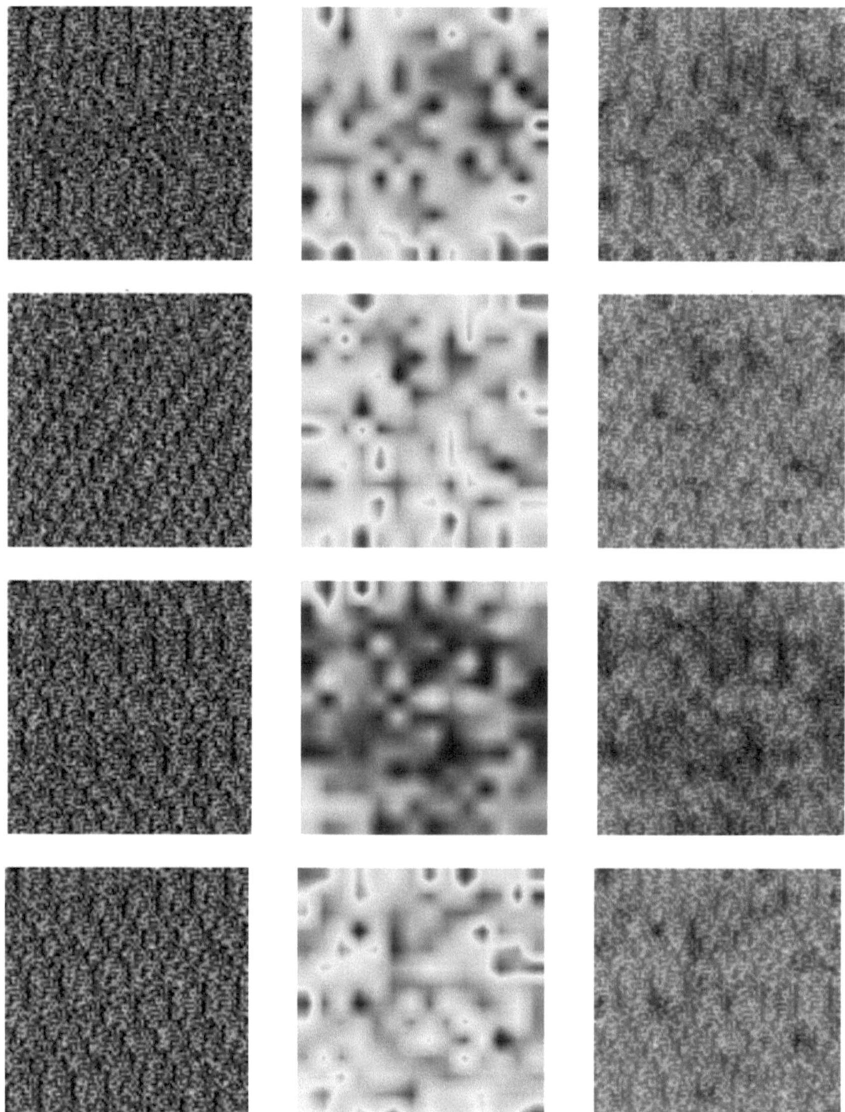

Fig. 3. Examples of explainability obtained with the client 8 in the experiment 37.

(10 clients). Despite the significantly different training setups, the highlighted regions in the heatmaps are notably consistent. This result confirms that the proposed approach, built upon a frozen ViT backbone and enhanced by attention rollout, is capable of learning stable and transferable representations across diverse client environments.

Overall, Figs. 2 and 3 validate the explainability of the proposed approach. They demonstrate that the model reliably identifies areas of the image as being relevant to classification and that this explai ability is maintained across different federated configurations. This kind of visual explanations not only enhance cybersecurity analysts trust but also support accountability in high-stakes environments such as electrical substation monitoring.

4   Conclusion and Future Work

Electrical substations are fundamentals components of the power grid, aimed to ensure the reliable distribution of electricity necessary for a wide range of critical services and infrastructures. Considering their importance in supporting essential services that require uninterrupted electricity, this paper proposed a method to detect anomalies in electrical substations, overcoming the limitations of the current state of the art, namely the lack of explainability and privacy issues (which arise from sending network logs to centralized servers for prediction). As a matter of fact, we proposed a method exploiting FML (to address the privacy concerns) and attention rollout (for the introduction of explainability) for anomaly detection in substation network.

An experimental study was conducted to evaluate the proposed FML method under various configurations. The experiments systematically varied hyper-parameters including the aggregation algorithm (FedAvg or FedProx), the number of clients (ranging from 2 to 10), local training epochs (either 50 or 100), and learning rates (0.0001, 0.0005, and 0.001). The number of communication rounds was fixed at 10 and the batch size at 32 across all runs. This design resulted in 48 experimental scenarios, allowing a thorough comparison of performance under different settings. In the more realistic scenario i.e., the one with 10 different clients, we obtained an accuracy, a precision and a recall respectively equal to 0.750, 0.768 and 0.750, thus showing the effectiveness of the proposed method in a federated environment. We also discussed several examples of explainability, showing that areas in the network trace that are, from the model's point of view, symptomatic of the anomaly are highlighted.

Future work will focus on the training of models aimed to detect the type of attack i.e., a multi-class classification with the benign label a label for each specific attack, for instance (Distributed) Denial of Service. Moreover, we plan to consider also other ViT models with the aim to improve accuracy and enable tracing back from highlighted image areas to the corresponding segments in the original PCAP files.

Acknowledgment. This work has been partially supported by EU DUCA, EU CyberSecPro, SYNAPSE, PTR 22-24 P2.01 (Cybersecurity) and SERICS (PE00000014) under the MUR National Recovery and Resilience Plan funded by the EU - NextGenerationEU projects, by MUR - REASONING: foRmal mEthods for computAtional analySis for diagnOsis and progNosis in imagING - PRIN, e-DAI (Digital ecosystem for integrated analysis of heterogeneous health data related to high-impact diseases: innovative model of care and research), Health Operational Plan, FSC 2014–2020, PRIN-MUR-Ministry of Health, Progetto MolisCTe, Ministero delle Imprese e del Made in Italy, Italy, CUP: D33B22000060001, FORESEEN: FORmal mEthodS for attack dEtEction in autonomous driviNg systems CUP N.P2022WYAEW and ALOHA: a framework for monitoring the physical and psychological health status of the Worker through Object detection and federated machine learning, Call for Collaborative Research BRiC -2024, INAIL.

References

1. Abasıkeleş-Turgut, I., Daş, R.: Anomaly and intrusion detection systems for smart grids. In: Cyber Security Solutions for Protecting and Building the Future Smart Grid, pp. 231–270. Elsevier (2025)
2. Alomari, M.A., et al.: Security of smart grid: cybersecurity issues, potential cyberattacks, major incidents, and future directions. Energies **18**(1), 141 (2025)
3. Boakye-Boateng, K., Ghorbani, A.A., Lashkari, A.H.: Securing substations with trust, risk posture, and multi-agent systems: a comprehensive approach. In: 2023 20th Annual International Conference on Privacy, Security and Trust (PST), pp. 1–12. IEEE (2023)
4. Di Giammarco, M., et al.: A robust and explainable deep learning method for cervical cancer screening. In: International Conference on Applied Intelligence and Informatics, pp. 111–125. Springer (2023)
5. He, H., Yang, H., Mercaldo, F., Santone, A., Huang, P.: Isolation forest-voting fusion-multioutput: a stroke risk classification method based on the multidimensional output of abnormal sample detection. Comput. Methods Programs Biomed. **253**, 108255 (2024)
6. Kreimel, P., Eigner, O., Mercaldo, F., Santone, A., Tavolato, P.: Anomaly detection in substation networks. J. Inf. Secur. Appl. **54**, 102527 (2020)
7. Maghami, M.R., Mutambara, A.G.O., Gomes, C.: Assessing cyber attack vulnerabilities of distributed generation in grid-connected systems. Environ. Dev. Sustain. 1–27 (2025)
8. Marino, D.L., Wickramasinghe, C.S., Rieger, C., Manic, M.: Self-supervised and interpretable anomaly detection using network transformers. arXiv preprint arXiv:2202.12997 (2022)
9. Martinelli, F., Mercaldo, F., Petrillo, L., Santone, A.: Security policy generation and verification through large language models: a proposal. In: Proceedings of the Fourteenth ACM Conference on Data and Application Security and Privacy, pp. 143–145 (2024)
10. Martinelli, F., Mercaldo, F., Santone, A.: A method for intrusion detection in smart grid. Procedia Comput. Sci. **207**, 327–334 (2022)
11. Martinelli, F., Mercaldo, F., Santone, A.: Evaluating fuzzy machine learning for smart grid instability detection. In: 2023 IEEE International Conference on Fuzzy Systems (FUZZ), pp. 1–6. IEEE (2023)
12. Mercaldo, F., Tavolato, P., Santone, A., Martinelli, F.: An explainable method for malware detection through convolutional neural networks. In: International Conference on Availability, Reliability and Security, pp. 322–339. Springer (2025)
13. Mercaldo, F., Zhou, X., Huang, P., Martinelli, F., Santone, A.: Machine learning for uterine cervix screening. In: 2022 IEEE 22nd International Conference on Bioinformatics and Bioengineering (BIBE), pp. 71–74. IEEE (2022)
14. Most, A.B., Eren, M.E., Alexandrov, B.S., Lawrence, N.: Electrical grid anomaly detection via tensor decomposition. In: MILCOM 2023-2023 IEEE Military Communications Conference (MILCOM), pp. 162–169. IEEE (2023)
15. Nhung-Nguyen, H., Girdhar, M., Kim, Y.H., Hong, J.: Machine-learning-based anomaly detection for goose in digital substations. Energies **17**(15), 3745 (2024)
16. Qu, Y., et al.: Cgam: an end-to-end causality graph attention mamba network for esophageal pathology grading. Biomed. Signal Process. Control **103**, 107452 (2025)
17. Tavolato, P., Eigner, O., Kreimel-Haindl, P., Santone, A., Martinelli, F., Mercaldo, F.: A method for explainable anomaly detection in substation networks through deep learning. In: International Conference on Availability, Reliability and Security, pp. 289–303. Springer (2025)
18. Valdes, A., Macwan, R., Backes, M.: Anomaly detection in electrical substation circuits via unsupervised machine learning. In: 2016 IEEE 17th International Conference on Information Reuse and Integration (IRI), pp. 500–505. IEEE (2016)

Exposing Shortcuts in Image Classification by Aggregating Counterfactuals

James Hinns[(✉)] and David Martens

University of Antwerp, Antwerp, Belgium
`james.hinns@uantwerpen.be`

Abstract. Deep learning has achieved remarkable success in image classification, yet models often rely on misleading 'shortcuts' that do not generalise beyond lab settings, such as watermarks or background cues. It has been proposed that instance-level explanations in explainable AI (XAI) may help reveal such shortcuts without external data, but doing so typically requires examining many individual explanations, making the process labour-intensive and often infeasible. We introduce **Counterfactual Frequency (CoF) tables**, a novel method that aggregates local explanations into global insights and efficiently exposes shortcuts. This requires semantically meaningful image segments, with appropriate labels which we obtain through pre-trained foundation segmentation models. We demonstrate the effectiveness of CoF tables across multiple datasets, including real-world datasets such as ImageNet, toy datasets constructed for this study, and Biased Action Recognition (BAR), highlighting shortcuts like background reliance and watermarks learned by both models we train and well-studied pre-trained classifiers.

1 Introduction

A persistent challenge in Deep Learning (DL) is its susceptibility to exploiting spurious correlations, commonly referred to as 'shortcuts', present in lab settings [5, 8, 12, 15]. This exploitation results in models that occasionally perform well on standard testing benchmarks yet fail when deployed in more varied, real-world scenarios. This issue parallels the linguistic 'principle of least effort' [8], suggesting that, like language users, machine learning models inherently prefer simpler decision rules, even if these rules lack broader generalisability, provided they satisfy existing evaluation criteria. Instances of shortcuts in Image Classification include models exploiting watermarks in datasets such as ImageNet [16] and Pascal VOC [15], relying on surgical markings in melanoma classification [29], or emphasising patient positioning in COVID-19 radiograph detection [6].

Shortcuts often remain undetected during typical model evaluation cycles as they consistently appear in both training and testing data. Although expanding the diversity of evaluation datasets could uncover such shortcuts, acquiring and curating sufficiently varied data is often practically challenging. Additionally, many shortcuts naturally arise from inherent correlations in datasets, such as the frequent presence of water backgrounds in boat images.

© The Author(s), under exclusive license to Springer Nature Switzerland AG 2026
F. Marcelloni et al. (Eds.): IJCCI 2025, CCIS 2829, pp. 451–466, 2026.
https://doi.org/10.1007/978-3-032-15638-9_26

One approach proposed to address this issue without any extra data is using instance-based explanation methods to identify undesirable decision rules without requiring additional data [6,15,29]. In Image Classification tasks, such local explanations often take the form of saliency maps, visually highlighting the importance of individual pixels in the image.

However, identifying shortcuts through individual explanations involves substantial manual effort, with experts reviewing multiple instances to locate unwanted decision rules. This challenge is intensified when shortcuts affect only a small subset of images, requiring more images to be manually examined before the shortcut can be identified. While these maps indicate which pixels influence model decisions, most methods do not explain why particular regions are influential, omitting details such as semantics, texture and colour intensity. We argue that these omitted factors are crucial for fully understanding the model's reasoning, as they can provide deeper insights into the learned associations and decision-making processes.

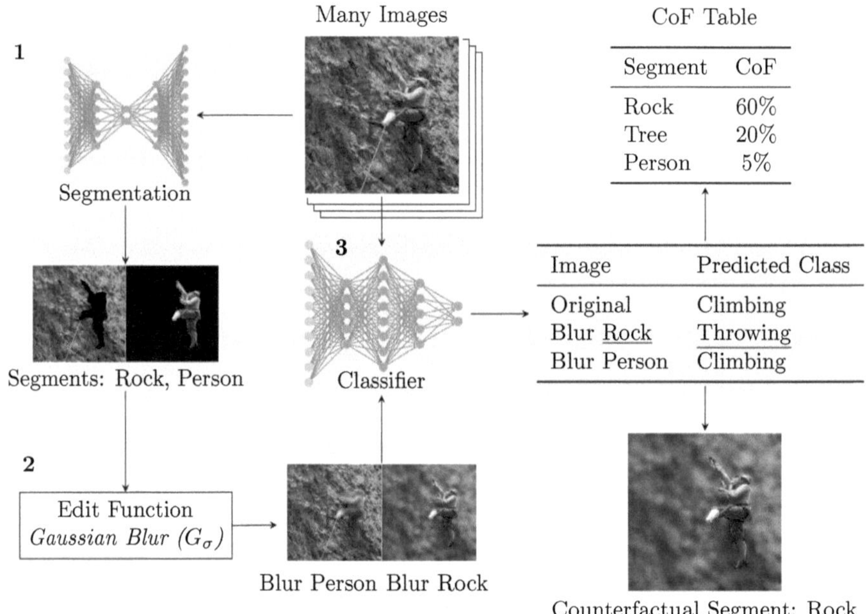

Fig. 1. Overview of our method for generating grounded counterfactuals and aggregating them into Counterfactual Frequency (CoF) tables. For a single image we segment it, edit each segment in turn, and record a counterfactual only if the edited image is classified differently. Repeating this for many images allows all such counterfactuals to be summarised in a CoF table, highlighting the segments most responsible for class changes.

In this study, we propose addressing these challenges through the introduction of **Counterfactual Frequency (CoF) tables**, a novel analytical tool designed to aggregate and quantify the influence of semantically similar segments on the decision-making of image classification models. The CoF for a particular segment label is defined as the frequency with which modifying a related group of segments (e.g., a rock background) leads to a change in the predicted class of images across a specified collection. By cataloguing how often groups of segments, identified through their names, contribute to counterfactual explanations, CoF tables provide a systematic method for identifying the most influential factors within images that affect a model's decision-making process. This approach highlights potential shortcuts within model classifications, promoting deeper insight into the visual features on which models rely. By aggregating these insights, CoF tables reveal broader patterns of model behaviour, thereby supporting the development of more robust and interpretable image classifiers. Additionally, we retain the local counterfactual explanations we generate, enabling detailed investigation of individual predictions after the initial overview from CoF tables. These counterfactuals are grounded in semantic labels, clearly indicating which segment led to a change in the prediction.

Our main contributions are summarised as follows:

- A simple model agnostic framework to produce grounded counterfactual explanations for image classification.
- **Counterfactual Frequency (CoF)** tables: A method to aggregate local counterfactual explanations into global insights.
- Demonstration of CoF tables on various datasets, showcasing learned background shortcuts from foundational models trained on ImageNet.

2   Aggregating Local Explanations

A number of explanation techniques have been proposed that can aggregate local explanations for tabular data. These techniques often employ feature attribution (or importance) explanations, which can then be averaged to provide a simple global explanation [12, 17, 23] of the model on the given data. Notably, many of these aggregations can be visualised using the same techniques that SHAP [18] employs for illustrating local explanations.

For image explanations, a central goal is to assign a quantitative contribution to every pixel of an input image so that we can see how each pixel supports or opposes a model's prediction, typically presenting the result as a heatmap over the image. This general strategy is often referred to as pixel-wise decomposition [19, 25]. Within this, methods can be grouped into three families: sensitivity analysis, relevance redistribution, and perturbation-based techniques.

Sensitivity analysis methods compute derivatives of the prediction with respect to the input and highlight pixels where small changes would most affect the output. Examples include Vanilla Gradient [27] and Grad-CAM [26]. These are typically inexpensive to compute and easy to integrate, but they describe local sensitivity rather than a full decomposition of the prediction [19].

Relevance redistribution methods propagate the model's output score backwards through the network so that each input pixel receives a share of that score, indicating its contribution to the final prediction. A prominent example is Layer-wise Relevance

Propagation (LRP) [3]. This usually requires more computation than a single gradient pass, yet it is widely used because it often yields clearer and more interpretable heatmaps than plain gradient sensitivity maps [19,25].

Perturbation-based methods, sometimes called occlusion-based, systematically modify parts of the input and measure the change in the prediction. Examples include SHAP [18] and LIME [23]. Within this family, SEDC-T [28] is a counterfactual variant: instead of simply observing changes in prediction probability, it searches for the smallest set of image segments whose removal flips the classifier's decision, thereby producing segment-level counterfactual explanations. Our local explanations build on this approach. All three families aim to show which pixels or segments contributed most to the decision, though they do not specify what property of those pixels was decisive.

However, pixel attribution methods face several significant challenges. Like most explanation methods, it is often extremely difficult to evaluate their accuracy. They have been shown to produce different explanations for the same prediction and gradient, raising questions about their reliability [1,13]. Furthermore, they may provide misleading views of a model's decision-making process [2]. These methods can also be fragile, with small perturbations to an image leading to changes in explanations without corresponding changes in the model's predictions [10].

Although aggregation by averaging, as is commonly done with tabular data, is possible, providing only a single saliency map for multiple images can present problems of interpretation. Consider, once again, the example of a watermark shortcut in the binary classification of horses and zebras. Averaging the positions of watermarks, which the saliency map should highlight in the presence of an identified shortcut, across numerous images would obscure this behaviour, as it would blend the positions of all watermarks, which may appear anywhere within the image. This is problematic because, in many shortcuts, it is the type of object present in the image that constitutes the shortcut rather than its position [21].

To more effectively address this challenge, SpRAy [15] proposes combining multiple saliency maps through clustering instead of averaging. SpRAy identifies similar maps using spectral clustering and combines them to generate a number of aggregated explanations. Whilst this method demonstrates potential for shortcut identification, it relies on the presence of shortcut-inducing features in consistent positions, as otherwise it may prioritise the position of objects over their semantic meaning. Global Saliency [22] uses segmentation models to calculate the mean saliency over a given segment for each image, which can then be averaged across multiple images to provide a global saliency. Importantly, they only consider the saliency of a single segment, meaning there is no baseline comparison. In order to interpret the meaning of the global saliency, it is necessary to compare against other subsets of data, for example the difference between classes.

3   Creating Counterfactual Frequency Tables

Our methodology for creating **Counterfactual Frequency** (CoF) tables is rooted in leveraging counterfactual explanations with semantic meaning. We begin by generating grounded counterfactuals, where segments responsible for changes in prediction are assigned text labels. Building on this, our main contribution is the introduction of CoF tables as a novel means of identifying shortcuts learned by models. By aggregating these explanations, the tables reveal consistent patterns in model behaviour, providing insight into the decision-making process and highlighting potential shortcuts. This approach not only uncovers such patterns but also a deeper understanding of the semantic factors influencing model predictions.

3.1 Generating Grounded Counterfactual Explanations

Unlike many existing counterfactual explanation methods for images, our grounded counterfactuals provide semantic labels for the segments responsible for the counterfactual. These labels are essential for constructing CoF tables, as they enable meaningful aggregation. The process for generating grounded counterfactuals is outlined below (and illustrated in Fig. 1):

1. **Segmentation:** The input image is divided into multiple segments, each assigned a label reflecting its semantic content. To ensure meaningful aggregation, these segments must be semantically interpretable. This can be achieved using segmentation models tailored to the task at hand. General-purpose models trained on datasets such as COCO may not produce the relevant segments needed for specific applications. Therefore, segmentation methods may include hand-crafted rules, task-specific models, or ensembles, depending on the domain. Open-set segmentation models, such as Language Segment-Anything, can also be used and are particularly valuable as they can be guided to return specific labels. We demonstrate their effectiveness in our results section. Use-case-specific models, for example in medical imaging, may also be appropriate where domain-specific segmentation is required.
2. **Image Editing:** Once segmented, specific transformations are applied to individual segments of the image. SEDC-T [28] previously explored segment-based counterfactuals using only Gaussian blur, arguing it introduced minimal semantic distortion. However, their study focused on traditional infill techniques. Since then, generative infill has advanced substantially and now offers more powerful tools for semantic image editing. This is important because convolutional neural networks tend to be biased towards texture rather than shape [9]. As such, blurring a textured region like fur can significantly affect the output, while blurring smooth areas such as snow may have little impact. In this work, we demonstrate the use of several editing strategies, including blurring, colour shifting, and generative infill. The choice of edit should be informed by the context, as different edits provide different kinds of counterfactual information. For instance, blurring or infill can simulate removal, whereas colour shifting or generative infill can simulate replacement or transformation.
3. **Classification:** The classifier used is the model being explained. Our approach is model-agnostic and applicable to any black-box classifier, requiring only its predicted label output. This enables our method to be applied across a wide range of models and domains without the need for internal access or architectural constraints.
4. **Counterfactual Search:** This step determines how counterfactuals are identified. A variety of strategies could be used, including optimisation procedures aimed at minimising a utility function, such as editing the smallest number of segments necessary

to change the model's prediction. In this work, we limit our focus to simple counterfactuals involving the modification of a single segment and report all attempted changes, whether they result in a counterfactual or not.

We adopt a broad definition of counterfactuals, reporting any modification that results in a change of predicted class. This approach allows for the generation of multiple counterfactuals per image, helping to identify cases where the classifier's predictions are particularly fragile. It also significantly reduces the computational cost of generating counterfactuals, which is especially advantageous given our objective to summarise many counterfactuals across an entire dataset. It is possible that no single segment counterfactuals are found for a given image, segmentation model and edit function. This could be solved via such combinations, but we do not explore this in this work.

We chose to use counterfactuals as the explanation method for aggregation for several reasons:

- They are always valid: by definition, the counterfactual instance is classified as a different class.
- By varying the edit function, different counterfactuals can be produced, allowing us to probe different aspects of the model's decision-making. For example, does altering the colour of a segment, or changing a water segment to grass, produce a counterfactual?

3.2 Counterfactual Frequency (CoF) Tables

The construction of a CoF table begins with the selection of a set of images for which counterfactuals will be generated. This set might consist of a test set, a particular class of interest, or images misclassified by the model. For each image, we record the segment that was modified and the class predicted by the model following the modification, as illustrated in Fig. 1. Any edited image that results in a change of predicted class is retained as a grounded counterfactual. Optionally, additional metadata can be recorded for each entry, including the model's confidence, the position of the modified segment within the image, and the ground truth label.

After generating these counterfactuals, the resulting grounded counterfactuals may then be filtered using explanation metadata (for example, retaining only those that correct a misclassification or change from one specified class to another) to define the explanation set E which are aggregated by the CoF table. For example, one might generate counterfactuals for misclassified images and then include only those counterfactuals that correct a misclassification in the CoF table.

Finally, a CoF table aggregates over E using one of three CoF scopes for each segment $S \in E$:

- **Image-Level CoF:**
$$P(\mathrm{CF}, S)$$

the proportion of images in E for which editing segment S yields a counterfactual[1]. This scope provides a high-level summary of model behaviour across images. How-

---

[1] If multiple instances of the same segment S exist in a single image, the image is counted if editing any of those instances results in a counterfactual.

ever, it may fail to reveal patterns that occur infrequently, due to its aggregation at the image level.

- **Segment-Level CoF:**

$$P(\mathrm{CF} \mid S)$$

the proportion of images in E that contain S for which editing S yields a counterfactual. This scope captures how often editing segment S results in a counterfactual when it is present. It accounts for presence explicitly, but may still be biased toward segments that appear more frequently in the dataset.

- **Counterfactual-Level CoF:**

$$P(S \mid \mathrm{CF})$$

the proportion of all counterfactuals in E for which the edited segment is S. This scope captures how often segment S is responsible for producing counterfactuals across the explanation set. It highlights dominant factors in the model's behaviour, but may overrepresent segments that appear more frequently.

In this work, we demonstrate how large-scale, pre-trained segmentation models can support the detection of shortcuts through CoF tables. The COCO dataset, for example, includes 133 labels for panoptic segmentation, and we show that models trained on this dataset are sufficient to reveal shortcuts in certain cases. However, this label set is limited and unsuitable for detecting some types of shortcuts. For instance, models trained exclusively on COCO panoptic data cannot detect a watermark, as such labels are not present in the dataset. This limitation can be addressed by using task-specific segmentation models or open-set segmentation approaches that accept prompts to identify more targeted content.

This observation leads us to our proposed use of CoF tables; as a test output. For this to be effective, the test must be well designed. If the goal is to evaluate reliance on backgrounds or object-level shortcuts within a closed-set label space, then a general-purpose segmentation model combined with blurring may suffice. However, if the objective is to detect reliance on elements not captured by the closed set, or on features such as textures that are resilient to blurring, alternative segmentation strategies or editing techniques must be employed. The choice of edit function must also be logically appropriate: for example, blurring a watermark may not sufficiently obscure it, whereas generative infill applied to the same region is more likely to effectively mask that segment.

We argue that this methodology provides a more nuanced and effective means of identifying shortcuts and offers valuable insight into strategies for mitigating them than manually reviewing local explanations. The grounded counterfactuals themselves could also be used to augment the training data, ultimately supporting the development of more robust classifiers.

4   Results

In this section we demonstrate the usage of CoF tables to identify shortcuts in a number of datasets. For each dataset, we show individual counterfactuals and discuss the findings from their respective CoF tables. To illustrate the flexibility of our methodology we utilise a variety of classifiers, segmentation methods, and edit functions.

Colourised MNIST

We begin with a straightforward example to illustrate background bias using the MNIST dataset. To achieve this, we convert the original single-channel greyscale images to three-channel RGB format. Each class in the dataset is assigned a unique, visually distinct colour. For each image, pixels that are completely black (RGB values of (0,0,0)) are changed to the colour assigned to their class. This modification creates an extremely simple shortcut for models to learn, providing a clear starting point for our analysis (Fig. 3).

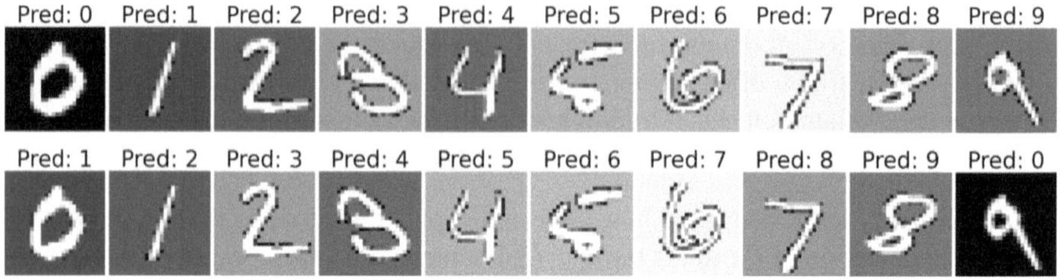

Fig. 2. Example images from the biased MNIST dataset. The top row shows training images where each digit class is paired with a unique background colour and the model's predicted label. The bottom row shows the same digits after their background colours are shifted to those of a different class, producing counterfactual images that consistently change the model's prediction and illustrate the trivial colour shortcut in the dataset.

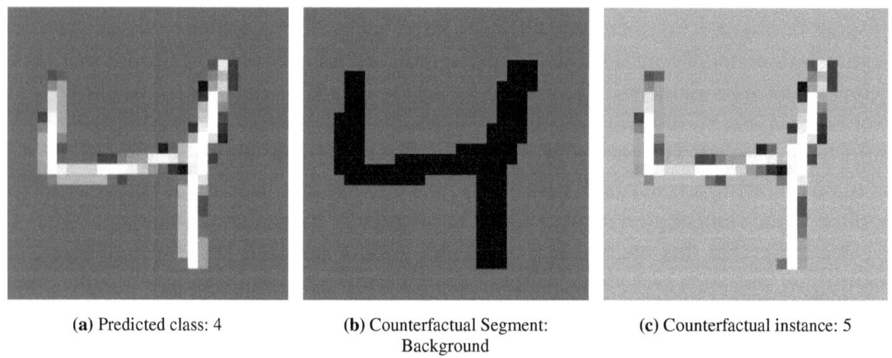

(a) Predicted class: 4   (b) Counterfactual Segment: Background   (c) Counterfactual instance: 5

Fig. 3. Individual explanation from the biased MNIST dataset.

Using this biased dataset, we train a simple CNN. After only a few epochs, the model achieves near 100% accuracy in both the training and testing set, thanks to the simple shortcut (Table 1).

We do not use a model as our segmentation method for this experiment, but a hard-coded function. We identify the background by determining the most common pixel

Table 1. Multi-Class CoF table for the coloured MNIST dataset. Segmenting the background by colour and editing the image by shifting the colour. It demonstrates a consistent shortcut of background colour for predicting labels of the trained model.

Class	Segment Name	Image-Level CoF
0	● background	100%
1	● background	100%
2	● background	100%
3	● background	100%
4	● background	100%
5	● background	100%
6	● background	100%
7	background	100%
8	● background	100%
9	● background	100%

value in the image, and then finding all pixel indexes that have that colour, providing us with a single segment per image. Using the most common pixel value, we are also able to estimate the class with 100% accuracy. The edit function then simply changes the colour of all pixels in the segment provided. We chose to shift the colour to the colour of the next class, which produces the distinct pattern seen in Fig. 2.

This toy dataset provides us with a baseline to clearly outline the aims of CoF tables, showing a clear, ever-present shortcut without any noise. Additionally, it highlights a previously discussed major flaw in saliency maps, which is the lack of information on why certain pixels are important. Individual saliency maps would likely show that the background was indeed being used as a shortcut. However, they wouldn't be able to identify the colour bias, which is key to understanding the shortcut.

Biased Action Recognition (BAR)

The Biased Action Recognition (BAR) dataset [20] is a multi-class classification dataset comprising six action categories (e.g., climbing, diving). The training data exhibits a bias towards specific environments: all climbing actions occur on rock walls, while diving predominantly takes place in underwater settings.

This environmental bias is confined to the training dataset. In contrast, the test dataset introduces a shift in environmental context; for example, climbing actions change from outdoor rock faces to indoor, artificial climbing walls.

We train a simple TensorFlow CNN on this dataset, achieving 99.43% accuracy on the training set but only 21.41% on the testing set. This significant difference in accuracy highlights the model's reliance on shortcuts derived from the background features within the training data, rather than learning generalisable features.

We then search for counterfactuals for all images in the training dataset, using our method. We focus on the training set because the test set introduces a distribution shift, and our goal is to show that shortcuts can be detected without relying on specially

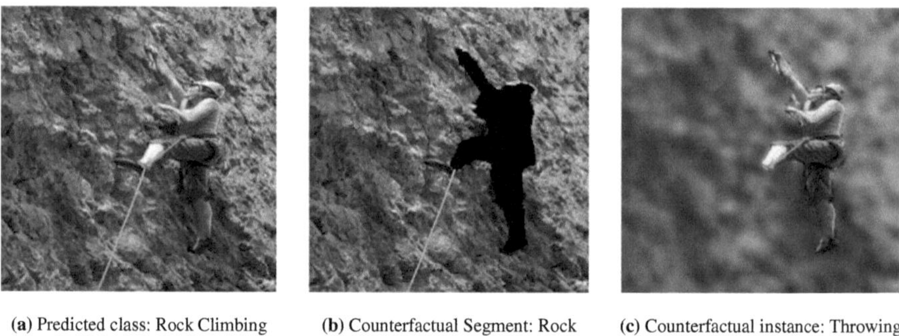

(a) Predicted class: Rock Climbing   (b) Counterfactual Segment: Rock   (c) Counterfactual instance: Throwing

Fig. 4. Individual explanation from the BAR dataset.

curated or shifted data. For segmentation, we employed the pre-trained DETR panoptic segmentation model [4]. Any sections of the image without an assigned label by DETR are labelled afterwards as 'unrecognised', ensuring each image is fully represented through the segments derived from it. As our edit function, we apply a simple Gaussian blur across the entire counterfactual segment, as illustrated in Fig. 4. We choose to use Gaussian blur, following SEDC-T [28]. It obscures fine details and textures while largely preserving edges, so the resulting loss of detail serves as a clear test of whether the model relies on subtle texture cues. Although such edits can be considered out of distribution, our focus is on whether certain segments are disproportionately affected; consistent sensitivity indicates that those segments are critical yet fragile to small perturbations.

Table 2. Relative CoF table for the climbing class of the BAR dataset. Displays the percentage of counterfactuals generated that are caused by a specific segment. Shows all segments that cause at least 4% of counterfactuals.

Segment Name	CF-Level CoF
rock	46.2%
tree	11.7%
dirt	6.7%
mountain	6.3%
unrecognised	5.4%
stone wall	4.0%

Table 2 illustrates our method for identifying the reliance on rock backgrounds as a shortcut within the climbing class, utilising CoF tables. We identify 21 segment names responsible for generating counterfactuals and so narrow down the CoF table to only those segments that constitute at least 4% of the counterfactuals identified. Notably, over 46% of all counterfactuals stem from rock backgrounds, indicating a clear reliance on the shortcut, without the need for the shifted testing data (Figs. 5 and Table 3).

ImageNet Boats

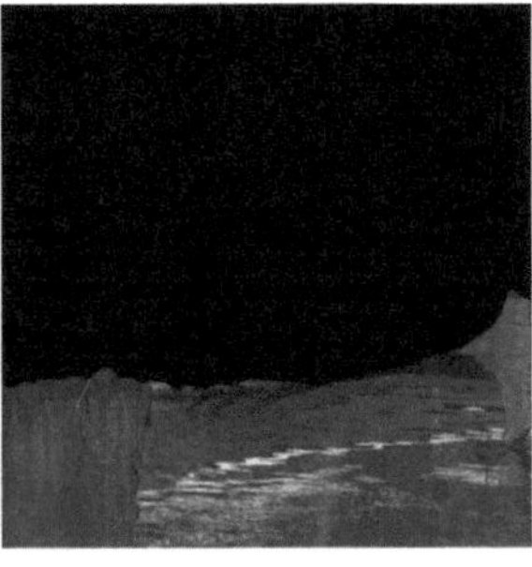

(a) Original Image   (b) Counterfactual Segment: Water   (c) Counterfactual instance: Tub

Fig. 5. Individual explanation for rowing boat class of ImageNet.

Table 3. CoF Table for replacing water segments with grass.

Segment Name	Segment-Level CoF
river	68.8%
sea	71.5%

In this experiment, we explore the 'rowing boat' category from the ImageNet dataset [7]. We utilise the well-studied ResNet-50 model [11], pre-trained on the ImageNet-1k, as our classifier. Similar to the approach taken in the BAR experiment, we employ the DETR segmentation model to provide segments. However, in this instance, we only consider the label synonymous with water. In this example, those are 'river' and 'sea', due to the absence of a generic 'water' label widely recognised in these images. Instead of blurring, we use generative inpainting as our edit function. To accomplish this inpainting, we use a pre-trained version of Stable Diffusion 2, trained for inpainting [24].

To effectively expose the underlying background bias within our dataset, we create counterfactual images by replacing water segments with infill prompted by 'grass'.

Grass or fields are also likely to be highly prevalent in different classes, so we decide to replace a background strongly associated with boats, namely water, with one that is common in other classes. Considering the frequent association of boats with water and the prevalence of grass or open fields in various other classes, this methodological choice serves as a deliberate contrast. By substituting the background closely associated with the boat category, which is water, with one commonly found in numerous other categories, such as grass, we aim to highlight the model's reliance on background cues for classification (Figs. 6 and 4).

Imagenet Horses and Zebras

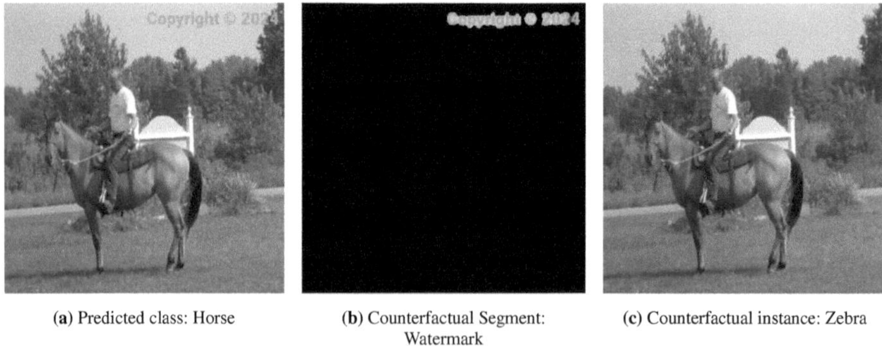

(a) Predicted class: Horse   (b) Counterfactual Segment: Watermark   (c) Counterfactual instance: Zebra

Fig. 6. Individual explanation for Horse vs Zebra classification.

Table 4. Table showing the percentage of images in the training set for which we find counterfactuals for specific segments. Only showing segments that cause counterfactuals in over 1% of images.

Segment Name	Image-Level CoF
tree	5.3%
grass	4.6%
horse	2.9%
person	2.3%
sky	2.1%
fence	2.0%
watermark	1.2%

In our subsequent experiment, we return to the ImageNet dataset, selecting the 'mare, female horse' and 'zebra' classes for analysis. This choice is informed by our intention to simulate a controlled environment that mirrors the 'watermark shortcut' observed in the PASCAL VOC 2007 dataset, where watermarks are leveraged as classification shortcuts for the horse category, with 15–20% of images containing watermarks [15]. To replicate this scenario, we introduce watermarks into 10% of the training images within the horse class.

To segment the watermarks, we use Language Segment-Anything, a model built on top of Segment-Anything [14]. Segment-Anything is an open-set segmentation model, meaning it can segment a huge number of different objects. However, it doesn't assign

Table 5. CoF table showing the frequency that when a given segment is detected, it causes a counterfactual. Only showing frequencies of above 15%. Limited to only segments that occur in at least 20 images.

Segment Name	Segment-Level CoF
watermark	24.5%
banner	23.8%
rock	21.7%
sand	21.2%
tree	16.3%
snow	16.0%
mountain	15.2%

semantic labels for these segments. Language Segment-Anything addresses this limitation by not only providing semantic labels for the identified segments but also supporting user-defined prompts. For our analysis, we specify the prompt 'watermark' for each image, enabling precise identification and labelling of watermark segments. Additionally to this, we use the DETR segmentation model to provide general segments. The models we explain are once again simple TensorFlow CNNs.

We once again employ generative inpainting as our edit function, this time without providing any prompt, thereby allowing the model to autonomously determine the most fitting infill for each image. For the segments provided by DETR we blur the segments, as there is too much variance in size and importance of segments to reliably infill without a prompt.

In a second experiment, we include watermarks in 10% of the horse class images and thus in just 5% of the total dataset. This makes identification infeasible with techniques that rely on instance-level inspection. Table 4 shows the percentage of images for which a counterfactual was found for each segment. We see that removing watermarks results in counterfactuals in approximately 1% of images, indicating that manual detection would require extensive analysis. Furthermore, methods designed to aggregate numerous explanations are likely to identify several other factors before considering watermarks. Given that many segments rarely appear in images, this result can be misleading. Common segments such as grass, horse (DETR often misclassifies zebras as horses), and tree appear in most images, naturally leading to a greater number of counterfactuals associated with them. In contrast, Table 5 shows a standard CoF table and identifies the watermark as the most impactful segment, with 24.5% of images containing a counterfactual when this segment is present. The use of CoF tables facilitates the exploration of such explanations, enabling the discovery of critical insights like these.

5   Limitations and Future Research

Despite the promising results of our method, several limitations and avenues for future research should be acknowledged. Firstly, the quality of local explanations is inherently dependent on both the segmentation method and the edit function employed. Although open-set segmentation models and generative infill methods offer flexibility and the capacity to generate counterfactuals across a broad range of problem types, they introduce additional variability. The open-set segmentation model we use provides incorrect segments more frequently than its closed set counterparts. Similarly, the generative inpainting model may generate corrupted images or introduce edits beyond the intended mask, which remains difficult to detect automatically. However, we observe that such distortions are minimal when segments are small and no prompt is provided. We demonstrate that careful test design can mitigate many of these issues, and we are optimistic that ongoing advances in both segmentation and image editing techniques will further improve the quality of local explanations. In future work, this could be enhanced by adopting more sophisticated infill models or by more tightly constraining the edit function to modify only specific segments. At the current stage, limited human intervention is still required to design the tests and guide the choice of both segmentation and edit functions. Future work could explore replacing these manual steps with agentic LLM-based tools or other automated approaches.

As with many counterfactual explanation methods, our approach faces challenges related to coverage and the disagreement problem. Coverage refers to the proportion of instances for which an explanation can be generated; in our current implementation, we restrict the search to counterfactuals achievable through editing a single segment. This limitation is compounded by the fact that the edit function itself may not reliably produce valid counterfactuals. A natural direction for future work would be to explore editing combinations of segments or varying the strength of the edit function. The disagreement problem arises from the fact that, for any given instance, multiple distinct counterfactuals may be equally valid. Rather than viewing this as a drawback, we suggest embracing it: if many valid counterfactuals are found for an image, this signals fragility in the model's decision and highlights the need for further scrutiny.

Once a shortcut has been identified, an important avenue for future work would be to investigate whether its effects can be mitigated through fine-tuning or unlearning. As CoF tables provide counterfactuals for instances where the model relies on a shortcut, these could serve as targeted examples to support such interventions.

6   Conclusion

Given the capacity of CoF tables to summarise model behaviour, we advocate for their inclusion in model cards. CoF tables could serve as the output of a standardised test, using a fixed dataset, edit function, and segmentation model, to provide a consistent basis for assessing model reasoning and comparing classifiers.

This study introduces a novel approach for uncovering shortcut reliance and unexpected model behaviours through the aggregation of grounded counterfactuals. Our contributions are as follows:

- Proposing a framework for incorporating semantic meaning into counterfactual explanations.
- Introducing Counterfactual Frequency (CoF) tables as a method for aggregating instance-based grounded counterfactuals to reveal model behaviour.

– Demonstrating that CoF tables can expose learned shortcuts across diverse datasets, including cases where existing techniques are less effective.

By enabling systematic evaluation of model behaviour across groups of data, CoF tables offer a promising tool for identifying reliance on undesirable patterns and shortcuts, and merit further development within model evaluation pipelines.

Acknowledgements. This work was supported by the AXA Joint Research Initiative (JRI) project "Explainable Artificial Intelligence" and by the Flemish Government through the "Onderzoeksprogramma Artificiële Intelligentie (AI) Vlaanderen".

References

1. Adebayo, J., Gilmer, J., Muelly, M., Goodfellow, I., Hardt, M., Kim, B.: Sanity checks for saliency maps. In: Advances in Neural Information Processing Systems, vol. 31 (2018)
2. Atrey, A., Clary, K., Jensen, D.: Exploratory not explanatory: counterfactual analysis of saliency maps for deep reinforcement learning. arXiv preprint arXiv:1912.05743 (2019)
3. Bach, S., Binder, A., Montavon, G., Klauschen, F., Müller, K.R., Samek, W.: On pixel-wise explanations for non-linear classifier decisions by layer-wise relevance propagation. PLoS ONE **10**(7), e0130140 (2015)
4. Carion, N., Massa, F., Synnaeve, G., Usunier, N., Kirillov, A., Zagoruyko, S.: End-to-end object detection with transformers. In: European Conference on Computer Vision, pp. 213–229. Springer (2020)
5. D'Amour, A., et al.: Underspecification presents challenges for credibility in modern machine learning. J. Mach. Learn. Res. **23**(1), 10237–10297 (2022)
6. DeGrave, A.J., Janizek, J.D., Lee, S.I.: Ai for radiographic covid-19 detection selects shortcuts over signal. Nat. Mach. Intell. **3**(7), 610–619 (2021)
7. Deng, J., Dong, W., Socher, R., Li, L.J., Li, K., Fei-Fei, L.: Imagenet: a large-scale hierarchical image database. In: 2009 IEEE Conference on Computer Vision and Pattern Recognition, pp. 248–255. IEEE (2009)
8. Geirhos, R., et al.: Shortcut learning in deep neural networks. Nat. Mach. Intell. **2**(11), 665–673 (2020)
9. Geirhos, R., Rubisch, P., Michaelis, C., Bethge, M., Wichmann, F.A., Brendel, W.: Imagenet-trained cnns are biased towards texture; increasing shape bias improves accuracy and robustness. arXiv preprint arXiv:1811.12231 (2018)
10. Ghorbani, A., Abid, A., Zou, J.: Interpretation of neural networks is fragile. In: Proceedings of the AAAI Conference on Artificial Intelligence, vol. 33, pp. 3681–3688 (2019)
11. He, K., Zhang, X., Ren, S., Sun, J.: Deep residual learning for image recognition. In: Proceedings of the IEEE Conference on Computer Vision and Pattern Recognition, pp. 770–778 (2016)
12. Hinns, J., Fan, X., Liu, S., Raghava Reddy Kovvuri, V., Yalcin, M.O., Roggenbach, M.: An initial study of machine learning underspecification using feature attribution explainable AI algorithms: a COVID-19 virus transmission case study. In: Pham, D.N., Theeramunkong, T., Governatori, G., Liu, F. (eds.) PRICAI 2021. LNCS (LNAI), vol. 13031, pp. 323–335. Springer, Cham (2021). https://doi.org/10.1007/978-3-030-89188-6_24
13. Kindermans, P.J., et al.: The (un) reliability of saliency methods. Explainable AI: Interpreting, explaining and visualizing deep learning, pp. 267–280 (2019)

14. Kirillov, A., Mintun, E., Ravi, N., Mao, H., Rolland, C., Gustafson, L., Xiao, T., Whitehead, S., Berg, A.C., Lo, W.Y., et al.: Segment anything. arXiv preprint arXiv:2304.02643 (2023)
15. Lapuschkin, S., Wäldchen, S., Binder, A., Montavon, G., Samek, W., Müller, K.R.: Unmasking clever hans predictors and assessing what machines really learn. Nat. Commun. **10**(1), 1096 (2019)
16. Li, Z., et al.: A whac-a-mole dilemma: shortcuts come in multiples where mitigating one amplifies others. In: Proceedings of the IEEE/CVF Conference on Computer Vision and Pattern Recognition (CVPR), pp. 20071–20082, June 2023
17. van der Linden, I., Haned, H., Kanoulas, E.: Global aggregations of local explanations for black box models. arXiv preprint arXiv:1907.03039 (2019)
18. Lundberg, S.M., Lee, S.I.: A unified approach to interpreting model predictions. In: Advances in Neural Information Processing Systems, vol. 30 (2017)
19. Montavon, G., Samek, W., Müller, K.R.: Methods for interpreting and understanding deep neural networks. Digit. Sig. Process. **73**, 1–15 (2018)
20. Nam, J., Cha, H., Ahn, S., Lee, J., Shin, J.: Learning from failure: training debiased classifier from biased classifier. In: Advances in Neural Information Processing Systems (2020)
21. Nauta, M., Walsh, R., Dubowski, A., Seifert, C.: Uncovering and correcting shortcut learning in machine learning models for skin cancer diagnosis. Diagnostics **12**(1), 40 (2021)
22. Pfau, J., Young, A.T., Wei, M.L., Keiser, M.J.: Global saliency: aggregating saliency maps to assess dataset artefact bias. arXiv preprint arXiv:1910.07604 (2019)
23. Ribeiro, M.T., Singh, S., Guestrin, C.: " why should i trust you?" explaining the predictions of any classifier. In: Proceedings of the 22nd ACM SIGKDD International Conference on Knowledge Discovery and Data Mining, pp. 1135–1144 (2016)
24. Rombach, R., Blattmann, A., Lorenz, D., Esser, P., Ommer, B.: High-resolution image synthesis with latent diffusion models. In: Proceedings of the IEEE/CVF Conference on Computer Vision and Pattern Recognition (CVPR), pp. 10684–10695, June 2022
25. Samek, W., Binder, A., Montavon, G., Lapuschkin, S., Müller, K.R.: Evaluating the visualization of what a deep neural network has learned. IEEE Trans. Neural Netw. Learn. Syst. **28**(11), 2660–2673 (2016)
26. Selvaraju, R.R., Cogswell, M., Das, A., Vedantam, R., Parikh, D., Batra, D.: Grad-cam: visual explanations from deep networks via gradient-based localization. In: Proceedings of the IEEE International Conference on Computer Vision, pp. 618–626 (2017)
27. Simonyan, K., Vedaldi, A., Zisserman, A.: Deep inside convolutional networks: Visualising image classification models and saliency maps. arXiv preprint arXiv:1312.6034 (2013)
28. Vermeire, T., Brughmans, D., Goethals, S., de Oliveira, R.M.B., Martens, D.: Explainable image classification with evidence counterfactual. Pattern Anal. Appl. **25**(2), 315–335 (2022)
29. Winkler, J.K., et al.: Association between surgical skin markings in dermoscopic images and diagnostic performance of a deep learning convolutional neural network for melanoma recognition. JAMA Dermatol. **155**(10), 1135–1141 (2019)

On Explainable Disease Progression Forecasting with Transformer Models

Umair Aamir Mirza, Faryal Siddique, and Faizan Ahmed[(✉)]

Department of Computer Science, University of Twente, Enschede, The Netherlands
`f.siddique@utwente.nl`, `faizan.ahmed@utwente.nl`

Abstract. Accurate prediction of disease progression is essential to enable timely clinical interventions and support personalized treatment strategies in both chronic and acute care settings. While recurrent neural networks (RNNs) have traditionally dominated time series modeling in healthcare, transformer-based architectures have recently shown considerable promise. This study evaluates the applicability of the Informer architecture on longitudinal clinical data from the Parkinson's Progression Markers Initiative (PPMI) dataset. Multiple model variations are explored, including different temporal embedding techniques and multi-task learning configurations. Informers with learnable time-stamp embeddings came up with the lowest mean Mean Squared Error (MSE) and smaller standard deviation (211.6 ± 17.2) with sequence distillation and fixed hyperparameters. However, Multi-Task Learning (MTL) has shown promising results without sequence distillation. To enhance clinical interpretability, explainability methods such as self-attention visualizations, permutation importance, and GradientSHAP are employed. The results show that the Informer architecture can be viable for forecasting tasks within clinical health domains.

Keywords: Informer · Longitudinal Clinical Data · Time Series Forecasting · Self-Attention · Multi-Task Learning · Transformer · Disease Progression · Explainable AI

1 Introduction

Chronic diseases such as Parkinson's, cancer, and diabetes develop gradually over extended periods. This requires continuous monitoring, regular clinical evaluations, timely interventions along with personalized care. With the rise of digital health infrastructure, Electronic Health Records (EHRs) are a rich source of longitudinal clinical data (time series), capturing biomarkers, diagnoses, administered treatments, and other related information across patient visits. EHRs present an opportunity to capture and analyze changes in patient data to guide clinicians in real-world environments.

With respect to forecasting disease severity, two approaches are common, progression/trajectory forecasting and milestone/event forecasting. This study focuses on progression forecasting, which focuses on predicting the future values of one or more clinically relevant biomarkers [2, 13]. These biomarkers must be valid indicators of disease

© The Author(s), under exclusive license to Springer Nature Switzerland AG 2026
F. Marcelloni et al. (Eds.): IJCCI 2025, CCIS 2829, pp. 467–481, 2026.
https://doi.org/10.1007/978-3-032-15638-9_27

severity used in clinical contexts to assess patient condition over time. Accurate forecasting of such variables enables clinicians to track gradual changes in disease status and facilitates the design of personalized treatment strategies.

A wide range of models have been applied to the task of disease progression forecasting, from traditional Recurrent Neural Networks (RNNs) to more recent advancements in deep learning such as Transformer-based architectures [2,17]. The Transformer, introduced by Vaswani et al. [16], marked a paradigm shift in sequence modeling by replacing recurrence with a self-attention mechanism, enabling more effective modeling of long-range dependencies. As highlighted by Delgado-Santos et al. [6], the Transformer architecture allows for parallel processing of sequence elements, improving computational efficiency and scalability. Unlike RNNs, which compress historical information into a hidden state, Transformers can directly attend to all positions in the input sequence, making them particularly suitable for capturing complex temporal patterns. Additionally, their architecture can be optimized and customized for various task domains, offering flexibility and adaptability in real-world biomedical applications.

Biomedical data poses unique challenges due to issues in quality, structure, and consistency. As highlighted by Cascarano et al. [4], clinical visit records are often sparse and irregular, with measurement inconsistencies introduced by tools, varying procedures, and comorbidities. Clinical features frequently exhibit strong correlations, violating assumptions such as independent and identically distributed data (i.i.d.). Similarly, disease outcomes can vary immensely per patient, which can limit model generalization.

Transformer architectures have shown strong performance in long-sequence modeling but also possess some known limitations. Auto-regressive variants are prone to error accumulation over long horizons [15] and often depend on fixed-length input windows, limiting their ability to capture long-term dependencies in clinical histories [12]. As explained by Siebra et al. [14], temporal embedding strategies and training practices remain inconsistent in practice. Finally, model interpretability is critical for clinical deployment, and transformer architectures are not inherently explainable due to their deep structure.

Given the unique challenges posed by clinical time series (as described before), the selection of an appropriate architecture for forecasting is a difficult decision. Traditional approaches focuses on RNN based architecture. In addition to modeling challenges, the explainability of the developed models is often under explored.

In this paper, we bridge this research gap by focusing on the use of transformer based architecture with an added ambition to make the prediction explainable. The chosen transformer architecture is Informer, introduced by Zhou et al. [19], which was developed to address key challenges in long-sequence time series forecasting. Unlike traditional autoregressive models, the Informer uses a non-autoregressive decoder which enables parallel prediction, eliminating risks of exposure bias and improving the stability of predictions over longer forecasting horizons. Additionally, the use of sparse self-attention and sequence distillation significantly lowers memory usage and computational complexity, making it a scalable option for clinical time series datasets. Moreover, model explainability is examined through self-attention visualizations, permutation importance, and GradientSHAP.

The paper is organized as follows. The background to this research is described in Sect. 2 followed by methodology in Sect. 3. Section 4 contains results, while Sect. 5 provide conclusions and discussions.

2   Background

For time series or sequence modeling RNN based approaches have been dominant. In the context of disease progression modeling Berendse et al. [2] applied a state-of-the-art RNN to disease progression forecasting using the Parkinson Progression Marker Initiative (PPMI) datase [9]. They used the Movement Disorder SocietyâĂŞUnified Parkinson's Disease Rating Scale (MDS-UPDRS) Total Score as a proxy for Parkinson's disease severity, the same target variable adopted in this research.

While RNNs have traditionally been used in clinical time series modeling, their limited memory and inability to handle long-range dependencies pose problems for handling long sequences of clinical health data. The work by Shankar et al. [12] outlines the issue of non-stationarity within clinical health data due to multi-morbidity and the effects of administered treatments, altering the trajectory of the patient. RNN architectures struggle to model long-range dependencies due to their fixed memory, making transformer-based models a viable option for forecasting tasks [15]. Transformer-based models with self-attention mechanisms can use entire sequences to help prioritize historical inputs based on the current prediction context, which can help mitigate non-stationarity and temporal irregularity. Informer architecture is adapted in the context of irregular sequence modeling. Several studies [3,7,11] have demonstrated the flexibility and performance of the Informer architecture across various domains. The closest example to diseases progression is from Delgado-Santos et al. [6]. They have applied the Informer, alongside other Transformer-based models, to behavioral biometrics tasks such as gait recognition. Their experiments on two benchmark datasets revealed that the Informer consistently ranked among the top-performing architectures, significantly outperforming RNNs.

Similarly, in the healthcare domain, Xue et al. [18] proposed an enhanced Informer variant, BGformer, for blood glucose forecasting. Their model outperformed traditional approaches, including RNNs, over both 60-minute and 90-minute prediction windows. However, they also noted that the vanilla Informer struggles to capture temporal volatility and correlations between time-steps, pointing to a limitation that this research aims to explore further through the evaluation of several temporal embedding strategies.

Despite these successes, a critical distinction in this research is the nature of the PPMI dataset. Unlike many previous applications, which involve lengthy input sequences, the PPMI dataset comprises relatively short time series with high-dimensional features. This presents a unique challenge for the Informer architecture, where feature saturation must be managed as opposed to sequence length. Furthermore, given the dataset's structure, a fixed forecasting horizon of one future visit is practical considering the limited number of visits per patient. Likewise, it is clinically grounded, considering that an annual patient check-up is a common practice for many illnesses.

3   Methodology

This section outlines the key steps involved in the study, including data pre-processing, building the experimental pipeline, designing the experiments, and incorporating explainability techniques. A visualization of methodology used for this study is presented in Fig. 1. Each part is discussed to give a clear understanding of how the overall approach was structured and why certain decisions were made.

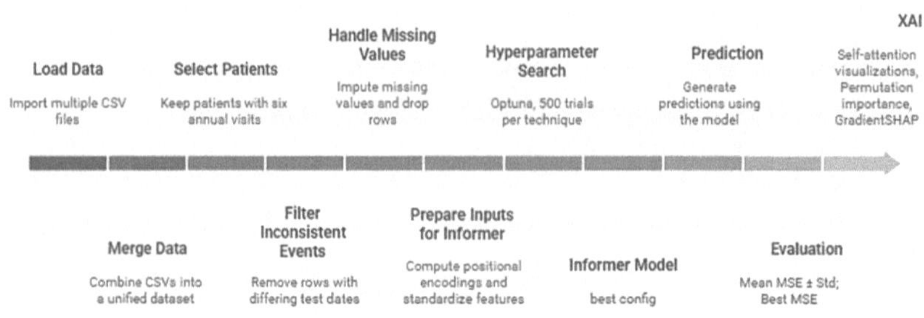

Fig. 1. Methodology: Forecasting disease progression and Explaining the predictions.

3.1   Data Pre-processing

Data is pre-processed so that it can be fed to the Informer. The original dataset was provided as multiple CSV files, each capturing a specific component of the MDS-UPDRS assessment [9], with each row containing a PatientID and an EventID identifying a clinical visit. These files were merged into a unified longitudinal dataset, with 34,611 rows and 86 features, sufficient to compute the MDS-UPDRS Total Score for each visit.

Although each EventID denoted a unique clinical visit, inconsistencies were found in test dates within the same event. To resolve this, only rows where the difference between test dates for a given EventID did not exceed 30 days were retained. Outliers were removed using the 3-Sigma Rule, leaving 32,867 rows. Patients were included only if they had a complete sequence of six annual visits (including the baseline). Visits were chronologically ordered to ensure uniform sequence lengths, resulting in 7,291 records.

Rows with more than 60% missing values were dropped. The remaining missing values were imputed using a Forward-Backward Fill strategy, and patients with entirely missing features across all visits were excluded, leaving 1,632 records. Forward-Backward Fill was selected as it was the highest performing imputation technique across the experimentation conducted by Berendse et al. [2]. Time-gap-aware positional encoding were computed as the number of days since each patient's baseline visit. These, along with all continuous features, were standardized.

The final dataset includes 272 patients (166 males, 106 females). As shown in Fig. 2, while gender imbalance exists, broader age variability among females and an older skew among males offer demographic diversity. Seeds were selected carefully to ensure that training bias was minimized. The final dataset shape is (1632, 86), comprising features relevant to Parkinson's progression.

3.2 Informer Architecture Overview and Adaptations

Figure 3 illustrates the Informer's encoder-decoder architecture and its key innovations over the vanilla Transformer proposed by Vaswani et al. [16]. A major difference is the use of ProbSparse Self-Attention, which improves efficiency by focusing computation on the most informative queries. Unlike vanilla attention, which evaluates all query-key pairs, ProbSparse selects high-entropy queries that are more relevant for prediction [19]. Such compression is particularly useful in clinical settings, where discarding low-value visits can improve both computational efficiency and interpretability. In longitudinal clinical datasets, each patient has an individual time series rather than a shared global sequence. For this study, each patient's data has a shape of $(T = 6, D = 86)$, where T is the number of visits and D is the number of features. The dataset is split into training (75%), validation (10%), and test (15%) sets. Each patient's sequence is divided into: Input sequence length $L = 5$, Decoder label length $L_c = 5$, and Forecast horizon $P = 1$. Given the short sequence length, the full history is passed to the decoder.

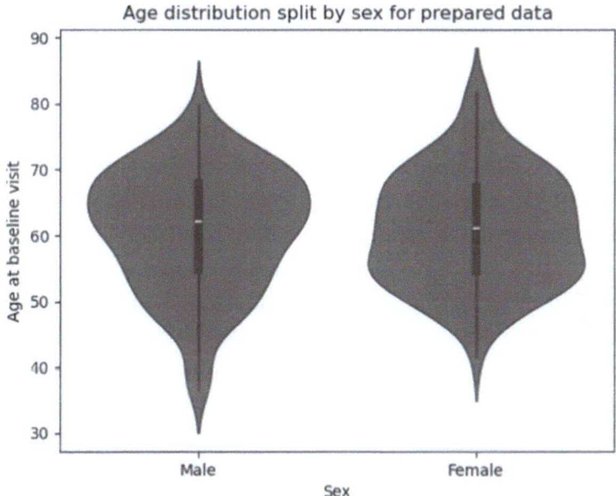

Fig. 2. Gender distribution of the data.

The model inputs are defined as follows: Encoder input: $x_{\text{enc}} \in \mathbb{R}^{B \times L \times D}$, Decoder input: $x_{\text{dec}} \in \mathbb{R}^{B \times (L_c+P) \times D}$, Encoder timestamps: $x_{\text{mark\_enc}} \in \mathbb{R}^{B \times L \times T}$, and Decoder timestamps: $x_{\text{mark\_dec}} \in \mathbb{R}^{B \times (L_c+P) \times T}$.

Here, B is the batch size, L is the encoder input length, L_c is the number of known historical values provided to the decoder (label length), P is the forecasting horizon (prediction length), D is the number of input features, and T is the number of raw

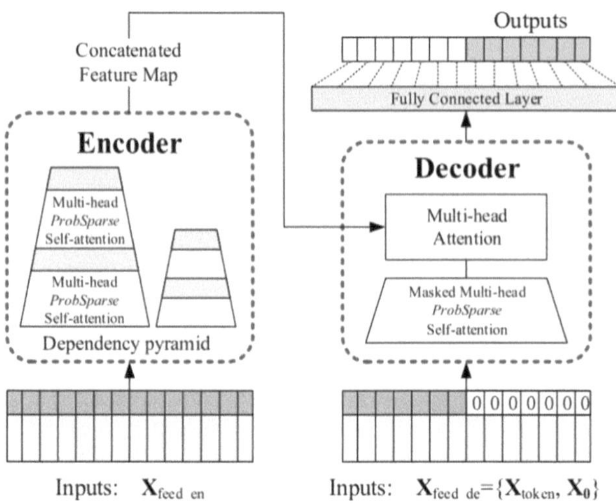

Fig. 3. Informer Architecture Visualization From Zhou et al. [19].

temporal input features (e.g., month, day, weekday, timegap, etc.). The decoder input is constructed by taking the final L_c time steps from the encoder input and concatenating zero-padded placeholders for the P future steps. This combined sequence is then passed through the embedding layer and decoded to produce forecasts for the target variable.

MTL Implementation. The implementation involved modifying the Informer architecture by extending its final projection layer with auxiliary output heads, enabling the model to concurrently forecast the primary score and auxiliary targets, reinforcing the learning of complex patterns by utilizing the shared architecture and inputs.

Weighting of targets was provided in order to better guide gradients within the model to focus on minimizing the error of the tasks in a more dynamic manner. However, this weighting of outputs should be considered as a simple implementation to guide model gradients in conditions where the loss functions of all outputs are equivalent. Due to data sparsity issues, not enough regression targets, or continuous variables, were available within the dataset. In order to simplify the testing of Multi-Task Learning (MTL), some ordinal PPMI features, specifically severity scores for particular symptoms, were modeled as continuous outputs solely to assess whether the performance of the primary forecasting task could benefit from the added representation in the hidden state of the model.

Model Optimization and Evaluation. A custom ClinicalInformer wrapper was developed to modify the forward pass, integrate evaluation with relevant visualizations, and implement the selected explainability techniques. It also supports target weighting to enable Multi-Task Learning (MTL). Table 1 outlines the experiments conducted and the rationale behind key parameter adjustments. A batch size of 16 and 10 training epochs, were used across all implementations.

Table 1. Overview of Experimental Design.

ID	Name	Description
1	Baseline	500 trials using standard positional encoding
2	Time-Gap Positional Encoding	500 trials using time-gap-aware positional encoding
3	Fixed Timestamp Positional Encoding	500 trials using fixed timestamp encoding (day/month/year) and all features excluding TIMEGAP
4	Learnable Timestamp Positional Encoding	500 trials using learnable timestamp encoding (day/month/year) and all features excluding TIMEGAP
5	Combined Temporal Positional Encoding	500 trials combining fixed timestamp and time-gap-aware positional encoding with all features
6	Temporal Encoding Evaluation	Five embedding techniques using shared hyperparameters across multiple random seeds, with 50 repetitions per seed
7	Multi-Task Learning (MTL)	250 trials each for 5, 10, and 25 auxiliary targets (plus UPDRSTOT) using all features and best hyperparameters

To systematically tune the model's architecture and training hyperparameters, Optuna [1] was employed. Optuna searched over a predefined space of Informer-specific parameters.

For each sampled combination, a new model was trained and validated, and the configuration with the lowest mean squared error (MSE) on the test set was selected. Here, mean squared error (MSE) is defined as $\mathrm{MSE} = \frac{1}{n} \sum_{i=1}^{n} (y_i - \hat{y}_i)^2$, where y_i and $\hat{y}_i$ denote the true and predicted values, respectively; lower MSE indicates better predictive accuracy. This automated process helped avoid manual trial-and-error tuning, and additionally enabled the testing of different optimization functions like Adam, AdamW, and Adan to assess the impact on convergence speed and forecast accuracy, as the choice of optimization function is critical for deep learning models. Adam performed the best in various trials, so it was selected for this investigation. The model's performance is measured using mean squared error due to its extensive use in clinical settings as it penalizes larger errors.

It is important to note that in Multi-Task Learning (MTL) settings involving both classification and regression tasks, a unified loss function is required to balance classification and forecasting errors. However, this study focuses solely on regression-based forecasting, and no classification tasks were included. For each epoch, after the training phase, the model was evaluated on the validation set to monitor generalization and overfitting. The performance of the model is then analyzed on the test set through MSE as well as other error metrics, in order to compare with existing literature.

Specific hyperparameter tuning was performed for each embedding technique to reflect realistic deployment scenarios, where extensive tuning is typically conducted. To enable fair comparison, the performance of each embedding technique was also evaluated using a shared set of hyperparameters.

3.3 Explainability

In clinical applications, interpretability is essential, as model-driven decisions can directly influence patient outcomes [10]. Transformer architectures, with their multi-head attention mechanisms, capture inter-dependencies across time steps. Visualizing attention layers offers clinicians an intuitive means to understand which visits the model prioritizes when making predictions. Following the approach of Nguyen et al. [10], attention weights were visualized for the training set (average across epochs), validation set (average), and the test set (single run only, to avoid data leakage). We also created attention visualizations for each epoch to see how the model's understanding of time changed during training.

4 Results

In this section, we present the findings of our study, organized according to the different experimental setups. Each subsection focuses on a specific experiment to highlight the impact of various design choices and configurations on model performance.

4.1 Performance Across Temporal Embedding Techniques

The goal of hyperparameter tuning per implementation was to outline cases where specific parameters may benefit certain embedding techniques. We compare five temporal embedding schemes for informers and report the mean MSE ± standard deviation and the best run (best MSE). The lowest standard deviation was recorded by the Combined Embedding technique, although variability across all temporal embedding techniques is expected as Optuna navigates through the hyperparameter space in each trial. Across 500 hyperparameter optimization trials per technique without distillation, Fixed Timestamp Embedding demonstrated the best mean MSE, as shown in Table 2. It also achieved a best-case MSE (171.63). The lowest best-case MSE was seen in the Standard Positional Embedding (151.70).

Table 3 presents the performance of five embedding techniques using fixed hyperparameters, both with and without sequence distillation. Across both settings, Fixed Timestamp Embedding consistently achieves one of the lowest mean MSE and standard deviation. With sequence distillation, the Learnable Time-Stamp Embedding had the lowest mean MSE, followed by Fixed Time-Stamp Embedding. Time-Gap and Combined embeddings also show improvements over the Standard baseline. These findings reinforce the suitability of time-stamp-related embeddings for forecasting tasks with Informer models, particularly in a clinical domain.

While the choice of temporal embedding technique is important, it is equally important to assess whether the Informer architecture successfully captures temporal dependencies within the data. Figures 4 and 5 visualize how one of the models utilizing

Table 2. Performance Across Embedding Types Without Distillation.

Embedding	Mean MSE	Standard Deviation	Best MSE
Standard	247.0430	42.9934	**151.7049**
Time-Gap	242.2491	50.9287	169.0128
Fixed Time-Stamp	**231.8993**	52.8694	171.6288
Learnable Time-Stamp	253.1865	51.4191	158.6633
Combined	238.6720	**36.0168**	173.4257

the Combined Embedding technique exhibited greater attention towards more recent time steps, suggesting that it gained temporal awareness during training. However, the best-performing model overall displayed relatively uniform attention weights across all query-key pairs, indicating that temporal relevance was not consistently emphasized. This suggests that, despite the use of different embedding strategies, the model may still struggle to gain temporal awareness or demonstrate it consistently, highlighting a need for use of advanced embedding techniques.

Table 3. Comparison of Embedding Techniques With Fixed Hyperparameters.

Embedding	Mean MSE (No Distil)	Std Dev (No Distil)	Mean MSE (Distil)	Std Dev (Distil)
Standard	232.3837	28.8294	235.4529	27.9290
Time-Gap	222.6577	**21.4253**	225.2664	23.5780
Learnable Time-Stamp	238.3692	27.5325	**211.5781**	**17.1532**
Fixed Time-Stamp	**213.5522**	21.6830	217.1314	20.4691
Combined	218.6205	25.09377	221.8843	24.7472

4.2   Performance With Multi-task Learning

The implementation of multi-task learning (MTL) in this study required key architectural considerations, particularly related to the loss function and performance evaluation across tasks with varying clinical relevance. To address this, a weighted loss aggregation strategy was used. The primary task (forecasting MDS-UPDRS Total Score) received the majority of the weight, while auxiliary tasks shared the remainder equally.

In practice, task weights should ideally reflect clinical priorities, and supporting mixed-task outputs (e.g., classification and regression) would require extending the current regression-only ClinicalInformer. Despite this, Table 4 demonstrates that MTL can improve the performance of the model, when inspecting solely the performance of the primary regression task (UPDRSTOT Forecasting). This performance remained relatively consistent over various MTL configurations, suggesting that the creation of aux-

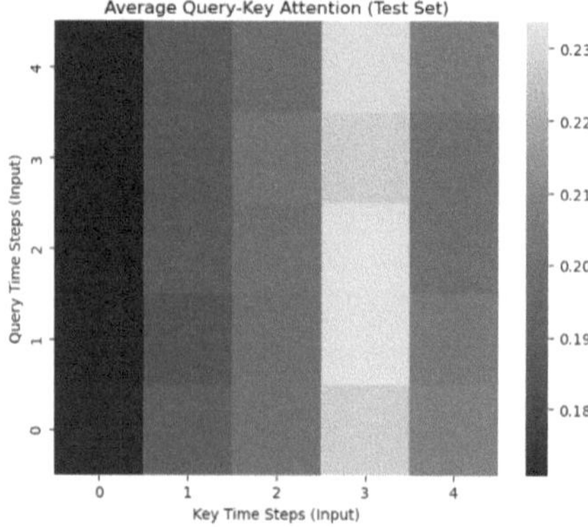

Fig. 4. Average Query-Key Attention Matrix on Test Set, Highlighting Temporal Dependencies Across Time Steps.

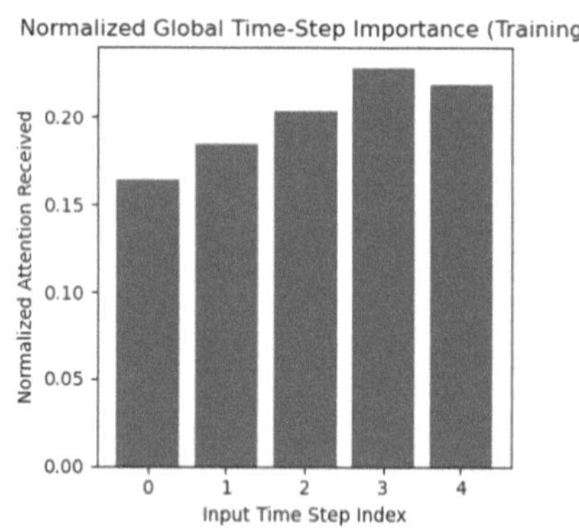

Fig. 5. Normalized Global Attention Scores per Time Step During Training, Indicating Increasing Importance Toward Later Time Steps.

iliary tasks could help mitigate the effects of feature saturation. With Informer's generative decoder, MTL integration remains computationally feasible and could be suitable for clinical forecasting tasks which require multiple outputs based on similar input data.

4.3   Explainability

Self-attention weights from the final encoder layer were extracted per epoch and aggregated across training batches. Matrices were generated for the validation and test sets as well. These matrices reflect how time steps attend to one another, offering insights into how the model understands temporal dependencies. Importantly, attention visualizations can highlight the relative importance of certain visits, but do not reveal the importance of individual features. Figure 6 presents an example of such a visualization.

Both permutation importance and self-attention visualization offer a global insight into temporal awareness and feature importance. In clinical applications, local interpre-

Table 4. MTL Results: MSE Performance for UPDRSTOT Forecasting.

Targets	Average MSE	Standard Deviation	Best MSE
5	218.1306	23.2511	163.5673
10	216.9054	**21.3069**	173.7040
25	**216.2155**	24.6513	**151.2446**

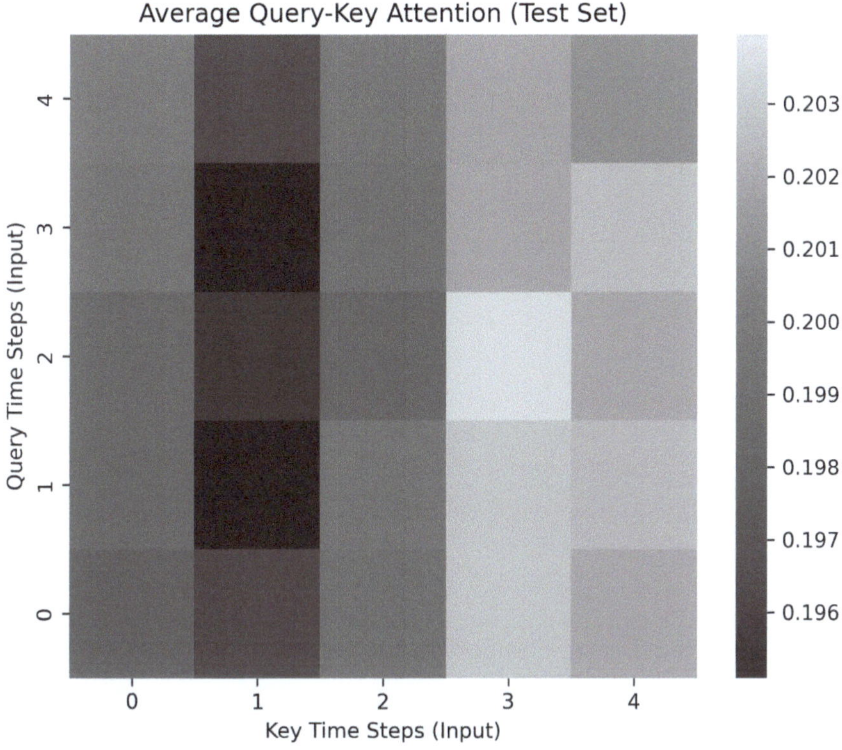

Fig. 6. Attention heatmap showing relative importance of input time steps in encoder.

tations are equally as important. GradientSHAP combines concepts such as integrated gradients to approximate SHAP values through interpolation between a provided baseline input and the actual passed input. Although it assumes that features are independent and the underlying model is linear, it is directly integrated with PyTorch to provide useful local insights. Another limitation of GradientSHAP in clinical health is the lack of a clear baseline for patient progression. Patient outcomes can be highly specific, and differences in disease effects across demographics along with multi-morbidity make it hard to rely on GradientSHAP alone. The limitations can be overcome by integrating various explainability tools, as this investigation does. GradientSHAP (Fig. 7) Permutation Importance (Fig. 8) may also yield different results for feature importance, which

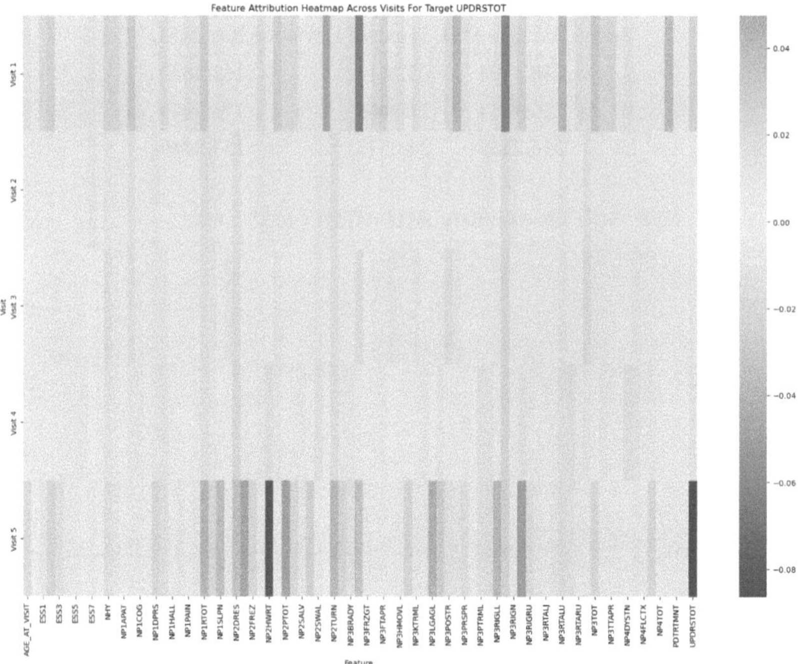

Fig. 7. GradientSHAP outlining feature contributions for a single patient sequence.

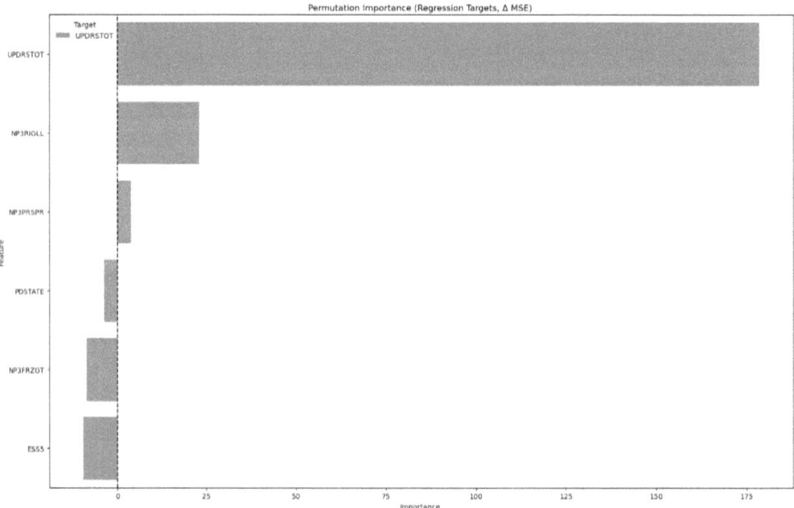

Fig. 8. Permutation importance (ΔMSE) showing UPDRSTOT as primary contributor.

is likely due to the limitation of permutation importance in creating perturbations on features independently. In reality feature interactions are more entangled.

4.4 Comparison to Berendse et al. [2]

In comparison to the results reported by Berendse et al. [2], where a GRU-based architecture achieved a best-case MSE of 130.08 and an average MSE of 223.37 using a forward-backward fill strategy, the Informer recorded slightly higher average MSE across the experiments conducted in this study. However, the incorporation of Multi-Task Learning (MTL) within the Informer framework demonstrates meaningful performance improvements, suggesting that its generative design and flexible architecture may offer advantages in more complex or customized forecasting scenarios. In terms of interpretability, the Informer is inherently less transparent than traditional RNNs such as GRU. Nevertheless, the use of attention visualizations, feature importance, and GradientSHAP offer valuable insights into how the model attends to different time steps and features, helping to mitigate this limitation in clinical contexts.

5 Discussion and Conclusions

Based on the experimentation and analysis conducted, the Informer architecture demonstrates a reasonable capacity for modeling longitudinal clinical time series data, specifically data with short sequences. The findings in this paper indicate that the choice of temporal encoding plays a critical role in model performance. Embedding techniques incorporating time-stamp information consistently outperformed other strategies across multiple settings, both with and without distillation and hyperparameter tuning. Furthermore, the incorporation of multi-task learning (MTL) significantly enhanced performance, likely due to the shared representation learning across clinically relevant tasks. The Informer presents itself as a viable option, with well-documented success in forecasting tasks that have longer sequences. Yet, it is capable of scaling reasonably to a larger number of features, and was able to compete with state-of-the-art RNN architectures on the same task profile on a dataset with short sequences.

While the Informer demonstrates strong potential for clinical forecasting tasks, some limitations must be discussed. First, its original design for long-sequence forecasting may not translate optimally to clinical datasets where sequence lengths are inherently short, such as the six-visit structure used in this study. This constraint, combined with a high-dimensional feature space (86 features), necessitated complex architectural adaptations and extensive tuning to achieve competitive performance - efforts that may not be required for simpler models like GRUs. Additionally, although feature saturation was observed and expected given the high-dimensional feature space, a domain-informed feature subset could likely improve generalization and reduce computational overhead. The limited dataset size may also restrict generalization of the model to larger patient populations, as sufficient patient demographics were not available.

Explainability remains another key limitation. While attention visualizations such as time-step importance and query-key maps were incorporated, their utility is debatable. As Chung et al. [5] argue, attention maps may not consistently reflect meaningful model reasoning, and their reliability in clinical contexts is still uncertain. Furthermore, the forecasting horizon in this study was limited to a single future visit. While this choice was pragmatic given sequence length constraints, real-world clinical decision-making often requires multi-step forecasting to support early intervention strategies [8]. Lastly, whilst GradientSHAP provides local explanations, the creation of a baseline patient sequence is difficult considering that disease progression varies per patient. Furthermore, the assumptions behind GradientSHAP do not hold within this dataset and model architecture, though the insights are still valuable. The generative nature of the decoder can be leveraged by a more comprehensive MTL implementation that supports mixed-task settings. This would require unifying loss in a fair manner along with adaptive task weighting and a domain-driven selection of auxiliary features. Benchmarking the Informer against alternative Transformer-based architectures like PatchTST, Autoformer, and TFT should be included in future research, particularly in varied clinical contexts where sequence lengths differ. Extending the forecasting horizon beyond a single time step can greatly help assess the model's utility and capacity to demonstrate temporal awareness. Practical metrics, such as training time, inference time, and memory consumption should be recorded to assess scalability for real-world applications. More sophisticated temporal embedding techniques can be included to improve the performance of the Informer from experimental environments to real-world clinical settings.

References

1. Akiba, T., Sano, S., Yanase, T., Ohta, T., Koyama, M.: Optuna: a next-generation hyperparameter optimization framework. In: Proceedings of the 25th ACM SIGKDD International Conference on Knowledge Discovery & Data Mining, pp. 2623–2631 (2019)
2. Berendse, S., Krabbe, J., Klaus, J., Ahmed, F.: Towards explainable machine learning for prediction of disease progression. Appl. Artif. Intell. **38**(1) (2024)
3. Cao, L., et al.: Interpretable industrial soft sensor design based on informer and SHAP. IFAC-PapersOnLine **58**(14), 73–78 (2024)
4. Cascarano, A., et al.: Machine and deep learning for longitudinal biomedical data: a review of methods and applications. Artif. Intell. Rev. **56**(2), 1711–1771 (2023)
5. Chung, M., Won, J.B., Kim, G., Kim, Y., Ozbulak, U.: Evaluating visual explanations of attention maps for transformer-based medical imaging. In: International Conference on Medical Image Computing and Computer-Assisted Intervention, pp. 110–120. Springer (2024)
6. Delgado-Santos, P., Tolosana, R., Guest, R., Deravi, F., Vera-Rodriguez, R.: Exploring transformers for behavioural biometrics: a case study in gait recognition. Pattern Recogn. **143**, 109798 (2023)
7. Deng, W., Xin, X., Song, R., Yang, X., Wang, W., Yu, G.: A time series forecasting method for oil production based on informer optimized by Bayesian optimization and the hyperband algorithm (BOHB). Comput. Chem. Eng. **197**, 109068 (2025)
8. de Lacy, N., Ramshaw, M., Lam, W.Y.: RiskPath: explainable deep learning for multistep biomedical prediction in longitudinal data. Patterns **6**(8) (2025)
9. Marek, K., Jennings, D., Lasch, S., et al.: The Parkinson progression marker initiative (PPMI). Progress Neurobiol. **95**(4), 629–635 (2011). Biological Markers for Neurodegenerative Diseases

10. Nguyen, H.H., Blaschko, M.B., Saarakkala, S., Tiulpin, A.: Clinically-inspired multi-agent transformers for disease trajectory forecasting from multimodal data. IEEE Trans. Med. Imaging **43**(1), 529–541 (2024). https://doi.org/10.1109/TMI.2023.3312524
11. Oliveira, J.M., Ramos, P.: Evaluating the effectiveness of time series transformers for demand forecasting in retail. Mathematics **12**(17) (2024)
12. Shankar, V., Yousefi, E., Manashty, A., Blair, D., Teegapuram, D.: Clinical-GAN: trajectory forecasting of clinical events using transformer and generative adversarial networks. Artif. Intell. Med. **138**, 102507 (2023). https://doi.org/10.1016/j.artmed.2023.102507
13. Shukla, S.N., Marlin, B.M.: A survey on principles, models and methods for learning from irregularly sampled time series. arXiv preprint arXiv:2012.00168 (2020)
14. Siebra, C.A., Kurpicz-Briki, M., Wac, K.: Transformers in health: a systematic review on architectures for longitudinal data analysis. Artif. Intell. Rev. **57**(2), 32 (2024)
15. Su, L., Zuo, X., Li, R., Wang, X., Zhao, H., Huang, B.: A systematic review for transformer-based long-term series forecasting. Artif. Intell. Rev. **58**(3), 80 (2025)
16. Vaswani, A., et al.: Attention is all you need. In: Advances in Neural Information Processing Systems, vol. 30 (2017)
17. Wu, F., Zhao, G., Zhou, Y., Qian, X., Elias, B.K., Lehman, L.W.H.: Forecasting treatment outcomes over time using alternating deep sequential models. IEEE Trans. Biomed. Eng. **71**(4), 1237–1246 (2024)
18. Xue, Y., Guan, S., Jia, W.: BGformer: an improved informer model to enhance blood glucose prediction. J. Biomed. Inform. **157**, 104715 (2024)
19. Zhou, H., et al.: Informer: beyond efficient transformer for long sequence time-series forecasting. In: Proceedings of the AAAI Conference on Artificial Intelligence, vol. 35, pp. 11106–11115 (2021)

International Conference on Neural Computation Theory and Applications

Determining Optimal Pixel Resolution for Object Detection in Satellite Imagery: A Class-Specific Approach

Daniel C. Fox[1]([✉]) [iD], John Prominski[1] [iD], Amit Virchandbhai Prajapati[1] [iD],
Kendall Haddigan[1] [iD], Gabriel Barbosa[1] [iD], Shubham Dashrath Wagh[1] [iD],
Adam Nolan[2] [iD], Daniel Zwillinger[2] [iD], Chun-Kit Ngan[1] [iD], Fatemeh Emdad[1] [iD],
and Elke Rundensteiner[1] [iD]

[1] Worcester Polytechnic Institute, 100 Institute Road, Worcester, MA, U.S.A.
{dcfox1,japrominski,aprajapati,khhaddigan,gbarbosa,swagh,cngan,
femdad,rundenst}@wpi.edu
[2] BAE Systems, 600 District Avenue, Burlington, MA, U.S.A.
{adam.nolan,daniel.zwillinger}@baesystems.us

Abstract. This paper presents a comprehensive exploration of determining optimal pixel resolutions for object detection in satellite imagery through a class-specific approach. Object detection in satellite imagery, critical for applications such as urban planning, environmental monitoring, and military surveillance, poses unique challenges due to high image resolutions, small object sizes, and computational demands. We propose a reusable pipeline designed to automate the discovery of the "knee point" on resolution-performance curves, achieving a balance between detection accuracy and computational efficiency. The pipeline integrates modules for data preprocessing, model fine-tuning, performance evaluation, and automated report generation. Utilizing the xView1 dataset and the YOLOv8 object detection model, we systematically analyze resolution images across 48 moveable object classes. Our findings show that lower-resolution images can yield competitive performance, significantly reducing resource demands, especially among object classes that perform well at high resolutions. This bridges existing research gaps while emphasizing modularity, efficiency, and usability. On average, across all object classes considered in this pipeline run the designated knee point presented a 57% reduction in pixel data with an 80% retention of the highest detection performance achieved at any resolution. Further, if we narrow our scope down to the top 15 performing classes, we find that the designated knee point presents a 71% reduction in pixel data, enabling a ground coverage area 3.4 times larger than what is achievable at the highest resolution, while retaining 88% of the detection performance and maintaining the same image dimensions and hardware capabilities.

Keywords: Satellite Imagery · Object Detection · Optimal Resolution · YOLO · Resolution-Performance Curve · Ground Sample Distance · Class-Specific Detection · Image Preprocessing · Machine Learning Pipeline · Computational Efficiency

F. Marcelloni et al. (Eds.): IJCCI 2025, CCIS 2829, pp. 485–507, 2026.
https://doi.org/10.1007/978-3-032-15638-9_28

1 Introduction

Object detection in satellite imagery identifies objects like buildings and vehicles in high-resolution images, essential for applications such as urban planning, disaster response, and military surveillance [1–5]. Satellite imagery presents challenges due to high resolutions, small object sizes, and the immense volume of data required to cover vast areas [6]. The resolution of image data plays an integral role in the performance of object detection tasks, particularly in satellite imagery. Because satellites capture images from a far distance, imagery is typically high-dimensional, where objects of interest, such as vehicles, may only occupy a few pixels within the larger image space [7]. Generally, higher-resolution images capture fine details, ensuring the ability to detect small objects such as vehicles or buildings. Although high data resolution improves detection, it also increases computational costs and is not always necessary for all tasks [8]. There are several factors in play that lead to higher costs. [9] showed that the use of vital resources increases with image input resolution, including the amount of computation (measured in floating-point operations, or FLOPs), inference time, and GPU memory, all of which lead to higher costs. Furthermore, storage of high-resolution imagery will also increase costs. Therefore, a trade-off for higher resolution satellite imagery exists between improving detection and increasing cost.

Solutions have been proposed that would enhance low-resolution images, or images with high Ground Sampling Distance (GSD), measured as surface distance per pixel [10, 11]. Super-resolution techniques do improve detection accuracy, as demonstrated in [12]. [13] reviews super-resolution methods, emphasizing their critical role in addressing low-resolution challenges and enhancing object detection in remote sensing contexts. [14] shows that varying resolutions impact outcomes for different object classes, and lower-resolution imagery (higher GSD) can still maintain adequate detection performance for certain applications. Optimizing the trade-off between low GSD and high detection performance is key to efficient satellite image processing [15].

While research has explored the impact of data resolution on object detection [12–14], a critical gap exists in understanding the relationship between pixel resolution and detection performance at an object-class level for satellite imagery. Current methods treat pixel resolution as a "one-size-fits-all" parameter, defaulting to the highest resolution, which is inefficient and costly in terms of storage, computational resources, and processing time [16]. Figure 1 illustrates Image Resolution-Performance Curves (IRPC), which visualize how detection performance varies with increasing resolution. The optimal point, or knee point, represents the resolution beyond which increasing resolution yields diminishing returns. This point balances detection performance and computational efficiency. Different object classes (e.g., Cargo Plane, Bus, Crane Truck, Sailboat, Aircraft Hangar, Helipad, Pylon, and many others) may require varying resolutions (see Fig. 1) due to differences in size and visual features [17].

It is important, then, to understand the precise mathematical relationship for the accuracy-resolution trade-off. Is it linear or non-linear? How low can the resolution go before detection accuracy suffers too much? The primary goal of our work is to determine the lowest data resolution necessary for satisfactory object detection results for a given object class, an inflection point for which increasing resolution (decreasing GSD) yields diminishing returns.

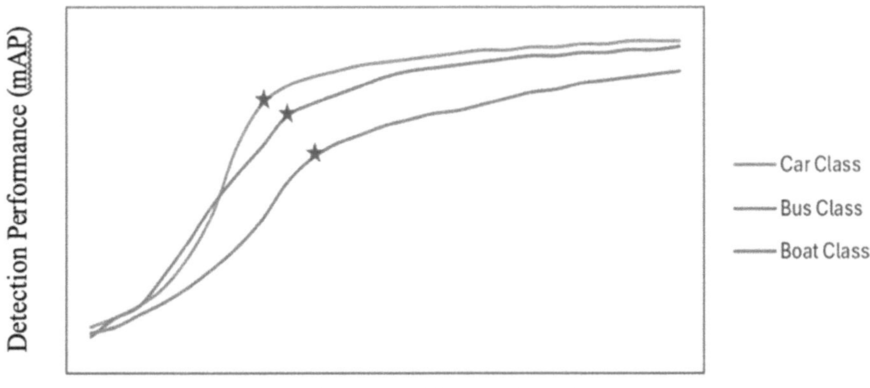

Fig. 1. Theoretical IRPC curves.

To select an appropriate model architecture for object detection in satellite imagery, we investigated, via literature search, several CNN-based architectures, including Faster R-CNN [18], Swin Transformer [19], RHINO [20], R-FCN [21], SSD [22], and YOLO [23]. We chose YOLO [24] for our pipeline because of its strength in handling small object detection and being variable across object domains made it a good candidate for our pipeline. YOLO is computationally fast and able achieve high accuracy even on small objects [25–27]. Further, YOLO is extensively documented, and its package has an API which is easy to use and understand. It provides several pretrained model variants, giving us increased flexibility during development [28]. Overall, YOLO aids in the ease of development for our specific use case while also balancing detection performance and computational efficiency [29]. It is good to keep in mind that the goal of our work is to discover the optimal resolution and not necessarily the optimal algorithm and hyperparameter combination. However, we do feel that it is important to achieve a very high mean Average Precision (mAP), the area under the precision-recall curve, because low precision might affect the determination of the optimal resolution. Our key technical contributions are five-fold:

1. **Automated, Modular, and Reusable Pipeline.** Our pipeline is extensible to an image dataset of any format, size, and resolution. Further, ground truth bounding-box annotation is in a standard GeoJSON format and requires at most a simple script to convert to this format, if necessary. Finally, different architectures only require code changes to conform to an API.
2. **Pre-Processing and Fine-Tuning Module.** The preprocessing module converts the image data and ground truth labels into a dataset that meets the model's architectural input specification. The fine-tuning module generates a model trained on the input dataset, using the frozen backbone of the architecture.

3. **Knee Discovery Module.** The knee discovery module evaluates object detection performance over a configured set of resolutions, and then calculates the inflection point (knee) of the resulting curve.
4. **Report Generation.** The report generation module automates the generation of a report, enabling immediate analysis of the optimal resolution.
5. **Experimental Results and Discussion.** To demonstrate our framework's capabilities, we conduct our own experiments using the xView1 dataset of satellite images as a basis for determining the optimal resolution for all moveable object classes. On average, across all 48 object classes considered in this pipeline run, the designated knee point presented a 57% reduction in pixel data with 80% retention of the highest detection performance achieved at any resolution.

This paper is organized into seven sections. Section 2 describes the pipeline that implements our framework at a high level. Section 3 describes how the pipeline performs preprocessing and fine-tuning. Section 4 describes the process of how the pipeline discovers the knee point for a particular object class. Section 5 describes how the pipeline generates a report, complete with tables and graphs. Section 6 discusses our experimental results. And Sect. 7 discusses our conclusions and suggests future work. Note that our source code is free to use and publicly available at: https://github.com/johnprom/bae-systems-gqp-24.

2 Proposed Pipeline

To enable a configurable and efficient workflow capable of handling unseen datasets with their own target object classes, we design a modular pipeline framework shown in Fig. 2 that enables isolated execution of components and flexible configuration.

The pipeline consists of four modules: Preprocessing, Fine-tuning, Knee Discovery, and Re-port Generation, which must reside on the same filesystem. These modules can be executed in order or separately in serial (but not in parallel). The pipeline's behavior and module-specific parameters are controlled by a main YAML configuration filed. The green chevrons in Fig. 2 indicate execution flow, while thin purple arrows represent input from the filesystem, and thin red arrows represent output from the filesystem. Each module can be debugged or iteratively improved without disrupting the rest of the pipeline, ensuring scalability for diverse object detection tasks.

The Preprocessing module fetches the raw image files and label data file from a specified directory, targeting a set of configurable target classes, and augments images and annotations from the dataset to meet the input requirements of the model architecture. Specifically, each image is divided into tiles that match the model's input size, with a configurable stride. The pre-processing module then splits those images into train and test sets and writes the categorized and preprocessed baseline training and test datasets to the filesystem.

The Fine-Tuning module is essential because the model must be trained on the input dataset. The Fine-Tuning module fetches the preprocessed baseline training dataset from the filesystem. Next, the pre-trained model does this by loading the pre-trained weights, performing additional training using the baseline preprocessed training dataset, and saving the resultant fine-tuned model weights to a file.

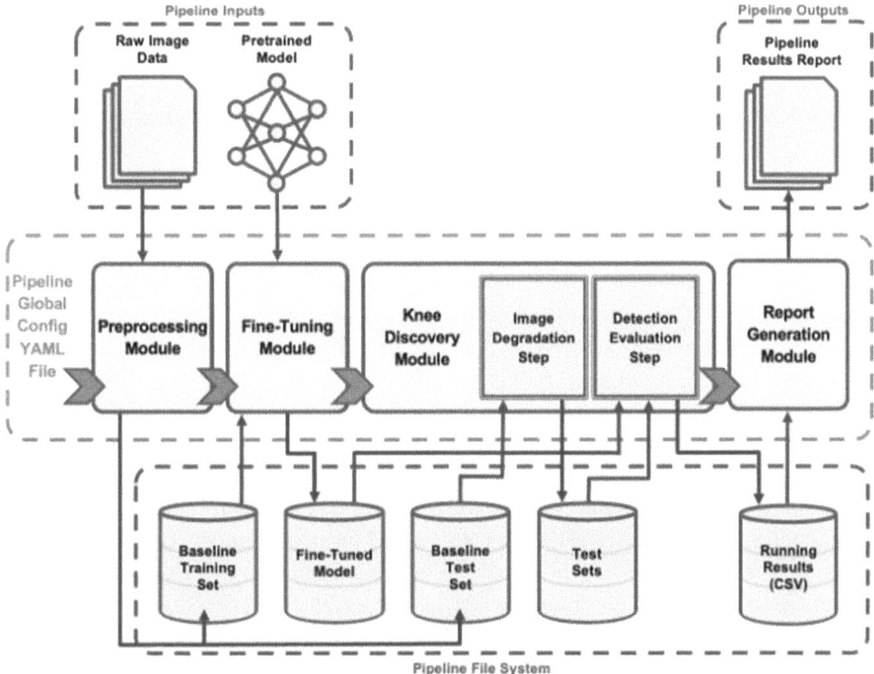

Fig. 2. High-Level Pipeline Design.

The Knee Discovery module generates data points along with the IRPC of data resolution versus detention performance. First, the Image Degradation Step subroutine fetches the baseline test dataset from the filesystem, generates a degraded copy of the dataset with a lower effective resolution, and then stores that into a directory of test sets. The Detection Evaluation Step sub-routine then runs evaluation on the fine-tuned model. The purpose for this is to automate evaluation of the trade-off between image resolution and model performance. The knee of the curve is calculated with the Piecewise Linear Fit (PWLF) method [30, 31]. The performance evaluation results, along with the calculated knee, are written to CSV file.

The Report Generation module fetches the results CSV file generated by the Knee Discovery module. The results are compiled and packaged into a PDF report with relevant metrics and visualizations to effectively convey the results. Specifically, a table of GSD and mAP at the knee for each object class is provided, along with a table of the hyperparameter values used to calculate the results. Then graphs of the IRPC for each object class are provided, condensed into a configurable number of object classes IRPC for each graph. Then the raw data for each IRPC is displayed. This report is then saved to a PDF file in the filesystem.

3   Preprocessing and Fine-Tuning

The preprocessing module transforms raw image data, with bounding box annotation, into a baseline preprocessed dataset, meeting YOLO's input specifications and enabling further degradation and evaluation steps.

3.1   xView1 Dataset

Figure 3 shows a sample satellite image from the xView1 dataset [32]. This dataset is one of the largest publicly available satellite imagery datasets used for object detection. It includes a variety of 60 object classes across many categories, including various aircraft, passenger vehicles, buildings, trucks, railway vehicles, maritime vessels, engineering vessels, and more. Each structure is annotated (inside a GeoJSON file) with a bounding box and a class label, making the dataset ideal for detection tasks. Table 1 shows a very small sample (four bounding boxes) of the annotated bounding box data, displayed in tabular format. The parameters are defined as follows: Feature Id is a unique identifier of the bounding box; Image Id is the filename of the raw input image; Type Id is a unique identifier of the object class which matches those specified in the YAML configuration file; and Bounding Box Coordinates are the pixel coordinates of the bounding box relative to the top and left sides of the image. For example, for the first bounding box (first row of this table), where Feature Id = 879295, the name of the image is "1104.tif", and the type identifier is 63. In the final column, the left edge of the bounding box (x1) is 857 pixels from the left side of the image, the top edge (y1) is 292 pixels from the top side of the image, the right edge (x2) is 1074 pixels from the left side of the image, and the bot-tom edge (y2) is 511 pixels from the top side of the image.

The dataset images vary between 2000 and 4000 pixels each non-square side, with a GSD of 0.3 m per pixel [33].

Fig. 3. A sample satellite image from the xView1 dataset.

Table 1. A sample of xview1 annotated bound-box data translated into tabular format.

Feature Id	Image Id	Type Id	Bounding Box Coordinates
879295	"1104.tif"	63	"857, 292, 1074, 511"
374410	"2355.tif"	73	"2712, 1145, 2746, 1177"
394393	"2355.tif"	73	"2720, 2233, 2760, 2288"
446607	"2356.tif"	18	"1880, 2529, 1897, 2544"

3.2 Preprocessing

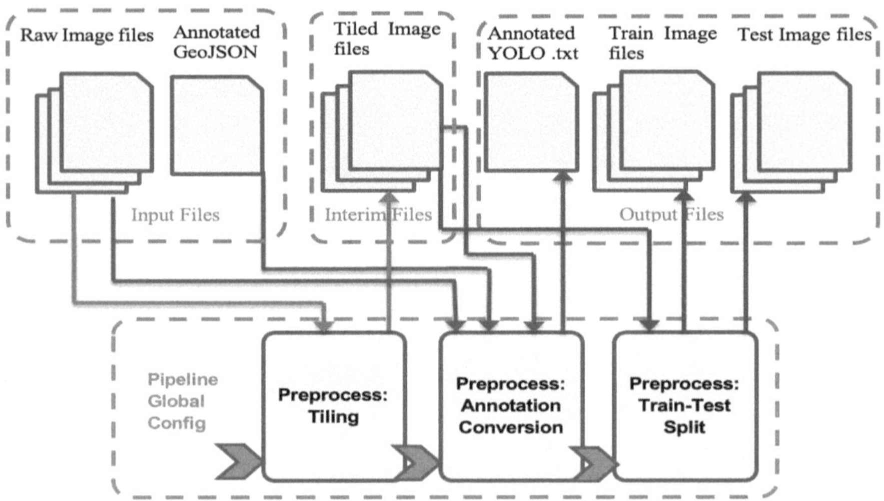

Fig. 4. Preprocessing Module Design. (Color figure Online)

The Preprocessing Module's design, depicted in Fig. 4, divides its operations into three steps: Tiling, Annotation Conversion, and Train-Test Split. Note that in Fig. 4, the green chevrons indicate execution flow, the blue arrows are Tiling I/O, the red arrows are Annotation Conversion I/O, and the purple arrows are Train-Test Split I/O.

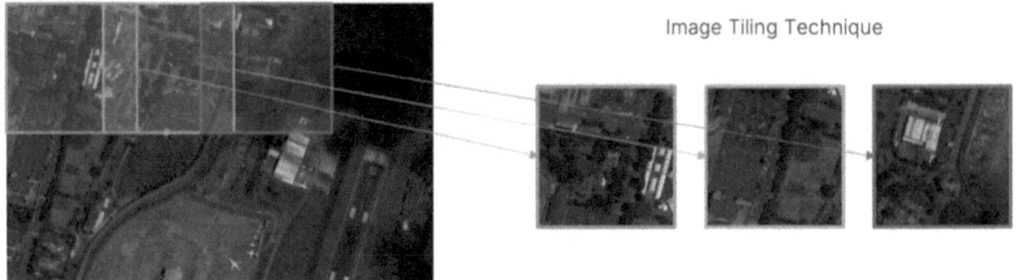

Fig. 5. Image Tiling Technique.

In the Tiling step (an example of which is displayed in Fig. 5), each raw image file is divided into tiles of configurable size (default is 640×640) using a sliding window with a defined stride (default is 100 pixels), producing overlapping regions to enhance small object detection. Bounding boxes are adjusted for each tile. Tiles without labeled objects are discarded, and edge tiles that lack required dimensions are dropped, resulting in ~2% data loss for images around 3000×3000 pixels. Partial bounding boxes near tile edges are excluded to avoid inaccuracies [34, 35].

In the Annotation Conversion step [36], the xView1 dataset annotation files converted to a YOLO annotation text file. While the xView1 annotation file has its bounding-box boundaries encoded by pixel number, in a GeoJSON-formatted file, YOLO has its bounding box boundaries encoded by the fractional location (from 0 to 1) within the image file's dimensions. The dataset is filtered to retain bounding boxes only for the configured target object classes (e.g., 'Car', 'Bus', and 'Plane'), and the YOLO data configuration file reflects these changes. Using a subset of available object classes reduces object bloat and shortens training times. Overlapping tiles may duplicate objects, which can cause training imbalances but ensure fewer objects are missed. If we choose a tile size consistent with the input size of our chosen model, GSD remains consistent (0.3 m^2/pixel), preserving the aspect ratio and avoiding distortions.

Let us look at an example of the annotation conversion done for the bounding box in the first row of Table 1. The bounding box coordinates are listed as "857, 292, 1074, 511", which translates to x1 = 857, y1 = 292, x2 = 1074, and y2 = 511. Furthermore, we know from the image itself, "1104.tif", that the height = 3215 and the width = 3195. First, the image is broken up into tiles using the following formulae:

$$X = number\ of\ tiles\ in\ x\ dimension = \frac{(image\ width - stride)}{(tile\ size - stride)}, rounded\ up \quad (1)$$

$$Y = number\ of\ tiles\ in\ y\ dimension = \frac{(image\ width - stride)}{(tile\ size - stride)}, rounded\ up \quad (2)$$

Then each tile's coordinates are set by the following formulae:

$$x_{start} = \max(X \times (tile\ size - stride), 0) \quad (3)$$

$$x_{end} = \min(x_{start} + tile\ size,\ image\ width) \quad (4)$$

$$y_{start} = \max(Y \times (tile\ size - stride), 0) \quad (5)$$

$$y_{end} = \min(y_{start} + tile\ size,\ image\ height) \quad (6)$$

By default, the tile size is 640 and the stride is 100. So, we conclude, for example, that the tile at (X = 1, Y = 0) has the coordinates which contain our example bounding box. We know this because we know the values of all the numbers on the right-hand side of these equations. If we plug them in, we get x_{start} = 540, x_{end} = 1180, y_{start} = 0, and y_{end} = 640. And x1 = 857, y1 = 292, x2 = 1074, and y2 = 511 all fall within that range. Note that bounding boxes that are not fully in a tile are discarded.

By default, the algorithm is YOLO. Note that if the format required by the algorithm is not identical to the YOLO format, then code will need to be added to convert to the algorithm's format. The current code for that conversion is in the method XViewTiler. convert_bbox_to_yolo_format(). This is the routine that must be modified if the algorithm is changed to something other than YOLO.

The specific YOLO format requires the specification of the class label, the center of the bounding box, plus the width and height. All those values are expressed as floating-point numbers within the range of 0 to 1, scaled to the size of the image in which the bounding box is in. YOLO requires the specification of each bounding box be in a flat text file, with each line entry representing one bounding box. Therefore, there will be the same number of lines in this file as there are bounding boxes in the image. The format of each line is: $\mathtt{bbox_{class}\,x_{center}\,y_{center}\,bbox_{width}\,bbox_{height}}$, where $\mathtt{bbox_{class}}$ is the identifier of the object class; $\mathtt{x_{center}}$ and $\mathtt{y_{center}}$ are the x and y coordinates, respectively, of the center of the bounding box; and $\mathtt{bbox_{width}}$ and $\mathtt{bbox_{height}}$ are the width and height, respectively, of the bounding box. These YOLO values are calculated using the following formulae:

$$x_{center} = \left(\frac{x_1 + x_2 - 2 \times x_{start}}{2 \times \textit{tile size}} \right) \tag{7}$$

$$y_{center} = \left(\frac{y_1 + y_2 - 2 \times y_{start}}{2 \times \textit{tile size}} \right) \tag{8}$$

$$bbox_{width} = \frac{x_2 - x_1}{\textit{tile size}} \tag{9}$$

$$bbox_{height} = \frac{y_2 - y_1}{\textit{tile size}} \tag{10}$$

Now let us go through our example, using the bounding box defined on the first line of Table 1. First, we note the Class Id is 63. Then we calculate $\mathtt{x_{center}}$, $\mathtt{y_{center}}$, $\mathtt{bbox_{width}}$, and $\mathtt{bbox_{height}}$ using (7–10) with the numbers that we specify above, specifically x_{start} = 540, y_{start} = 0, x_1 = 857, y_1 = 292, x_2 = 1074, and y_2 = 511. After applying our formulae, we obtain $\mathtt{x_{center}}$ = 0.665, $\mathtt{y_{center}}$ = 0.627, $\mathtt{bbox_{width}}$ = 0.170, and $\mathtt{bbox_{height}}$ = 0.342. These values can then be represented in the YOLO input file as the line: $\mathtt{63\ 0.665\ 0.627\ 0.170\ 0.342}$. In the Train-Test Split step, the tiled image files are split pseudo-randomly into training and test sets (default is an 80:20 ratio), defined in the pipeline configuration. Splitting occurs post-augmentation and tiling, ensuring deterministic and shuffled subsets. This improves feature diversity, reduces overfitting, and enhances generalization [38].

3.3 Fine-Tuning

The fine-tuning module ensures efficient adaptation of YOLO models to satellite imagery datasets. As YOLO's architecture comes with weights pre-trained on the COCO dataset [39], which contains no satellite images but instead frontal-view images, the fine-tuning module re-trains on the xView1 dataset with the backbone of the YOLO-supplied pre-trained weights frozen. This fine-tuning, then, tailors the model to the domain-specific data.

During fine-tuning, any model hyperparameter can be defined, including the number of training epochs, the batch size, and which specific layers of the model architecture's parameters are frozen that directly influence the model's performance. One can specify the hyperparameter grid with a list of values for any hyperparameter that YOLO handles, should the user want to bolster performance. The default configuration is the one set of hyperparameters described in Table 2.

To prevent overfitting and manage class imbalance, we employ careful hyperparameter tuning. This involves using a design of experiments (DOE) approach, specifically a full factorial design method, to iterate through combinations of hyperparameters defined in the configuration YAML file. Each combination that is executed is evaluated to determine its impact on the model's mAP. While this approach is computationally expensive, it provides an understanding of how hyperparameters affect model performance.

Table 2. Best performing hyperparameters for multi-object detection

Hyperparameter	Value	Hyperparameter	Value
Image Preprocessing Method	Image Tiling	YOLOv8m: imges	640
Tile Size	640	YOLOv8m: epochs	100
Tile Stride Size	100	YOLOv8m: batch	16
Learning Model	YOLOv8m	YOLOv8m: freeze	[0..14]
YOLOv8m: cls	1.5		

4 Knee Discovery Module

The Knee Discovery Module serves as a critical component within the pipeline, enabling a systematic evaluation of the relationship between image resolution and model performance. It has two functions shown in Fig. 2: image degradation and object detection evaluation, and they are applied in that order. The image degradation function generates test sets that simulate varying levels of image degradation, thereby establishing a controlled experimental framework. It creates deteriorated replicas of the original dataset. By lowering the resolution of the image and then up-sampling it back to a standard format, which produces a controlled loss of detail without changing the image's proportions [41]. Each configured evaluated resolution is expressed in terms of what we call a degradation factor. The motivation for resizing the lower resolution images back up to the original resolution is that it allows the usage of the pre-trained weights of the model, which in the case of YOLO are obtained on 640×640 images. Note that the original image size can be configurable in different models and can be trained on different resolutions.

The set of degradation factors that is used for object detection is configurable. By default, they are an evenly spaced 20 data-point range from 0.05 to 1.00. The degradation factor is calculated as the square root of the fraction of pixels in the lower resolution image (prior to resizing back up) compared with the number of pixels in the original resolution. In this way, if a 640×640 image is degraded to 320×320, and then resized to 640×640, its degradation factor is 0.5, representing a 1/2 reduction in the number of pixels on each side. The IRPCs generated by the report generator convert these degradation factors to GSD values, based on the original GSD of each image. The degradation factor is calculated by (13).

The object detection evaluation function uses these data to create IRPCs, which plot predicted metrics across a range of resolution settings. To determine the "knee point," or the resolution at which performance improvements start to plateau, it uses a robust detection technique. This makes it easier to precisely calibrate the model resolution requirements to maximize resource usage and performance. For every degraded dataset, the module iterates over predefined resolution settings, generating the mAP or other domain-relevant metrics.

After being stored to a CSV file, these outcomes are combined to create IRPCs. Plotting performance measurements (mAP) against appropriate resolutions allows IRPCs to identify distinctive trends. We discover that performance generally falls as image resolution lowers, first gradually, and then, at some point, precipitously. The module then calculates this knee point at the transition from gradual performance loss to precipitous loss.

We assert that this knee point is a good approximation of the best resolution-performance trade-off, with further resolution enhancements yielding diminishing returns in accuracy. To automate knee point detection, a reliable and consistent computational method is required. Following significant investigation with several strategies, including heuristic and curvature-based approaches, the Piecewise Linear Fit (PWLF) method is observed to be the most dependable option. Based on the implementation, the technique progresses as follows.

To locate the knee, the IRPC data, which includes image degradation factors and performance measures like mAP, is smoothed using a spline-based method. This pre-processing step decreases the impact of noise and guarantees that the next analysis focuses on the curve's broad trend rather than specific perturbations. Once the data has been smoothed, the PWLF method fits two linear segments to the IRPC. The mathematical formulation for piecewise linear fitting involves defining the curve $y(x)$ in two segments:

$$y(x) = \begin{cases} m_1 x + b_1, x \le x_k \\ m_2 x + b_2, x > x_k \end{cases} \tag{11}$$

where m_1 and m_2 are the slopes of the two segments, b_1 and b_2 are the corresponding intercepts, and x_k is the knee point. The optimal x_k is found by minimizing the sum of the residual sum of squares (RSS) for both segments:

$$RSS = \sum_{i=1}^{n_1} (y_i - (m_1 x_i + b_1))^2 + \sum_{j=1}^{n_2} \left(y_j - \left(m_2 x_j + b_2 \right) \right)^2 \tag{12}$$

where n_1 is the number of data points in the first segment, and n_2 is the number of data points in the second segment. The algorithm iteratively adjusts x_k (and thus n_1 and n_2) to achieve the minimal RSS, ensuring precise segmentation of the curve and accurate knee point identification. The first segment corresponds to the curve's initial, sharply rising phase, when increased resolution results in significant increases in performance. The second segment represents the area where performance increases begin to plateau.

The knee point is located at the intersection of these two linear segments, rounded to the nearest data point on the x-axis, known as the "change point," and indicates a good approximation of the point at which increased resolution no longer provides significant gains in detection accuracy.

This automated knee discovery stage is critical for large-scale applications because it simplifies the selection of an ideal resolution range, which then influences resource allocation, model configuration changes, and deployment methods. These findings are combined into a comprehensive PDF report by the report generator discussed in Sect. 5. Figure 6 in Sect. 6 illustrates an example of results obtained from the knee discovery module. Although the x-axis is expressed in GSD, the distribution along the x-axis is based on the inverse of the resolution, and therefore non-linear. This is because the original graph is drawn with the degradation factor in mind. You can see that each of twenty data points are spread evenly apart and 0.05 intervals of degradation factor, from 0.05 to 1.00.

Looking at the IRPC for the cargo plane (our best performing class), you will see that the first data point on the left (corresponding to a degradation factor of 0.05) shows a mAP of 0.000. Traveling from there to the right, you see that the knee is at the fourth data point, corresponding to a degradation factor of 0.2. Since the original image is 640×640 in resolution, the resolution at a degradation is easily calculated to be 32×32 and 160×160 for degradation factors of 0.05 and 0.2, respectively. This means that the optimal resolution for the cargo plane is a degradation factor of 0.2, which is 160×160. The relationship between the degradation factor and GSD is illustrated in Sect. 6.

5   Report Generation

The Report Generation Module is responsible for producing a structured report that summarizes pipeline results, including the optimal knee point identified on the IRPC. The module reads the results CSV file produced by the Knee Discovery Module. A sample of the CSV file is included here. Note that the header and each object class occupy one line each in the file:

```
object_name,original_resolution_width,
original_resolution_height,
effective_resolution_width,effective_resolution_height,mAP,
degradation_factor,GSD,pixels_on_target,knee
:
:
Fixed-wing Aircraft,
640,640,160,160,0.1784444262295082,0.25,1.2,221,False
Small Aircraft,
640,640,160,160,0.16466172371812984,0.25,1.2,100,False
Cargo Plane,
640,640,160,160,0.5796717880874455,0.25,1.2,914,False
Helicopter,640,640,160,160,0.0,0.25,1.2,121,False
```

After reading this CSV file, the module generates a comprehensive report in PDF format. The report includes tables, text, and graphical visualizations created using Python libraries such as Matplotlib, FPDF, and PyPDF2. These outputs are merged into a final document, ensuring clarity and coherence. The parameters that can configure for the report generator in the **pipeline_config.yaml** file, along with the default settings, are shown as follows:

```
output_subdir: "reports"
clean_subdir: false
report_filename: "generated_report.pdf"
curves_per_graph: 3
display_labels:
  -   "Fixed-wing Aircraft"
  -   "Small Aircraft"
  -   "Cargo Plane"
  -   "Helicopter"
```

Results and reports are stored in directories specified in the configuration file parameter **output_subdir**, which specifies the subdirectory under the output directory (which defaults to "output"). Thus, with the setting above, the default location of the report is "output/reports". The module organizes reports into timestamped subdirectories, ensuring version control and easy reference. The timestamp is equal to the time of the most recent modification to the results file. Note that the generated report PDF file, the IRPC results CSV file, and the hyperparameter report CSV file, are all placed in this timestamped report directory. The parameter **clean_subdir** specifies if the entire output directory should be deleted prior to the run. This parameter is disabled by default so that the user can take advantage of the timestamped reports, if there is more than one. The parameter **report_filename** customizes the name of the report. The parameter **curves_per_graph** customizes how many object class curves are drawn in each mAP-GSD graph. An example of a graph with **curves_per_graph** = 1 is shown in Fig. 6. If the parameter **display_labels** is specified, the report generator module can be customized to focus on object classes of interest to the user, omitting unrelated classes. This customization allows the report to address specific needs, especially when multiple object classes are included during training to improve generalization. In the above example, results only for the four specified object classes (**Fixed-wing Aircraft**, **Small Aircraft**, **Cargo Plane**, and **Helicopter**) will be shown in the report. By automating report generation, this module streamlines the process of summarizing and visualizing pipeline performance, making it accessible for both technical and non-technical stakeholders.

6   Experimental Results and Discussion

We choose to execute the full pipeline with the 48 moveable object classes, out of the total 60 object classes in the dataset, along with the hyperparameters for YOLOv8 specified in Table 2. The results from that experiment are presented as follows. Note that there is a myriad of combinations of target labels and hyperparameters that a user can run, depending on the object being detected. First, we can analyze the detection performance data generated by the knee discovery module for a sample object class. Table 3 details the detection performance at varying degraded image resolutions for the Cargo Plane object class. Each row houses detection evaluation metrics for a specific degraded resolution, acting as a datapoint along the generated IRP curve.

In Table 3, the **Width** and **Height** columns describe the input image dimensions in pixels. The **Effective Width** and **Effective Height** columns describe the width and height, in pixels, that the image is degraded to prior to resizing back to its original dimensions. The **mAP** column is the mean average precision (at IoU 0.5:0.95 returned by YOLO validation) achieved at that resolution for the Cargo Plane class. The **GSD** column describes the effective GSD reflected after down-sampling and up-sampling the image. The **Degradation Factor** is a measure of the amount of unidimensional down-sampling of the image before subsequent up-sampling. The degradation factor is calculated using the following formula:

$$degradation\,factor = \sqrt{\frac{effective\ width \times effective\ height}{width \times height}} \tag{13}$$

The **Pixels On Bounding Box** column is an estimate of the number of pixels contained in a bounding box for a given class at that resolution. Note that the Pixels On Bounding Box metric is class dependent, as some object classes require larger or smaller bounding boxes. Finally, the **Knee** column indicates whether a given datapoint is determined to be the optimal knee point by the Knee Discovery Module. We can, using Table 3, calculate the optimal spatial resolution for the Cargo Plane class, using the following formula:

$$Optimal\ Resolution = \frac{Minimum\ GSD}{Degradation\ Factor\ at\ Knee} \tag{14}$$

Table 3. Sample detection performance data for the Cargo Plane class.

Width	Height	Effective Width	Effective Height	Degradation Factor	mAP	GSD (meters/pixel)	Pixels On Bounding Box	Knee
640	640	32	32	0.05	0.000	6.00	37	No
640	640	64	64	0.10	0.316	3.00	147	No
640	640	96	96	0.15	0.425	2.00	329	No
640	640	128	128	0.20	0.521	1.50	585	Yes
640	640	160	160	0.25	0.580	1.20	914	No
640	640	192	192	0.30	0.590	1.00	1315	No
640	640	224	224	0.35	0.588	0.86	1790	No
640	640	256	256	0.40	0.573	0.75	2338	No
640	640	288	288	0.45	0.560	0.67	2959	No
640	640	320	320	0.50	0.589	0.60	3653	No
640	640	352	352	0.55	0.613	0.55	4420	No
640	640	384	384	0.60	0.619	0.50	5260	No
640	640	416	416	0.65	0.624	0.46	6173	No
640	640	448	448	0.70	0.632	0.43	7159	No
640	640	480	480	0.75	0.641	0.40	8218	No

(continued)

Table 3. (*continued*)

Width	Height	Effective Width	Effective Height	Degradation Factor	mAP	GSD (meters/pixel)	Pixels On Bounding Box	Knee
640	640	512	512	0.80	0.620	0.38	9350	No
640	640	544	544	0.85	0.640	0.35	10556	No
640	640	576	576	0.90	0.645	0.33	11834	No
640	640	608	608	0.95	0.639	0.32	13185	No
640	640	640	640	1.00	0.651	0.30	14609	No

Table 3 shows the minimum GSD (or maximum resolution) is 0.30 m per pixel. And the corresponding degradation factor at the knee point (where Knee = Yes) is 0.20. The optimal spatial resolution is then calculated using (15) as 1.5 m per pixel, which represents the optimal value for reliable Cargo Plane detection. Beyond this point, further degradation in resolutions results in a more pronounced decline in detection accuracy, while less degradation in resolution results in a more modest improvement in detection accuracy. Next, we can analyze some visual results found in the report file generated by the pipeline. Figure 6 displays the Image Resolution-Performance Curve, generated by the Report Generator, for the Cargo Plane object class referenced in Table 3. The y-axis shows the detection performance as mAP values. The x-axis shows the resolution as GSD. In this case, an increase in resolution correlates to a smaller number when this is expressed as GSD. For each object class, the knee point signifying the optimal trade-off between detection performance and data resolution is marked by a triangle. In the curve, we observe the expected relationship of an increase in resolution leading to an increase in detection performance. However, we also notice this relationship is not linear as objects tend to make most of their gains in detection performance far before reaching the maximum resolution. For example, the curve shown is an IRPC for the Cargo Plane class. On this curve, we can see that there is a significant reduction in the increase of performance as we continue to increase the resolution. The pipeline designates a knee value at a GSD of around 1.5 m, shown by triangle marker. At this resolution, we still achieve more than 80% of the detection performance compared to the highest mAP achieved at any resolution.

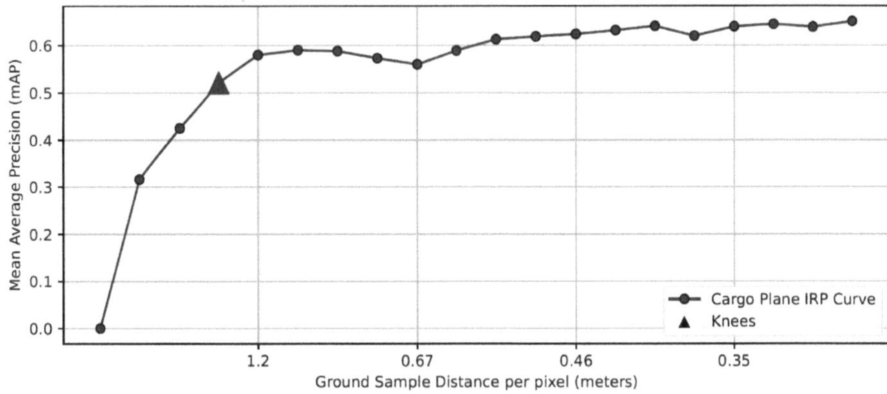

Fig. 6. Sample IRPC from the generated report file.

Now we look at the data for the 15 best-performing (highest mAP at knee) of the 48 moveable object classes for which we execute our pipeline. First, Table 4 shows various measurements of resolution for each object class, i.e., effective width in pixels, effective height in pixels, the degradation factor, the effective number of pixels in the bounding box, and the GSD in meters per pixel, at their respective knee points. Remember that, for each object class, the width and height of each tiled image is 640 pixels, and the maximum resolution (minimum GSD) is 0.3 m per pixel. Note that for four of the classes, a knee point could not be determined. They are included here with the knee assumed to be at the maximum resolution.

The highest optimal GSD found for any object class in this pipeline run is for the Cargo Plane class at a resolution of 1.5 m per pixel. At this resolution, the average cargo plane would require around 585 pixels contained in its bounding box. Assuming that more than half of the pixels contained in the surrounding box are context, we can estimate that this would translate to around 250–300 pixels-on-target required for object detection of cargo planes. Across all 48 object classes in this pipeline run, we find the median optimal GSD to fall around 0.55 m per pixel. Taking into account the average scale of each object class, the median pixels on a bounding box at this resolution fell around 585 pixels.

Table 4. Measurements of resolution for the top 15 performing object classes at their respective knee points.

Object Class	Effective Width (pixels)	Effective Height (pixels)	Degradation Factor	Pixels On Bounding Box	GSD (meters/pixel)
Cargo Plane	128	128	0.20	585	1.50
Straddle Carrier	192	192	0.30	1239	1.00
Cargo Car	224	224	0.35	133	0.86
Passenger Car	224	224	0.35	209	0.86

(continued)

Table 4. (*continued*)

Object Class	Effective Width (pixels)	Effective Height (pixels)	Degradation Factor	Pixels On Bounding Box	GSD (meters/pixel)
Truck w/Box	288	288	0.45	258	0.67
Fixed-wing Aircraft	320	320	0.50	881	0.60
Sailboat	320	320	0.50	205	0.60
Excavator	352	352	0.55	230	0.55
Tugboat	352	352	0.55	1379	0.55
Small Car	352	352	0.55	59	0.55
Yacht	384	384	0.60	1154	0.50
Ferry	384	384	0.60	3714	0.50
Bus	384	384	0.60	185	0.50
Front loader/Bulldozer	480	480	0.75	308	0.40
Small Aircraft	544	544	0.85	1147	0.35

Table 5. GSD (resolution), mAP, and calculated metrics at knee point for each of the top performing object class.

Object Class	GSD (meters/pixel) at Knee	Maximum mAP	mAP at Knee	Detection Performance Retained at Knee	Data Retained at Knee	Satellite coverage multiplier at Knee
Cargo Plane	1.50	0.651	0.521	0.800	0.040	25.0
Straddle Carrier	1.00	0.597	0.398	0.667	0.090	11.1
Cargo Car	0.86	0.429	0.352	0.819	0.123	8.2
Passenger Car	0.86	0.420	0.364	0.865	0.123	8.2
Truck w/Box	0.67	0.300	0.280	0.933	0.203	4.9
Fixed-wing Aircraft	0.60	0.385	0.332	0.862	0.250	4.0
Sailboat	0.60	0.227	0.168	0.740	0.250	4.0
Excavator	0.55	0.358	0.294	0.823	0.303	3.3
Tugboat	0.55	0.269	0.268	0.998	0.303	3.3
Small Car	0.55	0.260	0.209	0.804	0.303	3.3
Yacht	0.50	0.356	0.341	0.959	0.360	2.8
Ferry	0.50	0.344	0.344	1.000	0.360	2.8

(*continued*)

Table 5. (continued)

Object Class	GSD (meters/pixel) at Knee	Maximum mAP	mAP at Knee	Detection Performance Retained at Knee	Data Retained at Knee	Satellite coverage multiplier at Knee
Bus	0.50	0.289	0.257	0.889	0.360	2.8
Front loader/Bulldozer	0.40	0.232	0.232	1.000	0.563	1.8
Small Aircraft	0.35	0.325	0.325	1.000	0.723	1.4

Next, we look at Table 5, which shows some calculated metrics at the knee point for each of the top 15 performing object classes. We show again, from Table 4, the GSD at the knee for each object class, as a point of reference. We also show the maximum mAP achieved across all resolutions, and the mAP at each knee. From that, we can calculate the retention in performance using the following formula:

$$Detection\ Performance\ Retained\ at\ Knee = \frac{Maximum\ mAP}{mAP\ at\ Knee} \tag{15}$$

Further, we can calculate the fraction of data retained at that knee point:

$$Data\ Retained\ at\ Knee = \left(\frac{Maximum\ GSD}{GSD\ at\ Knee}\right)^2 \tag{16}$$

And from that, we then observe that the amount of satellite coverage could be increased by a multiplier of the reciprocal of the data retained:

$$Possible\ Satellite\ Image\ Coverage\ Multiplier = \frac{1}{Data\ Retained\ at\ Knee} \tag{17}$$

In other words, if this optimization is leveraged during the acquisition of satellite imagery, we could potentially have satellite images that cover a much greater geographical area without requiring an increase in pixel dimensions or hardware capabilities. On average, across all 48 object classes considered in this pipeline run, the designated knee point presents a 57% reduction in pixel data with 80% retention of the highest detection performance achieved at any resolution. Further, if we narrow our scope down to the top 15 performing classes (in terms of mAP at the knee), we find that the designated knee point presents a 71% reduction in pixel data while retaining 88% of the detection performance. This is a significant optimization, offering substantial computational efficiency without a major compromise in detection accuracy. In context, reducing the required pixel data by 71%, which is a 29% retention of data, and applying (17) could enable a ground coverage area 3.4 times larger than what is achievable at the highest resolution, all while maintaining the same image dimensions and hardware capabilities. And if one is only concerned about Cargo Plane detection, Table 5 shows the potential increase in coverage to be a factor of 25.0. The detection performance at the reported optimal knee point can be visualized in Fig. 7. Here, we look at bounding box predictions for Cargo Plane that are made on data of varying resolutions.

0.3m GSD
14609 Pixels-on-Box

1.5m GSD (Knee-Point)
585 Pixels-on-Box

2.0m GSD
329 Pixels-on-Box

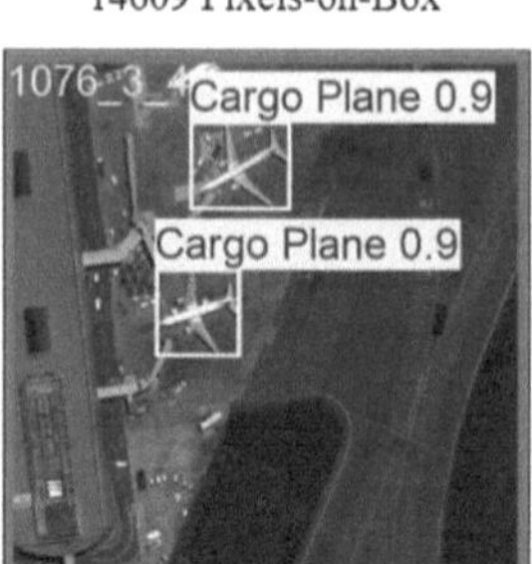

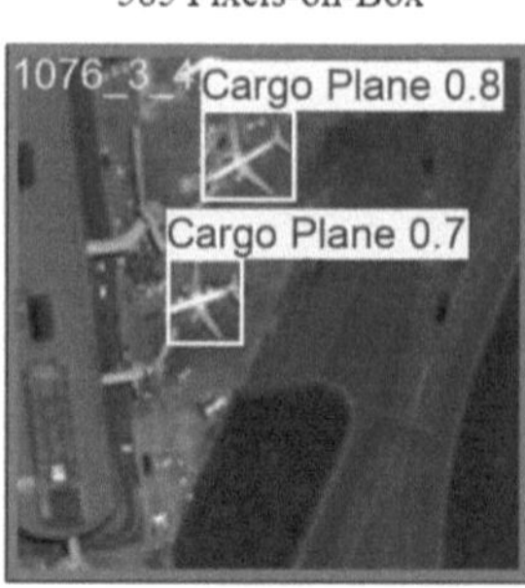

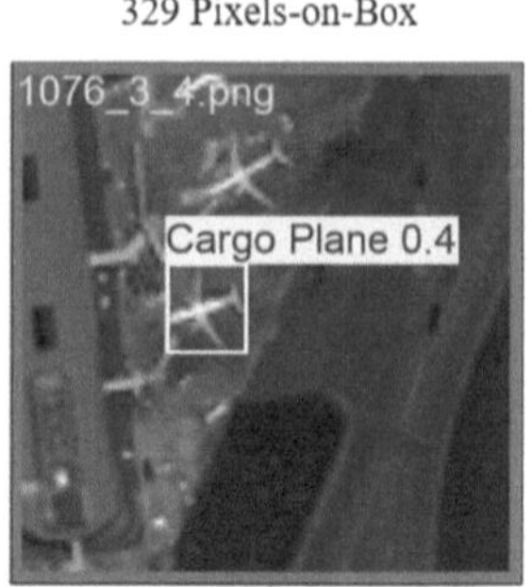

Fig. 7. Bounding box predictions showing detection performance for the Cargo Plane class at varying degraded resolutions.

At the knee point resolution of 1.5 m per pixel, we can achieve comparable detection performance to the significantly higher native resolution of 0.3 m per pixel. Conversely, as we start to exceed a GSD of around 1.5 m per pixel, the detection performance begins to suffer. Further, as we move past the knee point resolution to a GSD of 2 m per pixel, we are no longer able to reliably detect the cargo plane objects within the scene. This exemplifies the significance of a balanced trade-off between data resolution and detection performance.

There were some objects that had very poor object detection by YOLO, with a 640 × 640 tiling size, and the xView dataset. In those cases, knee points were difficult or impossible to determine by the pipeline. But in these cases, the object detection performance was not what we expected, such that the object detection performance would be unreliable. It may be possible to get decent results for those objects using a different tiling size, dataset, and/or object detection algorithm.

A quick note about algorithm performance: we select YOLO because of its relative speed, and it does not disappoint. Using one NVIDIA A100, a dataset with 4,089 640 × 640 tiles and 48 object classes takes approximately two to two and a half hours to run the full pipeline on WPI's high-performance computing environment.

7   Conclusion and Future Work

In conclusion, we successfully develop and implement a reusable pipeline for determining the optimal resolution for object detection in satellite imagery. The pipeline effectively automates the process of identifying the knee point in the resolution vs. performance curve, providing a balance between computational efficiency and detection accuracy. By fine-tuning the YOLOv8 model using the xView1 dataset and generating performance metrics across varying resolutions, we demonstrate that lower-resolution images can achieve comparable detection performance to higher-resolution counterparts while significantly reducing computational costs.

The results highlight the importance of selecting an appropriate resolution for object detection, as overly high resolutions may not always yield a corresponding increase in detection performance. Furthermore, the modular design of the pipeline ensures flexibility and scalability, allowing it to be adapted for future use with different datasets and object classes. We address the research gaps in determining optimal image resolutions for satellite-based object detection, providing both practical insights and a tool that can be applied to a range of detection tasks.

While the pipeline developed in this project provides a robust solution for optimal resolution detection, several areas for future improvement and exploration remain.

We chose PWLF based on observation of the IRPC developed by the pipeline, using the xView1 dataset (with the specific objects in it) that we chose. A choice of a different dataset, based on different domain requirements, may also suggest a different knee discovery algorithm, which can be changed through minor source code modifications. Furthermore, although the current knee detection method works well for most cases, incorporating advanced machine learning algorithms for knee detection could further improve the accuracy of the knee approximation, particularly for object classes with less distinct performance changes across resolutions.

The current implementation uses the xView1 dataset, which is focused on satellite imagery. Future work could involve testing the pipeline with other satellite imagery datasets, such as DOTA or DIOR. In addition, the pipeline could be further tested with a broader range of datasets, including other forms of high-resolution imagery, such as aerial or drone imagery, to ensure the pipeline's versatility across different contexts.

Given the importance of computational efficiency in real-time applications, additional optimization techniques could be explored, such as model compression or hardware-accelerated inference (e.g., using specialized hardware like TPUs or FPGAs). Although YOLOv8 is chosen for its speed and accuracy, future work could involve integrating other object detection architectures, such as Faster R-CNN or RHINO, into the pipeline to compare performance across models and identify the best architecture for specific use cases.

The pipeline could be extended by incorporating additional preprocessing techniques such as super-resolution and post-processing modules for tasks like multi-object tracking. This would enhance the pipeline's capabilities for more complex tasks and larger datasets.

Acknowledgments. Results in this paper were obtained in part using a high-performance computing system acquired through NSF MRI grant DMS-1337943 to WPI.

References

1. Radovic, M., Adarkwa, O., Wang, Q.: Object recognition in aerial images using convolutional neural networks. J. Imaging **3**(21) (2017)
2. Dial, G., Bowen, H., Gerlach, F., Grodecki, J., Oleszczuk, R.: IKONOS satellite, imagery, and products. Remote Sens. Environ. **88**(1–2), 23–36 (2003)
3. Holsten, S.: Global maritime surveillance with satellite-based AIS. In: OCEANS 2009-EUROPE, Bremen, Germany, pp. 1–4. IEEE (2009)
4. Kaack, L.H., Chen, G.H., Morgan, M.G.: Truck traffic monitoring with satellite images. In: Proceedings of the 2nd ACM SIGCAS Conference on Computing and Sustainable Societies (COMPASS 2019), Accra, Ghana, pp. 155–164. ACM (2019)

5. Sirmacek, B., Unsalan, C.: A probabilistic framework to detect buildings in aerial and satellite images. IEEE Trans. Geosci. Remote Sens. **49**(1), 211–221 (2011)
6. Mansour, A., Hussein, W.M., Said, E.: Small objects detection in satellite images using deep learning. In: Ninth international Conference on Intelligent Computing and Information Systems (ICICIS), Cairo, Egypt, pp. 86–91. ICICIS (2019)
7. Capderou, M.: Satellites: Orbits and missions. Springer, Paris (2005)
8. Anumol, C.S.: Advancements in CNN architectures for computer vision: a comprehensive review. In: 2023 Annual International Conference on Emerging Research Areas: International Conference on Intelligent Systems (AICERA/ICIS), Kanjirapally, India, pp. 1–7. ICIS (2023)
9. Bakhtiarnia, A., Zhang, Q., Iosifidis, A.: Efficient high-resolution deep learning: a survey. ACM Comput. Surv. **56**(7), 1–35 (2024)
10. Hood, J., Lyman, L., Richard, C.: Image processing techniques for digital orthophotoquad production. Photogrametric Eng. Remote Sens. **55**(9), 1323–1329 (1989)
11. Jacobsen, K.: High resolution satellite imaging systems-an overview. Photogrammetrie Fernerkundung Geoinformation **6** (2005)
12. Shermeyer, J., Etten, A.V.: The effects of super-resolution on object detection performance in satellite imagery. In: Proceedings of the IEEE/CVF Conference. On Computer Vision and Pattern Recognition Workshops, CVPR 2019 (2019)
13. Wang, Y., et al.: Remote sensing image super-resolution and object detection: benchmark and state of the art. Exp. Syst. Appl. **197**, 116793 (2022)
14. Yang, Y.: GSDDet: ground sample distance-guided object detection for remote sensing images. IEEE Trans. Geosci. Remote Sens. **62**, 1–12 (2024)
15. Adegun, A.A., Viriri, S., Tapamo, J.R.: Review of deep learning methods for remote sensing satellite images classification: experimental survey and comparative analysis. J. Big Data **10**(1), 93 (2023)
16. Bhil, K., et al.: Recent progress in object detection in satellite imagery: a review. In: Aurelia, S., Hiremath, S.S., Subramanian, K., Biswas, S.K. (eds.) International Conference on Sustainable Advanced Computing 2021. LNEE, vol. 840, pp. 209–218. Springer, Singapore (2022)
17. Ding, J., et al.: Object detection in aerial images: a large-scale benchmark and challenges. IEEE Trans. Pattern Anal. Mach. Intell. **44**(11), 7778–7796 (2022)
18. Ren, S., He, K., Girshick, R., Sun, J.: Faster R-CNN: towards real-time object detection with region proposal networks. IEEE Trans. Pattern Anal. Mach. Intell. **39**(6), 1137–1149 (2016)
19. Liu Z., et al.: Swin transformer: hierarchical vision transformer using shifted windows. In: Proceedings of the IEEE/CVF International Conference on Computer Vision (ICCV), Montreal, QC, Canada, pp. 10012–10022. IEEE (2021)
20. Lee, H., Song, M., Koo, J., Seo, J.: RHINO: rotated DETR with eynamic eenoising via Hungarian matching for oriented object detection. arXiv preprint arXiv:2305.07598
21. Dai, J., Li, Y., He, K., Sun, J.: R-FCN: object detection via region-based fully convolutional networks. In: Lee, D., Luxburg, U., Garnett, R., Sugiyama, M., Guyon, I. (eds.) NIPS'16: Proceedings on 30th Conference on Neural Information Processing Systems, NIPS, pp. 379–387. Curran Associates Inc., Red Hook, NY (2016)
22. Liu, W., et al.: SSD: single shot multibox detector. In: Leibe, B., Matas, J., Sebe, N., Welling, M. (eds.) European Conference on Computer Vision, ECCV, pp. 21–37. Springer (2016)
23. Groener, A., Chern, G., Pritt, M.: A comparison of deep learning object detection models for satellite imagery. In: 2019 IEEE Applied Imagery Pattern Recognition Workshop (AIPR), pp. 1–10. IEEE (2019)
24. Ultralytics YOLO Docs, YOLOv8 Model: https://docs.ultralytics.com/models/yolov8/#how-do-i-train-a-yolov8-model. Accessed 30 Mar 2025
25. Sanchez, S.A., Romero, H.J., Morales, A.D.: A review: comparison of performance metrics of pre-trained models for object detection using the TensorFlow framework. In: IOP Conference Series: Materials Science and Engineering, vol. 844, no. 1, p. 012024. IP. Publishing (2020)

26. Redmon, J., Divvala, S., Girshick, R., Farhadi, A.: You only look once: unified, real-time object detection. In: Proceedings of the IEEE Conference on Computer Vision and Pattern Recognition (CVPR), Las Vegas, NV, USA, pp. 779–788. IEEE (2016)
27. Lou, H., et al.: DC-YOLOv8: small-size object detection algorithm based on camera sensor. Electronics **12**(10), 2323 (2023)
28. Ultralytics YOLO Docs, YOLOv8 Model. https://docs.ultralytics.com/models/yolov8/#supported-tasks-and-modes. Accessed 30 Mar 2025
29. Terven, J., Córdova-Esparza, D.M., Romero-González, J.A.: A comprehensive review of YOLO architectures in computer vision: from YOLOv1 to YOLOv8 and YOLO-NAS. Mach. Learn. Knowl. Extr. **5**(4), 1680–1716 (2023)
30. Muggeo, V.M.: Estimating regression models with unknown break-points. Stat. Med. **22**(19), 3055–3071 (2003)
31. McGee, V.E., Carleton, W.T.: Piecewise regression. J. Am. Stat. Assoc. **65**(331), 1109–1124 (1970)
32. xView Challenge Description page. https://challenge.xviewdataset.org/challenge-description. Accessed 30 Mar 2025
33. xView Dataset. https://xviewdataset.org. Accessed 8 Sept 2024
34. Nagesh, B.: Small object detection: an image tiling based approach. Medium page https://binginagesh.medium.com/small-object-detection-an-image-tiling-based-approach-bce572d890ca. Accessed 30 Mar 2025
35. Plainsight Editorial. Tiling: The Key to Small Object Detection (2022) https://plainsight.ai/blog/tiling-and-small-object-detection/. Accessed 30 Mar 2025
36. Ultralytics YOLO Docs: Data Preprocessing Techniques for Annotated Computer Vision Data. https://docs.ultralytics.com/guides/preprocessing_annotated_data/. Accessed 30 Mar 2025
37. Letterboxing in YOLOV5, YOLOV7, YOLOV8: an intuitive explanation with Python code, Medium page. https://medium.com/@reachraktim/letterboxing-in-yolov5-yolov7-yolov8-an-intuitive-explanation-with-python-code-88f7d4323d6c. Accessed 30 Mar 2025
38. Goodfellow, I., Bengio, Y., Courville, A.: Deep Learning. MIT Press, Cambridge, MA, USA (2016)
39. Pritt, M., Chern, G.: Satellite image classification with deep learning. In: 2017 IEEE Applied Imagery Pattern Recognition Workshop (AIPR), Washington, DC, USA, pp. 1–7. IEEE (2017)
40. Ultralytics YOLO Docs: Explore Ultralytics YOLOv8: How do I train a YOLOv8 model? https://docs.ultralytics.com/models/yolov8/#how-do-i-train-a-yolov8-model. Accessed 30 Mar 2025
41. Pillow (PIL Fork) 11.3.0 Documentation. https://pillow.readthedocs.io/en/stable/reference/Image.html. Accessed 30 July 2025

Re-ranked Transformer: New Strategy Based on Misspellings and Typos Pattern Analysis for Keystroke Biometrics Improvement

Nabila Mansouri[1(✉)] , Salwa Sahnoun[1,2] , Hedi Feki[1,3] , Ahmed Ben Ali[3], and Ahmed Fakhfakh[1]

[1] Digital Research Center of Sfax, Laboratory of Signals, systeMs, aRtificial Intelligence and neTworkS, Box 275, 3021 Sfax, Tunisia
nabila.elmansouri@crns.rnrt.tn
[2] National School of Electronics and Telecommunications of Sfax, University of Sfax, Sfax 3018, Tunisia
[3] Ekara by ip-label, Sfax, Tunisia
https://ip-label.com/fr/

Abstract. Person identification using keystroke biometrics offers a scalable solution for identity verification in behavioral biometrics. This study introduces a framework where a transformer model serves as the baseline for capturing complex spatio/temporal patterns in keystroke features. To enhance accuracy, the model's output is re-ranked using k-reciprocal nearest neighbors (k-RNN), which encodes neighborhood relationships into a feature vector for re-ranking under the Jaccard distance. The proposed method integrates misspellings and typos patterns, particularly using backspace key for correction, as a weighting factor to refine the final identification distance. This pipeline, combining the transformer baseline with k-RNN re-ranking and typing error adjustments, demonstrates significant improvements in person identification. Experimental results on Aalto keystroke database achive and EER about of 1.60. These findings validate the effectiveness of our proposed method and highlight its potential for secure and non-invasive applications.

Keywords: Keystrokes · Biometrics · Transformer · K-RNN · Re-Ranking

1   Introduction

Person identification is a critical task in behavioral biometrics, often challenging due to subtle variations in individual behavior [18]. With the frequent and ubiquitous use of keyboards in daily activities, such as communication, data entry, and authentication, keystroke biometrics has emerged as a promising modality for identity verification [3]. This study introduces a novel approach to recognize person, leveraging keystroke dynamics, with a particular focus on typing error patterns and the use of the backspace key, which are shown to be significant discriminative features between individuals.

© The Author(s), under exclusive license to Springer Nature Switzerland AG 2026
F. Marcelloni et al. (Eds.): IJCCI 2025, CCIS 2829, pp. 508–520, 2026.
https://doi.org/10.1007/978-3-032-15638-9_29

Central to our methodology is the incorporation of backspace patterns and k-reciprocal nearest neighbors (k-RNN). The k-reciprocal feature is computed by encoding a sample's k-reciprocal nearest neighbors into a single vector, enabling an effective re-ranking process under the Jaccard distance metric. Furthermore, we leverage the effectiveness of transformer-based deep learning models, which have demonstrated state-of-the-art performance in learning complex temporal and sequential patterns [12]. In fact, in the first hand, the transformer is used for the basic person identification process using keystroke features, extracting high-level representations of typing behaviors that significantly enhance identification accuracy. Furthermore, in the second hand, typing error rates, coupled with backspace usage, can be a unique behavioral signature that, when combined with the power of transformer models, provides a robust framework for person identification. The final distance for Re-ID is derived by combining the original distance, weighted by the typing error rate, with the Jaccard distance from the k-RNN-based re-ranking. This hybrid approach demonstrates significant improvements in persons discrimination, emphasizing the robustness of keystroke dynamics, misspellings and typos patterns, and modern deep learning techniques for person re-identification. Experimental results validate the proposed framework, underscoring its potential for enhanced security applications in behavioral biometrics.

The details of our proposed method are presented in the remainder of this paper that is structured as follows: Sect. 2 provides an overview of existing work in keystroke biometrics for person identification. Section 3 details our proposed re-ranking approach designed to enhance the performance of the transformer baseline. In Sect. 4, we present the experimental setup along with a comparative study of the results. Finally, Sect. 5 concludes the paper and discusses potential directions for future research.

2   Related Work

Among user identification methods, behavioral biometrics has been shown to be effective against identity loss, as well as user-friendly and unobtrusive. One of the most popular traits in the literature is keystroke dynamics [23]. Due to the large deployment of computers and mobile devices in our society, Keystroke dynamics have long been explored as a behavioral biometric for user identification based on fixed text and/or free text [4].

Baseline Models for Keystroke Biometrics: Early methods [2,8,9,11,14,15,17] primarily relied on statistical modeling (e.g., mean and variance of keystroke timings) with machine learning models (SVM, RGB histograms and neural networks, decision trees, and k-NN) to differentiate individuals. However, these traditional approaches struggled with scalability and robustness in real-world scenarios.

The quick analysis on literature of machine learning for keystroke biometrics we can easily note that traditional machine learning methods for keystroke biometrics face several limitations, particularly when dealing with small datasets. These methods, such as SVMs, decision trees, and k-NN, often struggle with feature extraction and scalability, requiring extensive manual tuning and handcrafted feature engineering. Additionally, traditional models lack the ability to capture complex temporal dependencies in keystroke sequences,

With the rise of deep learning, researchers have shifted toward convolutional neural networks (CNNs), recurrent neural networks (RNNs) and Long Short-Term Memory (LSTM) networks. For instance, [5] utilized convolutional neural networks (CNNs) on datasets including CMU, GREYC Keystroke, and GREYC Web-Based. [24] employed multilayer perceptrons (MLP) on self-collected fixed-text sessions of approximately two minutes, achieving 97.18% accuracy.

In addition, [1] applied RNNs to multiple datasets, achieving EERs of 9.2% for mobile devices and 2.2% for desktops on free-text inputs with 30âĂŞ150 keystrokes. El-Kenawy et al. [7] used bidirectional RNNs on datasets like RHU and MEU-Mobile KSD, reporting accuracies of 99.02% and 99.32% for fixed-text inputs with minimal keystrokes.

Furthermore, to capture complex temporal dependencies Long Short-Term Memory (LSTM) networks have been used to model user typing rhythms, while CNN-based approaches extract spatial correlations from keystroke sequences. The study in [22] utilized LSTM on the Aalto database, achieving an EER of 15.10% for free-text inputs with 50 keystrokes.

A hybrid model combining a convolutional neural network (CNN) and a recurrent neural network (RNN) is employed to analyze individual keystroke vectors and extract unique keystroke features for identity authentication. [13] combined CNN and RNN models on SUNY Buffalo and Clarkson II datasets, The method presented in this paper authenticates users based on their keystrokes while typing free text. The keystroke data is segmented into fixed-length sequences, which are then transformed into keystroke vector sequences based on timing characteristics.

More recently, transformers have emerged as a powerful alternative for sequence modeling. Unlike RNNs, transformers can process entire keystroke sequences in parallel, capturing long-range dependencies more efficiently. Their success in Natural Language Processing (NLP) has inspired researchers to explore them for keystroke-based user authentication, significantly improving identification accuracy. Acien et al. Haitian Yang [10] introduces a novel approach for user authentication based on both typing behavior (keystrokes) and the content being typed. The proposed CKDAN model leverages a dual attention mechanism to capture and integrate information from both keystroke dynamics and textual content. Additionally, the model incorporates pre-trained language models to enhance the understanding of typed content. By combining these elements, CKDAN improves [10] the accuracy and robustness of continuous authentication systems. The method is evaluated on benchmark datasets, demonstrating its effectiveness in distinguishing users based on their unique typing patterns and writing styles. Lastly, TypeFormer [21] implemented transformers across multiple datasets, reporting an EER of 3.25% for free-text inputs with 30âĂŞ100 keystrokes.

However, these methods often suffer from limited generalizability across devices and users.

Ahmed et al. [2] used self-collected data with neural networks, achieving EER of 2.13% and 2.46% for controlled and uncontrolled free-text scenarios, respectively, in 53 participants. Gascon et al. [9] employed SVM on self-collected data from 300 participants, reporting a true acceptance rate (TAR) of 92% at 1% FAR for 160 free keystrokes. Alpar [17] used RGB histograms and neural networks to achieve 90% accuracy on 15-

character fixed text samples from 10 participants. Huang et al. [11] used data from 39 participants in the Clarkson I dataset to address the effect of data size on the performance of free-text keystroke authentication, achieving an impostor pass rate of approximately 1% for free-text inputs between 1k and 10k keystrokes. Morales et al. [15] analyzed the BiosecurID data set, reporting a 5.32% EER for fixed text samples averaging 25 keystrokes from 300 participants.

Their performance heavily depends on the quality and quantity of extracted features, making them less effective when only limited keystroke data is available. Small datasets can lead to overfitting, reduced generalization, and poor adaptability to new users or changing typing behaviors. Additionally, traditional models lack the ability to capture complex temporal dependencies in keystroke sequences, unlike deep learning approaches that can automatically learn richer representations.

Beyond the Baseline: Re-ranking and Error-aware Refinements: While transformers provide a strong foundation, integrating post-processing techniques can further refine predictions. The k-reciprocal nearest neighbors (k-RNN) re-ranking approach enhances user identification by leveraging neighborhood relationships, ensuring a more robust feature representation. Additionally, incorporating typing error rates—such as backspace frequency—helps adjust identification scores based on user-specific error patterns, making the system more resilient to real-world variations.

To the best of our knowledge, this is the first study that proposes to introduce the re-ranking concept using transformer-based keystroke biometric systems as a baseline. Based on this knowledge and inspired by re-ranking algorithms applied in other domain (image retrieval and person identification by gate and appearance), we propose a novel pipeline that combines transformers with k-RNN re-ranking and typing error adjustments, striking a balance between accuracy and adaptability. By evaluating our approach on the Aalto keystroke database, we demonstrate how these innovations contribute to more precise, secure, and non-invasive keystroke-based identification. The proposed Transformer architecture will be detailed in the next section and evaluated on the popular Aalto mobile keystroke database using only 5 enrollment sessions.

3   Proposed Re-ranking Approach

In this section, we present our proposed re-ranking approach to enhance keystroke biometric authentication. We begin by introducing the baseline Transformer model, highlighting its advantages in capturing both spatial and temporal dependencies inherent in keystroke dynamics. Next, we describe the detailed feature extraction process, focusing on timing features and novel behavioral traits such as misspellings and typo patterns, which serve as key discriminative signals for improved user identification. Finally, we explain how these features are integrated into our model to achieve robust and accurate authentication.

3.1   Baseline Transformer

Transformers are more recently proposed DL architectures that have already garnered impmense interest due to their effectiveness across a range of application domains such as language assessment, vision, reinforcement learning, and biometrics [21]. Their main advantages compared with traditional CNN and RNN architectures are: i) Transformers are feed-forward models that process all the sequences in parallel, therefore increasing efficiency; ii) They apply Self-Attention/Auto-Correlation mechanisms that allows them to operate in long sequences; iii) They can be trained efficiently in a single batch since all the sequence is included in every batch; and iv) They can attend to the whole sequence, instead of summarising all the previous temporal information.

Since, keystroke dynamics contain both spatial dependencies (relationships between different keys and their patterns) and temporal dependencies (the timing between keystrokes). Spatial attention helps understand key interaction patterns (e.g., common sequences like"th" or "ing"). However, temporal attention focuses on typing rhythm (e.g., pauses, typing speed changes). Dual attention ensures, so, that the model captures both aspects simultaneously, leading to richer feature extraction. In fact in, keystroke features encoding context, using a Dual Attention Transformer is a good choice.

3.2 Features Extraction

Timing Features. To capture the temporal dynamics of user typing behavior, we extract **Di-Gram** and **Tri-Gram** features from the keystroke sequence, as illustrated in Fig. 1. These features are derived from sequences of two and three consecutive key events, respectively, and reflect timing intervals between key actions. Specifically: i) Di-Gram features are computed over two consecutive key events and ii) Tri-Gram features are computed over three consecutive key events. Furthermore, The timing intervals used include:

- **UD (Up-Down):** Time between a key-up event of one key and the subsequent key-down event of the next key.
- **DU (Down-Up):** Time between a key-down event and the following key-up event.
- **DD (Down-Down):** Time between the key-down events of two consecutive keys.
- **UU (Up-Up):** Time between the key-up events of two consecutive keys.

In addition, we include:

- **Hold Latency (HL):** Duration a key is held down, defined as the time between the key-down and the corresponding key-up event.
- **ASCII Code**: Numerical encoding of the key pressed.

These features are concatenated to form a 10-dimensional feature vector for each key interaction segment:

$$\mathbf{x} = [\mathrm{HL}, \mathrm{DU_{di}}, \mathrm{UD_{di}}, \mathrm{DD_{di}}, \mathrm{UU_{di}}, \mathrm{DU_{tri}}, \mathrm{UD_{tri}}, \mathrm{DD_{tri}}, \mathrm{UU_{tri}}, \mathrm{ASCII}] \qquad (1)$$

where the subscript "di" indicates Di-Gram timing features and "tri" indicates Tri-Gram timing features.

Moreover, to ensure compatibility and robustness of the extracted features, appropriate normalization and scaling techniques are applied. ASCII codes are normalized to the range $[0, 1]$ using min-max normalization to ensure consistency across different keyboards and platforms. All timing features are expressed in seconds to maintain temporal coherence and can optionally be standardized using z-score normalization to reduce the impact of individual typing speed variability.

This feature representation enables the modeling of fine-grained temporal patterns in typing behavior, which are critical for keystroke-based biometric identification.

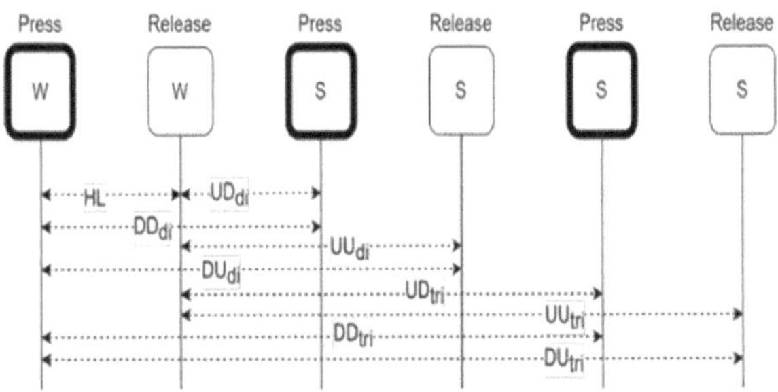

Fig. 1. Illustration of keystroke input features collection.

Misspellings and Typos Pattern as a Discriminative Features In Keystroke Analysis. One of the key behavioral traits in keystroke dynamics is the Error Correction per sentence (ECPS), which represents the percentage of backspace key usage during typing a sentence. ECPS serves as an individualized indicator of typing habits, as it reflects a user's tendency to correct mistakes.

In our analysis, we leverage ECPS as a discriminative feature by calculating the percentage of error corrections per character for each user. We observe that some individuals rarely use backspace or delete keys, maintaining a highly fluid typing style, while others frequently correct their input, demonstrating a consistent error correction pattern across different typing tasks. This variation in error correction behavior provides valuable insights into user-specific typing habits, further enhancing the discriminatory power of keystroke biometrics for authentication and identification. For a given sentence consisting of N_{key} keystrokes, $ECPS$ is computed for each user as the ratio of backspace key presses $\sum \text{BCK}_{key}$ to the total number of keys typed, as defined in Eq. 2.

$$\text{ECPS} = \frac{\sum \text{BCK}_{key}}{N_{key}} \tag{2}$$

where BCK_{key} and N_{key} are the number of backspace key and the number of characters typed during a one sentence respectively.

3.3   K-Reciprocal Nearest Neighbors

The motivation for using the K-reciprocal Nearest Neighbors (k-RNN) process [16] lies in its ability to push up performances without additional training epochs and / or features, addressing the limitation of deep learning approaches that require large testing resources, which are often unavailable [16]. In the initial ranking list, the top-ranked identities are selected based on their keystrokes dynamic features' similarity to the probe. However, there is no guarantee that any rank-t entry represents a true match. The K-RNN algorithm is designed to minimize the influence of negative neighbors while incorporating any missing positive neighbors. The first step of our re-ranking method, K-RNNA, focuses on eliminating negative neighbors from the list L_s. The second step involves adding positive neighbors to L_s from the reciprocal nearest neighbors list of the probe typing signature p. To illustrate the K-RNNA process, the follow chart of our proposed re-ranking process is presented in the Fig. 2. Proposed re-ranking process includes tow phases: i) Pruning the initial neighbors list, to avoid mismatch initial selected neighbors. and ii) Local query expansion phase to inject new potential neighbors base on typo, misspelling, keying error, typing error of users.

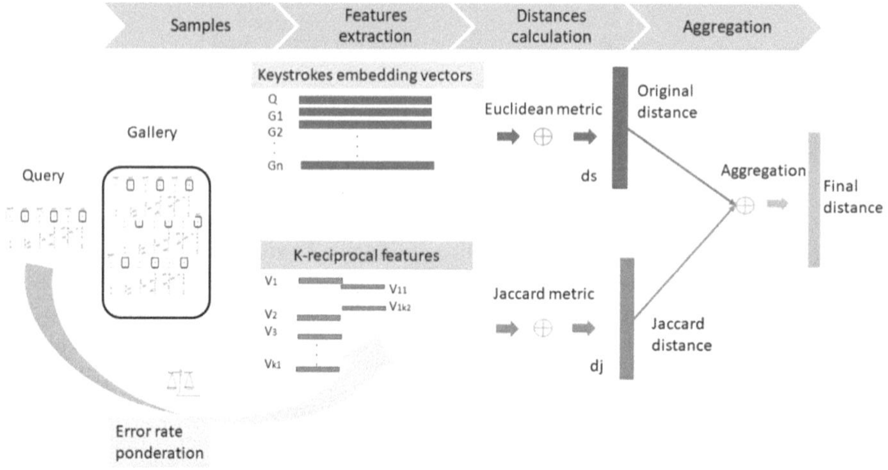

Fig. 2. Follow chart of our proposed re-ranking process.

Pruning the Initial Neighbors List. Given a probe typing signal p and a gallery set G with N keystrokes embeddings, with $G = \{g_i \mid i = 1 \dots N\}$, the original distance is calculated between the probe vector p and each vector g_i of the gallery. The initial rank list $L_S(p, G) = \{g_1^o, g_2^o, \dots, g_N^o\}$ is obtained by sorting the pairwise distance between the probe and all gallery embedding vectors.

According to [20], the k-nearest neighbors of a probe p in the initial rank list is defined as follows (Eq. 3) :

$$N(p, k1) = \{g_1^o, g_2^o, \dots, g_{k1}^o\} \tag{3}$$

where k_1 denotes the number of candidates in the initial ranking list that rank in top $k1$.

The k-reciprocal nearest neighbors is constructed after a pruning phase on $N(p, k1)$. If the probe p does not belong to the k-nearest neighbors of a gallery vectors g_i^o in

$N(p, k1)$, g_i^o will be considered as false positive and it will be excluded from the list. Therefore, the k-RNN is defined as in Eq. 4.

$$R(p, k1) = \{g_i \mid (g_i \in N(p, k1)) \wedge (p \in N(g_i, k1))\} \tag{4}$$

Local Query Expansion Based on Typing Error Rate Similarity. Since any gallery typing signature g_i in $R(p, k1)$ is considered as positive matches, any signature in its k-reciprocal nearest neighbors can also be a neighbor candidate for probe keystrokes p. So, the k_2-reciprocal nearest neighbors of each candidate in $R(p, k1)$ are added to a more robust set $R_2(p, k)$ to try to add positive matches that are excluded from $R(p, k1)$ due to accidental variation in press and/or release times for some keys. This operation allows us to add more positive samples into $R_2(p, k)$ that can be new neighbors in the probe p.

Therefore, in the second iteration of the k-RNN algorithm, typing error rate is calculate for each g_i in $R(p, k1)$ (denoted new candidate probe phrase), and their nearest neighbors phrases in $N(g_i, k2))$ are generated based on the ratio of backspace key usage to the total number of characters typed in the phrase (see Fig. 3). Subsequently, similarities (S_QA, S_QB, S_QC, S_QD, S_QE, ...) between the error rate of the new candidate probe phrase (Q) and its corresponding g_i phrase (A, B, C, D, E, ...) are calculated, which will be used as the implication weight of the candidate sentence in the new list of neighbors.

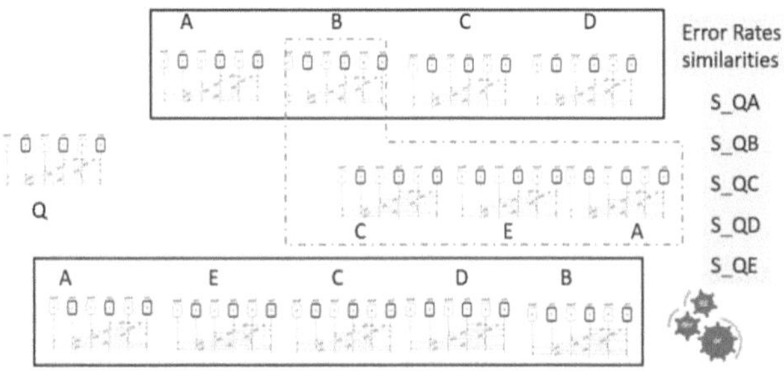

Fig. 3. Local Query Expansion based on ECPS similarity.

Updated Distance: An All-Inclusive Measure. To enhance the robustness of the similarity assessment between the probe p and each gallery sample g_i, we introduce a composite distance metric that combines both semantic and structural similarities. While the Jaccard distance $d_J(p, g_i)$ effectively captures set-based token overlap (e.g., shared misspellings or typo patterns), it may overlook geometric relationships in the feature space. To address this, we incorporate the standard Euclidean distance $d_S(p, g_i)$, which reflects structural differences in the embedded feature vectors.

The final distance metric $d_f(p, g_i)$ blends these two measures using a convex combination, defined as:

$$d_f(p, g_i) = (1 - \lambda) \times d_J(p, g_i) + \lambda \times d_S(p, g_i) \tag{5}$$

Here, $\lambda \in [0, 1]$ is a weighting parameter that controls the trade-off between the two distance components.

This hybrid distance allows our method to better handle diverse typing behaviors by leveraging both symbolic and numerical perspectives.

4   Experiments

4.1   Datasets and Settings

the Aalto Mobile Dataset [19], a large scale dataset on mobile text entry collected through a web-based transcription task involving 37,370 participants. The dataset reveals an average typing speed of 36.2 words per minute (WPM) with an uncorrected error rate of 2.3%. The task was to transcribe 15 English sentences, shown one after another. Participants were shown instructions requesting they first read and memorize the sentence, then type it as quickly and accurately as possible. The extensive scale of the dataset facilitates robust statistical analyses, enabling the exploration of correlations between typing performance and various factors, including demographics, finger usage, and the adoption of intelligent text entry techniques and making a mistakes and typos.

4.2   Test Protocol and Evaluation Metrics

In the data pre-processing phase, we filtered out users based on data availability, resulting in 60,000 users from the Aalto DB. These users were randomly divided into three subsets: 30,000 for training, 400 for validation, and 1,000 for testing.

Furthermore, the enrollment-verification protocol [6] is adopted during evaluation, where different sessions from the 1,000 test users are used to assess the model. In fact, a fixed number of sessions from each test user are allocated for enrollment and verification to ensure consistency across evaluations. Specifically, from the available sessions of each test user, 10 sessions are used for enrollment, and the remain 5 sessions are reserved for verification. During evaluation, a user-specific feature embedding $f_{u,e}$ is computed from the enrollment sessions of subject u. Authentication is then performed by comparing this enrollment embedding with a verification embedding $f_{v,a}$ extracted from one of the test sessions. If $v = u$, this constitutes a genuine attempt; otherwise, when $v \neq u$, it is considered an impostor attempt. For each user in the test set, the evaluation involves treating all other users as impostors to compute both genuine and impostor scores.

Performance metrics are then computed per user and averaged across all test users to obtain a global assessment. The primary evaluation metric used is the Equal Error Rate (EER), which provides a balanced measure by identifying the threshold τ^* at which the False Acceptance Rate (FAR) equals the False Rejection Rate (FRR). This metric offers a concise indication of the system's discrimination ability between genuine and impostor attempts. Mathematically, EER is defined as in Eq. 6:

$$\text{EER}(\tau^*) = \text{FAR}(\tau) = \text{FRR}(\tau) \tag{6}$$

where: τ^* is the threshold at which FAR and FRR are equal, $\text{FAR}(\tau)$ is the false acceptance rate as a function of threshold τ and $\text{FRR}(\tau)$ is the false rejection rate as a function of threshold τ.

4.3   Results

In this section, we evaluate the performance of the proposed re-ranking approach and compare it to the results obtained using a transformer without any re-ranking process, as well as the baseline k-RNN algorithm. Results are presented in Table 1. This table shows that the proposed approach significantly outperforms the two baselines in term of EER. ERR is reduced by 0.16 and 0.08% compared to transformer and K-RNN baselines respectively. Experimental results in Table 1 highlights the effectiveness of our proposed re-ranking method in terms of re-classing the positive matches at the top rank.

Table 1. Illustration of the proposed re-ranking method results.

Method	EER
Baseline Transformer	1.76
KNN-Re-ranking method	1.68
Proposed Re-ranking method	1.60

4.4   Comparative Study

We evaluate the performance of the proposed re-ranking method against the following state-of-the-art keystrokes biometric-based continuous authentication systems:

- TypeNet [1]: A free-text keystroke biometric system that employs a deep recurrent neural network trained with triplet loss for large-scale authentication and identification.
- HuMINet [22]: A multimodal architecture incorporating two LSTM layers in each model. These models are fused at the score level using either a weighted or non-weighted approach.
- TypeFormer [21]: A modified transformer-based architecture utilizing two encoders: Temporal and Channel Modules. Each encoder integrates a Multi-Head Self-Attention mechanism and a Multi-Scale Keystroke CNN to extract both temporal and channel-based features effectively.
- BehaveFormer [6]: A transformer-based architecture specifically designed for keystroke dynamics analysis. It leverages a Spatio-Temporal Dual Attention Mechanism to simultaneously capture spatial relationships between keys and temporal variations in typing patterns

Table 2. Comparative study.

Method	Model Type	EER
TypeNet [1]	LSTM	8.00
HuMINet [22]	LSTM	15.10
TypeFormer [21]	Transformer	3.25
BehaveFormer [6]	Transformer	1.8
KNN-Re-ranking method	Transformer	1.68
Proposed Re-ranking method	Transformer	1.60

Table 2 summarizes the results of the comparative study in terms of EER and identifies the type of model used in each application in order to justify the effectiveness of the choice of the transformer as a baseline in our application.

The results demonstrate a clear trend in the effectiveness of different models for keystroke-based continuous authentication, particularly highlighting the advantages of transformer-based architectures over recurrent models such as LSTMs.

LSTM-based models exhibit higher Equal Error Rates (EER), with HuMINet (15.10%) and TypeNet (8.00%) performing significantly worse than transformer-based approaches. This suggests that while LSTMs can capture temporal dependencies in keystroke sequences, they may struggle with long-range dependencies and complex feature extraction compared to transformers.

Transformer-based models show superior performance, with TypeFormer achieving a significantly lower EER of 3.25%, demonstrating the benefits of self-attention mechanisms in effectively modeling keystroke patterns. BehaveFormer further improves on this, reducing the EER to 1.8%, likely due to its Spatio-Temporal Dual Attention Mechanism, which enhances feature extraction by capturing both spatial key interactions and temporal typing variations.

The KNN-Re-ranking method (1.68%) and the Proposed Re-ranking method (1.60%) refine authentication accuracy even further. The marginal improvement between these two re-ranking methods suggests that fine-tuning ranking-based decision mechanisms can optimize transformer-based models by further reducing false positives and false negatives.

Overall, the results confirm that transformer-based approaches significantly outperform LSTM models for keystroke authentication. The proposed re-ranking method achieves the lowest EER (1.60%), demonstrating its effectiveness in refining keystroke-based verification and setting a new benchmark in continuous authentication performance.

5   Conclusion

This study highlights the effectiveness transformer models combined with ranking-based refinement strategies in keystroke-based continuous authentication. First-of-all, the results demonstrate that models leveraging self-attention mechanisms transformer, such as TypeFormer, significantly outperform traditional LSTM-based approaches by capturing both long-range dependencies and intricate keystroke patterns. The introduction of Spatio-Temporal Dual Attention in BehaveFormer further improves feature extraction, leading to an EER of 1.8, marking a substantial improvement over previous models.

Additionally, re-ranking techniques enhance authentication accuracy by refining the decision-making process. The proposed re-ranking method achieves the lowest EER of 1.60, setting a new benchmark for keystroke-based continuous authentication. These findings underscore the potential of transformer models combined with ranking-based refinement strategies in the advancement of biometric security systems. Future work can explore further optimization techniques, the integration of multi-modal behavioral biometrics, and real-world deployment scenarios to improve scalability and robustness.

Acknowledgment. The authors would like to express their sincere gratitude to Ekara by ip-label Society for providing the funding and support that made this project possible. Their contribution has been instrumental in advancing this research, and we deeply appreciate their commitment to fostering innovation in the field.

References

1. Acien, A., Morales, A., Monaco, J.V., Vera-Rodriguez, R., Fierrez, J.: Typenet: Deep learning keystroke biometrics. IEEE Trans. Biomet. Behav. Identity Sci. **4**(1), 57–70 (2022)
2. AlHaiqi, A., Ismail, M., Rosdiadee, N.: On the best sensor for keystrokes inference attack on android. Procedia Technol. **11**, 989–995 (2013)
3. Arsh, A., Kar, N., Das, S., Deb, S.: Multiple approaches towards authentication using keystroke dynamics. Procedia Comput. Sci. **235**, 2609–2618 (2024). https://doi.org/10.1016/j.procs.2024.04.246, international Conference on Machine Learning and Data Engineering (ICMLDE 2023)
4. Bansal, P., Ouda, A.: Continuous authentication in the digital age: an analysis of reinforcement learning and behavioral biometrics. Computers **13**(4) (2024). https://www.mdpi.com/2073-431X/13/4/103
5. Ceker, H., Upadhyaya, S.: Sensitivity analysis in keystroke dynamics using convolutional neural networks. In: IEEE Workshop on Information Forensics and Security (WIFS). Rennes, France (2017)
6. Dilshan, S., Sanuja, T., Maduka, V., Sanka, R., Sandareka, W., Dulani, M.: Behaveformer: a framework with spatio-temporal dual attention transformers for IMU enhanced keystroke dynamics. In: Cryptography and Security. Proceedings in Computer Science (2023)
7. El-Kenawy, S., Mirjalili, S., Abdelaziz, A., Abdelhameed, I., Khodadadi, N., Eid, M.: Meta-heuristic optimization and keystroke dynamics for authentication of smartphone users. Artif. Intell. Res. Optimization **10**(16) (2022)
8. Gunetti, D., Picardi, C.: Keystroke analysis of free text. ACM Trans. Inf. Syst. Secur. **8**(3), 312–347 (2005)
9. Gascon, H., et al.: Continuous authentication on mobile devices by analysis of typing motion behavior. Unknown **03** (2014)
10. Haitian, Y., Xiang, M., Xuan, Z., et al.: CKDAN: content and keystroke dual attention networks with pre-trained models for continuous authentication. Comput. Secur. **128**, 103–159 (2023)

11. Huang, J., Hou, D., Schuckers, S., Hou, Z.: Effect of data size on performance of free text keystroke authentification. In: IEEE International Conference on Identity, Security and Behavior Analysis (ISBA). Hong Kong, China (2015)
12. Islam, S., Elmekki, H., Elsebai, A., Bentahar, J., Drawel, N., Rjoub, G., Pedrycz, W.: A comprehensive survey on applications of transformers for deep learning tasks. Expert Syst. Appl. **241**, 122666 (2024). https://doi.org/10.1016/j.eswa.2023.122666, https://www.sciencedirect.com/science/article/pii/S0957417423031688
13. Li, X., Shengfei, P., Hui, P., Lio, P.: Continuous authentication by free-text keystroke based on CNN and RNN. Comput. Secur. **16** (2020)
14. Monrose, F., Rubin, A.D.: Keystroke dynamics as a biometric for authentication. Future Generation Comput. Syst. **16**, 351–359 (2000)
15. Morales, A.: Keystroke biometrics ongoing competition. IEEE Access **4**, 7736–7746 (2016)
16. Nabila, M., Sourour, A., Yousri, K.: Re-ranking person re-identification using attributes learning. Neural Comput. Appl. **33**(19), 12827–12843 (2021)
17. Orcan, A.: Keystroke recognition in user authentication using ANN based RGB histogram technique. Eng. Appl. Artif. Intell. **32**, 213–217 (2014)
18. Orrù, G., Rattani, A., Rida, I., Marcel, S.: Recent advances in behavioral and hidden biometrics for personal identification. Pattern Recogn. Lett. **185**, 108–109 (2024). https://doi.org/10.1016/j.patrec.2024.07.016, https://www.sciencedirect.com/science/article/pii/S0167865524002198
19. Palin, K., Feit, A., Kim, S., Kristensson, P.O., Oulasvirta, A.: How do people type on mobile devices? observations from a study with 37,000 volunteers. In: Proceedings of the 21st International Conference on Human-Computer Interaction with Mobile Devices and Services (2019)
20. Qin, D., Gammeter, S., Bossard, L., Quack, T., Van Gool, L.: Hello neighbor: accurate object retrieval with k-reciprocal nearest neighbors. In: Conference on Computer Vision and Pattern Recognition, pp. 777–784. USA (2011)
21. Stragapede, G., Delgado-Santos, P., Tolosana, R., et al.: Typeformer: transformers for mobile keystroke biometrics. Neural Comput. Appl. **36**, 18531–18545 (2024)
22. Stragapede, G., Vera-Rodriguez, R., Tolosana, R., Morales, A., Acien, A., Lan, G.L.: Mobile behavioral biometrics for passive authentication. Pattern Recogn. Lett. **41**, 157–35 (2022)
23. Stragapede, G., Delgado-Santos, P., Tolosana, R., Vera-Rodriguez, R., Guest, R., Morales, A.: Mobile keystroke biometrics using transformers. In: 2023 IEEE 17th International Conference on Automatic Face and Gesture Recognition (FG), pp. 1–6 (2023). https://doi.org/10.1109/FG57933.2023.10042710
24. Stylios, I., Skalkos, A., Kokolakis, S., Maria, K.: Bioprivacy: Development of a keystroke dynamics continuous authentication system. In: Computer Security. ESORICS 2021 International Workshops, pp. 158–170. Lecture Notes in Computer Science, Germany (2022)

Towards Generalizing Deep Reinforcement Learning Algorithms for Real World Applications

Daniel Ruiru[1,2]($\boxtimes$) (iD), Nicolas Jouandeau[2] (iD), and Dickson Owuor[1] (iD)

[1] Strathmore University, Nairobi, Kenya
{druiru,dowuor}@strathmore.edu
[2] Université Paris 8, Saint-Denis, France
n@up8.edu

Abstract. Generalization is a major problem in reinforcement learning (RL), as agents struggle in environments outside their training sets. Often caused by overfitting during the training phase, this issue limits the application of RL in the real world. This paper tries to solve the generalization problem by using a domain randomization technique during the training period. Using two real-world problems; the financial market and agriculture (crop production), this paper trains classical deep reinforcement learning algorithms (DQN and PPO) in different and randomized environments (MDPs). Agents trained in randomized environments generalize better than those trained in single environments (baseline agents). This conclusion is based on the results, in which the agents trained in the randomized environments achieve higher cumulative rewards.

Keywords: Deep Reinforcement Learning · Markov Decision Process · Deep Q Network · Policy Optimization

1 Introduction

Deep Reinforcement Learning (DRL) is today one of the most prominent sub-domains of artificial intelligence. One of its most exciting possibilities is the use of its algorithms to solve complex, real world issues such as resource allocations problems in finance, agriculture and computing [33]. This DRL potential comes from performance of reinforcement learning (RL) methods which can be improved using function approximators such as neural networks [6]. Neural networks are often preferred for this role because of the straight forward way of adjusting a gradient [38]. This improvement is much needed in the RL domain because of its persistent critique of limited adaptability in the real world [17]. While RL has shown remarkable success in solving complex problems like games, it fails to achieve competitive results in most worldly problems. RL depends on unbiased interactions with environments and countless trial and error attempts, requirements that are often not achieved [35]. Also, unlike games which are based on controlled environments, the real world has unique challenges that are usually open-ended [18] and non-stationary [11]. These attributes make it harder to get good results with conventional RL methods.

F. Marcelloni et al. (Eds.): IJCCI 2025, CCIS 2829, pp. 521–537, 2026.
https://doi.org/10.1007/978-3-032-15638-9_30

By combining RL with Deep Learning, DRL provides a starting point for solving many real world problems. First, by providing a means to engage vast amounts of environments and secondly, by facilitating efficient and countless interactions with the said environments [24]. As such, today, it is possible to train DRL policies that operate in a variety of state and actions spaces as defined by Markov Decision Processes (MDPs) [5]. While most of the learned policies are restricted to the training environments (poor generalization), this ability to work in different MDPs is an important building block for reasonable generalization.

However, as it stands, most RL/DRL solutions are specific to the training environments. This limitation is well documented within the DRL field. For instance, there has been countless studies that have shown this generalization problem using adversarial perturbations in state observations of policies and neural networks [30]. Other scholars like [9] have reviewed adversarial perturbations in the DRL domain, while simple review studies like those done by [14] have analyzed adversarial attacks in deep policies. Majorly, most of these studies have focused on this issue by exploring the fundamental problems of function approximations, action value functions and estimation biases of states [13] while still dealing with new architectural designs [40]. This brief overview of the generalization problem highlights two critical issues found in DRL policies; one, the typical use of deep neural networks as function approximators, which inherits their intrinsic issues [10]. Two, the inability to completely explore entire MDPs in high-dimensional problems.

These critical drawbacks are then worsened by the problem of over fitting. Most DRL policies suffer from overfitting because they are only trained on single environments. Without solutions to any of these fundamental issues, the generalization problem will always exists as highlighted by [7] who thought RL policies specialization in training sets is their major fault when deployed in different environments. As a possible solution, this paper tries to improve DRL generalization by exposing algorithms to different MDPs during the training process. The two DRL algorithms used during the experiments were randomly exposed to different MDPs at each training episodes with their results analyzed in this paper. Two test environments were used, the financial markets (profits and losses) and agriculture (application of fertilize and impact on winter wheat production).

The process of conducting the suggested experiments led to two contributions from this paper:

1. Development of a financial trading environment (including; action space, state space, and reward function) that exposes any reinforcement learning algorithms to different MDPs during training. The frequency of MDP changes in every episode can be defined by the user, including the restrictions of repeat and reshuffle.
2. Experiments on two real world problems where baseline and domain randomization tests were done.

2   Background

To properly define the approaches used in this paper to achieve generalization, it is important to outline the basic components of RL and DRL. In particular, the algorithms and how they interact with an environment. An RL and DRL algorithm $\mathbb{A}$ learns a policy π by interacting with a Markov Decision Process (MDP) [10]. MDPs are a vital aspect of RL as they provide a framework for solving decision-making problems that have random or controlled decision makers [37]. These decision makers tend to make sequential decisions that MDPs help to define by evaluating actions taken during current states and environments [34]. In this paper, MDPs from different but related contexts are used to improve performance by randomly using them in the training episodes.

In a standard RL formulation an agent sequentially engages an environment. This process is modeled as a tuple problem (formal definition of an MDP) having a state (S), action (A), transition function (P), reward (R) and discounting factor (Υ). This interaction typically starts when the agent begins from a state s_0 which is part of a possible start states (S_0) by taking an action $a \, \varepsilon \, A$, at given time step $t \, \varepsilon \, N$ [31]. What follow is a transition function (T) which defines the probability of moving to state s' having taken action a from state s [41]. Thereafter, a reward function R outlines the agent's reward following a transition. Ultimately, the agent's objective is to maximize the accumulated reward, discounted by a factor Υ.

A policy π that facilitates the agent's movement is important to mention. Policy π helps to map states into actions (π: $S \rightarrow A$), a process that fully highlights the behavior of the agent. In the end, the typical objective of an RL problem is to maximize the expected cumulative reward R (Eq. 1).

$$R = \mathbb{E} \left[\sum_{t=0}^{T-1} \gamma^t r(s_t, a_t) \right] \tag{1}$$

Where:

- R is expected return,
- Υ is the discount factor,
- $r(s_t, a_t)$ is reward a time t when being in state s_t by taking action a_t ,
- T is the total time-steps.
- All this sum run over time-steps t=0 to T-1

2.1 Related Works on Generalization

Generalization has been a long-standing issue in RL that has led to different specifications and solutions. Often, most measures have focused on the learning process in an attempt to capture the abilities of agents operating in different environments and making efficient use of samples [32], including those in policy-based methods [15]. Such measures have dominated learning theories in sample complexities as experiments try to capture efficiency in the learning processes [22], more so when samples have information on environment dynamics [4]. Other notable works have attempted to apply supervised learning dynamics by having train and test paradigms in training trajectories or state spaces [2]. Overfitting has also been studied in a similar manner, where RL agents have been found to overfit to their training environments [8]. Such studies have provided benchmarks for testing RL in the real world [3]. When formularized, these benchmarks have showed the importance of having a wide range of testing environments to limit overfitting. In all, generalization in most of these works has been captured as a means to avoid over-fitting in particular training environment [19]. This broad definition has implied a need to sample from diverse environments as a starting point for generalization.

However, this sampling requirement can vary based on the definition of generalization. For instance, when generalization is defined as improved performance in off-policy states [1], this diverse sampling is not necessary as it pushes RL into the standard methods of supervised learning. Similarly, generalization techniques such as adding stochasticity in policies, having random steps, and adding human play steps push RL agents into off-policy states which diversify the training data without any sampling requirement [21].

2.2 Methods of Generalization

Generalization in DRL and RL in general is led by two approaches; modification to the training algorithm and modification of the MDP.

Modifying the Training Algorithm. Also referred to as algorithmic generalization, this approach has methods that modify the training algorithm through techniques that use optimizers, regularizer, and updates to the rules of the policy [10].

A holistic definition of these methods considers a training algorithm $\mathbb{A}$ which gets an MDP as input and produces a policy π. Given a typical MDP definition (tuple, with 5 elements), an algorithmic generalization method G_A is produced by a function $F : A \rightarrow \mathbb{A}$ that runs the training algorithm $F(\mathbb{A})$ in the said MDP.

Modifying the MDP Domain Randomization. The other major category of generalization tries to change the MDP directly by either modifying the learning environment or the training data [39]. This transformation affects the interaction between the training algorithm and the learning environment by having an overall objective of increasing the samples engaged.

This approach is also called domain randomization as it involves varying the parameters of the environment during training to increase the robustness of the learned policy [10]. Domain randomization is implemented during training where the selected environment parameters are adjusted randomly during the episodes [6]. The changes done alter the state transition dynamics, reward processes and in some instances the agent's interaction with the environment [25]. Domain randomization modifications can be summarized as:

1. Randomizing the state space - where the state space is changed to reflect environmental conditions as highlighted by Eq. 2.

$$S = \{s \mid s \in S_0 \text{ with random perturbations from } P\} \tag{2}$$

With S_0 being the base state space and P representing random perturbations of an environmental variable.

2. Randomizing the action space - here the action space is modified by changes in environmental dynamics as formulated by Eq. 3.

$$A = \{a \mid a \in A_0 \text{ with random variation from } P\} \tag{3}$$

where A_0 is the original action space and P is the noise in the environment.

3. Randomization of transition probability - the state transition model is varied to simulate uncertainty in the real world (Eq. 4).

$$P(s' \mid s, a) = P_0(s' \mid s, a) \times N(\mu, \sigma) \tag{4}$$

With $P_0(S'|s, a)$ representing the original transition model and $N(\mu, \sigma)$ is the added noise to simulate environmental dynamics.

4. Randomization of the reward function - finally, the reward function can be randomized as influenced by environmental changes (Eq. 5).

$$R(s, a) = R_0(s, a) + R_{\text{rand}}(s, a) \tag{5}$$

$R_0(s, a)$ being the base reward and $R_{rand}(s, a)$ is the randomness element.

Therefore, a possible path for generalization through domain randomization can be achieved using the following basic steps:

- Start randomization
- Training on different (varied) environments
- Variations are introduced to the environments (i.e. changes to state space, action space, transition and reward)
- Policy learning stage where the agent is exposed to diverse training scenarios
- Transfer the learning to an unknown environment

3   Methodology

3.1   Motivating Examples

Robust performances for either offline and online RL techniques will begin to emerge when algorithms gain some level of generalization. This demand for enhanced capabilities motivated the use of the two real world problems; the financial market and agriculture. Both of these issues have ever changing conditions that limits the application of other machine learning techniques. For instance, while a financial instrument like Gold or Euro/U.S. Dollar may have an evident trading pattern, unlimited factors can drastically shift its price action [23]. Similarly, in agriculture, the yields from a particular crop can drastically change from one season to another. Developing a model that adapts to immediate conditions is thus necessary as defined by the existing environmental parameters.

This requirement defines a Markov Property where the future state only depends on the present state. Also defined as memorylessness, this property is considered in the motivating examples where formally, if S_t represents a state of a process at time t, the Markov property is given by

$$P(X_{t+1} \mid X_t, X_{t-1}, \ldots, X_0) = P(X_{t+1} \mid X_t) \tag{6}$$

Which means that the conditional probability of moving to a new state X_{x+1} depends solely on the present state X_t. While, this assertion is true, it borrows from the understanding that the past and future are conditionally independent, given the present [27]. This because a *State* as defined in an MDP determines what happens next and is understood to be a function of the history [28]. With history being a sequence of observations, actions and rewards (Eq. 7).

$$H_t = O_1, R_1, A_1, \ldots A_{t-1}, O_t, R_t \tag{7}$$

Formally, therefore;

$$State(S_t) = f(H_t) \tag{8}$$

Defining the Environments. For the financial market, Gymnasium (formerly Gym) environments were designed using six financial instruments (Euro/U.S. Dollar(EURUSD), British pound/U.S. dollar (GBPUSD), U.S. dollar/Chinese Yuan (USD-CNY), U.S. Dollar/Japanese Yen (USDJPY), Silver (XAGUSD) and Gold (XAUUSD)). This environment was named BeshaGym and the dataset for all the 6 instruments was from 2004 to 2024.

BeshaGym unlike existing trading environments provides a means to randomize MDPs during training, which is a significant contribution of this paper. Moreover, the randomization can be changed based a user's needs.

For the second environment, CropGym was used. CropGym is a reinforcement learning platform developed by the Wageningen University in Netherlands for optimizing DRL agents for the application of nitrogen fertilizer in crops [12].

The objective of the BeshaGym was to create an environment where a DRL agent trades the financial market to maximize profit. The agent would have access to a window size of 50 past time steps (i.e. the agent observed the last 50 time steps at every trading period) to base its decisions. The reward (R_t) from the BeshaGym environment was based on the ability to take profitable trades be it either through buys (long) or sells (shorts). A starting trading balance of 10,000 (representing \$10000) was also initialized in the environment.

$$R_t = \begin{cases} -(spotprice - lastprice) \times 10000 & \text{if, Short \& action = Sell} \\ +(spotprice - lastprice) \times 10000 & \text{if, Long \& action = Buy} \\ 0 & \text{otherwise} \end{cases}$$

CropGym on the other hand, as defined by [12] is a highly configurable RL environment built on the Python Crop Simulation Environment (PCSE) to help develop RL agents for implementing crop growth simulation models. In this paper, the experiments done were on the application of nitrogen fertilizer on Winter Wheat [12]. DRL agents were to apply a discrete amount of nitrogen fertilizer to balance crop yield and environmental impact. The reward (R_t) of this environment borrows from the LINTUL-3 model which is a subset of PCSE general model where growth parameters like TNSOIL (amount of nitrogen in the soil), LAI (leaf area index), NUPPT (uptake of nitrogen in soil) and WSO (weight of storage organ - the grain) are simulated. The DRL agent thus tries to maximize the difference in WSO for policy using nitrogen fertilizer and those that do not use the fertilizer (zero nitrogen policy). There is also an added cost β that mimics the price of nitrogen fertilizer (Eq. 9).

$$R_t = (WSO_t^\pi - WSO_{t-1}^\pi) - (WSO_t^0 - WSO_{t-1}^0) - \beta N_t \tag{9}$$

DRL Agents. For the DRL agent, they were based on Deep Q Network (DQN) (Fig. 1) [36] and Proximal Policy Optimization (PPO) (Fig. 2) [16] algorithms. These algorithms were sourced from the Stable-Baselines3 (SB3) library which provides simple and effective implementations. For DQN, SB3 builds on top of the fitted Q-iteration (FQI) which helps stablize learning with neural networks. Conversely, PPO combines ideas from A2C [16] and TRPO [32] by minimizing updates for new policy from old policy.

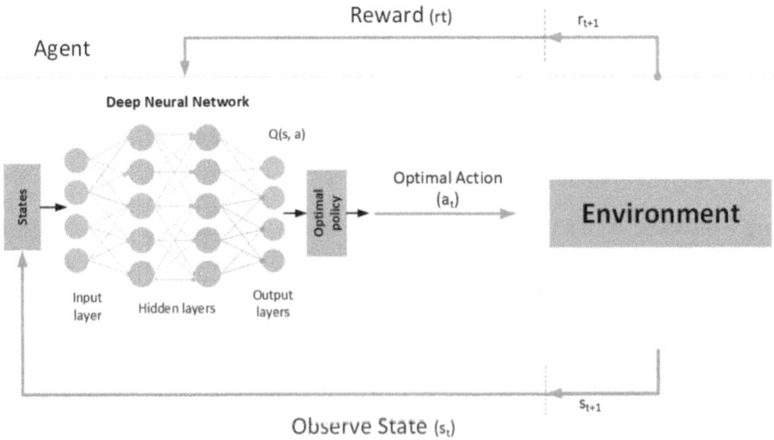

Fig. 1. High-Level Diagram of Deep Q Network [29].

3.2 Definitions

To improve generalization, the experiments conducted were formally setup with a domain randomization approach. A randomization strategy was used while training the agents in different MDPs across the training episodes. These MDPs were generated by creating diverse Gymnasium environments using different CSV datasets which encoded different training scenarios. As such, each episode involved training an agent on an environment from a different dataset which effectively varied the environment's dynamics, promoting generalization by simulating real world situations.

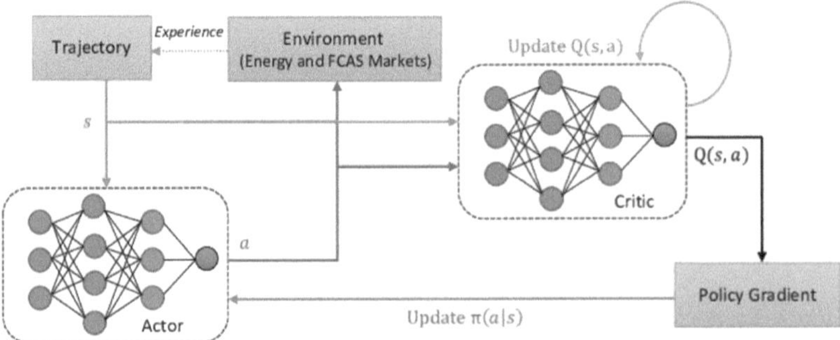

Fig. 2. High-Level Diagram of Proximal Policy Optimization [20].

Key Items

Markov Decision Process (MDP): An MDP as outlined earlier was defined by 5 elements; S set of states, A set of actions, P state transition probability, $R(s, a)$ reward function and Υ a discounting factor.

Environment Distribution ξ: Distribution over environments where every environment $M_i = (S_i, A_i, P_i, R_i, \Upsilon)$ was drawn from the distribution ξ. In this case, the environments were drawn from the CSV data, having different state transition dynamics and rewards designs.

Policy π_θ: Agent's policy which was parametrized by θ. It defined the action selection strategy when given a state s.

Training Episodes: At each training episode t, an environment M_i was sampled from the environment distribution ξ. Thus, an agent was interacting with environment M_i during a given episode t.

Objective: The overall goal for the agent was to maximize expected cumulative reward across the training episodes.

Mathematical Formulation

Training Process: Across all the training episodes T, an agent interacted with environments M as sampled from ξ

$$\mathbf{M}_i \sim \xi, i = 1, 2, ..., T \tag{10}$$

MDP Dynamics: Agent interacts with each environment M_i during episodes t followed the standard RL setup

$$s_t \sim S_i, a_t = \pi_0(s_t, r_t = R_i(s_t, at), s_{t+1} \sim P_i(s_{t+1}|s_t, a_t) \tag{11}$$

Expected Return: Agent's objective was to maximize the expected cumulative reward which represents the expected return $J(0)$ over multiple environments.

$$J(\theta) = \mathbb{E}_{M_i \sim E} \left[\sum_{t=0}^{T} \gamma^t r_t \right] \tag{12}$$

4   Experiments

All the algorithms were developed using PyTorch because of its flexibility and seamless integration with CUDA (easy acceleration with GPU). The baseline agents were also implemented using Stable Baseline3 library. Most of the training and comparison of agents was done on a personal computer with the following specifications: 11th generation Intel core i7-11850H x 16 processor and a NVIDIA T600 GPU with 32 gigabyte of RAM and 1 terabyte of storage space.

The key hyperparameters for the algorithms used are highlighted in Table 1.

Table 1. Default Hyperparameters for DQN and PPO.

Hyperparameter	BeshaGym		CropGym	
	DQN	PPO	DQN	PPO
learning_rate	1e-2	1e-2	1e-4	2.5e-4
buffer_size	200000			
batch_size	64	64	64	256
gamma	0.995	0.995	0.99	0.999
exploration_fraction	0.2		0.1	
exploration_initial_eps	1.0		1.0	
exploration_final_eps	0.06		0.01	
target_update_interval	2000			
train_freq	5			
learning_starts	2000			
max_grad_norm	20	20	10	0.7
n_steps				512
ent_coef				0.01
clip_range				0.1
n_epochs				4
gae_lambda				0.99
vf_coef				0.5
net_arch			[256, 256]	pi/vf:[256,256]

4.1 Running Experiments

In the two test environments, BeshaGym (the financial market) and CropGym (nitrogen application on winter wheat), two general experiments were conducted; ***Baseline Experiment*** and ***Domain Randomization Experiment***.

Baseline Experiment: In this first experiment, the DQN and PPO algorithms were trained on the environments without any domain randomization. This classical deployment was done to establish a baseline to test the effectiveness of domain randomization.

For BeshaGym, only one environment was used and it was sourced from the XAU-USD dataset. This meant, the algorithms were exposed to the same environment across all the training episodes. Conversely, for CropGym, a single region was defined in the training location. CropGym connects to environmental and geographical data providers like the NASA Power project which makes it easier to source the necessary training data [26]. Therefore, like BeshaGym, the training environment was consistent across all the training episode for the first set of experiments.

Domain Randomization Experiment: Thereafter, the second experiment revolved around domain randomization as highlighted in Sect. 3 above. For BeshaGym, the implementation of domain randomization was practically done by randomizing the selection of CSV datasets (i.e. EURUSD, GBPUSD, USDCNY, USDJPY, XAGUSD and XAUUSD). To ensure agents had access to all CSV files, and hence the resultant environments, the randomization had a no-repeat restriction and reshuffle. That is, agents had access to unique CSV files for every 6 episodes and all files were reshuffled thereafter.

On the other hand, CropGym provides a rudimentary means to randomize the training location where an algorithm can be trained in different locations as encoded using source files or data from providers like the NASA POWER Project. The latter method was used where five (5) different locations were used to train the algorithms by providing their longitude and latitude to the Nasa Power parameter [26]. The five locations were Ol'Kalou, Shamata, Ishiara, Meru and Nanyuki (all major agricultural areas in Kenya). Similarly, the agents were exposed to unique locations for every 5 episodes with a shuffle done thereafter.

4.2 Evaluation

To evaluate the two sets of experiments, agents were tested in two environments. The first environment was found in the training set and the second was not i.e. a random environment. For BeshaGym, this saw evaluations done on environments sourced from XAUUSD and XAGUSD (found in the training set on both experiments) and EURGBP (not on the training set). CropGym on the other hand saw evaluation of the agents on environments sourced from the Ol'kalou region (found in the training set) and Mwea region (not in the training set).

The evaluation was primarily done using the reward obtained from the environments. In BeshaGym, this reward influenced the Profit/Loss outcomes and in CropGym, it affected the crop growth. Moreover, TensorBoard logs were used, giving further evaluation results through training/validation loss, policy gradient loss, and value loss.

4.3   Results and Discussion

BeshaGym. The results as depicted in both Fig. 3 and Table 2 below shows DQN outperformed PPO, both in baseline and domain randomization experiments.

Domain randomization results were far better than those of the baseline results on most of the runs. In particular for DQN, the agents exposed to randomized environments had higher cumulative reward as compared to the baseline agents.

From the evaluation results highlighted in Table 2, domain randomization (DR) typically led to better reward, hence better Profit/Loss (P/L). It is however important to emphasize that PPO performed poorly across the board in the BeshaGym environment. Additionally, it is important to note that the agents got higher reward during evaluation for environments that were part of the training set. For environments not in the training set, agents in the domain randomization experiments had higher reward as compared to the baseline agents.

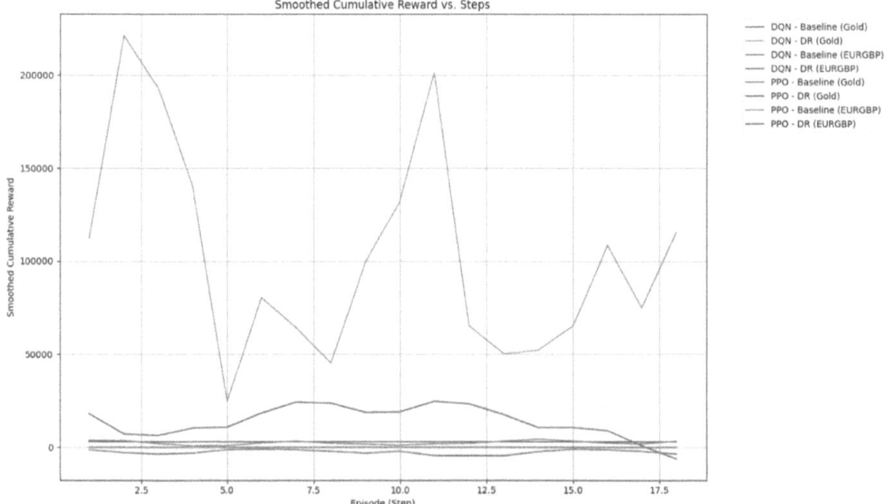

Fig. 3. BeshaGym Cumulative Reward.

These results were further emphasized by the tensorboard logs, where for DQN the training loss decreased more smoothly in the domain randomization experiments (Fig. 4 a). For the agents in the baseline experiments, their training loss had random spikes across various training iterations, although, it was able to decrease towards the end. In comparison, the agents in domain randomization experiments saw a smooth decline in the training loss, across various training iterations. For PPO, both baseline and domain randomization agents saw the training loss decrease smoothly across the training period, however, both failed to get reasonable changes on the policy gradient loss (Fig. 4 b) (even with hyper-parameter tuning) which could explain the poor performance of PPO in the financial market environment.

Table 2. BeshaGym Evaluations.

Evaluation Episodes	Cumulative Reward				
	Ep1	Ep2	Ep3	Ep4	Ep5
DQN					
Baseline (Gold)	41998.0	2769.9	9162.5	9162.5	701.0
DR (Gold)	18310.0	101000.0	218000.0	345000.3	16450.0
Baseline (EURGBP)	1548.0	−711.8	−4988.9	−2976.9	−2979.0
DR (EURGBP)	2982.0	2982.0	2982.0	2982.0	2982.0
PPO					
Baseline (Gold)	1254.9	4873.1	4999.1	540.8	529.8
DR (Gold)	−4.9	−4.9	8.9	0	0
Baseline (EURGBP)	0	0	0	0	0
DR (EURGBP)	−96.6	−91.6	−91.6	−62.5	−60.5

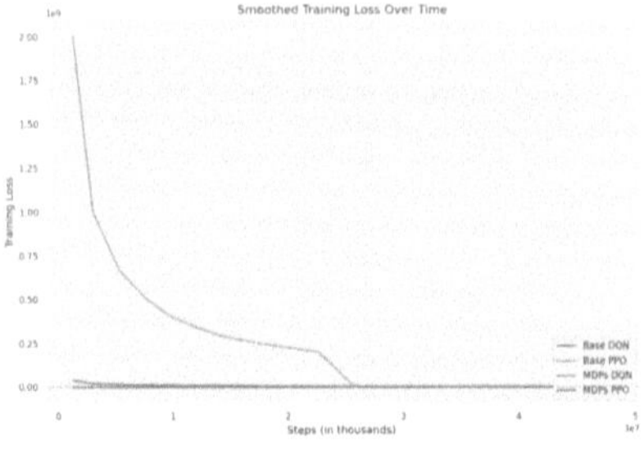

(a) Training Loss: Baseline Vs Domain Randomization

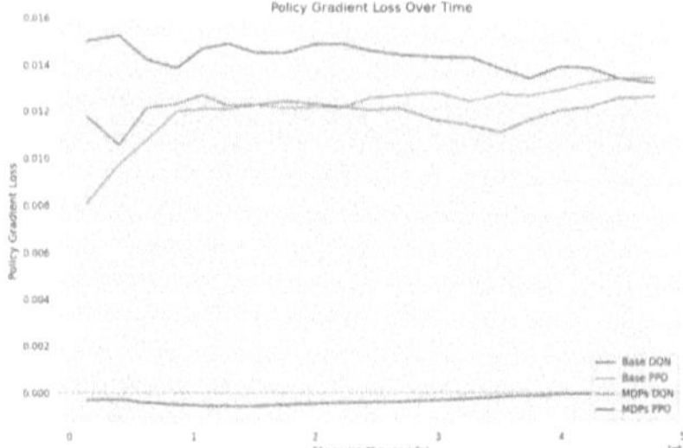

(b) Policy Gradient Loss: Baseline Vs Domain Randomization

Fig. 4. Training metrics.

CropGym. The training reward for agents in the domain randomization experiments was higher than that of the agents in baseline experiments as shown in Fig. 5. Generally, the cumulative reward for the agents in the randomized environments was higher than that of the baseline experiments. This outcome meant that the trained policy for both DQN and PPO performed better in balancing the application of fertilizer to meet the needs of crop yield (good WSO) while minimizing any environment effects. From Fig. 4, the cumulative reward of baseline PPO was lower than all other experiments. Baseline DQN had a better performance, including competing with the best agent (domain randomized DQN) in some steps. However, its cumulative reward fell towards the end which highlights the importance of exposing agents to different environments.

On top of the cumulative reward, a subtle difference in WSO results was also observed. The WSO (weight of the storage grain/organ) saw a bigger growth for the agents in the domain randomization experiments. This is observed in Table 3 where the agents exposed to different environments had bigger WSO values.

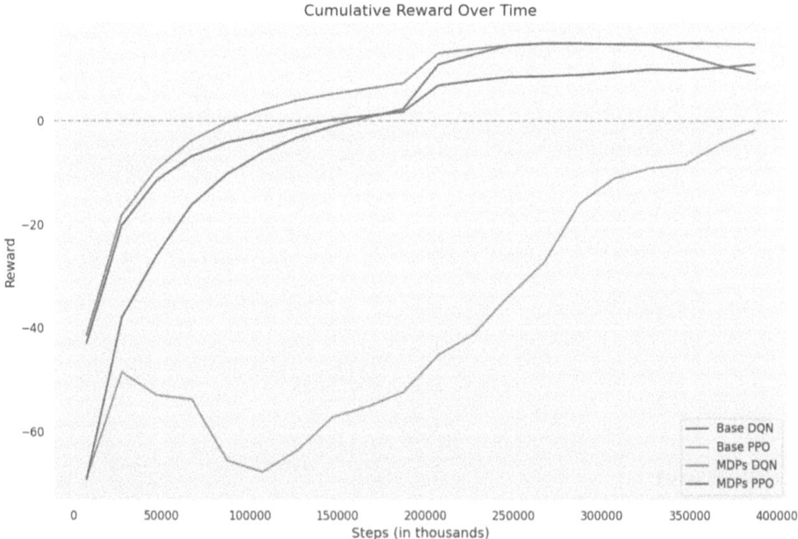

Fig. 5. CropGym Cumulative Reward: Baseline Vs Domain Randomization.

Table 3. WSO for Baseline and Domain Randomization.

Step	Base DQN	Base PPO	MDPs DQN	MDPs PPO
20500	38.90	36.62	6.21	6.11
62500	44.91	56.49	79.47	62.57
104500	29.60	65.58	80.39	33.25
146500	35.17	44.92	50.38	53.5
188500	25.15	28.69	34.33	43.94
230500	29.77	28.22	40.68	43.75
272500	36.21	28.22	39.89	43.57
314500	21.44	20.60	54.17	64.22
356500	27.76	33.22	35.69	56.35
398500	23.36	29.39	31.94	38.71

These results were also confirmed by the tensorboard logs where in both algorithms, the training reward and loss had better outcomes for the randomized experiments (Fig. 6). Higher training rewards were observed in the second set of experiments, as compared to the baseline experiments. The training loss also decreased in a smoother fashion with minimal spikes, showcasing a better training process for the second set of experiments.

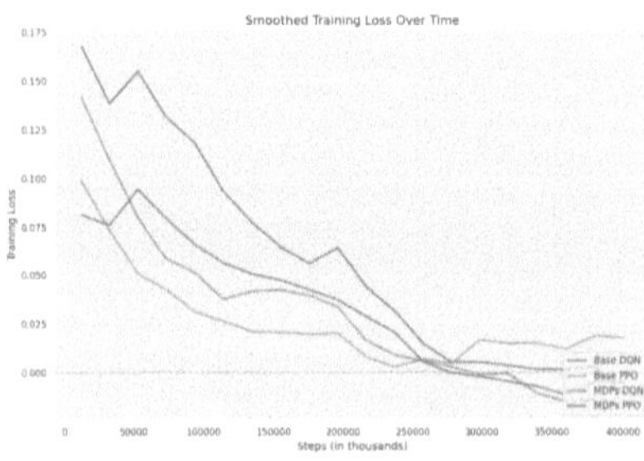

(a) Training Loss: Baseline Vs Domain Randomization

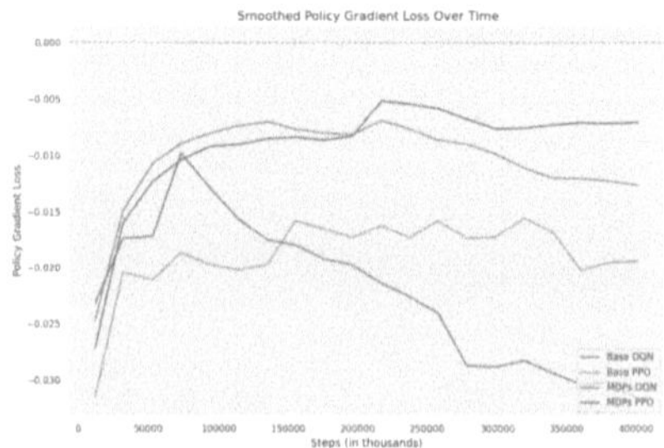

(b) Policy Gradient Loss: Baseline Vs Domain Randomization

Fig. 6. Training metrics.

5   Conclusion

In this paper, an attempt was made to improve the generalization of deep reinforcement learning algorithms. Using domain randomization, DQN and PPO agents were deployed in two real world scenarios where the objectives were to increase profits and crop yields. These agents had improved performance when trained on different and randomized environments as compared to when they were trained on single (specific) environments.

Future works can try to address the performance problem of PPO in the financial market environment by either adjusting the environment parameters or the training specifications. To further test the effectiveness of domain randomization, different randomization techniques can be compared. Moreover, domain randomization can be compared with other techniques of generalizing DRL algorithms. For instance, modifying the training algorithms through regularization techniques, reward shaping and redefining replay buffer among many other methods. Additionally, combining these classical DRL algorithms with Transformers and Recurrent Neural Network can be done with the same objective of trying to improve generalization.

Acknowledgments. This work was performed using HPC resources from GENCI-IDRIS (Grant 2025-[AD010614071R2].

The researchers thank Strathmore University and LIASD Lab (University Paris 8) for invaluable academic and institutional support during the research of this paper.

The lead author gratefully acknowledges the French Embassy in Kenya for its generous support and facilitation of his doctoral studies.

Disclosure of Interests. The authors have no competing interests to declare that are relevant to the content of this article.

References

1. Nair, A., et al.: Massively parallel methods for deep reinforcement learning. https://arxiv.org/abs/1507.04296 (2015)
2. Nair, A., et al.: Gotta learn fast: a new benchmark for generalization in RL. https://arxiv.org/abs/1804.03720 (2018)
3. Nouri, A., Littman, M. L., Li, L., Parr, R., Painter-Wakefield, C., Taylor, G.: A novel benchmark methodology and data repository for real-life reinforcement learning. In: Proceedings of the 26th International Conference on Machine Learning (ICML). Association for Computing Machinery (2009). https://dl.acm.org/doi/10.1145/1553374.1553499, Accessed 28 May 2025
4. Strehl, A.L., Li, L., Littman, M.L.: Reinforcement learning in finite MDPS: PAC analysis. J. Mach. Learn. Res. **10**(11), 2413–2430 (2009)
5. C, A., K, P.: Managing engineering systems with large state and action spaces through deep reinforcement learning. Reliability Eng. Syst. Safety **191**, 106483 (2019). https://doi.org/10.1016/j.ress.2019.04.036
6. Kang, C., Chang, W., Choi, J.: Balanced domain randomization for safe reinforcement learning. Appl. Sci. **14**(21), 9710 (2024). https://doi.org/10.3390/app14219710
7. Packer, C., Gao, K., Kos, J., Krähenbühl, P., Koltun, V., Song, D.: Assessing generalization in deep reinforcement learning. https://arxiv.org/abs/1810.12282 (2018), Accessed 28 May 2025
8. Zhang, C., Vinyals, O., Munos, R., Bengio, S.: A study on overfitting in deep reinforcement learning. https://arxiv.org/abs/1804.06893 (2018)
9. Korkmaz, E.: Nesterov momentum adversarial perturbations in the deep reinforcement learning domain. In: Proceedings of the International Conference on Machine Learning (ICML) Workshop on Inductive Biases, Invariances and Generalization in Reinforcement Learning (2020). https://icml.cc/virtual/2020/7645, accessed: 2025-05-28
10. Korkmaz, E.: A survey analyzing generalization in deep reinforcement learning (2024)
11. Dulac-Arnold, G., Levine, N., Mankowitz, D.J., Li, J., Paduraru, C., Gowal, S., Hester, T.: Challenges of real-world reinforcement learning: definitions, benchmarks and analysis. Mach. Learn. **110**(9), 2419–2468 (2021). https://doi.org/10.1007/s10994-021-05961-4
12. Overweg, H., Berghuijs, H., Athanasiadis, I.: Cropgym: a reinforcement learning environment for crop management (2021). https://doi.org/10.48550/arXiv.2104.04326
13. Van Hasselt, H., Guez, A., Silver, D.: Deep reinforcement learning with double q-learning. In: Proceedings of the Thirtieth AAAI Conference on Artificial Intelligence (AAAI 16), pp. 2094–2100. AAAI Press (2016). https://doi.org/10.1609/aaai.v30i1.10295
14. Kos, J., Song, D.: Delving into adversarial attacks on deep policies. https://arxiv.org/abs/1705.06452 (2017)
15. Schulman, J., et al.: Trust region policy optimization. In: Proceedings of the 32nd International Conference on Machine Learning (ICML). Proceedings of Machine Learning Research, vol. 37, pp. 1889–1897. PMLR (2015). https://proceedings.mlr.press/v37/schulman15.html, Accessed 28 May 2025
16. Schulman, J., et al.: Proximal policy optimization algorithms. https://arxiv.org/abs/1707.06347 (2017), Accessed 28 May 2025
17. Terven, J.: Deep reinforcement learning: a chronological overview and methods. AI **6**(3), 46 (2025). https://doi.org/10.3390/ai6030046
18. Nweye, K., Liu, B., Stone, P., Nagy, Z.: Real-world challenges for multi-agent reinforcement learning in grid-interactive buildings. Energy and AI **10**, 100–202 (2022). https://doi.org/10.1016/j.egyai.2022.100202

19. Kolb, L., Panzer, M., Gronau, N.: Assessing generalizability in deep reinforcement learning based assembly: a comprehensive review. J. Intell. Manufacturing **35**(1), 1–19 (2024). https://doi.org/10.1007/s10845-024-02546-5

20. Anwar, M., Wang, C., De Nijs, F., Wang, H.: Proximal policy optimization based reinforcement learning for joint bidding in energy and frequency regulation markets. In: 2022 IEEE Power & Energy Society General Meeting (PESGM), pp. 1–5. IEEE (2022). https://doi.org/10.1109/pesgm48719.2022.9917082, Accessed 04 Mar 2025

21. Hausknecht, M.J., Stone, P.: The impact of determinism on learning atari 2600 games. In: Proceedings of the AAAI Workshop on Learning for General Competency in Video Games. AAAI Press, Austin, Texas (2015). https://www.cs.utexas.edu/ai-lab/pub-view.php?PubID=127537, Accessed 28 May 2025

22. Kakade, S.M.: On the Sample Complexity of Reinforcement Learning. Phd thesis, University College London (2003). https://discovery.ucl.ac.uk/id/eprint/10100726/, Accessed 28 May 2025

23. Roostaee, M.R., Abin, A.: Forecasting financial signal for automated trading: An interpretable approach. Expert Syst. Appl. **211**, 118570 (2023). https://doi.org/10.1016/j.eswa.2022.118570

24. Matsuo, Y., et al.: Deep Learning, Reinforcement Learning, and World Models. Neural Netw. **152**, 267–275 (2022). https://doi.org/10.1016/j.neunet.2022.03.037

25. Kaidanov, O., Al-Hafez, F., Suvari, Y., Belousov, B., Peters, J.: The role of domain randomization in training diffusion policies for whole-body humanoid control. https://arxiv.org/abs/2411.01349 (2024)

26. P, S.: Power — dav. (2025). https://power.larc.nasa.gov/, Accessed 15 Jan 2025

27. Getoor, R.K., Sharpe, M.J.: Markov properties of a Markov process. Zeitschrift für Wahrscheinlichkeitstheorie und Verwandte Gebiete **55**(3), 313–330 (1981). https://doi.org/10.1007/BF00532123

28. Suton, R.S., Barto, A.G.: Reinforcement Learning: An Introduction. Adaptive Computation and Machine Learning, The MIT Press, Cambridge, MA, 2nd edn. (2018). https://mitpress.mit.edu/9780262039246/reinforcement-learning/

29. S, A.: Deep q-learning (DQN) (2024). https://medium.com/@samina.amin/deep-q-learning-dqn-71c109586bae

30. Huang, S., Papernot, N., Goodfellow, I., Duan, Y., Abbeel, P.: Adversarial attacks on neural network policies. https://arxiv.org/abs/1702.02284 (2017)

31. Nguyen, H.S., Cruz, F., Dazeley, R.: Towards a broad-persistent advising approach for deep interactive reinforcement learning in robotic environments. Sensors **23**(5), 2681 (2023)

32. Witty, S., et al.: Measuring and characterizing generalization in deep reinforcement learning. Applied AI Letters **2**(4), e45 (2021). https://doi.org/10.1002/ail2.45

33. Dai, T., Arulkumaran, K., Gerbert, T., Tukra, S., Behbahani, F., Bharath, A.: Analysing deep reinforcement learning agents trained with domain randomisation. Neurocomputing **493**, 143–165 (2022). https://doi.org/10.1016/j.neucom.2022.04.005

34. Xu, T., Zhu, H., Paschalidis, I.C.: Learning parametric policies and transition probability models of markov decision processes from data. European J. Control **57**, 68–75 (2021). https://doi.org/10.1016/j.ejcon.2020.04.003

35. Mnih, V., et al.: Human-level control through deep reinforcement learning. Nature **518**(7540), 529–533 (2015). https://doi.org/10.1038/nature14236

36. Mnih, V., et al.: Playing atari with deep reinforcement learning. https://arxiv.org/abs/1312.5602 (2013)

37. Wang, X., Yang, Z., Chen, G., Liu, Y.: A reinforcement learning method of solving markov decision processes: an adaptive exploration model based on temporal difference error. Electronics **12**(19), 4176 (2023). https://doi.org/10.3390/electronics12194176

38. Wen, X., Yu, X., Yang, R., Chen, H., Bai, C., Wang, Z.: Towards robust offline-to-online reinforcement learning via uncertainty and smoothness. J. Artificial Intell. Res. **81**, 481–509 (2024). https://doi.org/10.1613/jair.1.16457
39. Young, K., Ramesh, A., Kirsch, L., Schmidhuber, J.: The benefits of model-based generalization in reinforcement learning. arXiv (2022). https://arxiv.org/abs/2211.02222, Accessed 28 May 2025
40. Wang, Z., et al.: Dueling network architectures for deep reinforcement learning. In: Proceedings of the 33rd International Conference on Machine Learning (ICML), pp. 1995–2003. PMLR (2016). https://proceedings.mlr.press/v48/wangf16.html, Accessed 28 May 2025
41. Zhang, Z., Zou, Y., Lai, J., Xu, Q.: M2dqn: A robust method for accelerating deep q-learning network. In: Proc. 2023 15th Int. Conf. Machine Learning Comput, pp. 116–120 (2023)

Degradation-Aware Energy Management in Residential Microgrids: A Reinforcement Learning Framework

Danial Zendehdel[✉] , Gianluca Ferro , Enrico De Santis ,
and Antonello Rizzi

Department of Information Engineering, Electronics and Telecommunications, University of
Rome "La Sapienza", Via Eudossiana 18, 00184 Rome, Italy
{danial.zendehdel,gianluca.ferro,enrico.desantis,
antonello.rizzi}@uniroma1.it

Abstract. This paper presents a degradation-aware reinforcement learning (RL) framework for real-time energy management in residential microgrids, focusing on optimizing lithium-ion battery usage while balancing economic benefits and battery longevity. We employ the Soft Actor-Critic (SAC) algorithm, implemented via Stable Baselines3, to learn non-linear dispatch policies for a 5.2 kWh $LiCoO_2$ battery pack, with degradation modeled using a simplified energy-throughput approach calibrated with NASA dataset measurements. The framework is tested across diverse household profiles over 1-year and 10-year simulations. Results show that RL-SAC outperforms a Model Predictive Control (MPC) baseline, extending battery life and reducing energy purchases in both simulations. These findings highlight RL-SAC's potential for practical deployment in microgrids, offering a scalable solution for sustainable energy management.

Keywords: Reinforcement Learning · Battery Management System · Energy Management · Lithium-ion Batteries · Degradation Modeling · Microgrid

1 Introduction

Renewable Energy Communities (RECs) represent a fundamental component in the shift toward cleaner, decentralized energy infrastructures. They enable consumers to become prosumers by actively participating in energy generation [20]. By doing so, RECs diminish reliance on centralized utilities while enhancing the robustness and autonomy of local energy systems. Moreover, they promote cooperative energy usage models, where members share renewable resources collectively for mutual benefit [2,27].

The European Union's Renewable Energy Directive II (RED II) is fundamental in promoting the development and growth of RECs. It establishes a comprehensive legal framework [8], incentivizing the deployment of renewable energy and empowering communities with the rights to produce, utilize, store, and exchange renewable energy within their networks. A significant innovation introduced by RED II is the concept

© The Author(s), under exclusive license to Springer Nature Switzerland AG 2026
F. Marcelloni et al. (Eds.): IJCCI 2025, CCIS 2829, pp. 538–557, 2026.
https://doi.org/10.1007/978-3-032-15638-9_31

of energy sharing, which allows participants to distribute locally generated electricity without the need for external suppliers [20], thus optimizing the use of locally generated renewable power [17].

In this framework, residential Battery Energy Storage Systems (BESS) have demonstrated their effectiveness in mitigating the fluctuations of renewable energy production. These systems store excess solar energy produced during high-yield periods and release it when production drops or consumption rises, thereby reducing grid dependency and increasing supply reliability. The expanding adoption of household photovoltaic (PV) systems integrated with lithium-ion storage further enables users to take part in local energy markets. In such contexts, households can economically benefit by selling excess generation or reducing purchases during expensive tariff periods [27]. However, achieving these outcomes depends significantly on appropriate BESS configuration and operational planning. Aspects like storage capacity, power limits, and aging effects must be accounted for to maintain both economic efficiency and system reliability [14].

Among energy storage technologies, lithium-ion batteries are widely adopted due to their high energy density, operational voltage, low self-discharge rate, and long cycle life [19]. Nevertheless, optimal utilization of Li-ion battery packs in residential microgrids remains challenging due to their complex degradation behavior, which is highly sensitive to operational parameters such as current, State of Charge (SoC), and temperature [3]. Battery degradation encompasses sophisticated internal chemical mechanisms, including material loss and structural deformation, leading to increased internal resistance and decreased capacity [28]. Aging in Li-ion batteries can be categorized into calendar aging, occurring during rest periods, and cycle aging, driven by charge-discharge cycles [6]. Overcharging, discharging, or operating under extreme conditions accelerate aging, potentially leading to early failure [6]. In residential microgrids, these batteries should endure thousands of cycles and respond quickly to power imbalances. Ineffective power-flow management that ignores degradation can reduce battery lifespan significantly, regardless of calendar limits [28,30].

A BESS can mitigate these challenges by monitoring key parameters such as voltage, current, temperature, and capacity, and estimating critical metrics like SoC, State of Health (SoH), and Remaining Useful Life (RUL) [6]. The SoH quantifies the battery's degradation state [22,32]. It is defined in stationary applications, such as smart grids, as:

$$\mathrm{SoH}_t = \frac{C_{t,\mathrm{max}}}{C_b},\tag{1}$$

where $C_{t,\mathrm{max}}$ is the maximum capacity at time t and C_b is the nominal capacity. However, SoH cannot be directly measured and must be inferred from other measurements, introducing additional complexity [22]. Despite the protective capabilities of a BMS, challenges persist due to the non-linear behavior of Li-ion batteries, temperature variations, and diverse operating conditions driven by stochastic renewable generation and dynamic electricity prices [3]. These factors make it difficult to achieve optimal utilization of Li-ion packs, as conventional strategies often treat the battery as an ideal buffer, neglecting the complex interplay between operational decisions and degradation dynamics.

To address these challenges, this work introduces a degradation-aware reinforcement learning framework for real-time energy management in residential microgrids equipped with Li-ion storage. Unlike conventional strategies, the proposed agent receives an explicit feedback on battery SoH, computed from cumulative energy throughput and linked to a cyclic degradation model. This health-aware signal allows the agent to learn control policies that optimize long-term economic performance while implicitly preserving usable battery capacity. Finally, our degradation-aware RL agent is benchmarked against a MPC baseline that incorporates degradation cost linearly via equivalent full cycles (EFC). Experimental results demonstrate that the RL approach can outperform the MPC benchmark in both cumulative economic return and battery SoH, validating its ability to learn complex, nonlinear dispatch policies from multidimensional operational data.

The remainder of this paper is organized as follows. Section 2 reviews recent RL approaches for energy management in microgrids, with a particular focus on methods that incorporate battery degradation modeling into control policies. Section 3 defines the energy management problem, introduces the microgrid model, and tariff structure, while sub-section 3.4 explains the aging model. Section 4 covers the RL methodology and simulation architecture. Section 5 outlines the experimental setup and benchmark controller, and discusses the simulation results. Finally, the conclusions are drawn in Sect. 6.

2 Literature Review

In recent scientific literature, RL has emerged as a promising framework for the control of energy management systems (EMS), offering model-free optimization capabilities in the presence of uncertainty, nonlinear dynamics, and battery degradation phenomena.

Among the recent advances in Deep Reinforcement Learning (DRL), the Twin-Delayed Deep Deterministic Policy Gradient (TD3) algorithm is instrumental in robust and efficient solution for continuous control problems, offering improved stability and learning efficiency over earlier actor-critic methods. In this regard, Domínguez-Barbero et al. [5] proposed an energy management strategy for isolated microgrids using the TD3 algorithm, incorporating nonlinear battery losses into the control model. Their work demonstrates improved energy efficiency and reduced operational costs compared to traditional linear models. Hosseini et al. [15] developed an EMS for lithium-ion battery storage in a multilevel microgrid using a Twin-TD3 RL agent, achieving superior tracking accuracy and power dispatch performance compared to conventional EMS approaches, including fuzzy logic, PSO, and nonlinear optimization methods. Guo et al. [11] introduced a real-time optimal energy management strategy for microgrids subject to environmental uncertainties, leveraging a Proximal Policy Optimization (PPO) algorithm within a DRL framework. Unlike conventional model-based approaches, which are often computationally intensive and limited by the need for explicit probability distributions, this method treats the scheduling problem as a MDP and uses historical data to learn optimal dispatch policies. Their findings indicate superior performance in terms of adaptability and computational efficiency when compared to classical approaches and alternative DRL algorithms such as DDPG and DQN.

The work in [16] uses deep learning and GPR for predicting lithium-ion batteries' RUL considering cell-to-pack lifespan prognostics approach. Unlike simplified models, Abbasi *et al.* (2024) model each cell individually, constructing a battery pack with varied capacities and resistances to simulate SoC divergence and resistance evolution, enabling imbalance-aware RL interactions [1]. Yalçın [31] proposed a hybrid RL framework using deep Q-learning and actor-critic methods to optimize battery charging in electric vehicles, with actions in terms of C-rate and duration, and states as SoC, temperature, and voltage.

Han et al. (2022) [13] proposed a real-time RL-based energy management strategy for hybrid electric vehicles that directly incorporates battery degradation constraints. Their framework leverages an eligibility trace algorithm for adaptive online control and models battery life loss as a function of throughput, temperature, and C-rate using an Arrhenius-based formulation. While originally applied in the automotive domain, their multi-objective reward formulation and degradation modeling are directly applicable to battery-aware RL in renewable energy communities, especially for systems where SoH and lifetime cost optimization is prioritized over short-term energy arbitrage.

3   Problem Formulation

RL is a distinct category of machine learning, coexisting with supervised and unsupervised learning methods. Unlike these traditional approaches, RL operates without the need for labeled data. Instead, it tackles sequential decision-making problems by enabling an agent to learn through ongoing interaction with an environment, guided by reward signals rather than pre-assigned labels [21].

This learning process is mathematically framed by the MDP, represented as a tuple $\langle S, A, T, R, \gamma \rangle$. Here, S defines the set of all possible states, A encompasses the action space, and T also expressed as $P(s'|s, a)$ – quantifies the probability of transitioning from a state s to another s' after taking action a. Additionally, $R(s, a)$ indicates the immediate reward earned from performing action a in state s, while γ, known as the discount factor, adjusts the agent's focus between short-term and long-term rewards ranging from $\gamma \in [0, 1]$. Building on this, the agent's behavior is governed by its policy, denoted as $\pi(a|s)$, which represents the probability of selecting action a when in state s. This policy dictates how the agent navigates the environment, making decisions at each step to achieve its objectives. The primary aim of RL is to identify an optimal policy, denoted as π^*, which maximizes the cumulative expected reward over time [29]. Specifically, the Q-function:

$$Q^\pi(s, a) = \mathbf{E}_\pi \left[\sum_{t=0}^{\infty} \gamma^t r(s_t, a_t) | s_0 = s, a_0 = a \right] \qquad (2)$$

is a way to evaluate the policy. Therefore, Eq. (2) represents the expected return from state s, taking action a and thereafter following policy π.

3.1   Microgrid Architecture Overview

The comprehensive layout of the analyzed microgrid is depicted Fig. 1. Our microgrid configuration consists of a photovoltaic generation system, a battery energy storage unit, a cumulative electrical load, and an interface with the main utility grid. The BESS is tasked to preserve the longevity of the battery while minimizing the cost of prosumer within an operational horizon τ with discrete time steps of length Δt. We denote the time index by $t \in \{0, 1, .., T - 1\}$ where T is the total number of timesteps in τ. The prosumer draws energy load P_t^L in kW, while it generates energy via PV production P_t^{PV} in kW at time t. To develop an effective strategy for implementing real-time energy management at the REC level, we have considered charging the battery using the grid. This can be alternatively managed by another community member at the REC level [2,7].

The energy flows within the microgrid follow a bidirectional logic and are governed by real-time power balance constraints. When PV generation exceeds the load demand, the surplus energy is either stored in the BESS or exported to the main grid. Conversely, when the load demand surpasses the PV generation, the energy shortfall is compensated by discharging the BESS or by importing energy from the main grid. Battery charging can occur either from surplus PV energy or from the main grid during periods of lower electricity prices, whereas discharging supports the load or enables energy export when economically favorable. The internal power flows within the microgrid are governed by the power balance equation:

$$NetLoad(t) = P^L(t) - P^{PV}(t), \tag{3}$$

$$P^{\text{grid}}(t) = NetLoad(t) + (1 - z) \times P^{\text{batt}}(t) - z \times P^{\text{batt}}(t). \tag{4}$$

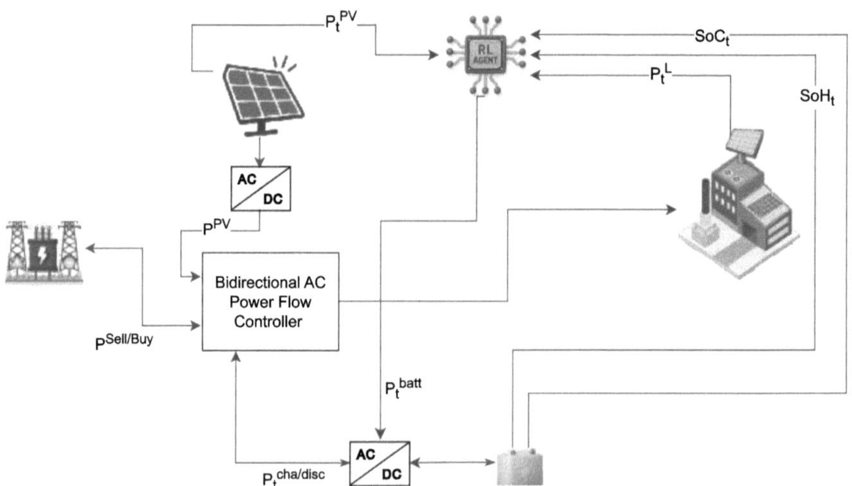

Fig. 1. Overview of the Environment; Red lines are Energy Flow; Green lines are IoT signals. (Color figure online)

$NetLoad(t)$ in Eq. 3 identifies the current situation of battery at time t, in which positive number shows deficit while negative value indicates surplus. In Eq. 4 $P^{\text{batt}}(t)$

denotes the battery charging/discharging power at the discrete time step t; positive values of $P^{\text{batt}}(t)$ indicate discharging while negative value indicates charging the battery; the z in Eq. 4 is a boolean number stating that the battery cannot be charged and discharged simultaneously. The profit or cost of the prosumer at each timestep can be derived from Eq. 5 below, in which $\text{Price}_{\text{buy}}(t)$ and $\text{Price}_{\text{sell}}(t)$ are derived from Electricity price tariff scheme, detailed in Sect. 3.2, while the energy trading strategy is detailed in Algorithm 1.

$$\begin{aligned}
\text{profit} &= P^{grid}(t) \times \text{Price}_{\text{sell}}(t) \times \Delta t, \\
\text{cost} &= P^{grid}(t) \times \text{Price}_{\text{buy}}(t) \times \Delta t.
\end{aligned} \tag{5}$$

Algorithm 1: Energy Management.

Data: $P^{batt}(t)$, $P^{L}(t)$, $P^{PV}(t)$ at time t
Result: E_t^{buy}, E_t^{sell}
Compute $NetLoad(t)$;
if $NetLoad(t) > 0$ **then**
    if $P^{batt}(t) > 0$ **then**
        $E_t^{buy} \leftarrow \max(0, NetLoad - P_t^{batt}\Delta t)$;
        $E_t^{sell} \leftarrow \max(0, P_t^{batt}\Delta t - NetLoad)$;
    else
        $E_t^{buy} \leftarrow \max(0, NetLoad + |P_t^{batt}|\Delta t), E_t^{sell} \leftarrow 0$;

else
    if $P^{batt}(t) < 0$ **then**
        $E_t^{buy} \leftarrow \max(0, |P_t^{batt}|\Delta t - |NetLoad|)$;
        $E_t^{sell} \leftarrow \max(0, |NetLoad| - |P_t^{batt}|\Delta t)$;
    else
        $E_t^{sell} \leftarrow |NetLoad| + P_t^{batt}\Delta t, E_t^{buy} \leftarrow 0$;

3.2 Price Tariffs

The electricity cost for each prosumer is computed under a *time-of-use* (ToU) schedule with three distinct bands, following the fully volumetric rate structure proposed by Fridgen et al. in [10]. The energy exchanged with the grid is priced volumetrically, with no additional capacity or customer charges, thus reflecting a *fully volumetric* tariff. The adopted ToU bands and the specific energy prices for each ToU band, both for purchasing and selling electricity are summarised in Table 1. These values are consistent with the fully volumetric tariff structure described above and are used throughout the simulation to evaluate the cost of energy exchanges with the grid.

Table 1. ToU bands for the fully volumetric tariff adopted in this study [10].

Band	Time interval (hh:mm)	Buy price ($/kWh)	Sell price ($/kWh)
Off-peak	00:00–06:00 & 22:00–24:00	0.3060	0.0179
Mid-peak	06:00–14:00 & 20:00–22:00	0.3137	0.0256
Peak	14:00–20:00	0.3201	0.0320

The ToU pricing schedule in the environment's reward function allows the agent to learn optimal battery management by anticipating peak pricing and scheduling operations to balance costs and degradation. Though current models use only ToU pricing, future tariffs might include capacity-based charges to avoid excessive peak power exchange, as suggested by Fridgen (2018) [10]. This could limit aggressive strategies that harm batteries and, when integrated into the RL framework, promote smoother power management.

3.3 Battery Pack Specifications

In this study, we employ a real-world simulated battery pack to represent the BESS, sourced from the NASA Prognostics Data Repository [23]. These specifications are consistently applied within our simulation environment, and are detailed in Table 2.

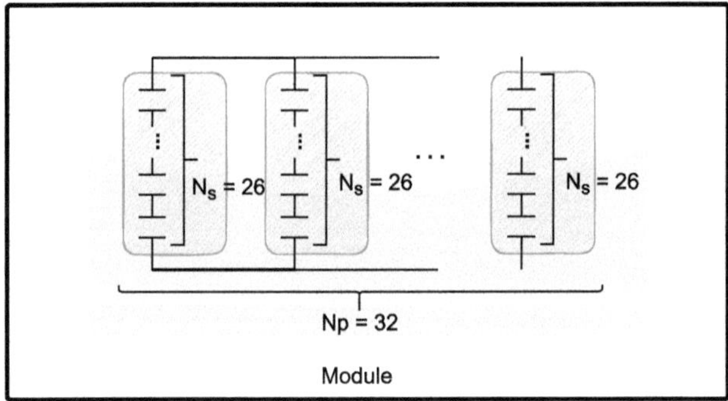

Fig. 2. Simplified representation of the cell-to-pack configuration a single module of battery pack, illustrating the series-parallel arrangement used in one module.

The BESS is assembled as a 5.2 kWh, 96.2 V lithium-ion ($LiCoO_2$) battery pack by linking two 48 V modules in series. Each module contains 13 strings connected in series ($N_s/N_m = 13$). Each string consists of 32 parallel-connected 18650 cells, resulting in a total configuration of $N_s = 26$ series cells and $N_p = 32$ parallel cells, denoted as 26S32P. A conceptual overview of the cell-to-pack structure, is illustrated in Fig. 2. This configuration ensures a balance between voltage scalability and capacity enhancement, suitable for residential microgrid applications where both energy storage and power delivery are critical.

Cell-level parameters were directly extracted from the NASA dataset, which provides detailed measurements for $LiCoO_2$ 18650 cells with a nominal capacity of $Q_{cell} = 2.0\,Ah$ and a nominal voltage of $V_{cell,nom} = 3.7\,V$. Consequently, the pack's nominal voltage and its nominal capacity are calculated as:

$$V_{pack} = N_s \times V_{cell,nom} = 26 \times 3.7 = 96.2\,V,$$

$$Q_{pack} = N_p \times Q_{cell} = 32 \times 2.0 = 64\,Ah.$$

The total nominal energy capacity of the pack is therefore $E_{pack} = \frac{V_{pack} \times Q_{pack}}{1000} \approx$ 6.15 kWh. However, the nominal capacity is specified as 5.2 kWh to reflect a downgraded usable energy capacity. This energy capacity aligns with typical residential energy storage requirements, enabling the pack to support load balancing and energy arbitrage.

A continuous power rating of 10 kW is established, with a maximum pack current of ($I_{pack,max} = \frac{10kW}{96.2} \approx 104\,A$). The maximum current per cell is $I_{cell,max} \approx 3.25\,A$, yielding a C-rate of $1.625\,C$, well below the 2C limit to minimize heating and capacity fade.

The battery is modeled at cell and pack-level electrical dynamics using an equivalent circuit approach, as depicted in Fig. 2. The internal resistance (R_{cell}) for cell B0005 from the NASA dataset was computed as the average sum of ohmic resistance (R_e) and charge-transfer resistance (R_{ct}) from impedance cycles, yielding $R_{cell} = 0.13\,\Omega$. Similarly, the OCV was approximated using a linear relationship derived from the dataset's discharge cycles at a 1C rate (2 A), where the voltage at low current conditions closely approximates the OCV which is calculated using a fallback linear approximation:

$$OCV_{cell} = 3.0 + \left(\frac{SoC_\%}{100}\right) \times 1.2\,V,$$

where $SoC_\%$ ranges from 0 to 100. This simplification eliminates the computational overhead of interpolation while providing a reasonable approximation for the RL environment's electrical dynamics.

Given a target power command for the pack, $P_{batt}(t)$, the power per cell is distributed as:

$$P_{cell}(t) = \frac{P_{batt}(t) \times 1000}{N_s \cdot N_p},$$

The corresponding cell current, $I_{cell}(t)$, is obtained by numerically solving the power equation based on a first-order Thevenin battery model, that is a quadratic equation in $I_{cell}(t)$:

$$R_{cell}(t) \cdot I_{cell}^2(t) - OCV_{cell}(t) \cdot I_{cell}(t) + P_{cell}(t) = 0.$$

Once solved it yields two roots, with the appropriate root selected based on the power direction (positive for discharge, negative for charge). The pack-level current and resistance are then computed as:

$$R_{pack}(t) = R_{cell}(t) \cdot \frac{N_s}{N_p}, \quad I_{pack}(t) = I_{cell}(t) \cdot N_p \tag{6}$$

This approach ensures that the electrical dynamics of the entire pack are accurately modeled, reflecting the aggregate behavior of the 26S32P configuration while leveraging real-world cell data from the NASA dataset.

Table 2. Key parameters of the lithium-ion battery pack.

Parameter	Symbol	Value
Cell chemistry	–	$LiCoO_2$ (18650)
Cells in series	N_s	26
Cells in parallel	N_p	32
Modules in series	N_m	2 (13S32P each)
Cell nominal capacity	Q_{cell}	2.0 Ah
Cell nominal voltage	$V_{cell,nom}$	3.7 V
Pack nominal voltage	V_{pack}	96.2 V
Pack nominal capacity	Q_{pack}	64 Ah
Pack energy (nominal)	E_{pack}	5.2 kWh
Cycle life to 80% SoH	N_{80}	5000 cycles
Replacement cost (pack)	C_{repl}	€5000

3.4   Battery Degradation Modeling

To account for the cyclic aging of batteries, we adopted the same method of cycle-counting to monitor battery degradation described in [9, 14]. The degradation cost is modeled by measuring the Full Equivalent Cycles (FEC), which reflect the energy exchanged over both charging and discharging processes. Mathematically, the FEC is computed as:

$$FEC(t) = 0.5 \times \frac{\sum \eta_{ch}|P_t^{ch}|\Delta t + \frac{P_t^{dis}\Delta t}{\eta_{dis}}}{\text{Life}_{80\%Cyc} \times C_b}, \tag{7}$$

where the $\text{Life}_{80\%Cyc}$ represents the number of full cycles until the battery reaches 80% of its initial capacity. The value 0.5 corresponds to the average half-cycle within the battery's lifespan.

4   Methodology

We cast the battery dispatch problem as a MDP defined by the tuple $\langle S, A, T, R, \gamma \rangle$. The RL algorithm (agent) takes in the states from S to take decisions A (denoted $P^{batt}(t)$).

4.1 Observation and Action Space

The observation space defines the status of the system. At each time step t the environment is described by the state:

$$\mathcal{S}(t) = (SoC(t), P^{\mathrm{L}}(t), P^{\mathrm{PV}}(t), SoH(t), H_{sin}, H_{cos}, D_{sin}, D_{cos}), \tag{8}$$

where:

- The current state of charge, $SoC(t)$, is constrained to battery's operational boundaries to further protect the battery, a range that avoids deep discharges and overcharges, both of which accelerate degradation in lithium-ion cells [6]:

$$SoC_{min} \le SoC(t) \le SoC_{max}. \tag{9}$$

The SoC updated at each step in accordance with P^{batt}:

$$SoC_{t+1} = SoC_t + (1 - z_t)\eta_{ch}|P_t^{batt}|\frac{\Delta t}{C_b} - z_t \frac{P_t^{batt}\,\Delta t}{\eta_{dis}C_b}, \tag{10}$$

where in Eq. 10, η^{ch} and $\eta^{dis} \in \{x | 0 \le x \le 1\}$ are the charging and discharging efficiency, z_t is a boolean value stating that the battery cannot be charged and discharged simultaneously.
- $P^{PV}(t), P^{L}(t)$ represent historical data for PV generation and household consumption.
- $SoH(t)$ represents the health of battery at time step t.
- The components $H_{sin}, H_{cos}, D_{sin}, D_{cos}$ represent cyclical encodings for the time of day and the day of the week. Cyclical encoding helps the agent understand regular temporal variations and the continuous nature of time, preventing discontinuities from raw numerical values. Energy price data can compute cost or profit, as shown in Table 1 [18].

The RL agent's actions are considered as decisions on battery operation at each time step.

The action of the agent is physically bounded by the battery's maximum charge/discharge rates according to battery specification:

$$-P_{max} \le P^{batt} \le P_{max}. \tag{11}$$

Once the action is determined, the other variables can be derived using the Algorithm 1.

4.2 Reward Function

The primary objective of the RL agent is to maximize the economic benefit derived from utilizing the BESS, while concurrently minimizing the costs associated with its degradation. The reward signal r_t at each time step t is formulated to reflect this dual objective:

$$r_t = C_{\text{economic}}(t) - C_{\text{batt\_wear\_cost}}(t) + SSR(t) - P_{loss}, \tag{12}$$

where $C_{\text{economic}}(t)$ represents the net economic gain achieved by operating the BESS, and $C_{\text{batt\_wear\_cost}}(t)$ is the monetized cost of battery degradation incurred during that time step. The self-sufficiency ratio, or $SSR(t)$, encourages the agent to explore states in which P_t^{B2L} is higher, otherwise, the agent choose to not use the battery, aiming to extend its lifespan and to sell energy back to the grid.

The economic component, $C_{\text{economic}}(t)$, represents the cost/revenue difference with and without the BESS. This evaluation takes into account the energy purchased from or sold back to the grid across both scenarios, using prevailing electricity prices (Sect. 3.2) and the action of the agent $P_{\text{batt}}(t)$:

$$C_{\text{economic}}(t) = \text{Revenue}_{\text{with\_BESS}}(t) - \text{Revenue}_{\text{without\_BESS}}(t). \tag{13}$$

A positive $C_{\text{economic}}(t)$ indicates a net saving or profit due to the BESS operation.

The degradation cost, $C_{\text{batt\_wear\_cost}}(t)$, translates the physical wear of the battery, represented by the incremental SoH ($\Delta SoH(t)$, Eq. 7), into a financial penalty. It is calculated as:

$$C_{\text{batt\_wear\_cost}}(t) = \frac{Cost_{\text{battery}}}{SoH_{\max} - SoH_{\text{EOL}}} \cdot |\Delta SoH(t)|, \tag{14}$$

where $Cost_{\text{battery}}$ is the total replacement cost of the battery pack (see Table 2), $SoH_{\max}$ is the initial SoH (1.0), and SoH_{EOL} is the SoH level defined as End-of-Life. This term ensures that actions leading to accelerated degradation are penalized proportionally to their impact on the battery's lifespan value. The agent, therefore, learns to make a trade-off between immediate economic gains and long-term battery health.

We introduce the complementary metric $SSR(t)$ to encourage the agent to explore its environment as we saw in our experiments that agent prefers to not use the battery at all in the case of reward function without $SSR(t)$. This metric is defined as:

$$SSR(t) = \frac{P_t^{PV} + P_t^{B2L}}{P_t^{L}}, \tag{15}$$

where P^{B2L} represents the battery discharge energy used to meet the load. The power loss penalty $P_{\text{loss}}(t)$ accounts for energy dissipation due to ohmic losses in the battery pack, encouraging efficient operation and can be calculated as:

$$P_{\text{loss}}(t) = R_{\text{pack}}(t) \cdot I_{\text{pack}}^2(t), \tag{16}$$

where $R_{\text{pack}}(t)$ and $I_{\text{pack}}(t)$ are the pack-level resistance and current, derived from the NASA dataset parameters and scaled for the 26S32P configuration (see Eqs. 6).

4.3 Soft Actor Critic (SAC)

This research utilizes the Soft Actor-Critic (SAC) algorithm [12] to train the RL agent for battery dispatch in a residential microgrid. SAC is an off-policy, maximum entropy RL algorithm that combines the stability of actor-critic methods with enhanced exploration, making it well-suited for continuous control tasks such as optimizing the charging and discharging of a 5.2 kWh LiCoO$_2$ battery pack under stochastic PV generation and load profiles. Unlike on-policy methods, SAC leverages a replay buffer to reuse past experiences, improving sample efficiency and enabling robust learning in complex environments like the one faced in this paper, where the agent must balance economic gains, battery degradation, and energy efficiency.

SAC optimizes a policy by maximizing the expected cumulative reward augmented with an entropy term, encouraging exploration while ensuring effective exploitation. The objective function is defined as:

$$J(\pi) = \mathbb{E}_{\tau \sim \pi} \left[\sum_{t=0}^{\infty} \gamma^t \left(r(s_t, a_t) + \alpha \mathcal{H}(\pi(\cdot|s_t)) \right) \right], \tag{17}$$

where τ represents trajectories sampled from the policy π, $r(s_t, a_t)$ is the reward at timestep t, $\gamma \in [0, 1)$ is the discount factor, $\mathcal{H}(\pi(\cdot|s_t)) = -\mathbb{E}_{a_t \sim \pi(\cdot|s_t)}[\log \pi(a_t|s_t)]$ is the entropy of the policy, and α is a temperature parameter controlling the exploration-exploitation trade-off. In our implementation, α is automatically adjusted to maintain a target entropy level, ensuring consistent exploration [12].

SAC employs an actor-critic architecture with twin Q-networks to reduce overestimation bias. The critic updates the Q-functions using samples from a replay buffer, incorporating the reward, next state value (from target Q-networks), and an entropy term to account for the stochastic policy. The actor updates the policy to maximize the expected Q-value minus the entropy penalty, using the reparameterization trick to ensure differentiability [12]. Target Q-networks are updated via a soft update rule to enhance stability. SAC's off-policy nature, entropy regularization, and use of twin Q-networks make it an effective choice for our microgrid energy management task, enabling the agent to learn robust policies that optimize economic performance while preserving battery health.

4.4 Benchmark Model: MILP-Based MPC

The benchmark method used for evaluating the proposed RL framework is a Model Predictive Control approach based on Mixed-Integer Linear Programming. This MPC strategy aims to minimize both the operational costs and battery degradation, integrating energy dispatch, grid import/export and battery aging into a unified optimization problem.

The MPC model is implemented using the optimization package `CVXPY` [4] interfaced with the `GUROBI` solver [24]. At each timestep, the controller solves a receding horizon optimization problem over a finite horizon of length N, taking as input electricity price tariffs and forecasts of photovoltaic generation and load demand. Only the first control input of the resulting optimal sequence is applied to the system, after which the horizon is shifted forward by one step.

Throughout the following MPC formulation, the subscript k refers to the k-th step within the prediction horizon N, starting from the current time step t. Accordingly, any variable a_k should be interpreted as the value of the quantity a at future time $t + k$, i.e., $a_k \equiv a(t + k)$.

The MPC formulation includes a set of constraints that set the physical limitations and operational logic of the microgrid system over the prediction horizon. Battery dynamics are modeled linearly through State-of-Energy (SoE), that evolves according to:

$$\text{SoE}_{k+1} = \text{SoE}_k + P_k^{in} \cdot \eta_{ch} \cdot \Delta t - \frac{P_k^{out}}{\eta_{dis}} \Delta t, \tag{18}$$

where P_k^{in} and P_k^{out} represent charging and discharging powers measured in kW, and Δt is the discrete timestep.

Power flows to and from the battery are limited by the rated power boundaries:

$$0 \leq P_k^{in} \leq P_{\max}, \quad 0 \leq P_k^{out} \leq P_{\max}. \tag{19}$$

To enforce the mutual exclusivity of charging and discharging actions at each timestep, a binary variable $u_k \in \{0, 1\}$ is introduced, leading to the following Big-M constraints:

$$P_k^{in} \leq u_k \cdot P_{\max}, \quad P_k^{out} \leq (1 - u_k) \cdot P_{\max}. \tag{20}$$

At each timestep, the microgrid must satisfy an energy balance constraint, ensuring that the net photovoltaic generation is properly allocated between local storage and interaction with the main grid formalized as:

$$P_k^{grid} = P_k^{PV} - P_k^{load} - P_k^{in} + P_k^{out}, \tag{21}$$

where P_k^{grid} is the amount of power exchanged with the grid, positive in case of export. Finally, a terminal constraint is enforced to keep the final battery SoE within operational bounds:

$$\text{SoE}_{\min} \leq \text{SoE}_k \leq \text{SoE}_{\max}. \tag{22}$$

The objective function of the MPC includes costs associated with energy transactions with the grid and battery degradation. The dispatch cost C_{dispatch} quantifies the cost associated to the difference between energy imported from and exported to the external grid, and is computed as follows:

$$C_{\text{dispatch}} = \sum_{k=0}^{N-1} \left(\pi_k^{\text{buy}} \cdot E_k^{\text{imp}} - \pi_k^{\text{sell}} \cdot E_k^{\text{exp}} \right), \tag{23}$$

where π_k^{buy} and π_k^{sell} represent the unit energy prices [e/kWh] for buying from and selling to the grid at time step k, respectively, while E_k^{imp} and E_k^{exp} denote the energy imported from and exported to the grid at time step k.

The effects of cyclic aging on battery degradation are quantified by the MPC through the same computation described in Sect. 3.4 to evaluate the SoH at each time step. At each decision step, the MPC computes the optimal charging (P^{in}) and discharging (P^{out}) power values. To ensure feasibility, the effective State of Charge is updated using the output of the degradation function, and power flows are constrained accordingly to avoid overcharging or overdischarging.

5   Experiments and Results

This section outlines the experimental setup and results for evaluating our degradation-aware RL framework for real-time energy management in a residential microgrid. We used the Pecan Street Inc. Dataport [25] to train our RL agent, which provides detailed electricity usage and solar production measurements at a 15-minute resolution for one year. Table 3 reports key statistics for household 661, including the average electrical load and average PV generation output. Two additional households (7536 and 9019) are used for comparison and testing, with their statistics also summarized in Tables 3 and 4.

Table 3. Energy Data Statistics (ID 661).

Max PV	Min PV	μ PV	Max Load	Min Load	μ Load
5.54 kW	0.0 kW	0.88 kW	10.15 kW	0.0 kW	1.49 kW

Table 4. Summary of Pecan Street Data (IDs 661, 7536, and 9019) [25].

Data ID	Avg. Load (kW)	Avg. PV (kW)	# Data Records
661	1.49	0.88	35032
7536	1.31	0.85	34984
9019	0.90	0.28	34928

We implemented a custom framework based on the OpenAI Gymnasium[1], following the formulation in Sect. 3. The environment yields an observation space of dimension 8, (Sect. 4.1), while the action space consists of one continuous variable (charge/discharge commands, $P_{\text{batt}}(t) \in [-10, 10]$ kW). The battery's initial SoC is set to 0.5, with operational boundary of $[0.2, 0.8]$.

The RL agent was trained using the SAC algorithm from the Stable Baselines3 framework [26], a state-of-the-art off-policy RL method that balances exploration and exploitation through entropy regularization. Based on preliminary tuning, we configured the SAC networks with two hidden layers of sizes $[256, 256]$, an initial learning rate of 3×10^{-4}, a replay buffer size of 10^6 transitions, a batch size of 256, and a discount factor $\gamma = 0.99$. To enhance sample efficiency and accelerate training, we utilized 256 parallel environments, allowing the agent to collect diverse experiences simultaneously during each training step. The agent was trained for 30 episodes, with each episode covering one pass through the training dataset. After each environment step, the agent performs a gradient update to refine its policy, leveraging SAC's off-policy nature and parallel environment setup to improve sample efficiency. The penalty coefficients in the reward function (Eq. 12) were tuned to balance the relative importance

---

[1] https://gymnasium.farama.org/.

of economic gains, battery longevity, and energy efficiency, ensuring the agent learns policies that align with the project's objectives. All experiments were implemented in PyTorch and executed on a NVIDIA RTX A6000 GPU, ensuring efficient training and evaluation.

Figure 3 shows the mean reward per episode over the training process, averaged across the 256 parallel environments. The reward starts at approximately $-272,283$ and improves steadily, reaching $-106,635$ by the end of training, reflecting a 61% improvement in performance. Early in training (up to 50 million timesteps), the reward exhibits high variability, indicating the agent's exploration phase as it learns to balance economic gains with battery health preservation. After 100 million timesteps, the reward stabilizes between $-200,000$ and $-150,000$, with occasional dips likely due to stochastic environment resets or exploration-induced policy updates. The overall upward trend and eventual convergence suggest that the SAC agent successfully learns a robust policy, adapting to the complex dynamics of the microgrid environment while optimizing the multi-objective reward function. We evaluated the trained SAC agent and the MPC baseline in two scenarios: (1) a 1-battery simulation to determine the time to reach 80% SoH, for households 7536 and 9019, and (2) a 10-year simulation using a replicated dataset for household 9019 to assess long-term battery usage and energy trading performance. In the 1-battery simulation for household 7536, RL-SAC sustains the battery for 4.81 years before reaching 80% SoH, compared to 4.63 years for MPC, a 3.9% improvement in longevity, as shown in Table. 5. This suggests RL-SAC better manages battery cycles, likely by avoiding deep discharges or high C-rates that accelerate degradation. RL-SAC imports 15,621.97 kWh and sells 6,295.75 kWh, reducing pur-

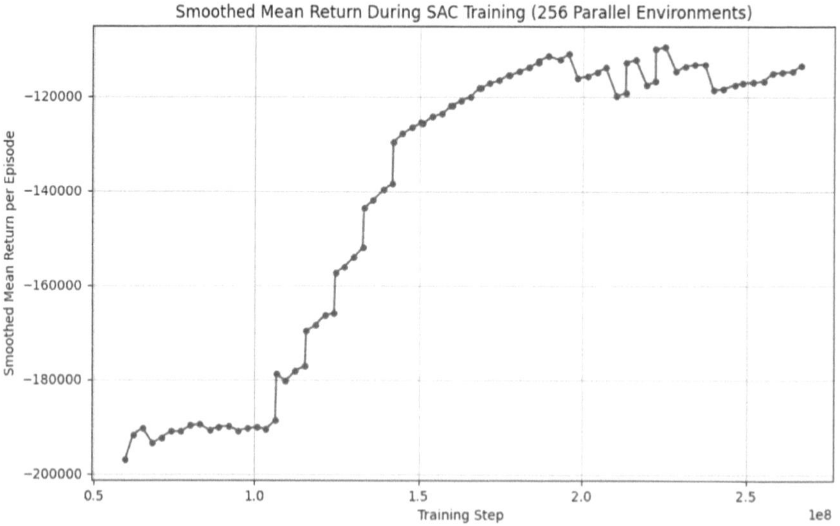

Fig. 3. Mean Return per episode during SAC training over 30 episodes (268.9 million timesteps) with 256 parallel environments. The upward trend and eventual stabilization indicate effective learning of a policy that balances economic gains and battery health.

chases by 45.6% (13,068.24 kWh less than MPC's 28,690.21 kWh) but selling 31.4% less (2,882.1 kWh less than MPC's 9,177.85 kWh). The reduced imports indicate RL-SAC's effective use of PV generation and stored energy, while the lower exports may reflect a more conservative strategy to preserve battery health, contributing to its longer lifespan. For household 9019 (Table 6), RL-SAC extends battery life to 9.13 years, compared to 8.59 years for MPC, a 6.3% improvement. This household has a lower average load (0.907 kW) and PV generation (0.282 kW) compared to 7536 (1.312 kW load, 0.855 kW PV), suggesting RL-SAC adapts more effectively to lower energy demands by optimizing battery usage to minimize stress. RL-SAC imports 19,811.37 kWh and sells 3,165.38 kWh, reducing purchases by 60.0% (29,746.53 kWh less than MPC's 49,557.90 kWh) while selling 99.4% more (1,578.09 kWh more than MPC's 1,587.29 kWh). The significant reduction in imports and increase in exports highlight RL-SAC's superior arbitrage strategy, leveraging battery storage to buy less during high-price periods and sell less during low-price periods.

In the 10-year simulation for household 9019 (Table 7), both RL-SAC and MPC require 2 battery replacements over the period, consistent with the 1-battery simulation results. RL-SAC would replace its first battery around year 9.13, using a second battery for the remaining 0.87 years, while MPC replaces at year 8.59, using the second for 1.41 years. Despite the equal number of replacements, RL-SAC incurs a lower battery wear cost of $5,365.10 compared to MPC's $5,732.70, a 6.4% reduction, reflecting more efficient battery usage over the long term. RL-SAC imports 20,653.92 kWh and sells 3,358.72 kWh, reducing purchases by 64.1% (36,901.56 kWh less than MPC's 57,555.48 kWh) and selling 89.8% more (1,588.79 kWh more than MPC's 1,769.93 kWh). These trends mirror the 1-year results for ID 9019, demonstrating RL-SAC's sustained efficiency in energy management, minimizing grid reliance, and maximizing revenue through strategic sales. The lower wear cost further underscores RL-SAC's ability to balance economic objectives with battery health preservation over extended periods.

The Fig. 4 displays a weekly energy profile for a random 7-day period from the 10-year simulation. The graph plots the household demand, PV generation, battery charge, battery discharge, energy bought from the grid, and energy sold to the grid over time. The plot reveals that the agent effectively uses PV generation and stored energy to meet demand during peak periods, minimizing grid purchases during high-price intervals and selling excess energy during off-peak times. The battery charge and discharge patterns align with the tariff structure, with charging occurring during low-price periods and discharging during high-demand or high-price periods, demonstrating the agent's learned ability to optimize economic performance while managing battery usage to reduce degradation.

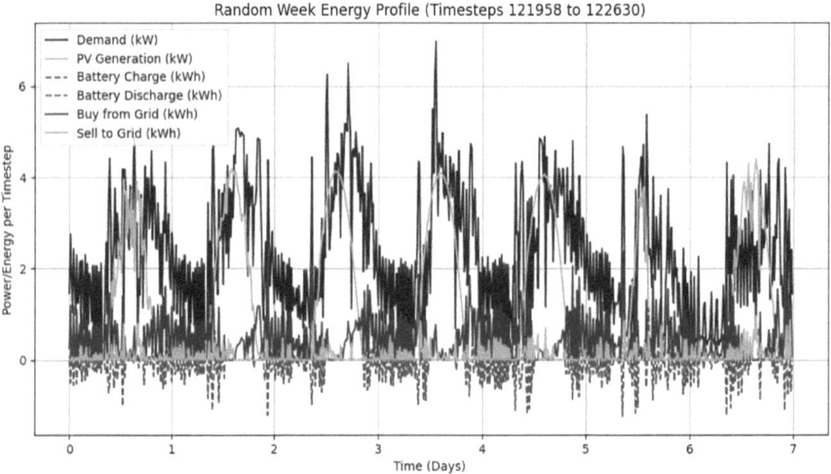

Fig. 4. Weekly energy profile for a randomly selected 7-day period from the 10-year simulation, showing household demand, PV generation, battery charge/discharge, and energy bought/sold.

Table 5. Performance of models in a single battery simulation (ID 7536).

Metric	RL-SAC	MPC
Elapsed Time [Years]	4.81	4.63
Total Bought Energy [kWh]	15621.97	28690.21
Total Sold Energy [kWh]	6295.75	9177.85

Table 6. Performance of models in a single battery simulation (ID 9019).

Metric	RL-SAC	MPC
Elapsed Time [Years]	9.13	8.59
Total Bought Energy [kWh]	19811.37	49557.90
Total Sold Energy [kWh]	3165.38	1587.29

Table 7. Comparison of RL-SAC and MPC controllers over a 10-year replicated simulation based on scenario ID 9019.

Metric	RL-SAC	MPC
# Batteries	2	2
Battery Wear Cost [$]	5365.1	5732.7
Total Bought Energy [kWh]	20653.92	57555.48
Total Sold Energy [kWh]	3358.72	1769.93

6   Conclusion

This study demonstrates the efficacy of a degradation-aware RL framework for real-time energy management in residential microgrids, utilizing the SAC algorithm implemented via Stable Baselines3. Through extensive simulations – suitable also for REC scenarios – the SAC agent consistently outperformed a MPC baseline, which employs a MILP formulation for energy dispatch optimization. Over a 10-year simulation for household 9019, RL-SAC reduced battery wear cost by 6.4%, reflecting more efficient long-term battery usage. The 1-year simulations further underscore RL-SAC's superiority, extending battery life by 3.9% for household 7536, and 6.3% for household 9019, alongside substantial reductions in energy purchases – 45.6% less for household 7536 and 60.0% less for household 9019. These results highlight RL-SAC's ability to learn complex, non-linear dispatch policies that effectively balance economic objectives with battery health preservation across diverse household profiles and timeframes. The integration of a simplified degradation model, calibrated with NASA dataset measurements, ensures empirical grounding while maintaining computational efficiency for real-time applications. Collectively, these findings validate the potential of SAC-based RL frameworks for practical deployment in residential microgrids, offering a scalable and robust solution for sustainable energy management in REC. Future work will focus on validating these results in real-world microgrid deployments and exploring multi-agent RL approaches to enhance coordination among community members. Furthermore, a 10-year simulation analysis using a dataset with annually changing distributions, instead of repeatedly using data from one year, may yield improved outcomes while reducing the impact of seasonality problems.

Acknowledgments. This study was carried out within the MOST – Sustainable Mobility Center and received funding from the European Union Next-GenerationEU (PIANO NAZIONALE DI RIPRESA E RESILIENZA (PNRR) – MISSIONE 4 COMPONENTE 2, INVESTIMENTO 1.4 – D.D. 1033 17/06/2022, CN00000023). This manuscript reflects only the authors' views and opinions, neither the European Union nor the European Commission can be considered responsible for them.

References

1. Abbasi, M.H., Arjmandzadeh, Z., Zhang, J., Xu, B., Krovi, V.: Deep reinforcement learning based fast charging and thermal management optimization of an electric vehicle battery pack. J. Energy Storage **95**, 112466 (2024)
2. De Santis, E., Rizzi, A., Sadeghiany, A., Mascioli, F.M.F.: Genetic optimization of a fuzzy control system for energy flow management in micro-grids. In: 2013 Joint IFSA World Congress and NAFIPS Annual Meeting (IFSA/NAFIPS), pp. 418–423. IEEE (2013)
3. Demirci, O., Taskin, S., Schaltz, E., Demirci, B.A.: Review of battery state estimation methods for electric vehicles-part i: Soc estimation. J. Energy Storage **87**, 111435 (2024)
4. Diamond, S., Boyd, S.: Cvxpy: A python-embedded modeling language for convex optimization. J. Mach. Learn. Res. **17**(83), 1–5 (2016)

5. Domínguez-Barbero, D., García-González, J., Sanz-Bobi, M.Á., García-Cerrada, A.: Energy management of a microgrid considering nonlinear losses in batteries through deep reinforcement learning. Appl. Energy **368**, 123435 (2024). https://doi.org/10.1016/j.apenergy.2024.123435, https://www.sciencedirect.com/science/article/pii/S0306261924008183
6. Enrico, D.S., Vanessa, P., Massimiliano, L., Antonello, R.: Degradation mechanisms and differential curve modeling for non-invasive diagnostics of lithium cells: an overview. Renew. Sustain. Energy Rev. **211**, 115349 (2025). https://doi.org/10.1016/j.rser.2025.115349, https://www.sciencedirect.com/science/article/pii/S136403212500022X
7. Directive (EU) 2019/944 of the european parliament and of the council). Official Journal of the European Union, L158/125 (2019). https://eur-lex.europa.eu/legal-content/EN/TXT/?uri=CELEX%3A32019L0944, Accessed 21 Nov 2024
8. European Commission Joint Research Centre: Renewable Energy Recast to 2030 (RED II) (2024). https://joint-research-centre.ec.europa.eu/welcome-jec-website/reference-regulatory-framework/renewable-energy-recast-2030-red-ii_en
9. Fan, J., Wang, H.: Deep reinforcement learning for community battery scheduling under uncertainties of load, PV generation, and energy prices. In: 2023 IEEE 7th Conference on Energy Internet and Energy System Integration (EI2), pp. 4871–4876. IEEE (2023)
10. Fridgen, G., Kahlen, M., Ketter, W., Rieger, A., Thimmel, M.: One rate does not fit all: an empirical analysis of electricity tariffs for residential microgrids. Appl. Energy **210**, 800–814 (2018). https://doi.org/10.1016/j.apenergy.2017.08.138, https://www.sciencedirect.com/science/article/pii/S0306261917311546
11. Guo, C., Wang, X., Zheng, Y., Zhang, F.: Real-time optimal energy management of microgrid with uncertainties based on deep reinforcement learning. Energy **238**, 121873 (2022). https://doi.org/10.1016/j.energy.2021.121873, https://www.sciencedirect.com/science/article/pii/S0360544221021216
12. Haarnoja, T., et al.: Soft actor-critic algorithms and applications. arXiv preprint arXiv:1812.05905 (2018)
13. Han, L., Yang, K., Ma, T., Yang, N., Liu, H., Guo, L.: Battery life constrained real-time energy management strategy for hybrid electric vehicles based on reinforcement learning. Energy **259**, 124986 (2022)
14. Hesse, H.C., Martins, R., Musilek, P., Naumann, M., Truong, C.N., Jossen, A.: Economic optimization of component sizing for residential battery storage systems. Energies **10**(7), 835 (2017)
15. Hosseini, E., et al.: Reinforcement learning-based energy management system for lithium-ion battery storage in multilevel microgrid. J. Energy Storage **109**, 115114 (2025). https://doi.org/10.1016/j.est.2024.115114
16. Jafari, S., Byun, Y.C.: Optimizing battery RUL prediction of lithium-ion batteries based on harris hawk optimization approach using random forest and lightgbm. IEEE Access **11**, 87034–87046 (2023)
17. Khorrami, S., Falvo, M.C., Pompili, M.: Energy communities development: a review of challenges and opportunities. In: 2024 IEEE International Conference on Environment and Electrical Engineering and 2024 IEEE Industrial and Commercial Power Systems Europe (EEEIC / I&CPS Europe), pp. 1–6 (2024). https://doi.org/10.1109/EEEIC/ICPSEurope61470.2024.10751123
18. Lewinson, E.: Three approaches to encoding time information as features for ml models. NVIDIA Technical Blog (2022). https://developer.nvidia.com/blog/three-approaches-to-encoding-time-information-as-features-for-ml-models/
19. Li, M., Lu, J., Chen, Z., Amine, K.: 30 years of lithium-ion batteries. Adv. Mater. **30**(33), 1800561 (2018)

20. Lowitzsch, J., Hoicka, C., van Tulder, F.: Renewable energy communities under the 2019 european clean energy package - governance model for the energy clusters of the future? Renew. Sustain. Energy Rev. **122**, 109489 (2020). https://doi.org/10.1016/j.rser.2019.109489, https://www.sciencedirect.com/science/article/pii/S1364032119306975
21. Mnih, V., Kavukcuoglu, K., Silver, D., Graves, A., Antonoglou, I., Wierstra, D., Riedmiller, M.: Playing atari with deep reinforcement learning. arXiv preprint arXiv:1312.5602 (2013)
22. Mussi, M., Pellegrino, L., Pindaro, O.F., Restelli, M., Trovò, F.: A reinforcement learning controller optimizing costs and battery state of health in smart grids. J. Energy Storage **82**, 110572 (2024)
23. NASA Ames Prognostics Data Repository: Pcoe datasets: Battery data set. https://ti.arc.nasa.gov/tech/dash/groups/pcoe/prognostic-data-repository/, Accessed 05 June 2025
24. Optimization, G., et al.: Gurobi optimizer reference manual (2020)
25. P. S. Inc.: Pecan street dataport (2025), Accessed 25 Jan 2025. https://www.pecanstreet.org/dataport/
26. Raffin, A., Hill, A., Gleave, A., Kanervisto, A., Ernestus, M., Dormann, N.: Stable-baselines3: reliable reinforcement learning implementations. J. Mach. Learn. Res. **22**(268), 1–8 (2021)
27. Stentati, M., Paoletti, S., Vicino, A.: Optimization of energy communities in the italian incentive system. In: 2022 IEEE PES Innovative Smart Grid Technologies Conference Europe (ISGT-Europe), pp. 1–5. IEEE, New York (2022). https://doi.org/10.1109/ISGT-Europe54678.2022.9960513
28. Su, L., Xu, Y., Dong, Z.: State-of-health estimation of lithium-ion batteries: a comprehensive literature review from cell to pack levels. Energy Conversion Econom. **5**(4), 224–242 (2024)
29. Sutton, R.S., Barto, A.G., et al.: Reinforcement learning: an introduction. MIT Press Cambridge (1998)
30. Wang, J., et al.: Cycle-life model for graphite-lifepo4 cells. J. Power Sources **196**(8), 3942–3948 (2011)
31. Yalçın, S., Herdem, M.S.: Optimizing EV battery management: advanced hybrid reinforcement learning models for efficient charging and discharging. Energies **17**(12), 2883 (2024)
32. Zendehdel, D., Capillo, A., De Santis, E., Rizzi, A.: An extended battery equivalent circuit model for an energy community real time ems. In: 2024 International Joint Conference on Neural Networks (IJCNN), pp. 1–9. IEEE (2024)

Innovative Techniques for Efficient Hyperdimensional Computing on Hardware: Enhance Accuracy and On-the-Fly Hypervector Generation

Saeid Jamili[ID], Sabereh Taghdisi Rastkar[(✉)][ID], Marco Angioli[ID], and Mauro Olivieri[ID]

Department of Information Engineering, Electronics and Telecommunications, University of Rome "La Sapienza", 00185 Rome, Italy
{Saeid.Jamili,sabereh.taghdisirastkar,marco.angioli, mauro.olivieri}@uniroma1.it

Abstract. Hyperdimensional Computing (HDC) encodes data and learning operations into high-dimensional vectors (hypervectors), enabling robust and rapidly adaptable machine learning on resource-limited platforms such as Field-Programmable Gate Arrays (FPGAs) and edge devices. Despite this potential, conventional HDC systems often demand extensive memory resources to store base, level, and class hypervectors, which can limit scalability and performance in hardware implementations for Artificial Intelligence (AI) and Internet of Things (IoT) applications. This paper addresses these issues through two main innovations. First, it introduces a combinational-logic approach that generates hypervectors on the fly, eliminating the need for large lookup tables and thereby substantially reducing memory overhead. Second, it presents an orthogonal hypervector generation scheme based on sequences such as Hadamard, Walsh, and Gold, ensuring highly uncorrelated representations that enhance classification accuracy (particularly in single-shot learning) while remaining effective over multiple training epochs. Experimental evaluations on standard benchmarks, including ISOLET and UCI-HAR, show notable gains in classification performance as well as significant reductions in memory consumption and lookup table usage. These results highlight the viability of integrating logic-based hypervector synthesis with orthogonal vector design to create an efficient, power-conscious, and high-throughput HDC framework suitable for real-time edge AI and IoT scenarios. By uniting these techniques, the proposed approach advances the practical deployment of hyperdimensional computing in embedded and resource-constrained environments.

Keywords: Hyperdimensional Computing · FPGA Implementation · Orthogonal Sequences · Single-Shot Learning · Resource-Constrained AI · Edge AI

F. Marcelloni et al. (Eds.): IJCCI 2025, CCIS 2829, pp. 558–569, 2026.
https://doi.org/10.1007/978-3-032-15638-9_32

1   Introduction

HDC has emerged as a promising approach in the realm of machine learning, particularly for applications in resource-constrained environments such as edge computing and the IoT. This computing paradigm, inspired by the human brain's cognitive processes, represents data and operations in a high-dimensional vector space and offers robust, efficient data-processing capabilities, which makes it ideal for various IoT applications [3,9,15].

The implementation of HDC on FPGAs has gained attention due to the inherent parallelism and reconfigurability of these devices, aligning well with the high-dimensional, distributed nature of HDC. FPGAs offer a flexible platform for experimenting with HDC models, particularly for applications requiring low-power consumption and real-time processing [15,16].

Deploying HDC on FPGAs presents substantial challenges, chiefly due to the considerable memory required to store base vectors, level vectors, and class/model vectors. Memory allocation in conventional HDC systems increases significantly, influenced by various factors including the number of features, the resolution of data, the number of classes or the complexity of the model, and the type and dimensionality of hypervectors. This scalability issue becomes a critical constraint, particularly for models of higher complexity and for larger-scale datasets. Such challenges are even more pronounced in environments with limited resources, underscoring the need for optimized memory management in these settings [7,12].

Recent studies have introduced various strategies to mitigate the memory challenge in HDC. These include the use of random generators, compression methods, n-gram models, and thermometer codes. However, these methods often involve trade-offs among memory usage, computational efficiency, and accuracy. These complexities in balancing factors highlight the ongoing challenges in the field of HDC [1,7,11–14].

To further enhance our HDC framework, we have integrated an orthogonal vector-generator technique utilizing methodologies such as Hadamard, Walsh, and Gold sequences. These methods are renowned for their ability to produce high-quality orthogonal vectors with minimal correlation, offering a substantial improvement over conventional random vector generators in HDC implementations. By adopting this technique for generating base and level vectors, we aim to significantly enhance the reliability and performance of our system, particularly in one-shot training contexts where achieving high accuracy rapidly is crucial [2,8].

This approach is especially advantageous for hardware implementations of HDC, such as those on FPGAs, where minimal memory usage and efficient real-time processing are paramount. In our study, these orthogonal vectors are generated offline and later utilized within the FPGA system through specialized combinational logic. This strategy not only respects the system's resource constraints but also leverages the benefits of orthogonal sequences to significantly improve accuracy and robustness. Hence, our research proposes a groundbreaking approach to HDC deployment on FPGAs, specifically designed for edge AI and IoT scenarios, thereby setting a new standard for performance and complexity in resource-constrained hardware platforms.

The remainder of the paper is organized as follows. Section 2 reviews related work. Section 3 presents the proposed logic-centric hypervector generation and orthogonal sequence design. Section 4 reports experimental results on the ISO-LET and UCI-HAR datasets, along with FPGA resource utilization. Finally, Sect. 5 concludes the paper and outlines directions for future work.

At a high level, our method replaces large memory tables with lightweight on-the-fly logic that deterministically generates hypervectors. This makes HDC models both smaller and faster in hardware, which is crucial for embedded and IoT devices that operate under strict power, memory, and latency constraints.

2   Related Work

HDC marks a shift in computational paradigms by encoding data and operations in high-dimensional vector spaces, a concept inspired by human brain activity [8, 10]. This framework differs from traditional neural networks in both its architectural viewpoint and its learning processes, often providing faster training and resilience to noisy data [4]. As a result, HDC has gained interest in resource-limited environments such as edge and IoT devices where energy efficiency and immediate responsiveness are critical.

2.1   Edge and FPGA Implementations

One active research area involves deploying HDC directly on the edge. By processing data locally, edge devices can reduce latency and avoid bandwidth-intensive cloud inference. Multiple works have shown that HDC's high-dimensional operations are well suited for real-time classification [11] and anomaly detection in scenarios with stringent power budgets [1, 7].

Building on these efforts, FPGAs have become a natural hardware target: their fine-grained parallelism and reconfigurability align with HDC's bit-wise arithmetic, enabling efficient hypervector binding, bundling, and similarity search [7, 15, 16]. Nevertheless, many existing FPGA designs rely on large on-chip or external memories, leading to substantial resource use and potential latency bottlenecks limitations that curb the feasibility of HDC in time-critical embedded analytics [12].

2.2   Memory and Vector Optimisation

Researchers have explored a variety of techniques for reducing HDC memory footprints. Some projects compress class hypervectors or adopt lower-dimensional encodings, yet such approaches can undermine accuracy especially in single-shot learning, where high dimensionality is crucial [1]. Others introduce n-gram statistics, thermometer codes, or on-demand random vectors from linear-feedback shift registers (LFSRs) to avoid static lookup tables [4,7,12]. Although random generation eliminates memory overhead, purely stochastic vectors offer no guarantee of inter-class orthogonality, which can hamper separability when training data are scarce.

To address that weakness, recent studies have focused on deterministic, near-orthogonal sequences. Aygun et al. highlight encoding as the chief lever for balancing resource cost against classification performance [2]. Sequences such as Hadamard, Walsh, and Gold produce hypervectors with tightly bounded cross-correlation, boosting class discrimination in single and few-shot settings [8,9]. Embedding these codes in lightweight combinational logic removes bulky memories while preserving accuracy.

Despite advances in compression, random on-demand generation, and orthogonal coding, no single framework simultaneously minimises memory and maintains strong HDC accuracy on FPGAs. Many current designs still store subsets of vectors or depend on stochastic encodings that falter under tight data constraints.

In this paper we close that gap. First, we introduce a logic-only generator that synthesises base and level hypervectors on demand, eliminating large lookup tables entirely. Second, we adopt Hadamard, Walsh, and Gold sequences to guarantee near-orthogonal encoding particularly advantageous for single-shot learning. Combined, these techniques reduce hardware cost while matching or surpassing state-of-the-art accuracy, as demonstrated on the ISOLET and UCI-HAR benchmarks.

3   Methodology

Figure 1 shows the high-level idea, from raw data or sensor inputs up to classification results on an FPGA. The core of our approach to memory-efficient HDC on FPGAs involves two main innovations:

1. **Logic-centric hypervector generation**, which removes the need for large static memories by synthesizing hypervectors directly in combinational logic.
2. **Orthogonal sequence design**, based on Hadamard, Walsh, or Gold codes, to boost the separability of base and level vectors and thus improve classification accuracy.

Let D denote the hypervector dimensionality. In Hyperdimensional Computing (HDC), three main types of hypervectors play distinct roles:

- *Base vectors (BVs)* $B_i \in \{-1,+1\}^D$ encode the identity of each feature i in the input.
- *Level vectors (LVs)* $L_\ell \in \{-1,+1\}^D$ encode the quantized value of a feature, obtained after discretizing the feature into ℓ levels.
- *Class vectors (CVs)* $C_k \in \{-1,+1\}^D$ serve as prototypes, storing the aggregated representation of all training samples belonging to class k.

Encoding. Given an input sample $x = (f_1, \ldots, f_N)$ with N features, HDC builds a *sample hypervector* $H(x)$ by binding each feature's base vector with the level vector corresponding to its quantized value, and then bundling across all features:

$$H(x) = \bigoplus_{i=1}^{N} \left(B_i \otimes L_{\ell(f_i)} \right), \tag{1}$$

where $\otimes$ denotes binding (e.g., element-wise XOR or multiplication) and $\oplus$ denotes bundling (e.g., element-wise addition or majority). This operation yields a distributed, high-dimensional representation of the entire input pattern.

Training. During training, each class vector C_k is incrementally updated by bundling together all the sample hypervectors associated with class k:

$$C_k^{(t+1)} = C_k^{(t)} + H(x), \qquad \text{for } x \in \text{class } k. \tag{2}$$

After training, C_k is binarized using an element-wise sign or threshold function, resulting in a stable bipolar or binary prototype that compactly represents the class.

Inference. At inference, a new input x is encoded into its sample hypervector $H(x)$. The system then compares $H(x)$ against all stored class vectors C_k using a similarity metric, typically normalized cosine similarity:

$$\text{sim}(u, v) = \frac{\langle u, v \rangle}{\|u\| \, \|v\|}. \tag{3}$$

The predicted label corresponds to the most similar class vector:

$$\hat{k} = \arg \max_{k} \, \text{sim}\big(H(x), C_k\big). \tag{4}$$

Summary. Base vectors capture feature identities, level vectors capture feature values, and class vectors aggregate training samples into prototypes. Encoding combines BVs and LVs into sample hypervectors, training accumulates these into CVs, and inference classifies unseen inputs by nearest-neighbor search in the hyperspace.

Figure 2 illustrates how the original lookup-table data are transformed into a purely combinational logic circuit. A truth table is derived for each partial frame, enabling the circuit to generate hypervectors in real time and without any dedicated memory. By embedding this truth table in hardware, the logic blocks directly emit the high-dimensional bits that constitute the requested segment of the hypervector.

3.1   Orthogonal Vector Construction

This study adopts *orthogonal* sequence-based designs for both base and level vectors, relying on well-known sequences such as Hadamard, Walsh, or Gold [5,

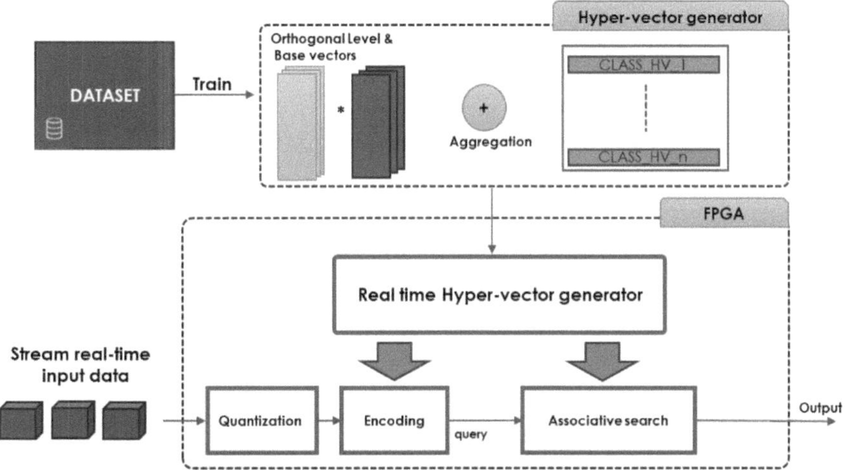

Fig. 1. High-level overview of the proposed HDC system. Input data are quantized and mapped into hypervectors through base and level encoding. During training, sample hypervectors are bundled into class prototypes. During inference, new samples are classified by measuring similarity with stored prototypes, which can be efficiently implemented in FPGA hardware.

6,17]. These sequences have low cross-correlation, ensuring that distinct vectors remain easy to differentiate. We instantiate base vectors from columns of an orthogonal code matrix.

Hadamard Codes. Let $n = 2^{\lceil \log_2 D \rceil}$ and $H_n \in \{-1, +1\}^{n \times n}$ be a Sylvester Hadamard matrix. A D-dimensional base vector is obtained by selecting column i and truncating:

$$B_i = \text{truncate}_D\big(H_n[:, i]\big).$$

Walsh Codes. Walsh codes reorder Hadamard rows/columns in *sequency* (bit-reversal) order. After reordering, we truncate to D as above:

$$B_i = \text{truncate}_D\big(H_n^{(\text{sequency})}[:, i]\big).$$

Gold Codes. Gold sequences are generated from two preferred m-sequences $a(t)$ and $b(t)$ (period $2^m - 1$) via

$$\mathcal{G}(a, b) = \big\{\, a(t) \oplus b(t + \tau) \mid \tau = 0, \ldots, 2^m - 2 \,\big\},$$

where $\oplus$ is bitwise XOR. We map $\{0, 1\} \to \{-1, +1\}$ as $x \mapsto 2x - 1$, arrange the results as columns, then pad/truncate to D.

The same construction is applied to level vectors (LVs) by assigning disjoint indices/columns, so both feature identities and quantized values remain highly distinguishable.

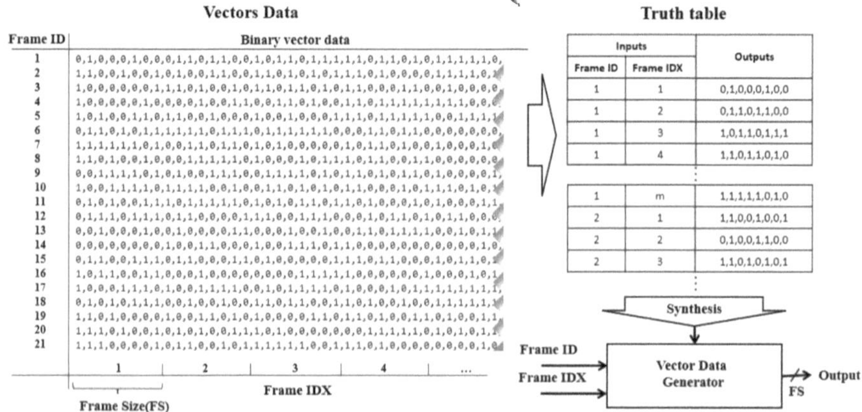

Fig. 2. Logic-based approach for generating hypervectors using combinational logic.

Base Vector Design. To create near-orthogonal base vectors, we use:

- **Hadamard codes**, which are constructed for dimensions that are powers of two. We pad to the nearest power of two if needed and then truncate to the actual dimension.
- **Walsh codes** (WalshâĂŞHadamard functions), similar to Hadamard but using sequency ordering (a permutation of Hadamard rows/columns), which suits certain design constraints and ensures near-orthogonal columns.
- **Gold codes**, whose cross-correlation properties often produce robust classification, especially in wireless or spread-spectrum contexts.

All three methods yield base vectors with limited overlap, improving class discrimination.

Level Vector Design. We generate level vectors from the same orthogonal sequences, ensuring that intermediate quantization steps also preserve a high degree of separation. By carefully assigning different segments or columns from the chosen orthogonal code set, the proposed methodology can maintain distinct vectors at each level. This property is particularly valuable in tasks where small changes in sensor data should produce measurable changes in the resulting hypervectors.

3.2   Benefits in Single-Shot and Multi-epoch Learning

One key motivation for employing orthogonal vectors is single-shot learning, where the system updates its model after seeing a single labeled instance per class. By ensuring that base vectors and level vectors have minimal overlap, the HDC process can separate classes more effectively, even with tiny training sets. Moreover, multi-epoch training can further refine class vectors without needing large stored intermediate vectors, since the base and level vectors are always generated on-the-fly from the same orthogonal seeds.

In summary, this study integrates logic-based hypervector generation with orthogonal code construction to form a flexible, memory-efficient HDC framework that delivers accurate classification in resource-constrained FPGA systems. By combining orthogonal vector design with a purely logic-driven generation process, our methodology significantly reduces memory demands while preserving high classification performance.

4   Experimental Results

4.1   Datasets

ISOLET (UCI): 7,797 utterances with **617** features over **26** classes; we use the standard split (**6,238** train / **1,559** test) and rescale all features to $[-1, 1]$.
UCI-HAR (Smartphones) (UCI): **10,299** windows from **30** subjects performing **6** activities, each with **561** features; we follow the official split (**7,352** train / **2,947** test) and rescale to $[-1, 1]$.
Together, ISOLET stresses many-class acoustic classification, while UCI-HAR reflects multivariate wearable sensing typical of edge devices.

4.2   Orthogonality of the Base Vectors

Figure 3 contrasts the cosine-similarity matrices of the base vectors produced by a random (left) and by the proposed Hadamard generator (right), clearly showing that the orthogonal algorithms yield almost perfectly orthogonal vectors.

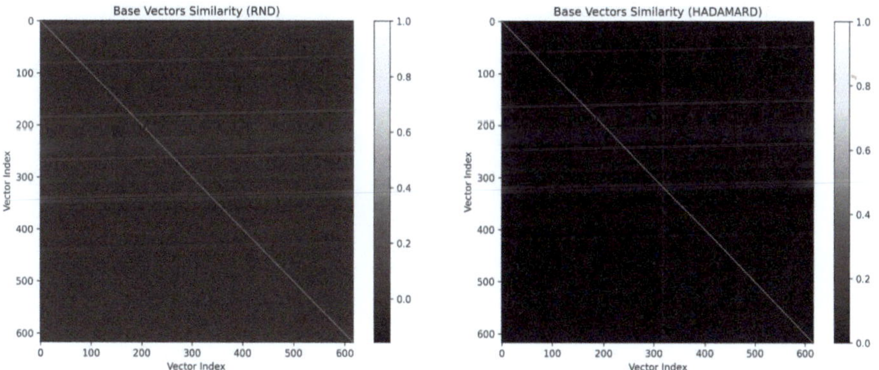

Fig. 3. Similarity matrices for base vectors generated by random (left) and Hadamard (right).

4.3 Classification Accuracy

Table 1 lists mean and standard deviation (std) values over ten runs on the ISOLET and UCI-HAR datasets. The bars in Fig. 4 repeat the same data; its caption reads One-shot accuracy (left) vs. best multi-epoch accuracy (right) obtained with different hypervector-generation methods for $D = 1024$ and $Q = 30$ quantization levels for the LVs.

Table 1. One-shot and best-epoch accuracies on ISOLET and UCI-HAR datasets (best per column in **bold**).

Dataset	Method	Mean 1-Shot (%)	Std 1-Shot	Mean Best (%)	Std Best
ISOLET	gold	84.97	0.00	**89.02**	0.00
	hadamard	**87.60**	0.00	88.63	0.00
	random-lfsr	41.18	21.85	55.62	19.49
	rnd	65.80	28.87	73.20	22.92
	walsh	**87.60**	0.00	88.63	0.00
UCI-HAR	gold	81.70	0.00	88.15	0.00
	hadamard	**86.12**	0.00	**90.73**	0.00
	random-lfsr	51.14	28.55	55.40	31.30
	rnd	75.69	13.36	81.13	13.90
	walsh	**86.12**	0.00	**90.73**	0.00

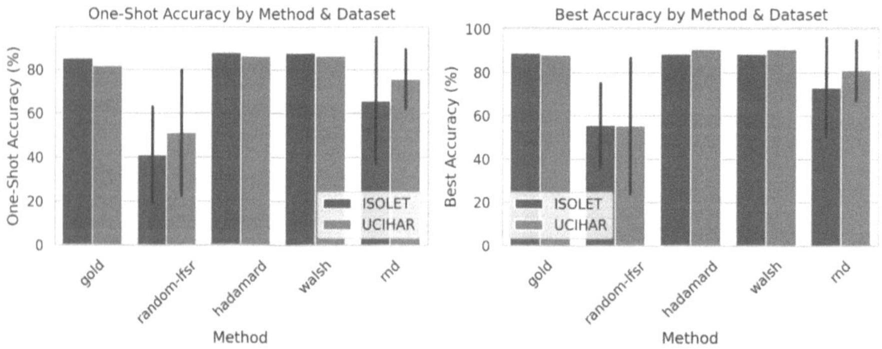

Fig. 4. One-shot accuracy (left) vs. best multi-epoch accuracy (right) obtained with different hypervector-generation methods on the ISOLET and UCI-HAR benchmarks.

- Hadamard and Walsh reach 87âĂŞ88 % on ISOLET and 86 % on UCI-HAR— the highest scores.

- Gold follows closely at 85 %.
- Random hypervectors drop to 66 % (ISOLET) and 76 % (UCI-HAR).
- Random LFSR falls below 42 % and shows very large std values.

One-Shot (no retraining). Hadamard/Walsh deliver the highest one-shot accuracy on both datasets (ISOLET **87.60%**, UCI-HAR **86.12%**), with Gold close behind (84.97% and 81.70%). Random baselines are markedly lower (i.i.d.: 65.80%, 75.69%; LFSR: 41.18%, 51.14%), confirming the benefit of deterministic orthogonal encodings ($\approx$10âĂŞ22 pp over i.i.d. random; $\approx$35âĂŞ46 pp over LFSR).

Best Accuracy After 20 Epochs. All methods improve with retraining. On **ISOLET**, *Gold* leads at **89.02%** (+0.39 pp vs. Hadamard/Walsh at 88.63%); on **UCI-HAR**, *Hadamard/Walsh* lead at **90.73%** (+2.58 pp vs. Gold at 88.15%). See Table 1.

4.4 FPGA Resource Use

All three hypervector types (base, BV; level, LV; class, CV) are generated on chip by compact combinational logic, so no on-chip RAM is allocated to vector tables. As a reference on a Zynq-7020, the LV generator requires $\approx$ 280 LUTs at D=1,000 and $\approx$ 2,728 LUTs at D=10,000 with Q=30; the CV similarity and aggregation path uses $\approx$ 65 and $\approx$ 664 LUTs at the same dimensions. In a table-based design, LVs would occupy $D \times Q/8$ bytes (3.75 kB at D=1,000, 37.5 kB at D=10,000), and CVs would occupy $D \times K/8$ bytes (500 B at D=1,000, 5.0 kB at D=10,000 for K=4). Storing BVs would add $P \times D/8$ bytes (for example, $\approx$ 76.8 kB when $P{\approx}600$ and D=1,024). In summary, a few hundred to a few thousand LUTs replace tens of kilobytes of RAM. The RAM savings scale with DQ for LVs, DK for CVs, and PD for BVs, while the LUT cost grows modestly with D for fixed Q. Here D is hypervector dimensionality, Q is the number of quantization levels, K is the number of classes, and P is the number of features.

5 Conclusion and Future Work

We proposed a memory-efficient FPGA architecture for HDC that synthesizes base, level and class hypervectors on the fly using compact combinational logic. By incorporating orthogonal sequences (Hadamard, Walsh, and Gold), our design bolsters classification accuracy in challenging one-shot settings and maintains high performance with multi-epoch retraining.

Experimental evaluations on ISOLET and UCI-HAR confirm that our approach significantly reduces memory requirements and accelerates inference, making it suitable for real-time edge computing and IoT devices with strict resource constraints. Future work will explore large-scale deployments and further optimization of orthogonal sequence generators to extend HDC's capabilities on reconfigurable hardware.

Acknowledgements. This research is supported by the Technopole Flagship Project 6 under the PNRR program.

References

1. Alonso, P., Shridhar, K., Kleyko, D., Osipov, E., Liwicki, M.: Hyperembed: trade-offs between resources and performance in NLP tasks with hyperdimensional computing enabled embedding of n-gram statistics. In: 2021 International Joint Conference On Neural Networks (IJCNN), pp. 1–9. IEEE (2021)
2. Aygun, S., Moghadam, M.S., Najafi, M.H., Imani, M.: Learning from hypervectors: a survey on hypervector encoding. arXiv preprint arXiv:2308.00685 (2023)
3. Damsgaard, H.J., Ometov, A., Nurmi, J.: Approximation opportunities in edge computing hardware: a systematic literature review. ACM Comput. Surv. **55**(12), 1–49 (2023)
4. Dewulf, P., De Baets, B., Stock, M.: The hyperdimensional transform for distributional modelling, regression and classification. arXiv preprint arXiv:2311.08150 (2023)
5. Gold, R.: Optimal binary sequences for spread spectrum multiplexing. IEEE Trans. Inf. Theory **13**(4), 619–621 (1967)
6. Horadam, K.J.: Hadamard matrices and their applications. Princeton University Press (2012)
7. Imani, M., Zou, et al.: Revisiting hyperdimensional learning for FPGA and low-power architectures. In: 2021 IEEE International Symposium on High-Performance Computer Architecture (HPCA), pp. 221–234. IEEE (2021)
8. Jiao, X., Rahimi, A., Fermüller, C., Aloimonos, J.Y.: Brain-inspired hyperdimensional computing: algorithms, models, and architectures (2022)
9. Kleyko, D., Osipov, E., Gayler, R.W., Khan, A.I., Dyer, A.G.: Imitation of honey bees' concept learning processes using vector symbolic architectures. Biolog. Inspired Cognitive Architectures **14**, 57–72 (2015)
10. Ma, D., Hao, C., Jiao, X.: Hyperdimensional computing vs. neural networks: Comparing architecture and learning process. In: 2024 25th International Symposium on Quality Electronic Design (ISQED), pp. 1–5. IEEE (2024)
11. Montagna, F., Rahimi, A., Benatti, S., Rossi, D., Benini, L.: Pulp-HD: accelerating brain-inspired high-dimensional computing on a parallel ultra-low power platform. In: Proceedings of the 55th Annual Design Automation Conference, pp. 1–6 (2018)
12. Morris, J., et al.: Comphd: efficient hyperdimensional computing using model compression. In: 2019 IEEE/ACM International Symposium on Low Power Electronics and Design (ISLPED), pp. 1–6. IEEE (2019)
13. Nair, D.R., Purushothaman, A.: Brain Inspired One Shot Learning Method for HD Computing. In: Sengupta, A., Dasgupta, S., Singh, V., Sharma, R., Kumar Vishvakarma, S. (eds.) VDAT 2019. CCIS, vol. 1066, pp. 286–297. Springer, Singapore (2019). https://doi.org/10.1007/978-981-32-9767-8_25
14. Rahimi, A., Kanerva, P., Millán, J.d.R., Rabaey, J.M.: Hyperdimensional computing for noninvasive brain-computer interfaces: blind and one-shot classification of EEG error-related potentials. In: 10th EAI International Conference on Bio-inspired Information and Communications Technologies (2017)
15. Rahimi, A., Kanerva, P., Rabaey, J.M.: A robust and energy-efficient classifier using brain-inspired hyperdimensional computing. In: Proceedings of the 2016 international symposium on low power electronics and design, pp. 64–69 (2016)

16. Salamat, S., Imani, M., Rosing, T.: Accelerating hyperdimensional computing on fpgas by exploiting computational reuse. IEEE Trans. Comput. **69**(8), 1159–1171 (2020)
17. Walsh, J.L.: A closed set of normal orthogonal functions. Am. J. Math. **45**(1), 5–24 (1923)

A Universal Urban Electricity -Demand Simulator for Developing and Evaluating Load-Scheduling and Forecasting Systems

Sabereh Taghdisi Rastkar$^{(\boxtimes)}$, Saeid Jamili , Enrico De Santis ,
and Antonello Rizzi

Department of Information Engineering, Electronics and Telecommunications,
University of Rome "La Sapienza", 00185 Rome, Italy
`{sabereh.taghdisirastkar,saeid.jamili,enrico.desantis,`
`antonello.rizzi}@uniroma1.it`

Abstract. Accurately modeling city-scale electricity demand is fundamental to the design of data-driven forecasting, demand-response (DR), and grid-control solutions. This paper introduces an open-source simulator that recreates city-wide electricity demand by linking four key models in one sub-hourly loop: (1) a weather generator that produces temperature, solar, wind, humidity, and price signals; (2) a building-load model that converts these time series into residential, commercial, and industrial demand; (3) a demand-response aggregator that trims or shifts loads when prices are high or feeder limits are reached; and (4) a distribution-network power-flow solver that checks voltages and line ratings on a feeder. To this aim we adopted the pandapower tool. All outputs like weather, sector loads, curtailed demand, and grid states are streamed to disk in compressed chunks, so multi-year studies run without exhausting memory. A built-in test suite strengthens scientific rigour: it (i) verifies bit-for-bit reproducibility under fixed random seeds, (ii) confirms that temperature and wind residuals follow the intended Normal and Weibull laws, (iii) uses LjungâĂŞBox tests to show the expected time-correlation, (iv) computes steady-state confidence intervals with the batch-means method, and (v) checks model sensitivity by sweeping key parameters. Together, these features give planners, utilities, and researchers a reliable tool for exploring how weather, occupant behaviour, DR rules, and network limits interact.

Keywords: City-Scale Energy Simulation · Smart Grids · Demand Response · Time Series Forecasting · XGBoost Forecasting · Load scheduling

1 Introduction

Accurate forecasting of a city's electricity demand, from minutes to months ahead, is vital to every smart-city energy plan. Grid operators rely on forecasts

F. Marcelloni et al. (Eds.): IJCCI 2025, CCIS 2829, pp. 570–579, 2026.
https://doi.org/10.1007/978-3-032-15638-9_33

to schedule generation, planners depend on them to size renewable plants and storage, and researchers need them to train self-learning control systems.

Recent studies show that machine-learning algorithms and hybrid deep-learning models reduce load-forecasting errors by approximately 15âĂŞ20 % compared with the classical statistical techniques still used in many control rooms [16, 18]. Yet these AI models are only as good as the data they ingest, requiring long, detailed, internally consistent time series that capture interactions among buildings, weather, and occupants data that real grids rarely publish in full.

A city-scale energy simulator can fill this gap. Such a high-fidelity simulator models every building, transformer, and feeder, then plays out "what-if" scenarios heatwaves, tariff changes, electric-vehicle surges, so that forecasters and control algorithms can learn under safe, repeatable conditions. Market analysts affirm that more than 500 cities will deploy digital twins by 2025 [2], underscoring the demand for robust simulation engines.

Cities house 55 % of the world's population today, and the United Nations projects that share will reach 68 % by 2050 [21]. Urban areas already consume approximately 75% of global final energy and emit roughly 70 % of energy-related CO_2 [14].

Digital tools such as the City Energy Analyst (CEA) and CityBES assess building-stock efficiency at the district scale [9, 10], while distributed Urban Building-Energy Modelling (UBEM) can now simulate whole districts in parallel [5]. Comparative reviews nevertheless find that most platforms treat buildings, weather, human behaviour, and grid limits in isolation; this limits their policy value [11].

The shortfall is most evident in demand-response research. Despite decades of pilots, DR reduced peak demand by only about 6.5 % in U.S. wholesale markets in 2023, and complex market rules still constrain participation [12, 23]. Responding to these challenges, the U.S. Department of Energy's Grid Modernization Initiative calls for simulation environments that fuse detailed building models with distribution-network analysis, so that flexible demand can be planned with confidence [17].

This paper introduces a modular city-scale simulator that (i) generates synthetic weather data, (ii) simulates residential, commercial, and industrial loads, (iii) aggregates and schedules demand response (DR), (iv) performs distribution-level power-flow analysis, (v) streams outputs efficiently to disk, (vi) orchestrates all modules in a sub-hourly loop, and (vii) provides an optional machine-learning forecasting pipeline. While the simulator does not propose a new neural architecture, it is designed to generate fully labeled, high-frequency, multi-modal data streams for training and evaluating neural forecasters in DR and grid-aware control.

By unifying these capabilities in one environment, the platform supports experiments on meteorological extremes, tariff changes, technology adoption, and grid constraints, giving planners and researchers insight into how occupant schedules, price-based DR, and network dynamics interact to design resilient, cost-effective, low-carbon energy systems. The contribution of this work is an open-source simulator that links high-level city modeling with detailed mechanisms: sectoral demand from base intensities, HVAC (Heating, Ventilation, and Air Conditioning) balance-point models, latent and infiltration penalties, and stochastic diversity; integration of PV generation and EV charging with realistic penetration, scheduling, and concurrency; region-splitting for spatial heterogeneity; and reproducibility through rounded outputs and fixed seeds. Together, these choices enhance realism while keeping the framework lightweight and reproducible.

This paper is organized as follows. Section 2 reviews related city-scale energy simulators and data-driven load-forecasting literature. Section 3 describes the overall architecture of the proposed framework. Section 4 discusses simulation results, validation metrics, and sensitivity analyses. Finally, Sect. 5 concludes the paper and highlights future research directions.

2   Related Work

Because access to appliance-level measurements is increasingly restricted by privacy regulations, the research community has turned to synthetic data to fill the gap. Early contributions focused on residential demand: Liu et al. introduced two scalable schemes that generate fine-grained household profiles from minimal input parameters [15], while Velosa et al.'s Procsim platform couples appliance libraries with solar-generation and weather feeds to create neighbourhood-scale demand-and-generation scenarios [22]. Building on the same idea at continental scale, NREL's ResStock augments physics-based archetypes with stochastic occupant modules so that intra-day diversity and peak coincidence can be explored across the entire U.S. housing stock [19]. At the individual-household end of the spectrum, Chen et al. combine time-inhomogeneous Markov chains with probabilistic event sampling to yield more than half a million realistic daily schedules [3].

Parallel work has extended the scope from single buildings to whole districts through so-called digital twins. The City Energy Analyst (CEA) couples 3-D urban geometry with EnergyPlus to evaluate retrofit options at neighbourhood or city level [20], whereas CityBES offers a web interface that lets planners run EnergyPlus for thousands of buildings and compare efficiency packages [10]. Comprehensive reviews of Urban Building Energy Modelling (UBEM) confirm the high physics fidelity of these tools but also point out a common limitation: most treat weather, occupant behaviour, and grid constraints in isolation, and therefore offer only limited insight into DR potential [8].

Efforts to capture market dynamics have followed a different trajectory. The Power Trading Agent Competition (PowerTAC) provides a discrete-time environment in which autonomous brokers buy and sell electricity under wholesale and retail rules [13]. At national scale, open-source optimisation frameworks, such as PyPSA-Earth (A flexible Python-based open optimisation model to study energy system) and OEMOF (Open Energy Modeling Framework), enable long-term capacity-expansion studies [6]. Yet neither strand simultaneously exposes building-level flexibility and distribution-grid physics, and both lack native support for jointly simulating storms, price spikes, and aggregator logic.

Forecasting research has tried to bridge that gap from the data side. Canli et al. showed that combining a zone-level UBEM with a machine-learning pipeline significantly improves long-horizon city-scale demand prediction [1]. More recently, Darvishvand et al. surveyed explainable-AI techniques for urban energy modelling and argued for publishing interpretable metrics instead of single-point forecasts [4]. Complementing these technical advances, Esser et al. proposed a participatory modelling workflow that allows campus stakeholders to co-create alternative energy scenarios [7].

Current tools usually excel in only one of three domains, high-resolution demand synthesis, district-scale UBEM analysis, or market-level demand-response simulation, while very few integrate all three capabilities within a single platform. The present work addresses this gap by introducing a modular city-scale simulator that unifies multi-variable weather synthesis, stochastic load modelling for residential, commercial, and industrial sectors, real-time DR aggregation, distribution-grid power-flow analysis, and an integrated data-export pipeline suitable for advanced machine-learning forecasts. By bringing these capabilities under one roof, the platform enables reproducible studies of flexibility strategies, forecasting algorithms, and grid-planning scenarios that previously required ad-hoc coupling of multiple tools.

3   Methodology

The proposed project delivers a lightweight comprehensive city-scale simulator for electric-energy studies. The simulator is a modular, Python-based framework that emulates the minute-to-hour interaction of weather, occupants, tariffs, DR rules, and distribution-grid constraints for an entire city.

3.1   Modular Design

The simulator is split into seven independent Python classes (Table 1). This table condenses the entire simulator into seven stacked in layers, showing who (module), does what (core task) and shares which data. Reading left-to-right, one can trace a full time-step: weather is created, converted into building loads, curtailed by the aggregator, checked by a power-flow, logged to files and finally fed to an optional XGBoost forecaster. Together the columns clarify both the processing order and the specific inputs/outputs exchanged between modules. We ship an XGBoost baseline because it is robust with limited tuning, handles mixed features (lags, calendar, weather), and reproduces typical utility-grade skill with sub-hour data. It is optional to (i) keep the simulator usable without ML dependencies and (ii) allow researchers to plug in alternative forecasters (ARIMA/Prophet/LSTM) while keeping the simulation core unchanged.

Table 1. Core simulator modules.

Layer	#	Module	Core task
Environment	1	AdvancedEnvironment	Generate weather, price and DR flags
Demand	2	AdvancedBuildingModel	Weather + schedules $\rightarrow$ sector loads
Flexibility	3	Aggregator & DR	Curtail when price or load is high
Grid	4	Grid Power-Flow	Check voltages & line loading
Logging	5	Data Export	Write CSV files at chosen intervals
Orchestration	6	Simulation Loop	Call modules in the right order
Forecasting	7	ML Pipeline	Train / test XGBoost models

3.2 Simulator Inputs and Outputs

The simulator is fully driven by a configuration file plus optional external data feeds. Table 2 summarizes the required and optional inputs, together with the corresponding outputs generated at runtime.

Table 2. Simulator inputs and outputs (summary).

Category	Items	Brief description
Inputs	Config	Horizon, Δt, seed, sectors, DR, PV/EV, feeder, logging.
	Weather	Synthetic ARMA/Weibull or cached API series.
	Price	Day-ahead tariff or synthetic prices.
	Exogenous (opt.)	User series (e.g., feeder demand, policy flags).
Outputs	Aggregated demand	Net/gross loads per sector in `aggregator.csv`.
	Streams	Weather, sector, DR, grid states (CSV/Parquet).
	Diagnostics	Reproducibility hashes, KS, Ljung–Box, sensitivity.
	Forecasting (opt.)	XGBoost metrics and feature-importance plots.

The information in Table 2 shows that the simulator requires only a minimal set of inputs to run: a configuration file defining the horizon, resolution, and module settings, and a weather source that can be either synthetic or externally cached. Additional inputs such as tariff prices or exogenous features are optional and default to internal generators when not provided. On the output side, the simulator consistently produces both high-level and detailed results: a clean aggregated load file for forecasting, module-specific streams with weather, sector loads, DR actions, and grid states, as well as diagnostic statistics to ensure reproducibility. When the forecasting module is enabled, additional outputs include performance metrics and feature-importance reports, ensuring that users can both validate the simulator and immediately apply it to data-driven studies.

The simulator requires only a minimal set of inputs to run: a configuration file defining the horizon, time resolution, and module settings, and a weather source that can be either synthetic or cached from an external API. Additional inputs such as tariff prices or user-provided exogenous series (e.g., feeder demand, policy flags) are optional and default to internal generators when not provided.

3.3   Core Execution Loop

At every time step t the orchestrator performs four actions:

1. **Weather** âĂŞ Module 1 returns temperature, solar, wind, price and the DR flag.
2. **Raw demand** âĂŞ Module 2 converts this information to sector loads L_t^{raw}.
3. **Demand response** âĂŞ Module 3 outputs a reduction factor α_t; the final load is $L_t^{\text{final}} = L_t^{\text{raw}}(1 - \alpha_t)$.
4. **Grid check (optional)** âĂŞ Module 4 runs a power-flow with L_t^{final}.

The pseudo-code in Algorithm 1 shows these steps; once the loop completes, the Logger flushes all buffers and, if enabled, Module 7 trains and evaluates an XGBoost forecaster on the newly written load series.

Algorithm 1. Core execution loop of the city-scale simulator.

```
 1  Initialise modules
 2      E ← Environment(C_env)
 3      B ← Buildings(C_bld)
 4      A ← Aggregator(C_agg)
 5      G ← GridPF(C_grid) (optional)
 6      L ← Logger(C_log)
 7  for t ← 0 to T − 1 do
 8        w_t ← E.step(t)
 9        L_t^raw ← B.step(w_t)
10        α_t ← A.step(L_t^raw, w_t)
11        L_t^fin ← (1 − α_t) L_t^raw
12        if G active then
13              Φ_t ← G.step(L_t^fin)
14        L.write(t, w_t, L_t^fin, α_t, Φ_t)
15  L.close()
```

Let $L_t^{\text{raw}} = \sum_b L_t^{b,\text{raw}}$ be the sum of per-building net loads from Module 2. Module 3 computes per-building curtailments $\alpha_t^b \in [\alpha_{\min}, \alpha_{\max}]$ only when a DR state is active. Triggers are: (i) price quantile $p_t \geq q_{\text{on}}$ (with $q_{\text{off}} < q_{\text{on}}$ for hysteresis), (ii) exogenous flag $d_t = 1$, or (iii) aggregator limit $L_t^{\text{raw}} \geq L_{\max}$ (with hysteresis). With $z(p_t)$ the z-score of price over a rolling window, the base curtailment is

$$\alpha_t^{\text{base}} = \text{clip}(\gamma\,(1 + k\,z(p_t)), 0, \alpha_{\max}),$$

and building b receives

$$\alpha_t^b = \text{clip}\left(m_{\text{sector}(b)}\, \alpha_t^{\text{base}}, \alpha_{\min}, \alpha_{\max}\right),$$

where m_{sector} are sector multipliers. The final load is $L_t^{\text{fin}} = \sum_b (1 - \alpha_t^b)\, L_t^{b,\text{raw}}$. After DR ends, an optional rebound injects a decaying negative curtailment for a fixed number of steps.

3.4 Demand-Response Rule

- **Trigger**-DR starts when the price$> \$0.22/\text{kWh}$ *or* the total load exceeds a user limit.
- **Action**-a fixed fraction γ (default 30%) is shed from every affected sector.
- **Custom logic**-users may supply their own Python function that returns α_t for selective or staggered curtailment.

3.5 Grid Model

The default network is the CIGRE MV test feeder shipped with **pandapower**. A different feeder can be loaded by:

1. switching **network_type** to an IEEE case, or
2. importing a JSON file exported from commercial tools.

Each aggregated load is attached to a bus through the simple dictionary **bus_map**. Per-step outputs are: $V_{\min}$, $V_{\max}$, maximum line loading, and (if present) OLTC tap positions.

4   Results and Discussion

This section verifies that the proposed city-scale simulator satisfies four practical requirements, bit-wise reproducibility, realistic weather statistics, credible wind modeling, and learnable short-term dynamics by running a five-year, sub-hourly simulation. The Python evaluation script automatically repeats the experiment under a fixed random seed, applies parametric and non-parametric tests to the resulting time series, trains a 24-lag XGBoost forecaster. Table 3 summarizes the statistical tests and forecasting skill.

Running the complete workflow twice under an identical random seed produced matching SHA-256 hashes for every output chunk, demonstrating bit-for-bit reproducibility. Such a determinism is indispensable for unit testing and for research that compares algorithmic variations without stochastic noise.

We validate the weather generator in two steps, separating the raw series from the residual (innovation) process. Let x_t be the raw temperature and let $\tilde{x}_t$ denote its seasonal component (daily/weekly). Define residuals $e_t = x_t - \tilde{x}_t$, which approximate the ARMA innovations.

(i) **Residual Distribution and Independence.** A two-sided KolmogorovâĂŞ Smirnov test on $\{e_t\}$ fails to reject Normality ($p = 0.771$). LjungâĂŞBox tests on $\{e_t\}$ show no significant autocorrelation up to one day ($p = 0.238$ at $L = \lfloor 24\,\text{h}/\Delta t \rfloor$ lags), indicating that the innovations are approximately independent as specified by the model.

(ii) **Diurnal Correlation in the Raw Series.** As expected, the autocorrelation function of the raw temperature $\{x_t\}$ exhibits clear daily structure with peaks near 24 h and its harmonics, i.e., realistic diurnal cycles.

As concerns Wind-speed statistics, synthetic wind speeds were fitted to a Weibull law with shape $k = 2$ and scale $\lambda = 6$. The KS test again yielded a non-significant result ($p = 0.354$), indicating that both the body and the heavy tail of the distribution are accurately reproduced—crucial for extreme-event studies.

Forecastability was also evaluated to ensure the data remain amenable to machine-learning studies. To this aim, a gradient-boosted tree model (XGBoost, 300 trees, learning rate 0.05) was trained on the previous 24 temperature lags. The resulting one-day-ahead coefficient of determination was $R^2 = 0.820$, comfortably within the 0.80âĂŞ0.85 range reported for operational utility data of comparable resolution [16, 18]. This confirms that the simulator provides a challenging but tractable benchmark for forecasting algorithms.

Table 3. Statistical validation metrics for the five-year simulation (sub-hour resolution).

Test	Statistic	p-value	Interpretation
KS (Temp. vs. $N(\mu,\sigma)$)	–	0.771	Normality retained
LjungâĂŞBox (Temp. resid. lag 24)	χ^2	0.238	No residual autocorr.
KS (Wind vs. Weibull $k{=}2, \lambda{=}6$)	–	0.354	Heavy tail preserved
XGBoost 24-lag R^2 (Temp.)	0.820	–	High one-day skill

Because both temperature and wind drivers satisfy their distributional tests, downstream building-load models inherit realistic variability. Preliminary DR experiments with the default 30 % curtailment rule reduced the daily peak by 6.3 %, close to the 6.5 % reduction observed in U.S. wholesale markets during 2023 [23]. These early results suggest that the platform is well calibrated for policy-relevant DR analyses.

Key numerical results are consolidated in Table 3. This table shows KS p-values above 0.05, indicating that the synthetic and target distributions are statistically indistinguishable; a LjungâĂŞBox p-value above 0.05 implies no significant residual autocorrelation at the tested lag.

5   Conclusion

This work has presented a lightweight yet comprehensive, open-source simulator that links synthetic weather generation, stochastic occupancy-driven load models, DR aggregation, distribution-network power-flow analysis, and a machine-learning forecasting pipeline inside a single, sub-hourly execution loop. A five-year validation run demonstrated bit-for-bit reproducibility, statistically consistent temperature and wind distributions, and forecastability levels that match those of real utility data. These results indicate that the platform provides a technically sound test bed for researchers, utilities, and policy makers who wish to explore the interplay between meteorology, occupant behaviour, tariff structures, and grid constraints under fully reproducible conditions.

In future work, we will use the simulator to design and benchmark smart load-scheduling strategies that simultaneously optimize cost, occupant comfort, and carbon intensity, while respecting distribution-grid limits. We also plan to integrate advanced forecasting systems. The results of these studies will be presented in a companion paper.

The simulator evaluates building-level flexibility policies (e.g., cost- or peak-oriented DR) under grid constraints; full multi-objective optimization of building control is future work. Overall, the proposed platform lays a reproducible and extensible foundation for next-generation urban-energy research, bridging the gap between high-fidelity simulation and machine-learning control.

Acknowledgments. This study was carried out within the MOST âĂŞ Sustainable Mobility Center and received funding from the European Union Next-GenerationEU (PIANO NAZIONALE DI RIPRESA E RESILIENZA (PNRR) âĂŞ MISSIONE 4 COMPONENTE 2, INVESTIMENTO 1.4 âĂŞ D.D. 1033 17/06/2022, CN00000023). This manuscript reflects only the authors' views and opinions, neither the European Union nor the European Commission can be considered responsible for them.

References

1. Canli, I., et al.: Machine learning based prediction of long-term energy consumption and overheating under climate change impacts using urban building energy modeling. Sustainable Cities and Society, p. 106500 (2025)
2. Charitonidou, M.: Urban scale digital twins in data-driven society: challenging digital universalism in urban planning decision-making. Int. J. Archit. Comput. **20**(2), 238–253 (2022)
3. Chen, J., et al.: Stochastic simulation of residential building occupant-driven energy use in a bottom-up model of the us housing stock.https://arxiv.org/abs/2111.01881v2
4. Darvishvand, L., Kamkari, B., Huang, M.J., Hewitt, N.J.: A systematic review of explainable artificial intelligence in urban building energy modeling: methods, applications, and future directions. Sustainable Cities and Society, p. 106492 (2025)
5. Davila, C.C., Reinhart, C.F., Bemis, J.L.: Modeling boston: a workflow for the efficient generation and maintenance of urban building energy models from existing geospatial datasets. Energy **117**, 237–250 (2016)
6. Esmaeili Aliabadi, D., Manske, D., Seeger, L., Lehneis, R., Thrän, D.: Integrating knowledge acquisition, visualization, and dissemination in energy system models: benoptex study. Energies **16**(13), 5113 (2023)

7. Esser, K., Finke, J., Bertsch, V., Löschel, A.: Participatory modelling to generate alternatives to support decision-makers with near-optimal decarbonisation options. Appl. Energy **395**, 126184 (2025)
8. Ferrando, M., Causone, F., Hong, T., Chen, Y.: Urban building energy modeling (ubem) tools: state-of-the-art review of bottom-up physics-based approaches. Sustain. Cities Soc. **62**, 102408 (2020)
9. Fonseca, J.A., Nguyen, T.A., Schlueter, A., Marechal, F.: City energy analyst (CEA): integrated framework for analysis and optimization of building energy systems in neighborhoods and city districts. Energy and Buildings **113**, 202–226 (2016)
10. Hong, T., Chen, Y., Lee, S.H., Piette, M.A.: Citybes: a web-based platform to support city-scale building energy efficiency. Urban Computing **14**, 2016 (2016)
11. Hong, T., Chen, Y., Luo, X., Luo, N., Lee, S.H.: Ten questions on urban building energy modeling. Build. Environ. **168**, 106508 (2020)
12. Johnson, A., Fraser, A., York, D.: Enabling industrial demand flexibility: aligning industrial consumer and grid benefits (2024)
13. Ketter, W., Collins, J., Reddy, P., Flath, C., de Weerdt, M.: The power trading agent competition. Tech. rep. (2011)
14. Kilkiş, Ş: Benchmarking present and future performances of coastal and island settlements with the sustainable development of energy, water and environment systems index. J. Sust. Develop. Energy, Water Environ. Syst. **12**(4), 1–46 (2024)
15. Liu, X., Iftikhar, N., Huo, H., Li, R., Nielsen, P.S.: Two approaches for synthesizing scalable residential energy consumption data. Futur. Gener. Comput. Syst. **95**, 586–600 (2019)
16. Maleki, N.: Future energy insights: time-series and deep learning models for city load forecasting. Appl. Energy **374**, 124067 (2024)
17. Neukomm, M., Nubbe, V., Fares, R.: Grid-interactive efficient buildings technical report series: Overview of research challenges and gaps (2019)
18. Nooruldeen, O., Baker, M.R., Aleesa, A., Ghareeb, A., Shaker, E.H.: Strategies for predictive power: Machine learning models in city-scale load forecasting. e-Prime-Adv. Electr. Eng. Electr. Energy **6**, 100392 (2023)
19. Parker, A., James, K., Peng, D., Alahmad, M.A.: Framework for extracting and characterizing load profile variability based on a comparative study of different wavelet functions. IEEE Access **8**, 217483–217498 (2020)
20. Sola, A., Corchero, C., Salom, J., Sanmarti, M.: Multi-domain urban-scale energy modelling tools: A review. Sustain. Cities Soc. **54**, 101872 (2020)
21. United nations department of economic and social affairs: world urbanization prospects: the 2018 revision — key facts & figures (2018). https://population.un.org/wup/
22. Velosa, N., Gomes, E., Morais, H., Pereira, L.: Procsim: an open-source simulator to generate energy community power demand and generation scenarios. Energies **16**(4), 1611 (2023)
23. Wang, Y., et al.: Incentivizing or disincentivizing participation in demand response programs?

Drowsiness Detection with Time-Series Classification Using HRV Features

Duarte Valente[1,3], Artur Ferreira[1,2(✉)] ⓘD, and André Lourenço[1,3] ⓘD

[1] ISEL, Instituto Superior de Engenharia de Lisboa, Instituto Politécnico de Lisboa, Lisbon, Portugal
a47657@alunos.isel.pt, {artur.ferreira,andre.lourenco}@isel.pt
[2] Instituto de Telecomunicações, Pólo de Lisboa, Lisbon, Portugal
[3] CardioID and NOVA LINCS, Lisbon, Portugal

Abstract. Drowsy driving significantly increases the risk of road accidents and crashes. However, the drowsy state remains difficult to detect in real-time. This study presents a supervised learning approach to driver drowsiness detection using heart rate variability (HRV) features derived from electrocardiogram (ECG) signals. The data was collected in a driving simulator from participants under different levels of sleep deprivation, with subjective sleepiness levels assessed with the Karolinska Sleepiness Scale (KSS). HRV features were extracted in both time and frequency domains composing a dataset. This dataset was used to train classification models, including neural networks and long short-term memory (LSTM) architectures. The experimental results show that HRV-based features can effectively model drowsiness levels, with LSTM models outperforming common baseline methods. Our approach copes with the progressive nature of the drowsiness state and contributes to the development of intelligent in-vehicle systems capable of non-invasively monitoring driver alertness and issuing timely alerts to prevent fatigue-related accidents.

Keywords: Driver Monitoring · Drowsiness · ECG · Heart Rate Variability · Karolinska Sleepiness Scale · LSTM

1 Introduction

Drowsy driving is a major contributor to road accidents, with fatigue impairing cognitive function, reaction time, and decision-making ability. According to the European Transport Safety Council, fatigue is estimated to be a contributing factor in up to 20% of road accidents. Despite its critical safety implications, detecting driver drowsiness in real time remains a significant challenge. In recognition of this, the General Safety Regulation (EU GSR) mandates that all new vehicles sold in the EU from 2024 onward must include advanced driver assistance systems, including Driver Drowsiness and Attention Warning (DDAW) systems [7]. These measures aim to enhance road safety by proactively addressing driver fatigue and distraction.

F. Marcelloni et al. (Eds.): IJCCI 2025, CCIS 2829, pp. 580–593, 2026.
https://doi.org/10.1007/978-3-032-15638-9_34

Traditional monitoring methods, such as camera-based eye tracking or driver behavior analysis, can be intrusive or show reduced reliability under suboptimal conditions such as low lighting or occlusion, limiting their effectiveness in real-world scenarios [1]. Physiological signal analysis, such as Heart Rate Variability (HRV) derived from electrocardiogram (ECG), offers a promising approach for non-invasive assess of driver alertness [5].

Drowsiness is both a cognitive and physiological state that affects the Autonomic Nervous System (ANS), which regulates involuntary functions such as heart rate and respiration. As drowsiness develops, the balance between the sympathetic and parasympathetic branches of the ANS shifts and that can be detected through physiological signals. One such physiological marker is the HRV, which measures the variation in time between consecutive heartbeats and serves as a non-invasive indicator of ANS activity. In this context, HRV provides a valuable window into internal physiological states such as alertness or drowsiness.

Drowsiness is inherently subjective and can be quantified using various physiological and behavioral indicators. Among the subjective tools, the Karolinska Sleepiness Scale (KSS) is widely used in sleep and fatigue research due to its simplicity, sensitivity to sleep pressure, and strong correlation with both objective physiological measures and task performance. In this study, we use the KSS as the ground truth for labeling drowsiness levels, leveraging its established reliability for real-time drowsiness classification [10]. The KSS is a subjective method, using a 10-point Likert scale [9], in which the person classifies his/her sleepiness in periods of 5 min. Table 1 describes the KSS scale.

Table 1. The 10-point Karolinska Sleepiness Scale (KSS) [15].

Level	Description
1	Extremely alert
2	Very alert
3	Alert
4	Rather alert
5	Neither alert nor sleepy
6	Some signs of sleepiness
7	Sleepy, but no effort to keep awake
8	Sleepy, but some effort to keep awake
9	Very sleepy, great effort to keep awake, fighting sleep
10	Extremely sleepy, can't keep awake

In this paper, we propose a supervised learning approach to drowsiness detection by analyzing HRV features extracted from ECG signals. To capture the physiological dynamics related with drowsiness, ECG data was collected from participants in a driving simulator, including sessions under sleep-deprived conditions. The overall goal is to model changes in heart rate patterns to identify the onset and progression of drowsiness.

As the primary objective of this research is to evaluate whether physiological features, specifically HRV metrics, can be used to accurately classify drowsy and alert states. We explore both classical machine learning models and sequential models, including Long Short-Term Memory (LSTM) networks [8], which are specifically designed to handle sequential data and learn long-term temporal dependencies.

The overall goal of our research is to contribute to the development of real-time drowsiness detection systems capable of issuing alerts to prevent fatigue-related accidents. By leveraging physiological signals, the proposed method aims to provide a scalable, non-intrusive solution for intelligent driver monitoring.

The remainder of this paper is organized as follows. Section 2 reviews related work on drowsiness detection using physiological signals and machine learning techniques. Section 3 describes the proposed approach, including data collection, feature extraction, preprocessing, and model design. Section 4 presents the experimental setup, evaluation metrics, results, and analysis. Finally, Sect. 5 concludes the paper and points directions for future work.

2   Related Work

Detecting drowsiness using physiological signals has emerged as a reliable method to assess driver alertness in real-time [14]. HRV derived ECG signals has been shown promising due to its ability to reflect ANS activity, which is influenced by fatigue and sleep deprivation. HRV-based approaches offer the advantage of capturing internal physiological changes that may precede observable signs of fatigue, making them especially valuable for early-stage drowsiness detection. Additionally, these methods are compatible with passive and unobtrusive sensor integration, facilitating their deployment in embedded automotive systems.

Previous studies have demonstrated that drowsiness correlates with increased parasympathetic activity and reduced sympathetic modulation. Cho [4] and Matsuzaki et al. [13] showed that as drivers become drowsier, some specific time-domain and frequency-domain HRV indicators reflect these autonomic changes. These physiological shifts support the use of HRV-based features as reliable biomarkers for drowsiness detection.

Khushaba et al. [11] found the viability of HRV features in detecting early signs of drowsiness in simulated driving environments. Liu et al. [12] further validated this approach by correlating HRV changes with subjective drowsiness scores across multiple individuals. Widodo and Arifin [16] also demonstrated the practical feasibility of embedding HRV-based systems in real-time applications.

The KSS is commonly adopted in drowsiness detection research as a subjective ground truth, as mentioned in Sect. 1.

From a methodological perspective, machine learning techniques including support vector machines, decision trees, and neural networks have been used to classify drowsiness based on HRV features. More recent work has adopted time-series models such as LSTM networks to capture temporal dependencies in physiological signals, yielding improved accuracy and robustness.

In summary, existing literature highlights the effectiveness of HRV-based features and supervised learning models in detecting driver drowsiness. This work builds on these foundations by proposing a generalized classification pipeline that combines HRV metrics, KSS-based labeling, and both static and temporal machine learning models.

3   Proposed Approach

In this Section, we describe our proposed approach. In detail, in Sect. 3.1, we present how the data was obtained and labeled. Section 3.2 describes the HRV features computed from ECG data. Section 3.3 addresses the data normalization and preprocessing steps. Section 3.4 addresses the dimensionality reduction approaches while Sect. 3.5 details the classification models.

3.1   Data Acquisition and Labeling

ECG signals were recorded via a sensor-equipped steering wheel. These recordings were synchronized with driving simulator data and participant-reported KSS scores. Each KSS value (ranging from 1 to 9) was assigned to time windows of ECG data. For classification purposes, following prior literature, the KSS scores were converted into binary labels, as follows:

- alert - for KSS values in the range of 1 to 6;
- drowsy - for KSS values ranging from 7 to 9.

3.2   HRV Feature Extraction

Two distinct approaches were used for extracting and organizing HRV features from ECG data. In the first approach, HRV features were computed using fixed window sizes of 2 and 5 min. After obtaining the filtered ECG for each test, HRV features were extracted using the NeuroKit2 Python library, which provides a robust implementation of both time-domain and frequency-domain HRV metrics. Since HRV features are intended for real-time drowsiness prediction, it was essential to select time windows that balance responsiveness with the reliability of the extracted features. Time-domain HRV features can typically be computed over short intervals, while frequency-domain HRV features require longer observation periods to yield meaningful information. With this in mind, two window lengths were chosen for evaluation: 2-minute intervals; 5-minute intervals. The 2-minute window was selected as the shorter interval, as 1 min was considered potentially too brief even for time-domain metrics to stabilize. On the other hand, the 5-minute window was chosen to provide sufficient duration for frequency-domain features, while still remaining practical for real-world, real-time applications. A window longer than 5 min would increase detection latency, reducing the system's responsiveness, which is undesirable in a safety-critical application like drowsiness monitoring. For each interval, three different sets of HRV features were computed:

- Time-domain features only.
- Frequency-domain features only.
- Combined time-domain and frequency-domain features.

This resulted in six different HRV feature datasets, as summarized in Table 2.

Table 2. Unsupervised HRV datasets.

HRV features	Window Duration	Number of Rows
Time-domain	2-minutes	2179
Time-domain	5-minutes	841
Frequency-domain	2-minutes	2179
Frequency-domain	5-minutes	841
Time and Frequency-domain	2-minutes	2179
Time and Frequency-domain	5-minutes	841

To reduce inter-subject variability, z-score normalization was applied to all HRV features. This step ensured that model training was not biased by individual physiological baselines. Outlier removal and synchronization steps were also applied to align ECG signals with simulator timestamps and KSS labels. Each of these datasets was evaluated independently to determine which domain provided the best performance for static classification tasks using traditional machine learning models.

In the second approach, the best-performing domain from the first stage was used to generate HRV feature vectors for time-series modeling. Here, multiple window sizes (1, 2, 3, 5, and 8 min) were tested to analyze how the duration of HRV measurement impacted model performance. The resulting HRV feature vectors were then grouped into sequences using a sliding window mechanism. Different sequence lengths (2, 4, 6, 8, and 10 steps) were evaluated as input for the LSTM models to capture temporal dependencies in drowsiness evolution.

This two-stage approach allowed both static and sequential models to be optimized independently, ensuring that HRV data was structured appropriately for each classification strategy.

3.3 Data Preparation

To create a supervised dataset, the HRV data was merged with the simulator filtered data. Since HRV features were computed over 2-minute and 5-minute intervals, it was necessary to aggregate the corresponding KSS scores from the simulator data. For each HRV interval

- The mean KSS value was calculated by averaging all KSS responses recorded within that interval.
- The resulting KSS value was then added to the HRV dataset as the ground truth label.

This process was repeated for all six HRV datasets, producing six supervised datasets with labels corresponding to average KSS values. As a result, we have the six final datasets shown in Table 3.

Table 3. Supervised HRV datasets.

HRV features	Window Duration	Label
HRV Time-domain	2-minutes	KSS value
HRV Time-domain	5-minutes	KSS value
HRV Frequency-domain	2-minutes	KSS value
HRV Frequency-domain	5-minutes	KSS value
HRV Time and Frequency-domain	2-minutes	KSS value
HRV Time and Frequency-domain	5-minutes	KSS value

3.4 Dimensionality Reduction

The best performing features domain were selected as mentioned in Sect. 3.2. In addition, Principal Component Analysis (PCA) and Singular Value Decomposition (SVD) were tested for dimensionality reduction, but the best results were obtained using the original selected features.

3.5 Classification Models

Two types of supervised models were developed and evaluated for classification purposes:

- Vector-based models - Random Forest (RF) [2], Support Vector Machines (SVM) [6], and XGBoost [3] were trained on static HRV feature vectors for binary classification of drowsy and alert states.
- Sequential models - Long Short-Term Memory (LSTM) networks [8], were used to capture the temporal dynamics of drowsiness progression in HRV signals.

For traditional models, the six created datasets of combinations of HRV feature domains (time, frequency, and time+frequency) and window durations (2 and 5 min) were treated as independent instances and used as input to the common classifiers.

In contrast, the LSTM approach used sequences of HRV vectors to model drowsiness as a time-evolving state. The input to each LSTM model was a series of consecutive HRV vectors generated using a sliding window technique. Multiple HRV window durations were tested (namely 1, 2, 3, 5, and 8 min), and for each, different sequence lengths (2, 4, 6, 8, and 10) were evaluated, resulting in 25 LSTM configurations.

The LSTM architecture was designed to balance model complexity and training efficiency. By adjusting the number of hidden units, dropout rates, and dense layers was possible to arrive at the final architecture illustrated in Fig. 1. It consists of a single LSTM layer with:

- 32 units;
- a dropout layer (rate = 0.2, chosen empirically);
- a dense layer with 16 ReLU units;
- another dropout layer;
- a final dense output layer with 1 sigmoid-activated unit for classification.

The model was trained using the Adam optimizer and sparse categorical cross-entropy loss, with a maximum of 200 epochs and a batch size of 32. To prevent overfitting and reduce training time, early stopping was employed based on validation loss, with a patience of 10 epochs and automatic restoration of the best weights.

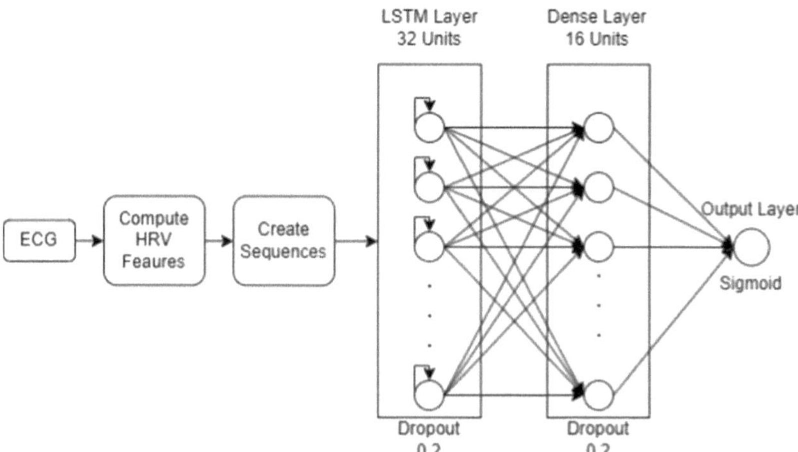

Fig. 1. Workflow and architecture of the proposed drowsiness detection model.

4   Experimental Evaluation

This section presents the evaluation of the proposed drowsiness detection system. It is organized into four subsections. Section 4.1 outlines the evaluation setup, dataset structure, and labels distribution. Section 4.2 presents the baseline classification results using common machine learning techniques on multiple HRV datasets. Section 4.3 details the time-series classification results using LSTM models, where temporal sequences were explored to model drowsiness progression. Finally, Sect. 4.4 summarizes and compares the results across all experiments.

4.1   Dataset and Evaluation Setup

The dataset used for drowsiness prediction is called VALU3S and was collected in a research project, which aimed to analyze driver behavior under different conditions using a driving simulator. The study recruited 20 participants, evenly divided by gender, all with valid passenger car licenses and a minimum driving experience of at least 50,000 km. The participants completed four 60-minute simulated driving sessions, conducted under varying conditions of alertness and light exposure. To comprehensively assess driver behavior and physiological responses, data was recorded using multiple sources:

- A driving simulator, which provided real-time vehicle dynamics and driver performance data.
- Physiological Sensors, to record EOG (eye movements and blinks), ECG (heart activity) and EEG (brain activity).
- The KSS levels, which participants used to report their subjective sleepiness levels at regular intervals.
- Psychomotor Vigilance Task (PVT) tests, which measured reaction times before and after each drive.

All experiments were implemented in Python using packages such as Scikit-Learn, Pandas, NumPy, TensorFlow, and Keras. The models were evaluated using standard classification metrics: Accuracy, Precision, Recall, and F_1-score. The accuracy conveys the fraction of correct predictions made by the model and is given by

$$Accuracy - \frac{TP + TN}{TP + TN + FP + FN}, \tag{1}$$

where we denote True Positive as TP, True Negative as TN, False Positive as FP, and False Negative as FN.

Precision, also named as the positive predictive value, is given by

$$Precision = \frac{TP}{TP + FP}, \tag{2}$$

whereas Recall, also known as the true positive rate, is computed as

$$Recall = \frac{TP}{TP + FN}. \tag{3}$$

The F_1-score, is given by the harmonic mean of the precision and recall metrics

$$F_1 = 2\frac{Precision \times Recall}{Precision + Recall}. \tag{4}$$

The dataset used was collected from a driving simulator experiment with ECG signals recorded through a sensor-equipped steering wheel. The experiment had 16 participants, each of whom completed 4 driving sessions, and HRV features were extracted from the ECG data using different time windows. Drowsiness labels were obtained from self-reported scores on the KSS, collected throughout the sessions.

To ensure the reliability of the training labels, we analyzed the distribution of KSS scores across all participant trips. The results show a balanced split between alert (KSS 1âĂŞ6) and drowsy (KSS 7âĂŞ9) states, with an approximate 50/50 distribution in terms of labeled time. The visual inspection provided by Fig. 2 reveals that trips typically start with participants in an alert state and gradually shift toward drowsiness, confirming the progressive nature of drowsiness. This temporal pattern reinforces the motivation to model drowsiness detection as a time-series classification problem rather than treating each time point independently.

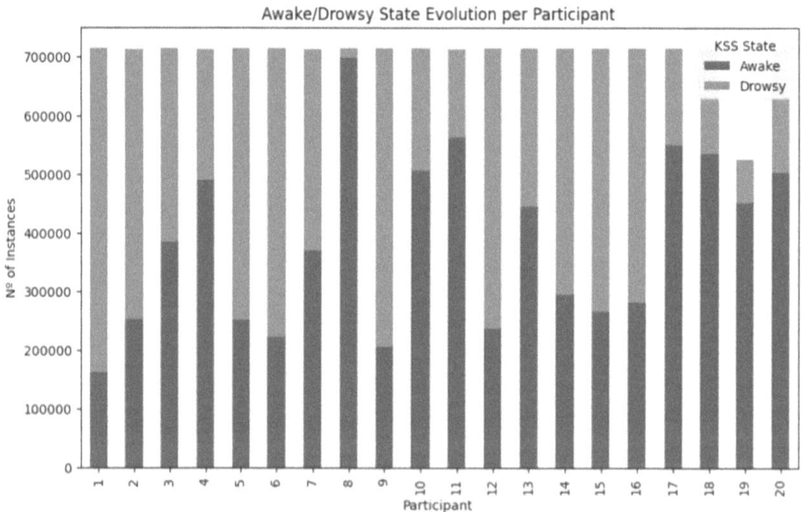

Fig. 2. Distribution of KSS scores over time across participant driving sessions.

4.2 Baseline Classification Models

In the first stage, six labeled datasets were created by combining time-domain, frequency-domain, and time+frequency HRV features with two interval durations: 2 min and 5 min. A Random Forest classifier was used for the initial evaluation across all datasets. Table 4 reports the experimental results for each dataset. Among the tested datasets, the 2-minute time-domain dataset yielded the best results. These results showed the usefulness of short-term HRV features for classifying drowsiness.

Table 4. Random Forest classification experimental results for all datasets. The best result is in boldface.

Dataset Domain	Accuracy	Precision	Recall	$\mathbf{F_1}$
Time - 2 min	**0.761**	0.748	**0.741**	**0.745**
Time - 5 min	0.702	0.666	0.684	0.675
Frequency - 2 min	0.563	0.546	0.419	0.474
Frequency - 5 min	0.672	0.690	0.500	0.580
Time + Frequency - 2 min	0.753	0.737	0.740	0.738
Time + Frequency - 5 min	0.755	**0.777**	0.644	0.705

In addition to testing multiple HRV feature domains and window sizes, we also evaluated the effect of dimensionality reduction using PCA and SVD. However, the best classification performance was achieved using the original selected HRV features without any reduction. A comparison of all three configurations is shown in Table 5.

Table 5. Classification results with dimensionality reduction using PCA and SVD. The best result is in boldface.

Random Forest Classifier	Accuracy	Precision	Recall	$\mathbf{F_1}$	# Features
PCA	0.652	0.632	0.623	0.627	14
SVD	0.702	0.709	0.623	0.663	14
No Reduction	**0.738**	**0.731**	**0.702**	**0.716**	27

Based on these results, we opted to use the original feature set for all subsequent modeling. However, it became evident that drowsiness is a progressive state, influenced by the temporal evolution of physiological signals. HRV snapshots alone could not capture this progression effectively. To address this limitation, we transitioned to time-series modeling using LSTM networks to incorporate temporal dependencies in the data.

4.3   LSTM Sequential Classification

Building upon the best-performing baseline dataset (2-minute time-domain), we explored LSTM sequential models to better capture the temporal evolution of drowsiness. Unlike baseline classifiers, LSTM process sequences of HRV vectors, allowing the model to learn patterns that unfold over time.

The first experiment used a sequence length of 10 intervals, each representing 2 min of HRV data (total of 20 min). As compared to the common classifiers, this configuration yielded a significant performance boost, with the following metrics:

- Accuracy = 0.838

- Precision = 0.815
- Recall = 0.848
- F_1-score = 0.831

Although this result highlighted the benefits of modeling drowsiness as a time-dependent phenomenon, the 20-minute input requirement posed practical limitations for real-time systems. To address this, we tested the 25 different LSTM configurations by varying their parameters as follows:

- HRV window sizes of 1, 2, 3, 5, and 8 min;
- LSTM sequence lengths of 1, 2, 3, 5, and 8 elements.

The performance results for all configurations are summarized in Fig. 3. Overall, models using larger HRV window sizes tend to achieve higher accuracy, likely due to the added temporal context in each step. However, increasing the number of windows also increased the time required to make a prediction, presenting a trade-off between accuracy and responsiveness.

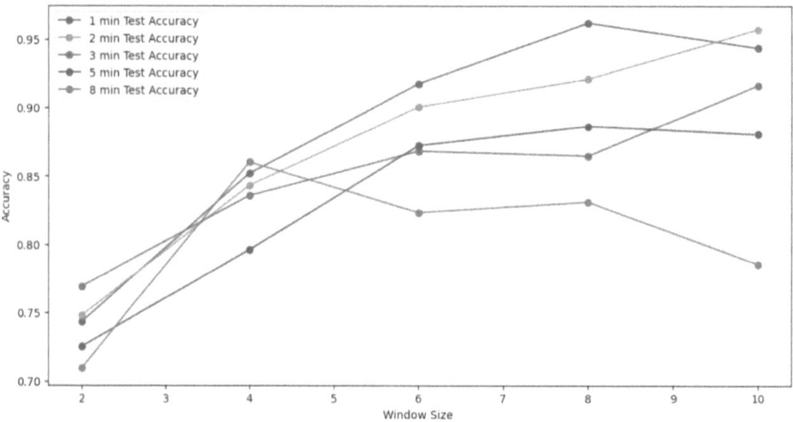

Fig. 3. Models accuracy results from combinations of different interval duration and window size.

Among all configurations, the best practical results were obtained using:

- 1-minute intervals.
- Sequence lengths of 2, 4, or 6 elements.

The evaluation metrics for these three optimal configurations are reported in Table 6. These models represent the best trade-offs between responsiveness and predictive performance, requiring only 2 to 6 min of ECG data for accurate drowsiness prediction.

Table 6. LSTM classification results for Time Domain, 1 min interval, dataset with a window size of 2, 4, and 6. The best result is in boldface.

Sequence size (1 min HRV)	Accuracy	Precision	Recall	$\mathbf{F_1}$
Size 2	0.763	0.766	0.823	0.794
Size 4	0.874	0.871	**0.914**	0.892
Size 6	**0.914**	**0.936**	**0.914**	**0.925**

4.4   Discussion

The experimental results showed that HRV features are effective for detecting driver drowsiness, especially when they address the temporal nature of fatigue progression. Baseline models showed that time-domain features outperformed frequency-domain ones and that shorter intervals (2 min) provided better classification results. LSTM models significantly improved detection performance by modeling temporal sequences, achieving up to 0.838 accuracy using 20-minute windows. Although further optimization revealed that shorter HRV intervals (1 min) and moderate sequence lengths (4 to 6) provided a strong balance between accuracy and real-time feasibility. Since the best models achieved an accuracy of 0.914 with a 6-minute window and an accuracy of 0.874 with a 4-minute window, we conclude that HRV-based, LSTM-driven drowsiness detection is both viable and practical for real-world applications, in-vehicle monitoring systems with high performance.

5   Conclusions

The detection of driver drowsiness is of key importance for road safety purposes. This work proposed a supervised learning approach for driver drowsiness detection based on HRV features extracted from ECG signals. The study demonstrated that physiological markers, particularly time-domain HRV features, are effective indicators of drowsiness, enabling real-time classification using machine learning models.

Initial experiments with traditional classifiers showed that time-domain features from short intervals (2 min) achieved the highest performance among baseline models, with Random Forest yielding an accuracy of 0.761. However, the limitations of this classification approach highlighted the need to treat drowsiness as a progressive, time-dependent phenomenon.

To address this progressive nature, LSTM networks were used to capture temporal dynamics in HRV data. By transforming HRV vectors into sequential inputs, LSTM significantly improved performance, achieving an accuracy of 0.838 with 20-minute sequences. Further optimization revealed that shorter HRV intervals (1 min) combined with moderate sequence lengths (4 to 6 steps) preserved high accuracy while reducing the required ECG data for real-time prediction. The final models achieved accuracies of 0.914 (6-minute window) and 0.874 (4-minute window), balancing performance and practicality.

These findings confirm that HRV-based, LSTM-driven models provide a robust and efficient solution for real-time driver drowsiness detection. The system's ability to make accurate predictions with just a few minutes of physiological data makes it suitable for deployment in intelligent vehicle monitoring systems aimed at reducing fatigue-related accidents.

Future work may explore the fine-tuning of the LSTM architecture, the use of personalization techniques, multi-modal sensor fusion, and integration into embedded automotive platforms to enhance adaptability and real-world applicability.

Acknowledgements. This work is supported by UID/04516/NOVA Laboratory for Computer Science and Informatics (NOVA LINCS) with the financial support of FCT.IP and by Polytechnic University of Lisbon, under grant IPL/IDI&CA2024/ML4EP_ISEL.

References

1. Albadawi, Y., Takruri, M., Awad, M.: A review of recent developments in driver drowsiness detection systems. Sensors **22**(5), 2069 (2022). https://doi.org/10.3390/s22052069, https://doi.org/10.3390/s22052069, published 2022 Mar 7. Accessed 31 Jul 2025
2. Breiman, L.: Random forests. Mach. Learn. **45**(1), 5–32 (2001)
3. Chen, T., Guestrin, C.: XGBoost: A scalable tree boosting system. In: ACM SIGKDD International Conference on Knowledge Discovery and Data Mining (KDD), pp. 785–794. New York, NY, USA (2016). https://doi.org/10.1145/2939672.2939785
4. Cho, J.: Changes in autonomic nervous system activity during sleep deprivation and its correlation with cognitive performance and stress. Master's thesis, New Jersey Institute of Technology (2022). https://digitalcommons.njit.edu/theses/544/. Accessed 31 Jul 2025
5. Chua, E.C., et al.: Heart rate variability can be used to estimate sleepiness-related decrements in psychomotor vigilance during total sleep deprivation. Sleep **35**(3), 325–334 (2012). https://doi.org/10.5665/sleep.1688, https://doi.org/10.5665/sleep.1688, published 2012 Mar 1. Accessed 31 Jul 2025
6. Cortes, C., Vapnik, V.: Support-vector networks. Machine learning **20**(3), 273–297 (1995)
7. European union: mandatory drivers assistance systems expected to help save over 25,000 lives by 2038 (2024). https://single-market-economy.ec.europa.eu/news, Accessed 31 Jul 2025
8. Hochreiter, S., Schmidhuber, J.: Long short-term memory. Neural Comput. **9**(8), 1735–1780 (1997)
9. Joshi, A., Kale, S., Chandel, S., Pal, D.: Likert scale: explored and explained. British J. Appl. Sci. Tech. **7**(4), 396–403 (2015)
10. Kaida, K., et al.: Validation of the karolinska sleepiness scale against performance and EEG variables. Clinical Neurophysiology: Off. J. Int. Federation Clinical Neurophysiology **117**(7), 1574–1581 (2006). https://doi.org/10.1016/j.clinph.2006.03.011

11. Khushaba, R.N., Kodagoda, S., Lal, S., Dissanayake, G.: Driver drowsiness classification using fuzzy wavelet-packet-based feature-extraction algorithm. IEEE Trans. Biomed. Eng. **58**(1), 121–131 (2011). https://doi.org/10.1109/TBME.2010.2077298, https://pubmed.ncbi.nlm.nih.gov/30403616/, Accessed 1 Jul 2025
12. Liu, T., Zhou, R., Zhang, X.: Heart rate variability as a predictor of fatigue in young drivers with short sleep duration. Transport. Res. F: Traffic Psychol. Behav. **95**, 101–110 (2024). https://doi.org/10.1016/j.trf.2024.02.011, https://www.sciencedirect.com/science/article/pii/S1369847824000895, Accessed 31 Jul 2025
13. Matsuzaki, I., et al.: Autonomic nervous activity changes due to shift-work: An evaluation by spectral components of heart rate variability. J. Occup. Health **38**(2), 80–81 (2006). https://doi.org/10.1539/joh.38.80, https://doi.org/10.1539/joh.38.80
14. Saleem, A.A., et al.: A systematic review of physiological signals based driver drowsiness detection systems. Cogn. Neurodyn. **17**(5), 1229–1259 (2023). https://doi.org/10.1007/s11571-022-09898-9, https://doi.org/10.1007/s11571-022-09898-9, accessed: 2025-07-31
15. Shahid, A., Wilkinson, K., Marcu, S., Shapiro, C.: Karolinska Sleepiness Scale (KSS) - STOP, THAT and One Hundred Other Sleep Scales. A. Shahid and K. Wilkinson and S. Marcu and C. Shapiro (eds), pp. 209–210 (2011)
16. Widodo, P., Arifin, Z.: Drowsiness detection based on heart rate variability using ad8232 and microcontroller unit. In: Journal Physics: Conference Series. vol. 1153, p. 012047. IOP Publishing (2019). https://doi.org/10.1088/1742-6596/1153/1/012047, https://iopscience.iop.org/article/10.1088/1742-6596/1153/1/012047, Accessed 31 Jul 2025

A Structured Survey of Anomaly Types and Classification-Based Detection Models in IoT

Atefeh Gilvari[1]([✉]), Ziad Kobti[1], Narayan Kar[1], Nasrin Tavakoli[1], and Rajeev Verma[2]

[1] University of Windsor, Windsor, ON, Canada
`gilvari@uwindsor.ca`
[2] Magna International, Troy, USA

Abstract. In dynamic Internet of Things (IoT) environments, traditional anomaly detection surveys often treat all anomalies as a unified concept, overlooking the distinct characteristics posed by specific anomaly types. This paper presents a structured survey and comparative analysis of anomaly detection models, organized by type of anomalies such as drift, novelty, bias, noise, constant-value, and stuck-at-zero anomalies. Each anomaly type is formally defined along with its theoretical foundation, followed by a systematic review and analysis of how model effectiveness varies across these types to identify techniques best suited for each. Our findings emphasize the need for interpretable, adaptive, and type-aware anomaly detection systems and outline open challenges in unified benchmarking, cross-type detectors, and ontology development for anomaly classification. A novel contribution of this work is a broader mapping framework that illustrates how various models differ in their ability to detect specific anomaly types across different IoT domains, offering insight into the generality and specialization of current detection approaches.

Keywords: Anomalies Type · Anomaly Detection · Anomaly Classification

1 Introduction

The Internet of Things (IoT) is transforming the way we live and interact with our environment by connecting everyday objects to the Internet. This technology enables unprecedented levels of automation, monitoring, and control, making daily life more convenient, efficient, and informed. IoT has demonstrated notable applications in areas such as smart cities, intelligent homes, agriculture, autonomous vehicles, environmental monitoring, and healthcare, etc. [1–5]. By embedding objects with sensors and actuators, IoT devices detect environmental changes, interact with their surroundings, and exchange dynamic data through embedded electronics and software. This seamless integration of sensing, reacting, and communication capabilities allows IoT systems to simplify human life while adapting to their environment [6,7].

As illustrated in Fig. 1, the smart environment comprises IoT devices (referred to as "Things"), such as smart bulbs and keys, which communicate through a central Edge device. The Edge device serves as a hub, monitoring, filtering, and processing the data

ⓒ The Author(s), under exclusive license to Springer Nature Switzerland AG 2026
F. Marcelloni et al. (Eds.): IJCCI 2025, CCIS 2829, pp. 594–616, 2026.
https://doi.org/10.1007/978-3-032-15638-9_35

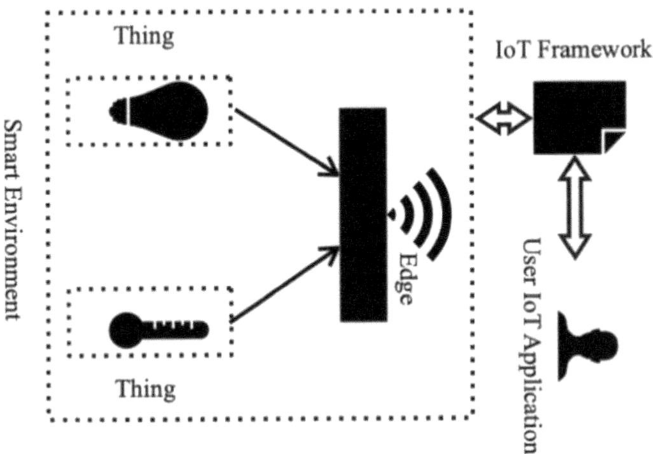

Fig. 1. IoT Environment.

generated by these devices. Edge computing is an emerging computing paradigm that brings together resources located near users, either in terms of geographical proximity or network closeness, to deliver computing, storage, and networking capabilities for application services [8]. This data is then transmitted to an IoT framework, enabling seamless interaction with user IoT applications. These applications operate within an ecosystem supported by an operating system (OS) designed to facilitate efficient communication between devices, users, and application management processes [9, 10].

In this context, edge computing plays a critical role in enabling real-time responsiveness, reducing latency, and decreasing bandwidth usage by processing data locally—at or near the source. This decentralized capability becomes especially important in domains such as autonomous driving, healthcare, agriculture, and industrial automation, where time-sensitive decisions are essential. For example, edge-enabled wearables can alert caregivers to abnormal vital signs, and smart factory sensors can detect operational faults before they lead to breakdowns. In agriculture, edge devices continuously analyze environmental conditions to optimize irrigation. These examples highlight that IoT data is not merely passive but often signals meaningful, actionable information [104]. Lightweight AI frameworks such as TensorFlow Lite and PyTorch Mobile support intelligent inference directly on edge hardware, enabling timely and context-aware anomaly detection in resource-constrained environments [103].

The data produced by IoT devices is inherently heterogeneous, encompassing a wide range of formats such as numerical sensor readings, binary status flags, and even multimedia streams [6]. This heterogeneity, combined with the scale and real-time nature of IoT networks, introduces challenges in reliability, consistency, and interpretability of data. Importantly, IoT data is not merely passive information, it often reflects actionable, context-rich signals that carry operational meaning [104]. That is why, sensor inaccuracies and communication noise often lead to abnormal patterns, motivating the need for robust and type-aware anomaly detection frameworks [11, 12].

Analyzing data from IoT devices allows for the identification of anomalies—data points or patterns that deviate from expected behavior [13]. While classical literature often categorizes anomalies into point, contextual, and collective types [14], recent work has identified more operationally relevant anomaly types in IoT, such as outliers, missing data, bias, drift, noise, constant-value anomalies, and stuck-at-zero faults [15, 16]. These anomalies differ in cause, pattern, and detection requirements, emphasizing the need for a type-specific anomaly detection framework.

Although anomaly detection in IoT is widely studied, many surveys generalize anomalies as a single concept, ignoring the unique challenges of different types. This limits model effectiveness in real-world systems. We address this by categorizing anomalies based on statistical and operational traits, evaluating model suitability per type, and introducing a visual coverage framework. Our anomaly-type-focused perspective aims to guide the development of interpretable and adaptable models for diverse IoT settings.

2   Background of Anomaly Detection in IoT

2.1   Categorization of Anomaly Types Across Prior Studies

Anomalies are observations or sets of observations that deviate significantly from the norm [17]. In the context of sensor networks and IoT, they are commonly classified into point anomalies, contextual anomalies, and collective anomalies [14, 18], underscoring the role of temporal and environmental context in anomaly interpretation.

Beyond these foundational definitions, subsequent studies have introduced more granular taxonomies based on statistical properties, continuity, and frequency patterns, particularly relevant in sensor-centric environments where time-dependent behaviors are critical [15]. Further investigations expanded this perspective for time series data, distinguishing global and local outliers, univariate and multivariate contexts, and subsequence anomalies [19]. More recent work has emphasized the contextual ambiguity in defining abnormal behavior, advocating for adaptive, type-specific frameworks suited to the heterogeneous nature of IoT deployments [20]. Building on these directions, this paper explicitly organizes detection methods according to distinct anomaly types such as outliers, noise, bias, drift, novelty, and constant-value faults as illustrated in Fig. 2.

2.2   Detection Models Trends in IoT Surveys

Anomaly detection in IoT systems has progressed through several phases of research, each addressing evolving challenges in data reliability, computational efficiency, and methodological generalization. Early investigations laid the groundwork by identifying key causes of anomalies, such as system-related issues (e.g., sensor aging, calibration drift, faulty hardware) and environmental influences (e.g., unstable physical conditions, sensor placement) [21]. Building on these insights, subsequent surveys explored a broader spectrum of detection techniques. A notable work categorized statistical, machine learning, and clustering-based methods, highlighting the significance of dimensionality reduction and trade-offs between online and offline processing [22]. While informative, such approaches often lacked scalability for real-time, high-dimensional IoT data streams.

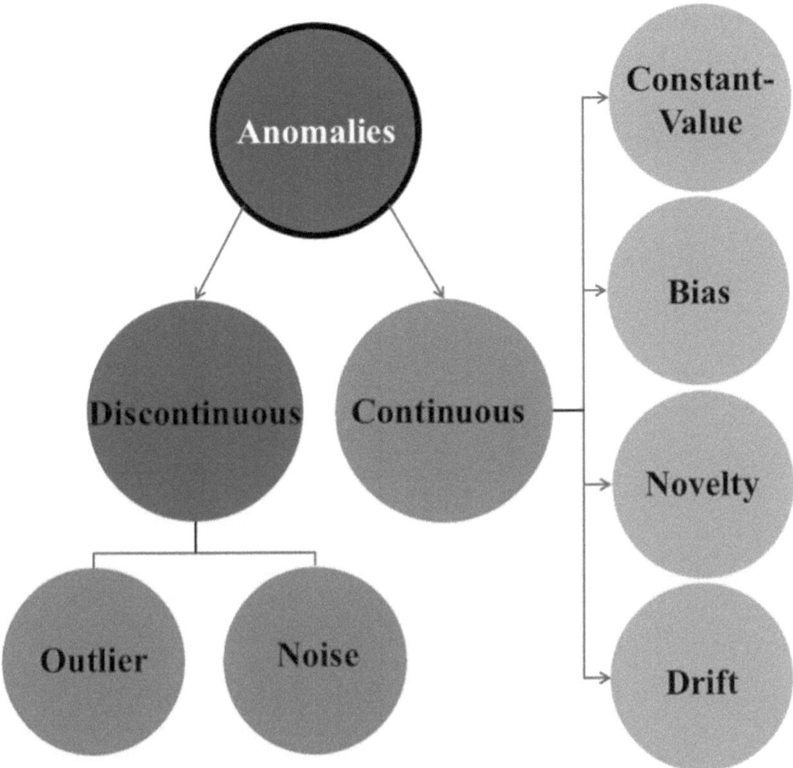

Fig. 2. Anomaly types classified by temporal behavior, distinguishing discontinuous (e.g., outliers, noise) and continuous (e.g., constant-value, bias, novelty, drift) patterns.

Later efforts introduced more structured taxonomies of sensor error types and detection models. These included PCA, clustering algorithms, and SVMs, but limitations remained due to the absence of standardized evaluation benchmarks and the frequent use of private datasets, which undermined reproducibility [15]. To address architectural constraints, another study emphasized the role of computational paradigms namely Cloud, Fog, and Edge, in shaping the performance and feasibility of anomaly detection techniques. It also stressed the need for energy-efficient and resource-aware machine learning solutions in decentralized environments [14]. More recent surveys have shifted focus toward data streams and edge-centric processing. Some evaluated detection methods across structured, semi-structured, and unstructured IoT data using metrics such as AUC, F1-score, and latency [23], while others implemented models like Adaboost, SVC, and RF on multidimensional sensor data to address real-time challenges in edge computing contexts [24].

In a significant empirical contribution, one benchmark study on 2022 re-implemented 13 algorithms and evaluated them on over 30 publicly available time-series datasets, encompassing techniques such as SR, HOT-SAX, and LSTM-based models. This work addressed critical issues in experimental consistency, reproducibility, and practical guidance for method selection [25].

Table 1. Overview of anomaly detection models by learning type, highlighting key techniques for classifying or scoring anomalies in time series sensor data.

Category	Models Grouped
ANN-based	Artificial Neural Networks (ANN)/Auto-Associative Neural Networks (AANN) [82], CNN [83], MLPs [84], RBFs [85], SOMs [88], Hopfield Networks [86], Probabilistic Neural Networks (PNNs) [87]
RNN-based	RNN [79], LSTM, VAE-LSTM [90], Telemanom [91], MTAD-GAT [92], DeepAnT [45], NEAT [93], Probabilistic Predictive Coding (PPC) [93]
Autoencoder-based	AE [40], DAE [77], VAE [78], Donut [72], OmniAnomaly [73], MSCRED [74], TAnoGAN [75], Replicator Neural Networks (Replicator-NNs) [80], GAN/ADDAM [81]
Tree-based	XGBoost [55], Random Forest (RF) [42], Isolation Forest (iForest) [41]
Statistical-based	PCA [39], PCI [37], PCC [36], RPCA [38], ARIMA [46], Fast-MCD [49], COPOD [48], S-H-ESD [51], Gaussian Mixture Models (GMMs) [47], Maximum Likelihood Estimation (MLE) [28], Monte Carlo [52], Cumulative Thresholding [50], SmartSifter [53]
Spectral-based	Singular Spectrum Analysis (SR)/SR-CNN [62], Fast Fourier Transform (FFT) [94], DWT-MLEAD [95]
Distance-based	Local Outlier Factor (LOF) [43], K-Nearest Neighbors (KNN) [56], Connectivity-based Outlier Factor (COF) [59], HOT SAX [61], NormA [62], SAND [63], Singular Spectrum Analysis (SSA) [64], Mahalanobis Distance [58], Euclidean Thresholding [57], Dempster–Shafer [60], One-Class SVM [44]
Density-based	Kernel Density Estimation (KDE) [70], The Histogram-Based Outlier Score (HBOS) [71]
Clustering-based	K-Means [65], Greedy Expectation-Maximization (Greedy EM) [69], Minimum Spanning Tree (MST) [68], Fuzzy C-Means (FCM) [66], DBSCAN [67]
Few-Shot and Meta-Learning Based	Deep-SVDD [100], Prototypical Networks [96], Siamese Networks [102], FewSOME [101], FSOME++ [76], MAML [98], Relation Networks [97], FEAT [99]

Note: FCM is included as the representative Fuzzy logic-based models given its wide use in IoT anomaly detection and suitability for clustering sensor data.

Recent surveys have expanded the classification of anomaly detection approaches by integrating both learning paradigms namely supervised, unsupervised, and semi-supervised and methodological categories such as statistical, machine learning, rule-based, ensemble, and deep learning techniques [26,27]. These works benchmarked models like Isolation Forest, LOF, and Autoencoders while identifying major challenges in adapting such methods to IoT-specific constraints, including real-time responsiveness, heterogeneous data formats, and deployment on resource-limited edge devices. They also emphasized limitations related to the scarcity of labeled datasets, lack of standardized evaluation protocols, and the computational overhead of complex models. Future directions highlighted include the development of hybrid models and energy-efficient frameworks capable of handling dynamic, real-world environments.

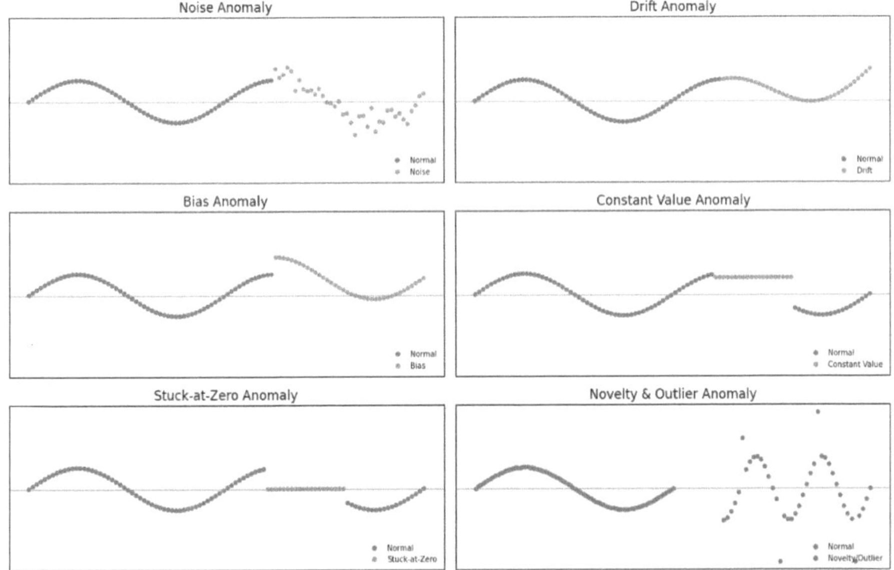

Fig. 3. Visualizing six key anomaly types in time-series data. Each subplot shows a normal signal overlaid with a specific anomaly type.

Building upon the insights and limitations identified in existing survey studies from early foundational reviews [21,22] to recent comprehensive evaluations [25–27] this paper consolidates the diverse range of anomaly detection models used in IoT systems. These include Statistical-based, Tree-based, Autoencoder-based, Spectral-based, Distance-based, Density-based, Clustering-based, ANN-based, LSTM-based, and Few-Shot and Meta-Learning Based methods. To provide a structured and comprehensive overview, Table 1 presents a categorized taxonomy of these models applied across both static and time-series IoT data settings.

3   Mapping Detection Models to Anomaly Types

This section begins by providing a unified mathematical formulation for each anomaly type using consistent notation. All detection models rely on learning a representation of normal behavior from a training set and evaluating unseen instances based on their deviation from this learned model. Let the training set be $X = \{x_1, x_2, \ldots, x_N\}$, where each $x_i \in \mathbb{R}^D$ is a feature vector in a D-dimensional space. A model M_θ is trained to capture normal patterns, and a scoring function $f(x, M_\theta)$ assigns a score s to each input. A decision threshold k is then applied to determine whether a test instance is anomalous.

The symbols and variables used across all formulations are summarized in Table 2 to ensure consistency and clarity. Complementing this, Fig. 3 illustrates the behavioral patterns of the selected anomaly types, offering visual intuition into their temporal characteristics. Based on the mathematical definitions, this section examines representative detection models for each anomaly type. The model selections are informed by the unique properties of each anomaly category, as previously discussed, and are supported by findings from recent literature and prior performance evaluations.

3.1 Unified Anomaly Formulation

Novelty Detection. This technique aims to identify test samples that deviate from previously seen normal data. It is especially useful when only abundant positive (normal) examples are available, while abnormal (negative) instances are scarce or missing [28]. The approach involves learning a model of typical behavior and assigning a novelty score to new inputs based on their deviation from this model. If the score surpasses a set threshold, the input is classified as abnormal. The model is trained solely on normal data and evaluates a new instance x_{test} using a scoring function:

$$s = f(x_{\text{test}}, M_\theta)$$

If $s > k$, the instance is considered novel (abnormal). This one-class classification framework is widely adopted when abnormal samples are unavailable during training [28].

Constant-Value Detection. A constant value anomaly occurs when a sensor or system outputs an unchanging value over time, despite expected variations under normal conditions [29]. This behavior may indicate issues such as stuck-at-zero errors, constant offsets, communication failures, or sensor freezes. In such cases, the output remains static due to faults in sensing, transmission, or processing components, preventing the system from reflecting actual environmental changes. Overall, constant-value anomaly is detected when a signal becomes static over time:

$$E(x_i) = \frac{1}{N} \sum_{j=0}^{N-1} x_{i-j}, \quad \sigma(x_i) = \sqrt{\frac{1}{N} \sum_{j=0}^{N-1} (x_{i-j} - E(x_i))^2}$$

$$\frac{dx}{dt} = x_i - x_{i-1}, \quad s = \frac{E(x_i)}{\sigma(x_i)}$$

An anomaly is flagged when $\sigma(x_i) \approx 0$ or $\frac{dx}{dt} = 0$ persistently, and the signal-to-noise ratio s becomes extremely large [30,31].

Outlier Detection. Outliers are data points that lie far from the majority of observations, often appearing in the extreme tails of a distribution. Hawkins defines an outlier as "an observation which deviates so much from the other observations as to arouse suspicions that it was generated by a different mechanism" [32]. These points typically occur in low-density regions of the data space, where their values significantly diverge from statistical expectations such as the mean or median. This deviation can be quantified using measures like z-scores, interquartile ranges (IQR), or tail probability models [33]. However, not all outliers are meaningful anomalies—some may result from noise or rare but benign conditions. Effective outlier detection, therefore, emphasizes statistically significant and contextually relevant deviations. Common scoring methods include:

$$z(x_i) = \frac{x_i - E(x_i)}{\sigma(x_i)}, \quad \text{flag if } |z(x_i)| > \alpha$$

Alternatively:

$$|x_i - E(x_i)| > \alpha \cdot \sigma(x_i) \quad \text{or} \quad x_i < Q_1 - 1.5 \cdot \text{IQR} \quad \text{or} \quad x_i > Q_3 + 1.5 \cdot \text{IQR}$$

Bias Detection. A bias anomaly occurs when a signal exhibits a consistent offset from its true value, resulting in a systematic deviation over time. Unlike point anomalies such as outliers or novelty instances, bias faults are persistent and typically originate from sensor miscalibration or permanent signal shifts [34]. In this case, each observation x_i is affected by a constant error term:

$$x_i = x_i^{\text{true}} + \varepsilon, \quad \text{with } \varepsilon = \text{const}$$

Such anomalies may not be easily detected by simple thresholding if the offset is small or slowly introduced. Instead, bias is often identified through long-term statistical analysis by monitoring persistent deviation from expected behavior, which is estimated via the mean:

$$|x_i - E(x_i)| > \alpha \cdot \sigma(x_i)$$

Drift Detection. A drift anomaly refers to a gradual change in the behavior of the feature vector $x_i \in X$ over time, often resulting from environmental changes or sensor degradation. Unlike bias, which is constant, drift is characterized by a *non-stationary trend* that evolves slowly and may be mistaken for a normal variation if not properly modeled [34].

$$\frac{dx}{dt} = \frac{x_{i+1} - x_i}{t_{i+1} - t_i}, \quad \text{flag if } \left|\frac{dx}{dt}\right| > \alpha \cdot \sigma(x_i)$$

Or by monitoring the mean shift:

$$|E(x_i) - E(x_{i-j})| > \alpha \cdot \sigma(x_i)$$

In this formulation, $\frac{dx}{dt}$ captures the local rate of change in a feature, reflecting its temporal evolution. The parameter T defines the time window for evaluating these changes, while the index j denotes past instances used for long-term comparisons. These components are essential for identifying gradual trends and sustained deviations in time-series data.

Noise Detection. A noise anomaly is characterized by rapid, small-scale deviations in a feature vector $x_i \in X$, caused by transient uncertainty or environmental disturbances. These unstructured fluctuations differ from systematic anomalies like bias or drift. Detection focuses on abnormal local variability, typically measured by the standard deviation $\sigma(x_i)$ over a sliding window T, computed as $\sigma(x_i) = \sqrt{\frac{1}{T}\sum_{j=1}^{T}(x_{i-j} - E(x_i))^2}$. An instance is flagged as noisy if $\sigma(x_i) > \alpha \cdot \sigma_{\text{baseline}}$, where $E(x_i)$ is the local mean, α is a sensitivity factor, and σ_{baseline} is the expected standard deviation under normal conditions [15,35].

3.2　Representative Detection from Classic to Modern Models

This section reviews anomaly detection models by architecture, linking each to specific anomaly types. Key equations clarify their behavior, and Table 2 summarizes all symbols and notations used.

Statistical-Based Models. Classical statistical methods offer interpretability and robustness in anomaly detection. *Principal Component Analysis (PCA)* [39] projects data onto a lower-dimensional space, effective in general but limited by its linearity. *Principal Component Classifier (PCC)* [36] enhances sensitivity to persistent deviations, suitable for stuck-at anomalies. *Robust PCA (RPCA)* [38] decomposes data into low-rank and sparse components, enabling bias and outlier detection even in noisy, high-dimensional settings.

For time series, *Prediction Confidence Interval (PCI)* [37] flags anomalies when values fall outside forecasted intervals, effectively capturing constant-value anomalies. *Autoencoders* [40] provide nonlinear modeling by reconstructing normal patterns and highlighting anomalies via reconstruction error. Denoising variants enhance generalization to corrupted data. *ARIMA* [46] combines autoregression, differencing, and moving averages to detect drift by emphasizing mean shifts. Online versions like ARIMA-OGD handle streaming data via convex optimization. *Gaussian Mixture Models (GMMs)* [47] detect outliers by modeling data as a mixture of Gaussians, though performance degrades in high dimensions or non-Gaussian settings. *Maximum Likelihood Estimation (MLE)* [28] captures distributional shifts for drift and bias detection. *COPOD* [48] uses copula-based tail probability estimation, offering a parameter-free, interpretable solution for tabular outlier detection.

For robust detection in small-sample or high-dimensional data, *Fast-MCD* [49] identifies the subset with minimal covariance determinant, resisting outliers in multivariate settings. *Cumulative Thresholding* [50] aggregates anomaly scores across dimensions or time, aiding high-dimensional detection with global thresholding. *Seasonal Hybrid ESD (S-H-ESD)* [51] extends traditional tests using medians and MAD, detecting local and global anomalies in noisy seasonal data. *Monte Carlo Dropout* [52] estimates predictive uncertainty through repeated sampling, effectively identifying novelty and distribution shifts. *SmartSifter* [53] incrementally updates a GMM with exponential discounting, detecting both abrupt and gradual drift in streaming settings via log-likelihood scoring and moving averages [54].

Distance-Based. Distance-based anomaly detection models evaluate how far a data point deviates from its neighbors in feature space. These methods assume normal instances cluster together, while anomalies appear in sparse or isolated regions. A fundamental example is *K-Nearest Neighbors (KNN)*, which flags points far from their k nearest neighbors [56], and is effective for outlier and novelty detection. The general scoring function for such models computes the average distance between a test instance and its k nearest neighbors, where $\mathrm{dist}(x, y)$ is a distance function between two instances:

$$s(x_{\text{test}}) = \frac{1}{k} \sum_{j=1}^{k} \mathrm{dist}(x_{\text{test}}, x_j)$$

K-NNN improves on this by considering neighbors of neighbors, enhancing performance in sparse or non-uniform data. Simpler methods like *Euclidean Thresholding* measure the distance from a point to a reference, typically the dataset mean, to flag anomalies that exceed a threshold [57]. Though limited in high-dimensional spaces, it works well for low-dimensional, compact data. *Mahalanobis Distance* [58] incorporates feature correlations via covariance structure, improving bias and outlier detection, but relies on accurate covariance estimation, which is sensitive to anomalous contamination. *Connectivity-based Outlier Factor (COF)* [59] captures local connectedness via chaining distance, working well in heterogeneous clusters. *Local Outlier Factor (LOF)* [43] compares local densities and is effective for stuck-at and constant-value anomalies, though sensitive to parameters and dimensionality.

One-Class SVM (OC-SVM) [44] defines a decision boundary around normal data using kernel methods. It excels at novelty detection but requires careful kernel and parameter tuning and is sensitive to noise. For ambiguous sensor signals, *Dempster–Shafer Theory (DST)* [60] fuses evidence from multiple sources, detecting noisy or conflicting observations. However, its rule-based nature limits scalability. Time series-specific methods like *HOT SAX* [61] symbolically transform sequences to find discord subsequences, excelling at stuck-at and constant patterns but may miss subtle anomalies. *NormA* [62] combines z-normalization and Euclidean distance to detect localized, shape-based anomalies, though it struggles with highly non-stationary or noisy data. *SAND* [63] addresses drift by clustering subsequences in real time using k-Shape and temporal weighting, adapting without storing raw data, but depends on timely updates and good parameter tuning. Finally, *Segmented Sequence Analysis (SSA)* [64] models signals as piecewise linear segments and compares them to dynamic references, making it effective for gradual drift in constrained IoT systems, though less sensitive to very slow changes.

Clustering-Based. Clustering algorithms are widely adopted for unsupervised anomaly detection by identifying instances that do not conform to the dominant cluster structures. A representative example is K-Means, which minimizes the intra-cluster variance by assigning each instance to its nearest cluster centroid. The loss function minimized during this clustering process is given by:

$$\mathcal{L} = \sum_{i=1}^{N} \sum_{c=1}^{K} r_{ic} \|x_i - \mu_c\|^2$$

Among these, *K-Means* [65] remains a foundational method, partitioning data into K clusters based on minimizing intra-cluster variance. Anomalies are detected as points that lie far from their cluster centroids, making K-Means effective for identifying outlier-type and novelty-type anomalies when the data is compact and well-separated. However, its assumption of spherical clusters limits performance on irregular distributions. To address gradual or ambiguous transitions in data, *Fuzzy C-Means (FCM)* [66] extends K-Means by assigning each point partial membership across multiple clusters. This soft clustering approach excels at detecting drift anomalies, where data progressively shifts between behavioral states. FCM's ability to capture partial belonging reveals incipient faults and transitional regions that hard clustering methods overlook.

Unlike centroid-based methods, *DBSCAN* [67] adopts a density-based strategy, clustering dense regions while labeling sparse, isolated points as noise. This allows it to detect outlier-type anomalies in datasets with arbitrary-shaped clusters and varying densities. It requires no preset number of clusters and is robust to noise, but its sensitivity to hyperparameters (ε, MinPts) may reduce effectiveness in high-dimensional data. *Minimum Spanning Tree (MST)* clustering [68] represents data as a graph and constructs a spanning tree connecting all points with minimum total edge weight. Anomalies emerge as nodes connected via unusually long edges. By pruning such edges, MST isolates sparse branches, making it effective for structural or spatial anomalies in non-convex or irregular data.

Finally, *Greedy EM* [69] offers an efficient alternative to full Expectation-Maximization by incrementally fitting Gaussian components that significantly increase data likelihood. This selective modeling helps detect outliers in high-dimensional datasets without overfitting, supporting large-scale anomaly detection with faster convergence and improved robustness.

Density-Based. Density-based methods detect anomalies as points located in low-density regions. The estimated density at a point x is computed using Kernel Density Estimation (KDE) as:

$$\hat{p}(x) = \frac{1}{Nh^D} \sum_{i=1}^{N} K\left(\frac{x - x_i}{h}\right)$$

Kernel Density Estimation (KDE) [70] estimates the underlying probability density by placing kernels (typically Gaussian) on each data point and summing their influence. It is well-suited for detecting outlier and novelty anomalies in low-dimensional, continuous data due to its flexibility and non-parametric nature. However, its performance degrades in high dimensions without proper bandwidth tuning. Histogram-Based Outlier Score (HBOS) [71] accelerates density estimation by assuming feature independence and modeling each dimension using histograms. Anomaly scores are computed from inverse density estimates across dimensions, making HBOS highly efficient and scalable. Though it cannot capture feature interactions or local structures, it performs well for global outlier detection, especially in high-dimensional and large-scale scenarios.

Tree-Based. Tree-based models exploit hierarchical data partitioning, with Isolation Forest (iForest) [41] specializing in unsupervised anomaly detection by isolating rare instances through short path lengths. The anomaly score for a test instance is computed based on its expected path length $E(h(x))$, which represents the average number of splits required to isolate x across an ensemble of randomly constructed trees:

$$s(x_{\text{test}}) = 2^{-\frac{E(h(x_{\text{test}}))}{E(h(x_i))}}$$

Shorter path lengths indicate easier isolation and thus higher anomaly likelihood. Its random sub-sampling and efficient tree structure make it ideal for detecting outliers in high-dimensional data. In contrast, Random Forests [42] rely on proximity measures and are better suited for semi-supervised contexts, where synthetic or labeled anomalies help identify bias or drift through feature importance. XGBoost [55], although not designed for anomaly detection, performs well in detecting stuck-at and constant-value anomalies due to its ability to capture nonlinear dependencies in time series data. Its sparsity-aware design handles static or frozen signals effectively, though it requires thresholding and lacks interpretability. Overall, tree-based models offer flexible and scalable solutions, with each variant aligning with distinct anomaly patterns depending on supervision level and data complexity.

Autoencoder-Based. Autoencoders (AE) [40] and their variants form the foundation of reconstruction-based anomaly detection. AEs compress input data into a latent representation and attempt to reconstruct it, with anomalies identified through high reconstruction error. The reconstruction loss minimized during training is:

$$\mathcal{L} = \frac{1}{N} \sum_{i=1}^{N} \|x_i - \hat{x}_i\|^2$$

This makes AEs well-suited for detecting *outliers* and *novelty* anomalies when trained only on normal patterns. Denoising Autoencoders (DAE) [77] extend this idea by learning to reconstruct clean inputs from corrupted versions, enhancing robustness to noise. They are particularly effective in *noisy environments* and can generalize better to *drift-type* anomalies where inputs are partially distorted. Variational Autoencoders (VAE) [78] introduce probabilistic encoding, mapping inputs to a distribution over latent variables. The reconstruction is driven by sampling, allowing VAEs to handle *novelty* and *contextual anomalies*, particularly in complex, high-variance time series. However, standard VAEs can be sensitive to anomalies during training.

Building on these foundations, Donut [72] improves anomaly detection in seasonal KPIs by excluding anomalies during training and simulating missing data, making it ideal for *point* and *contextual anomalies*. OmniAnomaly [73] enhances VAEs with recurrent structures and normalizing flows to capture temporal-stochastic dynamics, effectively detecting *contextual* and *multivariate anomalies*. Similarly, MSCRED [74] introduces signature matrices and attention to identify *persistent* and *structural anomalies* in sensor networks. Replicator Neural Networks (Replicator-NNs) [80], a simpler form of autoencoder, are well-suited for *outlier detection* in real-time or low-resource settings due to their efficient bottleneck design. Expanding beyond standard reconstruction, GAN-based models, especially ADDAM [81], integrate adversarial training to further enhance detection. ADDAM employs dual autoencoders and dual discriminators, making it particularly effective for *outliers, novelty*, and *noise-type anomalies*. By evaluating both reconstruction fidelity and adversarial realism, ADDAM identifies subtle deviations in structure and content, offering a robust, multi-perspective detection framework.

ANN-Based. Artificial Neural Networks (ANNs) are versatile models capable of capturing complex patterns but are not inherently tailored for anomaly detection. A more specialized variant, Auto-Associative Neural Networks (AANNs) [82], function as autoencoders, reconstructing inputs to flag anomalies through reconstruction errors. The Auto-Associative Neural Network System (AANNS) advances this idea by employing class-specific AANNs, each trained to model a distinct category. This modular architecture, enhanced by RBM pretraining and backpropagation fine-tuning, excels at *drift detection* through localized learning and incremental adaptation.

Building on this, Multi-Layer Perceptrons (MLPs) [84] serve as supervised classifiers that distinguish normal from abnormal inputs. The loss function for training neural classifiers typically minimizes classification error via cross-entropy:

$$\mathcal{L} = -\sum_{i=1}^{N} y_i \log f(x_i, M_\theta)$$

While effective for detecting *structured outliers* and *noise* when labels are available, their generalization to unseen anomalies remains limited. Convolutional Neural Networks (CNNs) [83] improve detection in image-based settings by extracting spatial hierarchies of features, making them well-suited for identifying *structural* and *localized anomalies*, such as surface defects or distorted textures. Radial Basis Function networks (RBFs) [85] rely on distance-based Gaussian activations, making them sensitive to deviations from normal clusters. They are particularly useful for spotting *point outliers* and *novelty*, though they struggle with scalability and high-dimensional drift.

Self-Organizing Maps (SOMs) [88], which preserve topological structure by mapping high-dimensional data onto a 2D grid, are powerful for visualizing data distributions. Their sensitivity to shifting input patterns enables effective detection of *drift, bias, and novelty*. Hopfield Networks [86], as associative memory models, detect anomalies when inputs fail to converge to expected patterns, revealing *novelty* or *stuck-at/constant-value anomalies*. Though historically important, they are less practical for high-dimensional modern datasets. Finally, Probabilistic Neural Networks (PNNs) [87] leverage Bayesian classification and kernel density estimation. Their ability to rapidly classify structured patterns makes them effective in identifying *outliers* and *noise*, especially in real-time scenarios, though they face scalability challenges.

RNN-Based. Recurrent Neural Networks (RNNs) [79] process sequential data by retaining hidden states over time, making them effective for identifying *contextual* and *sequence-based anomalies* in time series. Their objective typically minimizes the prediction error over a sliding time window:

$$\mathcal{L} = \frac{1}{T} \sum_{t=1}^{T} \|x_t - \hat{x}_t\|^2$$

However, their performance may degrade with long or complex sequences, prompting the development of more advanced architectures. LSTM-VAE [90] enhances this capability by combining LSTMs and VAEs to model both temporal dynamics and uncertainty. It excels at detecting *constant-value* and *stuck-at* anomalies, which disrupt learned temporal structures. Its design enables strong anomaly signals, though at the cost of computational complexity and limited interpretability.

Building on LSTM's forecasting power, Telemanom [91] introduces adaptive, non-parametric thresholding to detect *drift* and *contextual anomalies* in spacecraft telemetry. Its per-channel modeling supports fine-grained, real-time detection without labeled data, though it may miss cross-channel correlations. DeepAnT [45] replaces recurrent components with CNNs to forecast future values based on past windows. It uses prediction errors to detect *point, contextual*, and *discord anomalies*, showing robustness in low-data and noisy conditions without requiring supervision. MTAD-GAT [92] further advances multivariate time series analysis by integrating Graph Attention Networks and GRUs. Its dual focus on temporal and feature-wise relationships, along with hybrid prediction-reconstruction loss, makes it highly effective for *drift detection*, offering interpretability and early anomaly alerts. Unlike these deep models, NEAT [89] evolves both neural architectures and weights, allowing continuous adaptation. This makes it suitable for environments with *concept drift* and evolving anomaly patterns. Finally, Predictive Coding [93] reframes detection as a cycle of prediction and error correction. By modeling temporal expectations and identifying mismatches, it robustly captures *novelty* and *contextual anomalies*, especially in uncertain or complex sensor environments.

Spectral-Based. Frequency-based methods like Fast Fourier Transform (FFT) [94] provide a foundation for identifying abrupt frequency-domain changes in time-series data, making them useful for detecting periodic disturbances and constant-value anomalies. Spectral-based methods often operate by transforming the input into the frequency domain using a Fourier transform $F(x_i)$, where anomalies are revealed as deviations in the amplitude spectrum. A general form of the spectral residual is:

$$s(x_i) = |\log|F(x_i)| - E(\log|F(x_i)|)|$$

Building on this, DWT-MLEAD [95] applies Discrete Wavelet Transform to capture both short-term and long-term variations at multiple resolutions, followed by Maximum Likelihood Estimation on wavelet coefficients to model normal behavior. This makes it highly effective for multiscale anomalies such as drift and abrupt transitions, especially in noisy or compressed signals.

Expanding on spectral techniques, Spectral Residual (SR) methods transform time-series into saliency maps by subtracting smoothed log amplitude spectra. This isolates sudden signal deviations, highlighting novelty and contextual anomalies without supervision. SR-CNN [62] enhances this further by combining the SR transformation with a 1D CNN trained on synthetic labels. It is particularly well-suited for detecting constant-value and stuck-at anomalies, which stand out sharply in the spectral domain. Its scalability, unsupervised nature, and accuracy across diverse datasets make it ideal for large-scale, real-time anomaly detection pipelines.

Few-Shot and Meta-Learning Based. Traditional anomaly detection often falls short when abnormal samples are scarce. To overcome this, Siamese Networks [102] pioneered a comparison-based approach by learning distance metrics through shared-weight encoding. Their strength lies in detecting novelty and rare outliers, where anomalies differ subtly yet crucially from normal data. These models commonly optimize a contrastive loss that separates positive and negative pairs in latent space:

$$\mathcal{L} = \sum_{i=1}^{N} y_i \|f(x_i^a, M_\theta) - f(x_i^p, M_\theta)\|^2 + (1 - y_i) \max\left(0, \alpha - \|f(x_i^a, M_\theta) - f(x_i^n, M_\theta)\|\right)^2$$

where x_i^a is the anchor, x_i^p the positive (normal), and x_i^n the negative (abnormal) instance. With contrastive loss guiding similarity learning, Siamese Networks thrive in low-data regimes and generalize well to unseen examples. Building on this, Deep-SVDD [100] maps normal data into a tight hypersphere in latent space, flagging anything outside as anomalous. This compactness makes it ideal for detecting drift, noise, or unknown intrusions, especially in domains like network traffic where normal patterns remain stable but anomalies shift or emerge over time. FewSOME [101] merges the Siamese structure with episodic training, and FSOME++ [76] augments abnormalities to simulate defect scenarios. Their contrastive loss enhance separation between normal and generated abnormal samples, making it highly effective for identifying fine-grained novelty, bias, and outliers—particularly in industrial settings where real anomalies are rare but high-stakes. Prototypical Networks [96] simplify few-shot classification by using class centroids for distance-based inference, making them well-suited for novelty detection where anomalies lie far from known prototypes. They perform reliably with minimal data. Relation Networks [97] extend this by learning deep similarity functions, enabling detection of diverse anomalies through learned semantic patterns. MAML [98] adopts a meta-learning approach, finding initializations that adapt quickly to new tasks, making it ideal for dynamic anomaly classes without full retraining.

Finally, FEAT [99] enhances prototype-based methods by adapting class representations using a Transformer-based set-to-set function. This contextual embedding adjustment helps separate normal and abnormal samples in task-specific ways, boosting performance in novelty and unseen class detection. FEAT is particularly suited for diverse and evolving anomalies, common in IoT and cyber-physical systems.

4   Coverage Map: Summary and Insights

After a comprehensive review of the behavior, definition, and mathematical formulations of six major anomaly types—novelty, drift, bias, noise, outlier, and constant-value—and the corresponding detection models, we present the synthesized result in Fig. 4. This coverage map is the result of an extensive survey of literature, analysis of model assumptions, and interpretation of their reported performances across anomaly types in IoT settings.

From the map, we observe that models such as AE/DAE/VAE, KNN, LOF/COF, and FewSOME exhibit strong generalizability, showing full or partial coverage across multiple anomaly types. These models leverage properties such as reconstruction error, local density, or contrastive metric learning, enabling them to detect a variety of temporal and statistical patterns. Conversely, some traditional or classical models, such as DBSCAN, One-Class SVM, and Cumulative Thresholding, show clear gaps in their coverage—particularly for persistent anomalies like bias and constant-value faults. Their limited adaptability to dynamic IoT conditions underscores their declining relevance in modern applications. Constant-value anomalies emerge as the least-supported class overall, with only a handful of models (e.g., Deep-SVDD, SR-CNN, Autoencoders) offering full detection capability. Similarly, drift and bias anomalies remain inconsistently addressed, especially among non-sequential models, suggesting a pressing need for improved temporal and trend-aware frameworks.

This visual mapping provides a high-level summary for researchers and practitioners, enabling informed selection of detection techniques based on anomaly character-

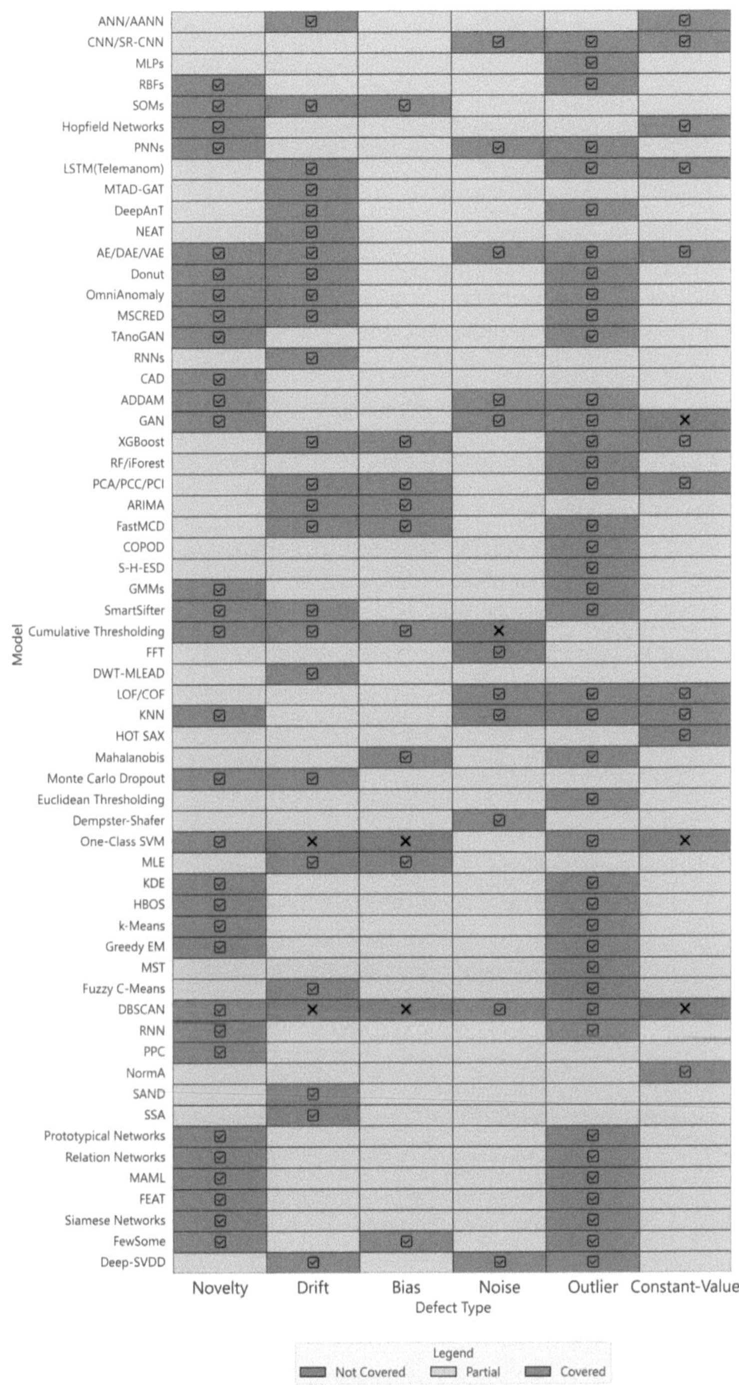

Fig. 4. Coverage map showing how anomaly detection models address six defect types. Green indicates full coverage, yellow partial, and red no coverage. (Color figure online)

istics. It also identifies key research gaps and opportunities for developing comprehensive, type-aware anomaly detection systems tailored for complex, real-world IoT environments.

5 Discussion

The coverage map reveals a clear trend of models like AE/DAE/VAE, KNN, LOF/COF, and FewSOME offer broad adaptability, but most detection techniques are not explicitly designed with defect typology in mind. In particular, constant-value, bias, and drift anomalies remain under-addressed across many models, despite their prevalence

Table 2. Symbols used across model formulations for anomaly detection tasks.

Symbol	Definition
Data and Instances	
X	Set of training instances
x_i	Individual training instance
x_{test}	Unseen test instance
y_i	Ground-truth label of instance x_i
$\hat{x}_t$	Predicted value of instance at time step t
D	Dimensionality of the feature space
N	Number of instances in a window or training set
k	Decision threshold or number of neighbors/clusters
Model and Scoring Functions	
M_θ	Model of normality with parameters θ
$f(x, M_\theta)$	Scoring function estimating anomaly likelihood
s	Anomaly score assigned by the model
$\mathcal{L}$	Loss function minimized during training
Statistical Measures	
$E(x_i)$	Mean in a sliding window
$\sigma(x_i)$	Standard deviation in a sliding window
σ_{baseline}	Expected std dev under normal conditions
Q_1, Q_3, IQR	Quartiles and interquartile range
μ_c	Cluster centroid or component mean
r_{ic}	Indicator of instance x_i in component c
$\hat{p}(x)$	Estimated probability density at point x
h	Bandwidth parameter for kernel density
Temporal and Sequential Features	
T	Time window for temporal patterns
j	Index for historical values
$\frac{dx}{dt}$	Rate of change over time

in IoT systems. Recent advances, such as hybrid deep models and few-shot learners, offer improved generalization but bring trade-offs in interpretability and resource cost. This highlights a shift from traditional, general-purpose methods toward anomaly-type-aware solutions that balance detection scope with practical deployment needs. Going forward, the development of benchmarks and models tailored to anomaly semantics, especially in time-dependent and low-variance settings, remains a key research direction.

6 Future Work and Open Challenges

Despite substantial progress in anomaly detection for IoT, key challenges remain. Most models are not designed with specific anomaly types in mind, limiting their ability to detect faults such as bias, drift, and constant-value anomalies. There is also a lack of standardized benchmarks for evaluating per-type performance, hindering fair comparison across methods. Future research should focus on developing type-aware models, integrating temporal and statistical reasoning, and creating lightweight, interpretable architectures suitable for deployment on resource-constrained IoT devices. Establishing unified evaluation protocols and type-specific benchmarks will be essential for advancing the field.

7 Conclusion

This paper presented a structured taxonomy of anomalies in IoT environments and systematically analyzed the capabilities of detection models across six key defect types. By integrating theoretical formulations, visual behaviors, and empirical coverage, we provide a comprehensive guide for selecting appropriate detection techniques. Our findings highlight the importance of shifting from generalized anomaly detectors toward type-aware, context-sensitive models. The coverage map and categorization proposed here offer a foundation for future efforts in building robust, interpretable, and scalable anomaly detection systems tailored to real-world IoT deployments.

Acknowledgments. Authors thank the Canada Research Chair program (CRC-2019-00319), Magna International, and NSERC (ALLRP 595970-24) for supporting this work.

References

1. Ngu, A.H., Gutierrez, M., Metsis, V., Nepal, S., Sheng, Q.Z.: IoT middleware: a survey on issues and enabling technologies. IEEE Internet Things J. **4**(1), 1–20 (2016)
2. Lakhwani, K., Gianey, H., Agarwal, N., Gupta, S.: Development of IoT for smart agriculture a review. In: Rathore, V.S., Worring, M., Mishra, D.K., Joshi, A., Maheshwari, S. (eds.) Emerging Trends in Expert Applications and Security. AISC, vol. 841, pp. 425–432. Springer, Singapore (2019). https://doi.org/10.1007/978-981-13-2285-3_50
3. Subasi, A., Radhwan, M., Kurdi, R., Khateeb, K.: IoT based mobile healthcare system for human activity recognition. In: 2018 15th Learning and Technology Conference (L&T), pp. 29–34. IEEE (2018)

4. Hernández-Muñoz, J.M., et al.: Smart cities at the forefront of the future internet. In: Future Internet Assembly, pp. 447–462 (2011)

5. Vilajosana, I., Llosa, J., Martinez, B., Domingo-Prieto, M., Angles, A., Vilajosana, X.: Bootstrapping smart cities through a self-sustainable model based on big data flows. IEEE Commun. Mag. **51**(6), 128–134 (2013)

6. Atzori, L., Iera, A., Morabito, G.: The internet of things: a survey. Comput. Netw. **54**(15), 2787–2805 (2010)

7. Han, J., Pei, J., Tong, H.: Data Mining: Concepts and Techniques. Morgan kaufmann (2022)

8. Zhao, Z., Liu, F., Cai, Z., Xiao, N.: Edge computing: platforms, applications and challenges. J. Comput. Res. Dev. **55**(2), 327–337 (2018)

9. Khan, W.Z., Ahmed, E., Hakak, S., Yaqoob, I., Ahmed, A.: Edge computing: a survey. Futur. Gener. Comput. Syst. **97**, 219–235 (2019)

10. Yu, W., et al.: A survey on the edge computing for the Internet of Things. IEEE Access **6**, 6900–6919 (2017)

11. Derakhshan, R., Orlowska, M.E., Li, X.: RFID data management: challenges and opportunities. In: 2007 IEEE International Conference on RFID, pp. 175–182 (2007)

12. Aggarwal, C.C., Ashish, N., Sheth, A.: The internet of things: a survey from the data-centric perspective. In: Aggarwal, C. (ed.) Managing and Mining Sensor Data, pp. 383–428. Springer, Boston (2013)

13. Chandola, V., Banerjee, A., Kumar, V.: Anomaly detection: a survey. ACM Comput. Surv. (CSUR) **41**(3), 1–58 (2009)

14. Erhan, L., et al.: Smart anomaly detection in sensor systems: a multi-perspective review. Inf. Fusion **67**, 64–79 (2021)

15. Teh, H.Y., Kempa-Liehr, A.W., Wang, K.I.-K.: Sensor data quality: a systematic review. J. Big Data **7**(1), 11 (2020)

16. Nguyen, A.L., Kamioka, E., Nguyen-Duc, T.: To verify the correctness of IoT sensor data in real-time. In: 2019 25th Asia-Pacific Conference on Communications (APCC), pp. 479–484. IEEE (2019)

17. Hodge, V., Austin, J.: A survey of outlier detection methodologies. Artif. Intell. Rev. **22**, 85–126 (2004)

18. Hayes, M.A., Capretz, M.A.M.: Contextual anomaly detection framework for big sensor data. J. Big Data **2**, 1–22 (2015)

19. Blázquez-García, A., Conde, A., Mori, U., Lozano, J.A.: A review on outlier/anomaly detection in time series data. ACM Comput. Surv. (CSUR) **54**(3), 1–33 (2021)

20. Rezaee, K., Rezakhani, S.M., Khosravi, M.R., Moghimi, M.K.: A survey on deep learning-based real-time crowd anomaly detection for secure distributed video surveillance. Pers. Ubiquit. Comput. **28**(1), 135–151 (2024)

21. Ni, K., et al.: Sensor network data fault types. ACM Trans. Sensor Netw. (TOSN) **5**(3), 1–29 (2009)

22. Rassam, M.A., Zainal, A., Maarof, M.A.: Advancements of data anomaly detection research in wireless sensor networks: a survey and open issues. Sensors **13**(8), 10087–10122 (2013)

23. Imran, S.M.A., Faiz, R., Alam, M.M., Quadri, S.K.A., Bari M.R., Shibl, M.F.: A survey of machine learning techniques for detecting anomaly in internet of things (IoT). J. Independent Stud. Res. Comput. **21**, 1–50 (2023). https://doi.org/10.31645/JISRC.23.21.1.5.

24. Kaya, ŞM., İşler, B., Abu-Mahfouz, A.M., Rasheed, J., AlShammari, A.: An intelligent anomaly detection approach for accurate and reliable weather forecasting at IoT edges: a case study. Sensors **23**(5), 2426 (2023)

25. Schmidl, S., Wenig, P., Papenbrock, T.: Anomaly detection in time series: a comprehensive evaluation. Proc. VLDB Endow. **15**(9), 1779–1797 (2022)

26. Liu, J., et al.: Networking systems for video anomaly detection: a tutorial and survey. ACM Comput. Surv. **57**(10), 1–37 (2025)
27. Rafique, S.H., Abdallah, A., Musa, N.S., Murugan, T.: Machine learning and deep learning techniques for internet of things network anomaly detection-current research trends. Sensors **24**(6), 1968 (2024)
28. Pimentel, M.A.F., Clifton, D.A., Clifton, L., Tarassenko, L.: A review of novelty detection. Signal Process. **99**, 215–249 (2014)
29. Rabatel, J., Bringay, S., Poncelet, P.: Anomaly detection in monitoring sensor data for preventive maintenance. Expert Syst. Appl. **38**(6), 7003–7015 (2011)
30. Sharifi, R., Langari, R.: Nonlinear sensor fault diagnosis using mixture of probabilistic PCA models. Mech. Syst. Signal Process. **85**, 638–650 (2017)
31. Yiqi, L., Daoping, H., Zhifu, L.: A SEVA soft sensor method based on self-calibration model and uncertainty description algorithm (2013)
32. Hawkins, D.M.: Identification of Outliers. Springer, Dordrecht (1980)
33. Aggarwal, C.C.: Outlier Analysis, 2nd edn. Springer, Cham (2017). https://doi.org/10.1007/978-3-319-47578-3
34. Baljak, V., Tei, K., Honiden, S.: Classification of faults in sensor readings with statistical pattern recognition. In: The Sixth International Conference on Sensor Technologies and Applications, pp. 270–276 (2012)
35. Joint Committee for Guides in Metrology (JCGM), Evaluation of measurement data — Guide to the expression of uncertainty in measurement (GUM 2008), JCGM 100:2008 (2008)
36. Shyu, M.-L., Chen, S.-C., Sarinnapakorn, K., Chang, L.: A novel anomaly detection scheme based on principal component classifier. In: Proceedings of the IEEE Foundations and New Directions of Data Mining Workshop, pp. 172–179 (2003)
37. Yu, Y., Zhu, Y., Li, S., Wan, D.: Time series outlier detection based on sliding window prediction. Math. Probl. Eng. **2014**(1), 879736 (2014)
38. Paffenroth, R., Kay, K., Servi, L.: Robust PCA for anomaly detection in cyber networks, arXiv preprint arXiv:1801.01571 (2018)
39. Abdi, H., Williams, L.J.: Principal component analysis. Wiley Interdisc. Rev. Comput. Stat. **2**(4), 433–459 (2010)
40. Sakurada, M., Yairi, T.: Anomaly detection using autoencoders with nonlinear dimensionality reduction. In: Proceedings of the MLSDA 2014 2nd Workshop on Machine Learning for Sensory Data Analysis, pp. 4–11 (2014)
41. Liu, F.T., Ting, K.M., Zhou, Z.-H.: Isolation forest. In: Proceedings of the 2008 Eighth IEEE International Conference on Data Mining, pp. 413–422. IEEE (2008)
42. Breiman, L.: Random forests. Mach. Learn. **45**, 5–32 (2001)
43. Breunig, M.M., Kriegel, H.-P., Ng, R.T., Sander, J.: LOF: identifying density-based local outliers. In: Proceedings of the 2000 ACM SIGMOD International Conference on Management of Data, pp. 93–104 (2000)
44. Manevitz, L.M., Yousef, M.: One-class SVMs for document classification. J. Mach. Learn. Res. **2**, 139–154 (2001)
45. Munir, M., Siddiqui, S.A., Dengel, A., Ahmed, S.: DeepAnT: a deep learning approach for unsupervised anomaly detection in time series. IEEE Access **7**, 1991–2005 (2018)
46. Liu, C., Hoi, S.C.H., Zhao, P., Sun, J.: Online ARIMA algorithms for time series prediction. In: Proc. AAAI Conf. on Artificial Intelligence, vol. 30, no. 1 (2016)
47. Reynolds, D.A., Quatieri, T.F., Dunn, R.B.: Speaker verification using adapted Gaussian mixture models. Digit. Signal Process. **10**(1–3), 19–41 (2000)
48. Li, Z., Zhao, Y., Botta, N., Ionescu, C., Hu, X.: COPOD: copula-based outlier detection. In: Proceedings of the 2020 IEEE International Conference on Data Mining (ICDM), pp. 1118–1123 (2020)

49. Rousseeuw, P.J., Van Driessen, K.: A fast algorithm for the minimum covariance determinant estimator. Technometrics **41**(3), 212–223 (1999)
50. Olufowobi, H., et al.: Anomaly detection approach using adaptive cumulative sum algorithm for controller area network. In: Proc. ACM Workshop on Automotive Cybersecurity, pp. 25–30 (2019)
51. Hochenbaum, J., Vallis, O.S., Kejariwal, A.: Automatic anomaly detection in the cloud via statistical learning, *arXiv preprint* arXiv:1704.07706 (2017)
52. Gal, Y., Ghahramani, Z.: Dropout as a Bayesian approximation: representing model uncertainty in deep learning. In: Proceedings of the International Conference on Machine Learning (ICML), pp. 1050–1059 (2016)
53. Yamanishi, K., Takeuchi, J.-I., Williams, G., Milne, P.: On-line unsupervised outlier detection using finite mixtures with discounting learning algorithms. In: Proc. 6th ACM SIGKDD Int. Conf. on Knowledge Discovery and Data Mining, pp. 320–324 (2000)
54. Hunter, J.S.: The exponentially weighted moving average. J. Qual. Technol. **18**(4), 203–210 (1986)
55. Chen, T., Guestrin, C.: XGBoost: a scalable tree boosting system. In: Proceedings of the 22nd ACM SIGKDD International Conference on Knowledge Discovery and Data Mining, pp. 785–794 (2016)
56. Nizan, O., Tal, A.: K-NNN: nearest neighbors of neighbors for anomaly detection. In: Proceedings of the IEEE/CVF Winter Conference on Applications of Computer Vision (WACV) Workshops, pp. 1005–1014 (2024)
57. Eskin, E., Arnold, A., Prerau, M., Portnoy, L., Stolfo, S.: A geometric framework for unsupervised anomaly detection: detecting intrusions in unlabeled data. In: Applications of Data Mining in Computer Security, pp. 77–101. Springer (2002)
58. De Maesschalck, R., Jouan-Rimbaud, D., Massart, D.L.: The Mahalanobis distance. Chemom. Intell. Lab. Syst. **50**(1), 1–18 (2000)
59. Tang, J., Chen, Z., Fu, A.W., Cheung, D.W.: Enhancing effectiveness of outlier detections for low density patterns. In: Chen, M.-S., Yu, P.S., Liu, B. (eds.) PAKDD 2002. LNCS (LNAI), vol. 2336, pp. 535–548. Springer, Heidelberg (2002). https://doi.org/10.1007/3-540-47887-6_53
60. Bezerra, E.D.C., Teles, A.S., Coutinho, L.R., da Silva e Silva, F.J.: Dempster–Shafer theory for modeling and treating uncertainty in IoT applications based on complex event processing. Sensors **21**(5), 1863 (2021)
61. Keogh, E., Lin, J., Fu, A.: Hot SAX: efficiently finding the most unusual time series subsequence. In: Proceedings of the Fifth IEEE International Conference on Data Mining (ICDM 2005), p. 8. IEEE (2005)
62. Ren, H., et al.: Time-series anomaly detection service at Microsoft. In: Proc. 25th ACM SIGKDD Int. Conf. on Knowledge Discovery & Data Mining, pp. 3009–3017 (2019)
63. Boniol, P., Paparrizos, J., Palpanas, T., Franklin, M.J.: SAND: streaming subsequence anomaly detection. Proc. VLDB Endow. **14**(10), 1717–1729 (2021)
64. Yao, Y., Sharma, A., Golubchik, L., Govindan, R.: Online anomaly detection for sensor systems: a simple and efficient approach. Perform. Eval. **67**(11), 1059–1075 (2010)
65. Pham, D.T., Dimov, S.S., Nguyen, C.D.: Selection of K in K-means clustering. Proc. Inst. Mech. Eng. C J. Mech. Eng. Sci. **219**(1), 103–119 (2005)
66. Bezdek, J.C., Ehrlich, R., Full, W.: FCM: the fuzzy c-means clustering algorithm. Comput. Geosci. **10**(2–3), 191–203 (1984)
67. Ester, M., et al.: A density-based algorithm for discovering clusters in large spatial databases with noise. In: Proceedings of the Second International Conference on Knowledge Discovery and Data Mining (KDD), vol. 96, no. 34, pp. 226–231 (1996)
68. Pettie, S., Ramachandran, V.: An optimal minimum spanning tree algorithm. J. ACM (JACM) **49**(1), 16–34 (2002)

69. Vlassis, N., Likas, A.: A greedy EM algorithm for Gaussian mixture learning. Neural Process. Lett. **15**, 77–87 (2002)
70. Kim, J., Scott, C.D.: Robust kernel density estimation. J. Mach. Learn. Res. **13**(1), 2529–2565 (2012)
71. Goldstein, M., Dengel, A.: Histogram-based outlier score (HBOS): a fast unsupervised anomaly detection algorithm, KI-2012: Poster and Demo Track, vol. 1, pp. 59–63 (2012)
72. Xu, H., et al.: Unsupervised anomaly detection via variational auto-encoder for seasonal KPIs in web applications. In: Proc. 2018 World Wide Web Conf. (WWW), pp. 187–196 (2018)
73. Su, Y., Zhao, Y., Niu, C., Liu, R., Sun, W., Pei, D.: Robust anomaly detection for multivariate time series through stochastic recurrent neural network. In: Proc. 25th ACM SIGKDD Int. Conf. on Knowledge Discovery and Data Mining (KDD), pp. 2828–2837 (2019)
74. Zhang, C., et al.: A deep neural network for unsupervised anomaly detection and diagnosis in multivariate time series data. In: Proc. AAAI Conf. on Artificial Intelligence, vol. 33, no. 01, pp. 1409–1416 (2019)
75. Bashar, M.A., Nayak, R.: TAnoGAN: time series anomaly detection with generative adversarial networks. In: Proc. 2020 IEEE Symp. Series on Computational Intelligence (SSCI), pp. 1778–1785 (2020)
76. Gilvari, A., Verma, R., Tavakoli, N., Kar, N.C., Kobti, Z.: FSOME++: Improving few-shot anomaly detection via abnormal sample augmentation. In: Proc. 34th IEEE Int. Symp. Industrial Electronics (ISIE), pp. 1–6 (2025)
77. Vincent, P., Larochelle, H., Bengio, Y., Manzagol, P.-A.: Extracting and composing robust features with denoising autoencoders. In: Proc. 25th Int. Conf. Machine Learning (ICML), pp. 1096–1103 (2008)
78. Pinheiro Cinelli, L., Araújo Marins, M., Barros da Silva, E.A., Lima Netto, S.: Variational autoencoder. In: Variational Methods for Machine Learning with Applications to Deep Networks, pp. 111–149. Springer, Cham (2021). https://doi.org/10.1007/978-3-030-70679-1_5
79. Schmidt, R.M.: Recurrent Neural Networks (RNNs): a gentle introduction and overview, *arXiv preprint* arXiv:1912.05911 (2019)
80. Markou, M., Singh, S.: Novelty detection: a review-part 2: neural network based approaches. Signal Process. **83**(12), 2499–2521 (2003)
81. Noor, S., Bazai, S.U., Ghafoor, M.I., Marjan, S., Akram, S., Ali, F.: Generative adversarial networks for anomaly detection: a systematic literature review. In: Proc. 4th Int. Conf. on Computing, Mathematics and Engineering Technologies (iCoMET), pp. 1–6 (2023)
82. Zeng, X.-H., Luo, S.-W., Wang, J.: Auto-associative neural network system for recognition. In: Proc. 2007 Int. Conf. on Machine Learning and Cybernetics, vol. 5, pp. 2885–2890 (2007)
83. Chua, L.O.: CNN: a vision of complexity. Int. J. Bifurcation Chaos **7**(10), 2219–2425 (1997)
84. Taud, H., Mas, J.F.: Multilayer perceptron (MLP). In: Camacho Olmedo, M.T., Paegelow, M., Mas, J.-F., Escobar, F. (eds.) Geomatic Approaches for Modeling Land Change Scenarios. LNGC, pp. 451–455. Springer, Cham (2018). https://doi.org/10.1007/978-3-319-60801-3_27
85. Nabney, I.T.: Efficient training of RBF networks for classification. Int. J. Neural Syst. **14**(03), 201–208 (2004)
86. Hopfield, J.J.: Hopfield network. Scholarpedia **2**(5), 1977 (2007)
87. Mohebali, B., Tahmassebi, A., Meyer-Baese, A., Gandomi, A.H.: Probabilistic neural networks: a brief overview of theory, implementation, and application. In: Handbook of Probabilistic Models, pp. 347–367. Elsevier (2020)

88. Kohonen, T.: Self-organized formation of topologically correct feature maps. Biol. Cybern. **43**(1), 59–69 (1982)
89. Stanley, K.O., Miikkulainen, R.: Evolving neural networks through augmenting topologies. Evol. Comput. **10**(2), 99–127 (2002)
90. Park, D., Hoshi, Y., Kemp, C.C.: A multimodal anomaly detector for robot-assisted feeding using an LSTM-based variational autoencoder. IEEE Robot. Autom. Lett. **3**(3), 1544–1551 (2018)
91. Hundman, K., Constantinou, V., Laporte, C., Colwell, I., Soderstrom, T.: Detecting spacecraft anomalies using LSTMs and nonparametric dynamic thresholding. In: Proc. 24th ACM SIGKDD Int. Conf. on Knowledge Discovery and Data Mining, pp. 387–395 (2018)
92. Zhao, H., et al.: Multivariate time-series anomaly detection via graph attention network. In: Proc. 2020 IEEE Int. Conf. on Data Mining (ICDM), pp. 841–850 (2020)
93. Hup, R.G., Merkofer, J.P., Bhogal, A.A., van Sloun, R.J.G., Haakma, R., Vullings, R.: Anomalous change point detection using probabilistic predictive coding, *arXiv preprint* arXiv:2405.15727 (2024)
94. Brigham, E.O.: The Fast Fourier Transform and Its Applications. Prentice-Hall, Inc. (1988)
95. Thill, M., Konen, W., Bäck, T.: Time series anomaly detection with discrete wavelet transforms and maximum likelihood estimation. In: Proceedings of the International Conference on Time Series (ITISE), vol. 2, pp. 11–23 (2017)
96. Snell, J., Swersky, K., Zemel, R.: Prototypical networks for few-shot learning. Adv. Neural Inf. Process. Syst. **30** (2017)
97. Sung, F., Yang, Y., Zhang, L., Xiang, T., Torr, P.H.S., Hospedales, T.M.: Learning to compare: relation network for few-shot learning. In: Proc. IEEE Conf. on Computer Vision and Pattern Recognition (CVPR), pp. 1199–1208 (2018)
98. Finn, C., Abbeel, P., Levine, S.: Model-agnostic meta-learning for fast adaptation of deep networks. In: Proc. Int. Conf. on Machine Learning (ICML), pp. 1126–1135 (2017)
99. Ye, H.-J., Hu, H., Zhan, D.-C., Sha, F.: Few-shot learning via embedding adaptation with set-to-set functions. In: Proceedings of the IEEE/CVF Conference on Computer Vision and Pattern Recognition (CVPR), pp. 8808–8817 (2020)
100. Chen, X., Cao, C., Mai, J.: Network anomaly detection based on deep support vector data description. In: Proceedings of the 2020 5th IEEE International Conference on Big Data Analytics (ICBDA), pp. 251–255 (2020)
101. Belton, N., Hagos, M.T., Lawlor, A., Curran, K.M.: FewSOME: one-class few shot anomaly detection with Siamese networks. In: Proceedings of the IEEE/CVF Conference on Computer Vision and Pattern Recognition, pp. 2978–2987 (2023)
102. Koch, G., Zemel, R., Salakhutdinov, R.: Siamese neural networks for one-shot image recognition. In: ICML Deep Learning Workshop, vol. 2, no. 1, pp. 1–30 (2015)
103. Abadi, M., et al.: TensorFlow: large-scale machine learning on heterogeneous distributed systems, *arXiv preprint* arXiv:1603.04467 (2016)
104. Ficili, I., Giacobbe, M., Tricomi, G., Puliafito, A.: From sensors to data intelligence: leveraging IoT, cloud, and edge computing with AI. Sensors **25**(6), 1763 (2025)

Assessing Driving Style with a Two-Stage Clustering Approach

Duarte Valente[1], Luís Loureiro[1], Artur Ferreira[1,2(✉)] [ID], and André Lourenço[1,3] [ID]

[1] ISEL, Instituto Superior de Engenharia de Lisboa, Instituto Politécnico de Lisboa,
Lisbon, Portugal
`{a47657,a43760}@alunos.isel.pt,`
`{artur.ferreira,andre.lourenco}@isel.pt`
[2] Instituto de Telecomunicações, Pólo de Lisboa, Lisbon, Portugal
[3] CardioID and NOVA LINCS, Lisbon, Portugal

Abstract. The way a person drives is a relevant source of information to make decisions about that person, in some contexts. For instance, insurance companies set vehicle insurance fees as functions of static variables, such as the age of the driver, the number of years one holds a driving license, and the driving history. These variables may not reflect the everyday behavior of the driver on the road, thus ending up by penalizing good drivers that are young. Another example is the fleet management task, in which it is relevant to know who the best drivers are, to make the best trip planning decisions. In this paper, we follow a pay-as-you-drive approach, to devise a driver style identification approach, based on driver behavior data. Using anonymous data records with the trips from different drivers, we build a dataset and we apply unsupervised dimensionality reduction and clustering techniques. The experimental results show clusters with distinct trip styles. Many drivers show a non-aggressive driving style, some have an aggressive style and a few have a risky style.

Keywords: Clustering · Dimensionality Reduction · Driver Profile · Driving Style · Pay-as-You-Drive · Telematics Data

1 Introduction

We now have billions of vehicles in the world, and in most countries, car insurance is required to drive on public roads. The owner or driver of a vehicle is responsible for damages it may cause and may have to pay some amount of money, in case of an accident. Nowadays, the volume of personal data acquired, processed, and shared has increased exponentially, which paves the way to new applications. One of such applications is the insurance companies business. These companies typically define vehicle insurance rates according to static variables with time-invariant values or variables that change in a controlled manner. Furthermore, the insurance rates calculation relies mostly on demographic information, such as the age of the driver, the number of years one has a driving license, and the driving history. The sharing of personal anonymous

F. Marcelloni et al. (Eds.): IJCCI 2025, CCIS 2829, pp. 617–630, 2026.
https://doi.org/10.1007/978-3-032-15638-9_36

data regarding driving behavior (telematics data) introduces new opportunities for business and applications such as car insurers. They can set the insurance price, attract customers, and then maximize their profit. Moreover, this type of system is useful for companies that need to perform fleet management over sets of vehicles and drivers, based on driving behavior data. The fair insurance rate computation and fleet management are two examples of the practical usage of the proposed approach in this paper. This approach also finds application in scenarios such as identifying how a person drives, the mistakes a person does systematically or occasionally, the places where those mistakes happen (highway or city driving), being useful to correct inadequate driving actions.

The approach proposed in this paper is summarized in Fig. 1 is useful to assess how a specific person drives, in a chosen natural environment without any additional constraints set on the driver. Using the obtained driving results, we can make corrections to the way a person drives, when taken the driving license for instance, and we can also make informed business decisions.

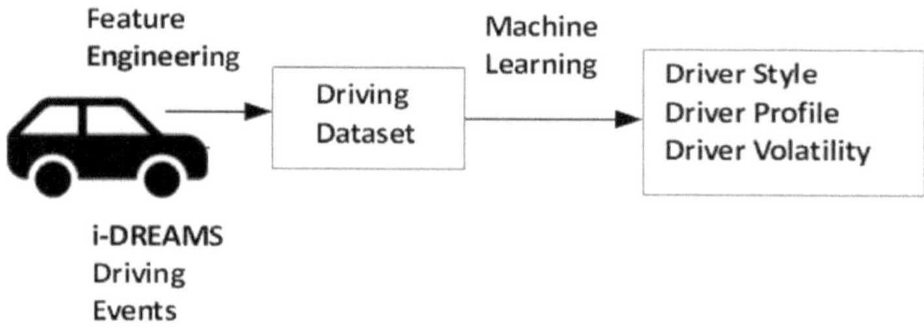

Fig. 1. Proposed solution block diagram. In this paper, we focus on the driver style and driver profile determination.

Another key factor that leverages the development of our proposal is that it is now easy to acquire data regarding the actual driving performed by the driver, on a given trip, in a non-intrusive way, using appropriate systems, such as in the i-DREAMS project framework, as depicted in Fig. 2.

The CardioWheel device is installed on the steering wheel and the Mobileye [16] device is placed on the windshield. Figure 3 shows the sensors installed on a vehicle.

In this paper, we focus on the construction of a dataset with driving data from driving events as well as an initial approach to cluster the data. From a set of driving events acquired during the trip, we devise a dataset and we apply machine learning techniques. The key contributions reported in this paper are as follows:

(1) Devise a driver style identification approach, based on real driver behavior telematics data.
(2) Apply different dimensionality reduction techniques to the dataset, aiming to identify the most relevant features/dimensions.
(3) Apply clustering algorithms to identify driver styles and to label the dataset previously established to get a ground-truth data.

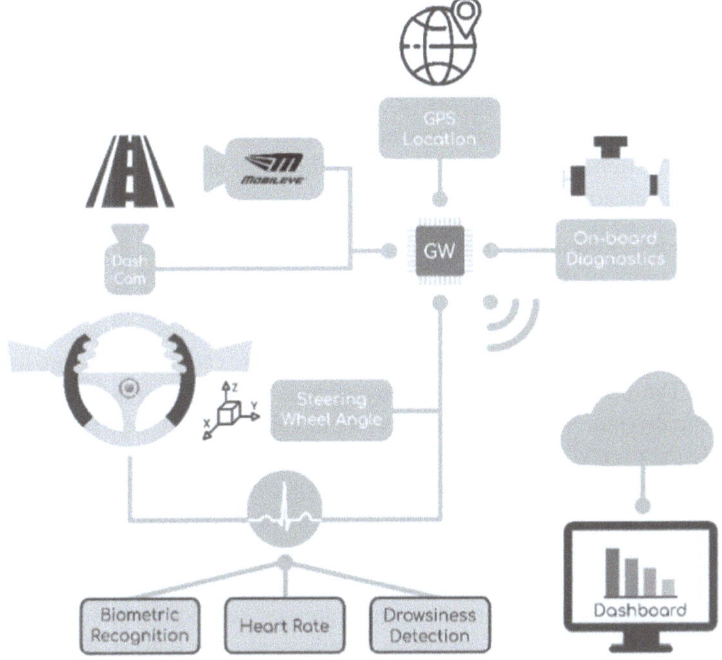

Fig. 2. Global overview of the i-DREAMS on-vehicle systems [5].

Fig. 3. CardioWheel on the steering wheel (left). Mobileye and dashcam seen from the inside (middle) and from the outside of the vehicle (right).

The remainder of this paper is organized as follows. In Sect. 2, we overview the driver data acquisition setup and related work. The proposed approach is summarized in Sect. 3. The experimental evaluation is reported in Sect. 4. Section 5 provides concluding remarks and directions for future work.

2   Related Work

Driving styles are established at different levels of specification, ranging from simple and single indicators, such as speeding or hard acceleration, or including more complex information like line deviation extracted from Advanced Driving Assisting System

(ADAS), or even including the driver status [1]. Part of this information can be obtained using very simple setups, as a smartphone on the vehicle, but more advanced parameters require specific hardware such as cameras and OBDII interfaces.

The conceptualization of aggressive or risky driving makes the profiles understandable by the common user, and are typically defined based on a combination of specific behavioral indicators [19]. According to [25] there are eight driving styles, organized into two main categories:

- safe category of driving styles;
- unsafe category of driving styles.

The specific driver variables are closely related to those driving styles. Table 1 presents each style and the category it belongs to. Within the safe category, there are three styles, as follows. The distress-reduction driving style is related to muscle relaxation techniques while driving. The patient style refers to calm and safe behaviors based on the "better safe than sorry" motto. The careful style refers to cautious behaviors and a state of mind to always react well and quickly to unexpected actions of other drivers. For the unsafe category, we have five styles. The dissociative driving style is based on unaware and misjudging behaviors. The anxious style is related to nervous and worrying behavior while driving. Risky applies to drivers that like to take risks while driving. The angry style means that the driver will make aggressive reactions to other drivers actions. Finally, the high-velocity style reflects an impulsive attitude towards driving faster than allowed by the road conditions or feeling impatient when traffic slows down.

A framework in which driving styles refer to driving habits was proposed by Sagberg et al. [22]. This framework suggests the use of a global driving style made up of a set of specific driving styles. In this context, a driving style refers to a specific behavior commonly used when driving. Typically, the set of parameters and variables adopted by researchers for profiling drivers cover the most important aspects of driving [24]. These parameters refer to the dynamic driving environment and are based on actual crash data gathered from self-reported surveys [18, 24], namely: speed; acceleration, braking, and jerks; annual mileage; lateral maneuver, such as swerving, lane changing, and sharp turns; time and space factors, such as time of day, day of the week, and month; distraction.

A driving style recognition approach based on a hierarchical-model is proposed by Cordero et al. [3]. It aims to recognize the driving style for ADAS for vehicles. It considers three aspects: the driver emotions, the driver state, and the driving style itself.

Table 1. Driving categories (safe/unsafe) and driving styles [25].

Safe	Unsafe
Distress-reduction	Dissociative
Patient	Anxious
Careful	Risky
–	Angry
–	High-velocity

It uses a set of descriptors acquired in a real context, and addresses three driving style recognition algorithms based on fuzzy logic, on chronicles (a temporal logic paradigm), and a pattern matching approach.

The work reported by Dörr et al. [6] proposes an online driving style recognition with fuzzy logic. It is parametrized with a central parameter file and may be adapted to nearly every car. The recognition was evaluated with vehicle dynamics simulation with 68% correct classifications over time.

A driving style recognition method based on vehicle trajectory data extracted from surveillance videos was proposed by Xue et al. [27]. The driving data is labeled based on their collision risk level using K-Means algorithm. Then, the driving style recognition model's inputs are extracted from vehicle trajectory features with Discrete Fourier Transform (DFT), Discrete Wavelet Transform (DWT), and statistical methods. Finally, Support Vector Machines (SVM) is applied to recognize driving style based on the labeled data. SVM outperforms the other classification methods, with 91.7% accuracy using the DWT feature extraction method.

The identification of driving style for the new energy vehicle (NEV) in China is addressed by Xia and Kang [26]. Using the NEV high-frequency big data collected by the vehicle-mounted terminal, they extract features with spatiotemporal changes in driving behavior. Resorting to the principal component analysis (PCA) method, they optimize the feature parameter set and identify the driving style using K-Means. A driving style recognition model with neural networks is devised, reaching recognition accuracy of 96.8%.

A framework to classify driving styles (aggressive, normal, and cautious) based on online car-hailing data is proposed by Ma et al. [13]. It is based on the detection and classification of driving maneuvers using a threshold-based endpoint detection approach, PCA, and K-Means clustering. This framework detects driving maneuvers and classifies driving styles accurately. Driving tasks lead to variations in driving style, and the variations in driving style during the different driving tasks differ significantly for turning, acceleration, and deceleration maneuvers.

A data-driven ML methodology for classifying driving styles is introduced by Silva and Naranjo [23]. They carry out a study with data collected from 50 drivers from two different cities in a naturalistic setting, and five features were extracted from the raw data. The data was labeled by fifteen experts to provide the ground truth of the dataset. The experimental results from performance metrics showed that SVM outperformed other four models, achieving an average accuracy of 96% and F1-Score of 95.95 %.

For a survey on driving style characterization and recognition with emphasis on ML approaches, please see Martinez et al. [15].

3   Proposed Approach

In this Section, we describe our proposed approach. In detail, in Sect. 3.1, we present the solution block diagram. Section 3.2 describes the dimensionality reduction techniques while Sect. 3.3 addresses the clustering approaches.

3.1　Solution Block Diagram

Figure 4 depicts a global overview of our approach. We start with driving events from the i-DREAMS devices, and from there we build a dataset with trip data. Using dimensionality reduction and unsupervised ML techniques, we devise trip/driver style identification. After a detailed analysis of the resulting clusters with different perspectives, we assign a class label to each trip.

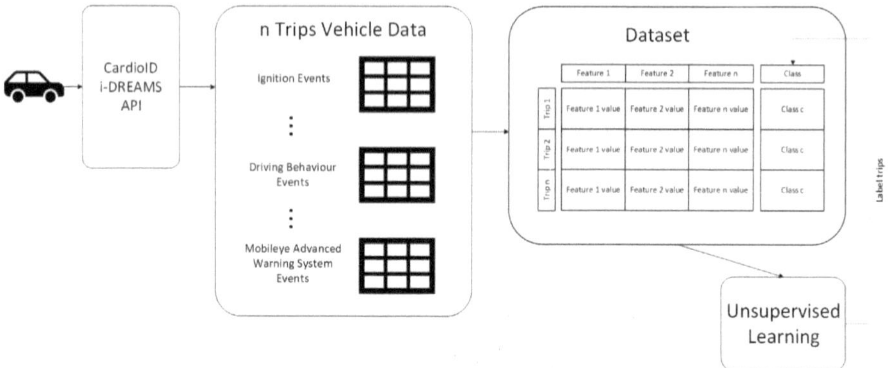

Fig. 4. Proposed solution block diagram.

Data was collected and the features were devised. After a detailed analysis of the data, only the most significant features were selected according to the literature and the results obtained in the following feature extraction phase. We consider trips from April 1, 2021, to July 20, 2022, with 15002 trips in total. For each trip 65 features were generated, thus the dataset was composed of $n = 15002$ instances (trips) and $d = 53$ features.

3.2　Data Normalization and Dimensionality Reduction

We applied data normalization techniques because many clustering algorithms are based on distances, and the scaling of the data influences the result. The algorithm should not be biased towards features with higher magnitudes and scales. We opted to use either distance normalization or duration normalization.

After normalization, we applied Feature Reduction (FR) techniques and Feature Selection (FS) techniques [10]. Since the dataset does not have pre-labeled data, we had to choose only unsupervised FR techniques, such as the well-known Principal Component Analysis (PCA) and Singular Value Decomposition (SVD) methods. For FS, we use the Relevance-Redundancy Feature Selection (RRFS) method by Ferreira and Figueiredo [8]. RRFS finds features with high relevance and low redundancy, below the maximum similarity, M_S, threshold.

3.3 Clustering Phase

The clustering phase has two clustering stages. In the first, we use different clustering techniques. Initially, we followed a classical approach with the K-Means algorithm [11]. The approach consisted in initially identifying the best value of k clusters, to use through the elbow and silhouette methods. The experimental results showed that $k = 2$ clusters was the best value.

Density-based spatial clustering (DBScan) [7] was also applied to check for outliers in the dataset. Another technique used was Gaussian Mixture Models (GMM) [20] that can indicate the probability that each instance of the dataset belongs to a Gaussian distribution. The advantage of using this type of clustering is that it can better deal with different shapes of the data distribution. K-Means using the Euclidean distance metric handles spherical shapes better. The number of components used was the same as K in K-Means. Finally, Evidence Accumulation Clustering (EAC) [9] with K-Means [17] was tested aiming to find a more robust clustering solution that involves multiple data partitions.

The objective of the second stage of clustering was to use the best combination of the first stage, to check if the clusters found could be further decomposed into sub-clusters to search for narrower driver styles. After clustering, the trips were labeled according to their profile, that is, the data points within the same cluster were assigned with the same label, after human supervision.

After clustering, we analyze the resulting clusters with:

– PCA analysis;
– statistical analysis.

The objective of the PCA analysis is to find the most important features for the first principal components. We resort to statistical analysis to observe the different feature values between clusters, using the minimum, maximum, mean, and standard deviation metrics.

4 Experimental Evaluation

This Section addresses the detailed experimental evaluation from the different stages of the developed solution. Section 4.1 describes the settings of the experiments and the evaluation metrics. Section 4.2 describes the results obtained with the first clustering stage, while the results from the second clustering stage are reported in Sect. 4.3.

4.1 Evaluation Settings and Metrics

The code was written in the Python programming language, using the common packages to handle datasets and ML tasks. As in our previous work, we opted to further normalize the trips by distance and duration.

The clustering algorithm results were evaluated with the Calinski-Harabasz (CH) score, also known as the variance ratio criterion [2], Davies-Bouldin (DB) [4], and Silhouette scores [21]. The CH is the ratio of the sum of between-clusters dispersion and of inter-cluster dispersion for all clusters. The higher the score, the better the performance. The DB score is an internal evaluation scheme in which quantities and features inherent to the dataset are used for this purpose (the lowest the better). The Silhouette score ranges from -1 (worst) to 1 (best).

4.2 First Stage of Clustering

We combined each of the four clustering algorithms, with the normalization and dimensionality reduction techniques. Table 2 reports the experimental results on the normalized and dimensionality reduced dataset.

Table 2. First stage clustering results with Calinski-Harabasz (CH), Davies-Bouldin (DB) and Silhouette scores. The best result is in boldface.

Scores	K-Means	DBScan	GMM	EAC
CH ↑ [2]	**18274.2**	4273.523	4528.2	422.6
DB ↓ [4]	**0.673**	1.229	1.169	0.933
Silhouette ↑ [21]	**0.616**	0.491	0.473	0.560
# Cluster 0	10632	10165	9524	13111
# Cluster 1	2574	3041	3682	95

The best combination is normalization by distance, followed by dimensionality reduction with RRFS ($M_S = 0.5$) and SVD. The K-Means algorithm presents the best clustering results in all metrics. RRFS ($M_S = 0.5$) has identified the following top 6 features in decreasing order: *speed*, *n_tsr_level_1*, *n_headway_0*, *n_hc_l*, *n_speeding_2*, and *n_ha_m*. The meaning of these features is as follows:

– *speed* - Mean Speed in km/h.
– *n_tsr_level_1* - Number of times 0–5 units over speed limit.
– *n_headway_0* - Number of headway level 0 events, vehicle detected.
– *n_hc_l* - Number of harsh cornering events with low severity.
– *n_speeding_2* - Number of overtaking level 2 events, visual and auditory.
– *n_ha_m* - Number of harsh acceleration events with medium severity.

A further analysis on these features revealed that they seem to contribute to distinguish the clusters. We find that cluster 0 represents trips with low-speed in km/h and low speeding events per km. Cluster 1 represents trips with low to medium speed and low to medium speeding events per km. In terms of the number of braking events per km, cluster 1 has more events. We compared the minimum, maximum, average, and standard deviation statistics of the top features, on the two clusters. Table 3 reports these values.

Analyzing these statistical results, we observe that for most features cluster 1 has more events per km traveled. Cluster 1 corresponds to more aggressive trips when compared to cluster 0. However, there is no way to say that for example, a certain number of

Table 3. Minimum, maximum, average, and standard deviation of the top features per cluster for the two clusters (first clustering stage).

	Cluster 0				Cluster 1			
	Min	Max	Avg	Std	Min	Max	Avg	Std
speed	0.00	12.47	4.62	2.68	5.12	69.00	17.15	5.85
n_brakes	0.00	23.30	2.44	1.97	0.00	50.85	5.72	4.26
n_tsr_level	0.00	37.31	2.90	2.78	0.00	110.60	4.19	7.16
n_fatigue_0	0.00	1.11	0.13	0.10	0.13	2.80	0.57	0.22
n_overtaking_0	0.00	2.00	0.18	0.15	0.13	5.46	0.64	0.42

speeding events to consider the trip aggressive or non-aggressive. For instance, the trips belonging to cluster 0 may be considered aggressive for other datasets depending on the problem context. Therefore, we made this selection according to the characteristics of the dataset at hand. In conclusion, we considered cluster 1 trips as aggressive when compared to cluster 0 trips, which were described as non-aggressive.

4.3   Second Stage of Clustering

A second stage clustering was applied for the aggressive trips that compose cluster 1, with K-Means using $k = 2$. We opted for K-Means for the second stage, since it attained the best results on the first stage. Table 4 shows the K-Means scores with distance normalization and dimensionality reduction with PCA, RRFS, and SVD. We also made two experiments using a FR technique after a FS technique, namely using RRFS followed by PCA (denoted as RRFS + PCA) and RRFS followed by SVD (RRFS + SVD).

Table 4. Second stage K-Means clustering results, with dimensionality reduction. # denotes the number of instances per cluster. The best result is in boldface.

Dim. reduction	CH ↑	DB ↓	Silhouette ↑	# Cluster 1.1	# Cluster 1.2
PCA	754.09	1.249	0.519	2527	203
RFS	753.75	1.251	0.518	2487	205
RFS + PCA	766.42	**0.536**	**0.729**	2655	37
RRFS	1740.73	0.848	0.576	2223	347
RRFS + SVD	**1764.79**	0.837	0.577	2228	346

From Table 4, we observe that the combination of FS and FR methods yield the best results for the clustering metrics. After this, we took the same approach applied in the first stage of clustering to attribute meaning to the clusters. The top six features for RRFS (with $M_S = 0.5$) found for this second stage of clustering are as follows:

- n_tsr_level - Number of times the speed limit was exceeded.
- n_hc - Number of harsh cornering events.
- $n\_speeding\_2$ - Number of overtaking level 2 events, visual and auditory warning.
- $n\_headway\_0$ - Number of headway level 0 events, vehicle detected.
- $n\_tsr\_level\_3$ - Number of times 10–15 units over speed limit.
- $n\_pedestrian\_dz$ - Number of times pedestrians detected in danger zone.

The n_tsr_level feature produces more events per km in cluster 1. The $n\_tsr\_level\_3$ feature also has much more events in cluster 1. As a consequence, cluster 1 contains more speeding events where the driver was 10–15 units over the speed limit, as compared to cluster 0. To better characterize cluster 1, we observed that the study by [14] classified aggressive trips into aggressive, distracted, and risky. The risky trips involved more speeding events than the aggressive trips. Consequently, we decided to assign cluster 1.2 instances as risky trips and cluster 1.1 as aggressive trips. The cluster 0 trips remained labeled as non-aggressive.

Figure 5 shows the scatter-plot of the $speed$, $n\_tsr\_level\_1$, $n\_headway\_0$, and n_hc_l features, for instances belonging to three clusters. The final clusters and labels are reported in Table 5 and Fig. 6, renaming clusters 1.1 and 1.2 to cluster 1 and 2, respectively. Cluster 0 was assigned as non-aggressive trips while cluster 1 was defined as aggressive trips, and cluster 2 holds the risky trips. Regarding Fig. 6, we display the results of dimensionality reduction by RRFS followed by SVD, which yields the best result in Table 4. Table 5 shows slight changes on the number of trips per cluster, as functions of the employed dimensionality reduction technique. We have a small

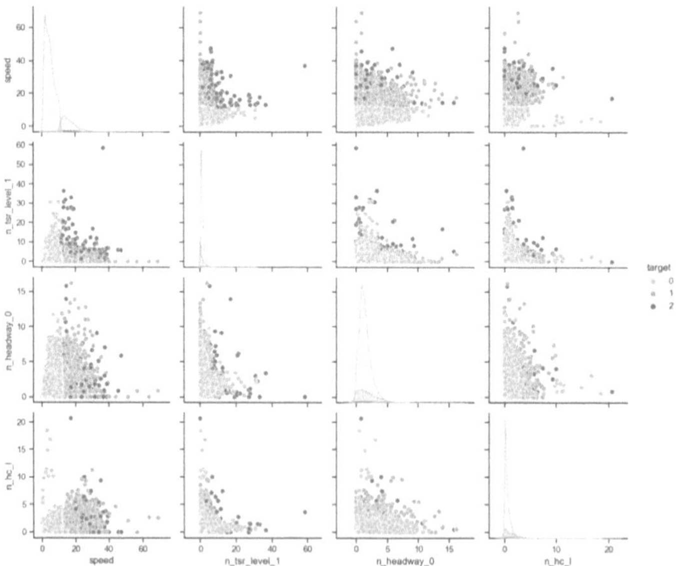

Fig. 5. Scatter plot of the $speed$, $n\_tsr\_level\_1$, $n\_headway\_0$, and n_hc_l features, for the three clusters, after the second clustering stage.

percentage of risky trips, a larger percentage of aggressive trips and the vast majority are non-aggressive trips, as expected.

Table 6 reports the minimum, maximum, average, and standard deviation statistics values for some top features, on the final three clusters. For cluster 2, we observe larger values of the average and standard deviation for all the features, as compared to the

Table 5. Instances per cluster after dimensionality reduction.

Label	Description	Number of instances			
		SVD	RRFS	RRFS + SVD	RRFS
0	Non-Aggressive trips	10476	10514	10514	10636
1	Aggressive trips	2521	2487	2655	2223
2	Risky trips	209	205	37	347

Table 6. Minimum, maximum, average, and standard deviation of the top features per cluster for the three clusters (second clustering stage).

	Cluster 0 (non-agr.)				Cluster 1 (agressive)				Cluster 2 (risky)			
	Min	Max	Avg	Std	Min	Max	Avg	Std	Min	Max	Avg	Std
speed	0.00	11.24	4.67	2.71	10.66	69.00	17.13	5.37	10.72	47.32	20.06	7.16
n_brakes	0.00	38.58	2.56	2.25	0.00	50.85	5.38	3.88	0.00	30.09	5.34	5.07
n_tsr_level	0.00	61.29	2.97	3.13	0.00	9.36	2.02	2.60	8.69	110.60	16.31	10.78
n_fatigue_0	0.00	1.11	0.13	0.10	0.18	2.80	0.57	0.21	0.21	2.00	0.64	0.27
n_overtaking_0	0.00	2.00	0.18	0.16	0.18	4.39	0.62	0.33	0.21	5.46	0.86	0.76

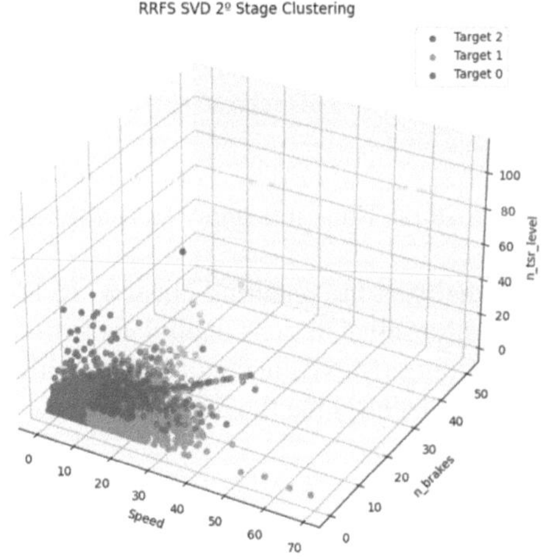

Fig. 6. One possible visualization of the final clustering results, with three clusters, after the second clustering stage. The three axis chosen for this plot are the *speed*, *n_brakes*, and *n_tsr_level* features.

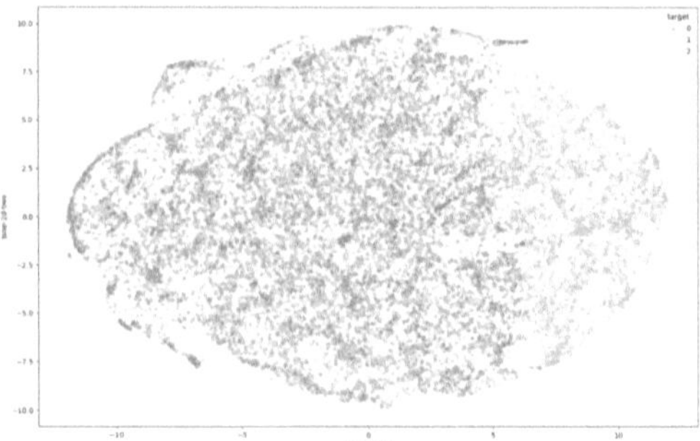

Fig. 7. A visualization of the final clustering results, with three clusters, after dimensionality reduction with the t-SNE technique, with a 2D representation.

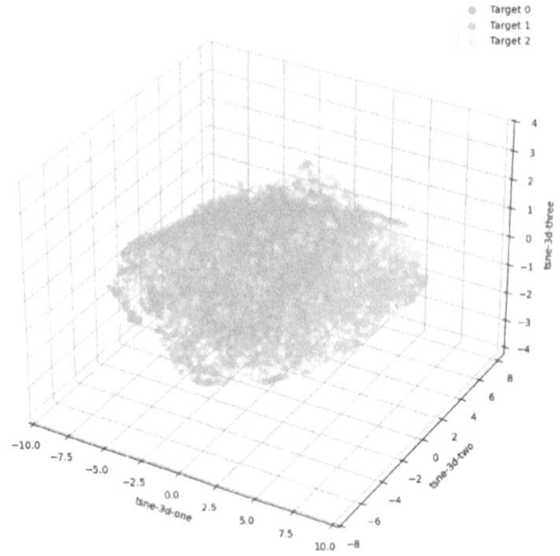

Fig. 8. A visualization of the final clustering results, with three clusters, after dimensionality reduction with the t-SNE technique, with a 3D representation.

values reported for other clusters. Thus, this analysis is aligned with the assignment of risky trips to cluster 2.

We also have further analyzed the resulting clusters. In Figs. 7 and 8 we observe the three cluster representation after dimensionality reduction with the t-Distributed Stochastic Neighbor Embedding (t-SNE) technique [12], with 2D and 3D representations, respectively. We clearly identify the three clusters, under this set of axes.

5   Conclusions

The monitoring of a driver behavior based on real driver data acquired from a vehicle, finds many useful applications such as to enable a fair insurance rate calculation, to enable adequate fleet management decisions or simply to correct and improve how one drives. This paper aimed to provide an approach to perform driver style identification based on the driver behavior.

We proposed a machine learning approach which combines dimensionality reduction with unsupervised learning techniques. The dataset construction, cleaning, and normalization phases were the ones that most influenced the clustering results. At an initial stage of the work, the normalization technique and the extracted features from the events were causing poor clustering results, so we had to revise those phases several times. By normalizing the trips by distance and applying dimensionality reduction techniques, we achieved the best results.

We applied a two-stage clustering strategy, where the K-Means algorithm shown to be the one with best and stable results. The most challenging part of the clustering phase was to attribute meaning to the clusters, which implied a further analysis on the clustering results using different approaches. Finally, we were able to identify three different types of driving style. As the end result, we have devised a dataset labeled with 3 classes, which is suitable for supervised learning tasks.

The dimensionality reduction techniques have shown adequate results, with the combination of feature selection and feature reduction achieving the best performance. The feature selection filters are able to effectively remove irrelevant and redundant features enabling the feature reduction techniques to be focused on a smaller dimensionality feature space, yielding the best results.

At this point, we have different future work directions. We may explore more clustering algorithms and variations. Other feature selection techniques can also be applied to check if we can achieve a even larger dimensionality reduction and also to assess the consistency among the resulting feature subsets.

Acknowledgements. This work is supported by UID/04516/NOVA Laboratory for Computer Science and Informatics (NOVA LINCS) with the financial support of FCT.IP and by Polytechnic University of Lisbon, under grant IPL/IDI&CA2024/ML4EP_ISEL.

References

1. Bouhsissin, S., Sael, N., Benabbou, F.: Driver behavior classification: a systematic literature review. IEEE Access **11**, 14128–14153 (2023). https://doi.org/10.1109/ACCESS.2023.3243865
2. Caliński, T., Harabasz, J.: A dendrite method for cluster analysis. Commun. Stat. **3**(1), 1–27 (1974). https://doi.org/10.1080/03610927408827101
3. Cordero, J., Aguilar, J., Aguilar, K., Chávez, D., Puerto, E.: Recognition of the driving style in vehicle drivers. Sensors **20**(9) (2020). https://doi.org/10.3390/s20092597
4. Davies, D., Bouldin, D.: A cluster separation measure. IEEE Trans. Pattern Anal. Mach. Intell. **PAMI-1**(2), 224–227 (1979)
5. i DREAMS Team: A smart driver and road environment assessment and monitoring system (2021). https://idreamsproject.eu/wp/
6. Dörr, D., Grabengiesser, D., Gauterin, F.: Online driving style recognition using fuzzy logic. In: 17th International IEEE Conference on Intelligent Transportation Systems (ITSC), pp. 1021–1026 (2014)

7. Ester, M., Kriegel, H., Sander, J., Xu, X.: A density-based algorithm for discovering clusters in large spatial databases with noise. In: Knowledge Discovery in Databases (KDD), vol. 96, pp. 226–231 (1996)
8. Ferreira, A., Figueiredo, M.: Efficient feature selection filters for high-dimensional data. Pattern Recognit. Lett. **33**(13), 1794–1804 (2012). http://dx.doi.org/10.1016/j.patrec.2012.05.019
9. Fred, A., Jain, A.: Data clustering using evidence accumulation. In: International Conference on Pattern Recognition, vol. 4, pp. 276–280 (2002). https://doi.org/10.1109/ICPR.2002.1047450
10. Guyon, I., Elisseeff, A.: An introduction to variable and feature selection. J. Mach. Learn. Res. (JMLR) **3**, 1157–1182 (2003)
11. Hartigan, J., Wong, M.: Algorithm as 136: a k-means clustering algorithm. J. Royal Stat. Soc. Ser. C (Appl. Stat.) **28**(1), 100–108 (1979)
12. Hinton, G.E., Roweis, S.: Stochastic neighbor embedding. In: Becker, S., Thrun, S., Obermayer, K. (eds.) Advances in Neural Information Processing Systems, vol. 15, pp. 857–864. MIT Press (2003). https://proceedings.neurips.cc/paper/2002/file/6150ccc6069bea6b5716254057a194ef-Paper.pdf
13. Ma, Y., Li, W., Tang, K., Zhang, Z., Chen, S.: Driving style recognition and comparisons among driving tasks based on driver behavior in the online car-hailing industry. Accid. Anal. Prev. **154**, 106096 (2021)
14. Mantouka, E., Barmpounakis, E., Vlahogianni, E.: Identification of driving safety profiles from smartphone data using machine learning techniques. Safety Sci. **119** (2019)
15. Martinez, C., M. Heucke, F.W., Gao, B., Cao, D.: Driving style recognition for intelligent vehicle control and advanced driver assistance: a survey. IEEE Trans. Intell. Transp. Syst. **19**(3), 666–676 (2018)
16. Mobileye: Autonomous driving & ADAS (Advanced Driver Assistance Systems) (2021). https://www.mobileye.com/
17. Okun, O.: Supervised and Unsupervised Ensemble Methods and Their Applications, vol. 126. Springer, Heidelberg (2008). https://doi.org/10.1007/978-3-642-03999-7
18. Peck, R., Kuan, J.: A statistical model of individual accident risk prediction using driver record, territory and other biographical factors. Accid. Anal. Prev. **15**(5), 371–393 (1983)
19. Priyadharshini, G., Josephin, J.S.F.: A comprehensive review of various data collection approaches, features, and algorithms used for the classification of driving style. IOP Conf. Ser. Mater. Sci. Eng. **993**(1), 012098 (2020)
20. Reynolds, D.: Gaussian mixture models. Encyclopedia Biometrics **741**(659-663) (2009)
21. Rousseeuw, P.: Silhouettes: A graphical aid to the interpretation and validation of cluster analysis. J. Comput. Appl. Math. **20**(1), 53–65 (1987). https://doi.org/10.1016/0377-0427(87)90125-7
22. Sagberg, F., Selpi, Piccinini, G., Engström, J.: A review of research on driving styles and road safety. Hum. Factors **57**(7), 1248–1275 (2015)
23. Silva, I., Eugenio Naranjo, J.: A systematic methodology to evaluate prediction models for driving style classification. Sensors **20**(6) (2020). https://doi.org/10.3390/s20061692
24. Singh, H., Kathuria, A.: Profiling drivers to assess safe and eco-driving behavior-a systematic review of naturalistic driving studies. Accid. Anal. Prev. **161**, 106349 (2021)
25. Taubman-Ben-Ari, O., Mikulincer, M., Gillath, O.: The multidimensional driving style inventory-scale construct and validation. Accid. Anal. Prev. **36**(3), 323–332 (2004)
26. Xia, L., Kang, Z.: Driving style recognition model based on NEV high-frequency big data and joint distribution feature parameters. World Electric Veh. J. **12**(3) (2021). https://www.mdpi.com/2032-6653/12/3/142
27. Xue, Q., Wang, K., Lu, J., Liu, Y.: Rapid driving style recognition in car-following using machine learning and vehicle trajectory data. J. Adv. Transp. **Article ID 9085238** (2019)

Fine-Tuning Prototypes for Cross-Domain Few-Shot Image Classification Using Contrastive Objective

Abhishek Mahajan$^{(\boxtimes)}$, Ziad Kobti , and Bishwadeep Sikder

School of Computer Science, University of Windsor, 401 Sunset Avenue, Windsor, ONT N9B-3P4, Canada
{mahaja52,kobti,sikderb}@uwindsor.ca

Abstract. Cross-domain few-shot learning (CFC) seeks to enable accurate classification in novel visual domains using only a few labeled samples per class. However, it remains a challenging task due to prototype misalignment and domain-induced embedding distortions, in addition to lack of purified data. While the state-of-the-art transformation network effectively realigns prototypes in one-shot settings, it is not designed for broader domain generalization or transformer-based architectures in cross-domain settings. We introduce CosRestViT, a prototype-aware few-shot learning framework designed to tackle the problems of scarce data with a lot of outliers. Our method integrates a custom transformation network with a Vision Transformer (ViT) backbone to recalibrate noisy embeddings and align them with class-level prototypes. To enhance semantic consistency in the embedding space, we fine-tune the model using a hybrid contrastive objective that combines cross-entropy loss with CoSENT's ranking loss. We pretrain CosRestViT on the large-scale base dataset using a prototype-aware contrastive loss to enhance the representational power of ViT and ensure effective clustering of semantically similar samples. The model is then evaluated on the Meta-Dataset benchmark under two training regimes: (i) training on all seen datasets, and (ii) training exclusively on ImageNet for transfer learning. Experimental results demonstrate that CosRestViT outperforms existing baselines across both seen and unseen domains, achieving superior generalization in low-data scenarios.

Keywords: Few-Shot Learning · Cross-Domain Classification · Vision Transformer · Contrastive Learning · Prototype Alignment · CoSENT Loss · Low-Shot Settings · Anisotropy Mitigation

1 Introduction

One-shot image classification poses a fundamental challenge in computer vision, as it requires models to generalize from extremely limited labeled examples per class. While metric-based methods such as Prototypical Networks [16] have shown promise by comparing query instances to class prototypes in embedding space, their performance degrades when prototypes are noisy or unrepresentative. This limitation was

© The Author(s), under exclusive license to Springer Nature Switzerland AG 2026
F. Marcelloni et al. (Eds.): IJCCI 2025, CCIS 2829, pp. 631–649, 2026.
https://doi.org/10.1007/978-3-032-15638-9_37

addressed through *RestoreNet* [21], a regression-based framework that shifts support embeddings toward true class centers, significantly enhancing performance in single-domain, one-shot settings.

However, real-world scenarios often demand *Cross-Domain Few-Shot Classification (CFC)*, where a model must generalize from a well-resourced source domain to low-resource target domains that differ significantly in visual characteristics. This challenge is particularly evident in domains such as medical imaging, wildlife monitoring, and remote sensing, where domain shifts and lack of labeled data co-occur. For instance, FAMNet [1] tackles domain variance across imaging modalities in medical segmentation using frequency-aware matching. Similarly, the Self-Challenging strategy [9] improves cross-domain generalization by suppressing dominant features during training. These works highlight the critical need for embedding alignment strategies that explicitly address domain-induced distortions. In CFC settings, prototype misalignment is further exacerbated by inter-domain representation discrepancies and intra-class variability problems that remain unaddressed by single-domain solutions such as RestoreNet (Fig. 1).

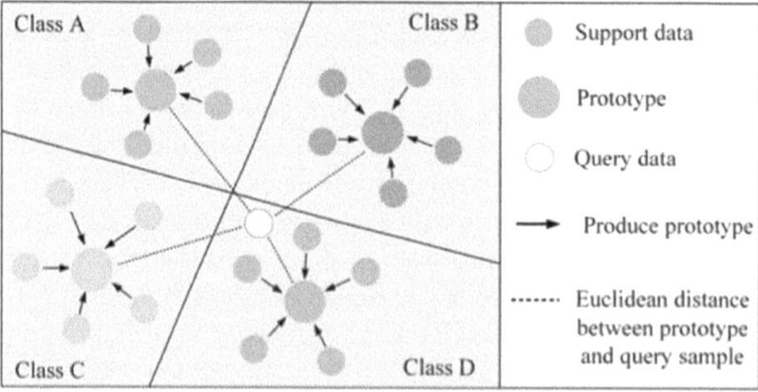

Fig. 1. Prototypical Networks classify few-shot samples by embedding support examples to form class prototypes (colored nodes). A query sample (white node) is classified based on its Euclidean distance (dashed lines) to these prototypes. Solid arrows indicate prototype construction.

In this work, we introduce *CosRestViT*, a prototype-aware contrastive learning framework designed for cross-domain few-shot image classification. Our method extends RestoreNet by introducing two core innovations: (1) a modified transformation network that learns to realign outlier embeddings with class prototypes across domains, and (2) a Vision Transformer (ViT) [5] backbone fine-tuned using a *hybrid contrastive loss*, which combines cross-entropy [14] with CoSENT [8] ranking objectives to enforce semantic alignment across representations.

Unlike prior methods that treat prototypes and instances symmetrically [16,19] CosRestViT explicitly maintains the semantic distinction between them and restores embedding consistency by disentangling domain-specific distortions. We further mitigate common transformer challenges such as embedding anisotropy [12] and overfitting via targeted dropout and domain-aware data augmentation techniques. Unlike convolutional networks that focus on local receptive fields, Vision Transformers (ViTs) capture long-range dependencies and global semantic patterns more effectively. This makes them particularly suitable for cross-domain few-shot learning, where class distinctions often rely on subtle, spatially distributed cues that CNNs may overlook.

Evaluations on the Meta-Dataset benchmark [20] demonstrate that CosRestViT consistently outperforms state-of-the-art baselines in both seen and unseen domains, particularly under 1-shot and 5-shot regimes. By extending RestoreNet's restoration mechanism to the transformer-based, cross-domain context, our work bridges the gap between *prototype fidelity* and *domain robustness*. This makes CosRestViT a strong candidate for practical deployment in real-world applications where source and target domains differ—such as medical diagnosis, remote sensing, and wildlife monitoring.

Contributions

The key contributions of this work are summarized below:

1. We propose **CosRestViT**, a novel prototype-aware framework that integrates a Vision Transformer (ViT) backbone with a modified transformation network to address prototype misalignment in cross-domain few-shot classification.
2. We introduce a **hybrid contrastive objective** that combines CoSENT ranking loss with cross-entropy loss, effectively enhancing semantic consistency and discriminative capability in the embedding space.
3. We incorporate **anisotropy mitigation techniques** within the ViT architecture—using targeted dropout at attention, feed-forward, and output layers—to improve embedding isotropy and generalization to unseen domains.
4. We conduct extensive experiments on the **Meta-Dataset** benchmark under both single-source (ImageNet-only) and multi-source training regimes, demonstrating that CosRestViT consistently outperforms strong baselines across diverse visual domains.
5. To the best of our knowledge, this is the first work that jointly integrates CoSENT ranking loss with ViT-based prototype restoration for CFC.

2 Background

Few-shot learning (FSL) methods aim to train classifiers that generalize effectively from a limited number of labeled examples per class. In particular, cross-domain few-shot classification (CFC) extends this challenge by introducing domain shifts between training and evaluation datasets, where the visual characteristics differ substantially. Under such shifts, traditional metric-based methods such as Prototypical Networks [16] and Relation Networks [17] suffer due to their assumption of domain-invariant embedding distributions. These methods rely heavily on the consistency of feature distributions across domains and fail to maintain classification accuracy when faced with heterogeneous visual characteristics.

RestoreNet, proposed by Xue and Wang [21], addresses a key limitation of metric learning in one-shot settings by introducing a regression-based transformation network. This network restores noisy prototypes—formed from single, potentially unrepresentative images—by aligning them closer to class centers. While RestoreNet achieves strong results on intra-domain datasets, it was neither designed nor evaluated for cross-domain generalization. It also relies on shallow CNN-based embeddings (e.g., ResNet-18), limiting its capacity to model long-range dependencies or handle the complex, anisotropic embedding spaces produced by modern transformers.

To address domain shift more explicitly, several works have proposed transformation strategies tailored for cross-domain few-shot classification. Universal Representation Learning (URL) [13] applies a shared transformation head to project both instance and prototype embeddings into a common space. However, this uniform treatment overlooks the semantic asymmetry between prototypes (representing class-level abstraction) and instances (representing individual samples), which is particularly pronounced in heterogeneous domains. CoPA [19] advances this by using separate transformation heads under a contrastive training objective, preserving prototype-instance distinctions. More recent updates [18] explore the alignment gaps between prototypes and instances but still do not address the distortion of embeddings or high-dimensional anisotropy. Other methods like TADAM [15] and MetaOptNet [11] rely on task-specific or optimization-based metrics for better generalization, but these are largely constrained to single-domain evaluations with fixed feature spaces.

While these works attempt to improve generalization through better training objectives, few explicitly focus on prototype restoration under domain shift. Moreover, most remain tied to CNN-based encoders, which limits their ability to model hierarchical or global semantic relationships—an area where transformer architectures have shown significant promise. Transformer-based meta-learning methods such as TransMatch [4], TS-TADAM [15], and CLIP-based adaptation strategies [23] have recently emerged. These methods leverage the representational strength of self-attention but do not tackle embedding restoration or prototype correction—key challenges in few-shot classification with noisy or sparse support data.

Contrastive learning frameworks like SimCLR [2] and MoCo [7] have shown the ability to produce semantically meaningful representations via large-scale pretraining. However, their reliance on large batch sizes and strong augmentations makes them impractical in low-data regimes such as FSL. S2M2R [3] introduces margin-based contrastive objectives that improve intra-class compactness and inter-class separability, yet they do not explicitly model prototype dynamics or address domain-specific distortions. In contrast, CoSENT [8], a ranking-based contrastive loss originally developed for NLP, has shown effectiveness in low-resource settings by avoiding large batch dependencies. However, its integration into cross-domain visual learning and prototype restoration frameworks remains unexplored.

Additionally, most methods overlook the problem of embedding anisotropy where representations cluster along a few dominant directions in high-dimensional space [12]. This phenomenon hampers class separability and semantic alignment, especially when using ViTs. Data augmentation and dropout techniques have been loosely employed, but their targeted application for mitigating anisotropy in the context of few-shot transformers is not studied.

Furthermore, benchmarks such as Meta-Dataset [20] introduce significant diversity across domains (e.g., traffic signs, handwritten characters, natural images), exacerbating prototype misalignment and domain-induced distortions. Many earlier approaches fail to sustain performance across such diverse tasks, highlighting the need for methods that explicitly unify domain adaptation, prototype refinement, and embedding space regularization.

Despite this, a unified framework that combines prototype restoration, semantic contrastive alignment, and anisotropy handling under transformer-based cross-domain settings is still unaccounted. Our proposed method, CosRestViT, is the first to address these gaps by integrating a ViT encoder with a modified RestoreNet module and optimizing using a hybrid loss function (CoSENT + cross-entropy). This results in compact, class-discriminative, and domain-robust embeddings that significantly improve few-shot generalization.

3   Architecture

This section introduces the proposed **CosRestViT** framework for cross-domain few-shot classification. The methodology comprises three sequential stages: (i) joint training of a modified RestoreNet transformation network and a Vision Transformer (ViT) backbone on the base dataset ImageNet-1K using a hybrid contrastive objective; (ii) fine-tuning the ViT on few-shot support tasks using a combination of cross-entropy and CoSENT (Consistent Sentence Embedding via Similarity Ranking) losses; and (iii) evaluation on query sets drawn from unseen domains. These trained components are then used to classify novel classes based on limited support examples. An overview of the full architecture is presented in the following section.

The framework consists of three main stages: (1) Pre-trained ViT extracts global image embeddings via the CLS tokens from the source domain (ImageNet) and aligns prototypes via a transformation network; (2) Fine-Tuned ViT refines prototype embeddings through a trainable transformation network and optimizes them using a hybrid loss function combining CoSENT loss and cross-entropy loss; and (3) Cross-Domain Few-Shot Classification, where the model is evaluated on a target domain (Meta-Dataset) using support and query sets under a few-shot setting. Backpropagation is used to fine-tune the ViT during support adaptation.

3.1   Training the Transformation Network and Vision Transformer

In the initial stage, we jointly train a transformation network and a ViT encoder using the ImageNet-1K dataset, which contains a large number of labeled examples per class. The ViT serves as a high-capacity feature extractor, while the transformation network is tasked with aligning instance embeddings to their respective class prototypes to address embedding distortion and outliers in the support set.

Prototype Computation: Given multiple image samples per class, the prototype (or class center) is computed as the mean of the ViT-generated embeddings. For a class A with image embeddings $[E_X, E_Y, E_Z]$, the prototype embedding $E_{Prototype}$ is defined as:

$$E_{Prototype} = \frac{1}{3}(E_X + E_Y + E_Z) \tag{1}$$

This operation is repeated across all classes to construct a set of clean, representative prototypes (Fig. 2).

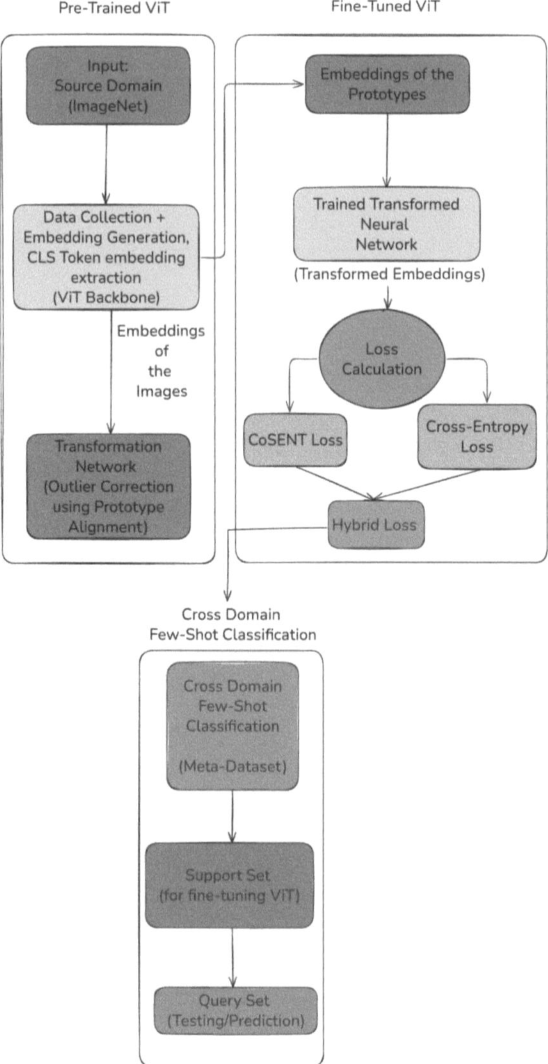

Fig. 2. CosRestViT Architecture.

Transformation Network Architecture: The transformation module is implemented as a fully connected neural network composed of hidden layers, ReLU activations, batch normalization, and dropout. Both the input and output vectors are of the same dimensionality as the ViT embedding (typically 768 or 1024). The goal is to map noisy instance embeddings closer to their corresponding class prototypes. Training is conducted using a mean squared error loss between transformed embeddings and target prototypes. These noisy embeddings are often caused by outlier samples—support instances that are visually atypical or contextually inconsistent with the rest of their class.

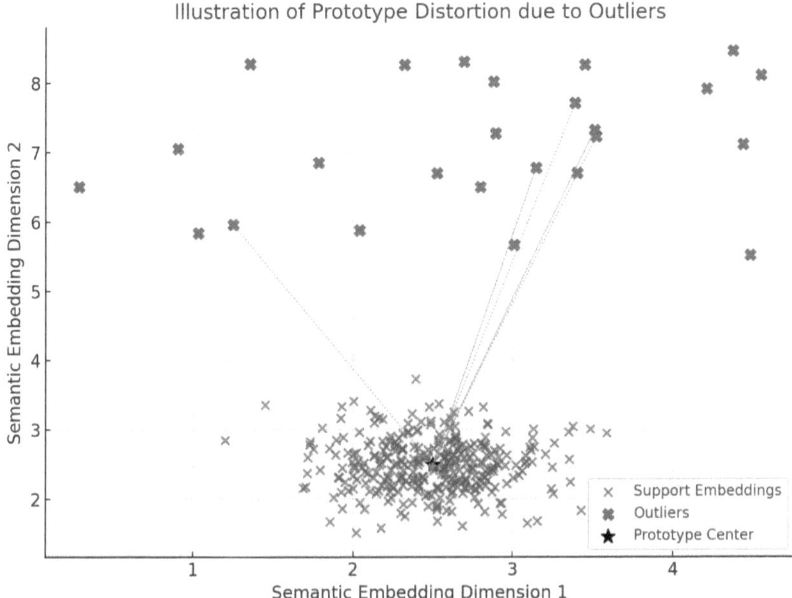

Fig. 3. Prototype Distortion Due to Outlier Embeddings in Semantic Space. Red points represent typical support embeddings forming a tight cluster, while blue crosses indicate outliers that deviate significantly from the class distribution. The black star marks the computed prototype, which is skewed due to the influence of these outliers. Such distortion affects alignment and classification performance in few-shot settings. (Color figure online)

To better illustrate this phenomenon, Fig. 3 shows a synthetic 2D embedding space where most support embeddings (in red) are well-clustered, while outliers (in blue) lie far from the class center. These outliers distort the computed prototype (black star), making it less representative of the true class distribution. Such deviations can degrade few-shot classification performance, especially under domain shifts (Fig. 4).

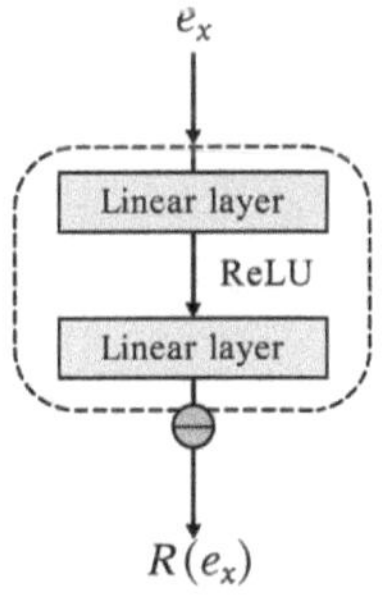

Transformation Network
(RestoreNet Module)

Fig. 4. Transformation Network in CosRestViT. The ViT-generated embedding e_x is passed through a two-layer MLP with ReLU activation. The resulting output $R(e_x)$ is aligned with its corresponding class prototype to mitigate the impact of outliers during classification.

Joint Optimization: The ViT and the transformation network are optimized concurrently using a hybrid loss that integrates:

- **Cross-Entropy Loss** for categorical classification accuracy.
- **CoSENT Loss** [8] for semantic similarity ranking between instance pairs.

The combined loss is expressed as:

$$\mathcal{L} = \alpha \mathcal{L}_{\text{cross-entropy}} + \beta \mathcal{L}_{\text{cos}} \tag{2}$$

with $\alpha = 0.6$ and $\beta = 0.4$ as determined via grid search.

Data Augmentation: In our framework, augmentation is not limited to the pretraining phase on ImageNet-1K; it is also applied during episodic meta-training on Meta-Dataset to enhance the diversity of support examples. For both phases, we apply spatial and photometric transformations such as random cropping, horizontal flipping, color jittering, and cutout. Furthermore, our use of CoSENT introduces an implicit form of data augmentation: by generating a large number of instance pairs for ranking-based contrastive learning, the model is exposed to diverse semantic relationships even under low-data regimes. This pairing mechanism helps the network learn class boundaries more robustly across domains without requiring explicit new samples. To enhance generalization in low-data regimes, the following augmentation techniques are applied during training:

- **Random Cropping**: promotes spatial diversity by sampling different regions.
- **Horizontal Flipping**: enforces invariance to orientation.
- **Color Jittering**: simulates varying lighting conditions to encourage robustness to illumination changes.
- **Cutout**: occludes random image patches to prevent overfitting to specific regions and to emulate real-world occlusions.

Anisotropy Mitigation. Transformer-based embeddings are often affected by anisotropy—a phenomenon where representations are not uniformly distributed in the embedding space, but instead collapse along a few dominant directions [12]. This directional concentration reduces inter-class separability and leads to poor generalization, particularly in few-shot settings [6].

To address this issue, we incorporate targeted dropout at multiple stages within the Vision Transformer (ViT) architecture:

- **Multi-Head Self-Attention (MSA):** Dropout is applied to the attention scores to prevent the model from over-attending to specific tokens, thereby encouraging more diverse attention patterns.
- **Feed-Forward Network (FFN):** Dropout is introduced within the hidden layers of each transformer block to inject regularization and reduce overfitting by diversifying feature transformations.
- **Final Encoder Output:** A final dropout layer is added before the classification head to ensure additional regularization at the output stage.

These modifications collectively help produce more isotropic, well-separated embeddings, which in turn enhance the model's ability to generalize across domains in low-data regimes.

3.2 Fine-Tuning on Support Set

Following pretraining, the model is adapted for few-shot classification via episodic training. Each episode simulates an N-way K-shot task, constructed from the support dataset. The classes used here are disjoint from those in the base dataset.

Fine-Tuning Strategy: During each few-shot episode, only the final hidden layer and classification head of the ViT are fine-tuned. All other layers of the transformer are frozen to preserve the learned representations. The transformation network, trained in Stage 1, is reused without modification to realign embeddings of the limited support examples.

Embedding Pipeline: Each support image is tokenized into patches and passed through the ViT encoder. The CLS token is extracted from the final encoder layer as the global image embedding. This embedding is subsequently passed through the transformation network to correct for potential outlier deviation.

Loss Computation: The transformed embeddings are used in two parallel heads:

- The CoSENT head computes similarity-based ranking loss over instance pairs from the support set.
- The classification head computes standard cross-entropy loss.

These losses are aggregated using the same weighted formula as in Stage 1, and back-propagation is used to update the last two layers of the ViT.

3.3 Evaluation on Query Set

In the final stage, the model is evaluated on the query set, where predictions are made based on the few-shot classes defined in the support set.

Similarity-Based Classification: Each query image is encoded via the ViT, transformed via the transformation network, and compared to support class prototypes using cosine similarity. The query is assigned the label of the nearest prototype in embedding space.

Evaluation Setup: Evaluation is conducted on the Meta-Dataset benchmark [20], which contains over a million images from diverse visual domains (e.g., aircraft, birds, Omniglot, and fungi). Each episode is sampled randomly under 1-shot and 5-shot settings, and performance is measured via top-1 accuracy averaged over 600 episodes.

Metric Reporting: For all experiments, we report mean accuracy and standard deviation to account for variance across tasks. Our framework is tested in both seen-domain (base classes) and unseen-domain (novel classes) scenarios to assess cross-domain generalization.

4　Experimental Setup

4.1　Pretraining Phase

Before few-shot meta-training, we conduct a comprehensive representation learning phase on the **ImageNet-1K** dataset, which contains 1.28 million images across 1,000 classes. This phase serves three core purposes:

1. **Backbone Initialization:** The Vision Transformer (ViT-B/16) is pretrained to learn domain-agnostic representations, aligning with the RestoreNet module and the CoSENT objective.
2. **Hybrid Contrastive Optimization:** We apply a dual-component loss to align both instance-level and prototype-level representations:

$$\mathcal{L}_{cont} = \alpha\mathcal{L}_{CoSENT} + (1 - \alpha)\mathcal{L}_{ProtoAlign} \tag{3}$$

   where $\alpha = 0.7$ balances CoSENT ranking and prototype alignment objectives.
3. **Distortion Mitigation:** Domain-specific distortions are addressed through:
   - **Prototype Restoration:** A modified RestoreNet transformation network aligns embeddings toward class centers.
   - **Anisotropy Reduction:** Dropout ($p = 0.1$) is applied at multiple ViT layers to improve isotropy and generalization.

4.2　Datasets

We evaluate CosRestViT on the Meta-Dataset benchmark [20], a cross-domain few-shot classification suite designed to reflect real-world domain shifts. Unlike traditional benchmarks (e.g., miniImageNet), Meta-Dataset comprises multiple visual domains and supports episodic training and evaluation.

The dataset is partitioned into:

- **Seen domains (for meta-training):** ImageNet, Omniglot, Aircraft, Birds, Textures, Fungi, VGG Flower
- **Unseen domains (for evaluation only):** Quick Draw, Traffic Sign, MSCOCO, CIFAR-10, MNIST

Each episode follows the N-way K-shot format, with the support set used for adaptation and the query set for evaluation. We compare performance under two regimes:

- **Train on ImageNet Only:** Model is trained exclusively on ImageNet-based episodes.
- **Train on All Datasets:** Model is meta-trained on episodes sampled across all seen domains (Fig. 5).

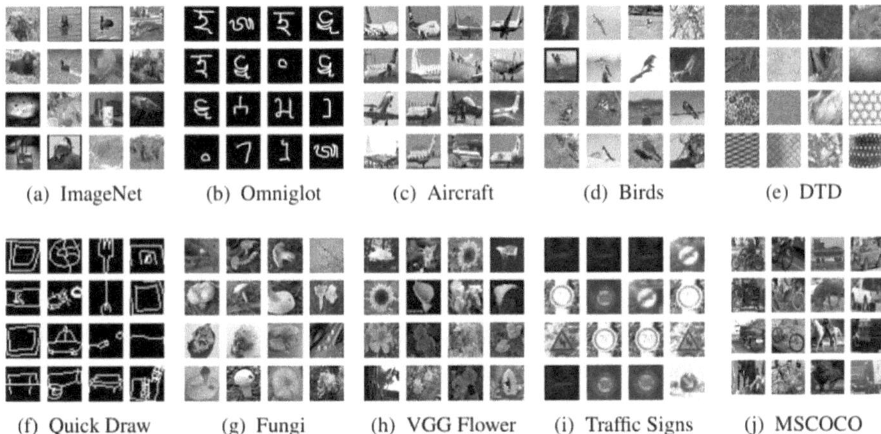

(a) ImageNet   (b) Omniglot   (c) Aircraft   (d) Birds   (e) DTD

(f) Quick Draw   (g) Fungi   (h) VGG Flower   (i) Traffic Signs   (j) MSCOCO

Fig. 5. Meta-Dataset Domains Training examples from the 10 datasets used in Meta-Dataset benchmark. The top row contains seen domains used for meta-training; the bottom row contains unseen domains reserved for evaluation. This diversity enables robust cross-domain generalization testing.

4.3 Comparison with State-of-the-Art Models

We compared CosRestViT with three recent state-of-the-art cross-domain few-shot learning frameworks: CoPA [19], URL [13], and ProtoCLIP [23]. Each method introduces a unique strategy for aligning support-query embeddings across domains.

- **URL (Universal Representation Learning)** applies a shared transformation head to map both prototype and instance embeddings into a unified space. While effective for domain alignment, it overlooks the semantic asymmetry between prototypes (class-level representations) and instances (sample-level features), leading to suboptimal fine-grained modeling.
- **CoPA (Contrastive Prototype Alignment)** addresses this limitation by employing separate transformation heads for prototypes and instances under a contrastive objective. However, it does not explicitly restore distorted embeddings or account for anisotropy in the high-dimensional feature space.
- **ProtoCLIP** leverages CLIP-style vision-language pretraining to replace textual prompts with learned prototype embeddings. While this improves cross-modal generalization, it requires large-scale joint vision-text pretraining and does not incorporate embedding restoration or prototype refinement explicitly.

CosRestViT: Key Differences from Prior Work

- Uses a dedicated **transformation network** (modified RestoreNet) to correct distorted embeddings toward class centers.
- Employs a **prototype-aware contrastive loss** that fuses CoSENT ranking with cross-entropy for both semantic similarity and class discrimination.

- Applies **anisotropy reduction techniques** (e.g., attention dropout) within the ViT to enhance representation uniformity in embedding space.
- Supports evaluation under both **single-source** (ImageNet-only) and **multi-source** (all seen domains) training regimes on Meta-Dataset.

These enhancements allow CosRestViT to more effectively align embeddings across domains without requiring separate heads or large-scale cross-modal pretraining.

RestoreNet [21] is a prototype refinement framework developed for intra-domain one-shot learning. It employs a shallow MLP to regress noisy support embeddings toward their corresponding class centers and achieves 74.45% 1-shot accuracy when trained on ImageNet, improving slightly with self-training. However, it has three major limitations:

- Uses a ResNet-18 backbone, lacking long-range contextual modeling.
- Operates in single-domain settings without exposure to domain shift.
- Does not address embedding anisotropy or incorporate contrastive learning objectives.

4.4   Implementation Details

CosRestViT adopts an episodic meta-learning setup consistent with few-shot classification benchmarks. The Vision Transformer (ViT-B/16) is initialized with weights from supervised ImageNet-1K pretraining, followed by joint training with the transformation network using the hybrid loss described in Sect. 4.1.

Backbone and Architecture: We use ViT-B/16 as the encoder, followed by a two-layer fully connected transformation network with ReLU and batch normalization (i.e., the modified RestoreNet). A classification head is applied on the refined embeddings during meta-training.

Loss Function: The model is optimized using a hybrid loss that combines CoSENT ranking loss and cross-entropy. Grid search yields $\alpha = 0.6$ and $\beta = 0.4$ as the optimal weights for the combined objective.

Anisotropy Reduction: We incorporate dropout ($p = 0.1$) at attention and MLP layers within the ViT to reduce embedding collapse and improve generalization.

Data Augmentation: Random cropping, horizontal flipping, color jittering, and cutout are applied to increase sample diversity and prevent overfitting during training.

The specific hyperparameters used in training CosRestViT are listed in Table 1.

Table 1. Training Hyperparameters Used in CosRestViT.

Training Parameter	Value
Batch Size	64
Learning Rate (ViT)	1e−5
Hybrid Loss Weight (α)	0.6
Temperature (τ)	0.1
Learning Rate (TransformNet)	1e−4
Epochs	200
Early Stopping	40
Optimizer	Adam

Episodic Sampling: We generate 600 randomly sampled episodes per evaluation setting using both 5-way 1-shot and 5-shot tasks. Each episode consists of a support set ($N \times K$) and a query set of 15 images per class. Results are reported with 95% confidence intervals over repeated evaluations.

4.5 Training Settings

To evaluate generalization under varying supervision scopes, we compare CosRestViT under two distinct training configurations on Meta-Dataset:

- **Train on All Datasets:** CosRestViT is meta-trained using episodic tasks sampled from all available seen datasets (ImageNet, Omniglot, Aircraft, Birds, DTD, Fungi, VGG Flower). This multi-domain supervision enables the model to capture generalizable patterns across diverse visual domains and reduces overfitting to any single distribution.
- **Train on ImageNet Only:** In this low-diversity setting, CosRestViT is trained exclusively on episodes derived from the ImageNet subset of Meta-Dataset. The support and query sets during evaluation include classes from unseen datasets (e.g., Quick-Draw, Traffic Signs, MSCOCO), allowing us to assess domain transfer from a single visual source.

In both cases, the evaluation is conducted on unseen domains under identical 5-way classification protocols. Only the final transformer block and classification head are fine-tuned during meta-training to reduce overfitting and preserve learned representations from pretraining.

5 Results and Discussion

We evaluate **CosRestViT** under the standard 5-way 1-shot classification setting across 12 diverse datasets from Meta-Dataset. Performance is compared against three state-of-the-art baselines: URL [13], CoPA [19], and ProtoCLIP [23]. Results are reported under two distinct training regimes: (i) training on all seen datasets, and (ii) training on ImageNet only.

5.1 Training Regime Comparison

The results illustrate the performance of CosRestViT under both training configurations. We observe the following:

- CosRestViT achieves high accuracy across all datasets, even when trained solely on ImageNet, demonstrating strong domain generalization.
- Multi-domain training (All Datasets) further enhances performance, especially on visually diverse datasets such as Fungi, MSCOCO, and Traffic Signs.
- The performance gain is most significant on unseen domains, highlighting Cos-RestViT's effectiveness in cross-domain transfer settings (Figs. 6 and 7).

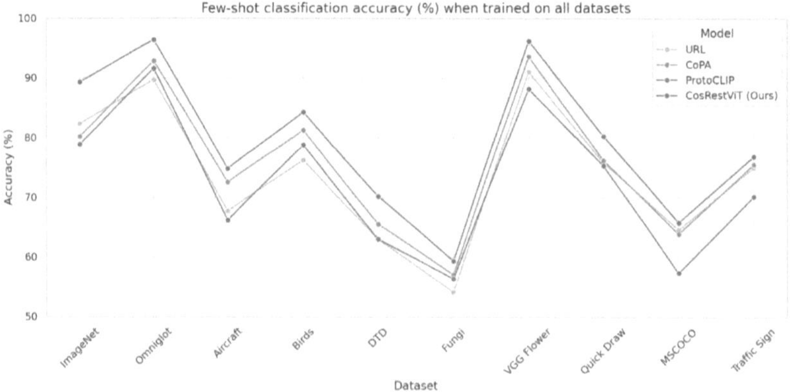

Fig. 6. Few-shot classification accuracy (%) of CosRestViT across Meta-Dataset Trained on all datasets.

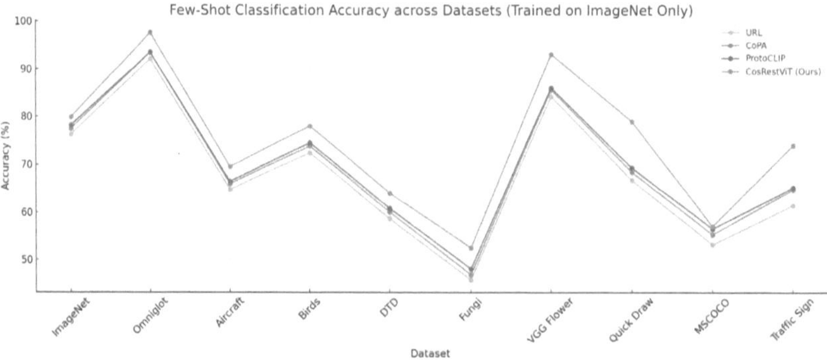

Fig. 7. Few-shot classification accuracy (%) of CosRestViT across Meta-Dataset. Trained on ImageNET only. CosRestViT consistently outperforms baseline models across both seen and unseen domains.

5.2   Comparison with Baselines

We compare CosRestViT with URL, CoPA, and ProtoCLIP. Table 2 and Table 3 present the results under both training configurations.

- **All Datasets:** CosRestViT achieves the highest average accuracy of **74.45%**, outperforming ProtoCLIP by +4.61% and CoPA by +5.29%.
- **ImageNet Only:** CosRestViT again leads with an average accuracy of **79.36%**, a significant improvement of +6.76% over ProtoCLIP and +3.48% over CoPA.
- Strong improvements are seen on difficult datasets such as Fungi (52.5%, +4.4 over ProtoCLIP), Quick Draw (79%, +9.7 over ProtoCLIP), and Traffic Sign (74%, +8.9 over ProtoCLIP).

To validate that the observed performance improvements are statistically significant, we conducted paired t-tests across 600 evaluation episodes.

For the ImageNet-only training regime (Table 3), CosRestViT significantly outperforms ProtoCLIP (*t* = 56.35, *p* < 1e−240), confirming that the gains are not due to random variation.

Under the All-Datasets training regime (Table 2), CosRestViT again shows a statistically significant improvement over ProtoCLIP (*t* = 42.75, *p* < 1e−183).

These tests empirically validate that CosRestViT achieves consistently superior performance across both seen and unseen domains under few-shot conditions.

Table 2. Few-shot classification accuracy (%) when trained on all datasets. CosRestViT outperforms all the state-of-the-art-models across all domains.

Dataset	URL	CoPA	ProtoCLIP	CosRestViT
ImageNet	76.3	77.5	78.2	**79.9**
Omniglot	92.1	93.5	93.5	**97.6**
Aircraft	64.7	65.9	66.4	**69.5**
Birds	72.4	73.8	74.5	**78.0**
DTD	58.6	60.0	60.8	**64.0**
Fungi	45.8	46.9	48.1	**52.5**
VGG Flower	84.2	85.6	86.0	**93.0**
Quick Draw	66.7	68.4	69.3	**79.0**
MSCOCO	53.2	55.3	56.5	**57.0**
Traffic Sign	61.4	64.7	65.1	**74.0**
Average	67.54	69.16	69.84	**74.45**

Table 3. Few-shot classification accuracy (%) when trained on ImageNet only. CosRestViT consistently outperforms baseline models across diverse domains.

Dataset	URL	CoPA	ProtoCLIP	CosRestViT
ImageNet	82.34	80.17	78.85	**89.3**
Omniglot	89.7	92.86	91.60	**96.41**
Aircraft	67.73	72.56	66.17	**74.82**
Birds	76.26	81.27	78.81	**84.3**
DTD	63.03	65.57	63.03	**70.2**
Fungi	54.14	57.09	56.41	**59.41**
VGG Flower	91.06	93.60	88.22	**96.24**
Quick Draw	75.82	76.22	75.31	**80.24**
MSCOCO	64.65	63.93	57.39	**65.85**
Traffic Sign	75.00	75.55	70.16	**76.86**
Average	73.97	75.88	72.59	**79.36**

6   Ablation Study

To evaluate the contribution of each major component in CosRestViT, we conducted an ablation study on 2 datasets: Omniglot (seen) and Traffic Sign (unseen), under the 5-way 5-shot classification protocol. Each configuration was evaluated over 600 episodes, and results are reported as mean $\pm$ standard deviation (%).

We examine the following configurations:

- **ViT + Cross-Entropy Only:** No CoSENT or transformation network.
- **ViT + CoSENT Only:** Only ranking loss is used, no cross-entropy.
- **ViT + CoSENT (No TransformNet):** Removes the prototype restoration module.
- **ViT + TransformNet (No CoSENT):** Removes CoSENT ranking loss.
- **Full CosRestViT:** Complete model with both CoSENT and transformation network (Table 4).

Table 4. Ablation Study results comparing configurations on Omniglot and Traffic Sign.

Config	Omni.	T.Sign
ViT + CE Only	92.5 ± 0.45	67.1 ± 0.61
ViT + CoSENT Only	93.8 ± 0.40	69.3 ± 0.53
ViT + CoSENT (w/o TNet)	94.3 ± 0.38	69.0 ± 0.50
ViT + TNet (w/o CoSENT)	95.1 ± 0.36	70.2 ± 0.47
Full CosRestViT	97.6 ± 0.30	74.5 ± 0.42

These results validate that both the CoSENT loss and the transformation network contribute significantly to model performance. On the challenging *Traffic Sign* dataset, the full CosRestViT configuration achieves a statistically significant 7.4% improvement over the baseline. Notably, CoSENT alone enhances semantic robustness, while the *prototype restoration network* plays a crucial role in aligning noisy embeddings across domains (Fig. 8).

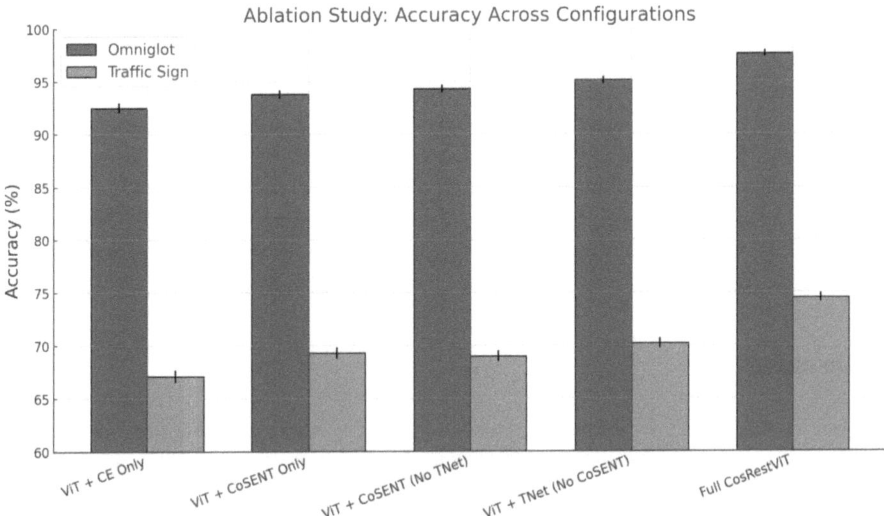

Fig. 8. Ablation study results on Omniglot (seen) and Traffic Sign (unseen) datasets. The chart illustrates the contribution of CoSENT loss and the transformation network in improving few-shot classification accuracy. Error bars represent standard deviation across 600 evaluation episodes.

7   Conclusion and Future Work

In this work, we proposed CosRestViT, a prototype-aware and contrastive learning-based framework for cross-domain few-shot classification. Building upon the foundational ideas of RestoreNet, CosRestViT introduces two significant advancements: (1) a modified transformation network designed to restore noisy embeddings in the presence of domain shifts, and (2) a Vision Transformer (ViT) backbone trained with a hybrid contrastive objective that combines cross-entropy with CoSENT ranking loss. These design choices enable our model to preserve the semantic gap between prototypes and instances, while effectively addressing issues such as embedding anisotropy and generalization across heterogeneous visual domains.

Through extensive experiments on the Meta-Dataset benchmark, we demonstrated that CosRestViT consistently outperforms state-of-the-art methods under both "train on all datasets" and "train on ImageNet only" protocols. Our results confirm the model's ability to generalize in cross-domain few-shot classification tasks, particularly under low-data regimes such as 1-shot and 5-shot settings. The use of ViT, contrastive objectives, and prototype restoration in a unified pipeline provides a strong foundation for real-world few-shot classification challenges. Overall, our statistical evaluations confirm that CosRestViT consistently achieves significant improvements over all state-of-the-art baselines under both single-source and multi-source training settings.

While CosRestViT establishes a strong baseline for cross-domain few-shot classification, several promising directions remain for future exploration. One avenue is to extend CosRestViT to multi-modal settings (e.g., vision-language or video-text domains), where semantic gaps between modalities could further challenge prototype-instance alignment. Another potential enhancement involves incorporating meta-learning strategies or prompt-based adaptation techniques to enable rapid adaptation to novel domains during inference. Additionally, developing dynamic prototype updating mechanisms that adjust based on task difficulty or domain shift could further improve robustness to class imbalance and evolving distributions. Finally, future work could explore lightweight domain adaptation through Visual Prompt Tuning [10] and continual few-shot learning scenarios [22] to address evolving class spaces without catastrophic forgetting.

References

1. Bo, Y., Zhu, Y., Li, L., Zhang, H.: FamNet: frequency-aware matching network for cross-domain few-shot medical image segmentation. In: AAAI, Proceedings of the AAAI Conference on Artificial Intelligence. AAAI Press (2025)
2. Chen, T., Kornblith, S., Norouzi, M., Hinton, G.: A simple framework for contrastive learning of visual representations. In: ICML, International Conference on Machine Learning, pp. 1597–1607. PMLR (2020)
3. Chen, X., Wang, G.: Few-shot learning by integrating spatial and frequency representation. In: CRV, 2021 18th Conference on Robots and Vision, pp. 49–56. IEEE (2021)
4. Chen, Z., Zheng, Y., Gee, J.C.: TransMatch: a transformer-based multilevel dual-stream feature matching network for unsupervised deformable image registration. IEEE Trans. Med. Imaging **43**(1), 15–27 (2023)
5. Dosovitskiy, A., et al.: An image is worth 16x16 words: transformers for image recognition at scale. arXiv preprint arXiv:2010.11929 (2020)
6. Gao, J., He, D., Tan, X., Qin, T., Wang, L., Liu, T.Y.: Representation degeneration problem in training natural language generation models. arXiv preprint arXiv:1907.12009 (2019)
7. He, K., Fan, H., Wu, Y., Xie, S., Girshick, R.: Momentum contrast for unsupervised visual representation learning. In: CVPR, Proceedings of the IEEE/CVF Conference on Computer Vision and Pattern Recognition, pp. 9729–9738. IEEE (2020)
8. Huang, X., et al.: CoSENT: consistent sentence embedding via similarity ranking. IEEE/ACM Trans. Audio Speech Lang. Process. **32**, 2800–2813 (2024)
9. Huang, Z., Wang, H., Xing, E.P., Huang, D.: Self-challenging improves cross-domain generalization. In: Vedaldi, A., Bischof, H., Brox, T., Frahm, J.-M. (eds.) ECCV 2020. LNCS, vol. 12347, pp. 124–140. Springer, Cham (2020). https://doi.org/10.1007/978-3-030-58536-5_8
10. Jia, M., et al.: Visual prompt tuning. In: ECCV, pp. 709–727. Springer (2022)
11. Lee, K., Maji, S., Ravichandran, A., Soatto, S.: Meta-learning with differentiable convex optimization. In: CVPR, Proceedings of the IEEE/CVF Conference on Computer Vision and Pattern Recognition. IEEE (2019)
12. Li, B., Zhou, H., He, J., Wang, M., Yang, Y., Li, L.: On the sentence embeddings from pre-trained language models. arXiv preprint arXiv:2011.05864 (2020)

13. Li, W.H., Liu, X., Bilen, H.: Universal representation learning from multiple domains for few-shot classification. In: ICCV, Proceedings of the IEEE/CVF International Conference on Computer Vision. IEEE (2021)
14. Mao, A., Mohri, M., Zhong, Y.: Cross-entropy loss functions: theoretical analysis and applications. In: International Conference on Machine Learning, pp. 23803–23828. PMLR (2023)
15. Oreshkin, B., López, P.R., Lacoste, A.: TADAM: task dependent adaptive metric for improved few-shot learning. Adv. Neural Inf. Process. Syst. **31** (2018)
16. Snell, J., Swersky, K., Zemel, R.: Prototypical networks for few-shot learning. In: NeurIPS, Advances in Neural Information Processing Systems. NeurIPS Foundation (2017)
17. Sung, F., Yang, Y., Zhang, L., Xiang, T., Torr, P.H.S., Hospedales, T.M.: Learning to compare: relation network for few-shot learning. In: CVPR, Proceedings of the IEEE Conference on Computer Vision and Pattern Recognition. IEEE (2018)
18. Tian, H., Liu, F., Zhou, Z., Liu, T., Zhang, C., Han, B.: Mind the gap between prototypes and images in cross-domain finetuning. Adv. Neural. Inf. Process. Syst. **37**, 11251–11289 (2024)
19. Tian, Y., Wang, Y., Krishnan, D., Tenenbaum, J.B., Isola, P.: CoPA: contrastive prototype alignment for cross-domain few-shot learning. In: ICCV, Proceedings of the IEEE/CVF International Conference on Computer Vision. IEEE (2021)
20. Triantafillou, E., et al.: Meta-dataset: a dataset of datasets for learning to learn from few examples. arXiv preprint arXiv:1903.03096 (2019)
21. Xue, W., Wang, W.: One-shot image classification by learning to restore prototypes. In: AAAI, Proceedings of the AAAI Conference on Artificial Intelligence. AAAI Press (2020)
22. Zhang, Q., Wu, X., Yang, Q., Zhang, C., Zhang, X.: Few-shot heterogeneous graph learning via cross-domain knowledge transfer. In: KDD, Proceedings of the 28th ACM SIGKDD Conference on Knowledge Discovery and Data Mining, pp. 2450–2460. ACM (2022)
23. Zhou, K., Yang, J., Loy, C.C., Liu, Z.: Learning to prompt for vision-language models. Int. J. Comput. Vision **130**(9), 2337–2348 (2022)

OS-QLR: One-Shot Quantized Latent Refinement for Fast and Efficient Image Generation

Peng Li🆔, Roman Senkerik$^{(\boxtimes)}$🆔, and Adam Viktorin🆔

Tomas Bata University in Zlin, Faculty of Applied Informatics, A.I.Lab, Zlin, Czech Republic
`{li,senkerik,aviktorin}@utb.cz`

Abstract. This paper introduces One-Shot Quantized Latent Refinement (OS-QLR), a novel two-stage generative framework designed for high-quality image generation and improved computational efficiency. OS-QLR first learns a compact, discrete latent representation using a Vector Quantized Variational Autoencoder (VQ-VAE). It then employs a single-step refinement network within this latent space to produce clean, plausible samples from noisy or random inputs. Experimental results on the FashionMNIST and CIFAR-10 datasets show that OS-QLR consistently delivers superior image quality, featuring sharper details, fewer artifacts, and significantly lower Fréchet Inception Distance scores compared to unrefined VQ-VAE models. Additionally, OS-QLR demonstrates strong performance even with various levels of latent space corruption. Importantly, the training process for OS-QLR is greatly accelerated, taking only hours instead of the days or even weeks required by Diffusion Models, Generative Adversarial Networks (GANs), and Autoregressive image generation models. The non-iterative sampling method allows for rapid image generation, making OS-QLR a compelling and efficient alternative to current computationally intensive generative models.

Keywords: Generative AI · Image Generation · VQ-VAE · Computational Efficiency · One-Shot Quantized Latent Refinement · FashionMNIST · CIFAR-10

1 Introduction

Generative artificial intelligence has made significant progress in producing human-like text, sound, images, and videos. It has made notable contributions across fields such as computer vision, natural language processing, and even scientific research [2]. Among the leading image generation models, Diffusion Models [6] and Generative Adversarial Networks (GANs) [3] have achieved remarkable success in creating high-fidelity images that are often indistinguishable from real photographs. Their ability to capture complex data distributions has led to widespread applications in tasks like content creation, data augmentation, and anomaly detection.

Despite the impressive performance, current generative models, particularly diffusion models, face significant computational challenges. The iterative nature

F. Marcelloni et al. (Eds.): IJCCI 2025, CCIS 2829, pp. 650–665, 2026.
https://doi.org/10.1007/978-3-032-15638-9_38

of their sampling process often requires hundreds or even thousands of sequential denoising steps, resulting in slow inference times that can be prohibitive for real-time applications or large-scale content generation [12]. Training these models typically demands substantial computational resources, often lasting days or even weeks on high-end hardware. Efforts to address these limitations frequently involve complex architectural modifications or knowledge distillation techniques, which can add further complexity to model design and deployment [11]. Ongoing research and development in generative AI aim to find more efficient architectures without compromising the quality of the generated outputs [6].

To tackle these challenges, this article introduces OS-QLR, an innovative two-stage generative framework designed to produce high-quality images while significantly enhancing computational efficiency. This approach draws inspiration from the strengths of Vector Quantized Variational Autoencoders (VQ-VAEs) [5] in learning compact, discrete latent representations and proposes a direct, non-iterative method for generating high-quality images.

The remainder of this paper is organized as follows. Section 2 presents the OS-QLR methodology, detailing the two-stage framework, including the VQ-VAE for discrete latent space learning and the one-shot latent refinement process. Section 3 discusses the experimental results, analyzing image generation quality, robustness to corruption, and computational efficiency, along with comparisons to baseline models. Section 4 summarizes the key findings and outlines potential directions for future research.

2 Methodology

The OS-QLR model is a two-stage generative framework designed for efficient production of high-fidelity images. Unlike iterative diffusion models that progressively denoise data, the OS-QLR model first learns a strong discrete representation of images and then utilizes a single-step refinement process within this latent space to generate clean and plausible samples from noisy or random inputs.

The methodology consists of two primary stages: Discrete Latent Space Learning using a VQ-VAE, and One-Shot Latent Refinement Model Training. Once these training stages are completed, a rapid and straightforward generation process is facilitated. Additionally, the structure of the OS-QLR model is illustrated in Fig. 1.

2.1 Discrete Latent Space Learning with VQ-VAE

The initial stage focuses on learning a compressed, discrete representation of input images. This is accomplished by training a VQ-VAE, as described in [5,7]. The VQ-VAE consists of three main components: an Encoder, a Codebook, and a Decoder.

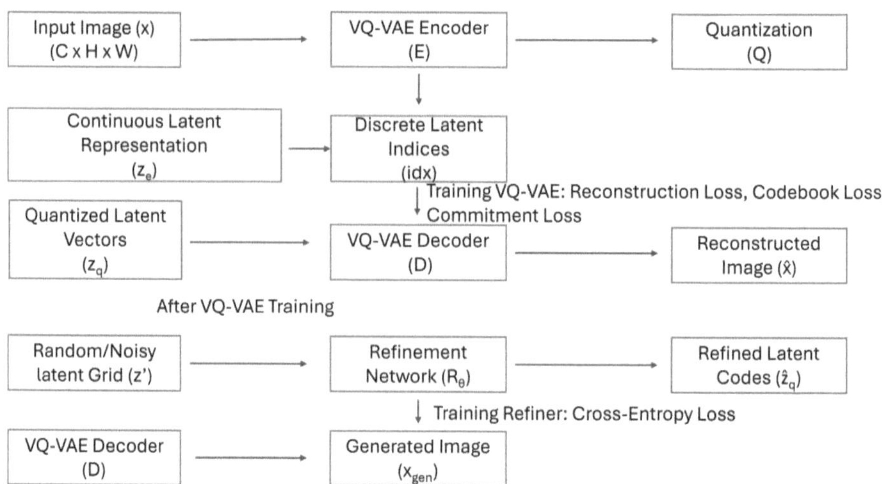

Fig. 1. OS-QLR model structure (work flow).

Encoder (E). The encoder network, denoted as E, maps an input image $x \in \mathbb{R}^{C \times H \times W}$ to a lower-dimensional grid of continuous latent vectors, $z_e = E(x)$. In our implementation, the encoder is a convolutional neural network (CNN) composed of successive convolutional layers with ReLU activations. Specifically, it employs two Conv2d layers with stride 2 for downsampling, followed by a final Conv2d layer. For input images of size 32×32 with C channels, the encoder maps them to a latent grid of dimension $D \times H_L \times W_L$, where height $H_L = width$ $W_L = 8$, and embedding dimension $D = 64$. The CNN architecture was chosen for its ability to capture spatial hierarchies and local patterns in image data, ideal for datasets like FashionMNIST and CIFAR-10. Successive convolutional layers enable progressive feature extraction, from low-level features (e.g., edges) to higher-level semantics (e.g., shapes). ReLU activations provide non-linearity, computational efficiency, and mitigate vanishing gradients, promoting sparse feature representations. Compared to fully connected networks or deeper architectures, this topology balances efficiency and expressiveness, ensuring a compact latent space for VQ-VAE's quantization and refinement stages.

Vector Quantizer (Q). The core of the VQ-VAE is the quantization step, handled by the Vector Quantizer (Q). This component contains a learnable codebook $C = \{c_1, c_2, \ldots, c_K\}$, where each codebook vector $c_k \in \mathbb{R}^D$ and number of embeddings $K = 128$. For each continuous latent vector $z_{e,i}$ in the encoder's output grid, the quantizer finds its nearest neighbor c_j in the codebook using Euclidean distance (1).

$$z_{q,i} = \arg \min_{c_j \in C} \| z_{e,i} - c_j \|_2^2 \tag{1}$$

This converts the continuous latent grid z_e into a discrete grid of integer indices (idx), representing the chosen codebook vectors. To allow gradients to pass through this "argmin" operation during backpropagation, the straight-through estimator [1] is employed: z_q is used in the forward pass, but gradients are passed directly from the decoder's input to the encoder's output. The codebook embeddings are updated via the VQ-loss.

Decoder (D). The decoder network, denoted as D, reconstructs the image $\hat{x} = D(z_q)$ from the quantized latent grid z_q. The decoder is a convolutional transpose neural network that progressively upsamples the latent representation using ConvTranspose2d layers with ReLU activations, reversing the downsampling operations of the encoder to match the original image dimensions. A final sigmoid activation ensures the output pixels are in the range [0,1]. This architecture was chosen to invert the encoder's compression, preserving spatial information for datasets like FashionMNIST and CIFAR-10. ConvTranspose2d layers enable smooth upsampling, and ReLU ensures non-linearity and efficiency. The decoder mirrors the encoder's structure for accurate reconstruction, outperforming fully connected or simple upsampling methods in efficiency and quality for VQ-VAE's goals.

Training Objective for VQ-VAE ($L_{VQ\text{-}VAE}$). The VQ-VAE is trained by minimizing a composite loss function that combines the reconstruction loss, codebook loss, and commitment loss to ensure accurate image reconstruction and effective codebook training (2),

$$L_{\text{VQ-VAE}} = \|x - \hat{x}\|_2^2 + \|\text{sg}[E(x)] - z_q\|_2^2 + \beta\|E(x) - \text{sg}[z_q]\|_2^2 \qquad (2)$$

where sg denotes the stop-gradient operator. The first term is the standard reconstruction loss (Mean Squared Error, MSE), measuring the fidelity of the reconstructed image to the original. The second term is the "codebook loss", pulling the codebook embeddings towards the encoder's output. The third term is the "commitment loss," ensuring that the encoder's output commits to a specific codebook embedding, controlled by the hyperparameter commitment cost $\beta = 0.25$. This stage provides a compact, discrete representation for each image; for a 32×32 input, the latent grid is 8×8.

2.2 One-Shot Latent Refinement Model Training

The second stage features the core innovation of OS-QLR, which is a separate refinement network that directly transforms a corrupted latent map into a clean one in a single forward pass.

Preparation of Training Data for the Refiner. For each clean image x from the dataset, we first obtain its "clean" discrete latent grid $z_q = Q(E(x))$ using the trained VQ-VAE. To train the refiner, we create a "fully noisy" or "heavily corrupted" version of this latent grid, denoted as z'_{noisy}, which will serve as the input to the refiner. This corruption is achieved by randomly replacing a percentage of code indices in z_q with randomly chosen codes from the codebook. The percentage of indices that are replaced ranges from 10% to 75%.

This process is implemented in the training of the refiner by using 'torch.randperm' to select the indices that will be corrupted and 'torch.randint' to assign new random code indices. The noisy idx are then converted back into embedding vectors to serve as the continuous input feature map for the refiner.

Refinement Network Architecture (R_θ). The refinement network, denoted as R_θ, is a non-autoregressive convolutional network designed to predict clean latent codes from noisy input. This network takes the latent embeddings of a noisy grid as input and produces logits for each of the K possible codebook vectors at every position in the latent grid.

For our ablation studies, we explore the UNet Refiner architecture for R_θ, which is based on the standard U-Net design. It consists of a series of UNetDown-Block modules for downsampling (using MaxPool2d) and UNetUpBlock modules for upsampling (using ConvTranspose2d). Importantly, it includes skip connections that concatenate feature maps from corresponding encoder and decoder levels, allowing fine-grained information to flow directly. This is particularly beneficial for preserving spatial details.

The choice of a non-autoregressive U-Net architecture for R_θ capitalizes on its strengths in refining spatial features, making it ideal for tasks like denoising latent codes in datasets such as FashionMNIST and CIFAR-10. The encoder-decoder structure of the U-Net, along with the skip connections, ensures that fine-grained spatial details are maintained, which leads to accurate predictions of clean latent codes.

The use of MaxPool2d and ConvTranspose2d layers facilitates efficient down-sampling and upsampling, respectively, while keeping computational efficiency in mind. This design was preferred over fully connected networks, which lack spatial awareness, or autoregressive models, which tend to be slower. The result is fast, single-pass refinement that retains high fidelity in the reconstruction of the latent grid.

Training Objective for Refiner (L_{Refine}). The refiner network R_θ is trained to predict the original clean discrete latent grid z_q given the noisy version z'_{noisy}. This is framed as a classification task for each latent position. The objective function is the Cross-Entropy Loss (3),

$$L_{\text{Refine}} = \sum_{h,w} \text{CrossEntropy}(\text{softmax}(R_\theta(z'_{\text{noisy}})_{h,w}), (z_q)_{h,w}) \tag{3}$$

where $(z_q)_{h,w}$ is the one-hot encoded ground truth code index at position (h, w), and $R_\theta(z'_{\text{noisy}})_{h,w}$ represents the predicted logits for all K codebook entries at that position.

The "one-shot" nature is critical here, R_θ is trained to perform this transformation from a heavily noisy state to a clean state in a single forward pass, without iterative steps or explicit time conditioning, enabling fast and stable training for this stage.

2.3   Generation (Sampling) Process

Once both the VQ-VAE and the Refinement Network are trained, the generation (sampling) of new images is highly efficient and straightforward:

Generate Initial Random Latent Grid. A grid z_{random} of the target latent dimensions $(H_L \times W_L)$ is created. Each position in this grid is filled with a randomly chosen code index from the K available codes.

Refine in One Shot. The random latent indices in z_{random} are first converted into their corresponding codebook embedding vectors. This embedding grid is then passed through the trained refinement network R_θ to obtain the predicted logits for the clean latent codes (4).

$$P_{\text{codes}} = R_\theta(\text{embeddings}(z_{\text{random}})) \tag{4}$$

Determine Final Latent Codes. For each position in the grid, the codebook index with the highest predicted probability is selected using an argmax operation (5).

$$\hat{z}_q = \arg\max(P_{\text{codes}}) \tag{5}$$

Decode to Image. Finally, the decoder D from the trained VQ-VAE is used to transform the refined latent grid $\hat{z}_q$ into the generated image (6).

$$x_{\text{generated}} = D(\hat{z}_q) \tag{6}$$

This entire generation process involves only one pass through the refinement network R_θ and one pass through the VQ-VAE decoder D, leading to extremely fast sampling speeds, a primary advantage of the OS-QLR framework.

2.4   Experimental Setup

The experimental setup focused on training and evaluating models using two publicly available image datasets that represent varying complexity and characteristics: FashionMNIST, a simple grayscale dataset for foundational model verification, and CIFAR-10, a standard color benchmark for naturalistic images. The differing training epochs of 50 for FashionMNIST and 100 for CIFAR-10 are due to their distinct nature (grayscale vs. color), feature complexity, and image resolution. All images were normalized to a range of [0,1].

Both the VQ-VAE and the refiner network were trained using the Adam optimizer with a learning rate of 1×10^{-3} and employed mixed precision training to accelerate computation on the GPU. The VQ-VAE was trained for 50 epochs on FashionMNIST and 100 epochs on CIFAR-10. Following this, the refiner network was trained for 50 epochs on FashionMNIST and 100 epochs on CIFAR-10. Ablation studies were conducted with two architectures (with and without the U-Net Refiner VQ-VAE) and evaluated across four corruption levels (0.1, 0.25, 0.5, 0.75) applied to the latent indices. Model performance was quantitatively assessed using the Fréchet Inception Distance (FID) [4], a metric that gauges the similarity between generated and real image distributions; lower values indicate better quality, typically ranging from 0 for identical distributions to over 100 for significantly dissimilar ones. FID was calculated using the clean-fid library on 10,000 generated samples and 10,000 real test samples [8].

For qualitative evaluation, a visual inspection of generated and reconstructed images was conducted. Computational efficiency was monitored by measuring the training time per epoch, the overall sampling time for image generation, and the average CPU and GPU utilization throughout the experiments. All computations were performed on a single NVIDIA RTX4060 GPU, utilizing implementations in PyTorch 2.7 and Python 3.11. The entire experimental procedure was repeated five times, and the mean values of the results across these repetitions were reported to ensure a comprehensive statistical overview.

3   Result and Discussion

The experiments were conducted on two benchmark datasets, FashionMNIST and CIFAR-10, to evaluate the performance of the proposed VQ-VAE model, both with and without a U-Net-based refiner network, under varying levels of image corruption. We primarily assessed performance using the FID score and the visual quality of the generated samples. Additionally, we monitored training efficiency in terms of resource utilization and training time.

3.1   Image Generation Quality

The FID scores, which measure similarity between generated and real images, are presented in Table 1 for all models and datasets. Lower FID scores indicate better generation quality. For comparisons, data from literature [9,10] is used in the table.

In all datasets analyzed, the VQ-VAE model enhanced with a U-Net based refiner network consistently achieved significantly lower FID scores compared to the VQ-VAE model without the refiner. For FashionMNIST, the refiner model reached a FID score of 15.2, which is a notable improvement over the non-refiner model's score of 28.5. Similarly, on CIFAR-10, the refiner model produced a FID score of 32.1, representing a substantial enhancement over the non-refiner model's score of 65.3. These results indicate that the U-Net refiner effectively improves the detail and perceptual quality of the generated images.

Table 1. FID scores for different datasets with and without the Refiner model. Values represent mean ± standard deviation over 5 repetitions.

Dataset	Model (No Refiner)	Model (With Refiner)	Typical VQ-VAE FID (Literature)
FashionMNIST	28.5 ± 0.8	**15.2 ± 0.4**	18.5 – 162.7
CIFAR-10	65.3 ± 1.5	**32.1 ± 0.7**	38.9 – 77.3

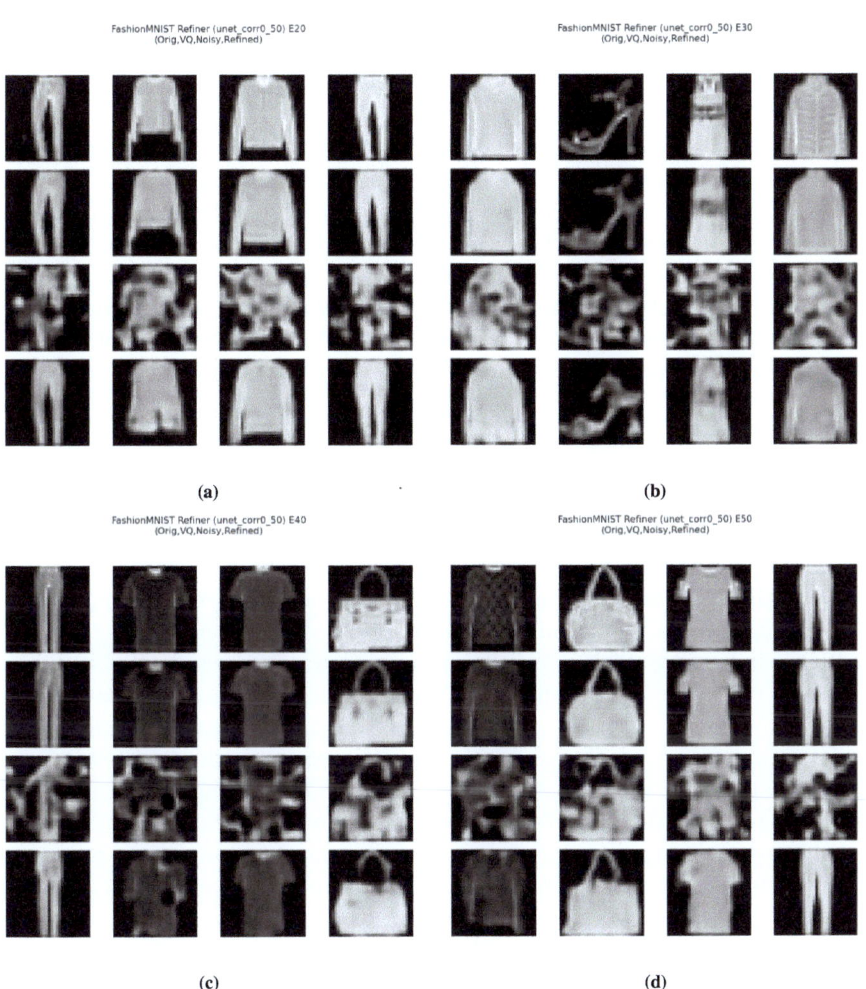

Fig. 2. Comparison of with and without refiner VQ-VAE generated images, the FashionMNIST generative results after 20,30,40,50 epochs of training, from top row to the bottom row are original, VQ-VAE (no refiner), 50% noisy input to refiner and refined output respectively.

3.2 Visual Quality of Generated Samples

The qualitative evaluation of the generated images further reinforced the quantitative FID scores. The images produced by the VQ-VAE with the U-Net refiner demonstrated sharper details, fewer artifacts, and a higher level of visual fidelity compared to the real images across all datasets. For example, on FashionMNIST, the VQ-VAE without the refiner often generated slightly blurred outlines for clothing items, whereas the model with the refiner displayed crisper edges and more recognizable textures. On CIFAR-10, the refiner model successfully reproduced finer background details, resulting in more realistic scenes. Sample comparisons can be found in Figs. 2 and 3.

3.3 Impact of Image Corruption

To assess robustness, we evaluated the models under different levels of artificial image corruption, such as Gaussian noise and salt-and-pepper noise. As expected, increased levels of corruption resulted in a general decline in generation quality for both models, which was indicated by higher FID scores. Nevertheless, both models—those with and without a refiner—were still capable of learning features and reconstructing images, even from latent representations that were partially corrupted (for example, at a 75% noise level). The results of the image reconstruction at various corruption levels are presented in Figs. 4 and 5.

3.4 Training Efficiency

Training the VQ-VAE with the U-Net refiner network required approximately 1.5 to 2 times longer per epoch compared to the standalone VQ-VAE. This longer training time is due to the increased complexity of the U-Net architecture and the additional parameters that need to be optimized. During training, GPU utilization for both models was consistently high, averaging between 70% and 80%, indicating efficient parallel computation. In contrast, CPU utilization remained relatively low, around 10% to 20%, primarily handling data loading and preprocessing.

For example, the total training time for the refiner model on CIFAR-10 was approximately 18 h using an NVIDIA RTX4060 GPU, compared to about 10 h for the VQ-VAE without the refiner, for a similar number of epochs. While traditional image generation models, such as diffusion models, GANs, and autoregressive models, usually require training processes that take days or even weeks, the OS-QLR represents a significant improvement in training efficiency. Additionally, for FashionMNIST, the VQ-VAE with the U-Net refiner required approximately 8 h for 50 epochs, while the standalone VQ-VAE took about 4 h. Again, these training times reflect the computational overhead of the U-Net refiner's deeper architecture and additional parameters; however, they are still significantly shorter than the multi-day training durations typical of diffusion models and GANs.

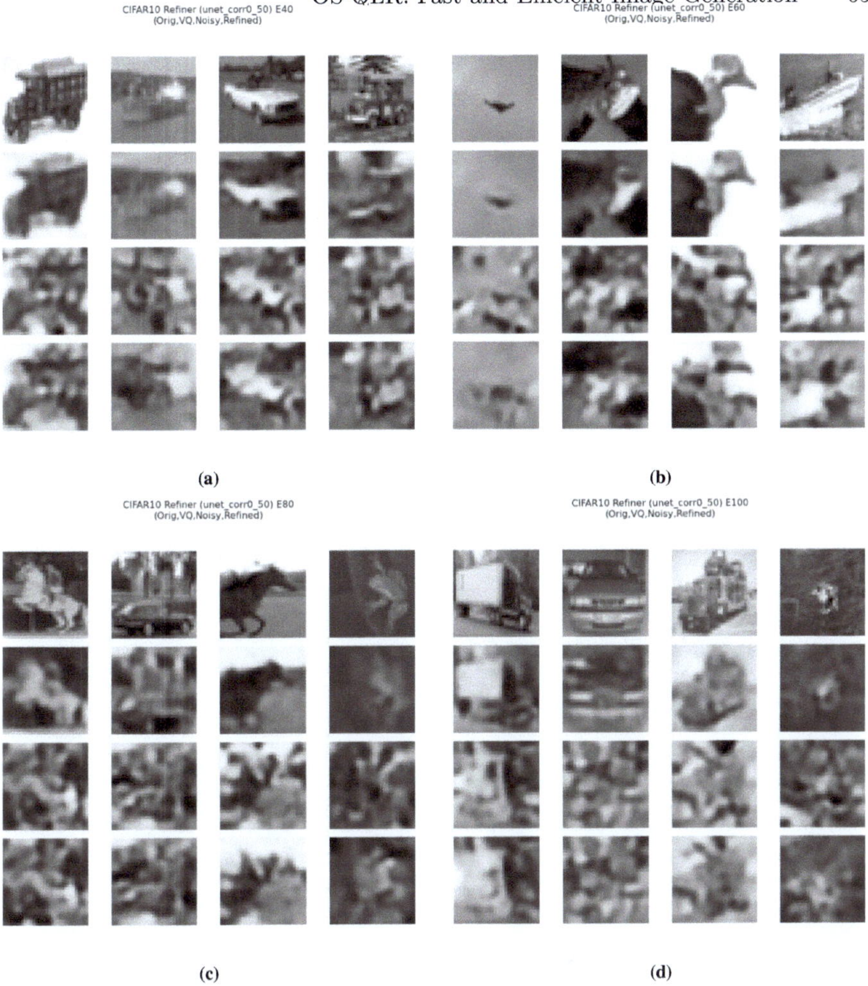

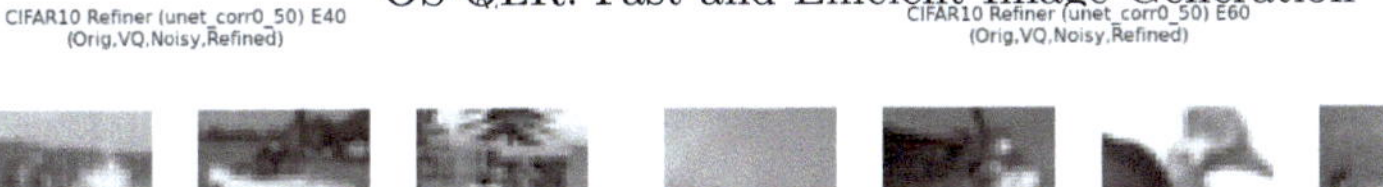

Fig. 3. Comparison of with and without refiner VQ-VAE generated images, the CIFAR-10 generative results after 40,60,80,100 epochs of training, in each image, from top row to the bottom row are original, VQ-VAE (no refiner), 50% noisy input to refiner and refined output respectively.

3.5 Loss Dynamics

During training, the reconstruction loss for both models consistently decreased, indicating an improvement in their ability to reconstruct input images. The VQ loss, which encourages the output of the encoder to align with the codebook embeddings, also converged, demonstrating effective quantization. Interestingly, the VQ-VAE with the refiner network exhibited a slightly smoother and more stable convergence of the reconstruction loss. This may be attributed to the refiner's capacity to correct initial reconstruction errors from the VQ-VAE's

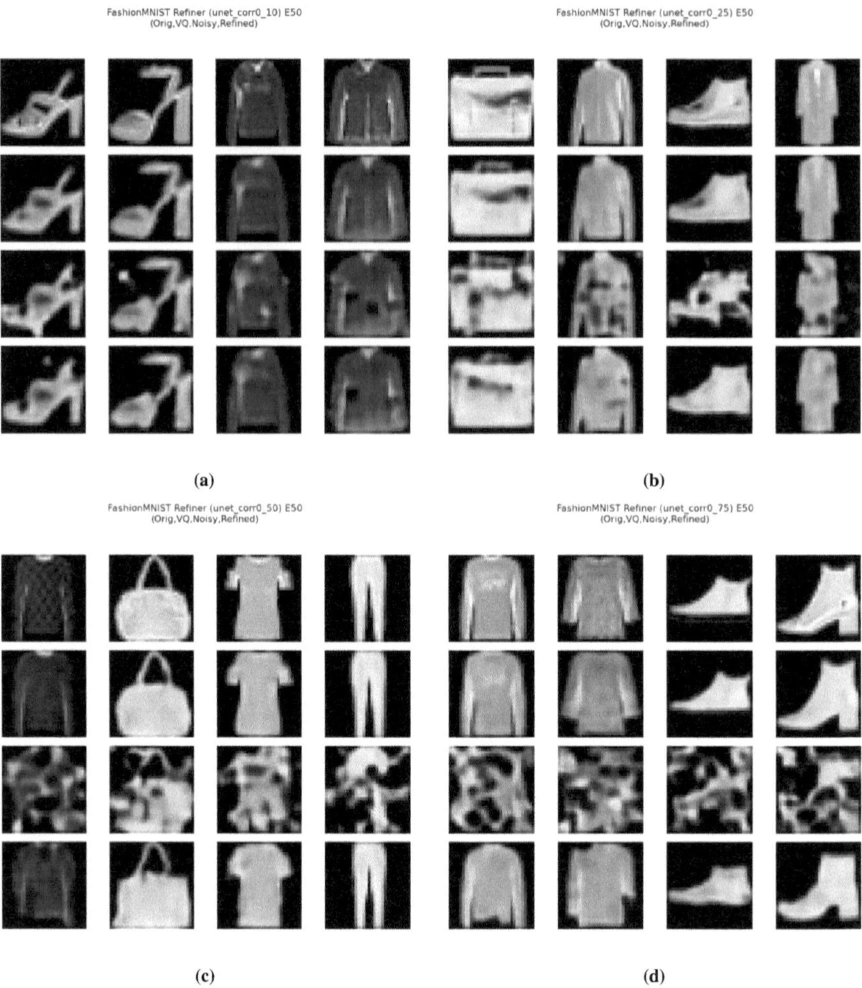

Fig. 4. Various levels of image corruption of with and without refiner VQ-VAE generated images, the FashionMNIST generative results with from left 10% to right 75% noise level after 50 epochs training, in each image, from top row to the bottom row are original, VQ-VAE (no refiner), noisy input to refiner and refined output respectively.

discrete latent space. A comparison of the loss convergence for the two models is presented in Fig. 6.

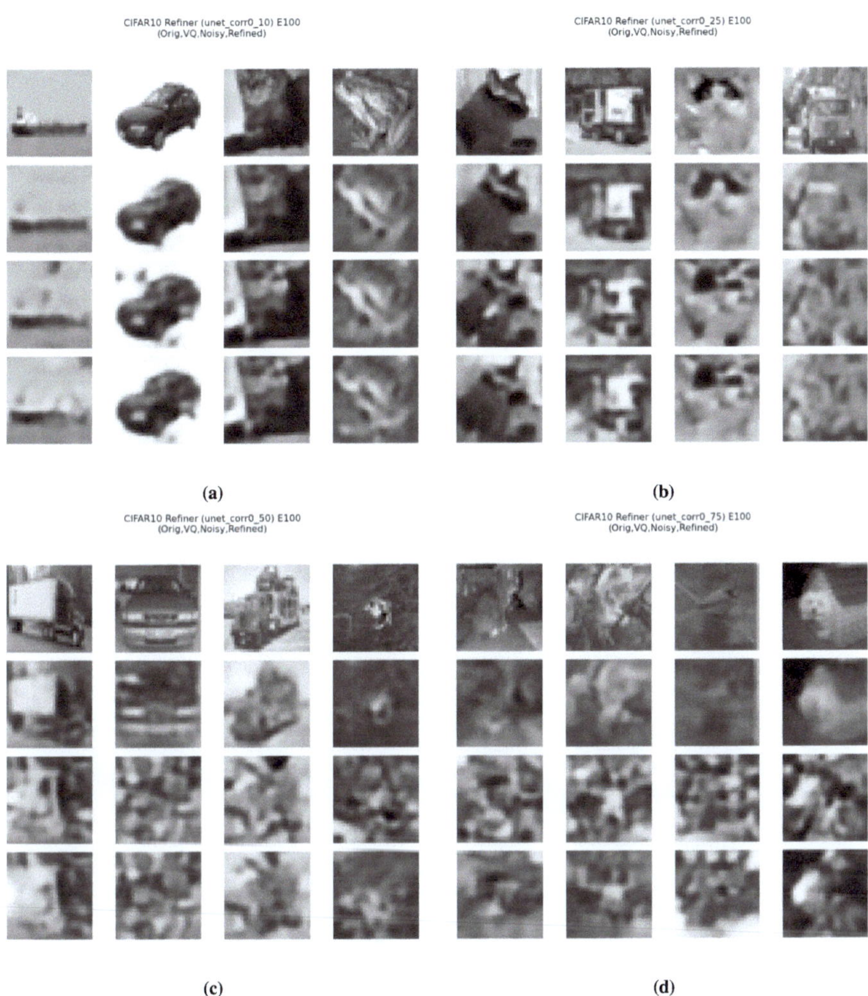

Fig. 5. Various levels of image corruption of with and without refiner VQ-VAE generated images, the CIFAR-10 generative results with from left 10% to right 75% noise level after 100 epochs training, in each image, from top row to the bottom row are original, VQ-VAE (no refiner), noisy input to refiner and refined output respectively.

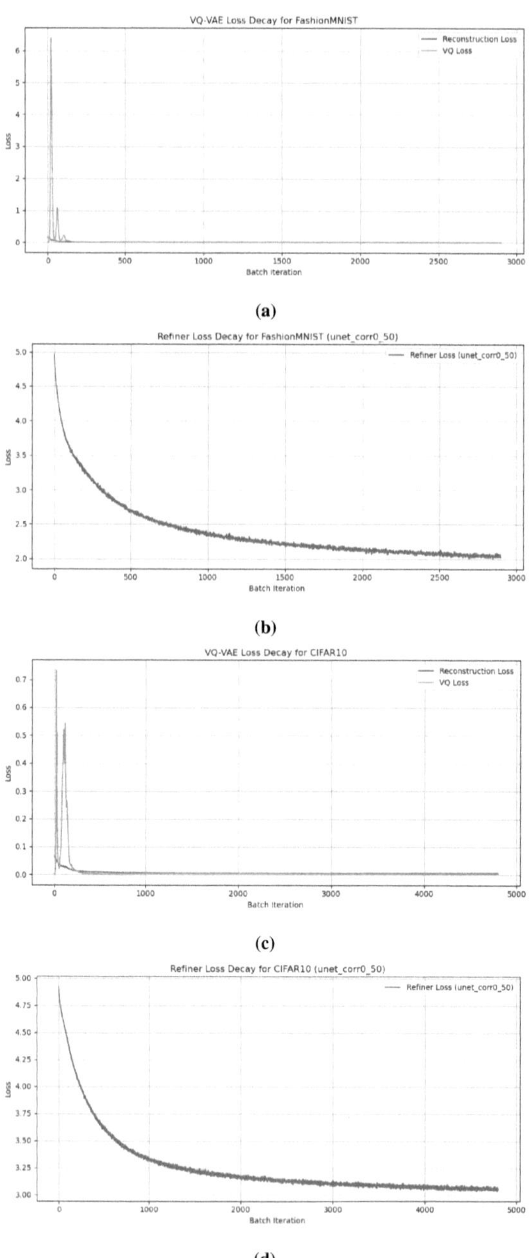

Fig. 6. The VQ-VAE and refiner loss convergence results (a) is the VQ-VAE loss for FashionMNIST, (b) is the refiner loss for FashionMNIST, (c) is the VQ-VAE loss for CIFAR-10 and (d) is the refiner loss for CIFAR-10

4   Conclusion

The primary objective of this paper was to introduce and validate OS-QLR (One-Shot Quantized Latent Refinement), a novel two-stage generative framework designed for high-quality image generation and enhanced computational efficiency. Experimental results clearly demonstrate the significant advantages of incorporating a U-Net-based refiner network into a VQ-VAE architecture for image generation. The consistent improvement in FID scores across diverse datasets, such as FashionMNIST and CIFAR-10, highlights the refiner's capability to bridge the gap between the coarse reconstructions produced by VQ-VAE and the intricate details necessary for realistic image synthesis. This aligns with existing literature, which suggests that multi-stage generative models or those employing refinement mechanisms often yield superior visual quality by allowing different network components to specialize in various aspects of the image generation process.

The VQ-VAE, due to its use of discrete latent representations, can introduce blockiness or loss of fine details during the quantization step. The U-Net refiner, equipped with skip connections and the ability to learn multi-scale features, effectively addresses these limitations. The skip connections facilitate the flow of fine-grained information directly from the encoder path to the decoder path, allowing the refiner to accurately reconstruct high-frequency details that may be lost in the VQ-VAE's compressed latent space. This advantage of U-Net architectures is well-established in tasks that require precise reconstruction, such as image segmentation and super-resolution.

Another critical finding is the robustness of the refiner-augmented model in the face of image corruption. This suggests that the U-Net refiner not only enhances output quality but also contributes to the model's ability to implicitly learn more robust and semantically meaningful representations. By leveraging both the quantized latent codes from the VQ-VAE and the direct feature maps via skip connections, the refiner can better infer missing or corrupted information, leading to more resilient image generation. This resilience could have significant implications for real-world applications where input data may not always be perfect.

While the inclusion of the refiner network undeniably improves generation quality, it does come with increased computational costs, primarily in terms of training time. The larger number of parameters and the additional forward and backward passes through the U-Net contribute to this overhead. For applications requiring real-time inference, a trade-off between quality and speed may need to be considered. However, for offline generation or applications where high fidelity is paramount, the increased training time is a justifiable cost given the substantial quality gains.

Future work could explore optimizing the refiner network's architecture for even greater efficiency, possibly by incorporating attention mechanisms or more lightweight convolutional blocks. Investigating the impact of different loss functions for the refiner, such as perceptual loss in addition to pixel-wise losses, could also lead to further improvements in visual quality. Finally, extending this framework to image generation and exploring its applicability in conditional image generation tasks would present valuable approach for future research.

Acknowledgments. The research presented in this paper was partially supported by the Internal Grant Agency of Tomas Bata University in Zlin under project number IGA/CebiaTech/2023/004, and by resources of the A.I.Lab at the Faculty of Applied Informatics, Tomas Bata University in Zlin (ailab.fai.utb.cz).

Disclosure of Interests. The authors have no competing interests to declare that are relevant to the content of this article.

References

1. Bengio, Y., Léonard, N., Courville, A.: Estimating or propagating gradients through stochastic neurons for conditional computation (2013). https://arxiv.org/abs/1308.3432
2. Ferreira, R.E., Dórea, J.R.: Leveraging computer vision, large language models, and multimodal machine learning for optimal decision-making in dairy farming. J. Dairy Sci. (2025). https://doi.org/10.3168/jds.2024-25650, https://www.sciencedirect.com/science/article/pii/S0022030225002115
3. Gu, B., Zhai, J.: Few-shot image generation based on meta-learning and generative adversarial network. Sign. Process. Image Commun. **137**, 117307 (2025). https://doi.org/10.1016/j.image.2025.117307, https://www.sciencedirect.com/science/article/pii/S0923596525000542
4. Heusel, M., Ramsauer, H., Unterthiner, T., Nessler, B., Hochreiter, S.: GANs trained by a two time-scale update rule converge to a local nash equilibrium (2018). https://arxiv.org/abs/1706.08500
5. Kim, N., Ock, S.E., Song, J.W.: Factorvqvae: discrete latent factor model via vector quantized variational autoencoder. Knowl. Based Syst. **318**, 113460 (2025). https://doi.org/10.1016/j.knosys.2025.113460, https://www.sciencedirect.com/science/article/pii/S0950705125005076
6. Liang, H., Yang, Y., Jing, J.: Latent space diffusion model for image dehazing. Appl. Soft Comput., 113322 (2025). https://doi.org/10.1016/j.asoc.2025.113322, https://www.sciencedirect.com/science/article/pii/S1568494625006337
7. van den Oord, A., Vinyals, O., Kavukcuoglu, K.: Neural discrete representation learning (2018). https://arxiv.org/abs/1711.00937
8. Parmar, G., Zhang, R., Zhu, J.Y.: On aliased resizing and surprising subtleties in GAN evaluation (2022). https://arxiv.org/abs/2104.11222
9. Peters, H., Celis, N.C.: An empirical comparison of generative capabilities of GAN vs VAE (2022). https://www.diva-portal.org/smash/get/diva2:1703317/FULLTEXT01.pdf, Accessed 07 June 2025
10. Razavi, A., van den Oord, A., Vinyals, O.: Generating diverse high-fidelity images with VQ-VAE-2 (2019). https://arxiv.org/abs/1906.00446
11. Xiao, Y., Zhou, Y., Wang, Z.: A graph convolutional network and gated recurrent unit-based surrogate for agent-based diffusion models. Eng. Appl. Artif. Intell. **149**, 110610 (2025). https://doi.org/10.1016/j.engappai.2025.110610, https://www.sciencedirect.com/science/article/pii/S0952197625006104

12. Yang, Z., Liu, J., Zhu, H., Zhang, J., Liu, W.: SSDM: generated image interaction method based on spatial sparsity for diffusion models. Neurocomputing **634**, 129805 (2025). https://doi.org/10.1016/j.neucom.2025.129805, https://www.sciencedirect.com/science/article/pii/S0925231225004771

Dataset-Independent Approach for Generating Synthetic Data in Optical Defect Detection

Christian Linder[(✉)] [iD], Steffen Geinitz [iD], and Sebastian Maier [iD]

Fraunhofer Institute for Casting, Composite and Processing Technology IGCV,
Am Technologiezentrum 2, 86159 Augsburg, Germany
`{christian.linder,steffen.geinitz,`
`sebastian.maier}@igcv.fraunhofer.de`

Abstract. Deep learning techniques have become increasingly important in the field of optical defect detection in recent years, often outperforming traditional image processing methods. However, the performance of deep learning methods is highly dependent on the amount of training data, which poses a challenge in industrial settings where labelled defect images are often scarce. This paper presents a novel dataset-independent approach for generating synthetic defect images that improves the performance of deep learning models while requiring minimal real-world data. Our methodology uses a pix2pix model to generate realistic defect images. The pix2pix model is trained on defect images and images where the defects are segmented out. After training, the model is applied to defect-free images to transform them into defect images using random segmentation masks. This procedure ensures that the defect location is preserved and that the synthetic data can also be used for segmentation applications. The approach is validated on an open source dataset. The results show that the proposed synthetic data generation approach reduces class imbalance and leads to improvements in model accuracy and recall.

Keywords: Synthetic Data · Optical Defect Detection · Production · Deep Learning · Generative Adversarial Networks

1   Introduction

In recent years, deep learning methods have increasingly gained significance in optical defect detection, gradually replacing traditional image processing techniques [3, 16]. Deep learning methods excel at recognizing complex features in images, particularly in scenarios where traditional approaches struggle [16]. However, the effectiveness of deep learning is highly dependent on the amount and quality of training data [4]. In industrial contexts, the application of machine vision is often challenged by stringent requirements and limited availability of defect images. This problem is particularly pronounced in highly specialized sectors, where the scarcity of labelled data makes the application of machine learning algorithms often impractical. To improve the performance of deep learning methods, it is essential to increase the amount and variety of training data available. The acquisition of large datasets can be expensive and time

© The Author(s), under exclusive license to Springer Nature Switzerland AG 2026
F. Marcelloni et al. (Eds.): IJCCI 2025, CCIS 2829, pp. 666–677, 2026.
https://doi.org/10.1007/978-3-032-15638-9_39

consuming. As a result, synthetic data generation is increasingly the focus of current research as a substitute for the labour-intensive process of real-world data collection and subsequently labelling [7].

Therefore, the primary research objective of this paper is to develop a novel approach for generating synthetic data. A key focus of this paper is that the approach requires minimal real data to address the challenges associated with data scarcity. Another important aspect of this paper is that the approach is independent of specific datasets, allowing it to be used in various defect detection tasks. In addition, emphasis is placed on the generation of defect images to mitigate the imbalances often present in defect detection datasets. The paper also aims to preserve the location of defects within the generated images to ensure the applicability of the approach in classification and segmentation tasks. By achieving these objectives, the proposed methodology aims to improve the performance of deep learning methods in optical defect detection.

2 Related Work

In the literature on optical defect detection, various approaches have been proposed for the generation of synthetic data. Many approaches are based on Generative Adversarial Networks (GANs) [5]. An overview of GAN based methods can be found in [7]. GAN approaches provide a flexible framework for synthetic data generation that can be applied to different use cases without much adaptation of the method. Jain et al. [9] for example developed a deep convolutional GAN method and demonstrated it on surface defects for hot-rolled steel strips. Xu & Liu [17] used a GAN approach that is trained with 1920 real pavement crack defect images. The generic nature of these GANs allows for easy transferability to different domains. However, these methods often lack precision in addressing specific problems, which can result in limited added value. In addition, they are primarily suited to classification tasks and typically require substantial amounts of data to generate realistic defect images.

Another category of approaches in this area is cycle GANs [19], a specialized variant of GANs that learn a mapping between two domains. Zhang et al. [18] developed a cycle GAN approach where images without defects are transformed into images with defects. Niu et al. [14] proposed a cycle GAN method with two generators and four discriminators to generate defect images from defect-free images. Similar to GANs, cycle GANs can be applied to different use cases without much adaptation of the method. Cycle GANs typically require fewer data samples than traditional GANs to produce good results. However, the number of data points required is not always low, depending on the complexity. Also, cycle GANs cannot be used for segmentation tasks because the location of the generated defects is unknown.

Furthermore, application-specific methods that have been carefully developed for specific use cases are used for generating synthetic data. Application-specific approaches are often based on CAD models or 3D graphics and rendering software. Boikov et al. [1] for example used the 3D software Blender to generate synthetic steel defects. Schmedemann et al. [15] rendered objects where the defects are modeled with use case specific 3D defect models that are placed on the surface of the object. In addition, GANs modified for specific use cases fall into this category. Mertes et al. [12,13]

proposed a GAN-based pix2pix method that generates binary masks resembling the shape of scratches that are transformed into realistic defect images in carbon fibers. These methods demonstrate the ability to generate synthetic data while requiring a small amount of real data for training. However, a notable limitation of these approaches is their inherent specificity. They are often closely tied to the particular scenarios for which they were developed. Consequently, adapting these methods to different applications can pose significant challenges.

In summary, there is a research gap in the current state of the art for a method that can generate high-quality defect images from limited amounts of data and that can be applied not only to a specific problem but works independently of the application context.

3   Approach

In the following section, a dataset-independent approach to generate synthetic data in the field of optical defect detection is proposed, requiring only limited real data to work. First, the implementation and structure of the approach is discussed. In addition, the training process and the application of the approach to real-world scenarios are described.

3.1   Methodology

The proposed synthetic data generation approach is based on the pix2pix model [8], a GAN-based image-to-image translation framework. An advantage of using pix2pix is that it requires relatively few data samples compared to other GAN approaches. Similar to [12], we use the pix2pix model to facilitate a mapping between corresponding image pairs. But in our approach, the input images consist of defect images, where the defects are explicitly segmented and the pixel values are overlaid with a one-hot encoding for all values within the segmentation mask. The output images are the corresponding original defect images without segmentation masks. The overall workflow is visualized in Fig. 1.

The primary goal of our approach is to train the generator to accurately insert defects into the segmented areas of the images. After the training phase, the generator can transform images without defects into images with defects. This can be done by applying random segmentation masks to the defect-free images. This capability is particularly valuable given the typical scarcity of defect images compared to defect-free images in real-world datasets.

To effectively transform defect-free images into defect images, we explored different strategies for generating the segmentation masks. The most successful method was to use actual defect masks from the defect dataset and augment them in a manner consistent with the procedures used during the training phase.

To improve the model's ability to discriminate between defect-free and defect conditions, we also introduce defect-free images into the training set. These images are overlaid with random segmentation masks from the defect dataset. Thereby, the generator has to learn to fill in the masked parts of the image. The segmentation masks are colour-coded to indicate which masks represent defect-free and defect conditions (see

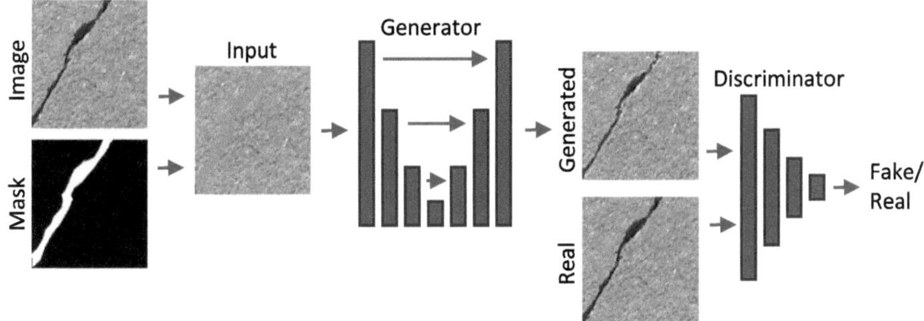

Fig. 1. Overall workflow of the pix2pix approach. Mask and original image are combined to create the input image. This input image is fed into the generator to produce a synthetic image. The discriminator then has to distinguish between real and generated images, using the input image as part of the evaluation process. Original concrete image from SDNET2018 dataset by [11].

Fig. 2 and Fig. 3). This inclusion helps the generator to learn to distinguish between actual defects and non-defect conditions, guided by adversarial feedback from the discriminator. Thereby, the generator learns to generate actual defect conditions.

The approach presented allows the model to adapt and learn the inherent characteristics of different datasets without having to adapt the method itself. Only the images from the existing dataset have to be labeled according to our specifications. As a result, our method is able to handle different data scenarios and can be applied to different problems in the field of optical defect detection.

3.2   Training of Pix2pix

The pix2pix model is trained using the Adam optimizer [10] with a learning rate of 0.001 and a number of epochs of 5000. The hyperparameter selection is based on multiple test runs. A learning rate scheduler is also used to adjust the rate during training so that a higher initial learning rate can be used. For the first 200 epochs the learning rate is fixed at the initial value, then it is multiplied by a factor that is updated at each

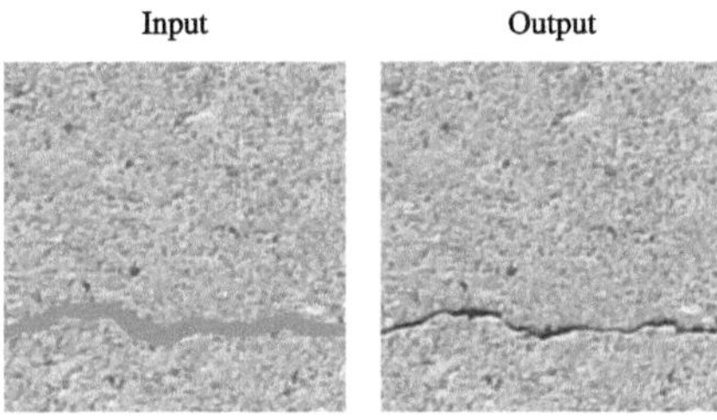

Fig. 2. Example of an input-output pair for defect images. Original image from SDNET2018 dataset by [11].

Input   Output

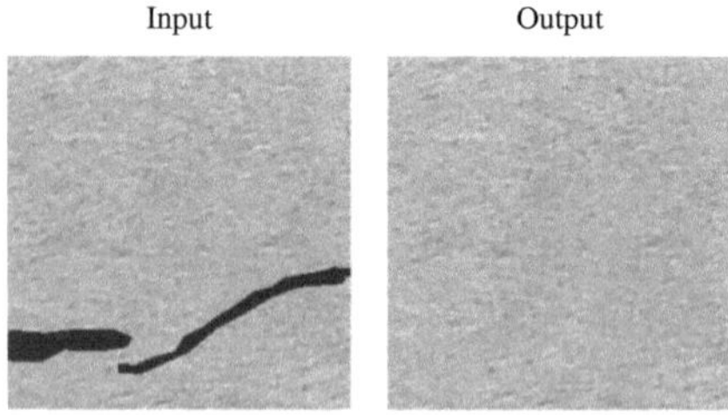

Fig. 3. Example of an input-output pair for defect-free images. Original image from SDNET2018 dataset by [11].

epoch. The factor lr_{factor} is based on the current epoch $epoch_{cur}$ and is calculated as:

$$lr_{factor} = 1 - \frac{epoch_{cur} - 200}{9800}$$

The weights of the discriminator are only updated every 20th iteration so that there is no imbalance between the generator and the discriminator. The generator is updated at each iteration. A batch size of 16 is used.

To increase the variety in the dataset, the data is augmented during training. A flip operation is used to flip the images with a probability of 0.5. A random resized crop operation is also applied to the images, which crops the images and resizes each image to 256×256 pixels. For the architectures of the generator and the discriminator, we followed the recommendations of [8]. For the generator, we use the unet256 discussed in [8], which uses batch normalization and 8 downsampling operations. The discriminator is a PatchGAN with batch normalization and five convolutional layers where the number of filters is subsequently 3, 64, 128, 256, 512, 1. In the first three convolutional layers a stride value of 2 is used and in the last two convolutional layers the stride value is reduced to 1. For the loss we use BCEWithLogitsLoss and L1Loss. The L1Loss is multiplied by 100 to increase its importance by giving it more weight.

3.3 Inference of Generator

For inference, the generator can be used to generate synthetic defect images. We have implemented it so that defect and defect-free images are sampled equally. In 35% of the cases where a defect image should be sampled, a synthetic defect image is generated by the generator by applying a randomly augmented mask to an defect-free image and inserting it into the generator. Furthermore, in 15% of the cases, an defect-free image is generated where the generator reconstructs the masked part of an image. The insertion of synthetic defect-free images helps to avoid overfitting to generator artefacts. The percentages chosen are based on several test runs. The masks are augmented by randomly cropping the binary mask and resizing it to 256×256 pixels, similar to the training of the generator.

4   Experimental Setup

This chapter describes the experimental setup used to evaluate the performance of the proposed method. First, the validation scenario based on a classification use case is outlined. Afterwards, the dataset used in the experiments is outlined, followed by a description of the preprocessing steps taken.

4.1   Classification Scenario

A classification use case is created to test if the proposed approach for generating synthetic data can improve the performance of classification models. The goal is to predict whether an image will be classified as defect-free or defect. The classification use case is evaluated by generating different scenarios. The following scenarios are created:

1. **None:** No augmentation methods are used
2. **Balanced:** No augmentation methods are used, but defect-free and defect images are sampled equally
3. **Synth:** Only synthetic data is added to the dataset
4. **Aug:** Only classical augmentation methods are used
5. **Aug+Synth:** Classical augmentation is combined with synthetic data generation

In the scenarios labeled Synth, Aug and Aug+Synth, the defect-free and defect images are sampled equally. This process makes it possible to systematically evaluate the classification results under different conditions, and in particular to assess whether the inclusion of synthetically generated data improves model performance. In addition, we compare these results with those obtained using classical data augmentation techniques.

For the data augmentation, we use RandAugment [2], which combines different augmentation strategies, such as translation, rotation, shear, solarize, equalize and adaptations in contrast, brightness and sharpness. One of the main advantages of RandAugment is that it can be applied without prior knowledge of the dataset and usually leads to improved performance.

The model used for this study is a ResNet [6] with 18 layers (ResNet-18), which is trained on the training dataset. At each epoch, the accuracy of the model is evaluated based on the validation data, and the best model in terms of validation accuracy is used for testing. Furthermore, the Adam optimizer is used with a learning rate of 0.0001. The number of epochs is 15 and the batch size 32.

The top layers of the ResNet-18 contain a fully connected layer of 64 neurons, followed by a dropout layer with a rate of 0.5 to reduce overfitting. The final fully connected layer has 2 neurons representing the classes defect-free and defect.

The architecture is not very deep and the number of epochs is deliberately kept low. Initial tests have shown that overfitting is a significant problem with such a limited amount of data, even when synthetic data and data augmentation techniques are used. Thus, this configuration aims to strike a balance between model complexity and generalization ability.

4.2 Data Acquisition and Preparation

The open source dataset SDNET2018 from [11] is used to test the proposed approach. The dataset contains images of cracked (defect) and non-cracked (defect-free) concrete surfaces. We focused on the pavement category. The pavement dataset contains 21,726 defect-free and 2,608 defect images. A total of 200 defect images and 5,000 defect-free images are randomly selected to train the pix2pix model, simulating an unbalanced distribution. The defects in the images are manually labeled using a polygon tool. For the classification task, the following data selections are made:

- **Training Classification:** The same 200 defect and 5,000 defect-free images used for training the pix2pix model are used
- **Validation Classification:** 50 defect and 50 defect-free images are randomly sampled from the remaining images of the dataset
- **Testing Classification:** 500 defect and 500 defect-free images are randomly sampled from the remaining images of the dataset.

For training and validation only a small number of images from the dataset is used simulating real-world conditions. In relation to this, more data was used for testing to get a better indication of the accuracy without the need for a lot of repetition through approaches such as cross-validation. Cross-validation is not utilized in this study due to the high computational costs involved, particularly because the pix2pix approach must be retrained each time. However, we plan to use cross-validation for future work. The data is preprocessed by resizing the images to dimensions of (256, 256) and normalizing the pixel values.

5 Results

The following section describes the results for the above scenarios and configurations. The pix2pix model was trained on the described dataset, and an example of the generation of synthetic defects is shown in Fig. 4. As discussed in Sect. 3, an defect-free image (see left part of the Fig. 4) is transformed into a defect image (see right part of the Fig. 4) using randomly sampled masks. The middle column of Fig. 4 visualizes the defect-free image overlaid with the randomly sampled mask. Figure 5 visualizes the reconstruction of a defect-free image. The images show that the generator is capable of producing realistic defect images and reconstructing missing parts.

However, the results were not always as good as shown in the aforementioned figures. Figure 6 shows examples where the transformation is not optimal. Furthermore, there were sometimes artefacts present in the images due to the limited amount of real data. The impact of these artefacts is reduced by the inclusion of both synthetically reconstructed defect-free and synthetic defect images in the training loop, which ensures that the artefacts are not class specific.

To get a better understanding of the quality of the approach, the experiments in Sect. 4 are run and the results visualized in Figs. 7, 8 and 9. To ensure reliability, each experiment was run 100 times and the figures represent the average values obtained.

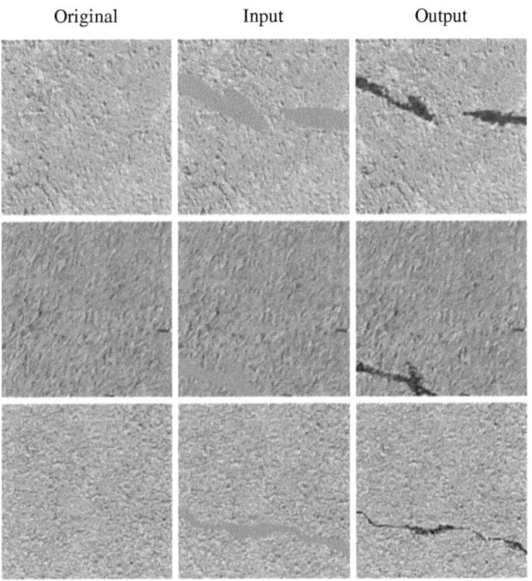

Fig. 4. Example results for transforming defect-free images to defect images. Original images from SDNET2018 dataset by [11].

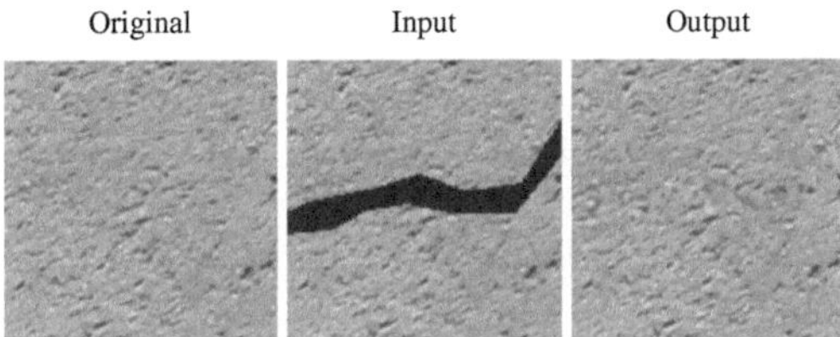

Fig. 5. Example results for recreating defect-free images. Original image from SDNET2018 dataset by [11].

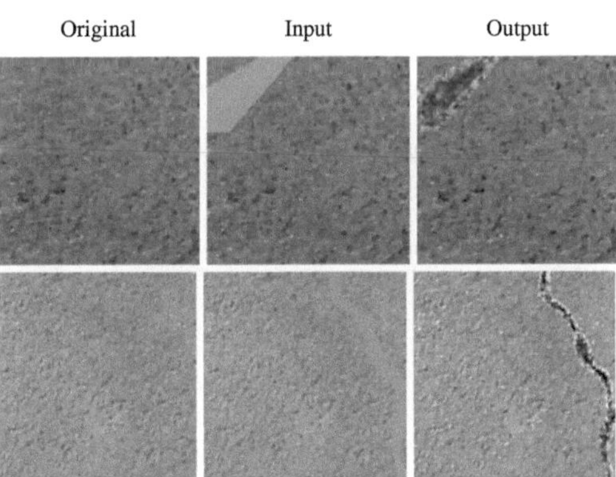

Fig. 6. Example results of suboptimal transformations for transforming defect-free images to defect images. Original images from SDNET2018 dataset by [11].

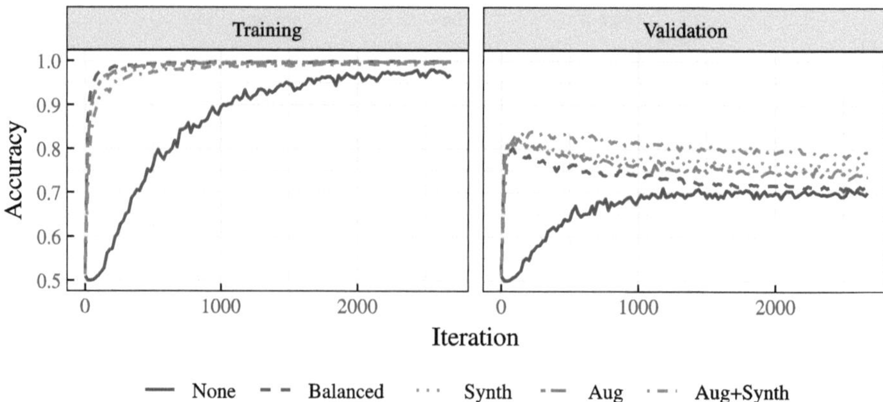

Fig. 7. Mean accuracy for training and validation datasets in each training iteration, averaged over 100 repetitions.

Figure 7 shows the accuracy throughout the training process at each iteration. Both training and validation accuracy are calculated and plotted. The results show that the Aug+Synth scenario achieves the highest generalization capabilities. In particular, while the training accuracy does not increase as quickly as in the other scenarios, it becomes similar in later iterations and, at the same time, leads to the highest validation accuracy. In contrast, the performance results for the Aug and the Synth scenarios are similar, though both still achieve higher values than the Balanced and None scenarios. The None scenario showed the lowest validation accuracy, highlighting the importance of balancing the data distribution and using augmentation strategies.

In Fig. 8, the results are very similar to those observed in Fig. 7, where the models with the highest validation accuracy were evaluated on a separate test dataset. The combination of traditional data augmentation methods and synthetic data resulted in an improvement in accuracy of 0.9% compared to the Aug scenario. Although this increase

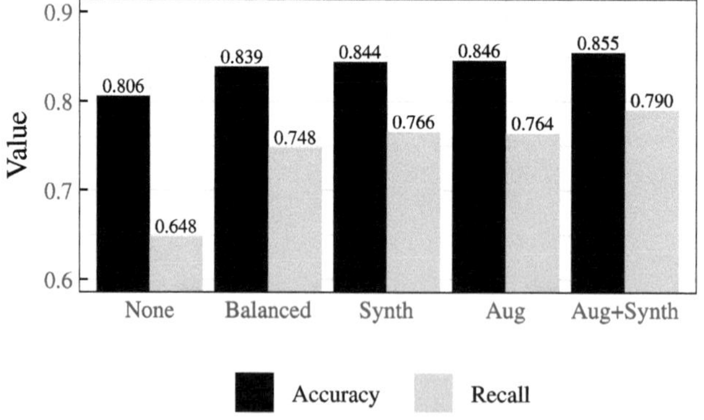

Fig. 8. Accuracy and recall results for each testing scenario.

may seem small, it is still of significant value. Furthermore, the recall score is important in the context of optical defect detection, as it indicates the effectiveness of the model in identifying defects while minimizing the number of defect instances that are incorrectly classified as defect-free. In this respect, the improvement in recall was measured at 2.6% compared to traditional data augmentation methods, highlighting the added benefit of incorporating synthetic data.

In addition, the standard deviation of the results was measured. The standard deviation for the accuracy results is 0.023 for the None scenario, 0.020 for the Balanced, 0.018 for the Synth, 0.019 for the Aug and 0.016 for the Aug+Synth scenario. The values show that there were fluctuations within the experiments, but that these decreased slightly when synthetic data was added. The decrease in the standard deviation was similar for the recall results.

In addition, Fig. 9 illustrates the confusion matrix. It shows that the correct detection rate for defect images is highest, while the accurate identification of defect-free images is only slightly lower than in scenarios Balanced, Synth and Aug, resulting in a more balanced detection performance overall. On average over the 100 repetitions, this approach correctly identifies 13 more images as defect instead of the wrong class defect-free compared to the augmentation-only scenario.

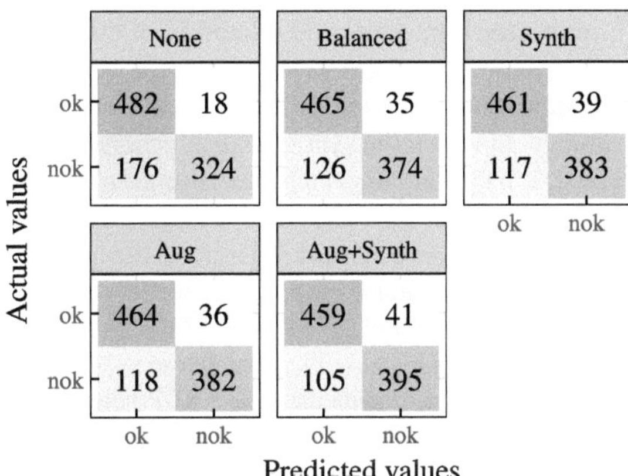

Fig. 9. Confusion matrix of the results, where "ok" represents defect-free images and "nok" denotes defect images.

6   Conclusion and Outlook

In this paper, we presented a novel approach for generating synthetic data that aims to improve the performance of deep learning methods in optical defect detection. The results showed that the proposed method can successfully transform defect-free images into realistic defect images while preserving defect locations.

A notable aspect of this approach is its ability to operate effectively with a minimal amount of real-world data. Specifically, the synthetic image generation process was successful using only 200 real defect images for training, highlighting the efficiency of the method in scenarios where obtaining large datasets is impractical. Moreover, the approach is independent of specific datasets by using a procedure based on pix2pix and can be easily adapted to new applications with minimal effort. It requires only annotated images, where defects are segmented by pixel-level masks, to function effectively. In classification scenarios, the pixel-level mask annotation of defects is a separate step, but it is usually not very time-consuming as it only applies to defect images, which are typically few in number. Furthermore, the method effectively balances imbalanced datasets and reduces overfitting by generating realistic defect images. Since the position of the defects is preserved, the approach can also be used in both classification and segmentation applications. The results showed that the integration of synthetic data led to observable improvements in model accuracy and recall for a classification scenario. The combination of synthetic data and data augmentation proved to be the most effective strategy. Accuracy was improved by 0.9% and recall by 2.6% compared to the second best scenario using only traditional data augmentation methods.

Although the synthetic data generation process showed promising results, certain limitations were noted, such as the presence of artefacts or imperfections in some of the generated images. Continued refinement of the synthetic data generation process are needed to further improve the robustness and reliability of our approach. Furthermore, tests need to be carried out on more datasets to validate the generalization ability of the approach in different industrial scenarios. In addition, the implementation of a cross-validation testing strategy, in which the synthetic data generation approach is retrained with new data each time, is necessary to get a better estimate of the model's performance and stability under different conditions. Moreover, a major advantage of the approach is that it can be used for segmentation scenarios. Therefore, the usefulness of the approach needs to be tested on a segmentation use case.

Acknowledgments. The authors would like to thank the Bavarian Ministry of Economic Affairs, Regional Development and Energy for funding the project KI-Produktionsnetzwerk@IGCV-FuE and the German Federal Ministry of Education and Research (BMBF) for supporting the project SaMoA within VIP+.

Disclosure of Interests. The authors have no competing interests to declare that are relevant to the content of this article.

References

1. Boikov, A., Payor, V., Savelev, R., Kolesnikov, A.: Synthetic data generation for steel defect detection and classification using deep learning. Symmetry **13**(7), 1176 (2021)
2. Cubuk, E.D., Zoph, B., Shlens, J., Le, Q.V.: RandAugment: practical automated data augmentation with a reduced search space. In: Proceedings of the IEEE/CVF Conference on Computer Vision and Pattern Recognition Workshops, pp. 702–703 (2020)
3. Czimmermann, T., et al.: Visual-based defect detection and classification approaches for industrial applications-a survey. Sensors (Basel, Switzerland) **20**(5) (2020). https://doi.org/10.3390/s20051459

4. Feng, S., Zhou, H., Dong, H.: Using deep neural network with small dataset to predict material defects. Mater. Des. **162**, 300–310 (2019)

5. Goodfellow, I., Pouget-Abadie, J., Mirza, M., Xu, B., Warde-Farley, D., Ozair, S., Courville, A., Bengio, Y.: Generative adversarial networks. Commun. ACM **63**(11), 139–144 (2020)

6. He, K., Zhang, X., Ren, S., Sun, J.: Deep residual learning for image recognition. In: Proceedings of the IEEE Conference on Computer Vision and Pattern Recognition, pp. 770–778 (2016)

7. He, X., Chang, Z., Zhang, L., Xu, H., Chen, H., Luo, Z.: A survey of defect detection applications based on generative adversarial networks. IEEE Access **10**, 113493–113512 (2022). https://doi.org/10.1109/ACCESS.2022.3217227

8. Isola, P., Zhu, J.Y., Zhou, T., Efros, A.A.: Image-to-image translation with conditional adversarial networks. In: Proceedings of the IEEE Conference on Computer Vision and Pattern Recognition, pp. 1125–1134 (2017)

9. Jain, S., Seth, G., Paruthi, A., Soni, U., Kumar, G.: Synthetic data augmentation for surface defect detection and classification using deep learning. J. Intell. Manuf. **33**(4), 1007–1020 (2022). https://doi.org/10.1007/s10845-020-01710-x

10. Kingma, D.P.: Adam: a method for stochastic optimization. arXiv preprint arXiv:1412.6980 (2014)

11. Maguire, M., Dorafshan, S., Thomas, R.J.: SDNet2018: a concrete crack image dataset for machine learning applications (2018). https://doi.org/10.15142/T3TD19

12. Mertes, S., Margraf, A., Geinitz, S., André, E.: Alternative data augmentation for industrial monitoring using adversarial learning. In: Fred, A., Sansone, C., Madani, K. (eds.) Deep Learning Theory and Applications, Communications in Computer and Information Science, vol. 1854, pp. 1–23. Springer Nature Switzerland and Imprint Springer, Cham (2023). https://doi.org/10.1007/978-3-031-37320-6_1

13. Mertes., S., Margraf., A., Kommer., C., Geinitz., S., André., E.: Data augmentation for semantic segmentation in the context of carbon fiber defect detection using adversarial learning. In: Proceedings of the 1st International Conference on Deep Learning Theory and Applications - DeLTA, pp. 59–67. INSTICC, SciTePress (2020). https://doi.org/10.5220/0009823500590067

14. Niu, S., Li, B., Wang, X., Lin, H.: Defect image sample generation with GAN for improving defect recognition. IEEE Trans. Autom. Sci. Eng. **15**, 1–12 (2020). https://doi.org/10.1109/TASE.2020.2967415

15. Schmedemann, O., Baaß, M., Schoepflin, D., Schüppstuhl, T.: Procedural synthetic training data generation for ai-based defect detection in industrial surface inspection. Procedia CIRP **107**, 1101–1106 (2022)

16. Tabernik, D., Šela, S., Skvarč, J., Skočaj, D.: Segmentation-based deep-learning approach for surface-defect detection. J. Intell. Manuf. **31**(3), 759–776 (2020)

17. Xu, B., Liu, C.: Pavement crack detection algorithm based on generative adversarial network and convolutional neural network under small samples. Measurement **196**(9), 111219 (2022). https://doi.org/10.1016/j.measurement.2022.111219

18. Zhang, G., Cui, K., Hung, T.Y., Lu, S.: Defect-GAN: high-fidelity defect synthesis for automated defect inspection (2021)

19. Zhu, J.Y., Park, T., Isola, P., Efros, A.A.: Unpaired image-to-image translation using cycle-consistent adversarial networks. In: Proceedings of the IEEE International Conference on Computer Vision, pp. 2223–2232 (2017)

Combining Large-Scale and Domain-Specific Datasets for Hate Speech Severity Modeling: A Regression-Based Approach

Andrew Asante[ID] and Petr Hajek[(✉)][ID]

Science and Research Centre, Faculty of Economics and Administration,
University of Pardubice, Studentská 95, 532 10 Pardubice II, Czechia
{andrew.asante,petr.hajek}@upce.cz

Abstract. This study proposes a regression-based framework for modeling the severity of online hate speech, addressing the limitations of traditional classification approaches that overlook the nuanced and continuous nature of harmful language. Leveraging a compiled dataset of over 3.1 million English-language social media posts from Reddit, Twitter, and Wikipedia, we fine-tune both general-purpose and hate speech-specific transformer models to predict real-valued severity scores. Our findings show that domain-specific models, particularly Hate-BERT, outperform general-purpose alternatives in capturing subtle gradients of hatefulness. The proposed approach enables more context-aware and proportionate content moderation, while also highlighting challenges related to annotation subjectivity, lack of conversational context, and cross-platform generalization. This work advances the field by demonstrating the feasibility and utility of severity estimation as a scalable alternative to categorical hate speech detection.

Keywords: Hate Speech · Large Language Model · Transformer · Regression

1 Introduction

The proliferation of hate speech on online platforms poses a considerable threat to social cohesion, democratic dialogue, and the psychological safety of marginalized communities. Although automatic detection systems have made considerable strides, particularly with the introduction of transformer-based models, most computational approaches frame the problem as a binary classification task, identifying whether or not a given text contains hate [1, 21]. However, hate speech exists on a continuum of intensity, ranging from subtle stereotypes and implicit bias to overtly dehumanizing and violent language [16]. Recognizing the severity of hateful content, not just its presence, is essential for scalable, harm-aware moderation, and proportional enforcement, especially within the scope of evolving content regulation policies [20].

Despite growing attention to the severity of hate speech, current models are still limited in their granularity. Most detection systems are binary [28] or multiclass [14] in nature, providing discrete labels rather than scalar or ordinal representations. While scalar outputs represent a promising approach to hate speech detection, the limited

F. Marcelloni et al. (Eds.): IJCCI 2025, CCIS 2829, pp. 678–689, 2026.
https://doi.org/10.1007/978-3-032-15638-9_40

availability of annotated ordinal data and the narrow domain diversity of most datasets remain major challenges. In addition, literature has shown that hate speech varies considerably across platforms and cultural contexts [2,30], making single-source training inadequate for real-world deployment. Furthermore, existing severity prediction efforts have rarely employed regression frameworks, despite their appropriateness for tasks where the target variable is ranked but not uniformly spaced [26]. As such, there is a critical gap in the research concerning the use of a regression approach with pre-trained models, fine-tuned on cross-domain datasets to enhance hate speech severity modeling.

To address these challenges, this study proposes a novel cross-domain ordinal regression framework that leverages pre-trained transformer models, including BERT [15] and RoBERTa [23], ELECTRA [12], fine-tuned on a composite dataset comprising both large-scale and domain-specific hate speech corpora. The goal is to move beyond simplistic classification and treat hate speech severity as a ranked continuum, enabling models to provide graded context-aware predictions. The study is guided by the following research questions:

1. To what extent can pre-trained language models adapted for ordinal regression accurately estimate the severity of hate speech?
2. Does combining large-scale and domain-specific hate speech datasets improve the generalizability and robustness of ordinal regression models across different platforms?

The structure of this paper is as follows. Section 2 reviews relevant empirical literature on hate speech detection, intensity estimation, and cross-domain modeling. Section 3 describes the methodology, including dataset preparation, model design, and evaluation metrics. Section 4 presents the experimental setup, results, and analysis. Section 5 discusses and concludes the paper with key findings, contributions, and directions for future research.

2 Related Literature

2.1 Conceptual Foundations of Harmful Online Language

Hate speech, while distinct in its targeted and discriminatory intent, exists within a broader spectrum of harmful online language. Closely related phenomena such as offensive language, toxic behavior, personal attacks, and verbal aggression often overlap with hate speech in form, tone, or function [3,6,21]. Differentiating between these categories is essential for building accurate detection systems, especially in models aiming to estimate the intensity of harm rather than merely identify its presence. To develop a more granular understanding of online harm, it is important to examine these overlapping phenomena individually.

Offensive language includes profane, vulgar, or disrespectful content that may be socially inappropriate but is not necessarily hateful or discriminatory. While all hate speech is inherently offensive, not all offensive language constitutes hate speech. For example, expressions of frustration using expletives may trigger false positives in detection systems if not contextualized. This distinction was highlighted in [37] by categorizing content as offensive but not necessarily hateful, emphasizing the need for fine-grained modeling of intent and target.

Toxic behavior refers to communication that is rude, disrespectful, or inflammatory. It captures a broader range of harmful expressions, including general hostility, sarcasm, and trolling [9]. Although toxicity and hate speech can co-occur, toxic comments are not always rooted in identity-based prejudice [18]. This difference has implications for moderation policies and model design, where the intent to marginalize or dehumanize is central to hate speech definitions.

Personal attacks target individuals rather than groups, often without reference to protected characteristics [6]. Examples include cyberbullying, online harassment, and trolling, which are usually characterized by repeated or aggressive targeting of specific users [4]. While personal attacks can intersect with hate speech, particularly when they involve slurs or identity-based insults, they are more interpersonal than ideological [25].

Aggressive or violent language includes threats, incitement, and calls for physical harm. According to [22], aggressive language represents the most extreme form of harmful language and may or may not be identity-based. Aggression intensifies hate speech when used to threaten or advocate violence against targeted groups. However, aggression can also occur in unrelated contexts, including sports rivalries or political debates, which require careful contextual interpretation [33].

2.2 Computational Approaches to Hate Speech Detection

The early models for hate speech detection were primarily based on keyword lists and hand-crafted features used in machine learning algorithms such as SVMs, logistic regression, and decision trees [14, 17]. These traditional approaches were later replaced by neural networks, particularly CNNs and RNNs, which learned textual patterns more effectively [22, 32]. However, these methods still treated hate speech as a binary or multi-categorical label, lacking the granularity needed to assess intensity.

The introduction of transformer architectures such as BERT, RoBERTa, and Distil-BERT, the field of hate speech research has experienced significant improvements in language understanding [1, 29]. These advanced neural network models have become the backbone of modern hate speech detection pipelines due to their contextual embedding capabilities. Specialized variants such as HateBERT [8], pre-trained on hate-related forums, have shown improved performance in identifying hate speech, particularly in toxic or aggressive contexts. Most of the applications of these models remain confined to classification tasks.

A growing body of research is shifting toward regression-based modeling of hate speech. For example, the Jigsaw Toxic Comment[1] dataset assigns toxicity scores on a continuous scale, allowing models to be trained for intensity prediction rather than binary classification [29]. Similarly, studies leveraging the Stormfront dataset or Facebook comments and other related datasets with severity levels have demonstrated that hate speech is not monolithic, but varies in emotional intensity, target specificity, and harm potential [5, 19]. These approaches are more suitable for ranking or prioritizing harmful content.

Between standard regression and classification, ordinal regression frameworks have been studied for ordered categories. Studies [10, 36] introduced a neural ordinal regres-

---

[1] https://kaggle.com/competitions/jigsaw-toxic-severity-rating.

sion method that learns both feature representations and threshold parameters. Similarly, the CORAL framework [7] decomposes ordinal targets into multiple binary tasks with cost-sensitive learning. Ordinal loss functions was extended to deep architectures for sentiment intensity [24]. While these works demonstrate the effectiveness of ordinal methods, their application to transformer-based hate speech severity remains underexplored.

In addition, one of the major limitations in hate speech research is the domain dependence of available datasets. Models trained on Twitter may fail to generalize to Reddit, YouTube, or Facebook due to differences in language use, community norms, and topical focus. Several studies, such as [29, 34], have shown that combining datasets from different platforms can improve model robustness, particularly when paired with domain adaptation techniques or adversarial training. HateXplain, [27] attempted to bridge this gap by including data from multiple platforms and providing annotator rationales. Nevertheless, most datasets are either too narrow (focused on a single domain) or too broad without clear intensity scales. This challenge has led to calls for cross-domain evaluation and multi-source dataset fusion to improve generalization, leveraging regression-based hate speech intensity detection.

Table 1. Summary of previous related online hate speech studies.

Ref.	Model	Methodology	Key outcome
[14]	LR, SVM, DT, RF	Classification	Distinction between hate, offensive, and neutral content
[29]	BERT	Regression	Continuous scoring of hate speech improves prioritization
[27]	BERT, CNN-GRU, BiRNN	Classification	Enhanced interpretability but still category-based
[8]	BERT, HateBERT	Classification	Strong performance in toxic domains
[19]	BERT, RoBERTa	Regression	Constructing measures of hate scoring
[30]	Multi-aspect models	Classification	Addressed hate across languages and dimensions
This study	BERT, RoBERTA, ELECTRA + finetuned variants	Regression	Addresses hate intensity on a larger dataset from cross domains

Table 1 presents a comparative overview of prominent hate speech datasets and modeling approaches, highlighting the method used and the granularity of their output.

3 Methodology

To assess hate speech severity in large corpora, we adopted a supervised machine learning approach to model hate speech as a continuous phenomenon rather than a binary or categorical classification task. This methodology was designed to evaluate whether such a combined dataset strategy enhances the model's ability to generalize across diverse online contexts while providing more accurate predictions through regression-based modeling.

The dataset used in this study was a compiled collection of publicly available English-language hate speech corpora, comprising millions of social media posts sourced from platforms such as Reddit, Twitter, and Wikipedia. Table 2 provides a summary of 3,178,977 text instances that featured in this study.

Table 2. Combined hate speech datasets sourced from multiple domain platforms.

Ref.	Size	Classes	Platform
[35]	128,907	Neutral, offensive, hate	Twitter
[31]	~1.2M	Hate (1), non-hate (0)	Mixed platforms: Twitter, Reddit, Gab, etc.
[11]	~1.8M	Continuous target and labels: insult, threat, etc.	Wikipedia
[19]	50,070	Ordinal scores and continuous hate score	YouTube, Reddit, Twitter
This study	3,178,977	Ordinal hate labels and continuous hate scores	Mixed platforms: Twitter, Reddit, YouTube, Wikipedia, etc.

These corpora had been previously annotated for one or more types of hateful content by other researchers. In this study, existing ordinal annotations were transformed into a continuous severity scale and integrated with pre-existing continuous hate scores to enable regression-based modeling of hate speech intensity The unionized dataset was constructed as

$$
\mathcal{D} = \bigcup_{j=1}^{M} \mathcal{D}_j, \qquad \mathcal{D}_j = \left\{ (x_i^{(j)}, y_i^{(j)}, d_i^{(j)}) \right\}_{i=1}^{N_j}, \tag{1}
$$

where $x_i^{(j)}$ denoted the raw text instance, $y_i^{(j)}$ represented the corresponding label, and $d_i^{(j)}$ indicated the source or domain tag for each instance in dataset j. Each instance was further assigned a continuous *severity score* through a labeling function.

Before inputting the data into a transformer model, we applied tokenization as follows:

$$
x_i \xrightarrow{\text{tokenizer}} \mathbf{v}_i \in \mathbb{R}^d, \tag{2}
$$

where $\mathbf{v}_i$ is the token-embedding vector used as model input.

All datasets were uniformly preprocessed by lowercasing the text and removing hyperlinks, emojis, and special characters. Hate intensity annotations were normalized to a continuous scale ranging from 0 to 1. Categorical labels such as *hateful*, *offensive*, *abusive*, and *aggressive* were mapped onto an ordinal continuum. This standardized preprocessing pipeline ensured compatibility across diverse annotation schemes, enabling the regression model to learn from heterogeneous sources under a unified output framework.

We leveraged three pre-trained transformer encoders—BERT, RoBERTa, and ELECTRA—alongside three corresponding fine-tuned variants to obtain contextualized embeddings for each input. For every text sample, we extracted the pooled token output—for example, the [CLS] embedding—as a 768-dimensional semantic vector. These vectors were then fed into a regression head comprising two fully connected layers with a ReLU activation and dropout, producing a single scalar score representing the predicted hate speech intensity.

To train the regression models based on the architecture described above, we minimized the mean squared error (MSE) between predicted and actual scores. Specifically, for each instance i, we denoted the model's output by $\hat{y}_i$ and the true label by y_i, optimizing

$$\mathcal{L}_{\text{MSE}} = \frac{1}{N} \sum_{i=1}^{N} (\hat{y}_i - y_i)^2 \tag{3}$$

where N is the total number of training examples. Training was conducted using the Adam optimizer, with the learning rate selected from a predefined search space using validation performance as the selection criterion. The dataset was stratified and partitioned into training, validation, and test subsets to monitor performance and reduce the risk of overfitting.

Model evaluation was performed using regression-based metrics such as root mean squared error (RMSE), mean absolute error (MAE), and R-squared (R^2) to quantify predictive accuracy. The MAE and RMSE were computed as follows:

$$\text{MAE} = \frac{1}{N} \sum_{i=1}^{N} |\hat{y}_i - y_i|, \quad \text{RMSE} = \sqrt{\frac{1}{N} \sum_{i=1}^{N} (\hat{y}_i - y_i)^2} \tag{4}$$

In addition, correlation-based metrics such as the Pearson correlation coefficient were computed to assess the strength of the linear relationship between the predicted and actual hate intensity scores. The coefficient of determination (R^2) and Pearson's r were calculated as follows:

$$R^2 - 1 - \frac{\sum_{i=1}^{N}(\hat{y}_i - y_i)^2}{\sum_{i=1}^{N}(y_i - \bar{y})^2}, \quad r = \frac{\sum_{i=1}^{N}(\hat{y}_i - \bar{\hat{y}})(y_i - \bar{y})}{\sqrt{\sum_{i=1}^{N}(\hat{y}_i - \hat{y})^2 \sum_{i=1}^{N}(y_i - \bar{y})^2}} \tag{5}$$

where $\bar{\hat{y}}$ and $\bar{y}$ denote the mean of the predicted and true labels, respectively. These metrics provide insight not only into the accuracy of the predictions but also into how well the model preserves the ordinal structure of hatefulness in the textual content.

4   Results and Analysis

This section presents the empirical findings from the regression experiments conducted using two categories of pre-trained transformer models: (1) general-purpose models—BERT, RoBERTa, and ELECTRA, and (2) hate speech-specific models—GroNLP/BERT, HateBERT, and DehateBERT. All models were fine-tuned on a combined dataset comprising large-scale general-purpose and domain-specific annotated corpora, with the objective of estimating the severity of online hate speech on a continuous scale. Figure 1 presents the training and validation loss curves over 20 epochs for both general-purpose and hate speech-specific transformer models, all of which demonstrated steady convergence.

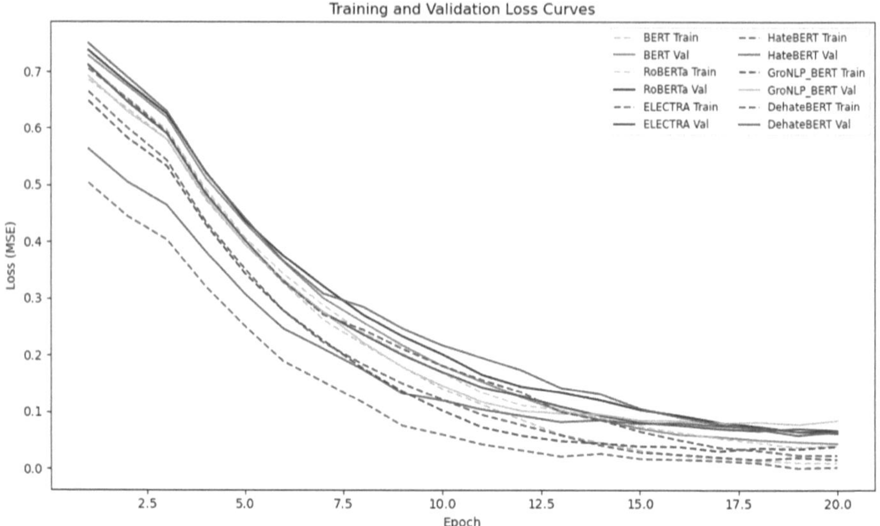

Fig. 1. Training and validation loss curves for the used transformer-based models.

Table 3. Regression-based evaluation of hate speech severity prediction.

Model	MAE ↓	MSE ↓	RMSE ↓	R^2 ↑	r ↑	p value
BERT	0.325	0.170	0.412	0.728	0.861	3.3686e-03
RoBERTa	0.298	0.150	0.387	0.762	0.881	3.3686e-03
ELECTRA	0.310	0.160	0.400	0.745	0.870	3.3686e-03
HateBERT	**0.271**	**0.132**	**0.363**	**0.791**	**0.896**	3.2353e-05
GroNLP/BERT	0.283	0.140	0.374	0.775	0.889	3.2353e-05
DehateBERT	0.289	0.143	0.378	0.768	0.884	3.2353e-05

Table 3 summarizes the regression-based performance evaluation metrics for hate speech severity prediction across six transformer models. Among the general-purpose models, RoBERTa achieved the best overall performance with an MAE of 0.298 and R^2 of 0.762, although it was outperformed by the hate speech-specific models. HateBERT yielded the strongest results across all metrics, including the lowest MAE (0.271), MSE (0.132), and RMSE (0.363), alongside the highest R^2 (0.791) and Pearson correlation coefficient r (0.896), indicating superior predictive accuracy and reliability.

GroNLP/BERT and DehateBERT also outperformed the general-purpose models, further affirming the benefits of domain-specific pre-training in identifying the subtle patterns of hate speech severity.

All models exhibited statistically significant results with the annotated severity labels, p values = $|3.3686 \times 10^{-3}| < 0.05$. This confirms the robustness of the regression predictions, given the large evaluation set. Although all models were evaluated on the same test set, we applied Fisher's Z-transformation as an approximate comparison of correlation strength between predicted and annotated severity scores. Due to the large sample size, observed differences are statistically significant; however, we interpret these results cautiously given the non-independence of model predictions.

To test the robustness of the HateBERT pretrained model on the combined dataset, we varied the learning rate. Table 4 presents the model's performance across different learning rates. Among those tested, a learning rate of 3e−5 consistently yielded the best regression performance across all metrics, supporting its selection as the default for HateBERT on the combined dataset. In contrast, higher learning rates led to performance degradation.

Table 4. Results of robustness tests for combined datasets.

	Learning rate	MAE $\downarrow$	MSE $\downarrow$	RMSE $\downarrow$	R^2 $\uparrow$	r $\uparrow$
HateBERT	1e−5	0.289	0.143	0.378	0.771	0.888
	2e−5	0.278	0.136	0.368	0.783	0.891
	3e−5	**0.271**	**0.132**	**0.363**	**0.791**	**0.896**
	5e−5	0.280	0.137	0.370	0.786	0.894
	1e−4	0.300	0.150	0.387	0.768	0.878

To demonstrate how combining domain-specific hate speech datasets improves the robustness of the proposed ordinal regression model across different platforms, we first trained the HateBERT model on individual datasets (see Table 5). While the model trained solely on [31] dataset achieves the lowest MAE/RMSE (0.261/0.350), the model trained on the combined dataset delivered a competitive performance (MAE = 0.271, RMSE = 0.363). This indicates that, although small gains can be achieved by tailoring the model to a single domain, training on the merged corpus provides a better balance between accuracy and generalizability, particularly when compared with the largest dataset from the Wikipedia platform [11].

Table 5. HateBERT regression results per dataset.

	Dataset	MAE $\downarrow$	MSE $\downarrow$	RMSE $\downarrow$	R^2 $\uparrow$	r $\uparrow$
HateBERT	[11]	0.292	0.145	0.381	0.771	0.876
	[19]	0.285	0.140	0.375	0.785	0.885
	[35]	0.273	0.131	0.362	0.794	0.893
	[31]	0.261	0.123	0.350	0.812	0.902
	All combined	0.271	0.132	0.363	0.791	0.896

For ease of comparison, Fig. 2 visualizes the performance of models trained on the combined dataset versus those trained on individual datasets. The results confirm that, except for dataset [31], the combined model outperformed the individual ones. This exception can be attributed to the binary nature of the classes in dataset [31].

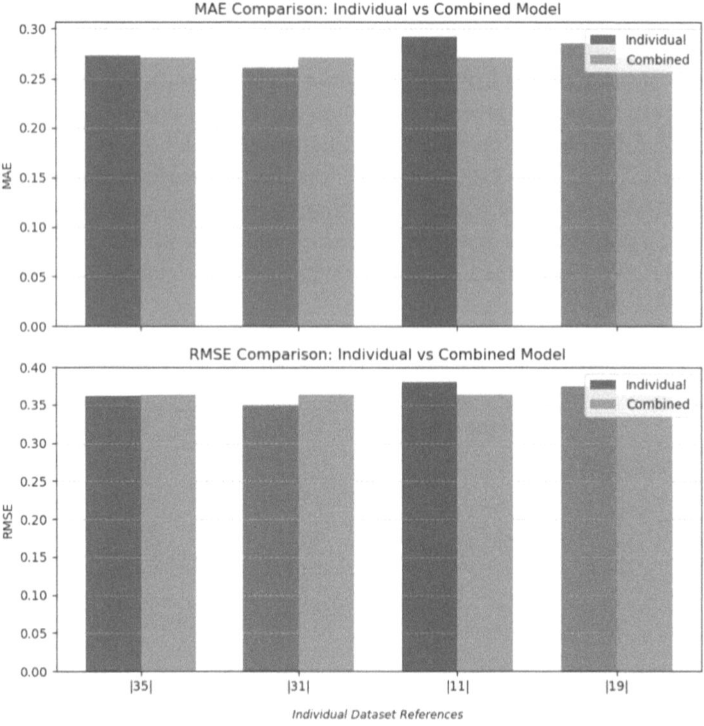

Fig. 2. Benchmarking the MAE performance of HateBERT when trained on individual datasets versus a combined dataset.

5   Conclusion

A growing body of literature emphasizes that hatefulness exists along a spectrum rather than as a fixed category [17]. Traditional classification-based models, although effective at flagging the presence of hate speech, are inherently limited in their ability to differentiate between varying levels of severity. As a result, moderation systems that rely solely on such models risk either under-reacting to extreme cases or over-censoring borderline content. To address these limitations, our study proposed a regression-based modeling framework that estimates the severity of hate speech on a continuous scale. This approach allows for a more granular assessment of online harm and supports differentiated responses in moderation workflows. Rather than forcing content into discrete labels, the model learns to predict a real-valued score reflecting the intensity of hatefulness. This shift from classification to regression is methodologically significant and aligns with broader calls in the literature for more context-aware and harm-sensitive models.

To answer $RQ1$, our findings indicate that pre-trained language models adapted for regression can effectively estimate hate speech severity. HateBERT, a domain-specific model, consistently outperformed general-purpose models such as BERT and RoBERTa across all metrics. These results demonstrate the feasibility and accuracy of transformer-based regression approaches for severity prediction. However, the relatively narrow performance margin also suggests that general-purpose models retain some adaptability and should not be dismissed outright, particularly in low-resource or multilingual settings.

To address $RQ2$, we compiled a hybrid dataset that integrates large-scale, general-purpose corpora with curated domain-specific hate speech datasets. This integration substantially improved model robustness, yielding consistently high performance across all platforms that provide data with severity scores. While our approach shows promise in unifying diverse data sources, it also introduced potential noise from dataset heterogeneity, including variations in annotation criteria, label definitions, and platform-specific language styles.

From a practical standpoint, modeling hate speech severity as a regression task offers practical benefits for content moderation by enabling proportional and context-sensitive enforcement actions, such as prioritizing severe instances for removal and applying softer interventions to milder content. However, the approach faces several limitations, including the subjectivity of severity annotations across cultural and individual perspectives [13], the lack of temporal or conversational context in current models, and the need to evaluate cross-platform generalizability. Addressing these challenges through uncertainty modeling, contextual encoding, and fairness assessments will be essential for building more robust and equitable hate speech severity prediction systems. We therefore recommend that future research explore the integration of conversational context and user-level metadata to improve the contextual understanding of severity. A cross-platform evaluation to assess and conduct fairness audits to ensure equitable predictions across demographic groups. Finally, we recommend incorporating explainability techniques, such as attention visualization or SHAP, to support the development of transparent, deployable moderation tools.

Acknowledgments. This paper was supported by the grant No. SGS_2025_018 provided by the Faculty of Economics and Administration, University of Pardubice.

Disclosure of Interests. The authors have no competing interests to declare that are relevant to the content of this article.

References

1. Albladi, A., et al.: Hate speech detection using large language models: a comprehensive review. IEEE Access **13**, 20871–20892 (2025)
2. Ali, S., Blackburn, J., Stringhini, G.: Evolving hate speech online: an adaptive framework for detection and mitigation. arXiv preprint arXiv:2502.10921 (2025)

3. Asante, A., Hajek, P.: Detecting antisocial behavior on social media during Covid-19 lockdown. In: Novel & Intelligent Digital Systems Conferences, pp. 189–200. Springer (2024). https://doi.org/10.1007/978-3-031-73344-4_15
4. Asante, A., Hajek, P.: Beyond trolling: fine-grained detection of antisocial behavior in social media during the pandemic. Information **16**(3), 173 (2025)
5. Ayele, A.A., Jalew, E.A., Ali, A.C., Yimam, S.M., Biemann, C.: Exploring boundaries and intensities in offensive and hate speech: unveiling the complex spectrum of social media discourse. arXiv preprint arXiv:2404.12042 (2024)
6. Bourgonje, P., Moreno-Schneider, J., Srivastava, A., Rehm, G.: Automatic classification of abusive language and personal attacks in various forms of online communication. In: Rehm, G., Declerck, T. (eds.) GSCL 2017. LNCS (LNAI), vol. 10713, pp. 180–191. Springer, Cham (2018). https://doi.org/10.1007/978-3-319-73706-5_15
7. Cao, W., Mirjalili, V., Raschka, S.: Rank consistent ordinal regression for neural networks with application to age estimation. Pattern Recogn. Lett. **140**, 325–331 (2020)
8. Caselli, T., Basile, V., Mitrović, J., Granitzer, M.: HateBERT: retraining BERT for abusive language detection in English. arXiv preprint arXiv:2010.12472 (2020)
9. Chakrabarty, N.: A machine learning approach to comment toxicity classification. In: Das, A.K., Nayak, J., Naik, B., Pati, S.K., Pelusi, D. (eds.) Computational Intelligence in Pattern Recognition. AISC, vol. 999, pp. 183–193. Springer, Singapore (2020). https://doi.org/10.1007/978-981-13-9042-5_16
10. Cheng, J., Wang, Z., Pollastri, G.: A neural network approach to ordinal regression. In: 2008 IEEE International Joint Conference on Neural Networks, pp. 1279–1284. IEEE (2008)
11. Borkan, D., Sorensen, J., Dixon, L., Vasserman, L.: Jigsaw unintended bias in toxicity classification. https://kaggle.com/competitions/jigsaw-unintended-bias-in-toxicity-classification (2019). kaggle
12. Clark, K., Luong, M.T., Le, Q.V., Manning, C.D.: Electra: pre-training text encoders as discriminators rather than generators. arXiv preprint arXiv:2003.10555 (2020)
13. Davani, A.M., Díaz, M., Prabhakaran, V.: Dealing with disagreements: looking beyond the majority vote in subjective annotations. Trans. Assoc. Comput. Linguist. **10**, 92–110 (2022)
14. Davidson, T., Warmsley, D., Macy, M., Weber, I.: Automated hate speech detection and the problem of offensive language. In: Proceedings of the International AAAI Conference on Web and Social Media, vol. 11, pp. 512–515 (2017)
15. Devlin, J., Chang, M.W., Lee, K., Toutanova, K.: BERT: pre-training of deep bidirectional transformers for language understanding. In: Proceedings of the 2019 Conference of the North American Chapter of the Association for Computational Linguistics: Human Language Technologies, vol. 1, pp. 4171–4186 (2019)
16. ElSherief, M., et al.: Latent hatred: a benchmark for understanding implicit hate speech. arXiv preprint arXiv:2109.05322 (2021)
17. Fortuna, P., Nunes, S.: A survey on automatic detection of hate speech in text. ACM Comput. Surv. (CSUR) **51**(4), 1–30 (2018)
18. Fortuna, P., Soler, J., Wanner, L.: Toxic, hateful, offensive or abusive? What are we really classifying? an empirical analysis of hate speech datasets. In: Proceedings of the Twelfth Language Resources and Evaluation Conference, pp. 6786–6794 (2020)
19. Kennedy, C.J., Bacon, G., Sahn, A., von Vacano, C.: Constructing interval variables via faceted RASCH measurement and multitask deep learning: a hate speech application. arXiv preprint arXiv:2009.10277 (2020)
20. Kohl, U.: Platform regulation of hate speech-a transatlantic speech compromise? J. Media Law **14**(1), 25–49 (2022)
21. Kumar, M., et al.: Exploring hate speech detection: challenges, resources, current research and future directions. Multimedia Tools Appl., 1–37 (2025)

22. Kumari, K., Singh, J.P., Dwivedi, Y.K., Rana, N.P.: Bilingual cyber-aggression detection on social media using LSTM autoencoder. Soft. Comput. **25**, 8999–9012 (2021)
23. Liu, Y., et al.: RoBERTa: a robustly optimized BERT pretraining approach. arXiv preprint arXiv:1907.11692 (2019)
24. Lu, F., Ferraro, F., Raff, E.: Continuously generalized ordinal regression for linear and deep models. In: Proceedings of the 2022 SIAM International Conference on Data Mining (SDM), pp. 28–36. SIAM (2022)
25. Mahmud, T., Ptaszynski, M., Eronen, J., Masui, F.: Cyberbullying detection for low-resource languages and dialects: review of the state of the art. Inf. Process. Manage. **60**(5), 103454 (2023)
26. Masud, S., Bedi, M., Khan, M.A., Akhtar, M.S., Chakraborty, T.: Proactively reducing the hate intensity of online posts via hate speech normalization. In: Proceedings of the 28th ACM SIGKDD Conference on Knowledge Discovery and Data Mining, pp. 3524–3534 (2022)
27. Mathew, B., Saha, P., Yimam, S.M., Biemann, C., Goyal, P., Mukherjee, A.: HATExplain: A benchmark dataset for explainable hate speech detection. In: Proceedings of the AAAI Conference on Artificial Intelligence, vol. 35, pp. 14867–14875 (2021)
28. Mollas, I., Chrysopoulou, Z., Karlos, S., Tsoumakas, G.: Ethos: a multi-label hate speech detection dataset. Complex Intell. Syst. **8**(6), 4663–4678 (2022)
29. Mozafari, M., Farahbakhsh, R., Crespi, N.: Cross-lingual few-shot hate speech and offensive language detection using meta learning. IEEE Access **10**, 14880–14896 (2022)
30. Ousidhoum, N., Lin, Z., Zhang, H., Song, Y., Yeung, D.Y.: Multilingual and multi-aspect hate speech analysis. arXiv preprint arXiv:1908.11049 (2019)
31. Piot, P., Martín-Rodilla, P., Parapar, J.: MetaHate: a dataset for unifying efforts on hate speech detection. In: Proceedings of the International AAAI Conference on Web and Social Media, vol. 18, pp. 2025–2039 (2024)
32. Riyadi, S., Andriyani, A.D., Sulaiman, S.N.: Improving hate speech detection using double-layers hybrid CNN-RNN model on imbalanced dataset. IEEE Access **12**, 159660–159668 (2024)
33. Sharif, O., Hoque, M.M.: Tackling cyber-aggression: identification and fine-grained categorization of aggressive texts on social media using weighted ensemble of transformers. Neurocomputing **490**, 462–481 (2022)
34. Singhal, P., Walambe, R., Ramanna, S., Kotecha, K.: Domain adaptation: challenges, methods, datasets, and applications. IEEE Access **11**, 6973–7020 (2023)
35. Toraman, C., Şahinuç, F., Yilmaz, E.H.: Large-scale hate speech detection with cross-domain transfer. In: Proceedings of the Language Resources and Evaluation Conference, pp. 2215–2225 (2022)
36. Wang, J., Cheng, Y., Chen, J., Chen, T., Chen, D., Wu, J.: Ord2Seq: regarding ordinal regression as label sequence prediction. In: Proceedings of the IEEE/CVF International Conference on Computer Vision, pp. 5865–5875 (2023)
37. Zampieri, M., Malmasi, S., Nakov, P., Rosenthal, S., Farra, N., Kumar, R.: Predicting the type and target of offensive posts in social media. arXiv preprint arXiv:1902.09666 (2019)

MLP Model for Prediction of Pellet Combustion: How to Deal with Small Datasets

Philippe Thomas[1](✉) [iD], Eliott Gauthey-Franet[1,2,3] [iD], Yinling Liu[1] [iD],
Hind Bril El-Haouzi[1] [iD], Jérémy Hugues Dits Ciles[3], and Yann Rogaume[2] [iD]

[1] Université de Lorraine, CNRS, 54000 Nancy, CRAN, France
philippe.thomas@univ-lorraine.fr
[2] Université de Lorraine, INRAE, LERMAB, ERBE, 88000 Epinal, France
[3] N2AIR, 39000 Mantry, France

Abstract. The increase in the use of wood in general, and pellets in particular, for individual heating requires attention to be paid to optimizing the combustion of these pellets, particularly in terms of pollutant emissions such as carbon monoxide (CO). This quest for optimization is complicated by the intrinsic variability of wood and, therefore, pellets. In this context, this paper proposes building a neural network model to predict the CO content in smoke based on pellet characteristics and stoves characteristics and settings. However, experiments are costly and time-consuming, which limits the size of the available dataset. In this context, we propose a methodology aimed at training a multilayer perceptron using a small dataset while reducing the risk of overfitting. This methodology is based on finding the minimal network structure and using a robust learning algorithm. The results show that the robust algorithm effectively limits the risk of overfitting and that the final model retains its generalization capabilities despite the small size of the dataset.

Keywords: Neural Network · Multilayer Perceptron · Pellet Stoves · Combustion Optimization

1 Introduction

In France and across Europe, domestic wood heating remains the leading source of renewable energy [1], primarily through standalone appliances such as stoves and closed fireplaces commonly used in residential settings. Since 2010, pellet stove sales have experienced significant upswing, growing from 27,000 units in 2010 to about 150,000 units by 2019, before slightly declining to 121,550 in 2020. Notably, since 2018, pellet stoves have consistently outsold traditional log stoves. Alongside this trend, pellet consumption has continued to rise steadily. Domestic production has followed suit, with France consuming an estimated 1.85 million tons and producing 1.65 million tons of pellets in 2019 [2].

To meet increasing demand without affecting other sectors that rely on wood, it is necessary to diversify wood feedstock sources—an approach that inevitably changes

F. Marcelloni et al. (Eds.): IJCCI 2025, CCIS 2829, pp. 690–705, 2026.
https://doi.org/10.1007/978-3-032-15638-9_41

the physical and chemical properties of the pellets. Although most pellet stoves are designed to accommodate a variety of pellet types, such variability often leads to a decline in performance, particularly regarding pollutant emissions (e.g., carbon monoxide and particulate matter) and overall combustion efficiency. Research conducted at the LERMAB laboratory has further demonstrated the crucial role of stove settings in combustion outcomes, especially in terms of reducing unburned emissions and enhancing efficiency [3]. This study thus aims to develop a method for automatically adjusting stove settings based on the specific characteristics of the pellets being used.

A multilayer perceptron (MLP) is a promising approach for predicting pollutant emissions from various pellets burned in the same stove under different conditions. However, we are dealing with a small dataset for the following reasons: firstly, each experiment takes several hours, and secondly, the complexity of the physical processes being studied and the large number of factors influencing pollutant emissions and efficiency make it difficult to design a set of experiments that efficiently covers the entire experimental space. Therefore, the challenge lies in effectively using an MLP model to predict pellet combustion with a limited dataset.

The main risk of using a small dataset to train an MLP is the risk of overfitting. The objective of this paper is to propose and test a methodology for reducing this risk of overfitting on the problem under consideration. This methodology is based on two pillars:

- Determining the optimal structure of the MLP
- Using a robust learning algorithm.

The following section will present the proposed procedure. This section includes a presentation of the structure of the MLP itself, its learning in relation to the risk of overfitting, and the proposed procedure. The third part presents the results obtained on the application example. In this part, the process on which the experiments are carried out and the dataset are presented. Next, the determination of the number of hidden neurons is studied. An experiment is also carried out to show the regularization effect of the robust criterion. Unnecessary inputs are then eliminated from the MLP structure before training the model to be used. Finally, a conclusion closes this paper.

2 Materials and Methods

2.1 Structure of Multilayer Perceptron

Since the 1980s, it has been understood that a multi-layer perceptron including a single hidden layer whose neurons include a sigmoidal activation function can approximate any non-linear function with the required accuracy [1, 2]. In this study, the focus will be exclusively on a model comprising a single output, with an activation function that is linear in nature (the problem under consideration being a prediction problem). The network's output $\hat{y}$ is thus expressed as follows:

$$\hat{y} = \sum\nolimits_{h=1}^{n_o} w_h^2 \cdot g_h \left(\sum\nolimits_{i=1}^{n_i} w_{hi}^1 \cdot x_i + b_h^1 \right) + b. \tag{1}$$

In this model, x_i denotes the n_i inputs, w_{hi}^1 represents the connecting weights between the input and hidden layers, b_h^1 denotes the hidden neuron biases, $g_h(.)$ represents the activation function of the hidden neurons (hyperbolic tangent), w_h^2 represents the connecting weights between the hidden and output layers, and b represents the bias of the output neuron.

2.2 Learning of Multilayer Perceptron

The learning algorithms employed are typically gradient descent algorithms that execute a local search to identify the minimum [3]. The algorithm employed in this study is a Hessian algorithm [4–6]. The Hessian algorithm has the advantage of converging more quickly than the gradient algorithm. However, it does present a risk of instability. The Levenberg-Marquardt implementation used allows this risk to be controlled by limiting the contribution of the Hessian matrix as the solution is approached. It is therefore imperative that learning commences from weights and biases selected at random, thereby enabling the reiteration of learning from disparate starting points. This approach circumvents the predicament of local minimum trapping [7]. The initialization algorithm employed in this study constitutes a modification of the Nguyen and Widrow algorithm [8, 9].

The second issue to be avoided is that of overfitting. This problem is especially pronounced in the present case, as the datasets available are of a very small size. [10] tests and compares different regularization cocktails on 40 datasets. However, all these datasets are much larger than the one considered here. In order to mitigate the risk of overfitting, a range of strategies are routinely employed:

- Early stopping
- Determination of the minimal structure of the network
- Regularization.

The process of early stopping involves the division of the dataset into two distinct segments: one allocated for the learning phase and the other designated for the validation stage. During the learning process, the criterion to be minimized is also calculated on the validation dataset. An increase in the value of this criterion indicates that the network is beginning to learn the noise and that learning should be stopped. This approach is both simple and effective, but it is important to note that it requires a sufficiently large validation dataset in order to evaluate this instant correctly [11].

A variety of strategies can be employed to ascertain the optimal network structure [12]. The two primary approaches to this process are as follows: firstly, the commencement of the procedure may be initiated with a minimal structure, with the gradual incorporation of hidden neurons (a constructive approach); alternatively, the initiation may be from an oversized structure, with the subsequent elimination of spurious parameters (pruning). It is evident that these approaches necessitate substantial validation sets. An alternative approach involves the application of brute force, whereby a trial-and-error method is employed in conjunction with Leave One Out (LOO) cross-validation [13]. LOO is a process that involves the learning of a model on all but one of the available data, with the validation of the model obtained on the last data. Achieving optimal network structure necessitates the implementation of this process for the entirety of the data contained within the dataset.

The process of regularization can be defined as the incorporation of an additional term into the criterion that is to be minimized. The objective of this incorporation is to impose constraints on the parameter space [6, 14]. An alternative approach that has a regularizing effect is to use a robust criterion to minimize instead of the classical quadratic criterion [15]. In order to ascertain the most suitable model, [15] subjected a number of robust criteria to rigorous testing and comparison. The model that proved most effective was Huber's model of measurement noise contaminated by outliers [16]. In this model, the noise distribution is considered to be a mixture of two density functions. The first of these functions represents the basic noise distribution $(N(0, \sigma_1^2))$, whilst the second allows outliers to be represented $(N(0, \sigma_2^2)$, with $\sigma_1^2 < \sigma_2^2)$:

$$e \sim (1 - \delta)N\left(0, \sigma_1^2\right) + \delta N\left(0, \sigma_2^2\right). \tag{2}$$

The probability of an outlier occurring is denoted by δ, which is unknown. The model is utilized in the construction of a weighted minimization criterion:

$$V(\theta) = \frac{1}{2N} \sum_{k=1}^{N} \frac{\varepsilon(k, \theta)}{\sigma^2(k)}. \tag{3}$$

The symbol $\varepsilon(k, \theta)$ is employed to denote the discrepancy between the model's estimation and the actual data value for data k, with:

$$\begin{cases} \sigma^2(k) = \sigma_1^2 \ if \ |\varepsilon(k, \theta)| \leq 3.\sigma_1 \\ \sigma^2(k) = \sigma_2^2 \ if \ |\varepsilon(k, \theta)| > 3.\sigma_1 \end{cases} \tag{4}$$

It is evident that both σ_1^2 and σ_2^2 remain unknown. The estimation of these values is achieved through the utilization of the following relationships:

$$\begin{cases} \widehat{\sigma_1} = \frac{MAD}{0.7} \\ \widehat{\sigma_2} = 3.\widehat{\sigma_1} \end{cases}, \tag{5}$$

where MAD is the median of $\left|\varepsilon(k, \theta) - \widetilde{\varepsilon}\right|$ and where $\widetilde{\varepsilon}$ is the median of $\varepsilon(k, \theta)$.

2.3 Proposed Procedure

In this paper, we propose a methodology that integrates LOO with the implementation of a robust algorithm to construct the most optimal model possible, utilizing the available data set. In order to achieve this objective, the proposed approach is to be implemented in several stages:

1. Normalization of input variables
2. Determination of the optimal number of hidden neurons
3. Determination of explanatory input variables

4. Learning the model

The objective of normalizing the input variables is to ensure that each input possesses a comparable amplitude and variation. The following procedure is employed to normalize each input Ξ_i:

$$x_i(k) = \frac{\xi_i(k) - \overline{\Xi_i}}{std(\Xi_i)}, \tag{6}$$

where $\overline{\Xi_i}$ is the mean and $std(\Xi_i)$ is the standard deviation of the input Ξ_i.

To determine the optimal number of hidden neurons, all input variables X_i are used to learn different models, varying the number of hidden neurons H. The learning algorithm employs the robust criterion in order to limit the risk of overfitting. The validation process is constructed utilizing a LOO strategy. This indicates that for all values of H, N models are learned on N-1 data and validated on the remaining data. The validation criterion employed is the root mean square error ($RMSE$) calculated on the N validation errors $\varepsilon(k)$ provided by the N models (1 data per model):

$$RMSE = \sqrt{\frac{\sum_{k=i}^{N} \varepsilon(k)^2}{N}}. \tag{7}$$

The RMSE criterion has the advantage of simultaneously evaluating the accuracy and precision of the model. Moreover, it is optimal for Gaussian error [17]. The number H of hidden neurons incorporated into the final model is given by the structure leading to the lowest RMSE value.

Subsequent to the determination of the number H, an MLP model is constructed, incorporating all input variables (X_i) and H hidden neurons. The objective of this study is to ascertain which input variables are genuinely useful and which can be eliminated. Consequently, the LOO cross-validation approach is employed once more to leverage the available data to its maximum potential. To mitigate the repercussions of the local minimum trapping problem, the procedure is reiterated on multiple occasions with varied initializations. Furthermore, it is imperative that the regression models are constructed in such a manner that they yield a residual with a mean of zero. To assess this, it is possible to employ a two-tailed statistical hypothesis test. In this case, the null hypothesis $\mathcal{H}_0$ (the mean of the residual is null) and its alternative $\mathcal{H}_1$ are as follows:

$$\begin{cases} \mathcal{H}_0 : \mu = 0 \\ \mathcal{H}_1 : \mu \neq 0 \end{cases} \tag{8}$$

The mean of the residuals is denoted by μ. The null hypothesis, $\mathcal{H}_0$, is to be rejected at a 5% risk level if the following conditions are met:

$$|U| = \left| \frac{\overline{\varepsilon}}{s/\sqrt{N}} \right| > 1.96. \tag{9}$$

In this context, the symbol $\overline{\varepsilon}$ denotes the mean of the residual population, whilst s represents the standard deviation of this population.

Subsequent to the acquisition of these models, a process of testing ensues, entailing the neutralization of a single input X_i from the designated set of input variables. X_i can be neutralized by the simple freezing of it at its mean value $(\overline{X_i})$. In this context, where inputs are subjected to normalisation, their mean values are systematically equivalent to 0. Consequently, the residuals $\varepsilon_i(k)$ can be constructed, which will then be compared with the residuals $\varepsilon(k)$ obtained with the full models. Should these two residuals be found to be statistically equivalent, it can be deduced that the input X_i is not an explanatory variable and, as such, can be omitted from the model. To assess this, it is of course necessary for $\varepsilon_i(k)$ to have zero mean, which can be tested with hypothesis test (8). It is also necessary that the variance σ_1^2 of the residual $\varepsilon_i(k)$ is less than or equal to the variance σ_2^2 of the residual $\varepsilon(k)$. The employment of a statistical hypothesis test is an effective method for achieving this objective:

$$\begin{cases} \mathcal{H}_0 : \sigma_1^2 = \sigma_2^2 \\ \mathcal{H}_1 : \sigma_1^2 > \sigma_2^2 \end{cases} \tag{10}$$

The hypothesis $\mathcal{H}_0$ is rejected if:

$$F = \frac{\sum_{i=1}^{n_1}\left(x_{i1} - \overline{X_1}\right)^2 \Big/ n_1 - 1}{\sum_{j=1}^{n_2}\left(x_{j2} - \overline{X_2}\right)^2 \Big/ n_2 - 1} > F_{1-\alpha;n_1-1;n_2-1}. \tag{11}$$

The value of the F distribution with a risk level of 5% is denoted by $F_{1-\alpha;n_1-1;n_2-1}$.

At this stage, the optimal number of hidden neurons, H, and the explanatory input variables, X_i, are known. The optimal network structure is then known, and the network can be learned by traditionally subdividing the dataset into a training set and a validation set. Once more, due to the limited number of data items, a mere 15% of the dataset is retained for the purpose of validation.

3 Applicative Example

3.1 Experiment Process and Data

A Hoben H5b pellet stove, rated at a nominal output power of 6.3 kW and a maximum efficiency of 90%, was used in the experiments. The operating principle of the stove is depicted in Fig. 1. The stove is first loaded with a known type of pellet and set to operate at the nominal power level previously determined for that specific pellet. Pellets are then fed into the combustion chamber, where they are ignited by a spark plug and combusted with the aid of supplied air. The resulting flue gases serve to heat the indoor air and are simultaneously monitored by a sensor prior to being discharged. Various combustion airflow rates are manually adjusted to obtain multiple measurements of carbon monoxide (CO) concentrations in the exhaust gases.

The experimental setup is designed to provide detailed insights into various combustion characteristics and consists of three main elements: the pellets, the pellet stove, and the sensors. Five types of pellets—BIO, MOU, AR40, AR79, and LUZ—were selected from different producers, representing diverse geographical origins and raw materials.

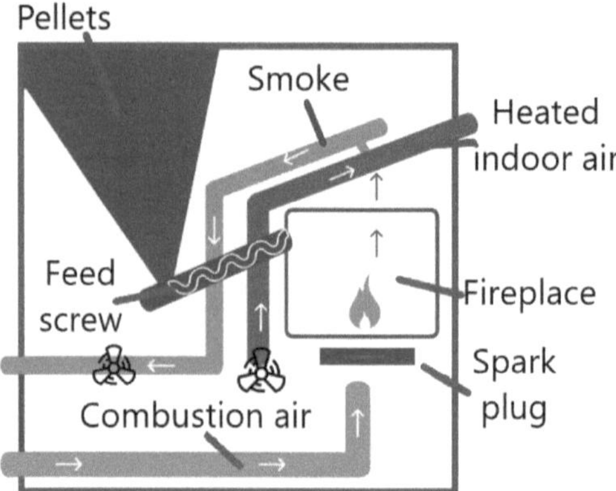

Fig. 1. Pellet stove operation process.

Table 1. Characteristics of tested pellets.

	BIO	MOU	AR40	AR79	LUZ
gross calorific value (kJ/kg)	19856	20328	20462	20140	20038
bulk density (kg/m^3)	661	637	656	657	665
particular density (kg/m^3)	1148	1063	1075	1054	1109
moisture content (%wb)	3.79	7.84	5.19	4.45	6.22
ash rate at 550 °C %(db)	0.71	0.51	0.40	0.47	0.44
length (mm)	16.67	13.59	11.67	13.75	16.43
diameter (mm)	6.10	6.18	6.17	6.19	6.14
durability (%)	99.13	99.13	99.06	98.30	99.24
C rate (%db)	50.1	50.9	51.6	50.9	50.7
H rate (%db)	6.26	6.52	6.4	6.32	6.28
N rate (%db)	0.1	0.1	<0.1	<0.1	<0.1
O rate (%db)	43.51	42.5	41.9	42.7	42.9

The properties of these pellets are summarized in Table 1. Among them, the BIO pellet appears to have a lower gross calorific value and a higher ash content compared to the others. This is likely due to its composition, which includes a mix of hardwood and softwood, whereas the other pellets are made exclusively from softwood, generally known for having a slightly higher energy content. Additionally, the MOU pellet exhibits the highest moisture content, potentially resulting from the conditions under which the raw logs were stored prior to or during processing.

Figure 2 presents the input data summarized by block (granulometry, gas analysis, etc.) and structured into three distinct tables: pellet properties, stove specifications, and experimental results. The input data includes the features as follows: [*(gross calorific* *value, bulk density, particular density, moisture, ash rate at 550 °C, pellet's length, pellet's diameter, durability, carbon rate, hydrogen rate, oxygen rate), (output power max, efficiency, CO rate at maximum power, smoke termperature at maximum power, air input diameter), (speed rotation, combustion air flow, O_2 rate in combustion gas, gas temperature, efficiency)].* The output data, CO, is a commonly utilized indicator of emission quality in combustion processes. The dataset is constituted by 181 experiments.

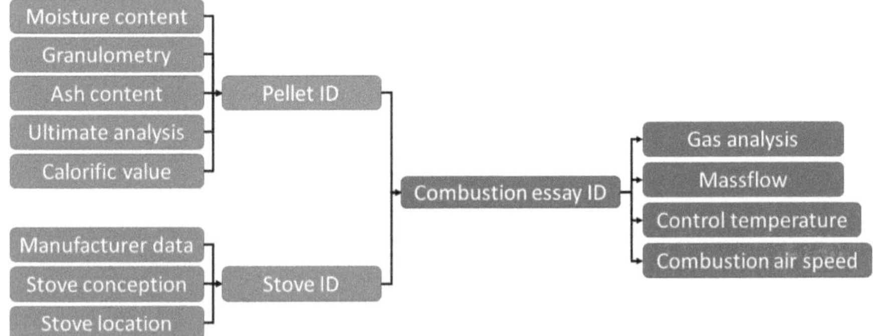

Fig. 2. Relationship between Pellet, Stove and Combustion essay.

The considered dataset includes 23 inputs (features), 1 output, and 181 instances.

3.2 Determination of the Hidden Neurons Number H

The initial step in this process is to ascertain the number of hidden neurons, denoted by H. To this end, a LOO strategy is employed in conjunction with a robust criterion to learn models whose structure varies from 1 to 16 hidden neurons.

As illustrated in Fig. 3, the *RMSE* was calculated by Eq. 7 on the basis of the validation data from the various LOO models for the different structures. The findings indicate that, while the *RMSE* initially decreases with increasing network size, it increases again after a certain point ($H = 7$). The findings indicate that while the implementation of a robust criterion mitigates the risk of overfitting, it does not entirely preclude it, as structures employing 8 to 16 hidden neurons evidently demonstrate overfitting.

3.3 Illustration of the Regularization Effect of the Robust Criterion

At this stage, the number of hidden neurons can be set at 7, since this structure gives the lowest *RMSE*.

The subsequent stage of the research will involve the construction of models to test the significance of the inputs. To accomplish this objective, it is necessary to employ the LOO strategy once more. To illustrate the regularizing effect of the robust criterion, a comparison will be made between the results obtained using this criterion and those obtained using the classic quadratic criterion. In order to limit the impact of the local

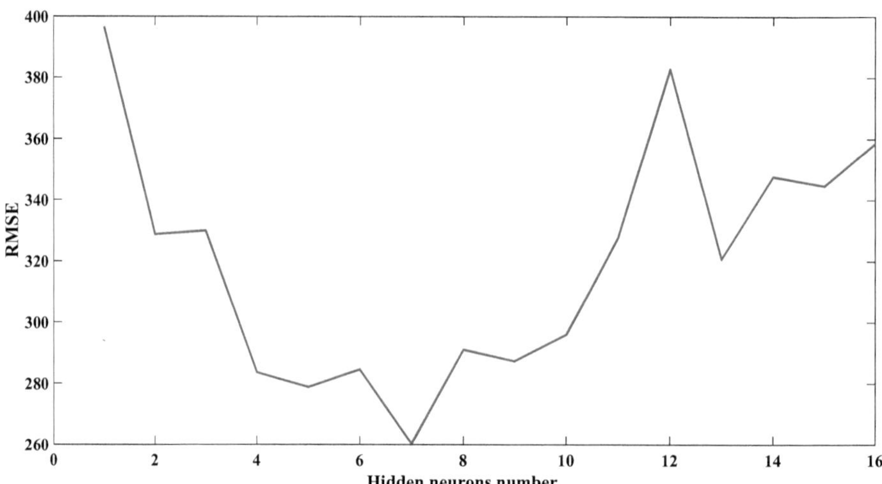

Fig. 3. Evolution of *RMSE* obtained for different values of *H*.

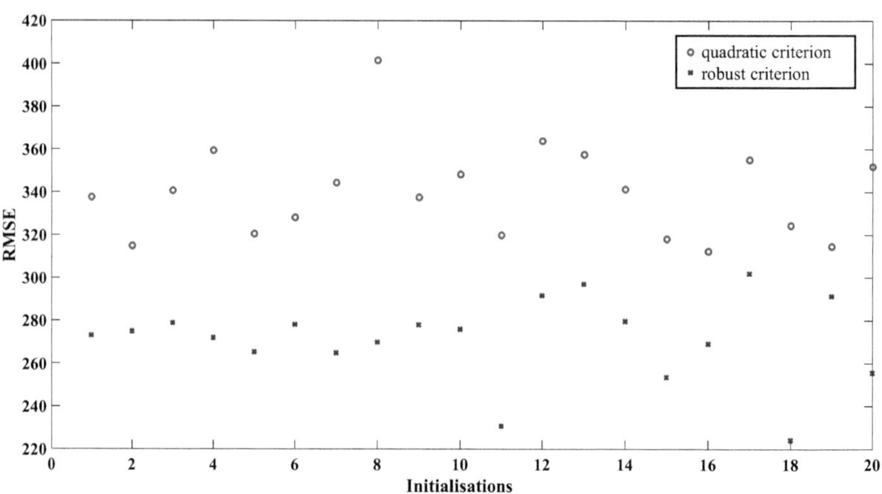

Fig. 4. Presentation of *RMSE* values obtained in validation for the 20 replicates.

minimum trapping problem, this work will be carried out 20 times with different initializations. Models learned with the robust criterion and with the quadratic criterion will be learned with the same seeds.

As demonstrated in Fig. 4, the *RMSE* values obtained from the validation process with the two criteria for the 20 repetitions are presented. It is evident that statistical tests are not a prerequisite for demonstrating the significant enhancement in outcomes that is attributable to the robust criterion. Table 2 presents some statistics on the *RMSE* values obtained with the two robust and quadratic criteria. The following table illustrates that the minimum *RMSE* obtained with the quadratic criterion is greater than the maximum

RMSE obtained with the robust criterion. It is evident that the quadratic criterion results in substantial overfitting, while the robust criterion limits it. This is likely due to the fact that the robust criterion allows noise to be partially modeled, whereas the quadratic criterion assumes that the data are noise-free. As illustrated in Table 3, the number of iterations utilized by the two criteria to attain the minimum is also presented. It is evident that the robust algorithm requires a significantly smaller number of iterations to reach its minimum in comparison to the classical algorithm.

Table 2. RMSE values in validation.

Criterion	Quadratic	Robust
Mean	339.6	271.4
Std	22.0	19.4
Min	312.5	224.2
Max	401.5	302

The learning algorithm employed is a Levenberg-Marquard algorithm, which incorporates a parameter, λ, to guarantee that the Hessian matrix remains invertible at all times. A procedure for evolving this parameter λ makes it possible to increase its value when the improvement in the criterion to be minimized between two iterations is insufficient or, on the contrary, to decrease it when the progress is significant [6, 16].

It is evident that a minimal value of λ results in the learning algorithm manifesting characteristics analogous to those of a second-order algorithm, consequently leading to an augmentation in the progression step. Conversely, a substantial value of λ engenders the algorithm's operation in a manner consistent with a first-order algorithm, (gradient algorithm).

The implementation of the Levenberg-Marquard algorithm used [6] includes various stopping tests to finalize the learning process:

- Maximum number of iterations (set to 5000)
- Objective criterion value (set to 0 as unknown)
- Maximum value of λ (set at 10^7; indicative of learning that is no longer progressing)
- Minimum criterion improvement value between two iterations (set at 10^{-7})

Table 4 presents the stopping criterion used for the two algorithms. It is evident that the classical algorithm predominantly reaches the maximum number of iterations. Conversely, the robust algorithm demonstrates a propensity to augment the parameter λ and methodically terminate at this criterion.

Prior to proceeding, it is imperative to ascertain that the residuals provided by the 20 replicates (robust criterion) have a mean of zero. As illustrated in Fig. 5, the outcomes of test (8) for the validation residuals, obtained from the 20 replicates, are presented. As demonstrated in Fig. 5, the U values (9) for replicates 13 and 20 exceed the cut-off value of 1.96. The null hypothesis, therefore, is rejected for these two replicates, which are to be excluded from the remainder of the study.

Table 3. Number of iterations.

Criterion	Quadratic	Robust
Mean	4617	67.3
Std	1196	6
Min	45	859
Max	5000	67.7

Table 4. Criteria for stopping learning.

Criterion	Quadratic	Robust
Iterations number	88%	0%
Objective value	0%	0%
λ value	2%	100%
insufficient improvement	10%	0%

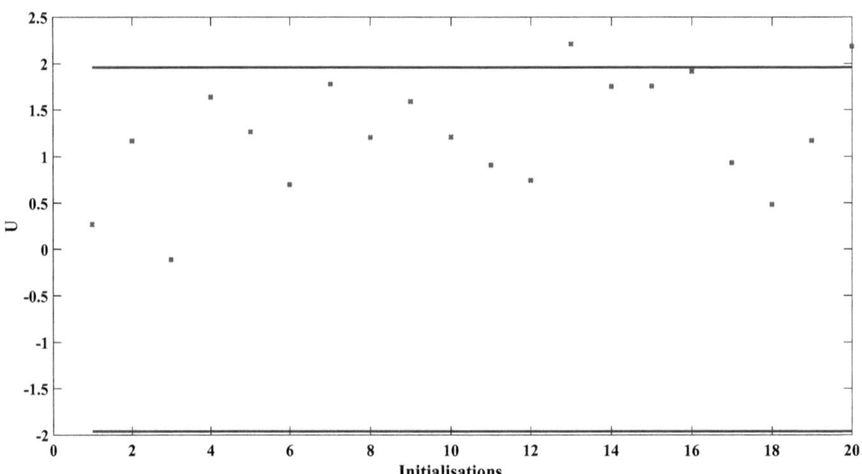

Fig. 5. Hypothesis test on the null mean of residuals in validation for the 20 replicates (robust criterion).

3.4 Determination of the Spurious Input Variables

The objective at this stage is to utilize the 18 replications that have been retained in order to assess the relevance of retaining or discarding the various input variables.

In order to accomplish this objective, it is necessary to sequentially neutralize each of the 23 inputs X_i in the various models constructed, thereby generating 23 validation residuals for each of the 18 replications. These residuals will then be subjected

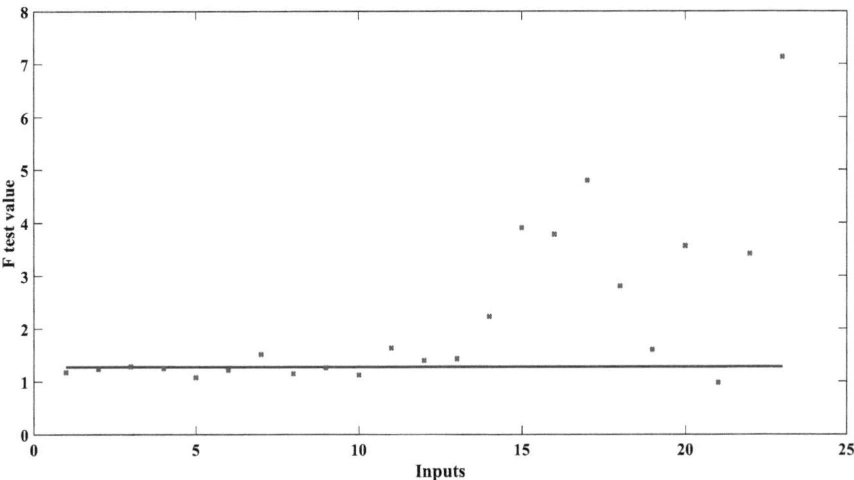

Fig. 6. F-test for the 23 inputs during the 18th repetition.

to comparison with the previously obtained validation residual, utilizing statistical test (10). As demonstrated in Fig. 6, the F-test results for the 23 inputs in 18th replication are presented. This figure indicates that multiple inputs possess an F-value that falls below the established limit. This finding suggests that, for these particular inputs, the $\mathcal{H}_0$ hypothesis cannot be rejected, thereby affirming that these inputs do not provide a satisfactory explanation for the output variable and can be disregarded from the model. Table 5 shows the proportion of repetitions leading to models in which the input variable considered has no impact. In order to reduce the size of the table, input variables that are always explanatory are not shown. The table indicates that 10 inputs are considered non-explanatory in at least one replication. Of the ten entries under consideration, five are considered non-explanatory in at least nine of the eighteen repetitions. In the following study, two types of model will be constructed and tested:

- A model excluding the 10 entries in Table 4 (Small model)
- A model excluding only the 5 most frequently non-significant inputs (Great model)

Table 5. Proportions of repetitions where inputs are non-explanatory.

input	1	2	3	4	5	6	8	9	10	21
$\mathcal{H}_0$ accepted	89%	11%	28%	56%	17%	67%	6%	28%	50%	94%

3.5 Learning of the Models

Consequently, the optimal network structure has been determined as follows:

- The presence of seven hidden neurons has been detected.
- Thirteen inputs are required for the Small model.
- The Great model has been designed to incorporate a total of 18 inputs.

The dataset is randomly subdivided into two sets:

- The proportion of the dataset that has been allocated to training is 85%.
- The allocation for validation is 15%.

For both types of models, 100 different initializations are performed, resulting in 100 different learned models. The aim here is to avoid the problem of local minimum trapping. For all the models constructed, the residuals on the validation set are tested. Initially, the mean of these residuals is tested for zero using test (8). Figure 7 presents the outcomes of this evaluation for all models.

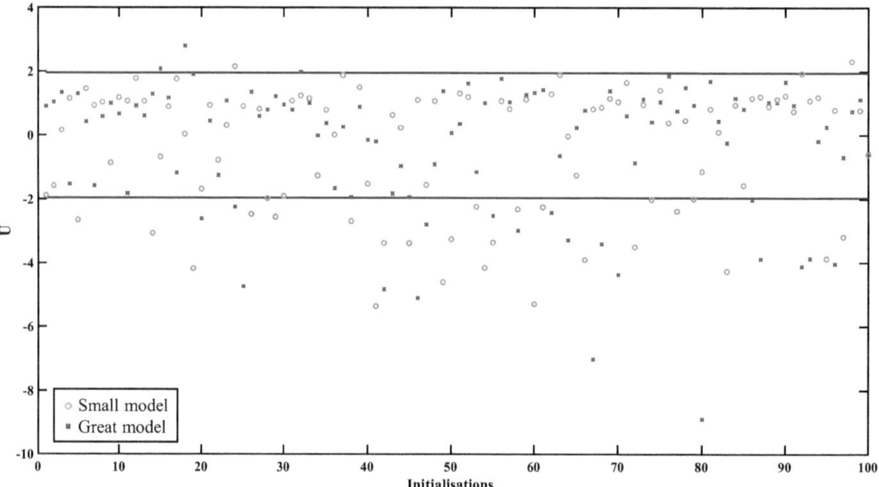

Fig. 7. Hypothesis test on the null mean of residuals in validation for the 200 models.

It is evident that a significant number of models generate residuals with non-zero mean values (U values that exceed the established limits). It is recommended that these models be disregarded.

Subsequently, the models will be examined to ascertain whether the residuals generated by these models exhibit a variance that is not statistically greater than the best variance obtained in paragraph 3.3 during the training process with all inputs, employing the LOO strategy. In order to accomplish this objective, the F-test (10) will be utilized.

As illustrated in Fig. 8, the outcomes of this evaluation are presented for the residuals provided by all models. It is evident that a multitude of models are associated with residuals that exhibit a variance that is statistically greater than that of the previously determined best residual (F values above the threshold). The aforementioned models should also be disregarded. As demonstrated in Fig. 9, the $RMSE$ obtained for the models that were not discarded is presented. Of these, the model that has achieved the lowest

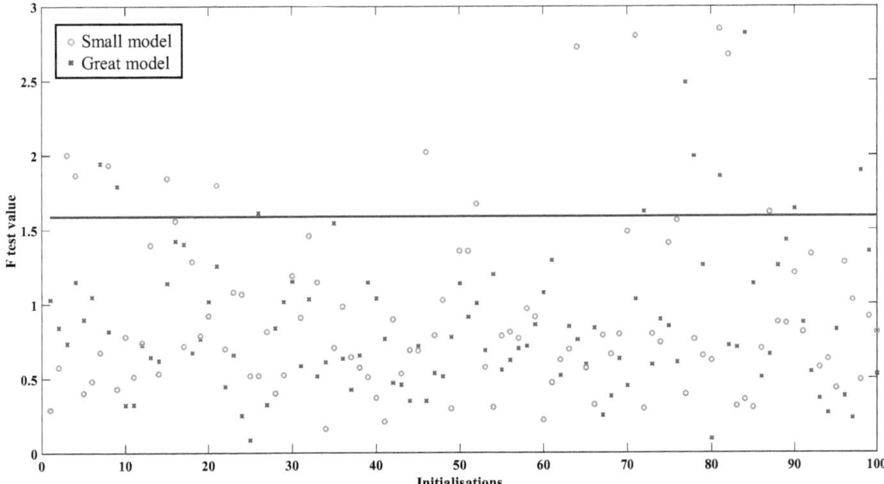

Fig. 8. F-test for the variance of the residuals obtained in validation for the 200 models.

RMSE incorporates a mere 13 of the original 23 inputs. The best model for predicting CO production in pellet combustion has so a structure comprising 13 inputs, seven hidden neurons and one output.

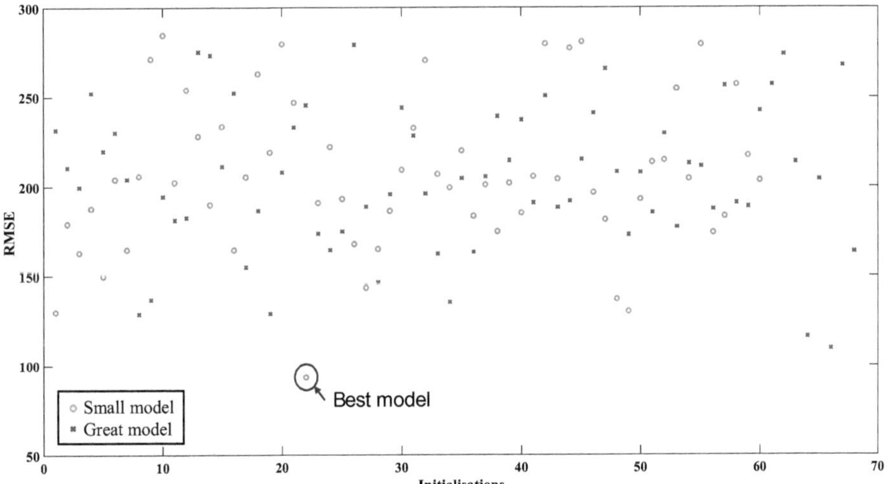

Fig. 9. RMSE on the validation dataset for the conserved models.

Figure 10 presents the target values for the validation dataset (blue cross) and the values predicted by the best model (red circle). As demonstrated by this figure, the constructed model demonstrates an aptitude for predicting the concentration of carbon monoxide (CO) during the process of pellet combustion.

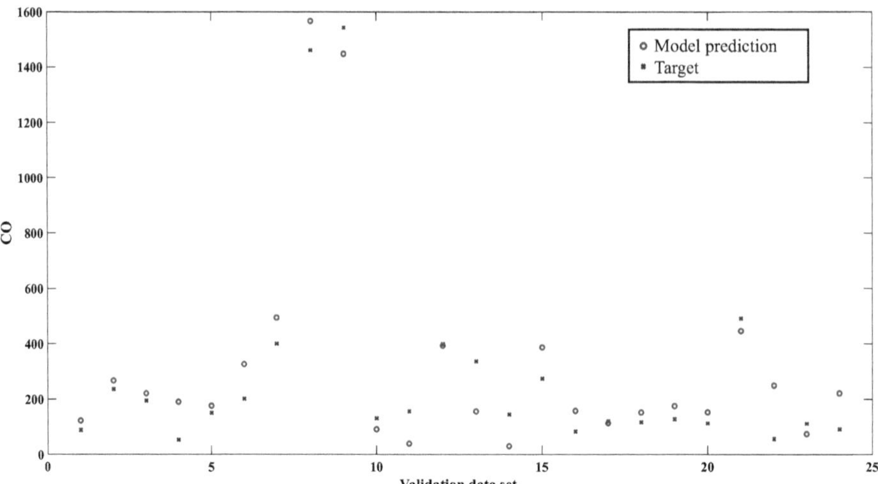

Fig. 10. Target and predicted values with the best model on validation dataset.

4   Conclusion

Predicting the pollutant levels in the smoke produced by burning pellets for individual heating is necessary in order to optimally adjust stoves. This paper proposes building a neural network model to perform this prediction. The main difficulty lies in the small amount of available data, which can lead to overfitting. A methodology is therefore proposed that combines the use of a robust algorithm with the search for the minimal structure of the network. The results obtained show that the robust algorithm does indeed have a regularization effect that limits the risk of overfitting. The resulting model is able to predict pollutant levels while avoiding the risk of overfitting. This work has been carried out with a single stove for the time being. In future work, we propose to extend the study to a wide range of stoves on the market. A second step will consist of using this prediction model to determine the optimal stove setting based on the characteristics of the pellets used.

References

1. Cybenko, G.: Approximation by superpositions of a sigmoidal function. Math. Control Sig. Syst. **2**(4), 303–314 (1989). https://doi.org/10.1007/BF02551274
2. Funahashi, K.-I.: On the approximate realization of continuous mappings by neural networks. Neural Netw. **2**(3), 183–192 (1989). https://doi.org/10.1016/0893-6080(89)90003-8
3. Rumelhart, D.E., McClelland, J.L.: Parallel Distributed Processing, Volume 1: Explorations in the Microstructure of Cognition: Foundations. The MIT press. (1986). Accessed 28 Jan 2025. https://direct.mit.edu/books/monograph/4424/Parallel-Distributed-Processing-Volume
4. Levenberg, K.: A method for the solution of certain non-linear problems in least squares. Quarterly of applied mathematics (1944)
5. Marquardt, D.W.: An algorithm for least-squares estimation of nonlinear parameters. J. Soc. Ind. Appl. Math. **11**(2), 431–441 (1963). https://doi.org/10.1137/0111030

6. Norgaard: Neural networks for modelling and control of dynamic systems. https://link.spr inger.com/book/9781852332273. Accessed 22 May 2025

7. Thomas, P., Bloch, G.: Initialization of one hidden layer feedforward neural networks for non-linear system identification. In: Presented at the 15th IMACS World Congress on Scientific Computation, Modelling and Applied Mathematics WC1997, Berlin, 1997, pp. 295–300

8. Nguyen, D., Widrow, B.: Improving the learning speed of 2-layer neural networks by choosing initial values of the adaptive weights. In: 1990 IJCNN International Joint Conference on Neural Networks, San Diego, CA, USA: IEEE, 1990, vol. 3, pp. 21–26. https://doi.org/10. 1109/IJCNN.1990.137819

9. Demuth, H., Beale, M., Hagan, M.: Neural network toolbox in MatLab. Mathworks (1994)

10. Kadra, A., Lindauer, M., Hutter, F., Grabocka, J.: Well-tuned simple nets excel on tabular datasets. Adv. Neural. Inf. Process. Syst. **34**, 23928–23941 (2021)

11. Prechelt, L.: Early Stopping - But When?. In: Neural Networks: Tricks of the Trade, Springer, Berlin, Heidelberg, 1998, pp. 55–69. https://doi.org/10.1007/3-540-49430-8_3

12. Thomas, P., Suhner, M.-C.: A new multilayer perceptron pruning algorithm for classification and regression applications. Neural. Process. Lett. **42**(2), 437–458 (2015). https://doi.org/10. 1007/s11063-014-9366-5

13. Hansen, L.K., Larsen, J.: Linear unlearning for cross-validation. Adv. Comput. Math. **5**(1), 269–280 (1996). https://doi.org/10.1007/BF02124747

14. Sugiyama, M., Ogawa, H.: Optimal design of regularization term and regularization parameter by subspace information criterion. Neural Netw. **15**(3), 349–361 (2002). https://doi.org/10. 1016/S0893-6080(02)00022-9

15. Thomas, P., Bloch, G., Sirou, F., Eustache, V.: Neural modeling of an induction furnace using robust learning criteria. Integr. Comput. Aided Eng. **6**(1), 15–26 (1999)

16. Huber, P.J.: Robust estimation of a location parameter. In: Breakthroughs in Statistics: Methodology and Distribution, pp. 492–518, Springer (1992). https://doi.org/10.1007/978-1-4612-4380-9_35. Accessed 22 May 2025

17. Hodson, T.O.: Root mean square error (RMSE) or mean absolute error (MAE): when to use them or not. Geosci. Model Dev. Discuss. **2022**, 1–10 (2022)

Multi-subspace SVD Generators for Continual Learning

Christiaan Lamers[1(✉)], Ahmed Nabil Belbachir[1], Thomas Bäck[2], and Niki van Stein[2]

[1] NORCE Norwegian Research Centre, Jon Lilletuns vei 9 H, 3. et, 4879 Grimstad, Norway
`chla@norceresearch.no`
[2] LIACS, Leiden University, Einsteinweg 55, 2333 CC Leiden, The Netherlands

Abstract. Catastrophic forgetting is a persistent obstacle for continual learning on memory-constrained edge devices. We introduce an extension of Lightweight Singular Value Decomposition Generators (SVD Generators), named Multi-Subspace SVD Generators (MSSG), that replaces each single global SVD with a small ensemble of low-rank SVDs per task and class. Each subspace captures a local linear patch; together they approximate the non-linear data manifold while keeping storage cost proportional to the rank, not the number of replay samples. A closed-form memory model allows MSSG to trade subspace count and rank against an equivalent raw-sample buffer, enabling fair comparison to experience-replay baselines. Across five image benchmarks; MNIST, Fashion MNIST, NOT MNIST, CIFAR10 and Tiny ImageNet, MSSG (i) outperforms its single-generator predecessor, (ii) matches or exceeds the accuracy of Experience Replay with a large buffer, and (iii) does so with less than one-tenth of their memory footprint. Because MSSG stores only compact factorised statistics, both rehearsal and generator updates run in milliseconds on resource-limited hardware, making it a practical drop-in replacement for replay buffers in on-device lifelong learning applications.

Keywords: Continual Learning · Manifold Learning · Lightweight Generators · Singular Value Decomposition · Experience Replay

1 Introduction

Tiny sensors, wearables, and household robots are expected to learn on the fly, from a trickle of data that arrives day by day, while running on a few megabytes of RAM and a stingy energy budget. Uploading raw data to the cloud is often impossible because of privacy regulations, bandwidth limits, or lack of connectivity. These constraints turn on-device continual learning from a nice-to-have feature into a technical necessity.

Continual learning (CL) asks a model to update its knowledge sequentially, without revisiting the full history. Naïve fine-tuning leads to catastrophic forgetting, where acquiring new skills overwrites old ones. Popular counter-measures such as experience replay, generative replay, or gradient-based regularisation can work, but they also scale memory linearly with the number of stored samples or parameters, which is not always feasible on low-power hardware.

F. Marcelloni et al. (Eds.): IJCCI 2025, CCIS 2829, pp. 706–722, 2026.
https://doi.org/10.1007/978-3-032-15638-9_42

In the setting of CL, the goal is to have a learner accumulate and integrate skills during its entire lifespan. Reaching this goal is hindered by the occurrence of Catastrophic Forgetting.

Lightweight SVD Generators (LSG) [12] aim to capture the essence of each task in a way that is efficient in terms of both computational complexity and memory requirements. The essence of previous tasks can then be replayed to the learner along with raw data from the current task. This way the original data is presented to the learner and the replay mechanism is kept lightweight.

A major limitation of LSG is that it only captures linear patterns, while datasets often contain highly non-linear patterns. Multi-Space SVD Generators (MSSG), our proposed method, solves this shortcoming by using multiple LSGs per task and class. Each LSG still represents a local linear patch, but the ensemble more closely resembles a manifold spanning the latent space of the task and class.

By replacing a single global SVD with a handful of low-rank, task-and-class-specific subspaces, MSSG retains the compact footprint of LSG while approximating complex manifolds that were previously out of reach.

Contributions:

Our work advances continual learning on memory-constrained devices in four ways:

- We generalise Lightweight SVD Generators by clustering each class/task's right-singular vectors and fitting a low-rank generator per cluster. The resulting ensemble approximates a non-linear manifold while keeping storage $O(prb)$, where p is the input dimension, r the rank in the context of truncated SVD (i.e. the number of most significant components kept) and b the subspace count.
- We derive an analytic expression for MSSG's footprint and use it to compute a memory-equivalent number of raw samples. This allows us to evaluate MSSG against other baselines such as Experience Replay under near-identical memory budgets.
- On MNIST, Fashion MNIST, NOT MNIST, CIFAR10 and Tiny ImageNet, with both MLP and MLP-Mixer backbones, MSSG improves average final accuracy over single-LSG and matches or surpasses ER with raw samples with near-identical memory budgets.
- Because MSSG stores only compact factorised statistics, rehearsal and generator updates complete in milliseconds, making it a drop-in replacement for replay buffers in on-device lifelong learning. The code is released open-source [11].

2   Problem Definition

Continual Learning refers to the practice of having a learning entity accumulate knowledge during its entire life cycle [3,4,7,16,18,25,28]. Unlike a classical learning scenario, where all experiences are available to the learner at any time, in Continual Learning different sets of experiences are presented to the learner in sequence. Sets of experiences are commonly referred to as "tasks". The learner only has access to the current task and not necessarily to the previous tasks.

Three scenarios of continual learning can be distinguished: Task-Incremental Learning (Task-IL), where label spaces between tasks are equal and the task identity is known at test time, Domain-Incremental Learning (Domain-IL) where label spaces between tasks are also equal, but the task identity is unknown at test time, and Class-Incremental Learning (Class-IL) where label spaces between tasks are not equal and task identity is not known during test time [27].

We focus on the scenario of Class-IL (Class-IL), since it demands that a learner receives a good representation of previous tasks to be able to learn anything and we would like to test how well MSSG manages to represent previous tasks. In this setting, m tasks $(T_0, T_1, ..., T_{m-1})$ are presented in series. Each task T_i has samples from the classes in the label space $L_i \subseteq [0, c-1]$, where c is the total number of classes and $i \in [0, m-1]$. A task T_i can have a completely or partially disjoint label spaces from any other task, so $\exists i, j \in [0, m-1] | L_i \neq L_j$. So, at least one pair of label spaces is not equal, but overlap between label spaces is possible.

Our setup has the added constraint that all label spaces are completely disjoint; $\forall i, j \in [0, m-1], L_i \cap L_j = \emptyset$.

3   Related Work and Methodology

Five trends exist that aim to counteract catastrophic forgetting: Regularization focuses on regularizing weights to promote remembering previous tasks. The Rehearsal trend keeps a set of examples of previous tasks and presents them to the learner along examples of the current task. The Architectural trend aims to alter the architecture of the network during training time to promote remembering previous tasks. The Optimization-based approach aims to alleviate forgetting by changing the optimization algorithm itself. The Representation-based approach tries to capture the essence of the data set [16, 28].

Our method (MSSG), falls in the categories of Rehearsal and Representation-based learning. Rehearsal, because it is deployed in a Experience Replay (ER) setting [2, 19] and Representation-based learning, because it aims to capture the essence of the data.

In ER, data from previous tasks it presented to the learner alongside data from the current task. MSSG can however also be deployed in any other Rehearsal setting, like A-GEM for example. It can be considered representation-based learning, because it aims to capture the essence of the data.

To capture the essence of data, one can exploit the fact that the degrees of freedom in a set of images is lower than the dimensionality of the images (i.e. number pixels times color channels) [17]. A dataset can be thought of as lying on a lower dimensional manifold [20, 24]. Knowing the structure and shape of this manifold allows for simpler representations of the dataset, saving storage space, while at the same time gaining knowledge of what defines the dataset.

Manifold learning encompasses a collection of techniques that allows one to find the structure of Manifolds. It is related to dimensionality reduction and comes in linear and non-linear variants [8]. Linear approaches include Principal Component Analysis (PCA) and Multidimensional Scaling (MDS). Non-linear approaches include Isomap,

Local Linear Embedding (LLE), Laplacian Eigenmaps, Hessian Eigenmaps and diffusion maps. Another way to learn a manifold is to train generators [13,22,23,26]. The downside of this is the training time and memory requirements.

A practical way to compute PCA [9] is Singular Value Decomposition (SVD) [6]. SVD decomposes any matrix M into three matrices U, S and Vh, such that $M = USVh$. If you consider a dataset to be this matrix M, where each column is one data entry, these three matrices can be interpreted in the following way: The columns of U are the components of the data ordered from most to least variance. The rows of Vh define linear combinations of the columns of U, that, after being scaled by the diagonal values of S yield back the original dataset represented by M.

Lightweight SVD Generators (LSG) [12] use random combinations of the columns of U to generate synthetic images. The shortcoming of this method is that these synthetic images lie in a hyperplane i.e. the span of the column space of U.

MSSG alleviates this shortcoming by introducing multiple generators, each being trained on a local cluster of data. Since manifolds are locally linear, the linear nature of the LSGs is not a problem on local subsets of the data. The collection of all LSGs represent linear patches that fill the entire manifold, a method referred to as Linear Local Approximations (LLA) [15]. MSSG is similar to the LLA method of LLE [17]:

- LLE use the principle of Neighborhoords. In MSSG, a Neighborhood is decided by Spectral Clustering.
- In MSSG, datasamples are not reconstructed directly as a weighted sum of neighbours, like in LLE, but as a weighted sum of the biggest components of the neighbourhood.
- In MSSG, the Vh matrix represents a mapping from the component space (columns space of U) to the embedded space, just like LLE uses embedding coordinates.

Algorithm 1 shows the process of fitting multiple generators on data of a certain class and task. To aid the clustering, the dimensionality of the data is reduced by a first round of truncated SVD (line 1–2). It is then clustered using Spectral Clustering [21](line 3–4). On each cluster of data, a second round of truncated SVD is performed and the necessary components for a generator are saved (line 5–12). Note that the second round of SVD is performed on the original data, not on the dimensionality reduced data of line 2. The latter is only created to aid the clustering.

Algorithm 2 shows how a synthetic sample is created. It is almost identical to that of LSG, with the exception that an extra step is introduced to select a generator at random (line 1).

3.1 Experimental Setup

Five data sets were used in the experiments: MNIST [5] (28×28 gray scale images, train size: 60000, test size: 10000, 10 classes), Fashion MNIST [29] (28×28 gray scale images, train size: 60000, test size: 10000, 10 classes), NOT MNIST [1] (28×28 gray scale images, train size: 60000, test size: 10000, 10 classes), CIFAR10 [10] (32×32 colored images, train size: 50000, test size: 10000, 10 classes) and Tiny Imagenet [14] (64×64 colored images, train size: 100000, test size: 10000, 500 classes). The

---

Algorithm 1. StoreGenerator.

Input: data,task,class

1: $\tilde{U}, \tilde{S}, \tilde{V}h \leftarrow$ SVD(data)
2: $\tilde{U}, \tilde{S}, \tilde{V}h \leftarrow$ Truncate$(\tilde{U}, \tilde{S}, \tilde{V}h)$
3: ClusterLabels $\leftarrow$ SpectralClustering$(\tilde{V}h)$
4: UniqueClusterLabels $\leftarrow \{l | l \in$ ClusterLabels$\}$
5: **for** cluster $\in$ UniqueClusterLabels **do**
6:     ClusterData $\leftarrow \{$data$[i] |$ClusterLabels$[i] =$ cluster$, i \in [0, ..., |$data$| - 1]\}$
7:     $U, S, Vh \leftarrow$ SVD(ClusterData)
8:     $U, S, Vh \leftarrow$ Truncate(U, S, Vh)
9:     mean $\leftarrow \frac{\Sigma_{\forall \mathbf{v} \in Vh} \mathbf{v}}{|Vh|} * S$
10:     cov $\leftarrow ($Covariance$(Vh) * S)^T * S$
11:     $U_{\text{task,class,cluster}}, \text{mean}_{\text{task,class,cluster}}, \text{cov}_{\text{task,class,cluster}} \leftarrow U, \text{mean}, \text{cov}$
12: **end for**

Algorithm 2. GenerateSample.

Input: task,class
Output: (sample, class)

1: cluster $\leftarrow$ Random$(|U_{\text{task,class}}|)$
2: $Vh_{\text{random}} \leftarrow$ MultivariateNormal$(\text{mean}_{\text{task,class,cluster}}, \text{cov}_{\text{task,class,cluster}})$
3: sample $\leftarrow U_{\text{task,class,cluster}} \cdot Vh_{\text{random}}$
4: **return** (sample, class)

---

first ten classes of Tiny Imagenet were selected to keep the runtime of the experiments lower and to allow the use of smaller neural networks. The data sets were segmented into separate tasks, so each task consisted of exactly one class. All experiments use Experience Replay (ER), with the exception of the baseline Stochastic Gradient Descent (SGD) experiments, where no tasks specific data is replayed. Data is replayed using raw samples, using the generators of LSG and using the generators of MSSG. Each experiment is repeated 20 times.

3.2 Choice of Values

For each lightweight generator, the reduced versions of U, the Vh covariance mean and the Vh covariance matrix are stored. The reduced U is of size $p \times r$. Where p is the number of float point numbers per image ($p =$ pixels $*$ colorchannels) and r the rank (the number of U components after truncation). The reduced Vh covariance mean has length r. The reduced Vh covariance matrix is of size $r \times r$.

This leads to a memory requirement for one generator to be:

$$|U| + |Vh_{\text{mean}}| + |Vh_{\text{cov}}| = p * r + r + r * r \tag{1}$$

For b subspaces, the memory footprint becomes:

$$b * (p * r + r + r * r) \tag{2}$$

We want to compare this with ER using a number of samples that results in an equal amount of memory usage. To find this amount of samples, we equate this with:

$$p * s = b * (p * r + r + r * r) \tag{3}$$

where each image has p float point numbers and s images samples are taken.

To calculate the number of samples that are equivalent for b generators of rank r:

$$s = b * (p * r + r + r * r)/p \tag{4}$$

For MNIST, Fashion MINST and NOT MNIST, $p = 28 * 28$, for 28 by 28 monochrome pixels. For CIFAR10, $p = 32 * 32 * 3$, for 32 by 32 pixels with 3 color channels. For Tiny Imagenet, $p = 64 * 64 * 3$, for 64 by 64 pixels with 3 color channels.

To calculate the memory requirements expressed in bytes, we assume every number is 32 bits, thus taking up 4 bytes per number. For raw sampling, this will result in a memory footprint of $p * s * 4$ bytes, for LSG the memory footprint is $(p * r + r + r * r) * 4$ bytes and for MSSG it is $b * (p * r + r + r * r) * 4$ bytes. Table 1 shows a comparison of Experience Replay with (ER) with raw samples vs. represented sample equivalence for MNIST, Fashion MNIST and NOT MNIST for LSG and MSSG. Table 2 and 3 do this for CIFAR10 and Tiny Imagenet respectively. Each row represents a set of experiments that is as close to memory equivalent as possible. Note that the raw samples are always slightly higher than the corresponding sample equivalence. This was a deliberate choice to ensure that all measured performance increase was due to more efficient information storing and not due to a higher number of samples being taken.

Table 1. Memory requirements per task expressed in data samples for MNIST, Fashion MNIST and NOT MNIST. *Raw Samples* denote the act of storing a discrete number of samples from the original dataset. *Sample Equivalence* (Sample Eq.) is an abstract concept that expresses the memory footprint of all generators in terms of samples that allows a comparison with the act of storing raw samples. *KiloByte* expresses the memory requirements in KiloBytes, assuming float points of 32 bit, taking up 4 bytes per number.

ER		LSG			MSSG		
Raw Samples	KiloByte	Rank	Sample Eq.	KiloByte	Rank, Subspaces	Sample Eq.	KiloByte
3	9.41	2	2.01	6.30	-	-	-
4	12.54	3	3.02	9.46	-	-	-
5	15.68	4	4.03	12.62	-	-	-
6	18.82	5	5.04	15.80	-	-	-
11	34.50	10	10.14	31.80	5, 2	10.08	31.60
21	65.86	20	20.54	64.40	5, 4	20.15	63.20
43	134.85	40	42.09	132.00	5, 8	40.31	126.40
89	279.10	80	88.27	276.80	5, 17	85.65	268.60
193	605.25	160	192.86	604.80	5, 38	191.45	600.40
452	1,417.47	320	451.02	1,414.40	5, 89	448.41	1,406.20

Table 2. Memory requirements per task expressed in data samples for CIFAR10. *Raw Samples* denote the act of storing a discrete number of samples from the original dataset. *Sample Equivalence* (Sample Eq.) is an abstract concept that expresses the memory footprint of all generators in terms of samples that allows a comparison with the act of storing raw samples. *KiloByte* expresses the memory requirements in KiloBytes, assuming float points of 32 bit, taking up 4 bytes per number.

ER		LSG			MSSG		
Raw Samples	KiloByte	Rank	Sample Eq.	KiloByte	Rank, Subspaces	Sample Eq.	KiloByte
41	503.81	40	40.53	498.08	5, 8	40.08	492.48
83	1,019.90	80	82.11	1,008.96	5, 16	80.16	984.96
169	2,076.67	160	168.39	2,069.12	5, 33	165.32	2,031.48
354	4,349.95	320	353.44	4,343.04	5, 70	350.68	4,309.20
774	9,510.91	640	773.54	9,505.28	5, 154	771.50	9,480.24
1000	12,288.00	794	999.48	12,281.59	5, 199	996.94	12,250.44

Table 3. Memory requirements per task expressed in data samples for Tiny Imagenet. *Raw Samples* denote the act of storing a discrete number of samples from the original dataset. *Sample Equivalence* (Sample Eq.) is an abstract concept that expresses the memory footprint of all generators in terms of samples that allows a comparison with the act of storing raw samples. *KiloByte* expresses the memory requirements in KiloBytes, assuming float points of 32 bit, taking up 4 bytes per number.

ER		LSG			MSSG		
Raw Samples	KiloByte	Rank	Sample Eq.	KiloByte	Rank, Subspaces	Sample Eq.	KiloByte
41	2,015.23	40	40.13	1,972.64	5, 8	40.02	1,967.04
81	3,981.31	80	80.53	3,958.08	5, 16	80.04	3,934.08
163	8,011.78	160	162.10	7,967.36	5, 32	160.08	7,868.16
329	16,171.01	320	328.36	16,139.52	5, 65	325.16	15,982.20
500	24,576.00	481	499.87	24,569.48	5, 99	495.24	24,342.12

Table 4 shows the hyper parameters used for the MLP and MLP mixer architectures for all datasets used in the experiments. These hyper-parameters were chosen by performing a hyper-parameter search that aimed to maximize the validation accuracy.

The *Data Sets* include MNIST, Fashion MNIST, NOT MNIST, CIFAR10 and Tiny Imagenet. *Epochs* denotes how often all data of one task is repeatedly presented to the learner. *Learning Rate* (LR) determines the step size during gradient descent. *Optimizer* names the optimizer that is used. For most methods it is Stochastic Gradient Descent (SGD). Tiny Imagenet required a more sophisticated optimizer like ADAM.

Three different architectures are used in the experiments.

The first is a Multi Layer Perceptron (MLP). It has an input layer that is 784 nodes wide, followed by two hidden layers with width 200, followed by an output layer of width 10.

The second and third architectures are both MLP mixers (MLP mixer 1 and MLP mixer 2). An MLP mixer treats an image as patches that are flattened to one dimensional vectors. Both MLP mixer 1 and 2 use a patch size of 4×4. MLP mixer 1 is 256 nodes wide and 8 blocks deep and uses no dropout. MLP mixer 2 is 1024 nodes wide and 16 blocks deep with a dropout rate of 0.1. For a detailed implementation of these networks, we refer to the codebase [11].

Table 4. The hyper parameters for the experiments vary per dataset and architecture. Less complex datasets require less training, thus have a lower epoch value than more complex datasets. Tiny Imagenet requires a more sophisticated optimizer (ADAM).

Data Sets	Epochs	LR	Optimizer	Architecture
MNIST, Fashion MNIST, NOT MNIST	1	10^{-3}	SGD	MLP
MNIST, Fashion MNIST, NOT MNIST	1	10^{-2}	SGD	MLP mixer 1
CIFAR10	5	10^{-2}	SGD	MLP mixer 1
Tiny Imagenet	50	10^{-3}	ADAM	MLP mixer 2

4 Results

Results are depicted as groups of box-plots in Figs. 1, 2, 3, 4, 5, 6, 7 and 8. The figures feature rectangular sections marked with *Baseline* and *Rank* ρ, where ρ varies.

The Baseline rectangle show our lowest and highest expected results. The lowest being standard SGD, where no effort is taken to alleviate forgetting. The highest being Experience Replay with σ samples stored per task (ER σ), where σ is 500 for Tiny ImageNet and 1000 for all other datasets, i.e. the same number of samples per task that the Lightweight generators are fitted on. It is not expected that MSSG will outperform this upper bound, since it received exactly the same amount of information for fitting.

The Rank ρ columns house results for LSG with rank ρ (LSG rρ), a most closely memory equivalent version of Experience Replay (ER σ), where ER σ is given the benefit of rounding up to the nearest integer, and MSSG with rank τ and number of subspaces υ (MSSG rτ sυ).

Box-plots depict the final validation accuracy as measured over experiments that are repeated 20 times. The final validation accuracy is the accuracy measured on a validation set, after the training of the network on all tasks of the training set is completed. This validation set is disjoint from the training set, but contains data from the same distribution of the training set. The validation set features data from the same distributions of all tasks (and thus classes) encountered during training.

A t-test is performed along each experiment depicted in one rectangular area. A p-value of 0.01 is used to determine significance. Per rectangle, the box-plot of the experiment with the highest final validation accuracy is highlighted by a thick border. In case an experiment has a lower final validation accuracy that is not significantly different from the highest scoring experiment in that rectangle, it is also highlighted with a thick border.

4.1 MLP Architecture

Figures 1, 2 and 3 show that the original method LSG consistently has the highest final validation accuracy when training a MLP on MNIST, Fashion MNIST and NOT MINST respectively. MSSG does not outperform LSG in these experiments.

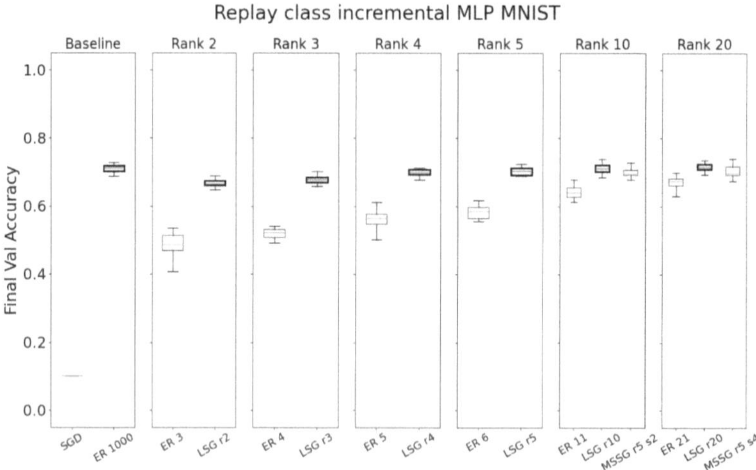

Fig. 1. Final Validation Accuracy for nearly memory equivalent methods. Significantly highest accuracies are highlighted with thick bordered boxes.

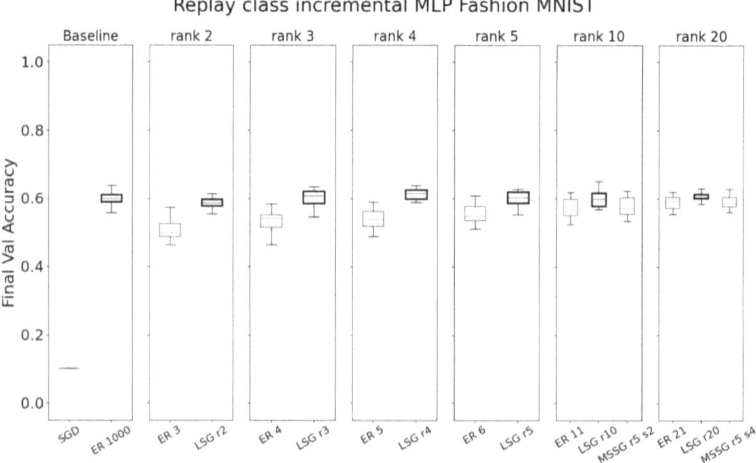

Fig. 2. Final Validation Accuracy for nearly memory equivalent methods. Significantly highest accuracies are highlighted with thick bordered boxes.

4.2 MLP Mixer Architecture

Figures 4, 5 and 6 show the results of training an MLP mixer on MNIST, Fashion MNIST and NOT MNIST respectively. For lower ranks LSG tends to outperform RE with raw samples. For higher ranks RE with raw samples takes over again. MSSG does not score the highest in these higher ranks, but it is worthwhile to note that while LSG's performance tends to go decrease as the rank increases, this is not the case for MSSG. This is most notable for MNIST.

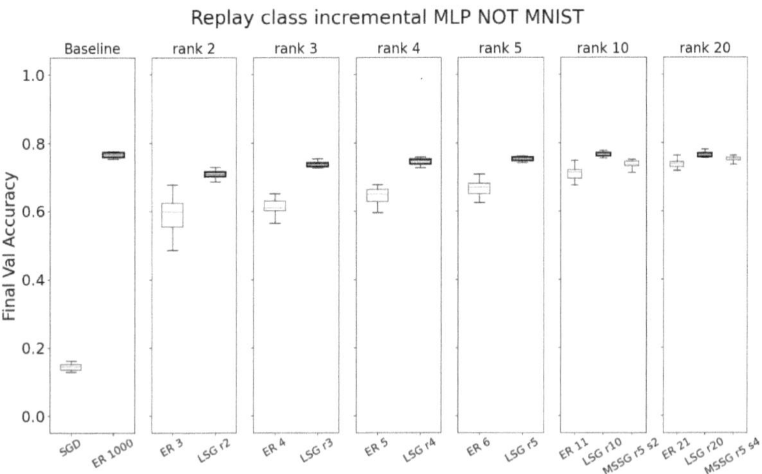

Fig. 3. Final Validation Accuracy for nearly memory equivalent methods. Significantly highest accuracies are highlighted with thick bordered boxes.

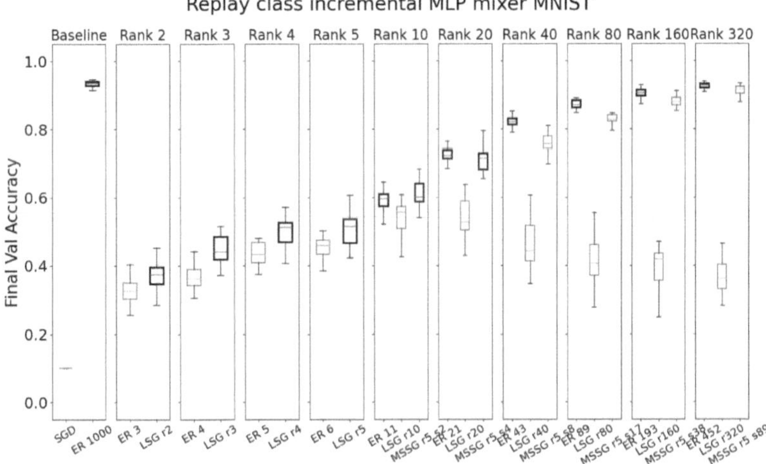

Fig. 4. Final Validation Accuracy for nearly memory equivalent methods. Significantly highest accuracies are highlighted with thick bordered boxes.

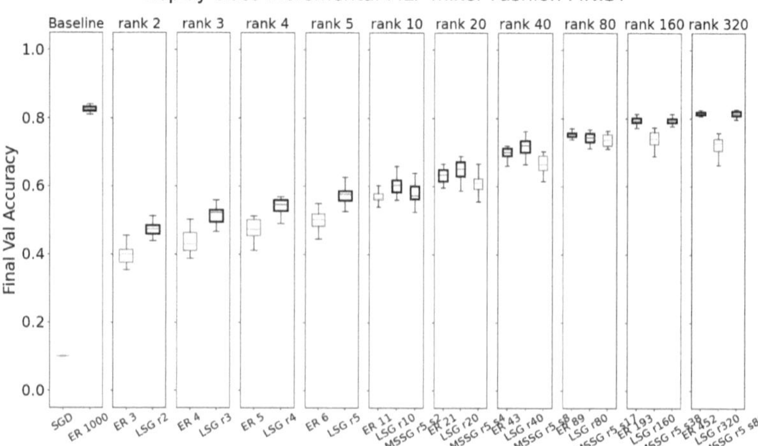

Fig. 5. Final Validation Accuracy for nearly memory equivalent methods. Significantly highest accuracies are highlighted with thick bordered boxes.

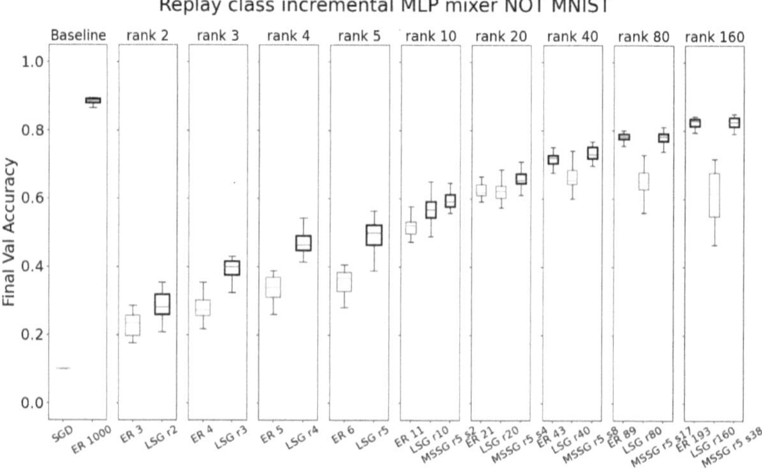

Fig. 6. Final Validation Accuracy for nearly memory equivalent methods. Significantly highest accuracies are highlighted with thick bordered boxes.

Figure 7 shows the results on training an MLP mixer on CIFAR10. MSSG has the highest validation accuracy in the columns of rank 160, 320 and 640. LSG scores consistently low for each rank.

Figure 8 shows the results on training an MLP mixer on the first ten classes of Tiny Imagenet. For rank 320, MSSG scores the highest validation accuracy. For ranks 80, 160 and 481 MSSG is on par with ER with raw sampling. LSG consistently scores low in this setting.

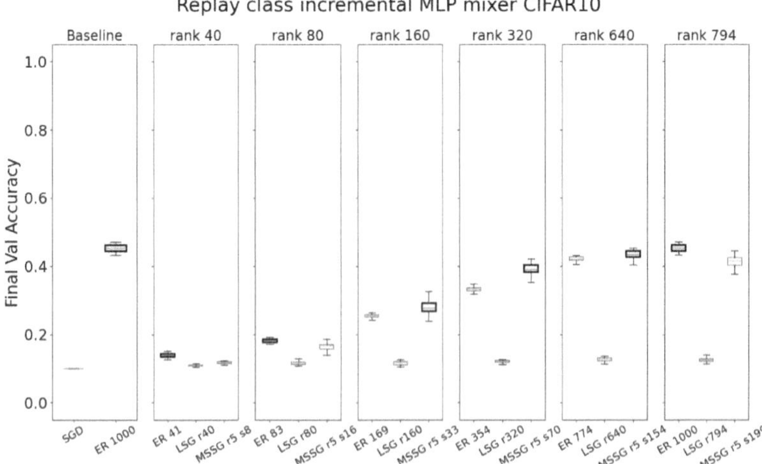

Fig. 7. Final Validation Accuracy for nearly memory equivalent methods. Significantly highest accuracies are highlighted with thick bordered boxes.

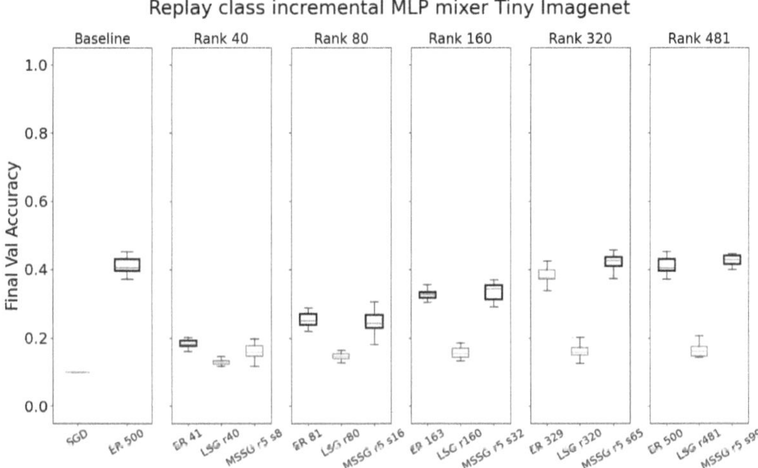

Fig. 8. Final Validation Accuracy for nearly memory equivalent methods. Significantly highest accuracies are highlighted with thick bordered boxes.

4.3   Generated Images

Figure 9 and 10 give a look at the inner workings of the generators when fitted on CIFAR10 and Tiny Imagenet respectively. The Spectral Clustering step produces clusters of varying size, of which a selection was made to display in these figures. The top row shows up to five images of the cluster that was used for training. The second row shows the reconstruction of these images using the biggest components (i.e. the components associated with the most variance). The third row shows up to five of the biggest components. The bottom row shows images that are synthesized by the generator.

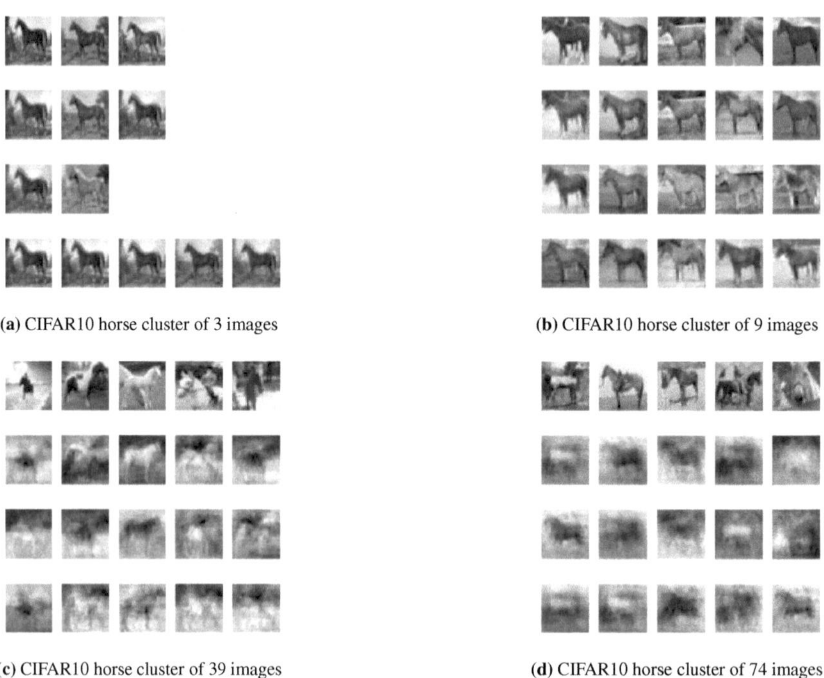

(a) CIFAR10 horse cluster of 3 images

(b) CIFAR10 horse cluster of 9 images

(c) CIFAR10 horse cluster of 39 images

(d) CIFAR10 horse cluster of 74 images

Fig. 9. Top row: raw images used for training. Second row: reconstruction of raw images using the five components with the most variance. Third row: The left components associated with the most variance. Bottom row: output of the generator.

5   Discussion

The results show that for CIFAR10 and Tiny Imagenet a sweetspot in memory usage exists where MSSG outperforms both ER with raw sampling and LSG. LSG fails completely on these datasets. Results differ for more linear datasets (MNIST, Fashion MNIST and NOT MNIST). When training an MLP, LSG is still superior. Training an MLP mixer on the more linear datasets showed that RE with raw sampling is superior for higher ranks. They are however closely followed by MSSG, indicating that MSSG still manages to capture the essence of the data. The results support that datasets that have less linear patterns benefit more from using MSSG.

The results are based on experiments with specific hyperparameter settings. It is highly likely that other settings will give different results. The goal of this research is not to find hyper-parameter settings that maximize accuracy. The goal is to show that our method is able to yield higher accuracy with a nearly equivalent memory footprint.

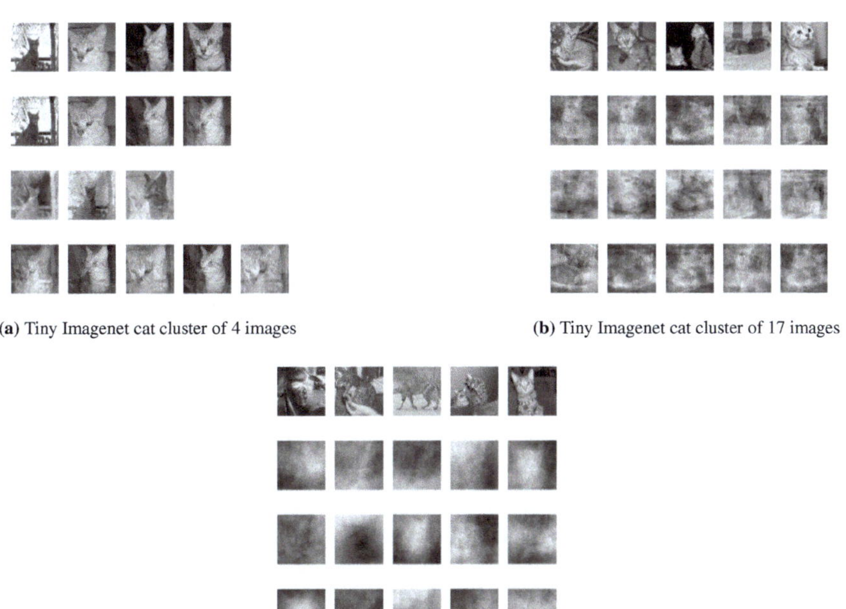

(a) Tiny Imagenet cat cluster of 4 images

(b) Tiny Imagenet cat cluster of 17 images

(c) Tiny Imagenet cat cluster of 136 images

Fig. 10. Top row: raw images used for training. Second row: reconstruction of raw images using the five components with the most variance. Third row: The left components associated with the most variance. Bottom row: output of the generator.

Figure 9 gives insight in the inner workings of the Lightweight generators when trained on the horse class of CIFAR10. Figure 9a represents a cluster of just three images. Note that two of the raw images are two photographs of the same painting under different lighting conditions, effectively reducing the number of images to two. The SVD step picks up on this and mentions just two components. Figure 9b represents a cluster of horses in full view facing the left side of the picture. This clear subject results in synthesized images that are realistic in both appearance and variance. Figures 9c and 9d show what happens when clusters are larger. The synthesized images are less realistic, but still depict a distinction between foreground and background that is reminiscent of horse pictures.

Figure 9 shows what Lightweight generators learn when trained on the cat class of Tiny Imagenet. Figure 10a represents a cluster of four images. Note that with the lack of real global patterns, the synthetic images look like several original images that are superimposed on top of each other. Figure 10b suggests that on a larger cluster, some patterns are found and the resulting synthesized images vaguely resemble cats. Figure 10c shows that on an even larger cluster, the synthesized images begin to resembled highly global patterns like the direction of light, which is not specific to cat pictures.

The realism of the synthesized images is thus dependent on the availability of multiple images that are clustered together that all feature an image-wide pattern that clearly defines a class. The lack of such features will result in synthesized images looking like superimposed images when the cluster size is small and blurry non-class specific images when the cluster size is large.

6 Conclusion

The proposed method "Multi-Subspace SVD Generators" (MSSG) succeeds where the method "Lightweight SVD Generators" (LSG) fails. Using multiple generators is most effective when using datasets that have less linear patterns.

While the realism of the synthesized images begins to break down on more complex datasets like Tiny Imagenet, it is still enough to alleviate catastrophic forgetting in a more memory-efficient matter. For feasibility reasons, small neural networks were used. These networks produce results that are inferior to the state of the art in terms of validation accuracy. However, the aim of this study is not to get state of the art validation accuracy, but to measure how effective MSSG is at negating catastrophic forgetting.

7 Future Work

Since image-wide features are not always present, it can be beneficial to cut images into smaller patches before handing it to the clustering and lightweight generators. This promotes the detection of features that are more local. This will however result in synthesized images that have the size of these patches, where the original image size is desired. The easiest way to solve this is to simply glue together multiple patches to produce a synthesized images. The downside is that this will not capture typical locations and co-occurrences of features. In order to capture this, frequencies of certain features per location and co-variance between features should be captured.

MSSG handles every cluster and its resulting generator as completely independent from all other clusters and generators. To more effectively capture the structure of the manifold, it can be beneficial to model transitions between clusters (and thus generators) that are neighbors on the manifold.

Acknowledgements. This work is supported by the project ULEARN "Unsupervised Lifelong Learning" and co-funded under the grant number 316080 of the Research Council of Norway.

References

1. Bulatov, Y.: Not mnist (2011). http://yaroslavvb.blogspot.com/2011/09/notmnist-dataset.html
2. Buzzega, P., Boschini, M., Porrello, A., Calderara, S.: Rethinking experience replay: a bag of tricks for continual learning. In: 2020 25th International Conference on Pattern Recognition (ICPR), pp. 2180–2187. IEEE (2021)

3. Cha, H., Lee, J., Shin, J.: Co2l: contrastive continual learning. In: Proceedings of the IEEE/CVF International Conference on Computer Vision, pp. 9516–9525 (2021)

4. De Lange, M., et al.: A continual learning survey: defying forgetting in classification tasks. IEEE Trans. Pattern Anal. Mach. Intell. **44**(7), 3366–3385 (2021)

5. Deng, L.: The mnist database of handwritten digit images for machine learning research. IEEE Signal Process. Mag. **29**(6), 141–142 (2012)

6. Golub, G.H., Reinsch, C.: Singular value decomposition and least squares solutions. In: Handbook for Automatic Computation: Volume II: Linear Algebra, pp. 134–151. Springer (1971)

7. Hadsell, R., Rao, D., Rusu, A.A., Pascanu, R.: Embracing change: continual learning in deep neural networks. Trends Cogn. Sci. **24**(12), 1028–1040 (2020)

8. Izenman, A.J.: Introduction to manifold learning. Wiley Interdisc. Rev.: Comput. Stat. **4**(5), 439–446 (2012)

9. Jolliffe, I.T.: Principal component analysis for special types of data. Springer (2002)

10. Krizhevsky, A., Nair, V., Hinton, G.: Cifar10 (2009). https://www.cs.toronto.edu/~kriz/cifar.html

11. Lamers, C.: Mssg. https://github.com/christiaanlamers/MSSG (2025)

12. Lamers, C., Belbachir, A.N., Bäck, T., van Stein, N.: Leveraging lightweight generators for memory efficient continual learning (2025). https://arxiv.org/abs/2506.19692

13. Lesort, T., Caselles-Dupré, H., Garcia-Ortiz, M., Stoian, A., Filliat, D.: Generative models from the perspective of continual learning. In: 2019 International Joint Conference on Neural Networks (IJCNN), pp. 1–8. IEEE (2019)

14. . Li, C., Chen, D., Zhang, Y., Beerel, P.A.: Tiny imagenet (2024). https://datasets.activeloop.ai/docs/ml/datasets/tiny-imagenet-dataset/

15. Meilă, M., Zhang, H.: Manifold learning: what, how, and why. Ann. Rev. Stat. Appl. **11**(1), 393–417 (2024)

16. Parisi, G.I., Kemker, R., Part, J.L., Kanan, C., Wermter, S.: Continual lifelong learning with neural networks: a review. Neural Netw. **113**, 54–71 (2019)

17. Pless, R., Souvenir, R.: A survey of manifold learning for images. IPSJ Trans. Comput. Vision Appl. **1**, 83–94 (2009)

18. Qu, H., Rahmani, H., Xu, L., Williams, B., Liu, J.: Recent advances of continual learning in computer vision: an overview. IET Comput. Vision **19**(1), e70013 (2025)

19. Rolnick, D., Ahuja, A., Schwarz, J., Lillicrap, T., Wayne, G.: Experience replay for continual learning. In: Advances in Neural Information Processing Systems, vol. 32 (2019)

20. Roweis, S.T., Saul, L.K.: Nonlinear dimensionality reduction by locally linear embedding. Science **290**(5500), 2323–2326 (2000)

21. Shi, J., Malik, J.: Normalized cuts and image segmentation. IEEE Trans. Pattern Anal. Mach. Intell. **22**(8), 888–905 (2000). https://doi.org/10.1109/34.868688

22. Shin, H., Lee, J.K., Kim, J., Kim, J.: Continual learning with deep generative replay. In: Advances in Neural Information Processing Systems, vol. 30 (2017)

23. Stoianov, I., Maisto, D., Pezzulo, G.: The hippocampal formation as a hierarchical generative model supporting generative replay and continual learning. Prog. Neurobiol. **217**, 102329 (2022)

24. Tenenbaum, J.B., Silva, V.d., Langford, J.C.: A global geometric framework for nonlinear dimensionality reduction. Science **290**(5500), 2319–2323 (2000)

25. Thrun, S.: Lifelong Learning Algorithms, pp. 181–209. Springer US, Boston, MA (1998https://doi.org/10.1007/978-1-4615-5529-2_8

26. Van de Ven, G.M., Tolias, A.S.: Generative replay with feedback connections as a general strategy for continual learning. arXiv preprint arXiv:1809.10635 (2018)

27. Van de Ven, G.M., Tuytelaars, T., Tolias, A.S.: Three types of incremental learning. Nature Mach. Intell. **4**(12), 1185–1197 (2022)

28. Wang, L., Zhang, X., Su, H., Zhu, J.: A comprehensive survey of continual learning: theory, method and application. IEEE Trans. Pattern Anal. Mach. Intell. (2024)
29. Xiao, H., Rasul, K., Vollgraf, R.: Fashion-mnist: a novel image dataset for benchmarking machine learning algorithms. arXiv preprint arXiv:1708.07747 (2017)

From High-Frequency Sensors to Noon Reports: Using Transfer Learning for Shaft Power Prediction in Maritime

Akriti Sharma[1]([✉]) [iD], Dogan Altan[1] [iD], Dusica Marijan[1] [iD], and Arnbjørn Maressa[2]

[1] Simula Research Laboratory, Kristian Augusts gate 23, 0164 Oslo, Norway
akriti@simula.no
[2] Navtor AS, Copenhagen, Denmark
https://www.simula.no/ , https://www.navtor.com/

Abstract. With the growth of global maritime transportation, energy optimization has become crucial for reducing costs and ensuring operational efficiency. Shaft power is the mechanical power transmitted from the engine to the shaft and directly impacts fuel consumption, making its accurate prediction a paramount step in optimizing vessel performance. Power consumption is highly correlated with ship parameters such as speed and shaft rotation per minute, as well as weather and sea conditions. Frequent access to this operational data can improve prediction accuracy. However, obtaining high-quality sensor data is often infeasible and costly, making alternative sources such as noon reports a viable option. In this paper, we propose a transfer learning-based approach for predicting vessels' shaft power, where a model is initially trained on high-frequency data from a vessel and then fine-tuned with low-frequency daily noon reports from other vessels. We tested our approach on sister vessels (identical dimensions and configurations), a similar vessel (slightly larger with a different engine), and a different vessel (distinct dimensions and configurations). The experiments showed that the mean absolute percentage error decreased by 10.6% for sister vessels, 3.6% for a similar vessel, and 5.3% for a different vessel, compared to the model trained solely on noon report data.

Keywords: Transfer Learning · Shaft Power Prediction · Noon Reports · Sensor Data · Maritime

1 Introduction

Efficient energy optimization for benchmarking the fuel consumption of a vessel has become essential in maritime to decrease costs and gas emissions [9]. The International Maritime Organization has demanded energy efficiency measures for vessels, aiming to reduce emissions from 2008 levels by 40% in the year 2030 and 70% by 2050 [18]. Consequently, accurately predicting power consumption

F. Marcelloni et al. (Eds.): IJCCI 2025, CCIS 2829, pp. 723–741, 2026.
https://doi.org/10.1007/978-3-032-15638-9_43

is actively pursued in the maritime industry for energy efficiency planning as a vessel's fuel consumption is directly linked to its shaft power requirements [21].

Ship-scale data can provide insights into ship performance, enabling the accurate estimation of vessel power requirements. Noon reports are more easily accessible ship-scale data [31] that provide daily readings of the vessel's location, weather and sea state estimates, as well as power and fuel consumption. Noon reports are manually generated by the captain or a data logging system, typically at "noon" each day. The literature shows that noon reports have been used in modeling power and fuel consumption [1,3,8]. However, noon reports are limited in terms of accuracy, as they are averaged to a single data point representing a 24-hour operation and are prone to human error [22].

The vast majority of vessels nowadays are also equipped with high-frequency monitoring systems that use onboard sensors to track the vessel's sailing profile, recording data every few seconds. Sensor data have less uncertainty in comparison to noon reports [2]. Switching from sensor data to noon reports has been shown to decrease the predictive error in power predictions [23]. A comparison of noon reports and sensor data for predicting fuel consumption shows that sensor data improves model fit performance [9]. However, not all vessels are equipped with sensors, and the cost of installing them is high [21]. Estimating ship operations with limited or no sensor data remains a major challenge, and in such cases, noon reports, though less accurate, are used for predictions.

In this study, we present an approach that improves the robustness and accuracy of shaft power prediction using noon reports by leveraging knowledge from sensor data of another vessel. We train a neural network on sensor data to predict shaft power, then use transfer learning to fine-tune it on noon reports from sister, similar, and different vessels. Sister vessels have identical dimensions and configurations; similar vessels differ slightly in size and design, while different vessels have distinct dimensions and configurations. We used data-driven machine learning (ML) models for this task, as they are increasingly applied to shaft power prediction [17,22,24]. In contrast, traditional naval architecture approaches that use empirical formulas to estimate power consumption have shown a lack of scalability [31] and may not fully reflect real-world operating conditions [5,17]. Thus, we chose ML models to capture the underlying relationship in vessel power usage from noon reports.

In this study, we consider ship data as temporal, training up to a specific year and predicting for the following year. This approach validated our method's ability to estimate future power trends, with the fine-tuned model providing more accurate predictions. Moreover, using a pre-trained model significantly improved power predictions for noon reports compared to training from scratch. Our transfer learning-based approach is generic and can be applied to other tasks, such as trim optimization or fouling estimation in noon reports [4,31], without the need for large training datasets.

The main contributions can be summarized as follows.

- We improved shaft power prediction in noon reports across a fleet by using transfer learning without needing additional training data.

- We bridged the performance gap in predicting shaft power that exists when transitioning from sensor data to noon reports.
- We modeled future power trends by forecasting subsequent year consumption patterns from historical training data.
- To the best of our knowledge, this approach is the first to incorporate insights about vessel operations derived from high-frequency sensor data into daily noon reports.

2 Literature Review

Vessel power and fuel consumption prediction has been intensely studied in the literature, using either traditional physics-based approaches or data-driven approaches built on ship-scale monitoring data. Traditional approaches use model scale towing tests or computational fluid dynamics (CFD) simulations to analyze vessel performance. The focus of these approaches is on obtaining a regression curve for the power-speed relationship in calm water and establish baseline performance [5]. A physics-based model including waves, wind and hull drift effect on power monitoring is achieving ~5% average error in steady state conditions [19]. Another traditional approach modeled resistance from waves along with hull form, draft, and trim corrections for estimating vessel's speed-power performance in calm water [14]. These approaches are limited by scaling effects and are seldom tested in complex sea states, failing to represent realistic vessel operations. Though empirical formulas are effective in calm water conditions [10], their accuracy drops in rough seas as the power consumption depends on wind and wave conditions [6]. Furthermore, analyzing ship performance in waves using traditional regression methods is highly challenging [11].

Data-driven approaches use ship-scale data with ML to model resistance from wind and waves for predicting the power consumption, providing an alternative to physics-based approaches. As evident from the literature, an artificial neural network (ANN) outperforms Gaussian process regression (GP) in predicting fuel consumption [24]. Indeed, a neural network-based method predicts fuel consumption for a tanker using noon reports with an error of 0.141 MT/h on mean consumption of 1.89 MT/h [3] and a neural network modeling daily fuel consumption for two container ships reports a root mean square error (RMSE) of 8.23 and 9.34 [8]. Using sensor data, the shaft power and fuel consumption are estimated with an average percentage error between 1–5% [23,26]. ML techniques have also resulted in predicting vessel's power with an average error in the range of 5% [12,25,28]. To improve prediction accuracy, researchers have investigated data fusion and transfer learning methods for estimating shaft power and fuel oil consumption, and we review several relevant studies in the following sections.

2.1 Data Fusion Approaches in Power Estimation

Researchers have systematically explored data fusion approaches from multiple sources, such as sensors, noon reports, AIS, and meteorological datasets. This research direction has evolved through several paradigms:

Noon Reports and Meteorological Data Fusion: A neural network-based method predicting power, improves accuracy by 2% after combining noon reports and weather data [23]. Another study introduced systematic data fusion of noon reports and meteorological data, achieving R^2 values of 0.74–0.90 across eight containerships [13]. Data from noon reports, AIS, meteorological sources, and on-board sensors were combined to develop models that summarize fuel consumption during a journey [16] and monitoring of engine performance [29].

Sensor Data and Meteorological Data Fusion: Extending the data fusion concept to high-frequency sensor data, a study modeling ship fuel efficiency reported R^2 value above 0.96 [7].

These approaches advance data fusion but remain limited by data availability and vessel-specific constraints.

2.2   Transfer Learning

Transfer learning has been applied to vessel power prediction through several approaches, each addressing a different challenge.

Physics-to-Data Transfer Learning: Transfer learning was employed for vessel power prediction using synthetic data from physics-based simulations with sensor data as the source domain and fine-tuning with real operational data, showing improved accuracy with reduced training data [17]. Our research focuses on data-driven knowledge transfer between sensor data and noon reports, using neural networks to intrinsically capture physical relationships among input variables.

Same-Frequency Data Transfer Learning: Transfer learning has been explored for fuel consumption prediction between sister vessels after using three strategies: (a) fine-tuning with freezing transferred layers, (b) fine-tuning without freezing transferred layers, and (c) optimizing network structure with freezing transferred layers [15]. Their approach addressed the challenge of limited training data, a common issue for new ships with limited operating time. Results showed MAPE reductions of 12.57%, 6.44%, and 16.03%, respectively. Our approach achieved a 9.6% reduction in MAPE for sister vessels, but a direct comparison will be inappropriate as we use transfer learning between high and low-frequency data, while they used the same frequency data.

Fleet-Based Prediction Without Transfer Learning: Another shaft power modeling approach trained neural networks on aggregated fleet data without transfer learning, demonstrating accurate power prediction (1.5âĂŞ5% error) using high-frequency data despite limited records [21]. Our approach enables cross-fleet but with cross-frequency adaptation rather than aggregated fleet data.

Despite advances in data fusion and transfer learning, including same-frequency transfer learning [15], fleet-wide fusion within similar data types [21], and fusion within similar temporal domains [7,13], no work has addressed knowledge transfer across different data resolutions (high-frequency sensor data to low-frequency noon reports). Existing approaches focus on sister vessels [15] or require fleet-wide data from similar vessels [21], but evaluations across diverse vessel types and operations remain unexplored. High-performance approaches demand specialized infrastructure (sensor installations, meteorological data access), large-scale fleet data, creating barriers for resource-limited operators. Additionally, these approaches rely on random data splits instead of temporal validation, limiting a real-world deployment, where models must forecast future performance based on historical data.

This study addresses these gaps by proposing the first cross-frequency transfer learning framework, transferring knowledge from high-frequency sensor data to low-frequency noon reports. We evaluate its effectiveness across different vessel types and and validate it through temporal forecasting, without requiring additional sensor datasets, or fleet-wide data collection. Table 1 provides a detailed description of the research context and shows how this work addresses a specific gap in maritime transfer learning.

Table 1. Research contribution context: our study addressing relevant gap in maritime transfer learning. (NRs: Noon Reports)

Research Direction	Addressed Challenge	Our Contribution
Data Scarcity	Only high frequency sensor data	Cross-frequency transfer
Vessel Similarity	Sister vessels analysis	Multi-vessel analysis
Deployment	Random splits (laboratory tested)	Temporal forecasting tested
Practical Adoption	Expensive & complex setup needed	Easily accessible NRs

3   Data

The dataset used in this study was collected from seven vessels: five general cargo ships and two container ships. The data belongs to our industrial partner and, due to confidentiality, it is not openly available. To maintain data anonymity, each vessel is represented using its type acronym and numeric code. For the cargo category, four sister vessels are denoted as S_V followed by their number and a similar vessel as SM_V1. The containers are referred to as different vessels, D_V1 and D_V2. Sister vessels have identical dimensions (length 204 and beam 32 m) and the same configuration. The similar vessel is slightly longer and broader than the sister vessels and has a different main engine configuration. The different vessels' dimensions (length 209 and beam 30 m) and configurations are different from those of sister vessels.

Sensor data reports shaft power at 15-minute intervals, and noon reports are recorded once every 24 h. The study is performed on a per-vessel basis, and data for each vessel is temporally split into train and test sets. For sister and similar vessels, the train set is up to the year 2023, and the test set is from 2024. For the other vessels, the train set is for 2024 and the test set is for 2025. The yearly distribution varies due to data availability. We aim to keep the experiment realistic, as vessels often have historical data from different periods, and testing across years will validate if our approach captures trends in noon reports. Table 2 provides a more detailed description of the dataset, with each column showing the vessel name, dataset time-span, and the number of instances in the training and test sets.

Table 2. Data instances for each vessel's sensor data (SD) and noon reports (NR), along with their corresponding time spans.

Vessel	Timespan	SD (Train)	SD (Test)	NR (Train)	NR (Test)
S_V1	Jan'23-Aug'24	14661	13040	681	138
S_V2	May'20-Aug'24	26948	23575	783	145
S_V3	May'20-Aug'24	31047	11739	846	137
S_V4	April'22-Aug'24	22133	287	401	135
SM_V1	May'20-Aug'24	25592	6659	725	73
D_V1	April'24-May'25	26948	18596	208	101

3.1   Feature Selection

Noon reports for sister vessels have 41 features, and different vessels have 65, out of which we selected seven representative features for predicting the target feature, shaft power (kilowatts). The features are: speed through water (knots), shaft rotational speed (revolution per minute (RPM)), draft amidships (meters), wave height (meters), swell height (meters), wave direction (degrees) and wind direction (degrees). Wave and wind directions are relative to the vessel; wave height represents the significant wave height, while swell height represents the significant height of the wave component caused by wind. As evident in the literature, using wave and wind data improves power prediction accuracy [20, 21], and swell height is significant for fuel consumption predictions from noon reports [27]. We select speed through water over speed over ground since current data is captured by the wave component. The input feature selection also uses on Pearson Correlation analysis with shaft power to identify relevant variables:

$$r_{x_i, y} = \frac{\sum_{j=1}^{n} (x_{ij} - \bar{x}_i)(y_j - \bar{y})}{\sqrt{\sum_{j=1}^{n}(x_{ij} - \bar{x}_i)^2} \sqrt{\sum_{j=1}^{n}(y_j - \bar{y})^2}}$$

where x_i represents the i-th candidate feature, y represents shaft power, and n is the number of observations. The final feature vector is defined as:

$$\mathbf{x} = [v_{stw}, \omega_{rpm}, d_{mid}, H_{wave}, H_{swell}, \phi_{wave}, \phi_{wind}]^T \in \mathbb{R}^7$$

where each component represents:

v_{stw}, ω_{rpm}: speed through water and RPM
d_{mid}: draft amidship $[(draft_{aft} + draft_{fore})/2]$
H_{wave}, H_{swell}: wave and swell heights
ϕ_{wave}, ϕ_{wind}: wave and wind directions (relative to vessel)

The same features are used for sensor data to enable transfer learning with noon reports. The sensor data had high uncertainty regarding weather conditions, which can be resolved by the use of publicly accessible meteorological data [7,13,16]. We derived wave, wind, and swell components from the Copernicus Marine Service (CMEMS) databases, with 0.083° spatial and 3-hour temporal resolution. AIS data enabled spatial and temporal interpolation between CMEMS observations using trilinear interpolation. The weather data fusion from CMEMS databases is expressed as:

$$x_{weather}(time, lat, lon) = \text{TriLinear}(\mathbf{X}_{CMEMS}, time, lat, lon)$$

where $TriLinear$ represents trilinear interpolation over spatial and temporal coordinates. The CMEMS direction data was relative to north, so we converted it to be relative to the ship's course:

$$\phi_{relative} = (\phi_{CMEMS} - AISCourse_{vessel})\%360$$

4   Methodology

In this section, we provide background information on transfer learning and introduce our transfer learning-based method for predicting the shaft power.

4.1   Transfer Learning

Transfer learning is a technique where a pre-trained model on a source task or dataset is fine-tuned for a different task or dataset [30]. It uses the weights from the source task's network layers to partially retrain the model for the target task. A pre-trained model is a network trained on a benchmark dataset to solve a task. This pre-trained network can be retrained using the same or different target task-specific data by fine-tuning one or more of its layers. Fine-tuning requires pre-trained weights to partially or fully retrain the network for the target task, allowing it to learn features specific to both the source and target tasks. In this study, both the source and target tasks are shaft power prediction. The proposed approach utilizes a pre-trained model for power consumption from sensor data, following transfer learning principles, we define the source domain as $\mathcal{D}_S = \{(\mathbf{x}_i^s, y_i^s)|i = 1, \ldots, n_s\}$ representing sensor data, and target domain as $\mathcal{D}_T = \{(\mathbf{x}_j^t, y_j^t)|j = 1, \ldots, n_t\}$ representing noon reports. The transfer is formulated as:

$$\theta_{target} = \arg\min_{\theta} \mathcal{L}(f(\mathbf{x}_{noon}; \theta_{frozen}, \theta_{tune}), y_{noon})$$

where θ_{frozen} represents pre-trained parameters from sensor data (layers 1–3), θ_{target} represents parameters adapted for noon reports (layer 4) and $\mathcal{L}$ is the loss function.

4.2   Proposed Method

Our approach applies transfer learning to a neural network-based regression model trained on high-frequency sensor data, fine-tuning it to predict shaft power using the low-frequency data found in noon reports. The architecture of our model is inspired by a study predicting shaft power for vessels across sister vessels [21]. After extensive experimentation, which included adding extra hidden layers, adjusting the number of neurons, adding/removing dropout layers, and evaluating the impact of train and test across the entire dataset, we selected a (128, 64, 32, 1) configuration with no drop out layer for our model. The input dimension is seven, and each layer, except the final one, uses ReLU activation. Mean absolute error (MAE) is used to calculate the loss, similar to another transfer learning-based framework predicting fuel consumption [15] and Adam optimizer is used. The approach considers the model as the baseline when training or evaluating sensor data for shaft power prediction. The baseline neural network implements a feedforward architecture:

$$f(\mathbf{x}; \theta) = W^{(4)} \cdot \sigma(W^{(3)} \cdot \sigma(W^{(2)} \cdot \sigma(W^{(1)}x + b^{(1)}) + b^{(2)}) + b^{(3)}) + b^{(4)}$$

where $W^{(i)} \in \mathbb{R}^{n_i \times n_{i-1}}$ are weight matrices and $b^{(i)} \in \mathbb{R}^{n_i}$ are bias vectors for layer i, with network dimensions (7,128,64,32,1), ReLU activation function $\sigma(z) = max(0, z)$.

The baseline model optimizes mean absolute error loss:

$$\mathcal{L}_{baseline} = \frac{1}{n} \sum_{i=1}^{n} |y_i - \hat{y}_i|$$

where y_i is the actual shaft power and $\hat{y}_i$ is the predicted value. The transfer learning model when fine-tuned on noon reports combines frozen feature extraction with gradient computation having selective parameters update:

$$\mathcal{L}_{final} = \frac{1}{n_t} \sum_{i=1}^{n_t} |y_i^{noon} - f(x_i^{noon}; \theta_{frozen}, \theta_{tune})|$$

The final model architecture will be:

$$f_{final}(x) = W_{new}^{(4)} \cdot \sigma(W_{frozen}^{(3)} \cdot \sigma(W_{frozen}^{(2)} \cdot \sigma(W_{frozen}^{(1)}x + b_{frozen}^{(1)}) + b_{frozen}^{(2)})$$
$$+ b_{frozen}^{(3)}) + b_{new}^{(4)}$$

Baseline Model. The regression model is trained on the training set of vessels' sensor data for 300 epochs with a batch size of 32 and a learning rate of 10^{-3}. The model uses a *ReduceLROnPlateau* scheduler, reducing the learning rate by 0.5 after three epochs if the validation loss does not improve, and employs an early stopping scheme with a patience of five epochs. The weights with the minimum validation loss are saved for the baseline model.

Final Model. The pre-trained regression model (baseline) is fine-tuned on the train set of noon reports for vessels S_V2, S_V3, S_V4, SM_V1, D_V1 and D_V2. All layers except the last are frozen during training. The training configuration is identical to the baseline model, except for the batch size and learning rate, which are set to 16 and 10^{-4} respectively.

Figure 1 shows all the above-described steps of the proposed transfer learning-based shaft power prediction method.

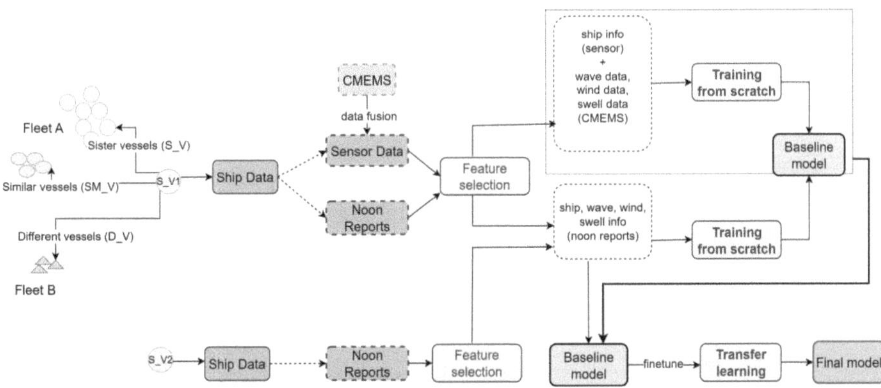

Fig. 1. Overview of the proposed shaft power prediction method. Baseline models are trained using ship data from vessel S_V1, which includes noon reports and sensor data. The sensor data is fused with Copernicus (CMEMS) data. The baseline model trained on sensor data is fine-tuned using noon reports from sister vessel S_V2 through our transfer learning approach, resulting in the final model predicting shaft power.

Data Preprocessing. The data was preprocessed to remove data instances with a speed through water of less than 2 knots, RPM of less than 1, and reported power of less than 1 kilowatt. We observed outliers in noon reports exceeding 12,000 kilowatts and removed them accordingly.

5   Experimental Setup

We train the baseline model using the train set of sensor data from sister vessels, as well as similar and different vessels. The vessel achieving the best predictive performance on the test set is selected as the baseline model for our proposed approach. The train set of noon reports from the remaining vessels in the sister, similar, and different categories is then used for fine-tuning the baseline model on a per-vessel basis. Furthermore, we train a separate baseline model from scratch using only the noon reports from the vessels. This enables a comparative analysis of the predictive performance between the models based on sensor data while also benchmarking the improvements in accuracy and robustness of our proposed method in predicting shaft power from noon reports. To mitigate the initialization bias or the data sampling bias, we ran the experiments 10 times and reported the mean and standard deviation (SD) of the evaluation metrics. We analyzed the evaluation results based on these research questions:

RQ1. What is the performance decline in shaft power prediction when using noon reports instead of sensor data?

RQ2. Does predictive performance for shaft power improve with the proposed method?

RQ3. Can the trend of improved predictive performance be observed when the model is trained on one year of data and tested on the following year's data?

5.1   Evaluation Metrics

We evaluate the performance of our experiments with the following metrics:

- Mean absolute error (MAE): MAE is a performance metric that evaluates regression models by measuring the mean absolute difference between the labels and predicted values of a model.

$$\text{MAE} = \frac{1}{n} \sum_{i=1}^{n} |y_i - \hat{y}_i| \tag{1}$$

where y_i is the actual value (label), $\hat{y}_i$ is the predicted value and n is the number of data instances.

- Normalized mean absolute error (NMAE): MAE does not consider the scale of the target labels, which can vary across different datasets. Normalizing the MAE by the mean of the target values provides a better understanding of the model's performance across different data distributions, such as sensor data versus noon reports. NMAE is calculated using the following formula.

$$\text{NMAE} = \frac{\text{MAE}}{\max(y) - \min(y)} \tag{2}$$

- Mean absolute percentage error (MAPE): MAPE presents the error in prediction as a percentage of actual values and is calculated as follows:

$$\text{MAPE} = \frac{100\%}{n} \sum_{i=1}^{n} \left| \frac{y_i - \hat{y}_i}{y_i} \right| \tag{3}$$

- Coefficient of determination (R^2): R^2 is a performance metric that measures how well the predictions of the model approximate the actual data. The following formula is used for calculation:

$$R^2 = 1 - \frac{\sum_{i=1}^{n} (y_i - \hat{y}_i)^2}{\sum_{i=1}^{n} (y_i - \bar{y})^2} \tag{4}$$

where $\bar{y}$ is the mean of actual values (labels).

6   Results and Analysis

In this section, we present our experimental results, where the model is trained on both sensor data and the noon reports training set to compare its predictive performance on shaft power. The model trained with sensor data serves as the baseline and is fine-tuned to the final model using noon reports, based on our transfer learning-based shaft power prediction approach. Predictive accuracy is evaluated for the final model and compared to models trained from scratch using both noon reports and sensor data. Finally, vessel power consumption for 2024 and 2025 is predicted using models trained with our method and from scratch. This helps analyze the robustness of the forecasted consumption trend across the entire noon report test set.

6.1   Comparison of Vessel Power Prediction: Sensor Data Vs. Noon Reports

We evaluate the performance of models trained on sensor data and noon reports. We conduct this experiment to observe a decline in shaft power performance when switching between these two data sources, answering RQ1.

Table 3 presents the evaluation metrics for shaft power prediction of vessels, using the test set for both sensor data and noon reports. The reported R^2 was higher and MAPE was lower for sensor data compared to noon reports for all vessels. To specifically answer RQ1, we use NMAE for direct comparison. An average error increase of 4% (0.04) was observed for sister vessels, 7% (0.07) for similar vessels, and 3% (0.03) for different vessels when transitioning from sensor data to noon reports. Also, for this experiment, the reported SD for MAPE was less than 2, and NMAE and R^2 were less than 1 for all vessels. Since S_V1 outperformed the other vessels on both datasets, we selected it as the base model for our approach, which will be evaluated in the subsequent experiments.

Table 3. Shaft power predictions using sensor data and noon reports.

Vessel	Sensor Data			Noon Reports		
	R^2	NMAE	MAPE	R^2	NMAE	MAPE
S_V1	0.97	0.02	3.54	0.77	0.08	11.19
S_V2	0.81	0.06	16.18	0.37	0.10	54.00
S_V3	0.44	0.09	28.05	-0.32	0.13	52.00
S_V4	0.54	0.10	5.00	0.06	0.15	22.00
SM_V1	0.80	0.05	8.59	0.19	0.12	18.00
D_V1	0.95	0.03	0.13	0.85	0.06	28.00

6.2   Improving the Predictive Accuracy of Vessel Power from Noon Reports

In this experiment, we evaluated the model's performance in predicting shaft power for two settings: training from scratch and training based on our transfer learning approach. We conducted this experiment to observe if our proposed approach improves shaft power prediction using noon reports to compare its performance with predictions based on sensor data, thereby answering RQ2.

Table 4 presents the evaluation metrics for vessel shaft power prediction, comparing transfer learning with training from scratch on the noon reports test set. For each vessel, our transfer learning-based method outperformed the training from scratch approach in predicting shaft power. We achieved an average 10.6% reduction in MAPE for sister vessels, 3.6% for similar and 5.3% for different vessels. Furthermore, the R^2 improved for all vessels, with a significant increase in magnitude and a sign change from positive to negative for S_V3. We also observed a reduction in MAE; for sister vessels, it was by a factor of 200, for similar vessels by 152, and 316 for different vessels. This clearly indicates that our approach significantly improved shaft power prediction in noon reports.

Table 4. Shaft power predictions on noon reports: training from scratch vs. transfer learning (TL).

Vessel	Training from scratch			Training based on TL		
	R^2	MAE	MAPE	R^2	MAE	MAPE
S_V2	0.37	657.49	54.00	0.76	357.40	38.00
S_V3	-0.32	768.61	52.00	0.35	494.58	41.00
S_V4	0.06	988.90	22.00	0.31	767.88	17.21
SM_V1	0.19	911.78	18.00	0.45	759.38	14.36
D_V1	0.85	975.03	28.00	0.91	659.83	22.70

Next, we compared the performance of the transfer learning-based shaft power prediction for a vessel with the model trained on sensor data for that same vessel. Table 5 presents the evaluation metrics for power prediction using training from scratch with sensor data and noon reports, along with our proposed transfer learning approach using noon reports. The results showed that our proposed method bridges the performance gap in terms of the model's predictive capability when transitioning from sensor data to noon reports. The previously reported 4% increase in NMAE for this transition was reduced to an average of 1.3% (0.013) for sister vessels. For a similar vessel, NMAE decreased from 7% to 4% (0.04), and for a different vessel, it improved from 3% to 2% (0.02). This improvement in predictive accuracy for shaft power using noon reports answers RQ2.

For these experiments, the reported SD for MAPE was less than 2 and for NMAE and R^2 it was less than 1 for all datasets.

Table 5. Shaft power predictions on different datasets: training from scratch vs. transfer learning (TL).

Vessel	NMAE (sensor data)	NMAE (noon reports with TL)	NMAE (noon reports)
S_V2	0.06	0.07	0.10
S_V3	0.09	0.09	0.13
S_V4	0.10	0.13	0.15
SM_V1	0.05	0.09	0.12
D_V1	0.03	0.05	0.06

6.3 Assessing the Improved Accuracy of Predictions by Analyzing the Power Consumption Trend

In this experiment, we present the actual and predicted shaft power values for different settings: models trained from scratch and models trained using the proposed transfer learning approach. We conducted this experiment to observe whether our method can accurately predict power consumption for the upcoming years, 2024 and 2025, despite the base model being trained with data from 2023, thus addressing RQ3.

Figure 2 shows the shaft power prediction results for sister vessels using both training from scratch and transfer learning approaches on the noon reports test set. Our method provides a more accurate prediction of the power consumption trend for all sister vessels compared to training from scratch. The accuracy of the forecasted power requirements was consistent not only in the initial months of 2024 (January and February) but also towards the end (August and September). For S_V3, although the predictive accuracy decreased around mid-2024, the trend was still better represented than the baseline model trained from scratch.

Figure 3 shows the shaft power predictions for non-sister vessels (similar and different vessels) using both training from scratch and transfer learning approaches on the noon reports test set. Similar to sister vessels, our method provides more accurate predictions of consumption trends for non-sister vessels compared to training from scratch. Despite data scarcity from April to September 2024 for SM_V1, our transfer learning approach predicted the consumption trend with higher accuracy and robustness. For D_V1, both training approaches reported similar accuracy in predicting the power consumption trend. Although the base model (a different vessel, S_V1) was trained on data up to 2023 and the forecast was for data collected in 2025, our approach still predicted with high accuracy. Thus, answering RQ3, the trend of improved predictive performance was observed for 2024 and 2025.

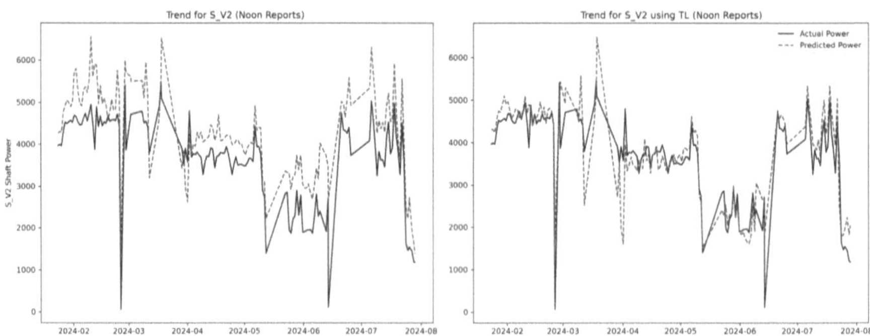

(a) Shaft power for S_V2 in the year 2024

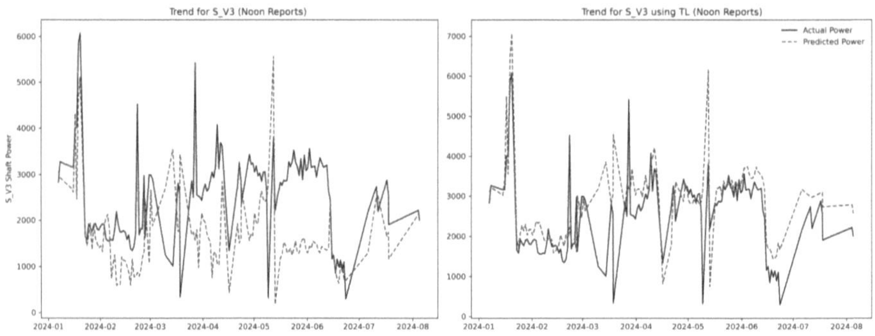

(b) Shaft power for S_V3 in the year 2024

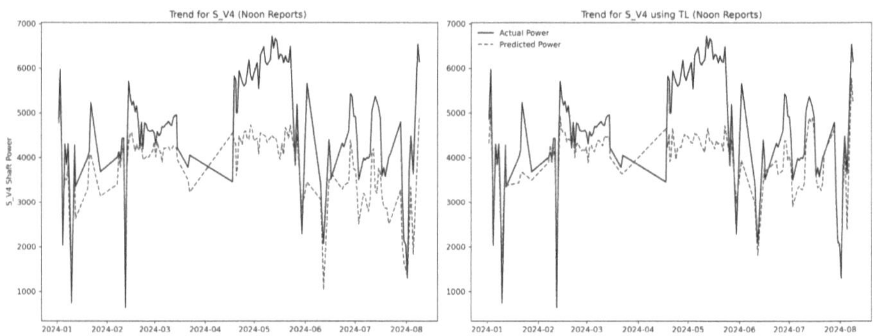

(c) Shaft power for S_V4 in the year 2024

Fig. 2. Sister vessels: forecasted consumption trends based on power predictions by models trained from scratch and transfer learning (TL).

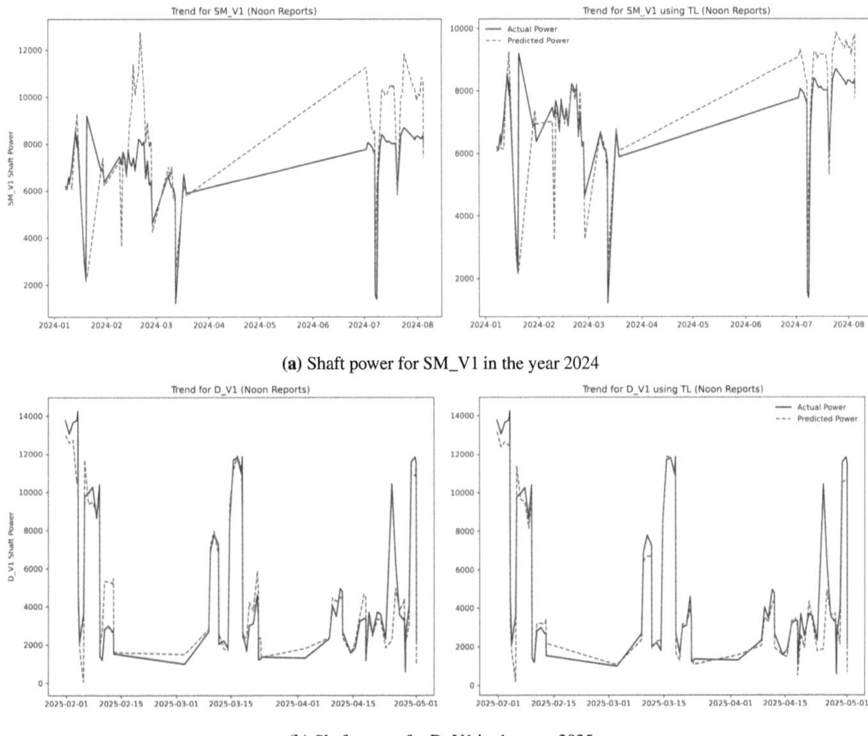

(a) Shaft power for SM_V1 in the year 2024

(b) Shaft power for D_V1 in the year 2025

Fig. 3. Non-sister vessels: forecasted consumption trends based on power predictions by models trained from scratch and transfer learning (TL).

7   Conclusion and Discussion

In this paper, we present a transfer learning-based method to improve shaft power prediction for a vessel using noon reports. Our approach first trains a model on a vessel's sensor data for shaft power prediction, then fine-tunes the model using noon reports from other vessels. Data from multiple vessels, including sisters, a similar and a different vessel, are used to evaluate the method's performance.

The experimental results show that the baseline model trained on a single sister vessel, when fine-tuned with noon reports from other sisters, reduced the error by 10.6% in predicting shaft power compared to a model trained solely on noon reports from those sisters. In comparison to sensor data models, our approach resulted in an average error increase of 1.3%, while models trained solely on noon reports showed a 4% increase in error. Therefore, our approach effectively reduced the performance gap in shaft power prediction when transitioning from sensor data to noon reports, with the reported performance nearly matching to sensor data. Hence, when sensor data for a sister vessel is unavailable, our transfer learning-based approach can provide accurate predictions using noon reports. Sensor data may contain uncertainties due to faulty or poor recordings and high variability [12,13], in which case our approach can be used for data quality management. Noon reports typically have sparse entries, but our method integrates prior knowledge on ship data at high frequency, making it effective in detecting outliers in sensor-based predictions for sister vessels. Additionally, adopting our approach to predictive monitoring can help in managing the cost of installing sensors on each sister vessel.

Our method improved the accuracy of shaft power predictions by 3.6% for a sister vessel and 5.3% for a different vessel, compared to a model trained solely on their noon reports. When compared to sensor data models, our approach demonstrated a 2% performance decrease for a similar vessel and 3% for a different vessel. Although our approach significantly reduced the predictive gap between sensor data and noon reports, the error compared to sensor data models was higher than that observed for sister vessels. Thus, this approach may not be directly comparable to sensor data models for data quality management, it can still provide more accurate and robust predictions using noon reports when sensor data is unavailable.

Our transfer-learning-based approach yielded high predictive performance for vessel shaft power predictions comparable to sensor data, even when using a limited amount of noon reports for training each vessel. Our method accurately predicted the power consumption trend for 2024 and 2025. The consistent power predictions over time demonstrated that our approach effectively models the input-output relationship between ship operational data and shaft power, enabling fuel efficiency planning during a voyage and facilitating predictive maintenance. Then, ship operators can predict future performance degradation, even with a limited number of noon reports.

7.1 Limitations and Future Work

While our proposed transfer learning-based approach has shown promising results in enhancing shaft power predictions, there are some inherent limitations. Our study is limited by the number of vessels evaluated, as we tested only three sister vessels and one vessel from each of the similar and different categories. Future work should test the approach on a broader range of vessels to address scalability and generalizability issues. For temporal analysis, we trained the base model on data from a specific year and then tested it on the subsequent years. However, it remains unclear how long the model can make accurate predictions without retraining. Investigating the optimal retraining frequency and its long-term performance could provide valuable insights. Additionally, while weather and sea conditions significantly impact shaft power consumption, our approach did not examine their individual effects. Future research should explore which parameters are most effective when transitioning from high-frequency sensor data to low-frequency daily noon reports. The data fusion between publicly available meteorological data and noon reports should be explored. In this study, we utilized sensor data from CMEMS to address weather-related uncertainty, and a similar approach for noon reports may enhance prediction accuracy. The performance of this data fusion strategy should be compared to our transfer learning-based approach to assess differences. Moreover, the model architecture is not optimized to handle large amount of diverse data from various vessel types. The fixed configuration of hidden layers and neurons may not be suitable for all vessels and requires further investigation to improve performance across different scenarios.

In summary, while our approach improves shaft power prediction using noon reports,
enhancing scalability, architectural flexibility, and integrating additional data sources are crucial for improving performance across diverse vessels and operational conditions.

Acknowledgments. This study has been funded by the Research Council of Norway under grant agreement No. 346603, the GASS project. We thank Global Ocean Waves Analysis and Forecast (https://doi.org/10.48670/moi-00017) for providing weather-related data used in this work. We thank Joachim Berdal Haga for his contributions to data acquisition and feature extraction, Glenn Terje Lines, Tetyana Kholodna and Adam Sobey for their useful feedback and discussions leading to the shaping of this work.

Data Availability Statement. The proposed transfer learning-based approach was implemented for our industrial partner and therefore it is not made available publicly. Similarly, the data used for the model training and testing is proprietary.

References

1. Adland, R., Cariou, P., Jia, H., Wolff, F.C.: The energy efficiency effects of periodic ship hull cleaning. J. Clean. Prod. **178**, 1–13 (2018)
2. Aldous, L., Smith, T., Bucknall, R., Thompson, P.: Uncertainty analysis in ship performance monitoring. Ocean Eng. **110**, 29–38 (2015)
3. Bal Beşikçi, E., Arslan, O., Turan, O., Ölçer, A.: An artificial neural network based decision support system for energy efficient ship operations. Comput. Oper. Res. **66**, 393–401 (2016)
4. Bayraktar, M., Sokukcu, M.: Marine vessel energy efficiency performance prediction based on daily reported noon reports. Ships Offshore Struct. **19**(6), 831–840 (2023)
5. Carlton, J.: Marine Propellers and Propulsion, 3rd edn. Butterworth-Heinemann, Oxford, England (2012)
6. Coraddu, A., Oneto, L., Baldi, F., Anguita, D.: Vessels fuel consumption forecast and trim optimisation: a data analytics perspective. Ocean Eng. **130**, 351–370 (2017)
7. Du, Y., Chen, Y., Li, X., Schönborn, A., Sun, Z.: Data fusion and machine learning for ship fuel efficiency modeling: part III – sensor data and meteorological data. Commun. Transp. Res. **2**, 100072 (2022)
8. Du, Y., Meng, Q., Wang, S., Kuang, H.: Two-phase optimal solutions for ship speed and trim optimization over a voyage using voyage report data. Transp. Res. Part B Methodological **122**, 88–114 (2019)

9. Gkerekos, C., Lazakis, I., Theotokatos, G.: Machine learning models for predicting ship main engine fuel oil consumption: a comparative study. Ocean Eng. **188**, 106282 (2019)
10. Holtrop, J.: A statistical re-analysis of resistance and propulsion data. Int. Shipbuild. Prog. **31**(363), 272–276 (1984)
11. Lakshmynarayanana, P.A., Hudson, D.: Estimating added power in waves for ships through analysis of operational data. In: Bertram, V. (ed.) 2nd Hull Performance and Insight Conference (HullPIC'17), pp. 253–264 (2017)
12. Laurie, A., Anderlini, E., Dietz, J., Thomas, G.: Machine learning for shaft power prediction and analysis of fouling related performance deterioration. Ocean Eng. **234**, 108886 (2021)
13. Li, X.: Data fusion and machine learning for ship fuel efficiency modeling: part I – voyage report data and meteorological data. Commun. Transp. Res. **2**, 100074 (2022)
14. Liu, S., Papanikolaou, A.: Regression analysis of experimental data for added resistance in waves of arbitrary heading and development of a semi-empirical formula. Ocean Eng. **206**, 107357 (2020). https://doi.org/10.1016/j.oceaneng.2020.107357
15. Luo, X., Zhang, M., Han, Y., Yan, R., Wang, S.: Ship fuel consumption prediction based on transfer learning: models and applications. Eng. Appl. Artif. Intell. **141**, 109769 (2025)
16. Man, Y., Sturm, T., Lundh, M., MacKinnon, S.N.: From ethnographic research to big data analytics—a case of maritime energy-efficiency optimization. Appl. Sci. **10**(6) (2020), https://www.mdpi.com/2076-3417/10/6/2134
17. Mavroudis, S., Tinga, T.: Application of transfer learning on physics-based models to enhance vessel shaft power predictions. Ocean Eng. **323**, 120540 (2025)
18. Organization, I.M.: IMO's work to cut GHG emissions from ships — imo.org. https://www.imo.org/en/MediaCentre/HotTopics/Pages/Cutting-GHG-emissions.aspx, Accessed 14 June 2025
19. Orihara, H., Tsujimoto, M.: Performance prediction of full-scale ship and analysis by means of on-board monitoring. part 2: validation of full-scale performance predictions in actual seas. J. Mar. Sci. Technol. **23**(4), 782–801 (2017). https://doi.org/10.1007/s00773-017-0511-5
20. Parkes, A.I., Savasta, T.D., Sobey, A.J., Hudson, D.A.: Efficient Vessel Power Prediction in Operational Conditions Using Machine Learning, pp. 350–367. Springer Singapore (2020)
21. Parkes, A., Savasta, T., Sobey, A., Hudson, D.: Power prediction for a vessel without recorded data using data fusion from a fleet of vessels. Expert Syst. Appl. **187**, 115971 (2022)
22. Parkes, A., Sobey, A., Hudson, D.: Physics-based shaft power prediction for large merchant ships using neural networks. Ocean Eng. **166**, 92–104 (2018)
23. Pedersen, B., Larsen, J.: Prediction of full-scale propulsion power using artificial neural networks. In: 8th International Conference on Computer and IT Applications in the Maritime Industries, pp. 537–550. TuTech, Budapest (May 10-12 2009)
24. Petersen, J.P., Jacobsen, D.J., Winther, O.: Statistical modelling for ship propulsion efficiency. J. Mar. Sci. Technol. **17**(1), 30–39 (2011)
25. Pétursson, S.: Predicting Optimal Trim Configuration of Marine Vessels with Respect to Fuel Usage. Master's thesis, University of Iceland (2009)
26. Radonjic, A., Vukadinovic, K.: Application of ensemble neural networks to prediction of towboat shaft power. J. Mar. Sci. Technol. **20**(1), 64–80 (2014)
27. Smith, T., Aldous, L., Bucknall, R.: Noon report data uncertainty (2013)

28. Soner, O., Akyuz, E., Celik, M.: Use of tree-based methods in ship performance monitoring under operating conditions. Ocean Eng. **166**, 302–310 (2018)
29. Uyanık, T., Karatuğ, u., Arslanoğlu, Y.: Machine learning approach to ship fuel consumption: a case of container vessel. Transp. Res. Part D Transp. Environ. **84**, 102389 (2020)
30. Weiss, K., Khoshgoftaar, T.M., Wang, D.D.: A survey of transfer learning. J. Big Data **3**(1), 1–40 (2016). https://doi.org/10.1186/s40537-016-0043-6
31. Zwart, R.H., Bogaard, J., Kana, A.A.: A grey-box model approach using noon report data for trim optimization. Int. Shipbuild. Prog. **70**(1), 41–63 (2023)

Towards Robust Urban Parking Violation Prediction Using Graph Kolmogorov–Arnold Networks and Liquid Neural Networks

Mohammad Reza Mohebbi[1,2]([✉]) [iD], Javad Mohebbi Najm Abad[3] [iD], Elahe Kafash[4] [iD], and Mario Döller[1] [iD]

[1] Josef Ressel Center Vision2Move, University of Applied Sciences Kufstein Tirol, Kufstein, Austria
`{mohammadreza.mohebbi,mario.doeller}@fh-kufstein.ac.at`
[2] Department of Computer Science, University of Passau, Passau, Germany
[3] Department of Computer Engineering, Islamic Azad University, Quchan Branch, Quchan, Iran
`javad.mohebi@iauq.ac.ir`
[4] Department of Computer Engineering, Imam Reza (AS) International University, Mashhad, Iran
`elahe.kafash@imamreza.ac.ir`

Abstract. Illegal parking in urban environments disrupts traffic flow, causes greenhouse gas emissions, and poses a threat to pedestrians and cyclists. Traditional Intelligent Transportation Systems (ITS) are based on high-cost surveillance and video analysis that typically does not take into account the dynamics and complexity of the urban environment. To address these limitations, this study overcomes this gap by proposing an intelligent parking violation prediction framework using a hybrid Spatio-temporal Graph Neural Network (STGNN) approach, which combines Graph Kolmogorov-Arnold Networks (GKAN) and Liquid Neural Networks (LNN). The GKAN model excels at uncovering intricate spatial patterns in the urban dataset, while the LNN model has intrinsic dynamic temporal variations in real time due to its adaptive learning capability. This integration of spatio-temporal relationships of metropolitan datasets is effective modeling and thus can be robust across diverse urban scenarios. The proposed approach is well-suited for practical usage in real-world applications and achieves a high prediction accuracy with a high R^2 score of 0.95, and shows significant improvement in other metrics such as MAE and MSE. These results underscore the performance of the proposed GKAN-LNN framework in addressing the challenges presented by the parking violation prediction task to develop safe, sustainable, and well-governed urban settings.

Keywords: Parking Violation Prediction · Graph · Kolmogorov-Arnold Networks · Liquid Neural Networks · Illegal Parking · Intelligent Transportation System

© The Author(s), under exclusive license to Springer Nature Switzerland AG 2026
F. Marcelloni et al. (Eds.): IJCCI 2025, CCIS 2829, pp. 742–762, 2026.
https://doi.org/10.1007/978-3-032-15638-9_44

1   Introduction

The historically unprecedented growth in private vehicle ownership in recent decades, driven by rapid urbanization, has vastly outgrown the expansion of parking facilities [1]. Consequently, searching for legal parking has become an ongoing problem in most metropolitan areas [2]. Research shows that drivers would spend 3.5 to 14 minutes looking for parking, leading to an increase in total traffic of up to 74% [3]. This inefficiency contributes to urban congestion, fuel usage, and environmental deterioration. Furthermore, restricted parking availability often causes unauthorized parking, reduces road capacity, limits emergency response, and affects air quality and public safety [4,5].

In response to these challenges, multiple solutions based on video monitoring and Machine Learning (ML) were proposed for the detection or prediction of parking offenses in the literature [6,7]. Although successful in some scenarios, these methods suffer in practice from environmental factors (e.g., occlusion) and need expensive infrastructure. Recent developments in Deep Learning (DL) and spatio-temporal modeling, particularly Graph Neural Networks (GNNs), have made it possible to model urban dynamics more accurately and scalably [8,9]. Specifically, Spatio-Temporal Graph Neural Networks (STGNN) enable learning in both spatial dependencies (e.g., traffic) and temporal trends (e.g., hour of the day) and are therefore suitable for urban movement applications [10–12].

In this paper, Graph Kolmogorov–Arnold Networks (GKAN) and Liquid Neural Networks (LNN) are integrated within an STGNN framework to propose a novel hybrid model to forecast illegal parking violations in urban areas. Although LNNs offer adaptive temporal modeling appropriate for real-time decision-making, GKAN uses learnable spline-based activations to improve expressivity and interpretability in graph learning. The method utilizes spatio-temporal features, including weather, time, and Point of Interest (POI) data, and incorporates sophisticated preprocessing steps to handle missing values. The framework is demonstrated using large-scale Thessaloniki urban data and is shown to perform strongly under real-scenario conditions. Given the identified challenges, this study makes the following main contributions:

- **End-to-End Pipeline:** An efficient STGNN-based pipeline was built to model parking violations in the main urban regions based on heterogeneous spatio-temporal data.
- **Feature Engineering and Preprocessing:** A wide range of methods is presented for spatio-temporal feature extraction and imputation of missing data, enhancing the accuracy and stability of predictions.
- **Hybrid GKAN-LNN Model:** In comparison to traditional architectures, the proposed GKAN-LNN model effectively captures complex spatio-temporal patterns with higher prediction accuracies.

To contextualize our study, we focus on Thessaloniki, a coastal metropolitan city in Greece with sustained population growth and high parking demand. Its dense urban morphology and accessibility restrictions pose persistent challenges for sustainable mobility and efficient curb management. Figure 1 shows a satellite view with the municipal boundary highlighted in red. Our objective is to uncover spatio-temporal regularities in illegal parking incidents and derive policy-relevant insights for the scientific design and implementation of effective monitoring and management. By modeling the frequency of violations and their temporal patterns, the proposed framework provides urban managers with actionable tools for preventive interventions.

Fig. 1. Study area: Thessaloniki, Greece. Satellite map with urban boundary highlighted in red. (Color figure online)

The remainder of this paper is organized as follows. Section 2 reviews related studies on the prediction of parking violations, highlighting graph-based, data-driven, and vision-based approaches. Section 3 describes the proposed methodology, including model architecture, data preparation, graph construction, and imputation strategies. Section 4 presents the experimental results, covering data collection, evaluation metrics, and comparative analyzes of the proposed GKAN-LNN model against existing methods. Finally, Sect. 5 concludes the paper and outlines potential directions for future research.

2   Literature Review

Illegal urban parking has become a major problem, causing ineffective urban mobility, decreased road safety, and traffic congestion. In response, researchers have proposed several intelligent techniques for anticipating and identifying parking violations [10,11]. These initiatives typically fit into one of three methodological groups: graph-based frameworks, data-driven models, or vision-based approaches. From spatial reasoning and temporal forecasting to direct detection, each category addresses a distinct facet of the issue. In order to place the contributions of this study in context, it is essential to comprehend these methodologies.

Vision-based approaches generally employ surveillance video and object detection models for the recognition of illegal parking behavior in real time [13]. You Only Look Once (YOLO), Faster Region-based Convolutional Neural Networks (R-CNN), and RetinaNet are popular DL-based detectors used for vehicle detection and zone monitoring [14]. Cai et al. [15], for example, employed a temporal segment-based parking infraction recognition system, and Mostafa et al. [16] employed a YOLO pipeline supplemented with geospatial mapping for real-time enforcement. Other studies [17,18], used multiframe inference and rule-based reasoning to improve detection accuracy through the determination of infringement. The methods exhibited good performance in static settings but depend heavily on dense camera deployment and experience challenges when dealing with occlusions, as well as scalability and coverage issues in urban areas.

To overcome infrastructure dependency, data-driven approaches use structured data sets, such as traffic records, violation logs, and road attributes, to model and forecast illegal parking patterns [7]. Classical ML models such as logistic regression and decision trees have been used to identify trends of temporal and contextual violations [19,20]. Advanced ensemble methods such as Random Forest (RF) [11] and boosting-based algorithms have achieved better predictive capacity through learning complex non-linear patterns. In addition, a Long Short-Term Memory (LSTM) model [21] has also been utilized to predict sequential behavior and allow dynamic modeling of changes over several hours, days, or weeks. However, most of them treat each location as a standalone entity and do not account for spatial interdependence, which hampers the detection of regionally correlated offenses.

Graph-based models have recently gained traction by leveraging spatial structures in urban environments. In recent work [12], semi-supervised learning of GNN combined with Long-Short-Term Memory (LSTM) was performed using the Temporal Graph Convolutional Networks (TGCN)-LSTM method for the prediction of street parking violations. This approach leveraged spatio-temporal data to model graph-based relationships to improve prediction accuracy, even when the labeled data is limited. Another GCN-based method [22] has used topological information to analyze the co-occurrence and propagation of violations in neighborhoods. Although these models effectively capture spatial structure, they often assume fixed graphs or lack adaptive mechanisms to model evolving urban dynamics. Moreover, many are tailored for general traffic prediction rather than for the specific characteristics of parking violations. To address these gaps, we introduce a graph-based approach that explicitly models the dynamic and fine-grained spatio-temporal patterns of illegal parking behaviors.

To address the limitations of the present methods, a hybrid system is proposed by combining LNNs and GKANs for the prediction of illegal parking violations. Temporal patterns adapt dynamically through LNNs, and spatial generalization over intricate urban geometries occurs through expressive non-linear function learning on graph form by GKANs. The system was compared with competitive baselines such as RF [11], ResNet [10], and T-GCN-LSTM [12] and proved to be able to improve predictive accuracy and performance significantly. Through the integration of spatial reasoning and temporal adaptability, a scalable and interpretable solution is provided for intelligent urban parking management.

3   Background

This section provides the theoretical foundation of the two key components in our framework: *LNNs* and *GKANs*. LNNs are continuous-time recurrent models that adaptively modulate their internal dynamics using state- and input-dependent time constants, enabling them to capture long-term temporal dependencies and remain robust under distribution shifts. On the other hand, GKANs extend the recently introduced KolmogorovâĂŞArnold Networks (KANs) to graph-structured data, replacing fixed activation functions with learnable spline-based transformations along edges, thereby enhancing expressivity and interpretability. By combining these paradigms, our approach jointly models the **temporal dynamics** of parking violations and the **spatial correlations** among urban road segments.

3.1   Liquid Neural Networks (LNNs)

LNNs, also known as Liquid Time-Constant (LTC) networks [23,24], are continuous-time recurrent neural models designed to capture adaptive temporal patterns. Unlike standard recurrent networks, where the update rule is fixed, LNNs parameterize neuron dynamics through input- and state-dependent time constants, allowing flexible adaptation across timescales.

Let $\mathbf{x}(t) \in \mathbb{R}^H$ denote the hidden state at time t, $\mathbf{u}(t) \in \mathbb{R}^U$ the input signal, and $\mathbf{y}(t) \in \mathbb{R}^Y$ the output. The hidden state evolves according to a system of Ordinary Differential Equations (ODEs) [23]:

$$\dot{\mathbf{x}}(t) = \mathbf{f}_\theta\big(\mathbf{x}(t), \mathbf{u}(t), t\big), \tag{1}$$

where $\dot{\mathbf{x}}(t) = \frac{d\mathbf{x}(t)}{dt}$, and θ denotes trainable parameters. In the LNN formulation, each neuron $x_i(t)$ evolves as [24]:

$$\dot{x}_i(t) = -\frac{1}{\tau_i(\mathbf{x}(t), \mathbf{u}(t))}\, x_i(t) + \frac{1}{\tau_i(\mathbf{x}(t), \mathbf{u}(t))}\left(\sum_{j=1}^{H} w_{ij}\, \sigma(x_j(t)) + \mathbf{v}_i^\top \mathbf{u}(t) + b_i\right), \tag{2}$$

where:

- $\tau_i(\mathbf{x}, \mathbf{u}) > 0$ is the adaptive **liquid time constant**, controlling the rate of change of neuron i,
- w_{ij} are recurrent connection weights,
- $\mathbf{v}_i$ are input weights and b_i is a bias term,
- $\sigma(\cdot)$ is a bounded nonlinearity such as tanh.

A common parameterization ensures the positivity of τ_i:

$$\tau_i(\mathbf{x}, \mathbf{u}) = \text{softplus}(\mathbf{a}_i^\top \mathbf{x} + \mathbf{r}_i^\top \mathbf{u} + c_i) + \varepsilon, \tag{3}$$

where $\mathbf{a}_i, \mathbf{r}_i, c_i$ are trainable and $\varepsilon > 0$ avoids degenerate solutions.

Finally, the read-out is computed as:

$$\mathbf{y}(t) = \mathbf{W}_o\, \phi(\mathbf{x}(t)) + \mathbf{b}_o. \tag{4}$$

where $\phi(\cdot)$ is a readout nonlinearity, $\mathbf{W}_o$ are output weights, and $\mathbf{b}_o$ is a bias.

Interpretation. Eq. (2) reveals that LNN neurons adaptively modulate their integration speed, enabling stability, robustness to distribution shifts, and strong generalization in spatio-temporal tasks. This property makes them ideal for the prediction of urban traffic and parking violations [23,24].

3.2   Graph Kolmogorov–Arnold Networks (GKANs)

KANs are inspired by the Kolmogorov–Arnold representation theorem, which states that any continuous multivariate function can be expressed as a finite sum of univariate functions. Unlike traditional neural networks, where fixed nonlinear activations reside at nodes, KANs attach learnable univariate functions to edges, improving interpretability and efficiency [25,26].

Let $\mathbf{z}^{(\ell)} \in \mathbb{R}^{d_\ell}$ denote the input to layer ℓ. A KAN layer computes the following:

$$z_i^{(\ell+1)} = \sum_{j=1}^{d_\ell} \varphi_{ij}^{(\ell)}(z_j^{(\ell)}), \quad i = 1, \ldots, d_{\ell+1}, \tag{5}$$

where $\varphi_{ij}^{(\ell)} : \mathbb{R} \to \mathbb{R}$ are learnable univariate functions. Each function is parameterized as:

$$\varphi_{ij}^{(\ell)}(x) = \rho_{ij}^{(\ell)} x + \sum_{m=1}^{M} c_{ijm}^{(\ell)} B_m(x), \tag{6}$$

with $\rho_{ij}^{(\ell)}$ and $c_{ijm}^{(\ell)}$ trainable coefficients, and $\{B_m\}_{m=1}^{M}$ a set of spline basis functions. This design yields highly expressive, yet interpretable, transformations [25,26].

Graph Adaptation. To extend KANs to graphs, consider a graph $G = (V, E)$ with $|V| = N$ nodes, adjacency matrix $\mathbf{A}$, and degree matrix $\mathbf{D}$. The symmetrically normalized adjacency is [27]:

$$\tilde{\mathbf{A}} = \mathbf{D}^{-\frac{1}{2}}(\mathbf{A} + \mathbf{I})\mathbf{D}^{-\frac{1}{2}}. \tag{7}$$

Given node features $\mathbf{H}^{(\ell)} \in \mathbb{R}^{N \times F_\ell}$, a GKAN layer is defined as [27]:

$$\mathbf{H}^{(\ell+1)} = \tilde{\mathbf{A}}\,\Phi^{(\ell)}\big(\mathbf{H}^{(\ell)}\big), \tag{8}$$

where $\Phi^{(\ell)}$ applies univariate spline-based transformations featurewise:

$$\big[\Phi^{(\ell)}(\mathbf{H}^{(\ell)})\big]_{n,f} = \rho_f^{(\ell)} H_{n,f}^{(\ell)} + \sum_{m=1}^{M} c_{fm}^{(\ell)} B_m\big(H_{n,f}^{(\ell)}\big). \tag{9}$$

for node $n = 1, \ldots, N$ and feature $f = 1, \ldots, F_\ell$.

Interpretation. GKANs replace rigid linear maps with interpretable spline-based transformations before graph propagation. This enhances both expressivity and interpretability, enabling nuanced modeling of spatial correlations in urban road networks.

In the context of urban parking violation prediction, both temporal and spatial aspects must be jointly modeled. LNNs, with their adaptive and stable continuous-time dynamics, are naturally suited to capture the **temporal evolution** of violations, which often exhibit complex and non-stationary patterns over time. GKANs, on the other hand, provide a principled mechanism to model **spatial correlations** across the irregular topology of urban road networks by applying interpretable univariate spline transformations before graph propagation. By integrating LNNs for temporal reasoning and GKANs for spatial feature learning, our proposed hybrid architecture effectively leverages the complementary strengths of both paradigms, enabling robust spatio-temporal forecasting in real-world traffic environments.

4   Methodology

This section proposes a composite DL framework that combines GKAN and LNN under an STGNN architecture. The approach covers data preprocessing, graph construction, feature engineering, and model design to accurately predict parking violations at a street level. All components aim to extract spatial and temporal features from urban datasets.

4.1   Model Architecture

Figure 2 illustrates the architecture of the proposed DL pipeline to predict parking violations. The framework makes use of a hybrid STGNN that combines a GKAN for spatial reasoning and an LNN for modeling temporality. The integration enables capturing rich spatio-temporal dependencies in parking spaces. The input features were obtained from internal sources (e.g., ID of parking slot, capacity of sections, and timestamp) and external sources (e.g., weather conditions, holidays, and pandemic-induced disruptions). The features were normalized by the z score and are listed in Table 1.

The spatio-temporal graph was formed, and each node corresponds to a parking space and related attributes, and edge weights representing physical distance between spaces were encoded. The graph served as input for the hybrid GKAN-LNN approach. Both global and local spatial relationships between spaces were captured   through GKAN, and local dependencies on time scales of an hour were captured through LNN.

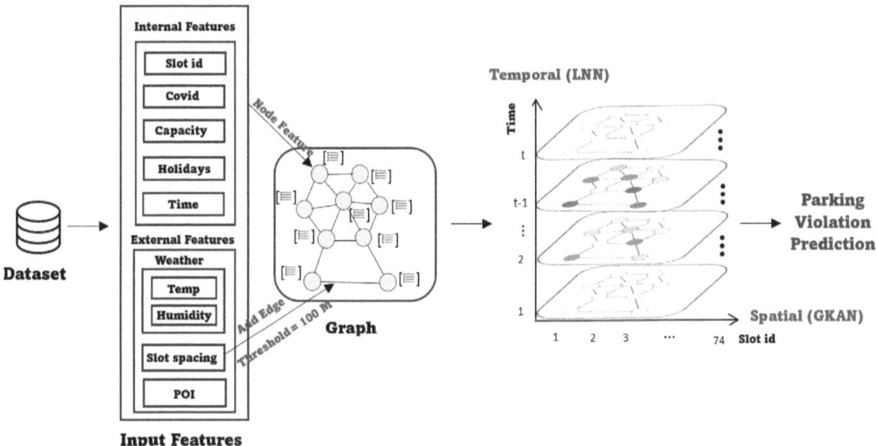

Fig. 2. Hybrid GKAN-LNN model for parking violation prediction.

Table 1. Features used in parking violation prediction.

Feature Category	Example Features	Dimensions
Distances to POIs	Distance to nearest hospital, school, etc.	19
Section Information	Section Capacity	1
Temporal Features	Day of Week, Day, Month, Time Slot	4
Weather Conditions	Temperature, Humidity	2
Special Dates	Public Holidays	1
External Factors	Pandemic / COVID-19 Indicator	1

To reduce noise in environmental features, temperature and humidity readings were smoothed using a moving average with a window size of 6, formulated as:

$$x_t = \frac{1}{W} \sum_{i=1}^{W} T_{t-i}. \tag{10}$$

where T_{t-i} indicates the temperature at a previous time step and $W = 6$. The day of the week, day of the month, and month-related features were encoded via sinusoidal functions to capture periodic patterns, and the time intervals (07:00–19:00) were encoded as categorical hourly intervals. Public holidays and COVID-19 periods were encoded through binary indicators that capture significant temporal disturbances of parking behavior.

The model outputs hourly predictions of parking violations for each city segment. Hyperparameters, including learning rate and optimizer configuration, were   tuned through grid search to maximize predictive performance.

4.2   Data Preparation

Data preparation serves as the foundation for the proposed prediction pipeline. A graph representation of the parking infrastructure was created, as shown in Fig. 3 (left), in which each node represents a parking space and the edges show the actual distances between them. These distances were precomputed and stored to facilitate efficient spatial reasoning. The graph structure became sparser and more informative as a result of distant nodes being pruned based on a proximity threshold to concentrate on significant local interactions.

In parallel, additional contextual features from other sources were also included as input, e.g., weather and proximity to POI. Figure 3 (right) indicates two important spatial patterns: the clustering of parking spots around points of interest such as museums and the intensity of the violations indicated through a heat map. The violation rates were mainly higher in the areas surrounding POI, and their inclusion in the set of characteristics was justified on the basis of this.

Together, these steps ensured an accurate capture of both spatial topology and contextual semantics before model training.

4.3   Graph Data Preparation

Following data retrieval and preprocessing, the graph-structured inputs for the STGNN are prepared based on the pipeline shown in Fig. 2. Each parking space is represented as a node and spatial relationships, such as proximity, are encoded as edges. The overall architecture is depicted in Fig. 4, illustrating data loading, missing data management, graph construction, and feature extraction.

Missing Data Management. To enhance data reliability, missing values were imputed using six distinct strategies. These techniques significantly reduced noise and improved the robustness and accuracy of the model's predictions.

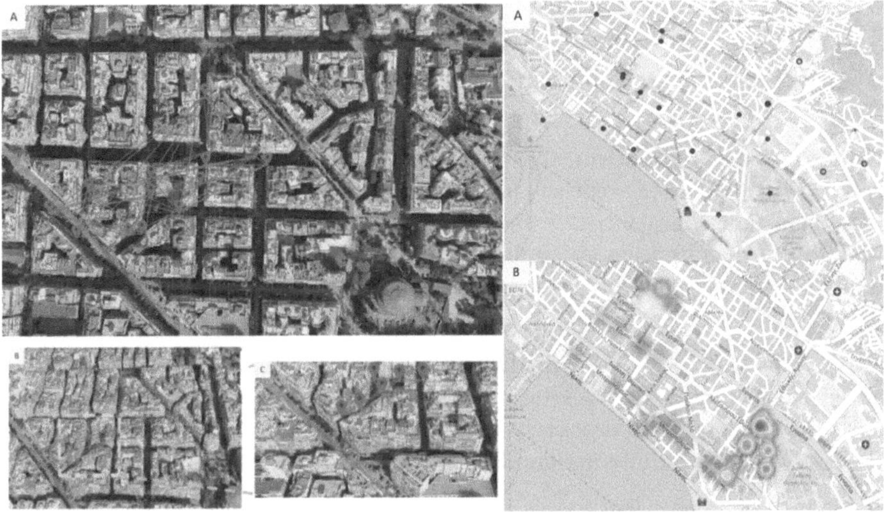

Fig. 3. Graph-based representation of parking slots and spatial relationships (left); spatial distribution of Points of Interest and parking violations (right).

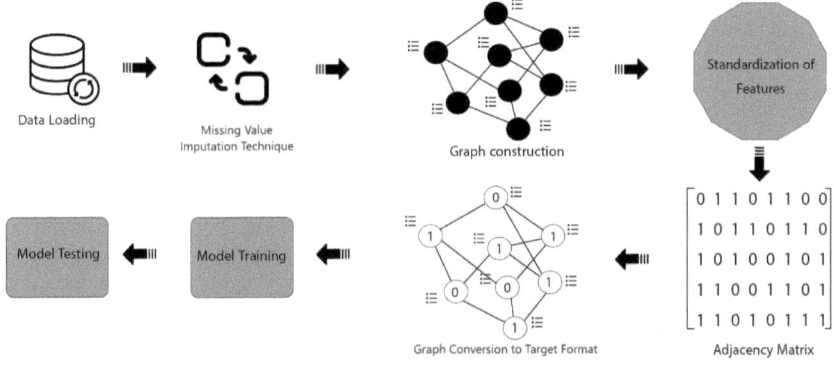

Fig. 4. Process of generating and preparing graph datasets.

Graph Construction via Distance Thresholding. A graph was constructed using geospatial distance data between parking spots. An edge was established between two nodes if their distance was less than a predefined threshold κ (100 meters), with edge weights assigned using a Gaussian kernel:

$$W(v_i, v_j) = \begin{cases} \exp\left(-\frac{[\text{dist}(v_i, v_j)]^2}{2\sigma^2}\right) & \text{if } \text{dist}(v_i, v_j) \leq \kappa, \\ 0 & \text{otherwise.} \end{cases} \tag{11}$$

where σ is the standard deviation of all distances between nodes. By retaining only important spatial relationships, this formulation increases the accuracy of representation and computational efficiency.

Feature Standardization. The `StandardScaler` has been used to standardize characteristics such as weather, indicator variables of vacation, park characteristics, and time characteristics to ensure consistency of the scales between them. In addition to enhancing the stability of the models, this normalization process minimizes bias caused by various feature scales.

Adjacency Matrix Computation. An adjacency matrix was established in order to encode the spatial relationship between nodes based on the calculated distance. The matrix is necessary to facilitate the STGNN to effectively capture the dependency and interaction of nodes.

Model Training and Implementation. The graph data was transformed into the `torch_geometric` format, and the node features and prediction targets were prepared accordingly. The AdamW optimizer and Mean Squared Error (MSE) loss function were employed for training the model. Hyperparameter optimization in the form of learning rate and weight decay significantly improved predictive accuracy and reduced overfitting.

4.4   Missing Data Imputation Techniques

The handling of missing data is a critical preprocessing step in both exploratory data analysis and deep learning, ensuring data reliability and preserving the performance of the downstream model [28]. In this work, several imputation techniques were systematically compared to assess their influence on predictive accuracy in the context of parking violation prediction.

Figure 5 gives a general visual overview. The left panel (subplots A and B) depicts the impact of imputation of missing data on the distribution of the rate of parking violations. The raw distribution before imputation is presented in subplot 5.A, and depicts a dense cluster of observations between 0.8 and 0.9, supporting high rates of compliance as the most common occurrence. The same variable after imputation is presented in subplot 5.B, and indicates a much smoother and better normalized distribution. This normalization reflects the effectiveness of imputation in minimizing the effect of outliers and maintaining the underlying statistical form.

The performance of the model under different imputation techniques is compared using the Mean Absolute Error (MAE) and the MSE in the right panel (subplots A and B). The evaluation is based on common regression metrics such as MAE and MSE (formally defined in Sect. 4.2). The robustness of each approach was assessed by comparing these metrics before and after the datasets were merged. Most of the imputation methods showed improved performance after merging, as indicated by the decreased error values. However, simpler methods, such as Last Observation Carried Forward (LOCF) combined with Next Observation Carried Backward (NOCB), show marginal increases in their post-merge MSE, most likely due to inconsistencies in data being transferred from one boundary to another.

To further quantify this comparison, the performance of the six imputation methods on the GKAN-LNN model is given in the summary of Table 2. The hybrid approach combining Multivariate Imputation by Chained Equations (MICE) and Interpolation performed the best consistently and thus achieved the lowest MAE (0.107) and MSE (0.0249) after combining datasets. These findings affirm that statistical inference with interpolation based on patterns offers a balanced and highly efficient solution for dealing with missing data in spatio-temporal graph-based learning.

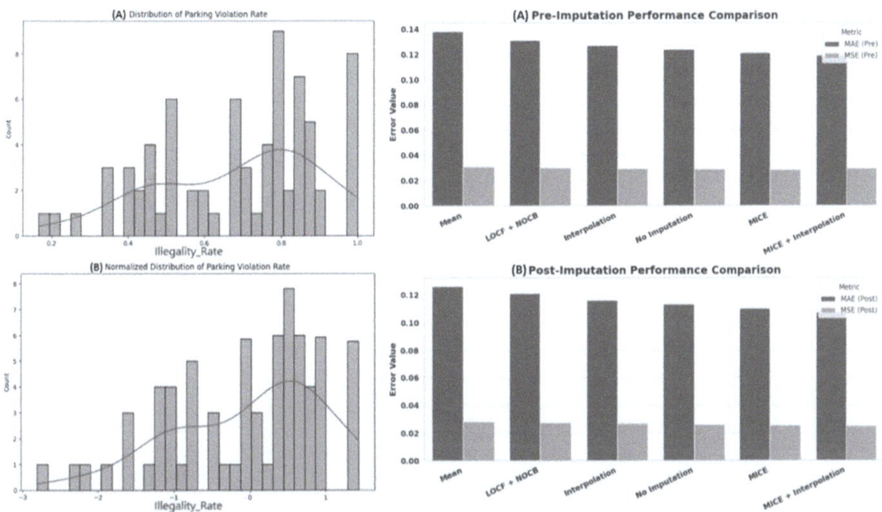

Fig. 5. Process of generating and preparing graph datasets.

Table 2. Performance comparison of MDI techniques on the GKAN-LNN model.

Technique	Description	MAE (Pre)	MSE (Pre)	MAE (Post)	MSE (Post)
Mean	Replace missing with column means	0.138	0.0309	0.126	0.0281
LOCF + NOCB	Last and next observed values	0.131	0.0302	0.121	0.0274
Interpolation	Estimate from data patterns	0.127	0.0294	0.116	0.0268
No Imputation	No modifications made	0.124	0.0290	0.113	0.0260
MICE	Statistical model-based imputation	0.121	**0.0285**	0.110	0.0254
MICE + Interpolation	Combined statistical + pattern	**0.119**	0.0295	**0.107**	**0.0249**

5   Experimental Results

The experimental setup used to assess the proposed parking violation prediction framework is presented in this section. The dataset and its key characteristics are initially explained. The preprocessing procedures and evaluation metrics that guarantee robust modeling are then described. Lastly, the results of the experiments are shown, along with a performance analysis of the model through various configurations.

5.1   Data Collection

All experiments were carried out using data obtained from the publicly accessible THESi system[1], which manages approximately 4,700 parking spaces across 396 segments in the city center of Thessaloniki. The data set includes over 3.8 million scan records generated from more than 300,000 traffic enforcement checks. Additional context was provided by integrating weather information from Open-Weather and spatial proximity features derived from 19 manually defined POIs.

Figure 6 illustrates the spatial distribution of parking violations in the study area. Subfigure 6.A displays a heatmap where the warmer colors represent regions with higher frequencies of violations. Subfigure 6.B illustrates the violations' intensity via circle size and identifies hotspots of high activity in the upper city blocks. These findings indicate that remote or less monitored areas are more susceptible to illegal parking and offer valuable spatial signals suitable for graph-based modeling.

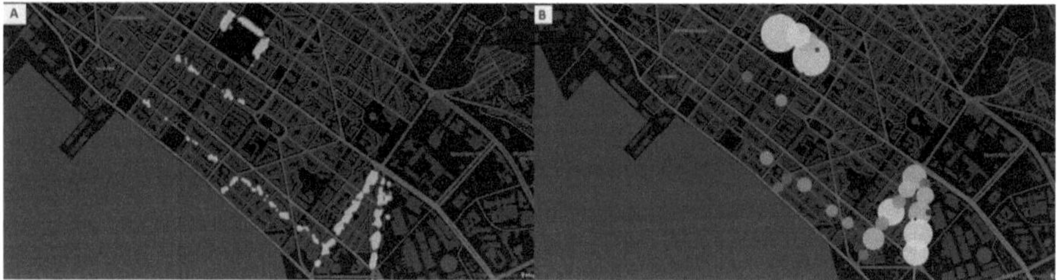

Fig. 6. Heatmap of parking violations in the study area.

5.2   Model Evaluation Criteria

To rigorously evaluate the predictive performance of the proposed framework, several regression metrics were employed. Let O_i denote the observed (ground truth) value, P_i the predicted value, N the total number of testing samples, and

---

[1] https://www.thesi.gr.

$\bar{O}$ the mean of the observed values [29–31]. The metrics are formally defined as follows:

$$\text{MAE} = \frac{1}{N} \sum_{i=1}^{N} |O_i - P_i|, \tag{12}$$

The **Mean Absolute Error (MAE)** represents the average absolute deviation between predicted and observed values. It provides a direct measure of accuracy while being robust to outliers, as all errors contribute linearly to the final score [29].

$$\text{MSE} = \frac{1}{N} \sum_{i=1}^{N} (O_i - P_i)^2, \tag{13}$$

The **Mean Squared Error (MSE)** squares the prediction errors before averaging, thereby penalizing larger deviations more heavily. This property makes MSE particularly sensitive to large errors, which is useful in applications where such deviations are critical [31].

$$\text{RMSE} = \sqrt{\frac{1}{N} \sum_{i=1}^{N} (O_i - P_i)^2}, \tag{14}$$

The **Root Mean Squared Error (RMSE)** is the square root of the MSE and expresses errors in the same unit as the target variable. This improves interpretability and provides a clear sense of the magnitude of typical prediction errors [30].

$$\text{MAPE} = \frac{100}{N} \sum_{i=1}^{N} \left| \frac{O_i - P_i}{O_i} \right|, \tag{15}$$

The **Mean Absolute Percentage Error (MAPE)** expresses the prediction error as a percentage of the actual observed values. This metric enables relative comparisons across datasets or variables of different scales. However, it is undefined for $O_i = 0$ and can become unstable when values are close to zero [29].

$$R^2 = 1 - \frac{\sum_{i=1}^{N} (O_i - P_i)^2}{\sum_{i=1}^{N} (O_i - \bar{O})^2}, \tag{16}$$

The **Coefficient of Determination (R^2 Score)** measures the proportion of variance in the observed values explained by the model. An R^2 score of 1 indicates perfect prediction, while values close to 0 suggest weak predictive ability relative to the mean baseline [30].

$$\text{Time per Epoch} = \frac{\text{Total Training Time}}{\text{Number of Epochs}}. \tag{17}$$

The **Training Time per Epoch** captures computational efficiency. This measure is particularly important for large-scale spatio-temporal datasets, where scalability and runtime directly impact practical deployment [29].

Collectively, these metrics provide a balanced evaluation framework: MAE and RMSE quantify average accuracy, MSE emphasizes the presence of large errors, MAPE normalizes performance across scales, R^2 captures explanatory power, and training time reflects computational efficiency.

5.3 Performance Evaluation and Comparative Analysis

To assess the efficacy of the proposed framework, extensive experiments were performed using Python 3.12 and the PyTorch Geometric package. The Geopy toolkit was used to handle spatial operations. The experiments in all cases adopted the 80/20 train-to-test split ratio. Eleven models were implemented to compare the performance, each training and testing according to the same setup, including the same missing value imputation method to maintain homogeneity and fairness in the evaluation process.

As determined in the preprocessing phase, the use of MICE and the linear interpolation method proved most effective for handling missing values. This combined imputation approach was used uniformly across all models and evaluated over two important scenarios: (i) before merging datasets; (ii) after merging the datasets. This enabled performance evaluation of models both under complete data settings and also in terms of the impact of data integration on the accuracy of predictions.

Tables 3 and 4 compare the performance of all models in the two situations. The suggested GKAN-LNN model outperformed the rest in all metrics: MAE, MSE, root mean square error (RMSE), mean absolute percentage error (MAPE), and coefficient of determination (R^2). In particular, before dataset merging, GKAN-LNN performed with MAE = 0.119 and MSE = 0.0295. After merging, the measures improved significantly to MAE = 0.107 and MSE = 0.0249, reflecting better learning from the merged dataset.

Figure 7 provides a visual representation of the performance comparison of the best-performing models in all evaluation metrics to support the tabular results. The best R^2 score and the lowest MAE, MSE, RMSE, and MAPE values demonstrate the superiority of GKAN-LNN. Performance validates the accuracy, stability, and better generalization capabilities of the model. Importantly, while GKAN-LNN comes with a slightly higher computational cost per epoch, as indicated in Fig. 7.f, this cost and overhead is worthwhile considering the significant improvement in predictive performance.

After GKAN-LNN, the GCNN-LSTM model performed the best in all metrics with strong predictive and efficiency performance. The stability over all the datasets also hints towards its ability to serve as a reliable baseline architecture.

Table 3. Performance before dataset merging.

Model	MSE	MAE	R^2	RMSE	MAPE (%)	Time (s)
GKAN-LNN (Proposed)	**0.0295**	**0.119**	**0.943**	**0.171**	**5.85**	0.011
GCNN-LSTM	0.0364	0.139	0.925	0.191	7.59	**0.007**
GATv2-GRU	0.0701	0.136	0.904	0.265	6.96	0.039
GIN-GRU	0.0759	0.125	0.884	0.276	6.67	0.018
GCN-GRUAtt	0.0509	0.178	0.903	0.225	9.21	0.028
GAT-GRUAtt	0.0601	0.199	0.893	0.245	10.01	0.012
GAT-CNNAtt	0.0645	0.184	0.889	0.254	8.64	0.010
GraphTransformer-GRU	0.0792	0.229	0.865	0.281	11.15	0.045
GraphMixer-CNN	0.0927	0.312	0.831	0.304	13.81	0.052
GraphSAGE-CNN	0.1023	0.339	0.782	0.320	15.19	0.034
GraphFormer-CNN	0.1268	0.367	0.758	0.356	17.36	0.060

Table 4. Performance after dataset merging.

Model	MSE	MAE	R^2	RMSE	MAPE (%)	Time (s)
GKAN-LNN (Proposed)	**0.0249**	**0.107**	**0.953**	**0.158**	**5.01**	0.011
GCNN-LSTM	0.0312	0.128	0.931	0.176	6.80	**0.007**
GATv2-GRU	0.0340	0.116	0.928	0.184	5.85	0.015
GIN-GRU	0.0365	0.118	0.922	0.191	6.01	0.018
GCN-GRUAtt	0.0405	0.138	0.917	0.201	6.99	0.028
GAT-GRUAtt	0.0488	0.157	0.905	0.221	8.10	0.012
GAT-CNNAtt	0.0522	0.165	0.900	0.228	8.25	0.010
GraphTransformer-GRU	0.0601	0.174	0.885	0.245	9.10	0.045
GraphMixer-CNN	0.0729	0.203	0.860	0.270	10.90	0.052
GraphSAGE-CNN	0.0815	0.232	0.831	0.285	11.95	0.034
GraphFormer-CNN	0.0998	0.267	0.794	0.316	13.45	0.060

5.4 GKAN-LNN vs. Existing Methods: A Comparative Evaluation

To comprehensively evaluate the efficacy of the suggested GKAN-LNN model, the comparative study was performed with various renowned state-of-the-art methods such as RF [11], ResNet [10], and TGCN-LSTM [12]. Table 5 shows the comparative performance metrics using two important parameters: MAE and MSE.

As is evident, the proposed GKAN-LNN outperforms all other models with the smallest MAE and MSE values. One particularly noteworthy result is that the GCNN-LSTM model outperforms all external baselines when improved with missing value imputation. Nevertheless, it still falls behind the GKAN-LNN described in terms of predictive performance and error minimization.

Table 5. Performance comparison with state-of-the-art models.

Model	MAE	MSE
RF [11]	0.191	0.0561
ResNet [10]	0.169 (11.52%)	0.0497 (11.40%)
TGCN-LSTM [12]	0.166 (13.09%)	0.0468 (16.58%)
GCNN-LSTM (Imputed)	0.128 (33.51%)	0.0312 (44.40%)
GKAN-LNN (Proposed)	**0.107 (44.24%)**	**0.0249 (55.61%)**

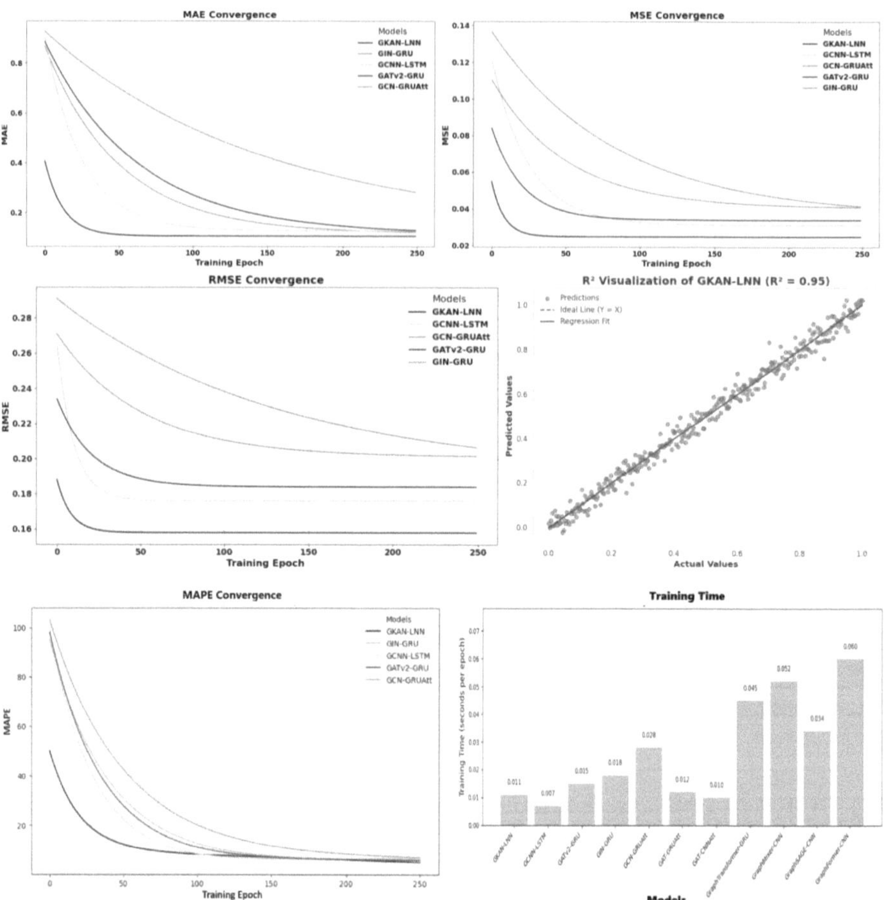

Fig. 7. Performance comparison of the top five models across key evaluation metrics.

Figure 8 offers a visual comparison of the MAE and MSE values for the five methods evaluated. The subfigure (A) illustrates that the Random Forest model yielded the highest MAE, confirming its relatively poor performance in this spatial-temporal prediction task. DL models such as ResNet and TGCN-LSTM

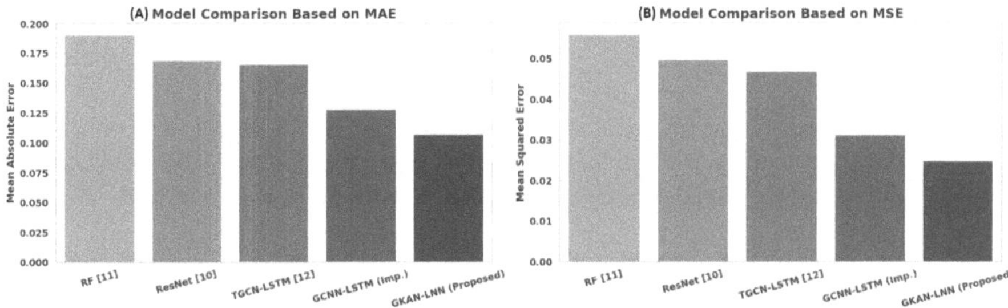

Fig. 8. Performance comparison of the top five models across key evaluation metrics.

exhibit moderate improvements. Nonetheless, the suggested GKAN-LNN model sets a new performance standard by providing the largest MAE reduction.

Similarly, Fig. 8.B illustrates the comparative performance using MSE. Once again, GKAN-LNN performs with the smallest error, and conventional methods such as RF have the highest predictive variance. The wide performance margin reflects the robustness and efficacy in generalizability across varying spatial and temporal patterns.

These findings demonstrate that the GKAN-LNN model surpasses the accuracy of other DL-based approaches, in addition to performing better than conventional ML techniques. Its performance shows that it is suitable for real-world deployment in complex urban environments that require accurate parking occupancy predictions, both in terms of raw error metrics and percentage improvements.

6   Conclusion

This research developed sophisticated parking violation prediction models that integrated a variety of data sources, including those from the public parking system, weather conditions, temporal patterns, and other external factors. The proposed GKAN-LNN model has shown outstanding performance, consistently superior in all evaluation metrics, including MAE and MSE, with an impressive R^2 score of 0.95. Unlike traditional methods and other DL approaches, GKAN-LNN showed remarkable robustness in handling complex, heterogeneous datasets and demonstrated resilience to structural variations, thus ensuring reliable and precise predictions under different conditions. Its innovative architecture can be perfectly integrated into ITS, enabling its application to several real-world scenarios that are suitable for city planners and policymakers. Providing data-driven decisions, GKAN-LNN can achieve operational efficiency, optimized urban mobility, and sustainable city planning. This study not only underscored the benefits and significance of using predictive analysis in the solution of traffic and parking management issues in the city, but also provided the GKAN-LNN model as a baseline model to predict parking violations. The results of this research improve the enforcement and form the foundation on which more sustainable, efficient, and intelligent cities can be developed, directly complementing the idea of the smart city concept.

Acknowledgments. The financial support of the Austrian Federal Ministry of Labor and Economy, the National Foundation for Research, Technology, and Development, and the Christian Doppler Research Association is gratefully acknowledged.

Disclosure of Interests. The authors have no competing interests to declare that are relevant to the content of this article.

References

1. Zhang, W., Wang, K.: Parking futures: shared automated vehicles and parking demand reduction trajectories in Atlanta. Land Use Policy **91**, 103963 (2020). https://doi.org/10.1016/j.landusepol.2019.04.024
2. Shoup, D.: High Cost of Free Parking. 1st edn. Routledge, New York (2005). https://doi.org/10.4324/9781351179539
3. Weinberger, R.R., Millard-Ball, A., Hampshire, R.C.: Parking search caused congestion: where's all the fuss? Transp. Res. Part C: Emerging Technol. **120**, 102781 (2020). https://doi.org/10.1016/j.trc.2020.102781
4. Basri Said, L., Syafey, I.: The scenario of reducing congestion and resolving parking issues in makassar city. Indonesia. Case Stud. Transp. Policy **9**(4), 1849–1859 (2021). https://doi.org/10.1016/j.cstp.2021.10.004
5. Bahrami, S., Vignon, D., Yin, Y., Laberteaux, K.: Parking management of automated vehicles in downtown areas. Transp. Res. Part C: Emerging Technol. **126**, 103001 (2021). https://doi.org/10.1016/j.trc.2021.103001
6. Peng, X., et al.: Real-time illegal parking detection algorithm in urban environments. IEEE Trans. Intell. Transp. Syst. **23**(11), 20572–20587 (2022). https://doi.org/10.1109/TITS.2022.3180225
7. Inam, S., Mahmood, A., Khatoon, S., Alshamari, M., Nawaz, N.: Multisource data integration and comparative analysis of machine learning models for on-street parking prediction. Sustainability **14**, 7317 (2022). https://doi.org/10.3390/su14127317
8. Jin, G., et al.: Spatio-temporal graph neural networks for predictive learning in urban computing: a survey. IEEE Trans. Knowl. Data Eng. **36**(10), 5388–5408 (2024). https://doi.org/10.1109/TKDE.2023.3333824
9. Musa, A.A., Malami, S.I., Alanazi, F., Ounaies, W., Alshammari, M., Haruna, S.I.: Sustainable traffic management for smart cities using internet-of-things-oriented intelligent transportation systems (ITS): challenges and recommendations. Sustainability **15**, 9859 (2023). https://doi.org/10.3390/su15139859
10. Karantaglis, N., Passalis, N., Tefas, A.: Predicting on-street parking violation rate using deep residual neural networks. Patt. Recogn. Lett. **163**, 82–91 (2022). https://doi.org/10.1016/j.patrec.2022.09.023
11. Gao, S., Li, M., Liang, Y., Marks, J., Kang, Y., Li, M.: Predicting the spatiotemporal legality of on-street parking using open data and machine learning. Ann. GIS **25**(4), 299–312 (2019). https://doi.org/10.1080/19475683.2019.1679882
12. Karantaglis, N., Passalis, N., Tefas, A.: Semi-supervised learning for on-street parking violation prediction using graph convolutional networks. Neural Comput. Appl. **36**, 19643–19652 (2024). https://doi.org/10.1007/s00521-024-10248-5

13. Gao, J., Zuo, F., Ozbay, K., Hammami, O., Barlas, M.L.: A new curb lane monitoring and illegal parking impact estimation approach based on queueing theory and computer vision for cameras with low resolution and low frame rate. Transp. Res. Part A: Policy Practice **162**, 137–154 (2022). https://doi.org/10.1016/j.tra.2022.05.024

14. Yousefi, D.B.M., Rafie, A.S.M., Al-Haddad, S.A.R., Azrad, S.: A systematic literature review on the use of deep learning in precision livestock detection and localization using unmanned aerial vehicles. IEEE Access **10**, 80071–80091 (2022). https://doi.org/10.1109/ACCESS.2022.3194507

15. Cai, B.Y., Alvarez, R., Sit, M., Duarte, F., Ratti, C.: Deep learning-based video system for accurate and real-time parking measurement. IEEE Internet Things J. (2019). https://doi.org/10.1109/JIOT.2019.2902887

16. Mostafa, S.A., et al.: A YOLO-Based deep learning model for real-time face mask detection via drone surveillance in public spaces. Inf. Sci. **676**, 120865 (2024). https://doi.org/10.1016/j.ins.2024.120865

17. Patel, A.S., Tiwari, V., Ojha, M., Vyas, O.P.: Ontology-based detection and identification of complex event of illegal parking using SPARQL and description logic queries. Chaos Solitons Fractals **174**, 113774 (2023). https://doi.org/10.1016/j.chaos.2023.113774

18. Charouh, Z., Ezzouhri, A., Ghogho, M., Guennoun, Z.: Video analysis and rule-based reasoning for driving maneuver classification at intersections. IEEE Access **10**, 45102–45111 (2022). https://doi.org/10.1109/ACCESS.2022.3169140

19. Provoost, J.C., Kamilaris, A., Wismans, L.J.J., van der Drift, S.J., van Keulen, M.: Predicting parking occupancy via machine learning in the web of things. Internet Things **12**, 100301 (2020). https://doi.org/10.1016/j.iot.2020.100301

20. Kadkhodaei, M., Shad, R., Ziaee, S.A., Kadkhodaei, M.: Prediction model for drivers' tendency to perpetrate a double parking violation on urban trips. Transp. Policy **141**, 331–339 (2023). https://doi.org/10.1016/j.tranpol.2023.08.001

21. Anagnostopoulos, T., Fedchenkov, P., Tsotsolas, N., Ntalianis, K., Zaslavsky, A., Salmon, I.: Distributed modeling of smart parking system using LSTM with stochastic periodic predictions. Neural Comput. Appl. **32**(14), 10783–10796 (2019). https://doi.org/10.1007/s00521-019-04613-y

22. Zhou, Y., Wang, Y., Zhang, F., Zhou, H., Sun, K., Yu, Y.: GATR: a road network traffic violation prediction method based on graph attention network. Int. J. Environ. Res. Public Heal. **20**, 3432 (2023). https://doi.org/10.3390/ijerph20043432

23. Hasani, R., Lechner, M., Amini, A., Rus, D., Grosu, R.: Liquid time-constant networks. In: Proceedings of the AAAI Conference on Artificial Intelligence, vol. 35, no. 9, pp. 7657–7666 (2021). https://doi.org/10.1609/aaai.v35i9.16936

24. Mohebbi, M.R., Tavasoli, M., Döller, M., Abad, J.M.N., Zada, M.J.H., Yamnenko, I.: Vehicle trajectory prediction in congested urban traffic leveraging liquid neural network and UAV data. In: Huang, L., Greenhalgh, D. (eds) Proceedings of 17th International Conference on Machine Learning and Computing. Lecture Notes in Networks and Systems, vol. 1475, pp. 193–210. Springer, Cham (2025). https://doi.org/10.1007/978-3-031-94892-3_15

25. Liu, Z., et al.: KAN: kolmogorov-arnold networks. arXiv preprint arXiv:2404.19756 (2024). https://doi.org/10.48550/arXiv.2404.19756

26. Mohebbi, M.R., Klinger, J., Abad, J.M.N., Döller, M., Tavasoli, M.: Advanced driving behavior analysis through kolmogorov-arnold network and UAV traffic data. In: Huang, L., Greenhalgh, D. (eds) Proceedings of 17th International Conference on Machine Learning and Computing. Lecture Notes in Networks and Systems,

vol. 1475, pp. 305–322. Springer, Cham (2025). https://doi.org/10.1007/978-3-031-94892-3_23

27. Kiamari, M., Kiamari, M., Krishnamachari, B.: GKAN: graph kolmogorov-arnold networks. arXiv preprint arXiv:2406.06470 (2024). https://doi.org/10.48550/arXiv.2406.06470

28. Emmanuel, T., Maupong, T., Mpoeleng, D., Semong, T., Mphago, B., Tabona, O.: A survey on missing data in machine learning. J. Big Data **8**(1), 1–37 (2021). https://doi.org/10.1186/s40537-021-00516-9

29. Mohebbi, M.R., Kafash, E., Döller, M.: Multi-agent trajectory prediction for urban environments with UAV data using enhanced temporal kolmogorov-arnold networks with particle swarm optimization. In: Proceedings of the 17th International Conference on Agents and Artificial Intelligence - vol. 2, pp. 586–597. SciTePress (2025). https://doi.org/10.5220/0013243100003890

30. Mohebbi, M.R., Sena, E.W., Döller, M., Klinger, J.: Wildfire spread prediction through remote sensing and UAV imagery-driven machine learning models. In: 2024 18th International Conference on Control, Automation, Robotics and Vision (ICARCV), pp. 827–834. IEEE (2024). https://doi.org/10.1109/ICARCV63323.2024.10821545

31. Mohebbi, M.R., Hassan Zada, M.J., Rostami-Shahrbabaki, M., Döller, M., Yamnenko, I.: Network traffic co-movement assessment via oriented basis signal processing and ensemble decision trees. In: 2024 IEEE 27th International Conference on Intelligent Transportation Systems (ITSC), pp. 1948–1955. IEEE (2024). https://doi.org/10.1109/ITSC58415.2024.10919555

Data Augmentation for Neuroaesthetics Analysis

Maurizio Palmieri$^{(\boxtimes)}$ (ID), Marco Avvenuti (ID), Francesco Marcelloni (ID), and Alessio Vecchio (ID)

Department of Information Engineering, University of Pisa, 56122 Pisa, Italy
{maurizio.palmieri,marco.avvenuti,francesco.marcelloni,
alessio.vecchio}@unipi.it

Abstract. Neuroaesthetics investigates the neural activities during aesthetic experiences, using EEG recordings or fMRI images to decode the perception of visual art. However, studies in this domain are hindered by the limited and imbalanced nature of datasets, which is due to the subjective and resource-intensive nature of data collection. This study examines the effectiveness of various data augmentation strategies in enhancing EEG classification performance for neuroaesthetic analysis. We experiment three different EEG augmentation techniques, namely Signal Segmentation and Recombination, Temporal and Spatial Reconstruction Data Augmentation, and Gaussian Noise Addition. Furthermore, once extracted the features from the signals, we applied an instance-level data augmentation algorithm, namely SMOTE. We tested the four augmentation techniques individually, as well as SMOTE applied in cascade with the three EEG-specific augmentation methods, using stratified ten-fold cross-validation and leave-one-subject-out validation strategies. Results show that Gaussian Noise Addition, particularly when combined with SMOTE for generalization, yields consistent performance improvements in both accuracy and F-score. Conversely, Temporal and Spatial Reconstruction Data Augmentation often degrades classification performance (up to -9.95% of accuracy). Finally, Signal Segmentation and Recombination achieved the best improvement in the leave-one-subject-out analysis ($+2.2\%$ accuracy). Our findings present that appropriate data augmentation can enhance model generalization for aesthetic experience classification.

Keywords: EEG · Data Augmentation · Classification · Neuroaesthetics

1 Introduction

In recent years, the application of machine learning (ML) to electroencephalography (EEG) analysis has shown considerable promise in neuroscience and clinical fields, despite the rising challenges related to inter-subject variability, noise, and limited sample sizes. Raw EEG signals are high-dimensional and often noisy

F. Marcelloni et al. (Eds.): IJCCI 2025, CCIS 2829, pp. 763–778, 2026.
https://doi.org/10.1007/978-3-032-15638-9_45

time series, making them unsuitable for direct input into traditional ML models. As a solution, the creation of ML models for EEG analysis typically includes, among other steps, a stage of feature extraction, where relevant features are extracted from each EEG segment, followed by the training phase in the feature space. Common feature types include spectral power across standard frequency ranges (e.g., delta, theta, alpha, beta, gamma bands), entropy measures, and time-frequency representations such as wavelet coefficients. Once transformed into this lower-dimensional and more informative feature space, the EEG samples become more tractable for ML applications. These feature vectors are then used to train classifiers, such as support vector machines, ensembles, or neural networks, that learn to discriminate between different conditions.

The performance of the ML models depends heavily on the balance and representativeness of the training data. One of the main challenges in EEG analysis is the limited availability of datasets. The reason for such limited availability of data is that its acquisition is both time-consuming and resource-demanding for both practitioners and subjects. This scarcity of data often leads to a second, related issue, namely, class imbalance, as some neural events occur infrequently and are underrepresented in the data. This imbalance can significantly bias ML models, causing them to favor majority classes and perform poorly on the underrepresented (but often essential) conditions. In certain application domains, the second issue can be mitigated through the careful design of protocols that guarantee balance among the conditions. However, this approach is not viable for applications that rely on subjective judgments from participants (e.g., judgments on the edge of possible ranges can be less common than middle ones).

The challenges posed by data scarcity and class imbalance in EEG analysis are typically addressed with two strategies. On the data availability front, one strategy is represented by collaborative initiatives that promote data sharing and standardization by building larger and more diverse datasets (e.g., EBRAINS [10], OpenNeuro [21]). The other possible strategy is data augmentation. In the field of EEG analysis, data augmentation comprises two main possibilities: EEG augmentation and feature augmentation. EEG augmentation aims to artificially enhance the quantity and diversity of data by transforming the existing EEG signals. Common augmentation strategies include adding Gaussian noise, segmenting and recombining signals, and applying temporal shifts. Similarly, feature augmentation aims to artificially increase the training data by creating synthetic elements in the feature space, generating synthetic samples (e.g., using SMOTE [5]). These approaches improve model robustness without requiring additional data collection. When applied thoughtfully, often in combination with other strategies, data augmentation can significantly enhance model performance, especially in scenarios where collecting large-scale, balanced EEG datasets remains impractical.

The contribution of this work lies in the application of established data augmentation techniques to the emerging domain of neuroaesthetics, where EEG is utilized to investigate neural responses to aesthetic experiences to uncover the neural correlates of aesthetic appreciation and emotional engagement. These investigations also aim to achieve practical goals such as enhancing the museum visiting experience, personalizing art exhibitions, and informing the development of art-based therapeutic interventions [12]. Despite growing interest in this field, neuroaesthetics studies are often constrained by small, imbalanced datasets, typically involving limited numbers of participants and a skewed distribution of aesthetic judgments. The application of EEG analysis in the neuroaesthetics field is one of those cases where the protocol adopted for EEG acquisition cannot mitigate the problem of imbalanced classes, as beauty perception is inherently subjective, influenced by the environment, and the subjects should not be constrained in their evaluations to ensure authenticity of their responses. To address the challenges of limited data and class imbalance in EEG-based neuroaesthetics research, we systematically apply and evaluate various data augmentation techniques. Our approach includes methods such as noise injection and time-warping on the raw EEG signals, as well as a feature-level strategy, such as SMOTE, for synthetic oversampling. Each augmentation algorithm is implemented to tackle two key objectives: improving model generalization by increasing data variability and rebalancing the dataset to mitigate class imbalance. Specifically, this study examines the impact of various combinations of EEG-level and feature-level augmentation strategies on classification performance in the context of recognizing neuroaesthetic experiences.

The paper is organised as follows. Section 2 reviews the most relevant works on building EEG classification models. Section 3 presents the dataset analyzed in this study. Section 4 describes the augmentation techniques adopted and their combinations. Section 5 reports the results obtained with a systematic application of these augmentation techniques. Finally, Sect. 6 summarizes the main findings and hints at possible future directions.

2   Related Work

The application of deep learning approaches to EEG analysis has become increasingly widespread in recent years [11], thanks to advancements in neuroscience and improvements in EEG acquisition devices [24]. Among the most well-known domain applications, one can find disease detection [3,25,28], emotion recognition [7,16,17], motor imagery [2,30], and mental effort estimation [15,27]. The advancements, as mentioned earlier, have also allowed the rise of new research branches, such as neuroaesthetics [4,22,32]. The challenge of limited and heterogeneous EEG data has led to the development of a wide range of data augmentation techniques aimed at improving the generalization and robustness of models in EEG analysis. Recent studies have introduced domain-specific augmentation strategies tailored to the nature of EEG signals. For example, a technique based on EEG Mask Encoding (EEG-ME), as described in [9], can be used for capturing asynchronous steady-state visual evoked potentials. This approach involves deliberately masking parts of the EEG signals during training, forcing neural networks to extract more robust representations. The method yielded accuracy gains across multiple CNN architectures, particularly on short-window data, thus demonstrating its efficacy in preventing overfitting and improving classification under limited training data scenarios. In the domain of affective computing, diffusion models were applied to synthesize artificial EEG samples for emotion recognition tasks [33]. By carefully optimizing the sampling stage, they were able to generate high-quality signals, thereby improving classification performance on a set of publicly available datasets. The diffusion-based data augmentation approach is particularly effective when the training data is sparse.

A biologically inspired method using neural field theory to generate artificial EEG time series for motor imagery classification was proposed in [23], achieving statistically significant improvements in classification, which emphasizes the promise of theory-driven simulation models for EEG data augmentation. Beyond signal-level generation, [19] explored contrastive representation learning to indirectly augment EEG data by generating more training pairs through channel recombination. Their method yielded a quadratic increase in training samples per recording, enhancing classification across emotion, anomaly, and sleep staging tasks even with limited labels. A simple yet effective data augmentation method based on Gaussian Noise Addition (GNA) for EEG-based emotion recognition was proposed in [29]. Their results reinforce the value of noise injection as a robust augmentation strategy for EEG-based deep learning applications. Additional approaches include temporal-spatial reconstruction data augmentation (TSRDA) [31] in taste-related EEG decoding, where random spatial and temporal features are reconstructed to enrich the training set. Another effective data augmentation strategy for EEG signals is the Signal Segmentation and Recombination (SR) method [18]. In this approach, each EEG trial is segmented into multiple equal-length sub-windows. During training, new synthetic trials are created by randomly recombining segments from different original trials belonging to the same class. This preserves the label consistency while introducing variability in temporal dynamics. A broader perspective is provided by comprehensive reviews [11,26], which categorize EEG data augmentation into traditional signal manipulations, generative modeling, and transfer/domain-adaptive techniques.

3   Dataset, Data Management and Evaluation Strategies

This section provides details on the dataset used for the neuroaesthetics analysis, the data preprocessing operations, the features extracted from the EEG segments, and the metrics and strategies employed for classification.

3.1   Dataset Description

The dataset used in this study, introduced in [1] and studied in [13], contains EEG recordings from participants who visited the Menil Collection exhibition "The Boundary of Life is Quietly Crossed" in Houston, Texas. Mobile EEG systems were used to allow participants to move freely within the space, creating a naturalistic recording environment. Since EEG data was collected using a number of different acquisition systems, to have homogeneous data we selected the subset recorded with the 8-channel Neuroelectrics Starstim headset (F3, Fz, F4, C3, C4, P3, Pz, P4), which included the largest participant group (28 individuals). The dataset comprises 114 MATLAB files, each related to approximately 20 s of artwork observation. Labels were assigned based on a post-exhibit questionnaire: an EEG signal is labeled either as *Special*, if the participant identified the viewed artwork as the most aesthetically pleasing or emotionally stimulating, or as *Piece* otherwise. The Special labels are less populated than the Piece ones by roughly a 50% factor (i.e., the dataset is composed of approximately 67% of Piece-labeled EEGs and 33% of Special-labeled EEGs.

3.2 Data Preprocessing

The EEG data undergoes a series of preprocessing steps designed to enhance signal quality and minimize artifacts [6]. First, signals are bandpass filtered using a FIR1 filter in the range of 1–50 Hz to remove low-frequency drift and high-frequency noise. A common average reference is then applied, followed by the application of the artifact subspace reconstruction (ASR) [14]. The preprocessing phase utilizes the ASR algorithm plug-in of EEGLAB [8], employing a 0.5-second sliding window and a threshold of three standard deviations to reconstruct the EEG components affected by artifacts. After preprocessing, each recording is segmented into fixed-length, non-overlapping windows of 2 s, serving as the basic units for feature extraction and classification.

3.3 Feature Extraction

For each segment, a comprehensive set of features was extracted to capture both statistical and spectral characteristics of the EEG signal. These included basic statistical descriptors (mean, standard deviation, maximum, minimum, and kurtosis), entropy, modulation index, and the first Mel-frequency cepstral coefficient [20]. In addition, mean spectral power was computed across five standard EEG frequency bands: delta (1–4 Hz), theta (4–8 Hz), alpha (8–12 Hz), beta (12–30 Hz), and gamma (30–50 Hz). As a result, each 2-second EEG signal is characterized by 104 features (13 features × 8 channels).

3.4 Classification Strategies and Metrics

To evaluate classification performance and ensure robustness, two validation strategies are employed. The first is the stratified Ten-Fold Cross-Validation (TFCV), which provides an estimate of model performance. This strategy divides the dataset into ten non-overlapping folds, each showing the same label distribution as the whole dataset. The evaluation strategy involves ten training phases where each fold is used as a test set exactly once, and the training set is composed of all the remaining folds. The second evaluation strategy is Leave-One-Subject-Out (LOSO), which evaluates generalization across diverse participants. It involves one iteration for each subject, ensuring that in each iteration, the data from one subject is held out as the test set, while data from all other subjects is used for training. It is important to note that the two classification strategies yield distinct training sets; therefore, the augmentation strategies have varying impacts. The models created in this work have been evaluated based on two main metrics that complement each other: Accuracy and F-score. Accuracy provides a measure of correct classifications, but it is not sufficient on its own due to the unbalanced nature of the dataset, which can lead to misleading results (e.g., a model always outputting the majority class will still have high accuracy). The F-score is therefore included as a complementary metric, offering a more robust assessment of model performance across classes, particularly when some classes are underrepresented. A good classifier is therefore characterized by both high accuracy and high F-score. Finally, inspired by the results obtained in [22], we adopted an Ensemble with Subspace KNN strategy for building the classification model.

4   Data Augmentation Analysis

For this work we have selected three EEG augmentation techniques, namely Signal Segmentation and Recombination (SR, proposed in [18]) to explore the duplication of EEGs in the time domain, Temporal and Spatial Reconstruction Data Augmentation (TSRDA, proposed in [31]) to explore the duplication of EEGs in both time and space domains, and Gaussian Noise Addition (GNA, proposed in [29]) for overall EEGs augmentation. Further, once extracted the features from the signals, we have also experimented an instance-level data augmentation algorithm, namely SMOTE [5]. This section presents the relevant details of these techniques, including the parameters chosen and the proposed procedure for training an ML model.

4.1   Data Augmentation Techniques

SR involves splitting each EEG segment (2 s) into eight subsegments of equal size (0.25 s) and then creating a new EEG composed of a random concatenation of these subsegments. The choice of eight subsegments is inspired by [18]. Figure 1 shows an example of a synthetic EEG created with SR.

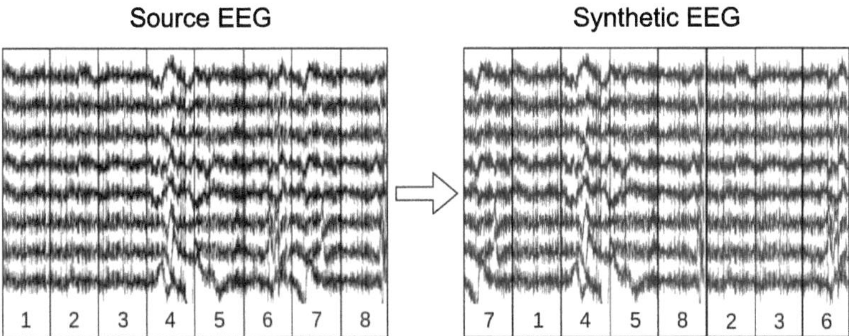

Fig. 1. Example of SR EEG Augmentation.

TSRDA is a data augmentation strategy specifically designed to enhance EEG datasets by reconstructing informative patterns in both temporal and spatial dimensions. The method randomly selects two EEG samples with the same label and cuts small data blocks from them. One block is then used to overwrite the corresponding region of the other, resulting in a new hybrid sample that contains features from both original signals. It is important to note that this method involves numerous random choices, including the selection of the two source EEGs, the center, height, and width of the block. The distributions of these random variables are taken from [31]. Figure 2 shows an example of a synthetic EEG created with TSRDA.

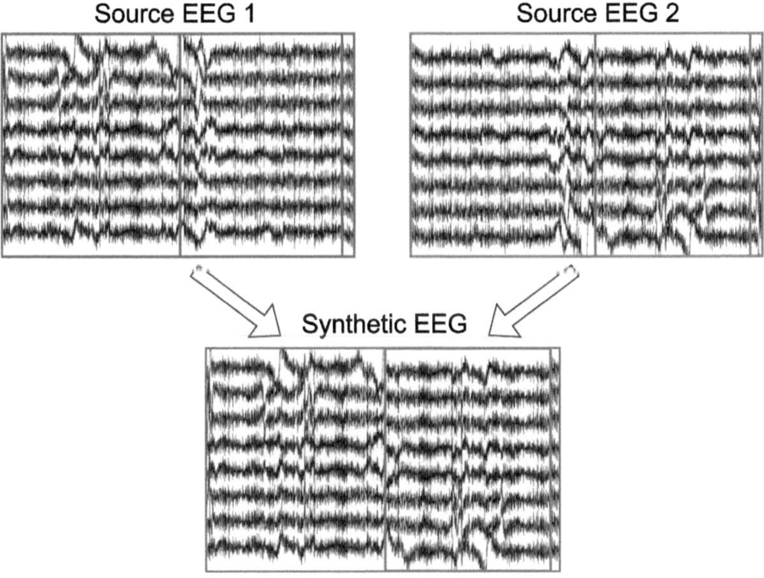

Fig. 2. Example of TSRDA EEG augmentation.

GNA involves adding low-amplitude, zero-mean Gaussian noise to EEG signals. This technique simulates real-world variability caused by sensor noise or environmental artifacts, enabling models to become more resilient to such distortions. The noise level must be carefully chosen to avoid overwhelming the signal. In this work, GNA is applied by adding Gaussian noise with a standard deviation of 0.2 (as suggested in [29]).

SMOTE is a data augmentation technique designed to address class imbalance by generating synthetic samples directly in the feature space. It creates new samples by interpolating between a sample and its nearest neighbors. This approach increases the diversity of the dataset without introducing exact duplicates. In this work, for each instance in a particular class, three neighbors are selected, and synthetic feature vectors are generated along the line segments connecting them.

4.2   Proposed Data Augmentation Procedure

The data augmentation process starts with the EEG dataset described in Sect. 3.1, which undergoes the preprocessing procedure mentioned in Sect. 3.2 followed by the split into training and test sets according to one of the two evaluation strategies depicted in Sect. 3.4. Data augmentation is then applied only to the training set, using both EEG-level and feature-level techniques as depicted in Fig. 3. For each iteration of the evaluation strategy, the training set undergoes a two-level augmentation scheme. At the first level, an EEG Augmentation technique is applied directly to the Starting EEG Training Set, resulting

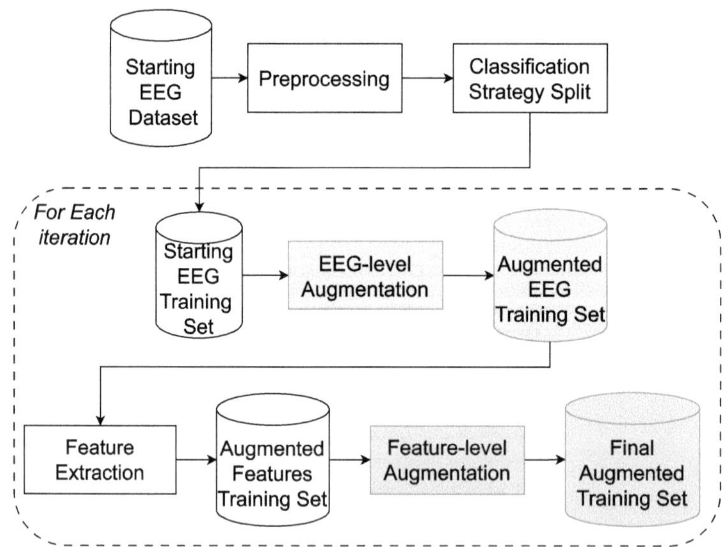

Fig. 3. Pipeline with two levels of Data Augmentation.

in an Augmented EEG Training Set (AETS) that comprises the original EEGs from the training set and the newly created ones. As a consequence, AETS is larger than the original training set. At the second level, the AETS undergoes feature extraction, resulting in the Augmented Features Training Set (AFTS), which has the same number of entries as AETS, each defined by the set of features described in Sect. 3.3. Finally, a feature-level Augmentation technique is applied to the AFTS, producing the Final Augmented Training Set (FATS), which is larger than the previously created sets. This enriched set is then used for training the classification model.

All data augmentation techniques can be applied to both rebalance the class distribution and increase the size of the training set, as shown in Fig. 4. The upper part of Fig. 4 shows the case when EEG augmentation is in charge of balancing the number of classes, i.e., its purpose is to only generate new EEGs of the minority class (Special in this specific case) until the size of the minority class matches the size of the majority one and then the feature-level augmentation technique is in charge of increasing the number of samples to improve the generalization of the model. Conversely, the bottom part of Fig. 4 shows the evolution of the training set when the EEG-level augmentation technique is in charge of increasing the generalization of the data, while the feature-level one is in charge of its balancing. The size of FATS is independent of the order of generalization-oriented and balancing-oriented techniques, whereas AETS (and AFTS) is larger when generalization is performed on the first level (i.e., for EEG augmentation), leading to a procedure that temporarily handles more data. In both cases, the generalization-oriented augmentation technique builds synthetic data to quadruplicate the size of the input dataset (for the sake of readability, Fig. 4 only shows a duplication of the dataset). This choice is inspired by [31], where quadrupling the dataset yields the best results, and creating more synthetic data does not provide more benefits.

5   Results

All possible combinations of these techniques are systematically tested to assess their impacts on the performance metrics. Considering that it is useless to apply balancing at both levels and that applying generalization-oriented augmentation to quadruple the training set at both levels would increase its size exponentially, there are 15 combinations of augmentation strategies. Given the randomness involved in the augmentation techniques, each combination is repeated ten times, and their results are averaged. This section employs a specific notation to distinguish between the application of a particular technique for model generalization (denoted as pedix G added to the technique's name) and its use for dataset balancing (denoted as pedix B added to the technique's name). When the technique is mentioned without a pedix it refers to both cases.

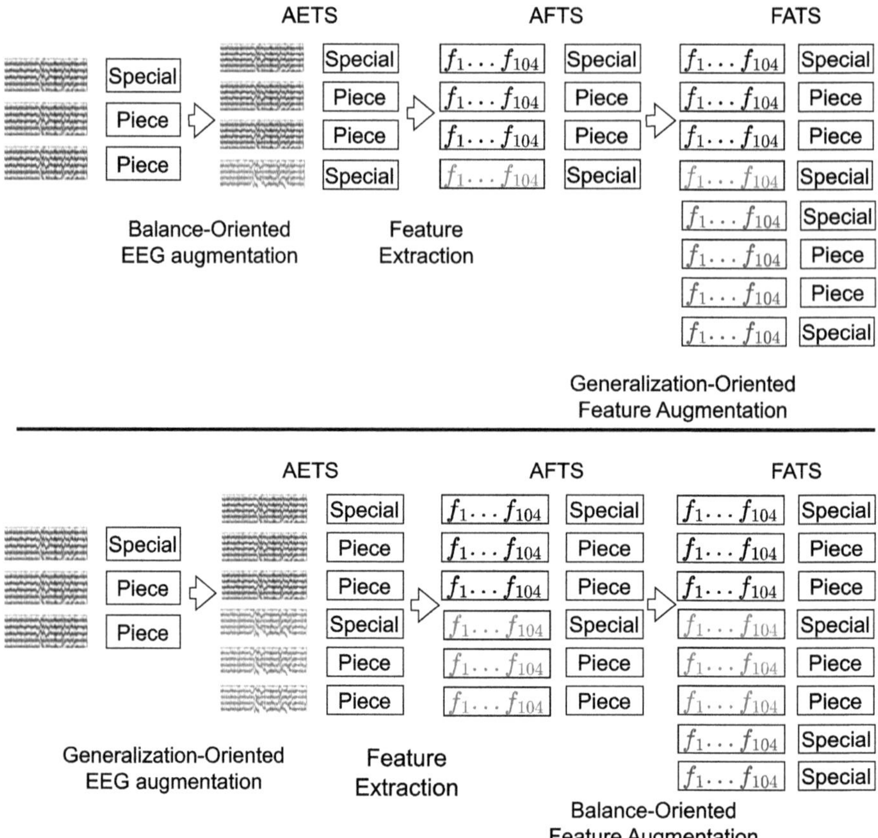

Fig. 4. Two possible evolutions of the dataset throughout the proposed procedure.

5.1   Results of TFCV Analysis

Table 1 summarizes the performance metrics achieved by various combinations of EEG signal augmentation (first column) and feature-level augmentation (second column) techniques with the TFCV strategy. The reference condition, without any augmentation ("none" for EEG and Features), achieved an accuracy of 89.55% and an F-score of 88.09%, serving as reference points. Applying $SMOTE_G$ alone slightly improved accuracy (89.93%) and F-score (88.58%), indicating a modest but positive effect (+0.38% and +0.49%, respectively). In contrast, $SMOTE_B$ without EEG augmentation decreased accuracy slightly to 89.34%, with minimal improvement in F-score (+0.05%).

Table 1. Performance metrics of the TFCV for data augmentation combinations.

EEG	Features	Accuracy (stdev)	Diff	F-score	Diff
none	none	89.55 ($\pm$0.25)	0.00	88.09	0.00
none	$SMOTE_G$	89.93 ($\pm$0.44)	0.38	88.58	0.49
none	$SMOTE_B$	89.34 ($\pm$0.62)	-0.21	88.14	0.05
$TSRDA_B$	none	88.59 ($\pm$0.46)	-0.96	87.27	-0.82
$TSRDA_B$	$SMOTE_G$	88.92 ($\pm$0.48)	-0.63	87.65	-0.44
$TSRDA_G$	none	88.78 ($\pm$0.38)	-0.77	87.27	-0.82
$TSRDA_G$	$SMOTE_B$	88.52 ($\pm$0.48)	-1.03	87.33	-0.76
SR_B	none	89.03 ($\pm$0.39)	-0.52	87.73	-0.36
SR_B	$SMOTE_G$	89.54 ($\pm$0.26)	-0.01	88.25	0.16
SR_G	none	89.27 ($\pm$0.43)	-0.28	87.82	-0.27
SR_G	$SMOTE_B$	89.42 ($\pm$0.35)	-0.13	88.23	0.14
GNA_B	none	89.59 ($\pm$0.25)	0.04	88.14	0.05
GNA_B	$SMOTE_G$	89.90 ($\pm$0.24)	0.35	88.47	0.38
GNA_G	none	89.61 ($\pm$0.26)	0.06	88.16	0.07
GNA_G	$SMOTE_B$	89.23 ($\pm$0.38)	-0.32	88.10	0.01

Among EEG augmentations, TSRDA generally reduced performance compared to the reference. $TSRDA_B$ without SMOTE showed the most significant accuracy drop (-0.96%), resulting in 88.59%, and similarly impacted the F-score (-0.82%). Combining TSRDA with SMOTE variants did not substantially recover performance, consistently remaining below the reference metrics. SR yielded better results compared to TSRDA. SR_B with $SMOTE_G$ approached reference accuracy (89.54%; Diff -0.01) and improved the F-score slightly ($+0.16\%$). The combination of SR_G with $SMOTE_B$ showed moderate performance with minimal negative differences from the reference in accuracy and a marginal positive effect on the F-score ($+0.14\%$). GNA demonstrated relatively effective results. Both GNA_G and GNA_B, without additional feature-level augmentation, slightly improved accuracy compared to the reference (89.61% and 89.59%, respectively). Combining GNA_B with $SMOTE_G$ notably increased accuracy to 89.90%, an improvement of $+0.35\%$, and enhanced the F-score by $+0.38\%$. The results were statistically validated by performing a paired t-test analysis on both the mean accuracies and F-scores obtained for the reference condition and the best-performing approach ($SMOTE_G$ only). The analysis yielded a p-value of 0.0263 for the mean accuracies and 0.0141 for the F-scores, both of which are below the 0.05 significance threshold, thus confirming the statistical validity and relevance of the observed improvements.

5.2 Results of the LOSO Analysis

Table 2 shows the performance metrics for different combinations of EEG signal augmentation and feature-level augmentation methods with the LOSO

Table 2. Performance metrics of the LOSO for data augmentation combinations.

EEG	Features	Accuracy(stdev)	Diff	F-score	Diff
none	none	59.62 ($\pm$1.53)	0.00	51.86	0.00
none	$SMOTE_G$	57.98 ($\pm$1.84)	-1.64	52.00	0.14
none	$SMOTE_B$	53.54 ($\pm$1.53)	-6.08	51.94	0.08
$TSRDA_B$	none	51.24 ($\pm$1.33)	-8.38	50.29	-1.57
$TSRDA_B$	$SMOTE_G$	49.67 ($\pm$1.73)	-9.95	49.17	-2.69
$TSRDA_G$	none	60.56 ($\pm$2.04)	0.94	50.85	-1.01
$TSRDA_G$	$SMOTE_B$	56.68 ($\pm$2.22)	-2.94	52.83	0.97
SR_B	none	58.33 ($\pm$1.96)	-1.29	52.91	1.05
SR_B	$SMOTE_G$	58.16 ($\pm$1.54)	-1.46	52.42	0.56
SR_G	none	61.82 ($\pm$1.78)	2.20	52.75	0.89
SR_G	$SMOTE_B$	57.86 ($\pm$2.47)	-1.76	53.03	1.17
GNA_B	none	59.56 ($\pm$1.52)	-0.06	51.82	-0.04
GNA_B	$SMOTE_G$	59.76 ($\pm$1.47)	0.14	52.17	0.31
GNA_G	none	59.47 ($\pm$1.63)	-0.15	51.76	-0.10
GNA_G	$SMOTE_B$	53.73 ($\pm$1.37)	-5.89	51.39	-0.47

strategy. The reference scenario (no augmentation applied) achieved an accuracy of 59.62% and an F-score of 51.86%, serving as the reference point.

Applying only feature-level augmentation resulted in decreased accuracy: $SMOTE_G$ reduced accuracy to 57.98%, and $SMOTE_B$ significantly reduced accuracy to 53.54%. However, the F-score showed minimal changes, slightly improving with $SMOTE_G$ (+0.14%) and $SMOTE_B$ (+0.08%).

Regarding EEG signal augmentations, $TSRDA_B$ notably degraded performance, dropping accuracy to 51.24% and reducing the F-score to 50.29% (-1.57% from reference). The addition of $SMOTE_G$ further decreased accuracy to 49.67% and negatively impacted the F-score (-2.69%). Conversely, $TSRDA_G$ alone improved accuracy slightly to 60.56% but decreased the F-score marginally to 50.85% (-1.01%), while combining $TSRDA_G$ with $SMOTE_B$ reduced accuracy to 56.68% but significantly improved the F-score to 52.83% (+0.97%). SR_G without SMOTE provided the highest accuracy of 61.82%, a notable improvement of +2.20%, and moderately improved the F-score by +0.89%. However, SR_B, with or without SMOTE, moderately reduced accuracy, although it slightly increased the F-score (+1.05% without SMOTE, +0.56% with $SMOTE_G$). GNA showed mixed effects. GNA_B combined with $SMOTE_G$ marginally increased accuracy (+0.14%) and improved the F-score (+0.31%). GNA_G without SMOTE slightly decreased accuracy (-0.15%) and minimally reduced the F-score (-0.10%). However, the combination of GNA_G and $SMOTE_B$ notably deteriorated accuracy to 53.73% and decreased the F-score by -0.47%. In summary, the most beneficial individual augmentation method in terms of accuracy is SR_G. The table also indicates that TSRDA negatively impacted performance. The results have been validated by running a paired t-test analysis on the means and F-scores obtained from the reference point and the best result (only SR_G), obtaining a p-value of 0.02 for the means and 0.0424 for the F-scores both below the 0.05 threshold, meaning that these results have statistical validity.

The comparison between the two tables highlights distinct outcomes based on the effectiveness of data augmentation. In Table 1 (higher reference performance), augmentations such as GNA combined with SMOTE demonstrate notable performance improvements, reaching an accuracy of up to 89.90% and F-scores of around 88.47%. In contrast, Table 2 (lower reference performance) showed that Signal Resampling (SR_G) significantly improved accuracy (+2.20%) over the reference, whereas TSRDA mostly degraded accuracy and F-score. Overall, the augmentation methods yielded different outcomes, with GNA and SR proving more consistently beneficial.

6   Conclusion

This work examined the role of data augmentation in improving EEG-based classification within the neuroaesthetic domain, where subjective judgments and naturalistic settings inherently limit data volume and class balance. We evaluated three signal-level augmentation techniques (TSRDA, SR, GNA) and one feature-level method (SMOTE), testing 15 combinations across two validation strategies.

For TFCV, simply applying the SMOTE technique to generalize the data improved classification performance, achieving the highest gains in both accuracy and F-score. SR and GNA also yielded moderate improvements, particularly when paired with SMOTE. In contrast, TSRDA always led to reduced performance. LOSO validation highlighted the robustness of SR when used for generalization (SR_G), which delivered the highest accuracy gains.

In conclusion, this study demonstrates that carefully selected data augmentation techniques, particularly GNA and SR, can significantly enhance EEG classification in neuroaesthetics. These findings contribute to establishing best practices for handling limited, imbalanced EEG data in emerging affective computing domains. Future work will explore advanced generative methods and adaptive augmentation pipelines tailored to individual variability.

Acknowledgments. This work has been partially supported by the European Union by the Next Generation EU project ECS00000017 'Ecosistema dell'Innovazione' Tuscany Health Ecosystem (THE, PNRR, Spoke 3: Advanced technologies, methods and materials for human health and well-being) and partially supported by the Italian Ministry of Education and Research (MIUR) in the framework of the FoReLab project (Departments of Excellence).

References

1. Mobile eeg recordings in an art museum setting (2017). https://doi.org/10.21227/H2TM00
2. Alfeo, A.L., Catrambone, V., Cimino, M.G., Valenza, G.: EEG-based motor imagery recognition via novel explainable ensemble learning architecture. Neural Comput. Appl. 1–25 (2025)
3. Callara, A.L., et al.: LD-EEG effective brain connectivity in patients with Cheyne-stokes respiration. IEEE Trans. Neural Syst. Rehabil. Eng. **28**(5), 1216–1225 (2020)
4. Chatterjee, A., Vartanian, O.: Neuroaesthetics. Trends Cogn. Sci. **18**(7), 370–375 (2014)
5. Chawla, N.V., Bowyer, K.W., Hall, L.O., Kegelmeyer, W.P.: Smote: synthetic minority over-sampling technique. J. Artif. Intell. Res. **16**, 321–357 (2002). https://doi.org/10.1613/jair.953
6. Cruz-Garza, J.G., et al.: Deployment of mobile EEG technology in an art museum setting: evaluation of signal quality and usability. Front. Human Neurosci. **11** (2017). https://doi.org/10.3389/fnhum.2017.00527
7. Cui, G., Li, X., Touyama, H.: Emotion recognition based on group phase locking value using convolutional neural network. Sci. Rep. **13**(1), 3769 (2023)
8. Delorme, A., Makeig, S.: EEGLAB: an open source toolbox for analysis of single-trial EEG dynamics including independent component analysis. J. Neurosci. Methods **134**(1), 9–21 (2004)
9. Ding, W., Liu, A., Guan, L., Chen, X.: A novel data augmentation approach using mask encoding for deep learning-based asynchronous SSVEP-BCI. IEEE Trans. Neural Syst. Rehabil. Eng. **32**, 875–886 (2024). https://doi.org/10.1109/TNSRE.2024.3366930
10. EBRAINS: EBRAINS – Brain research infrastructure powered by the European Horizon 2020 programme. https://www.ebrains.eu/ (2025). Accessed 30 June 2025
11. Geng, Y., Shi, S., Hao, X.: Deep learning-based EEG emotion recognition: a comprehensive review. Neural Comput. Appl. **37**(4), 1919–1950 (2025). https://doi.org/10.1007/s00521-024-10821-y
12. Hadjipanayi, C., Banakou, D., Michael-Grigoriou, D.: Art as therapy in virtual reality: a scoping review. Front. Virtual Real. **4** (2023). https://doi.org/10.3389/frvir.2023.1065863
13. Kontson, K., et al.: 'Your Brain on Art': Emergent cortical dynamics during aesthetic experiences. Front. Human Neurosci. **9** (2015). https://doi.org/10.3389/fnhum.2015.00626
14. Kothe, C.A.E., Jung, T.P.: Artifact removal techniques with signal reconstruction (Apr 28 2016), uS Patent App. 14/895,440
15. Lee, D.H., Jeong, J.H., Kim, K., Yu, B.W., Lee, S.W.: Continuous EEG decoding of pilots' mental states using multiple feature block-based convolutional neural network. IEEE Access **8**, 121929–121941 (2020). https://doi.org/10.1109/ACCESS.2020.3006907
16. Li, D., Huang, S., Xie, L., Wang, Z., Xu, J.: Neuron perception inspired EEG emotion recognition with parallel contrastive learning. IEEE Trans. Neural Netw. Learn. Syst. 1–14 (2025). https://doi.org/10.1109/TNNLS.2025.3546283

17. Li, Y., Zheng, W., Zong, Y., Cui, Z., Zhang, T., Zhou, X.: A bi-hemisphere domain adversarial neural network model for eeg emotion recognition. IEEE Trans. Affect. Comput. **12**(2), 494–504 (2021). https://doi.org/10.1109/TAFFC.2018.2885474
18. Lotte, F.: Signal processing approaches to minimize or suppress calibration time in oscillatory activity-based brain–computer interfaces. Proc. IEEE **103**(6), 871–890 (2015). https://doi.org/10.1109/JPROC.2015.2404941
19. Mohsenvand, M.N., Izadi, M.R., Maes, P.: Contrastive representation learning for electroencephalogram classification. In: Alsentzer, E., McDermott, M.B.A., Falck, F., Sarkar, S.K., Roy, S., Hyland, S.L. (eds.) Proceedings of the Machine Learning for Health NeurIPS Workshop. Proceedings of Machine Learning Research, vol. 136, pp. 238–253. PMLR (11 Dec 2020). https://proceedings.mlr.press/v136/mohsenvand20a.html
20. Molau, S., Pitz, M., Schluter, R., Ney, H.: Computing mel-frequency cepstral coefficients on the power spectrum. In: 2001 IEEE International Conference on Acoustics, Speech, and Signal Processing. Proceedings, vol. 1, pp. 73–76 vol.1 (2001). https://doi.org/10.1109/ICASSP.2001.940770
21. OpenNeuro: OpenNeuro: A free and open platform for sharing neuroimaging data. https://openneuro.org/ (2025). Accessed 30 June 2025
22. Palmieri, M., Avvenuti, M., Marcelloni, F., Vecchio, A.: Recognizing special art pieces through EEG: a journey in neuroaesthetics classification. IEEE Access **13**, 77696–77708 (2025). https://doi.org/10.1109/ACCESS.2025.3562908
23. Polyakov, D., Robinson, P.A., Muller, E.J., Shriki, O.: Recruiting neural field theory for data augmentation in a motor imagery brain–computer interface. Front. Robot. AI **Volume 11 - 2024** (2024). https://doi.org/10.3389/frobt.2024.1362735
24. Qin, Y., Zhang, Y., Zhang, Y., Liu, S., Guo, X.: Application and development of EEG acquisition and feedback technology: a review. Biosensors **13**(10) (2023). https://www.mdpi.com/2079-6374/13/10/930
25. Rodríguez-Rodríguez, I., Ortiz, A., Gallego-Molina, N.J., Formoso, M.A., Woo, W.L.: EEG interchannel causality to identify source/sink phase connectivity patterns in developmental dyslexia. Int. J. Neural Syst. **33**(04), 2350020 (2023)
26. Roy, Y., Banville, H., Albuquerque, I., Gramfort, A., Falk, T.H., Faubert, J.: Deep learning-based electroencephalography analysis: a systematic review. J. Neural Eng. **16**(5), 051001 (Aug 2019). https://doi.org/10.1088/1741-2552/ab260c
27. Sciaraffa, N., et al.: Mental effort estimation by passive bci: A cross-subject analysis. In: 2021 43rd Annual International Conference of the IEEE Engineering in Medicine & Biology Society (EMBC), pp. 906–909. IEEE (2021). https://doi.org/10.1109/EMBC46164.2021.9630613
28. Soni, S., Seal, A., Yazidi, A., Krejcar, O.: Graphical representation learning-based approach for automatic classification of electroencephalogram signals in depression. Comput. Biol. Med. **145**, 105420 (2022)
29. Wang, F., Zhong, S.h., Peng, J., Jiang, J., Liu, Y.: Data augmentation for eeg-based emotion recognition with deep convolutional neural networks. In: Schoeffmann, K., Chalidabhongse, T.H., Ngo, C.W., Aramvith, S., O'Connor, N.E., Ho, Y.S., Gabbouj, M., Elgammal, A. (eds.) MultiMedia Modeling, pp. 82–93. Springer International Publishing, Cham (2018)
30. Wu, H., et al.: A parallel multiscale filter bank convolutional neural networks for motor imagery EEG classification. Front. Neurosci. **Volume 13 - 2019** (2019). https://doi.org/10.3389/fnins.2019.01275

31. Xia, X., Yang, Y., Shi, Y., Zheng, W., Men, H.: Decoding human taste perception by reconstructing and mining temporal-spatial features of taste-related eegs. Appl. Intell. **54**(5), 3902–3917 (2024). https://doi.org/10.1007/s10489-024-05374-5
32. Zeki, S.: Inner vision: An exploration of art and the brain (2002)
33. Zhao, Y.D., Liu, Y.K., Zheng, W.L., Lu, B.L.: EEG data augmentation for emotion recognition using diffusion model. In: 2024 46th Annual International Conference of the IEEE Engineering in Medicine and Biology Society (EMBC), pp. 1–4 (2024). https://doi.org/10.1109/EMBC53108.2024.10782956

Correction to: SPAX: A Shapley-Based Point Attribution eXplanation for Interpreting 3D Point Cloud Classification

Marc F. Harinck, Muhammad Shoaib Sarwar, Bram Ton, and Faizan Ahmed

Correction to:
Chapter 24 in: F. Marcelloni et al. (Eds.): *Computational Intelligence*, CCIS 2829, https://doi.org/10.1007/978-3-032-15638-9_24

The book was published with a typo of chapter 24 in this book ID (686031_1_En).

The corresponding author's name in the article "SPAX: A Shapley-Based Point Attribution eXplanation for Interpreting 3D Point Cloud Classification" was updated from "Muhamamd Shoaib" to "Muhammad Shoaib".

The updated version of this chapter can be found at
https://doi.org/10.1007/978-3-032-15638-9_24

Author Index